T0199408

Dynamics of Complex Systems

Studies in Nonlinearity

Series Editor: *Robert L. Devaney*

Yaneer Bar-Yam

Dynamics of
Complex Systems

Routledge
Taylor & Francis Group

LONDON AND NEW YORK

First published 1997 by Westview Press

Published 2018 by Routledge
52 Vanderbilt Avenue, New York, NY 10017
2 Park Square, Milton Park, Abingdon, Oxon OX14 4RN

Routledge is an imprint of the Taylor & Francis Group, an informa business

Copyright © 1997 by Yaneer Bar-Yam

All rights reserved. No part of this book may be reprinted or reproduced or utilised in any form or by any electronic, mechanical, or other means, now known or hereafter invented, including photocopying and recording, or in any information storage or retrieval system, without permission in writing from the publishers.

Notice:
Product or corporate names may be trademarks or registered trademarks, and are used only for identification and explanation without intent to infringe.

Library of Congress Cataloging-in-Publication Data
Bar-Yam, Yaneer.
    Dynamics of complex systems / Yaneer Bar-Yam.
        p.   cm.
    Includes index.
    ISBN 0-201-55748-7
    1. Biomathematics.   2. System theory.   I. Title.
QH323.5.B358   1997
570'.15' 1—DC21   96-52033
                                                                                CIP

ISBN 13: 978-0-367-00510-8 (hbk)
ISBN 13: 978-0-367-15497-4 (pbk)

This book is dedicated with love to my family

Zvi, Miriam, Aureet and Sageet

Naomi
and our children
Shlomiya, Yavni, Maayan and Taeer

Aureet's memory is a blessing.

Contents

Preface

"Complex" is a word of the times, as in the often-quoted "growing complexity of life." Science has begun to try to understand complexity in nature, a counterpoint to the traditional scientific objective of understanding the fundamental simplicity of laws of nature. It is believed, however, that even in the study of complexity there exist simple and therefore comprehensible laws. The field of study of complex systems holds that the dynamics of complex systems are founded on universal principles that may be used to describe disparate problems ranging from particle physics to the economics of societies. A corollary is that transferring ideas and results from investigators in hitherto disparate areas will cross-fertilize and lead to important new results.

In this text we introduce several of the problems of science that embody the concept of complex dynamical systems. Each is an active area of research that is at the forefront of science. Our presentation does not try to provide a comprehensive review of the research literature available in each area. Instead we use each problem as an opportunity for discussing fundamental issues that are shared among all areas and therefore can be said to unify the study of complex systems.

We do not expect it to be possible to provide a succinct definition of a complex system. Instead, we give examples of such systems and provide the elements of a definition. It is helpful to begin by describing some of the attributes that characterize complex systems. Complex systems contain a large number of mutually interacting parts. Even a few interacting objects can behave in complex ways. However, the complex systems that we are interested in have more than just a few parts. And yet there is generally a limit to the number of parts that we are interested in. If there are too many parts, even if these parts are strongly interacting, the properties of the system become the domain of conventional thermodynamics—a uniform material.

Thus far we have defined complex systems as being within the mesoscopic domain—containing more than a few, and less than too many parts. However, the mesoscopic regime describes any physical system on a particular length scale, and this is too broad a definition for our purposes. Another characteristic of most complex dynamical systems is that they are in some sense purposive. This means that the dynamics of the system has a definable objective or function. There often is some sense in which the systems are engineered. We address this topic directly when we discuss and contrast self-organization and organization by design.

A central goal of this text is to develop models and modeling techniques that are useful when applied to all complex systems. For this we will adopt both analytic tools and computer simulation. Among the analytic techniques are statistical mechanics and stochastic dynamics. Among the computer simulation techniques are cellular automata and Monte Carlo. Since analytic treatments do not yield complete theories of complex systems, computer simulations play a key role in our understanding of how these systems work.

The human brain is an important example of a complex system formed out of its component neurons. Computers might similarly be understood as complex interacting systems of transistors. Our brains are well suited for understanding complex systems, but not for simulating them. Why are computers better suited to simulations of complex systems? One could point to the need for precision that is the traditional domain of the computer. However, a better reason would be the difficulty the brain has in keeping track of many and arbitrary interacting objects or events—we can typically remember seven independent pieces of information at once. The reasons for this are an important part of the design of the brain that make it powerful for other purposes. The architecture of the brain will be discussed beginning in Chapter 2.

The study of the dynamics of complex systems creates a host of new interdisciplinary fields. It not only breaks down barriers between physics, chemistry and biology, but also between these disciplines and the so-called soft sciences of psychology, sociology, economics, and anthropology. As this breakdown occurs it becomes necessary to introduce or adopt a new vocabulary. Included in this new vocabulary are words that have been considered taboo in one area while being extensively used in another. These must be adopted and adapted to make them part of the interdisciplinary discourse. One example is the word "mind." While the field of biology studies the brain, the field of psychology considers the mind. However, as the study of neural networks progresses, it is anticipated that the function of the neural network will become identified with the concept of mind.

Another area in which science has traditionally been mute is in the concept of meaning or purpose. The field of science traditionally has no concept of values or valuation. Its objective is to describe natural phenomena without assigning positive or negative connotation to the description. However, the description of complex systems requires a notion of purpose, since the systems are generally purposive. Within the context of purpose there may be a concept of value and valuation. If, as we will attempt to do, we address society or civilization as a complex system and identify its purpose, then value and valuation may also become a concept that attains scientific significance. There are even further possibilities of identifying value, since the very concept of complexity allows us to identify value with complexity through its difficulty of replacement. As is usual with any scientific advance, there are both dangers and opportunities with such developments.

Finally, it is curious that the origin and fate of the universe has become an accepted subject of scientific discourse—cosmology and the big bang theory—while the fate of humankind is generally the subject of religion and science fiction. There are exceptions to this rule, particularly surrounding the field of ecology—limits to population growth, global warming—however, this is only a limited selection of topics that could be addressed. Overcoming this limitation may be only a matter of having the appropriate tools. Developing the tools to address questions about the dynamics of human civilization is appropriate within the study of complex systems. It should also be recognized that as science expands to address these issues, science itself will change as it redefines and changes other fields.

Different fields are often distinguished more by the type of questions they ask than the systems they study. A significant effort has been made in this text to articulate questions, though not always to provide complete answers, since questions that define the field of complex systems will inspire more progress than answers at this early stage in the development of the field.

Like other fields, the field of complex systems has many aspects, and any text must make choices about which material to include. We have suggested that complex systems have more than a few parts and less than too many of them. There are two approaches to this intermediate regime. The first is to consider systems with more than a few parts, but still a denumerable number—denumerable, that is, by a single person in a reasonable amount of time. The second is to consider many parts, but just fewer than too many. In the first approach the main task is to describe the behavior of a particular system and its mechanism of operation—the function of a neural network of a few to a few hundred neurons, a few-celled organism, a small protein, a few people, etc. This is done by describing completely the role of each of the parts. In the second approach, the precise number of parts is not essential, and the main task is a statistical study of a collection of systems that differ from each other but share the same structure—an ensemble of systems. This approach treats general properties of proteins, neural networks, societies, etc. In this text, we adopt the second approach. However, an interesting twist to our discussion is that we will show that any complex system requires a description as a particular few-part system. A complementary volume to the present one would consider examples of systems with only a few parts and analyze their function with a view toward extracting general principles. These principles would complement the seemingly more general analysis of the statistical approach.

The order of presentation of the topics in this text is a matter of taste. Many of the chapters are self-contained discussions of a particular system or question. The first chapter contains material that provides a foundation for the rest. Part of the role of this chapter is the introduction of "simple" models upon which the remainder of the text is based. Another role is the review of concepts and techniques that will be used in later chapters so that the text is more self-contained. Because of the interdisciplinary nature of the subject matter, the first chapter is considered to have particular importance. Some of the material should be familiar to most graduate students, while other material is found only in the professional literature. For example, basic probability theory is reviewed, as well as the concepts and properties of cellular automata. The purpose is to enable this text to be read by students and researchers with a variety of backgrounds. However, it should be apparent that digesting the variety of concepts after only a brief presentation is a difficult task. Additional sources of material are listed at the end of this text.

Throughout the book, we have sought to limit advanced formal discussions to a minimum. When possible, we select models that can be described with a simpler formalism than must be used to treat the most general case possible. Where additional layers of formalism are particularly appropriate, reference is made to other literature. Simulations are described at a level of detail that, in most cases, should enable the student to perform and expand upon the simulations described. The graphical display of such simulations should be used as an integral part of exposure to the dynamics of these systems. Such displays are generally effective in developing an intuition about what are the important or relevant properties of these systems.

Acknowledgments

This book is a composite of many ideas and reflects the efforts of many individuals that would be impossible to acknowledge. My personal efforts to compose this body of knowledge into a coherent framework for future study are also indebted to many who contributed to my own development. It is the earliest teachers, who we can no longer identify by memory, who should be acknowledged at the completion of a major effort. They and the teachers I remember from elementary school through graduate school, especially my thesis advisor John Joannopoulos, have my deepest gratitude. Consistent with their dedication, may this be a reward for their efforts.

The study of complex systems is a new endeavor, and I am grateful to a few colleagues and teachers who have inspired me to pursue this path. Charles Bennett through a few joint car trips opened my mind to the possibilities of this field and the paths less trodden that lead to it. Tom Malone, through his course on networked corporations, not only contributed significant concepts to the last chapter of this book, but also motivated the creation of my course on the dynamics of complex systems.

There are colleagues and students who have inspired or contributed to my understanding of various aspects of material covered in this text. Some of this contribution arises from reading and commenting on various aspects of this text, or through discussions of the material that eventually made its way here. In some cases the discussions were originally on unrelated matters, but because they were eventually connected to these subjects, they are here acknowledged. Roughly divided by area in correspondence with the order they appear in the text these include: Glasses—David Adler; Cellular Automata—Gerard Vichniac, Tom Toffoli, Norman Margolus, Mike Biafore, Eytan Domany, Danny Kandel; Computation—Jeff Siskind; Multigrid—Achi Brandt, Shlomi Taasan, Sorin Costiner; Neural Networks—John Hopfield, Sageet Bar-Yam, Tom Kincaid, Paul Appelbaum, Charles Yang, Reza Sadr-Lahijany, Jason Redi, Lee-Peng Lee, Hua Yang, Jerome Kagan, Ernest Hartmann; Protein Folding—Elisha Haas, Charles DeLisi, Temple Smith, Robert Davenport, David Mukamel, Mehran Kardar; Polymer Dynamics—Yitzhak Rabin, Mark Smith, Boris Ostrovsky, Gavin Crooks, Eliana DeBernardez-Clark; Evolution—Alan Perelson, Derren Pierre, Daniel Goldman, Stuart Kauffman, Les Kaufman; Developmental Biology—Irving Epstein, Lee Segel, Ainat Rogel, Evelyn Fox Keller; Complexity—Charles Bennett, Michael Werman, Michel Baranger; Human Economies and Societies—Tom Malone, Harry Bloom, Benjamin Samuels, Kosta Tsipis, Jonathan King.

A special acknowledgment is necessary to the students of my course from Boston University and MIT. Among them are students whose projects became incorporated in parts of this text and are mentioned above. The interest that my colleagues have shown by attending and participating in the course has brightened it for me and their contributions are meaningful: Lewis Lipsitz, Michel Baranger, Paul Barbone, George Wyner, Alice Davidson, Ed Siegel, Michael Werman, Larry Rudolf and Mehran Kardar.

Among the readers of this text I am particularly indebted to the detailed comments of Bruce Boghosian, and the supportive comments of the series editor Bob Devaney. I am also indebted to the support of Charles Cantor and Jerome Kagan.

I would like to acknowledge the constructive efforts of the editors at Addison-Wesley starting from the initial contact with Jack Repcheck and continuing with Jeff Robbins. I thank Lynne Reed for coordinating production, and at Carlisle Communications: Susan Steines, Bev Kraus, Faye Schilling, and Kathy Davis.

The software used for the text, graphs, figures and simulations of this book, includes: Microsoft Excel and Word, Deneba Canvas, Wolfram's Mathematica, and Symantec C. The hardware includes: Macintosh Quadra, and IBM RISC workstations.

The contributions of my family, to whom this book is dedicated, cannot be described in a few words.

תושלב"ע

Yaneer Bar-Yam
Newton, Massachusetts, June 1997

0

Overview:
The Dynamics of Complex Systems —
Examples, Questions, Methods and Concepts

0.1 The Field of Complex Systems

The study of complex systems in a unified framework has become recognized in recent years as a new scientific discipline, the ultimate of interdisciplinary fields. It is strongly rooted in the advances that have been made in diverse fields ranging from physics to anthropology, from which it draws inspiration and to which it is relevant.

Many of the systems that surround us are complex. The goal of understanding their properties motivates much if not all of scientific inquiry. Despite the great complexity and variety of systems, universal laws and phenomena are essential to our inquiry and to our understanding. The idea that all matter is formed out of the same building blocks is one of the original concepts of science. The modern manifestation of this concept—atoms and their constituent particles—is essential to our recognition of the commonality among systems in science. The universality of constituents complements the universality of mechanical laws (classical or quantum) that govern their motion. In biology, the common molecular and cellular mechanisms of a large variety of organisms form the basis of our studies. However, even more universal than the constituents are the dynamic processes of variation and selection that in some manner cause organisms to evolve. Thus, all scientific endeavor is based, to a greater or lesser degree, on the existence of universality, which manifests itself in diverse ways. In this context, the study of complex systems as a new endeavor strives to increase our ability to understand the universality that arises when systems are highly complex.

A dictionary definition of the word "complex" is: "consisting of interconnected or interwoven parts." Why is the nature of a complex system inherently related to its parts? Simple systems are also formed out of parts. To explain the difference between simple and complex systems, the terms "interconnected" or "interwoven" are somehow essential. Qualitatively, to understand the behavior of a complex system we must understand not only the behavior of the parts but how they act together to form the behavior of the whole. It is because we cannot describe the whole without describing each part, and because each part must be described in relation to other parts, that complex systems are difficult to understand. This is relevant to another definition of "complex": "not easy to understand or analyze." These qualitative ideas about what a complex system is can be made more quantitative. Articulating them in a clear way is

1

both essential and fruitful in pointing the way toward progress in understanding the universal properties of these systems.

For many years, professional specialization has led science to progressive isolation of individual disciplines. How is it possible that well-separated fields such as molecular biology and economics can suddenly become unified in a single discipline? How does the study of complex systems in general pertain to the detailed efforts devoted to the study of particular complex systems? In this regard one must be careful to acknowledge that there is always a dichotomy between universality and specificity. A study of universal principles does not replace detailed description of particular complex systems. However, universal principles and tools guide and simplify our inquiries into the study of specifics. For the study of complex systems, universal simplifications are particularly important. Sometimes universal principles are intuitively appreciated without being explicitly stated. However, a careful articulation of such principles can enable us to approach particular systems with a systematic guidance that is often absent in the study of complex systems.

A pictorial way of illustrating the relationship of the field of complex systems to the many other fields of science is indicated in Fig. 0.1.1. This figure shows the conventional view of science as progressively separating into disparate disciplines in order to gain knowledge about the ever larger complexity of systems. It also illustrates the view of the field of complex systems, which suggests that all complex systems have universal properties. Because each field develops tools for addressing the complexity of the systems in their domain, many of these tools can be adapted for more general use by recognizing their universal applicability. Hence the motivation for cross-disciplinary fertilization in the study of complex systems.

In Sections 0.2–0.4 we initiate our study of complex systems by discussing examples, questions and methods that are relevant to the study of complex systems. Our purpose is to introduce the field without a strong bias as to conclusions, so that the student can develop independent perspectives that may be useful in this new field—opening the way to his or her own contributions to the study of complex systems. In Section 0.5 we introduce two key concepts—emergence and complexity—that will arise through our study of complex systems in this text.

0.2   Examples

0.2.1 *A few examples*

What are complex systems and what properties characterize them? It is helpful to start by making a list of some examples of complex systems. Take a few minutes to make your own list. Consider actual systems rather than mathematical models (we will consider mathematical models later). Make a list of some simple things to contrast them with.

Examples of Complex Systems

Governments

Families

The human body—physiological perspective

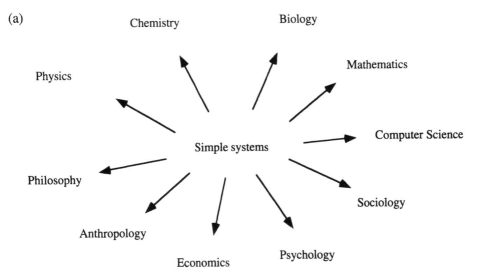

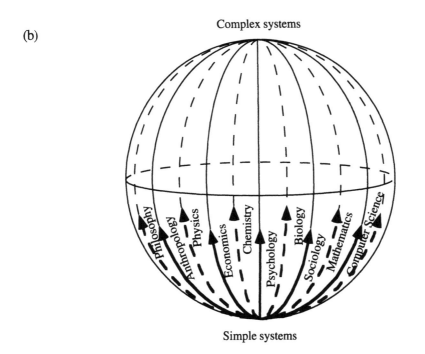

Figure 0.1.1 Conceptual illustration of the space of scientific inquiry. (a) is the conventional view where disciplines diverge as knowledge increases because of the increasing complexity of the various systems being studied. In this view all knowledge is specific and knowledge is gained by providing more and more details. (b) illustrates the view of the field of complex systems where complex systems have universal properties. By considering the common properties of complex systems, one can approach the specifics of particular complex systems from the top of the sphere as well as from the bottom.

A person—psychosocial perspective

The brain

The ecosystem of the world

Subworld ecosystems: desert, rain forest, ocean

Weather

A corporation

A computer

Examples of Simple Systems

An oscillator

A pendulum

A spinning wheel

An orbiting planet

The purpose of thinking about examples is to develop a first understanding of the question, What makes systems complex? To begin to address this question we can start describing systems we know intuitively as complex and see what properties they share. We try this with the first two examples listed above as complex systems.

Government

- It has many different functions: military, immigration, taxation, income distribution, transportation, regulation. Each function is itself complex.
- There are different levels and types of government: local, state and federal; town meeting, council, mayoral. There are also various governmental forms in different countries.

Family

- It is a set of individuals.
- Each individual has a relationship with the other individuals.
- There is an interplay between the relationship and the qualities of the individual.
- The family has to interact with the outside world.
- There are different kinds of families: nuclear family, extended family, etc.

These descriptions focus on function and structure and diverse manifestation. We can also consider the role that time plays in complex systems. Among the properties of complex systems are change, growth and death, possibly some form of life cycle. Combining time and the environment, we would point to the ability of complex systems to adapt.

One of the issues that we will need to address is whether there are different categories of complex systems. For example, we might contrast the systems we just described with complex physical systems: hydrodynamics (fluid flow, weather), glasses, composite materials, earthquakes. In what way are these systems similar to or different from the biological or social complex systems? Can we assign function and discuss structure in the same way?

0.2.2 *Central properties of complex systems*

After beginning to describe complex systems, a second step is to identify commonalities. We might make a list of some of the characteristics of complex systems and assign each of them some measure or attribute that can provide a first method of classification or description.

- Elements (and their number)
- Interactions (and their strength)
- Formation/Operation (and their time scales)
- Diversity/Variability
- Environment (and its demands)
- Activity(ies) (and its[their] objective[s])

This is a first step toward quantifying the properties of complex systems. Quantifying the last three in the list requires some method of counting possibilities. The problem of counting possibilities is central to the discussion of quantitative complexity.

0.2.3 *Emergence: From elements and parts to complex systems*

There are two approaches to organizing the properties of complex systems that will serve as the foundation of our discussions. The first of these is the relationship between elements, parts and the whole. Since there is only one property of the complex system that we know for sure — that it is complex—the primary question we can ask about this relationship is how the complexity of the whole is related to the complexity of the parts. As we will see, this question is a compelling question for our understanding of complex systems.

From the examples we have indicated above, it is apparent that parts of a complex system are often complex systems themselves. This is reasonable, because when the parts of a system are complex, it seems intuitive that a collection of them would also be complex. However, this is not the only possibility.

Can we describe a system composed of simple parts where the collective behavior is complex? This is an important possibility, called emergent complexity. Any complex system formed out of atoms is an example. The idea of emergent complexity is that the behaviors of many simple parts interact in such a way that the behavior of the whole is complex. Elements are those parts of a complex system that may be considered simple when describing the behavior of the whole.

Can we describe a system composed of complex parts where the collective behavior is simple? This is also possible, and it is called emergent simplicity. A useful example is a planet orbiting around a star. The behavior of the planet is quite simple, even if the planet is the Earth, with many complex systems upon it. This example illustrates the possibility that the collective system has a behavior at a different scale than its parts. On the smaller scale the system may behave in a complex way, but on the larger scale all the complex details may not be relevant.

0.2.4 *What is complexity?*

The second approach to the study of complex systems begins from an understanding of the relationship of systems to their descriptions. The central issue is defining quantitatively what we mean by complexity. What, after all, do we mean when we say that a system is complex? Better yet, what do we mean when we say that one system is more complex than another? Is there a way to identify the complexity of one system and to compare it with the complexity of another system? To develop a quantitative understanding of complexity we will use tools of both statistical physics and computer science—information theory and computation theory. According to this understanding, complexity is the amount of information necessary to describe a system. However, in order to arrive at a consistent definition, care must be taken to specify the level of detail provided in the description.

One of our targets is to understand how this concept of complexity is related to emergence—emergent complexity and emergent simplicity. Can we understand why information-based complexity is related to the description of elements, and how their behavior gives rise to the collective complexity of the whole system?

Section 0.5 of this overview discusses further the concepts of emergence and complexity, providing a simplified preview of the more complete discussions later in this text.

0.3   Questions

This text is structured around four questions related to the characterization of complex systems:

1. Space: What are the characteristics of the structure of complex systems? Many complex systems have substructure that extends all the way to the size of the system itself. Why is there substructure?

2. Time: How long do dynamical processes take in complex systems? Many complex systems have specific responses to changes in their environment that require changing their internal structure. How can a complex structure respond in a reasonable amount of time?

3. Self-organization and/versus organization by design: How do complex systems come into existence? What are the dynamical processes that can give rise to complex systems? Many complex systems undergo guided developmental processes as part of their formation. How are developmental processes guided?

4. Complexity: What is complexity? Complex systems have varying degrees of complexity. How do we characterize/distinguish the varying degrees of complexity?

Chapter 1 of this text plays a special role. Its ten sections introduce mathematical tools. These tools and their related concepts are integral to our understanding of complex system behavior. The main part of this book consists of eight chapters, 2–9. These chapters are paired. Each pair discusses one of the above four questions in the context of a particular complex system. Chapters 2 and 3 discuss the role of substructure in the context of neural networks. Chapters 4 and 5 discuss the time scale of dynamics in the context of protein folding. Chapters 6 and 7 discuss the mechanisms of organization of complex systems in the context of living organisms. Chapters 8 and 9 discuss complexity in the context of human civilization. In each case the first of the pair of chapters discusses more general issues and models. The second tends to be more specialized to the system that is under discussion. There is also a pattern to the degree of analytic, simulation or qualitative treatments. In general, the first of the two chapters is more analytic, while the second relies more on simulations or qualitative treatments. Each chapter has at least some discussion of qualitative concepts in addition to the formal quantitative discussion.

Another way to regard the text is to distinguish between the two approaches summarized above. The first deals with elements and interactions. The second deals with descriptions and information. Ultimately, our objective is to relate them, but we do so using questions that progress gradually from the elements and interactions to the descriptions and information. The former dominates in earlier chapters, while the latter is important for Chapter 6 and becomes dominant in Chapters 8 and 9.

While the discussion in each chapter is presented in the context of a specific complex system, our focus is on complex systems in general. Thus, we do not attempt (nor would it be possible) to review the entire fields of neural networks, protein folding, evolution, developmental biology and social and economic sciences. Since we are interested in universal aspects of these systems, the topics we cover need not be the issues of contemporary importance in the study of these systems. Our approach is to motivate a question of interest in the context of complex systems using a particular complex system, then to step back and adopt a method of study that has relevance to all complex systems. Researchers interested in a particular complex system are as likely to find a discussion of interest to them in any one of the chapters, and should not focus on the chapter with the particular complex system in its title.

We note that the text is interrupted by questions that are, with few exceptions, solved in the text. They are given as questions to promote independent thought about the study of complex systems. Some of them develop further the analysis of a system through analytic work or through simulations. Others are designed for conceptual development. With few exceptions they should be considered integral to the text, and even if they are not solved by the reader, the solutions should be read.

Question 0.3.1 Consider a few complex systems. Make a list of their elements, interactions between these elements, the mechanism by which the system is formed and the activities in which the system is engaged.

Solution 0.3.1 The following table indicates properties of the systems that we will be discussing most intensively in this text. ∎

System	Element	Interaction	Formation	Activity
Proteins	Amino Acids	Bonds	Protein folding	Enzymatic activity
Nervous system Neural networks	Neurons	Synapses	Learning	Behavior Thought
Physiology	Cells	Chemical messengers Physical support	Developmental biology	Movement Physiological functions
Life	Organisms	Reproduction Competition Predation Communication	Evolution	Survival Reproduction Consumption Excretion
Human economies and societies	Human Beings Technology	Communication Confrontation Cooperation	Social evolution	Same as Life? Exploration?

Table 0.3.1: Complex Systems and Some Attributes

0.4   Methods

When we think about methodology, we must keep purpose in mind. Our purpose in studying complex systems is to extract general principles. General principles can take many forms. Most principles are articulated as relationships between properties—when a system has the property x, then it has the property y. When possible, relationships should be quantitative and expressed as equations. In order to explore such relationships, we must construct and study mathematical models. Asking why the property x is related to the property y requires an understanding of alternatives. What else is possible? As a bonus, when we are able to generate systems with various properties, we may also be able to use them for practical applications.

All approaches that are used for the study of simple systems can be applied to the study of complex systems. However, it is important to recognize features of conventional approaches that may hamper progress in the study of complex systems. Both experimental and theoretical methods have been developed to overcome these difficulties. In this text we introduce and use methods of analysis and simulation that are particularly suited to the study of complex systems. These methods avoid standard simplifying assumptions, but use other simplifications that are better suited to our objectives. We discuss some of these in the following paragraphs.

- Don't take it apart. Since interactions between parts of a complex system are essential to understanding its behavior, looking at parts by themselves is not sufficient. It is necessary to look at parts in the context of the whole. Similarly, a complex system interacts with its environment, and this environmental influence is important in describing the behavior of the system. Experimental tools have been developed for studying systems *in situ* or *in vivo*—in context. Theoretical analytic methods such as the mean field approach enable parts of a system to be studied in context. Computer simulations that treat a system in its entirety also avoid such problems.

- Don't assume smoothness. Much of the quantitative study of simple systems makes use of differential equations. Differential equations, like the wave equation, assume that a system is essentially uniform and that local details don't matter for the behavior of a system on larger scales. These assumptions are not generally valid for complex systems. Alternate static models such as fractals, and dynamical models including iterative maps and cellular automata may be used instead.

- Don't assume that only a few parameters are important. The behavior of complex systems depends on many independent pieces of information. Developing an understanding of them requires us to build mental models. However, we can only have "in mind" 7±2 independent things at once. Analytic approaches, such as scaling and renormalization, have been developed to identify the few relevant parameters when this is possible. Information-based approaches consider the collection of all parameters as the object of study. Computer simulations keep track of many parameters and may be used in the study of dynamical processes.

There are also tools needed for communication of the results of studies. Conventional manuscripts and oral presentations are now being augmented by video and interactive media. Such novel approaches can increase the effectiveness of communication, particularly of the results of computer simulations. However, we should avoid the "cute picture" syndrome, where pictures are presented without accompanying discussion or analysis.

In this text, we introduce and use a variety of analytic and computer simulation methods to address the questions listed in the previous section. As mentioned in the preface, there are two general methods for studying complex systems. In the first, a specific system is selected and each of the parts as well as their interactions are identified and described. Subsequently, the objective is to show how the behavior of the whole emerges from them. The second approach considers a class of systems (ensemble), where the essential characteristics of the class are described, and statistical analysis is used to obtain properties and behaviors of the systems. In this text we focus on the latter approach.

0.5 Concepts: Emergence and Complexity

The objectives of the field of complex systems are built on fundamental concepts—emergence, complexity—about which there are common misconceptions that are addressed in this section and throughout the book. Once understood, these concepts reveal the context in which universal properties of complex systems arise and specific universal phenomena, such as the evolution of biological systems, can be better understood.

A complex system is a system formed out of many components whose behavior is emergent, that is, the behavior of the system cannot be simply inferred from the behavior of its components. The amount of information necessary to describe the behavior of such a system is a measure of its complexity. In the following sections we discuss these concepts in greater detail.

0.5.1 *Emergence*

It is impossible to understand complex systems without recognizing that simple atoms must somehow, in large numbers, give rise to complex collective behaviors. How and when this occurs is the simplest and yet the most profound problem that the study of complex systems faces. The problem can be approached first by developing an understanding of the term "emergence." For many, the concept of emergent behavior means that the behavior is not captured by the behavior of the parts. This is a serious misunderstanding. It arises because the collective behavior is not readily understood from the behavior of the parts. The collective behavior is, however, contained in the behavior of the parts if they are studied in the context in which they are found. To explain this, we discuss examples of emergent properties that illustrate the difference between local emergence—where collective behavior appears in a small part of the system—and global emergence—where collective behavior pertains to the system as a whole. It is the latter which is particularly relevant to the study of complex systems.

We can speak about emergence when we consider a collection of elements and the properties of the collective behavior of these elements. In conventional physics, the main arena for the study of such properties is thermodynamics and statistical mechanics. The easiest thermodynamic system to think about is a gas of particles. Two emergent properties of a gas are its pressure and temperature. The reason they are emergent is that they do not naturally arise out of the description of an individual particle. We generally describe a particle by specifying its position and velocity. Pressure and temperature become relevant only when we have many particles together. While these are emergent properties, the way they are emergent is very limited. We call them local emergent properties. The pressure and temperature is a local property of the gas. We can take a very small sample of the gas away from the rest and still define and measure the (same) pressure and temperature. Such properties, called intensive in physics, are local emergent properties. Other examples from physics of locally emergent behavior are collective modes of excitation such as sound waves, or light propagation in a medium. Phase transitions (e.g., solid to liquid) also represent a collective dynamics that is visible on a macroscopic scale, but can be seen in a microscopic sample as well.

Another example of a local emergent property is the formation of water from atoms of hydrogen and oxygen. The properties of water are not apparent in the properties of gasses of oxygen or hydrogen. Neither does an isolated water molecule reveal most properties of water. However, a microscopic amount of water is sufficient.

In the study of complex systems we are particularly interested in global emergent properties. Such properties depend on the entire system. The mathematical treatment of global emergent properties requires some effort. This is one reason that emergence is not well appreciated or understood. We will discuss global emergence by summarizing the results of a classic mathematical treatment, and then discuss it in a more general manner that can be readily appreciated and is useful for semiquantitative analyses.

The classic analysis of global emergent behavior is that of an associative memory in a simple model of neural networks known as the Hopfield or attractor network. The analogy to a neural network is useful in order to be concrete and relate this model to known concepts. However, this is more generally a model of any system formed from simple elements whose states are correlated. Without such correlations, emergent behavior is impossible. Yet if all elements are correlated in a simple way, then local emergent behavior is the outcome. Thus a model must be sufficiently rich in order to capture the phenomenon of global emergent behavior. One of the important qualities of the attractor network is that it displays global emergence in a particularly elegant manner. The following few paragraphs summarize the operation of the attractor network as an associative memory.

The Hopfield network has simple binary elements that are either ON or OFF. The binary elements are an abstraction of the firing or quiescent state of neurons. The elements interact with each other to create correlations in the firing patterns. The interactions represent the role of synapses in a neural network. The network can work as a memory. Given a set of preselected patterns, it is possible to set the interactions so that these patterns are self-consistent states of the network—the network is stable when it is in these firing patterns. Even if we change some of the neurons, the original pattern will be recovered. This is an associative memory.

Assume for the moment that the pattern of firing represents a sentence, such as "To be or not to be, that is the question." We can recover the complete sentence by presenting only part of it to the network "To be or not to be, that" might be enough. We could use any part to retrieve the whole, such as, "to be, that is the question." This kind of memory is to be contrasted with a computer memory, which works by assigning an address to each storage location. To access the information stored in a particular location we need to know the address. In the neural network memory, we specify part of what is located there, rather than the analogous address: Hamlet, by William Shakespeare, act 3, scene 1, line 64.

More central to our discussion, however, is that in a computer memory a particular bit of information is stored in a particular switch. By contrast, the network does not have its memory in a neuron. Instead the memory is in the synapses. In the model, there are synapses between each neuron and every other neuron. If we remove a small part of the network and look at its properties, then the number of synapses that a neuron is left with in this small part is only a small fraction of the number of synapses it started with. If there are more than a few patterns stored, then when we cut out the small part of the network it loses the ability to remember any of the patterns, even the part which would be represented by the neurons contained in this part.

This kind of behavior characterizes emergent properties. We see that emergent properties cannot be studied by physically taking a system apart and looking at the parts (reductionism). They can, however, be studied by looking at each of the parts in the context of the system as a whole. This is the nature of emergence and an indication of how it can be studied and understood.

The above discussion reflects the analysis of a relatively simple mathematical model of emergent behavior. We can, however, provide a more qualitative discussion that serves as a guide for thinking about diverse complex systems. This discussion focuses on the properties of a system when part of it is removed. Our discussion of local emergent properties suggested that taking a small part out of a large system would cause little change in the properties of the small part, or the properties of the large part. On the other hand, when a system has a global emergent property, the behavior of the small part is different in isolation than when it is part of the larger system.

If we think about the system as a whole, rather than the small part of the system, we can identify the system that has a global emergent property as being formed out of interdependent parts. The term "interdependent" is used here instead of the terms "interconnected" or "interwoven" used in the dictionary definition of "complex" quoted in Section 0.1, because neither of the latter terms pertain directly to the influence one part has on another, which is essential to the properties of a dynamic system. "Interdependent" is also distinct from "interacting," because even strong interactions do not necessarily imply interdependence of behavior. This is clear from the macroscopic properties of simple solids.

Thus, we can characterize complex systems through the effect of removal of part of the system. There are two natural possibilities. The first is that properties of the part are affected, but the rest is not affected. The second is that properties of the rest are affected by the removal of a part. It is the latter that is most appealing as a model of a truly complex system. Such a system has a collective behavior that is dependent on the behavior of all of its parts. This concept becomes more precise when we connect it to a quantitative measure of complexity.

0.5.2 *Complexity*

The second concept that is central to complex systems is a quantitative measure of how complex a system is. Loosely speaking, the complexity of a system is the amount of information needed in order to describe it. The complexity depends on the level of detail required in the description. A more formal definition can be understood in a simple way. If we have a system that could have many possible states, but we would like to specify which state it is actually in, then the number of binary digits (bits) we need to specify this particular state is related to the number of states that are possible. If we call the number of states Ω then the number of bits of information needed is

$$I = \log_2(\Omega) \tag{0.5.1}$$

To understand this we must realize that to specify which state the system is in, we must enumerate the states. Representing each state uniquely requires as many numbers as there are states. Thus the number of states of the representation must be the same as the number of states of the system. For a string of N bits there are 2^N possible states and thus we must have

$$\Omega = 2^N \tag{0.5.2}$$

which implies that N is the same as I above. Even if we use a descriptive English text instead of numbers, there must be the same number of possible descriptions as there are states, and the information content must be the same. When the number of possible valid English sentences is properly accounted for, it turns out that the best estimate of the amount of information in English is about 1 bit per character. This means that the information content of this sentence is about 120 bits, and that of this book is about 3×10^6 bits.

For a microstate of a physical system, where we specify the positions and momenta of each of the particles, this can be recognized as proportional to the entropy of the system, which is defined as

$$S = k \ln(\Omega) = k \ln(2)I \qquad (0.5.3)$$

where $k = 1.38 \times 10^{-23}$ Joule/°Kelvin is the Boltzmann constant which is relevant to our conventional choice of units. Using measured entropies we find that entropies of order 10 bits per atom are typical. The reason k is so small is that the quantities of matter we typically consider are in units of Avogandro's number (moles) and the number of bits per mole is 6.02×10^{23} times as large. Thus, the information in a piece of material is of order 10^{24} bits.

There is one point about Eq. (0.5.3) that may require some clarification. The positions and momenta of particles are real numbers whose specification might require infinitely many bits. Why isn't the information necessary to specify the microstate of a system infinite? The answer to this question comes from quantum physics, which is responsible for giving a unique value to the entropy and thus the information needed to specify a state of the system. It does this in two ways. First, it tells us that microscopic states are indistinguishable unless they differ by a discrete amount in position and momentum—a quantum difference given by Planck's constant h. Second, it indicates that particles like nuclei or atoms in their ground state are uniquely specified by this state, and are indistinguishable from each other. There is no additional information necessary to specify their internal structure. Under standard conditions, essentially all nuclei are in their lowest energy state.

The relationship of entropy and information is not accidental, of course, but it is the source of much confusion. The confusion arises because the entropy of a physical system is largest when it is in equilibrium. This suggests that the most complex system is a system in equilibrium. This is counter to our usual understanding of complex systems. Equilibrium systems have no spatial structure and do not change over time. Complex systems have substantial internal structure and this structure changes over time.

The problem is that we have used the definition of the information necessary to specify the microscopic state (microstate) of the system rather than the macroscopic state (macrostate) of the system. We need to consider the information necessary to describe the macrostate of the system in order to define what we mean by complexity. One of the important points to realize is that in order for the macrostate of the system to require a lot of information to describe it, there must be correlations in the microstate of the system. It is only when many microscopic atoms move in a coherent fashion that we can see this motion on a macroscopic scale. However, if many microscopic atoms move together, the system must be far from equilibrium and the microscopic information (entropy) must be lower than that of an equilibrium system.

It is helpful, even essential, to define a complexity profile which is a function of the scale of observation. To obtain the complexity profile, we observe the system at a particular length (or time) scale, ignoring all finer-scale details. Then we consider how much information is necessary to describe the observations on this scale. This solves the problem of distinguishing between a microscopic and a macroscopic description. Moreover, for different choices of scale, it explicitly captures the dependence of the complexity on the level of detail that is required in the description.

The complexity profile must be a monotonically falling function of the scale. This is because the information needed to describe a system on a larger scale must be a subset of the information needed to describe the system on a smaller scale—any finer-scale description contains the coarser-scale description. The complexity profile characterizes the properties of a complex system. If we wish to point to a particular number for the complexity of a system, it is natural to consider the complexity as the value of the complexity profile at a scale that is slightly smaller than the size of the system itself. The behavior at this scale includes the movement of the system through space, and dynamical changes of the system that are essentially the size of the system as a whole. The Earth orbiting the sun is a useful example.

We can make a direct connection between this definition of complexity and the discussion of the formation of a complex system out of parts. The complexity of the parts of the system are described by the complexity profile of the system evaluated on the scale of the parts. When the behavior of the system depends on the behavior of the parts, the complexity of the whole must involve a description of the parts, thus it is large. The smaller the parts that must be described to describe the behavior of the whole, the larger the complexity of the entire system.

0.6   For the Instructor

This text is designed for use in an introductory graduate-level course, to present various concepts and methodologies of the study of complex systems and to begin to develop a common language for researchers in this new field. It has been used for a one-semester course, but the amount of material is large, and it is better to spread the material over two semesters. A two-semester course also provides more opportunities for including various other approaches to the study of complex systems, which are as valuable as the ones that are covered here and may be more familiar to the instructor.

Consistent with the objective and purpose of the field, students attending such a course tend to have a wide variety of backgrounds and interests. While this is a positive development, it causes difficulties for the syllabus and framework of the course.

One approach to a course syllabus is to include the introductory material given in Chapter 1 as an integral part of the course. It is better to interleave the later chapters with the relevant materials from Chapter 1. Such a course might proceed: 1.1–1.6; 2; 3; 4; 1.7; 5; 6; 7; 1.8–1.10; 8; 9. Including the materials of Chapter 1 allows the discussion of important mathematical methods, and addresses the diverse backgrounds of the students. Even if the introductory chapter is covered quickly (e.g., in a one-semester course), this establishes a common base of knowledge for the remainder of the course. If a high-speed approach is taken, it must be emphasized to the students that this material serves only to expose them to concepts that they are unfamiliar with, and to review concepts for those with prior knowledge of the topics covered. Unfortunately, many students are not willing to sit through such an extensive (and intense) introduction.

A second approach begins from Chapter 2 and introduces the material from Chapter 1 only as needed. The chapters that are the most technically difficult, and rely the most on Chapter 1, are Chapters 4 and 5. Thus, for a one-semester course, the subject of protein folding (Chapters 4 and 5) could be skipped. Then much of the introductory material can be omitted, with the exception of a discussion of the last part of Section 1.3, and some introduction to the subject of entropy and information either through thermodynamics (Section 1.3) or information theory (Section 1.8), preferably both. Then Chapters 2 and 3 can be covered first, followed by Chapters 6–9, with selected material introduced from Chapter 1 as is appropriate for the background of the students.

There are two additional recommendations. First, it is better to run this course as a project-based course rather than using graded homework. The varied backgrounds of students make it difficult to select and fairly grade the problems. Projects for individuals or small groups of students can be tailored to their knowledge and interests. There are many new areas of inquiry, so that projects may approach research-level contributions and be exciting for the students. Unfortunately, this means that students may not devote sufficient effort to the study of course material, and rely largely upon exposure in lectures. There is no optimal solution to this problem. Second, if it is possible, a seminar series with lecturers who work in the field should be an integral part of the course. This provides additional exposure to the varied approaches to the study of complex systems that it is not possible for a single lecturer or text to provide.

1

Introduction and Preliminaries

Conceptual Outline

█ 1.1 █ A deceptively simple model of the dynamics of a system is a deterministic iterative map applied to a single real variable. We characterize the dynamics by looking at its limiting behavior and the approach to this limiting behavior. Fixed points that attract or repel the dynamics, and cycles, are conventional limiting behaviors of a simple dynamic system. However, changing a parameter in a quadratic iterative map causes it to undergo a sequence of cycle doublings (bifurcations) until it reaches a regime of chaotic behavior which cannot be characterized in this way. This deterministic chaos reveals the potential importance of the influence of fine-scale details on large-scale behavior in the dynamics of systems.

█ 1.2 █ A system that is subject to complex (external) influences has a dynamics that may be modeled statistically. The statistical treatment simplifies the complex unpredictable stochastic dynamics of a single system, to the simple predictable dynamics of an ensemble of systems subject to all possible influences. A random walk on a line is the prototype stochastic process. Over time, the random influence causes the ensemble of walkers to spread in space and form a Gaussian distribution. When there is a bias in the random walk, the walkers have a constant velocity superimposed on the spreading of the distribution.

█ 1.3 █ While the microscopic dynamics of physical systems is rapid and complex, the macroscopic behavior of many materials is simple, even static. Before we can understand how complex systems have complex behaviors, we must understand why materials can be simple. The origin of simplicity is an averaging over the fast microscopic dynamics on the time scale of macroscopic observations (the ergodic theorem) and an averaging over microscopic spatial variations. The averaging can be performed theoretically using an ensemble representation of the physical system that assumes all microscopic states are realized. Using this as an assumption, a statistical treatment of microscopic states describes the macroscopic equilibrium behavior of systems. The final part of Section 1.3 introduces concepts that play a central role in the rest of the book. It discusses the differences between equilibrium and complex systems. Equilibrium systems are divisible and satisfy the ergodic theorem. Complex systems are composed out of interdependent parts and violate the ergodic theorem. They have many degrees of freedom whose time dependence is very slow on a microscopic scale.

■ 1.4 ■ To understand the separation of time scales between fast and slow degrees of freedom, a two-well system is a useful model. The description of a particle traveling in two wells can be simplified to the dynamics of a two-state (binary variable) system. The fast dynamics of the motion within a well is averaged by assuming that the system visits all states, represented as an ensemble. After taking the average, the dynamics of hopping between the wells is represented explicitly by the dynamics of a binary variable. The hopping rate depends exponentially on the ratio of the energy barrier and the temperature. When the temperature is low enough, the hopping is frozen. Even though the two wells are not in equilibrium with each other, equilibrium continues to hold within a well. The cooling of a two-state system serves as a simple model of a glass transition, where many microscopic degrees of freedom become frozen at the glass transition temperature.

■ 1.5 ■ Cellular automata are a general approach to modeling the dynamics of spatially distributed systems. Expanding the notion of an iterative map of a single variable, the variables that are updated are distributed on a lattice in space. The influence between variables is assumed to rely upon local interactions, and is homogeneous. Space and time are both discretized, and the variables are often simplified to include only a few possible states at each site. Various cellular automata can be designed to model key properties of physical and biological systems.

■ 1.6 ■ The equilibrium state of spatially distributed systems can be modeled by fields that are treated using statistical ensembles. The simplest is the Ising model, which captures the simple cooperative behavior found in magnets and many other systems. Cooperative behavior is a mechanism by which microscopic fast degrees of freedom can become slow collective degrees of freedom that violate the ergodic theorem and are visible macroscopically. Macroscopic phase transitions are the dynamics of the cooperative degrees of freedom. Cooperative behavior of many interacting elements is an important aspect of the behavior of complex systems. This should be contrasted to the two-state model (Section 1.4), where the slow dynamics occurs microscopically.

■ 1.7 ■ Computer simulations of models such as molecular dynamics or cellular automata provide important tools for the study of complex systems. Monte Carlo simulations enable the study of ensemble averages without necessarily describing the dynamics of a system. However, they can also be used to study random-walk dynamics. Minimization methods that use iterative progress to find a local minimum are often an important aspect of computer simulations. Simulated annealing is a method that can help find low energy states on complex energy surfaces.

■ 1.8 ■ We have treated systems using models without acknowledging explicitly that our objective is to describe them. All our efforts are designed to map a system onto a description of the system. For complex systems the description must be quite long, and the study of descriptions becomes essential. With this recognition, we turn to information theory. The information contained in a communication, typically a string of characters, may be defined quantitatively as the logarithm of the number of possible messages. When different messages have distinct probabilities P in an ensemble, then the information can be identified as $-\ln(P)$ and the average information is defined accordingly. Long messages can be modeled using the same concepts as a random walk, and we can use such models to estimate the information contained in human languages such as English.

■ **1.9** ■ In order to understand the relationship of information to systems, we must also understand what we can infer from information that is provided. The theory of logic is concerned with inference. It is directly linked to computation theory, which is concerned with the possible (deterministic) operations that can be performed on a string of characters. All operations on character strings can be constructed out of elementary logical (Boolean) operations on binary variables. Using Turing's model of computation, it is further shown that all computations can be performed by a universal Turing machine, as long as its input character string is suitably constructed. Computation theory is also related to our concern with the dynamics of physical systems because it explores the set of possible outcomes of discrete deterministic dynamic systems.

■ **1.10** ■ We return to issues of structure on microscopic and macroscopic scales by studying fractals that are self-similar geometric objects that embody the concept of progressively increasing structure on finer and finer length scales. A general approach to the scale dependence of system properties is described by scaling theory. The renormalization group methodology enables the study of scaling properties by relating a model of a system on one scale with a model of the system on another scale. Its use is illustrated by application to the Ising model (Section 1.6), and to the bifurcation route to chaos (Section 1.1). Renormalization helps us understand the basic concept of modeling systems, and formalizes the distinction between relevant and irrelevant microscopic parameters. Relevant parameters are the microscopic parameters that can affect the macroscopic behavior. The concept of universality is the notion that a whole class of microscopic models will give rise to the same macroscopic behavior, because many parameters are irrelevant. A conceptually related computational technique, the multigrid method, is based upon representing a problem on multiple scales.

The study of complex systems begins from a set of models that capture aspects of the dynamics of simple or complex systems. These models should be sufficiently general to encompass a wide range of possibilities but have sufficient structure to capture interesting features. An exciting bonus is that even the apparently simple models discussed in this chapter introduce features that are not typically treated in the conventional science of simple systems, but are appropriate introductions to the dynamics of complex systems. Our treatment of dynamics will often consider discrete rather than continuous time. Analytic treatments are often convenient to formulate in continuous variables and differential equations; however, computer simulations are often best formulated in discrete space-time variables with well-defined intervals. Moreover, the assumption of a smooth continuum at small scales is not usually a convenient starting point for the study of complex systems. We are also generally interested not only in one example of a system but rather in a class of systems that differ from each other but share a characteristic structure. The elements of such a class of systems are collectively known as an ensemble. As we introduce and study mathematical models, we should recognize that our primary objective is to represent properties of real systems. We must therefore develop an understanding of the nature of models and modeling, and how they can pertain to either simple or complex systems.

1.1   Iterative Maps (and Chaos)

An iterative map f is a function that evolves the state of a system s in discrete time

$$s(t) = f(s(t - \delta t)) \tag{1.1.1}$$

where $s(t)$ describes the state of the system at time t. For convenience we will generally measure time in units of δt which then has the value 1, and time takes integral values starting from the initial condition at $t = 0$.

Many of the complex systems we will consider in this text are of the form of Eq. (1.1.1), if we allow s to be a general variable of arbitrary dimension. The generality of iterative maps is discussed at the end of this section. We start by considering several examples of iterative maps where s is a single variable. We discuss briefly the binary variable case, $s = \pm 1$. Then we discuss in greater detail two types of maps with s a real variable, $s \in \Re$, linear maps and quadratic maps. The quadratic iterative map is a simple model that can display complex dynamics. We assume that an iterative map may be started at any initial condition allowed by a specified domain of its system variable.

1.1.1 *Binary iterative maps*

There are only a few binary iterative maps. Question 1.1.1 is a complete enumeration of them.*

Question 1.1.1 Enumerate all possible iterative maps where the system is described by a single binary variable, $s = \pm 1$.

Solution 1.1.1 There are only four possibilities:

$$s(t) = 1$$
$$s(t) = -1$$
$$s(t) = s(t - 1) \tag{1.1.2}$$
$$s(t) = -s(t - 1)$$

---

*Questions are an integral part of the text. They are designed to promote independent thought. The reader is encouraged to read the question, contemplate or work out an answer and then read the solution provided in the text. The continuation of the text assumes that solutions to questions have been read.

It is instructive to consider these possibilities in some detail. The main reason there are so few possibilities is that the form of the iterative map we are using depends, at most, on the value of the system in the previous time. The first two examples are constants and don't even depend on the value of the system at the previous time. The third map can only be distinguished from the first two by observation of its behavior when presented with two different initial conditions.

The last of the four maps is the only map that has any sustained dynamics. It cycles between two values in perpetuity. We can think about this as representing an oscillator. ∎

Question 1.1.2

a. In what way can the map $s(t) = -s(t - 1)$ represent a physical oscillator?
b. How can we think of the static map, $s(t) = s(t - 1)$, as an oscillator?
c. Can we do the same for the constant maps $s(t) = 1$ and $s(t) = -1$?

Solution 1.1.2 (*a*) By looking at the oscillator displacement with a strobe at half-cycle intervals, our measured values can be represented by this map. (*b*) By looking at an oscillator with a strobe at cycle intervals. (*c*) You might think we could, by picking a definite starting phase of the strobe with respect to the oscillator. However, the constant map ignores the first value, the oscillator does not. ∎

1.1.2 *Linear iterative maps: free motion, oscillation, decay and growth*

The simplest example of an iterative map with s real, $s \in \Re$, is a constant map:

$$s(t) = s_0 \tag{1.1.3}$$

No matter what the initial value, this system always takes the particular value s_0. The constant map may seem trivial, however it will be useful to compare the constant map with the next class of maps.

A linear iterative map with unit coefficient is a model of free motion or propagation in space:

$$s(t) = s(t - 1) + v \tag{1.1.4}$$

at successive times the values of s are separated by v, which plays the role of the velocity.

Question 1.1.3 Consider the case of zero velocity

$$s(t) = s(t - 1) \tag{1.1.5}$$

How is this different from the constant map?

Solution 1.1.3 The two maps differ in their dependence on the initial value. ∎

Runaway growth or decay is a multiplicative iterative map:

$$s(t) = gs(t - 1) \tag{1.1.6}$$

We can generate the values of this iterative map at all times by using the equivalent expression

$$s(t) = g^t s_0 = e^{\ln(g)t} s_0 \tag{1.1.7}$$

which is exponential growth or decay. The iterative map can be thought of as a sequence of snapshots of Eq. (1.1.7) at integral time. $g = 1$ reduces this map to the previous case.

Question 1.1.4 We have seen the case of free motion, and now jumped to the case of growth. What happened to accelerated motion? Usually we would consider accelerated motion as the next step after motion with a constant velocity. How can we write accelerated motion as an iterative map?

Solution 1.1.4 The description of accelerated motion requires two variables: position and velocity. The iterative map would look like:

$$
\begin{aligned}
x(t) &= x(t - 1) + v(t - 1) \\
v(t) &= v(t - 1) + a
\end{aligned} \tag{1.1.8}
$$

This is a two-variable iterative map. To write this in the notation of Eq. (1.1.1) we would define s as a vector $s(t) = (x(t), v(t))$. ∎

Question 1.1.5 What happens in the rightmost exponential expression in Eq. (1.1.7) when g is negative?

Solution 1.1.5 The logarithm of a negative number results in a phase $i\pi$. The term $i\pi t$ in the exponent alternates sign every time step as one would expect from Eq. (1.1.6). ∎

At this point, it is convenient to introduce two graphical methods for describing an iterative map. The first is the usual way of plotting the value of s as a function of time. This is shown in the left panels of Fig. 1.1.1. The second type of plot, shown in the right panels, has a different purpose. This is a plot of the iterative relation $s(t)$ as a function of $s(t - 1)$. On the same axis we also draw the line for the identity map $s(t) = s(t - 1)$. These two plots enable us to graphically obtain the successive values of s as follows. Pick a starting value of s, which we can call $s(0)$. Mark this value on the abscissa. Mark the point on the graph of $s(t)$ that corresponds to the point whose abscissa is $s(0)$, i.e., the point $(s(0), s(1))$. Draw a horizontal line to intersect the identity map. The intersection point is $(s(1), s(1))$. Draw a vertical line back to the iterative map. This is the point $(s(1), s(2))$. Successive values of $s(t)$ are obtained by iterating this graphical procedure. A few examples are plotted in the right panels of Fig. 1.1.1.

In order to discuss the iterative maps it is helpful to recognize several features of these maps. First, intersection points of the identity map and the iterative map are the fixed points of the iterative map:

$$s_0 = f(s_0) \tag{1.1.9}$$

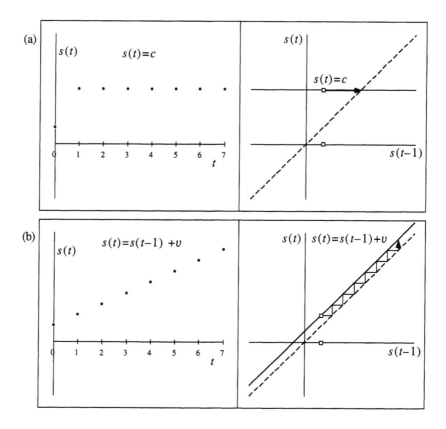

Figure 1.1.1 The left panels show the time-dependent value of the system variable $s(t)$ resulting from iterative maps. The first panel (a) shows the result of iterating the constant map; (b) shows the result of adding v to the previous value during each time interval; (c)–(f) show the result of multiplying by a constant g, where each figure shows the behavior for a different range of g values: (c) $g > 1$, (d) $0 < g < 1$, (e) $-1 < g < 0$, and (f) $g < -1$. The right panels are a different way of showing graphically the results of iterations and are constructed as follows. First plot the function $f(s)$ (solid line), where $s(t) = f(s(t-1))$. This can be thought of as plotting $s(t)$ vs. $s(t-1)$. Second, plot the identity map $s(t) = s(t-1)$ (dashed line). Mark the initial value $s(0)$ on the horizontal axis, and the point on the graph of $s(t)$ that corresponds to the point whose abscissa is $s(0)$, i.e. the point $(s(0), s(1))$. These are shown as squares. From the point $(s(0), s(1))$ draw a horizontal line to intersect the identity map. The intersection point is $(s(1), s(1))$. Draw a vertical line back to the iterative map. This is the point $(s(1), s(2))$. Successive values of $s(t)$ are obtained by iterating this graphical procedure. ∎

Fixed points, not surprisingly, play an important role in iterative maps. They help us describe the state and behavior of the system after many iterations. There are two kinds of fixed points—stable and unstable. Stable fixed points are characterized by "attracting" the result of iteration of points that are nearby. More precisely, there exists

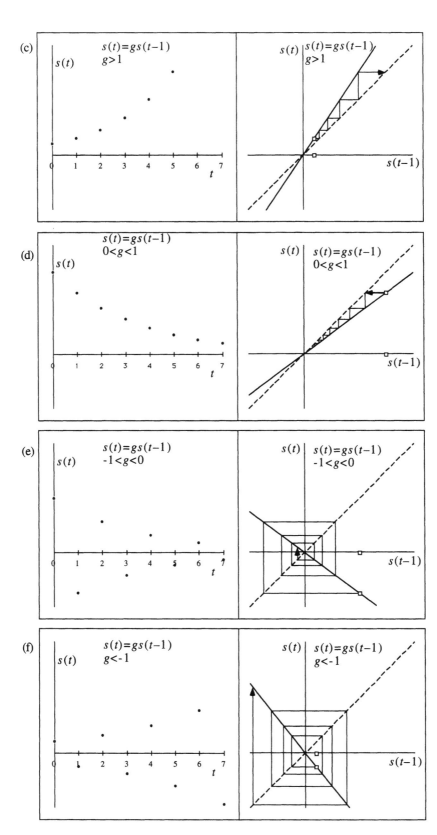

a neighborhood of points of s_0 such that for any s in this neighborhood the sequence of points

$$\{s, f(s), f^2(s), f^3(s), \ldots\} \tag{1.1.10}$$

converges to s_0. We are using the notation $f^2(s) = f(f(s))$ for the second iteration, and similar notation for higher iterations. This sequence is just the time series of the iterative map for the initial condition s. Unstable fixed points have the opposite behavior, in that iteration causes the system to leave the neighborhood of s_0. The two types of fixed points are also called attracting and repelling fixed points.

The family of multiplicative iterative maps in Eq. (1.1.6) all have a fixed point at $s_0 = 0$. Graphically from the figures, or analytically from Eq. (1.1.7), we see that the fixed point is stable for $|g| < 1$ and is unstable for $|g| > 1$. There is also distinct behavior of the system depending on whether g is positive or negative. For $g < 0$ the iterations alternate from one side to the other of the fixed point, whether it is attracted to or repelled from the fixed point. Specifically, if $s < s_0$ then $f(s) > s_0$ and vice versa, or $\text{sign}(s - s_0) = -\text{sign}(f(s) - s_0)$. For $g > 0$ the iteration does not alternate.

Question **1.1.6** Consider the iterative map.

$$s(t) = gs(t - 1) + v \tag{1.1.11}$$

convince yourself that v does not affect the nature of the fixed point, only shifts its position.

Question **1.1.7** Consider an arbitrary iterative map of the form Eq. (1.1.1), with a fixed point s_0 (Eq. (1.1.9)). If the iterative map can be expanded in a Taylor series around s_0 show that the first derivative

$$g = \left. \frac{df(s)}{ds} \right|_{s_0} \tag{1.1.12}$$

characterizes the fixed point as follows:

For $|g| < 1$, s_0 is an attracting fixed point.

For $|g| > 1$, s_0 is a repelling fixed point.

For $g < 0$, iterations alternate sides in a sufficiently small neighborhood of s_0.

For $g > 0$, iterations remain on one side in a sufficiently small neighborhood of s_0.

Extra credit: Prove the same theorem for a differentiable function (no Taylor expansion needed) using the mean value theorem.

Solution 1.1.7 If the iterative map can be expanded in a Taylor series we write that

$$f(s) = f(s_0) + g(s - s_0) + h(s - s_0)^2 + \ldots \tag{1.1.13}$$

where g is the first derivative at s_0, and h is one-half of the second derivative at s_0. Since s_0 is a fixed point $f(s_0) = s_0$ we can rewrite this as:

$$\frac{f(s) - s_0}{s - s_0} = g + h(s - s_0) + \ldots \qquad (1.1.14)$$

If we did not have any higher-order terms beyond g, then by inspection each of the four conditions that we have to prove would follow from this expression without restrictions on s. For example, if $|g| > 1$, then taking the magnitude of both sides shows that $f(s) - s_0$ is larger than $s - s_0$ and the iterations take the point s away from s_0. If $g > 0$, then this expression says that $f(s)$ stays on the same side of s_0. The other conditions follow similarly.

To generalize this argument to include the higher-order terms of the expansion, we must guarantee that whichever domain g is in ($g > 1$, $0 < g < 1$, $-1 < g < 0$, or $g < -1$), the same is also true of the whole right side. For a Taylor expansion, by choosing a small enough neighborhood $|s - s_0| < \delta$, we can guarantee the higher-order terms are less than any number ε we choose. We choose ε to be half of the minimum of $|g - 1|$, $|g - 0|$ and $|g + 1|$. Then $g + \varepsilon$ is in the same domain as g. This provides the desired guarantee and the proof is complete.

We have proven that in the vicinity of a fixed point the iterative map may be completely characterized by its first-order expansion (with the exception of the special points $g = \pm 1, 0$). ∎

Thus far we have not considered the special cases $g = \pm 1, 0$. The special cases $g = 0$ and $g = 1$ have already been treated as simpler iterative maps. When $g = 0$, the fixed point at $s = 0$ is so attractive that it is the result of any iteration. When $g = 1$ all points are fixed points.

The new special case $g = -1$ has a different significance. In this case all points alternate between positive and negative values, repeating every other iteration. Such repetition is a generalization of the fixed point. Whereas in the fixed-point case we repeat every iteration, here we repeat after every two iterations. This is called a 2-cycle, and we can immediately consider the more general case of an n-cycle. In this terminology a fixed point is a 1-cycle. One way to describe an n-cycle is to say that iterating n times gives back the same result, or equivalently, that a new iterative map which is the nth fold composition of the original map $h = f^n$ has a fixed point. This description would include also fixed points of f and all points that are m-cycles, where m is a divisor of n. These are excluded from the definition of the n-cycles. While we have introduced cycles using a map where all points are 2-cycles, more general iterative maps have specific sets of points that are n-cycles. The set of points of an n-cycle is called an orbit. There are a variety of properties of fixed points and cycles that can be proven for an arbitrary map. One of these is discussed in Question 1.1.8.

Question 1.1.8 Prove that there is a fixed point between any two points of a 2-cycle if the iterating function f is continuous.

Solution 1.1.8 Let the 2-cycle be written as

$$s_2 = f(s_1)$$
$$s_1 = f(s_2)$$

(1.1.15)

Consider the function $h(s) = f(s) - s$, $h(s_1)$ and $h(s_2)$ have opposite signs and therefore there must be an s_0 between s_1 and s_2 such that $h(s_0) = 0$—the fixed point. ∎

We can also generalize the definition of attracting and repelling fixed points to consider attracting and repelling n-cycles. Attraction and repulsion for the cycle is equivalent to the attraction and repulsion of the fixed point of f^n.

1.1.3 *Quadratic iterative maps: cycles and chaos*

The next iterative map we will consider describes the effect of nonlinearity (self-action):

$$s(t) = as(t-1)(1 - s(t-1))$$

(1.1.16)

or equivalently

$$f(s) = as(1 - s)$$

(1.1.17)

This map has played a significant role in development of the theory of dynamical systems because even though it looks quite innocent, it has a dynamical behavior that is not described in the conventional science of simple systems. Instead, Eq. (1.1.16) is the basis of significant work on chaotic behavior, and the transition of behavior from simple to chaotic. We have chosen this form of quadratic map because it simplifies somewhat the discussion. Question 1.1.11 describes the relationship between this family of quadratic maps, parameterized by a, and what might otherwise appear to be a different family of quadratic maps.

We will focus on a values in the range $4 > a > 0$. For this range, any value of s in the interval $s \in [0,1]$ stays within this interval. The minimum value $f(s) = 0$ occurs for $s = 0, 1$ and the maximal value occurs for $s = 1/2$. For all values of a there is a fixed point at $s = 0$ and there can be at most two fixed points, since a quadratic can only intersect a line (Eq. (1.1.9)) in two points.

Taking the first derivative of the iterative map gives

$$\frac{df}{ds} = a(1 - 2s)$$

(1.1.18)

At $s = 0$ the derivative is a which, by Question 1.1.7, shows that $s = 0$ is a stable fixed point for $a < 1$ and an unstable fixed point for $a > 1$. The switching of the stability of the fixed point at $s = 0$ coincides with the introduction of a second fixed point in the interval $[0,1]$ (when the slope at $s = 0$ is greater than one, $f(s) > s$ for small s, and since

$f(1) = 0$, we have that $f(s_1) = s_1$ for some s_1 in $[0,1]$ by the same construction as in Question 1.1.8). We find s_1 by solving the equation

$$s_1 = as_1(1-s_1) \tag{1.1.19}$$

$$s_1 = (a-1)/a \tag{1.1.20}$$

Substituting this into Eq. (1.1.18) gives

$$\left.\frac{df}{ds}\right|_{s_1} = 2-a \tag{1.1.21}$$

This shows that for $1 < a < 3$, the new fixed point is stable by Question 1.1.7. Moreover, the derivative is positive for $1 < a < 2$, so s_1 is stable and convergence is from one side. The derivative is negative for $2 < a < 3$, so s_1 is stable and alternating.

Fig. 1.1.2(a)–(c) shows the three cases: $a = 0.5$, $a = 1.5$ and $a = 2.8$. For $a = 0.5$, starting from anywhere within $[0,1]$ leads to convergence to $s = 0$. When $s(0) > 0.5$ the first iteration takes the system to $s(1) < 0.5$. The closer we start to $s(0) = 1$ the closer to $s = 0$ we get in the first jump. At $s(0) = 1$ the convergence to 0 occurs in the first jump. A similar behavior would be found for any value of $0 < a < 1$. For $a = 1.5$ the behavior is more complicated. Except for the points $s = 0,1$, the convergence is always to the fixed point $s_1 = (a-1)/a$ between 0 and 1. For $a = 2.8$ the iterations converge to the same point; however, the convergence is alternating. Because there can be at most two fixed points for the quadratic map, one might think that this behavior would be all that would happen for $1 < a < 4$. One would be wrong. The first indication that this is not the case is the instability of the fixed point at s_1 starting from $a = 3$.

What happens for $a > 3$? Both of the fixed points that we have found, and the only ones that can exist for the quadratic map, are now unstable. We know that the iteration of the map has to go somewhere, and only within $[0,1]$. The only possibility, within our experience, is that there is an attracting n-cycle to which the fixed points are unstable. Let us then consider the map $f^2(s)$ whose fixed points are 2-cycles of the original map. $f^2(s)$ is shown in the right panels of Fig. 1.1.2 for increasing values of a. The fixed points of $f(s)$ are also fixed points of $f^2(s)$. However, we see that two additional fixed points exist for $a > 3$. We can also show analytically that two fixed points are introduced at exactly $a = 3$:

$$f^2(s) = a^2 s(1-s)(1-as(1-s)) \tag{1.1.22}$$

To find the fixed point we solve:

$$s = a^2 s(1-s)(1-as(1-s)) \tag{1.1.23}$$

We already know two solutions of this quartic equation—the fixed points of the map f. One of these at $s = 0$ is obvious. Dividing by s we have a cubic equation:

$$a^3 s^3 - 2a^3 s^2 + a^2(1+a)s + (1-a^2) = 0 \tag{1.1.24}$$

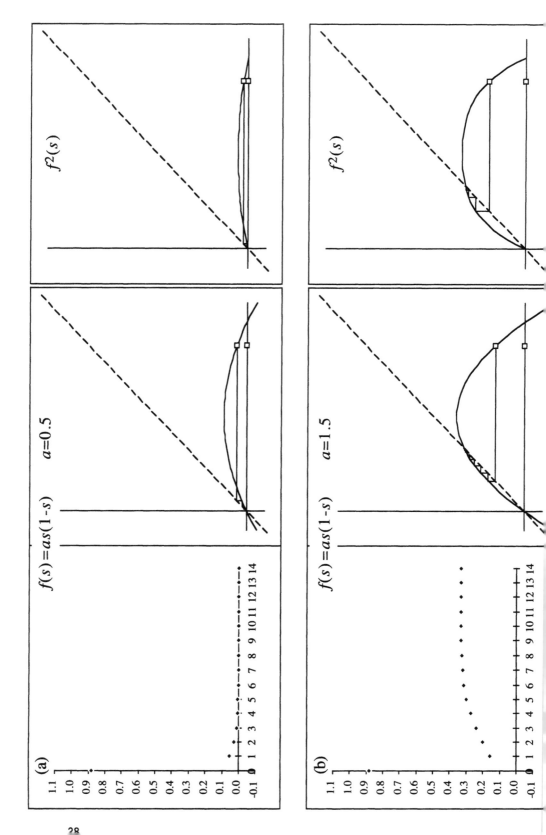

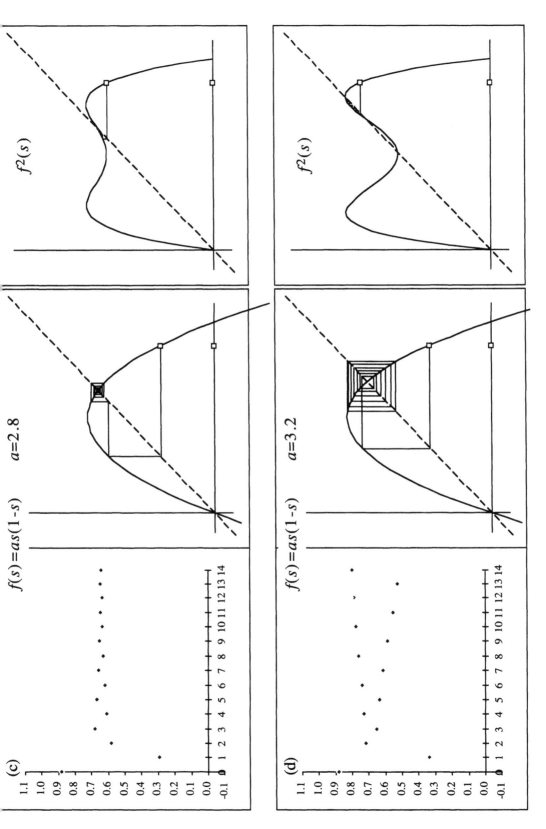

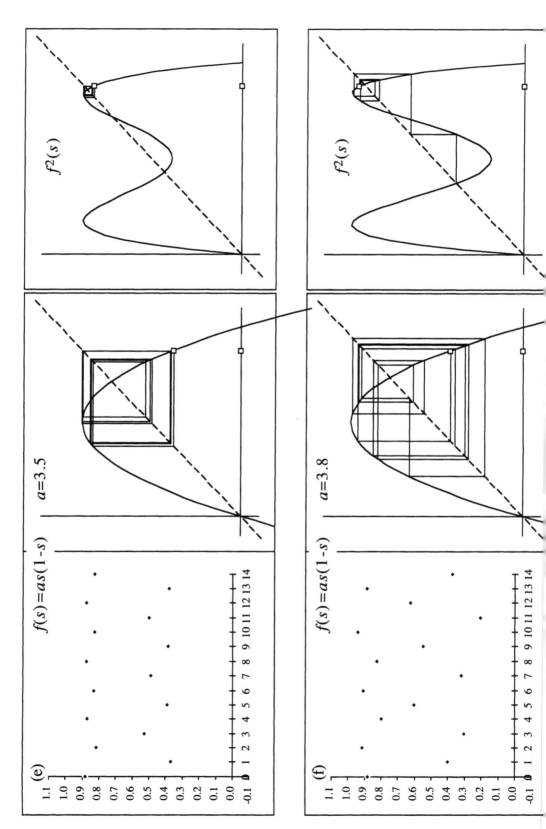

Figure 1.1.2 (pp. 28-30) Plots of the result of iterating the quadratic map $f(s) = as(1 - s)$ for different values of a. The left and center panels are similar to the left and right panels of Fig. 1.1.1. The left panels plot $s(t)$. The center panels describe the iteration of the map $f(s)$ on axes corresponding to $s(t)$ and $s(t - 1)$. The right panels are similar to the center panels but are for the function $f^2(s)$. The different values of a are indicated on the panels and show the changes from (a) convergence to $s = 0$ for $a = 0.5$, (b) convergence to $s = (a - 1)/a$ for $a = 1.5$, (c) alternating convergence to $s = (a - 1)/a$ for $a = 2.8$, (d) bifurcation — convergence to a 2-cycle for $a = 3.2$, (e) second bifurcation — convergence to a 4-cycle for $a = 3.5$, (f) chaotic behavior for $a = 3.8$. ∎

We can reduce the equation to a quadratic by dividing by $(s - s_1)$ as follows (we simplify the algebra by dividing by $a(s - s_1) = (as - (a - 1))$):

$$
\begin{array}{r}
a^2 s^2 - a(a+1)s + (a+1) \\[4pt]
\hline
(as - (a-1)) \overline{\big)\, a^3 s^3 - 2a^3 s^2 + a^2(1+a)s + (1-a^2)} \\
a^3 s^3 - (a-1)a^2 s^2 \\[2pt]
\hline
-(a+1)a^2 s^2 + a^2(1+a)s + (1-a^2) \\
-(a+1)a^2 s^2 + a(1+a)(a-1)s \\[2pt]
\hline
+ a(1+a)s + (1-a^2)
\end{array}
\tag{1.1.25}
$$

Now we can obtain the roots to the quadratic:

$$
a^2 s^2 - a(a+1)s + (a+1) = 0
\tag{1.1.26}
$$

$$
s_2 = \frac{(a+1) \pm \sqrt{(a+1)(a-3)}}{2a}
\tag{1.1.27}
$$

This has two solutions (as it must for a 2-cycle) for $a < -1$ or for $a > 3$. The former case is not of interest to us since we have assumed $0 < a < 4$. The latter case is the two roots that are promised. Notice that for exactly $a = 3$ the two roots that are the new 2-cycle are the same as the fixed point we have already found s_1. The 2-cycle splits off from the fixed point at $a = 3$ when the fixed point becomes unstable. The two attracting points continue to separate as a increases. For $a > 3$ we expect that the result of iteration eventually settles down to the 2-cycle. The system state alternates between the two roots Eq. (1.1.27). This is shown in Fig. 1.1.2(d).

As we continue to increase a beyond 3, the 2-cycle will itself become unstable at a point that can be calculated by setting

$$
\left. \frac{df^2}{ds} \right|_{s_2} = -1
\tag{1.1.28}
$$

to be $a = 1 + \sqrt{6} = 3.44949$. At this value of a the 2-cycle splits into a 4-cycle (Fig. 1.1.2(e)). Each of the fixed points of $f^2(s)$ simultaneously split into 2-cycles that together form a 4-cycle for the original map.

Question 1.1.9 Show that when f has a 2-cycle, both of the fixed points of f^2 must split simultaneously.

Solution 1.1.9 The split occurs when the fixed points become unstable—the derivative of f^2 equals -1. We can show that the derivative is equal at the two fixed points of Eq. (1.1.27), which we call $s_2^{\pm}$:

$$\left.\frac{df^2}{ds}\right|_{s_2} = \left.\frac{df(f(s))}{ds}\right|_{s_2} = \left.\frac{df(s)}{ds}\right|_{f(s_2)} \left.\frac{df(s)}{ds}\right|_{s_2} \qquad (1.1.29)$$

where we have made use of the chain rule. Since $f(s_2^+) = s_2^-$ and vice versa, we have shown this expression is the same whether $s_2 = s_2^+$ or $s_2 = s_2^-$.

Note: This can be generalized to show that the derivative of f^k is the same at all of its k fixed points corresponding to a k-cycle of f. ■

The process of taking an n-cycle into a $2n$-cycle is called bifurcation. Bifurcation continues to replace the limiting behavior of the iterative map with progressively longer cycles of length 2^k. The bifurcations can be simulated. They occur at smaller and smaller intervals and there is a limit point to the bifurcations at $a_c = 3.56994567$. Fig. 1.1.3 shows the values that are reached by the iterative map at long times—the stable cycles—as a function of $a < a_c$. We will discuss an algebraic treatment of the bifurcation regime in Section 1.10.

Beyond the bifurcation regime $a > a_c$ (Fig. 1.1.2(f)) the behavior of the iterative map can no longer be described using simple cycles that attract the iterations. The behavior in this regime has been identified with chaos. Chaos has been characterized in many ways, but one property is quite generally agreed upon—the inherent lack of predictability of the system dynamics. This is often expressed more precisely by describing the sensitivity of the system's fate to the initial conditions. A possible definition is: There exists a distance d such that for any neighborhood V of any point s it is possible to find a point s' within the neighborhood and a number of iterations k so that $f^k(s')$ is further than d away from $f^k(s)$. This means that arbitrarily close to any point is a point that will be displaced a significant distance away by iteration. Qualitatively, there are two missing aspects of this definition, first that the points that move far away must not be too unlikely (otherwise the system is essentially predictable) and second that d is not too small (in which case the divergence of the dynamics may not be significant).

If we look at the definition of chaotic behavior, we see that the concept of scale plays an important role. A small distance between s and s' turns into a large distance between $f^k(s)$ and $f^k(s')$. Thus a fine-scale difference eventually becomes a large-scale difference. This is the essence of chaos as a model of complex system behavior. To understand it more fully, we can think about the state variable s not as one real variable,

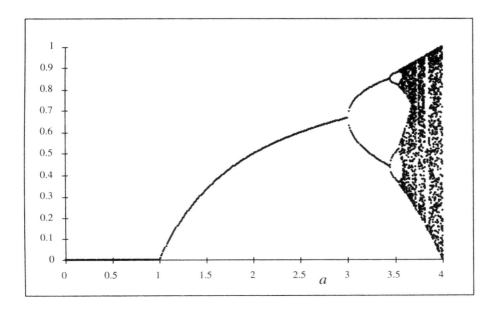

Figure 1.1.3 A plot of values of s visited by the quadratic map $f(s) = as(1 - s)$ after many iterations as a function of a, including stable points, cycles and chaotic behavior. The different regimes are readily apparent. For $a < 1$ the stable point is $s = 0$. For $1 < a < 3$ the stable point is at $s_0 = (a - 1)/a$. For $3 < a < a_c$ with $a_c = 3.56994567$, there is a bifurcation cascade with 2-cycles then 4-cycles, etc. 2^k-cycles for all values of k appear in progressively narrower regions of a. Beyond 4-cycles they cannot be seen in this plot. For $a > a_c$ there is chaotic behavior. There are regions of s values that are not visited and regions that are visited in the long time behavior of the quadratic map in the chaotic regime which this figure does not fully illustrate. ∎

but as an infinite sequence of binary variables that form its binary representation $s = 0.r_1 r_2 r_3 r_4 ...$ Each of these binary variables represents the state of the system—the value of some quantity we can measure about the system—on a particular length scale. The higher order bits represent the larger scales and the lower order ones represent the finer scales. Chaotic behavior implies that the state of the first few binary variables, $r_1 r_2$, at a particular time are determined by the value of fine scale variables at an earlier time. The farther back in time we look, the finer scale variables we have to consider in order to know the present values of $r_1 r_2$. Because many different variables are relevant to the behavior of the system, we say that the system has a complex behavior. We will return to these issues in Chapter 8.

The influence of fine length scales on coarse ones makes iterative maps difficult to simulate by computer. Computer representations of real numbers always have finite precision. This must be taken into account if simulations of iterative maps or chaotic complex systems are performed.

Another significant point about the iterative map as a model of a complex system is that there is nothing outside of the system that is influencing it. All of the information we need to describe the behavior is contained in the precise value of s. The complex behavior arises from the way the different parts of the system—the fine and course scales—affect each other.

Question 1.1.10: Why isn't the iterative map in the chaotic regime equivalent to picking a number at random?

Solution 1.1.10: We can still predict the behavior of the iterative map over a few iterations. It is only when we iterate long enough that the map becomes unpredictable. More specifically, the continuity of the function $f(s)$ guarantees that for s and s' close together $f(s)$ and $f(s')$ will also be close together. Specifically, given an ε it is possible to find a δ such that for $|s - s'| < \delta$, $|f(s) - f(s')| < \varepsilon$. For the family of functions we have been considering, we only need to set $\delta < \varepsilon/a$, since then we have:

$$|f(s) - f(s')| = a|s(1-s) - s'(1-s')| = a|s-s'||1-(s+s')| < a|s-s'| < \varepsilon$$

$$(1.1.30)$$

Thus if we fix the number of cycles to be k, we can always find two points close enough so that $|f^k(s') - f^k(s)| < \varepsilon$ by setting $|s - s'| < \varepsilon/a^k$. ∎

The tuning of the parameter a leading from simple convergent behavior through cycle bifurcation to chaos has been identified as a universal description of the appearance of chaotic behavior from simple behavior of many systems. How do we take a complicated real system and map it onto a discrete time iterative map? We must define a system variable and then take snapshots of it at fixed intervals (or at least well-defined intervals). The snapshots correspond to an iterative map. Often there is a natural choice for the interval that simplifies the iterative behavior. We can then check to see if there is bifurcation and chaos in the real system when parameters that control the system behavior are varied.

One of the earliest examples of the application of iterative maps is to the study of heart attacks. Heart attacks occur in many different ways. One kind of heart attack is known as fibrillation. Fibrillation is characterized by chaotic and ineffective heart muscle contractions. It has been suggested that bifurcation may be observed in heartbeats as a period doubling (two heartbeats that are inequivalent). If correct, this may serve as a warning that the heart structure, due to various changes in heart tissue parameters, may be approaching fibrillation. Another system where more detailed studies have suggested that bifurcation occurs as a route to chaotic behavior is that of turbulent flows in hydrodynamic systems. A subtlety in the application of the ideas of bifurcation and chaos to physical systems is that physical systems are better modeled as having an increasing number of degrees of freedom at finer scales. This is to be contrasted with a system modeled by a single real number, which has the same number of degrees of freedom (represented by the binary variables above) at each length scale.

1.1.4 *Are all dynamical systems iterative maps?*

How general is the iterative map as a tool for describing the dynamics of systems? There are three apparent limitations of iterative maps that we will consider modifying later, Eq. (1.1.1):

a. describes the homogeneous evolution of a system since f itself does not depend on time,

b. describes a system where the state of the system at time t depends only on the state of the system at time $t - \delta t$, and

c. describes a deterministic evolution of a system.

We can, however, bypass these limitations and keep the same form of the iterative map if we are willing to let s describe not just the present state of the system but also

a. the state of the system and all other factors that might affect its evolution in time,

b. the state of the system at the present time and sufficiently many previous times, and

c. the probability that the system is in a particular state.

Taking these caveats together, all of the systems we will consider are iterative maps, which therefore appear to be quite general. Generality, however, can be quite useless, since we want to discard as much information as possible when describing a system.

Another way to argue the generality of the iterative map is through the laws of classical or quantum dynamics. If we consider s to be a variable that describes the positions and velocities of all particles in a system, all closed systems described by classical mechanics can be described as deterministic iterative maps. Quantum evolution of a closed system may also be described by an iterative map if s describes the wave function of the system. However, our intent is not necessarily to describe microscopic dynamics, but rather the dynamics of variables that we consider to be relevant in describing a system. In this case we are not always guaranteed that a deterministic iterative map is sufficient. We will discuss relevant generalizations, first to stochastic maps, in Section 1.2.

Extra Credit Question 1.1.11 Show that the system of quadratic iterative maps

$$s(t) = s(t-1)^2 + k \qquad (1.1.31)$$

is essentially equivalent in its dynamical properties to the iterative maps we have considered in Eq. (1.1.16).

Solution 1.1.11 Two iterative maps are equivalent in their properties if we can perform a time-independent one-to-one map of the time-dependent system states from one case to the other. We will attempt to transform the family of quadratic maps given in this problem to the one of Eq. (1.1.16) using a linear map valid at all times

$$s(t) = ms'(t) + b \qquad (1.1.32)$$

By direct substitution this leads to:

$$ms'(t)+b=(ms'(t-1)+b)^2 +k \qquad (1.1.33)$$

We must now choose the values of m and b so as to obtain the form of Eq. (1.1.16).

$$s'(t)=ms'(t-1)(s'(t-1)+\frac{2b}{m})+\frac{1}{m}(b^2 +k-b) \qquad (1.1.34)$$

For a correct placement of minus signs in the parenthesis we need:

$$s'(t)=(-m)s'(t-1)(\left(-\frac{2b}{m}\right)-s'(t-1))+\frac{1}{m}(b^2 +k-b) \qquad (1.1.35)$$

or

$$b^2 -b+k=0 \qquad (1.1.36)$$

$$\frac{2b}{m}=-1 \qquad (1.1.37)$$

giving

$$b=(1\pm\sqrt{1-4k})/2 \qquad (1.1.38)$$

$$a=-m=2b=(1\pm\sqrt{1-4k}) \qquad (1.1.39)$$

We see that for $k< 1/4$ we have two solutions. These solutions give all possible (positive and negative) values of a.

What about $k > 1/4$? It turns out that this case is not very interesting compared to the rich behavior for $k < 1/4$, since there are no finite fixed points, and therefore by Question 1.1.8 no 2-cycles (it is not hard to generalize this to n-cycles). To confirm this, verify that iterations diverge to $+\infty$ from any initial condition.

Note: The system of equations of this question are the ones extensively analyzed by Devaney in his excellent textbook *A First Course in Chaotic Dynamical Systems.* ∎

Extra Credit Question 1.1.12 You are given a problem to solve which when reduced to mathematical form looks like

$$s= f_c(s) \qquad (1.1.40)$$

where f is a complicated function that depends on a parameter c. You know that there is a solution of this equation in the vicinity of s_0. To solve this equation you try to iterate it (Newton's method) and it works, since you find that $f^k(s_0)$ converges nicely to a solution. Now, however, you realize that you need to solve this problem for a slightly different value of the parameter c, and when you try to iterate the equation you can't get the value of s to converge. Instead the values start to oscillate and then behave in a completely erratic way. Suggest a solution for this problem and see if it works for the function $f_c(s) = cs(1 - s)$, $c = 3.8$, $s_0 = 0.5$. A solution is given in stages (a) - (c) below.

Solution 1.1.12(a) A common resolution of this problem is to consider iterating the function:

$$h_c(s) = \alpha s + (1 - \alpha)f_c(s) \qquad (1.1.41)$$

where we can adjust α to obtain rapid convergence. Note that solutions of

$$s = h_c(s) \qquad (1.1.42)$$

are the same as solutions of the original problem.

Question 1.1.12(b) Explain why this could work.

Solution 1.1.12(b) The derivative of this function at a fixed point can be controlled by the value of α. It is a linear interpolation between the fixed point derivative of f_c and 1. If the fixed point is unstable and oscillating, the derivative of f_c must be less than -1 and the interpolation should help.

We can also explain this result without appealing to our work on iterative maps by noting that if the iteration is causing us to overshoot the mark, it makes sense to mix the value s we start from with the value we get from $f_c(s)$ to get a better estimate.

Question 1.1.12(c) Explain how to pick α.

Solution 1.1.12(c) If the solution is oscillating, then it makes sense to assume that the fixed point is in between successive values and the distance is revealed by how much further it gets each time; i.e., we assume that the iteration is essentially a linear map near the fixed point and we adjust α so that we compensate exactly for the overshoot of f_c.

Using two trial iterations, a linear approximation to f_c at s_0 looks like:

$$\begin{aligned} s_2 &= f_c(s_1) \approx g(s_1 - s_0) + s_0 \\ s_3 &= f_c(s_2) \approx g(s_2 - s_0) + s_0 \end{aligned} \qquad (1.1.43)$$

Adopting the linear approximation as a definition of g we have:

$$g \equiv (s_3 - s_2)/(s_2 - s_1) \qquad (1.1.44)$$

Set up α so that the first iteration of the modified system will take you to the desired answer:

$$s_0 = \alpha s_1 + (1 - \alpha)f_c(s_1) \qquad (1.1.45)$$

or

$$s_0 - s_1 = (1 - \alpha)(f_c(s_1) - s_1) = (1 - \alpha)(s_2 - s_1) \qquad (1.1.46)$$

$$(1 - \alpha) = (s_0 - s_1)/(s_2 - s_1) \qquad (1.1.47)$$

To eliminate the unknown s_0 we use Eq. (1.1.43) to obtain:

$$(s_2 - s_1) = g(s_1 - s_0) + (s_0 - s_1) \tag{1.1.48}$$

$$(s_0 - s_1) = (s_2 - s_1)/(1-g) \tag{1.1.49}$$

or

$$1 - \alpha = 1/(1-g) \tag{1.1.50}$$

$$\alpha = -g/(1-g) = (s_2 - s_3)/(2s_2 - s_1 - s_3) \tag{1.1.51}$$

It is easy to check, using the formula in terms of g, that the modified iteration has a zero derivative at s_0 when we use the approximate linear forms for f_c. This means we have the best convergence possible using the information from two iterations of f_c. We then use the value of α to iterate to convergence. Try it! ∎

1.2   Stochastic Iterative Maps

Many of the systems we would like to consider are described by system variables whose value at the next time step we cannot predict with complete certainty. The uncertainty may arise from many sources, including the existence of interactions and parameters that are too complicated or not very relevant to our problem. We are then faced with describing a system in which the outcome of an iteration is probabilistic and not deterministic. Such systems are called stochastic systems. There are several ways to describe such systems mathematically. One of them is to consider the outcome of a particular update to be selected from a set of possible values. The probability of each of the possible values must be specified. This description is not really a model of a single system, because each realization of the system will do something different. Instead, this is a model of a collection of systems—an ensemble. Our task is to study the properties of this ensemble.

A stochastic system is generally described by the time evolution of random variables. We begin the discussion by defining a random variable. A random variable s is defined by its probability distribution $P_s(s')$, which describes the likelihood that s has the value s'. If s is a continuous variable, then $P_s(s')ds'$ is the probability that s resides between s' and $s' + ds'$. Note that the subscript is the variable name rather than an index. For example, s might be a binary variable that can have the value $+1$ or -1. $P_s(1)$ is the probability that $s = 1$ and $P_s(-1)$ is the probability that $s = -1$. If s is the outcome of an unbiased coin toss, with heads called 1 and tails called -1, both of these values are $1/2$. When no confusion can arise, the notation $P_s(s')$ is abbreviated to $P(s)$, where s may be either the variable or the value. The sum over all possible values of the probability must be 1.

$$\sum_{s'} P_s(s') = 1 \tag{1.2.1}$$

In the discussion of a system described by random variables, we often would like to know the average value of some quantity $Q(s)$ that depends in a definite way on the value of the stochastic variable s. This average is given by:

$$<Q(s)> = \sum_{s'} P_s(s') Q(s') \tag{1.2.2}$$

Note that the average is a linear operation.

We now consider the case of a time-dependent random variable. Rather than describing the time dependence of the variable $s(t)$, we describe the time dependence of the probability distribution $P_s(s';t)$. Similar to the iterative map, we can consider the case where the outcome only depends on the value of the system variable at a previous time, and the transition probabilities do not depend explicitly on time. Such systems are called Markov chains. The transition probabilities from a state at a particular time to the next discrete time are written:

$$P_s(s'(t) | s'(t-1)) \tag{1.2.3}$$

P_s is used as the notation for the transition probability, since it is also the probability distribution of s at time t, given a particular value $s'(t-1)$ at the previous time. The use of a time index for the arguments illustrates the use of the transition probability. $P_s(1|1)$ is the probability that when $s=1$ at time $t-1$ then $s=1$ at time t. $P_s(-1|1)$ is the probability that when $s=1$ at time $t-1$ then $s=-1$ at time t. The transition probabilities, along with the initial probability distribution of the system $P_s(s'; t=0)$, determine the time-dependent ensemble that we are interested in. Assuming that we don't lose systems on the way, the transition probabilities of Eq. (1.2.3) must satisfy:

$$\sum_{s''} P_s(s'' | s') = 1 \tag{1.2.4}$$

This states that no matter what the value of the system variable is at a particular time, it must reach some value at the next time.

The stochastic system described by transition probabilities can be written as an iterative map on the probability distribution $P(s)$

$$P_s(s';t) = \sum_{s''} P_s(s' | s'') P_s(s''; t-1) \tag{1.2.5}$$

It may be more intuitive to write this using the notation

$$P_s(s'(t);t) = \sum_{s'(t-1)} P_s(s'(t) | s'(t-1)) P_s(s'(t-1); t-1) \tag{1.2.6}$$

in which case it may be sufficient, though hazardous, to write the abbreviated form

$$P(s(t)) = \sum_{s(t-1)} P(s(t) | s(t-1)) P(s(t-1)) \tag{1.2.7}$$

It is important to recognize that the time evolution equation for the probability is linear. The linear evolution of this system (Eq. (1.2.5)) guarantees that superposition applies. If we start with an initial distribution $P(s;0) = \frac{1}{2}P^1(s;0) + \frac{1}{2}P^2(s;0)$ at time $t = 0$, then we could find the result at time t by separately looking at the evolution of each of the probabilities $P^1(s;0)$ and $P^2(s;0)$. Explicitly we can write $P(s;t) = \frac{1}{2}P^1(s;t) + \frac{1}{2}P^2(s;t)$. The meaning of this equation should be well noted. The right side of the equation is the sum of the evolved probabilities $P^1(s;t)$ and $P^2(s;t)$. This linearity is a direct consequence of the independence of different members of the ensemble and says nothing about the complexity of the dynamics.

We note that ultimately we are interested in the behavior of a particular system $s(t)$ that only has one value of s at every time t. The ensemble describes how many such systems will behave. Analytically it is easier to describe the ensemble as a whole, however, simulations may also be used to observe the behavior of a single system.

1.2.1 *Random walk*

Stochastic systems with only one binary variable might seem to be trivial, but we will devote quite a bit of attention to this problem. We begin by considering the simplest possible binary stochastic system. This is the system which corresponds to a coin toss. Ideally, for each toss there is equal probability of heads ($s = +1$) or tails ($s = -1$), and there is no memory from one toss to the next. The ensemble at each time is independent of time and has an equal probability of ± 1:

$$P(s;t) = \frac{1}{2}\delta_{s,1} + \frac{1}{2}\delta_{s,-1} \tag{1.2.8}$$

where the discrete delta function is defined by

$$\delta_{i,j} = \begin{cases} 1 & i = j \\ 0 & i \neq j \end{cases} \tag{1.2.9}$$

Since Eq. (1.2.8) is independent of what happens at all previous times, the evolution of the state variable is given by the same expression

$$P(s'|s) = \frac{1}{2}\delta_{s',1} + \frac{1}{2}\delta_{s',-1} \tag{1.2.10}$$

We can illustrate the evaluation of the average of a function of s at time t:

$$<Q(s)>_t = \sum_{s'=\pm 1} Q(s')P_s(s';t) = \sum_{s'=\pm 1} Q(s')\left(\frac{1}{2}\delta_{s',1} + \frac{1}{2}\delta_{s',-1}\right) = \frac{1}{2}\sum_{s'=\pm 1} Q(s') \tag{1.2.11}$$

For example, if we just take $Q(s)$ to be s itself we have the average of the system variable:

$$<s>_t = \frac{1}{2}\sum_{s'=\pm 1} s' = 0 \tag{1.2.12}$$

Question 1.2.1 Will you win more fair coin tosses if (*a*) you pick heads every time, or if (*b*) you alternate heads and tails, or if (*c*) you pick heads or tails at random or if (*d*) you pick heads and tails by some other system? Explain why.

Solution 1.2.1 In general, we cannot predict the number of coin tosses that will be won, we can only estimate it based on the chance of winning. Assuming a fair coin means that this is the best that can be done. Any of the possibilities (*a*)–(*c*) give the same chance of winning. In none of these ways of gambling does the choice you make correlate with the result of the coin toss. The only system (*d*) that can help is if you have some information about what the result of the toss will be, like betting on the known result *after* the coin is tossed. A way to write this formally is to write the probability distribution of the choice that you are making. This choice is also a stochastic process. Calling the choice $c(t)$, the four possibilities mentioned are:

(*a*) $\quad P(c;t) = \delta_{c,1}$ (1.2.13)

(*b*) $\quad P(c;t) = \frac{1+(-1)^t}{2}\delta_{c,1} + \frac{1-(-1)^t}{2}\delta_{c,-1} = \text{mod}_2(t)\delta_{c,1} + \text{mod}_2(t+1)\delta_{c,-1}$ (1.2.14)

(*c*) $\quad P(c;t) = \frac{1}{2}\delta_{c,1} + \frac{1}{2}\delta_{c,-1}$ (1.2.15)

(*d*) $\quad P(c;t) = \delta_{c,s(t)}$ (1.2.16)

It is sufficient to show that the average probability of winning is the same in each of (*a*)–(*c*) and is just 1/2. We follow through the manipulations in order to illustrate some concepts in the treatment of more than one stochastic variable. We have to sum over the probabilities of each of the possible values of the coin toss and each of the values of the choices, adding up the probability that they coincide at a particular time t:

$$<\delta_{c,s}> = \sum_{s'}\sum_{c'}\delta_{c',s'}P_s(s';t)P_c(c',t)$$ (1.2.17)

This expression assumes that the values of the coin toss and the value of the choice are independent, so that the joint probability of having a particular value of s and a particular value of c is the product of the probabilities of each of the variables independently:

$$P_{s,c}(s',c';t) = P_s(s';t)P_c(c';t)$$ (1.2.18)

—the probabilities-of-independent-variables factor. This is valid in cases (*a*)–(*c*) and not in case (*d*), where the probability of c occurring is explicitly a function of the value of s.

We evaluate the probability of winning in each case (*a*) through (*c*) using

$$< \delta_{c,s} > = \sum_{s'} \sum_{c'} \delta_{c',s'} (\tfrac{1}{2} \delta_{s',1} + \tfrac{1}{2} \delta_{s',-1}) P_c(c';t)$$
$$= \sum_{c'} (\tfrac{1}{2} \delta_{c',1} + \tfrac{1}{2} \delta_{c',-1}) P_c(c';t) \qquad (1.2.19)$$
$$= \sum_{c'} (\tfrac{1}{2} P_c(1;t) + \tfrac{1}{2} P_c(-1;t)) = \tfrac{1}{2}$$

where the last equality follows from the normalization of the probability (the sum over all possibilities must be 1, Eq. (1.2.1)) and does not depend at all on the distribution. This shows that the independence of the variables guarantees that the probability of a win is just 1/2.

For the last case (d) the trivial answer, that a win is guaranteed by this method of gambling, can be arrived at formally by evaluating

$$< \delta_{c,s} > = \sum_{s'} \sum_{c'} \delta_{c',s'} P_{s,c}(s',c';t) \qquad (1.2.20)$$

The value of s at time t is independent of the value of c, but the value of c depends on the value of s. The joint probability $P_{s,c}(s',c';t)$ may be written as the product of the probability of a particular value of $s = s'$ times the conditional probability $P_c(c'|s';t)$ of a particular value of $c = c'$ given the assumed value of s:

$$< \delta_{c,s} > = \sum_{s'} \sum_{c'} \delta_{c',s'} P_s(s';t) P_c(c'|s';t)$$
$$= \sum_{s'} \sum_{c'} \delta_{c',s'} P_s(s';t) \delta_{c',s'} = \sum_{s'} P_s(s';t) = 1 \qquad (1.2.21) \ \blacksquare$$

The next step in our analysis of the binary stochastic system is to consider the behavior of the sum of $s(t)$ over a particular number of time steps. This sum is the difference between the total number of heads and the total number of tails. It is equivalent to asking how much you will win or lose if you gamble an equal amount of money on each coin toss after a certain number of bets. This problem is known as a random walk, and we will define it as a consideration of the state variable

$$d(t) = \sum_{t'=1}^{t} s(t') \qquad (1.2.22)$$

The way to write the evolution of the state variable is:

$$P(d'|d) = \tfrac{1}{2} \delta_{d',d+1} + \tfrac{1}{2} \delta_{d',d-1} \qquad (1.2.23)$$

Thus a random walk considers a state variable d that can take integer values $d \in \{ \ldots, -1, 0, 1, \ldots \}$. At every time step, $d(t)$ can only move to a value one higher or one lower than where it is. We assume that the probability of a step to the right (higher) is equal to that of a step to the left (lower). For convenience, we assume (with no loss of generality) that the system starts at position $d(0) = 0$. This is built into Eq. (1.2.22). Because of the symmetry of the system under a shift of the origin, this is equivalent to considering any other starting point. Once we solve for the probability distribution of d at time t, because of superposition we can also find the result of evolving any initial probability distribution $P(d;t=0)$.

We can picture the random walk as that of a drunk who has difficulty consistently moving forward. Our model of this walk assumes that the drunk is equally likely to take a step forward or backward. Starting at position 0, he moves to either $+1$ or -1. Let's say it was $+1$. Next he moves to $+2$ or back to 0. Let's say it was 0. Next to $+1$ or -1. Let's say it was $+1$. Next to $+2$ or 0. Let's say $+2$. Next to $+3$ or $+1$. Let's say $+1$. And so on.

What is the value of system variable $d(t)$ at time t? This is equivalent to asking how far has the walk progressed after t steps. Of course there is no way to know how far a particular system goes without watching it. The average distance over the ensemble of systems is the average over all possible values of $s(t)$. This average is given by applying Eq. (1.2.2) or Eq. (1.2.11) to all of the variables $s(t)$:

$$
<d(t)> = \frac{1}{2} \sum_{s(t)=\pm 1} K \frac{1}{2} \sum_{s(3)=\pm 1} \frac{1}{2} \sum_{s(2)=\pm 1} \frac{1}{2} \sum_{s(1)=\pm 1} d(t)
$$

$$
= \sum_{t'=1}^{t} <s(t')> = 0
$$

(1.2.24)

The average is written out explicitly on the first line using Eq. (1.2.11). The second line expression can be arrived at either directly or from the linearity of the average. The final answer is clear, since it is equally likely for the walker to move to the right as to the left.

We can also ask what is a typical distance traveled by a particular walker. By typical distance we mean how far from the starting point. This can either be defined by the average absolute value of the distance, or as is more commonly accepted, the root mean square (RMS) distance:

$$
\sigma(t) = \sqrt{<d(t)^2>}
$$

(1.2.25)

$$
<d(t)^2> = <\left(\sum_{t'=1}^{t} s(t')\right)^2> = <\sum_{t',t''=1}^{t} s(t')s(t'')> = \sum_{t',t''=1}^{t} <s(t')s(t'')>
$$

(1.2.26)

To evaluate the average of the product of the two steps, we treat differently the case in which they are the same step and when they are different steps. When the two steps are the same one we use $s(t) = \pm 1$ to obtain:

$$
<s(t)^2> = <1> = 1
$$

(1.2.27)

Which follows from the normalization of the probability (or is obvious). To evaluate the average of the product of two steps at different times we need the joint probability of $s(t)$ and $s(t')$. This is the probability that each of them will take a particular value. Because we have assumed that the steps are *independent*, the joint probability is the product of the probabilities for each one separately:

$$P(s(t),s(t')) = P(s(t))P(s(t')) \qquad t \neq t' \qquad (1.2.28)$$

so that for example there is 1/4 chance that $s(t) = +1$ and $s(t) = -1$. The independence of the two steps leads the average of the product of the two steps to factor:

$$
\begin{aligned}
<s(t)s(t')> &= \sum_{s(t),s(t')} P(s(t),s(t'))s(t)s(t') \\
&= \sum_{s(t),s(t')} P(s(t))P(s(t'))s(t)s(t') \qquad t \neq t' \qquad (1.2.29) \\
&= <s(t)><s(t')> = 0
\end{aligned}
$$

This is zero, since either of the averages are zero. We have the combined result:

$$<s(t)s(t')> = \delta_{t,t'} \qquad (1.2.30)$$

and finally:

$$<d(t)^2> = \sum_{t',t''=1}^{t} <s(t')s(t'')> = \sum_{t',t''=1}^{t} \delta_{t',t''} = \sum_{t'=1}^{t} 1 = t \qquad (1.2.31)$$

This gives the classic and important result that a random walk travels a typical distance that grows as the square root of the number of steps taken: $\sigma(t) = \sqrt{t}$.

We can now consider more completely the probability distribution of the position of the walker at time t. The probability distribution at $t = 0$ may be written:

$$P(d;0) = \delta_{d,0} \qquad (1.2.32)$$

After the first time step the probability distribution changes to

$$P(d;1) = \frac{1}{2}\delta_{d,1} + \frac{1}{2}\delta_{d,-1} \qquad (1.2.33)$$

this results from the definition $d(1) = s(1)$. After the second step $d(2) = s(1) + s(2)$ it is:

$$P(d;2) = \frac{1}{4}\delta_{d,2} + \frac{1}{2}\delta_{d,0} + \frac{1}{4}\delta_{d,-2} \qquad (1.2.34)$$

More generally it is not difficult to see that the probabilities are given by normalized binomial coefficients, since the number of ones chosen out of t steps is equivalent to the number of powers of x in $(1 + x)^t$. To reach a position d after t steps we must take $(t + d)/2$ steps to the right and $(t - d)/2$ steps to the left. The sum of these is the number of steps t and their difference is d. Since each choice has 1/2 probability we have:

$$P(d,t)=\frac{1}{2^t}\binom{t}{(d+t)/2}\delta_{t,d}^{oddeven}=\frac{1}{2^t}\frac{t!}{[(d+t)/2]![(t-d)/2]!}\delta_{t,d}^{oddeven}$$

$$\delta_{t,d}^{oddeven}=\frac{(1+(-1)^{t+d})}{2}$$

(1.2.35)

where the unusual delta function imposes the condition that d takes only odd or only even values depending on whether t is odd or even.

Let us now consider what happens after a long time. The probability distribution spreads out, and a single step is a small distance compared to the typical distance traveled. We can consider s and t to be continuous variables where both conditions $d, t \gg 1$ are satisfied. Moreover, we can also consider $|d| \ll t$, because the chance that all steps will be taken in one direction becomes very small. This enables us to use Sterling's approximation to the factorial

$$x! \sim \sqrt{2\pi x}\, e^{-x} x^x$$

$$\ln(x!) \sim x(\ln x - 1) + \ln(\sqrt{2\pi x})$$

(1.2.36)

For large t it also makes sense not to restrict d to be either odd or even. In order to allow both, we, in effect, interpolate and then take only half of the probability we have in Eq. (1.2.35). This leads to the expression:

$$P(d,t)=\frac{\sqrt{t}}{\sqrt{2\pi(t-d)(t+d)}2^t}\frac{t^t e^{-t}}{[(d+t)/2]^{[(d+t)/2]}[(t-d)/2]^{[(t-d)/2]}e^{-(d+t)/2-(t-d)/2}}$$

$$=\frac{(2\pi t(1-x^2))^{-1/2}}{(1+x)^{[(1+x)t/2]}(1-x)^{[(1-x)t/2]}}$$

(1.2.37)

where we have defined $x = d / t$. To approximate this expression it is easier to consider it in logarithmic form:

$$\ln(P(d,t))=-(t/2)[(1+x)\ln(1+x)+(1-x)\ln(1-x)]-(1/2)\ln(2\pi t(1-x^2))$$

$$\approx -(t/2)[(1+x)(x-x^2/2+K)+(1-x)(-x-x^2/2+K)]-(1/2)\ln(2\pi t+K)$$

$$=-tx^2/2-\ln(\sqrt{2\pi t})$$

(1.2.38)

or exponentiating:

$$P(d,t)=\frac{1}{\sqrt{2\pi t}}e^{-d^2/2t}=\frac{1}{\sqrt{2\pi}\,\sigma}e^{-d^2/2\sigma^2}$$

(1.2.39)

The prefactor of the exponential, $1/\sqrt{2\pi}\sigma$, originates from the factor $\sqrt{2\pi x}$ in Eq. (1.2.36). It is independent of d and takes care of the normalization of the probability. The result is a Gaussian distribution. Questions 1.2.2–1.2.5 investigate higher-order corrections to the Gaussian distribution.

Question 1.2.2 In order to obtain a correction to the Gaussian distribution we must add a correction term to Sterling's approximation:

$$x! \sim \sqrt{2\pi x}\, e^{-x} x^x (1 + \frac{1}{12x} + ...)$$

$$\ln(x!) \sim x(\ln x - 1) + \ln(\sqrt{2\pi x}) + \ln(1 + \frac{1}{12x} + ...) \qquad (1.2.40)$$

Using this expression, find the first correction term to Eq. (1.2.37).

Solution 1.2.2 The correction term in Sterling's approximation contributes a factor to Eq. (1.2.37) which is (for convenience we write here $c = 1/12$):

$$\frac{(1+c/t)}{(1+2c/(t+d))(1+2c/(t-d))} = (1 - \frac{3c}{t} + ...) = (1 - \frac{1}{4t} + ...) \qquad (1.2.41)$$

where we have only kept the largest correction term, neglecting d compared to t. Note that the correction term vanishes as t becomes large. ∎

Question 1.2.3 Keeping additional terms of the expansion in Eq. (1.2.38), and the result of Question 1.2.2, find the first order correction terms to the Gaussian distribution.

Solution 1.2.3 Correction terms in Eq. (1.2.38) arise from several places. We want to keep all terms that are of order $1/t$. To do this we must keep in mind that a typical distance traveled is $d \sim \sqrt{t}$, so that $x \sim 1/\sqrt{t}$. The next terms are obtained from:

$$\begin{aligned}
\ln(P(d,t)) &= -(t/2)[(1+x)\ln(1+x)+(1-x)\ln(1-x)] \\
&\quad -(1/2)\ln(2\pi t(1-x^2))+\ln(1-1/4t) \\
&\approx -(t/2)[(1+x)(x-\tfrac{1}{2}x^2+\tfrac{1}{3}x^3-\tfrac{1}{4}x^4 K\,) \\
&\quad +(1-x)(-x-\tfrac{1}{2}x^2-\tfrac{1}{3}x^3-\tfrac{1}{4}x^4 K\,)] \\
&\quad -\ln(\sqrt{2\pi t})-(1/2)\ln(1-x^2)+\ln(1-1/4t) \\
&\approx -(t/2)[(x+x^2-\tfrac{1}{2}x^2-\tfrac{1}{3}x^3+\tfrac{1}{3}x^3+\tfrac{1}{4}x^4-\tfrac{1}{4}x^4 K\,) \\
&\quad +(-x+x^2-\tfrac{1}{2}x^2+\tfrac{1}{2}x^3-\tfrac{1}{3}x^3+\tfrac{1}{3}x^4-\tfrac{1}{4}x^4 K\,)] \\
&\quad -\ln(\sqrt{2\pi t})+(x^2/2+...)+(-1/4t+...) \\
&= -tx^2/2-\ln(\sqrt{2\pi t})-tx^4/12+x^2/2-1/4t
\end{aligned} \qquad (1.2.42)$$

This gives us a distribution:

$$P(d,t) = \sqrt{\frac{1}{2\pi t}} e^{-d^2/2t} e^{-d^4/12t^3 + d^2/2t^2 - 1/4t} \qquad (1.2.43) \quad \blacksquare$$

Question 1.2.4 What is the size of the additional factor? Estimate the size of this term as t becomes large.

Solution 1.2.4 The typical value of the variable d is its root mean square value $\sigma = \sqrt{t}$. At this value the additional term gives a factor

$$e^{1/6t} \qquad (1.2.44)$$

which approaches 1 as time increases. $\quad \blacksquare$

Question 1.2.5 What is the fraction error that we will make if we neglect this term after one hundred steps? After ten thousand steps?

Solution 1.2.5 After one hundred time steps the walker has traveled a typical distance of ten steps. We generally approximate the probability of arriving at this distance using Eq. (1.2.39). The fractional error in the probability of arriving at this distance according to Eq. (1.2.44) is $1 - e^{1/6t} \approx -1/6t = -0.00167$. So already at a distance of ten steps the error is less than 0.2%.

It is much less likely for the walker to arrive at the distance $2\sigma = 20$. The ratio of the probability to arrive at 20 compared to 10 is $e^{-2}/e^{-0.5} \sim 0.22$. If we want to know the error of this smaller probability case we would write $(1 - e^{-16/12t + 4/2t - 1/4t}) = (1 - e^{5/12t}) \approx -0.0042$, which is a larger but still small error.

After ten thousand steps the errors are smaller than the errors at one hundred steps by a factor of one hundred. $\quad \blacksquare$

1.2.2 Generalized random walk and the central limit theorem

We can generalize the random walk by allowing a variety of steps from the current location of the walker to sites nearby, not only to the adjacent sites and not only to integer locations. If we restrict ourselves to steps that on average are balanced left and right and are not too long ranged, we can show that all such systems have the same behavior as the simplest random walk at long enough times (and characteristically not even for very long times). This is the content of the central limit theorem. It says that summing any set of independent random variables eventually leads to a Gaussian distribution of probabilities, which is the same distribution as the one we arrived at for the random walk. The reason that the same distribution arises is that successive iteration of the probability update equation, Eq. (1.2.7), smoothes out the distribution, and the only relevant information that survives is the width of the distribution which is given by $\sigma(t)$. The proof given below makes use of a Fourier transform and can be skipped by readers who are not well acquainted with transforms. In the next section we will also include a bias in the random walk. For long times this can be described as an average motion superimposed on the unbiased random walk. We start with the unbiased random walk.

Each step of the random walk is described by the state variable $s(t)$ at time t. The probability of a particular step size is an unspecified function that is independent of time:

$$P(s;t) = f(s) \tag{1.2.45}$$

We treat the case of integer values of s. The continuum case is Question 1.2.6. The absence of bias in the random walk is described by setting the average displacement in a single step to zero:

$$<s> = \sum_s sf(s) = 0 \tag{1.2.46}$$

The statement above that each step is not too long ranged, is mathematically just that the mean square displacement in a single step has a well-defined value (i.e., is not infinite):

$$<s^2> = \sum_s s^2 f(s) = \sigma_0^2 \tag{1.2.47}$$

Eqs. (1.2.45)–(1.2.47) hold at all times.

We can still evaluate the average of $d(t)$ and the RMS value of $d(t)$ directly using the linearity of the average:

$$<d(t)> = <\sum_{t'=1}^{t} s(t')> = t<s> = 0 \tag{1.2.48}$$

$$<d(t)^2> = <\left(\sum_{t'=1}^{t} s(t')\right)^2> = \sum_{t',t''=1}^{t} <s(t')s(t'')> \tag{1.2.49}$$

Since $s(t')$ and $s(t'')$ are independent for $t' \neq t''$, as in Eq. (1.2.29), the average factors:

$$<s(t')s(t'')> = <s(t')><s(t'')> = 0 \qquad t' \neq t'' \tag{1.2.50}$$

Thus, all terms $t' \neq t''$ are zero by Eq. (1.2.46). We have:

$$<d(t)^2> = \sum_{t'=1}^{t} <s(t')^2> = t\sigma_0^2 \tag{1.2.51}$$

This means that the typical value of $d(t)$ is $\sigma_0\sqrt{t}$.

To obtain the full distribution of the random walk state variable $d(t)$ we have to sum the stochastic variables $s(t)$. Since $d(t) = d(t-1) + s(t)$ the probability of transition from $d(t-1)$ to $d(t)$ is $f(d(t) - d(t-1))$ or:

$$P(d'|d) = f(d' - d) \tag{1.2.52}$$

We can now write the time evolution equation and iterate it t times to get $P(d;t)$.

$$P(d;t) = \sum_{d'} P(d|d')P(d';t-1) = \sum_{d'} f(d-d')P(d';t-1) \tag{1.2.53}$$

This is a convolution, so the most convenient way to effect a t fold iteration is in Fourier space. The Fourier representation of the probability and transition functions for integral d is:

$$\tilde{P}(k;t) \equiv \sum_{d} e^{-ikd} P(d;t)$$
$$\tilde{f}(k) \equiv \sum_{s} e^{-iks} f(s) \tag{1.2.54}$$

We use a Fourier series because of the restriction to integer values of d. Once we solve the problem using the Fourier representation, the probability distribution is recovered from the inverse formula:

$$P(d;t) = \frac{1}{2\pi} \int_{-\pi}^{\pi} dk\, e^{ikd} \tilde{P}(k;t) \tag{1.2.55}$$

which is proved

$$\frac{1}{2\pi} \int_{-\pi}^{\pi} dk\, e^{ikd} \tilde{P}(k;t) = \frac{1}{2\pi} \int_{-\pi}^{\pi} dk\, e^{ikd} \sum_{d'} e^{-ikd'} P(d';t)$$
$$= \frac{1}{2\pi} \sum_{d'} P(d';t) \int_{-\pi}^{\pi} dk\, e^{ik(d-d')} = \sum_{d'} P(d';t)\delta_{d,d'} = P(d;t) \tag{1.2.56}$$

using the expression:

$$\delta_{d,d'} = \frac{1}{2\pi} \int_{-\pi}^{\pi} dk\, e^{ik(d-d')} \tag{1.2.57}$$

Applying Eq. (1.2.54) to Eq. (1.2.53):

$$\tilde{P}(k;t) = \sum_{d} e^{-ikd} \sum_{d'} f(d-d')P(d';t-1)$$
$$= \sum_{d'} \sum_{d} e^{-ik(d-d')} e^{-ikd'} f(d-d')P(d';t-1)$$
$$= \sum_{d'} \sum_{d''} e^{-ikd''} e^{-ikd'} f(d'')P(d';t-1)$$
$$= \sum_{d''} e^{-ikd''} f(d'') \sum_{d'} e^{-ikd'} P(d';t-1) = \tilde{f}(k)\tilde{P}(k;t-1) \tag{1.2.58}$$

we can iterate the equation to obtain:

$$\tilde{P}(k;t) = \tilde{f}(k)\tilde{P}(k;t-1) = \tilde{f}(k)^t \tag{1.2.59}$$

where we use the definition $d(1) = s(1)$ that ensures that $P(d;1) = P(s;1) = f(d)$.

For large t the walker has traveled a large distance, so we are interested in variations of the probability $P(d;t)$ over large distances. Thus, in Fourier space we are concerned with small values of k. To simplify Eq. (1.2.59) for large t we expand $\tilde{f}(k)$ near $k = 0$. From Eq. (1.2.54) we can directly evaluate the derivatives of $\tilde{f}(k)$ at $k = 0$ in terms of averages:

$$\left.\frac{d^n \tilde{f}(k)}{d^n k}\right|_{k=0} = \sum_s (-is)^n f(s) = (-i)^n < s^n > \tag{1.2.60}$$

We can use this expression to evaluate the terms of a Taylor expansion of $\tilde{f}(k)$:

$$\tilde{f}(k) = \tilde{f}(0) + \left.\frac{\partial \tilde{f}(k)}{\partial k}\right|_{k=0} k + \frac{1}{2}\left.\frac{\partial^2 \tilde{f}(k)}{\partial k^2}\right|_{k=0} k^2 + \mathrm{K} \tag{1.2.61}$$

$$\tilde{f}(k) = <1> - i <s> k - \frac{1}{2} <s^2> k^2 + \mathrm{K} \tag{1.2.62}$$

Using the normalization of the probability ($<1> = 1$), and Eqs. (1.2.46) and (1.2.47), gives us:

$$\tilde{P}(k;t) = \left(1 - \tfrac{1}{2}\sigma_0^2 k^2 + \mathrm{K}\right)^t \tag{1.2.63}$$

We must now remember that a typical value of $d(t)$, from its RMS value, is $\sigma_0 \sqrt{t}$. By the properties of the Fourier transform, this implies that a typical value of k that we must consider in Eq. (1.2.63) varies with time as $1/\sqrt{t}$. The next term in the expansion, cubic in k, would give rise to a term that is smaller by this factor, and therefore becomes unimportant at long times. If we write $k = q/\sqrt{t}$, then it becomes clearer how to write Eq. (1.2.63) using a limiting expression for large t:

$$\tilde{P}(k;t) = \left(1 - \frac{1}{2}\frac{\sigma_0^2 q^2}{t} + \mathrm{K}\right)^t \sim e^{-\sigma_0^2 q^2/2} = e^{-t\sigma_0^2 k^2/2} \tag{1.2.64}$$

This Gaussian, when Fourier transformed back to an expression in d, gives us a Gaussian as follows:

$$P(d;t) = \frac{1}{2\pi}\int_{-\pi}^{\pi} dk e^{ikd} e^{-t\sigma_0^2 k^2/2} \cong \frac{1}{2\pi}\int_{-\infty}^{\infty} dk e^{ikd} e^{-t\sigma_0^2 k^2/2} \tag{1.2.65}$$

We have extended the integral because the decaying exponential becomes narrow as t increases. The integral is performed by completing the square in the exponent, giving:

$$= \frac{1}{2\pi} \int_{-\infty}^{\infty} dk e^{-d^2/2t\sigma_0^2} e^{-(t\sigma_0^2 k^2 - 2ikd - d^2/t\sigma_0^2)/2} = \frac{1}{\sqrt{2\pi t\sigma_0^2}} e^{-d^2/2t\sigma_0^2} \tag{1.2.66}$$

or equivalently:

$$P(d;t) = \frac{1}{\sqrt{2\pi\sigma(t)^2}} e^{-d^2/2\sigma(t)^2} \tag{1.2.67}$$

which is the same as Eq. (1.2.39).

Question 1.2.6 Prove the central limit theorem when s takes a continuum of values.

Solution 1.2.6 The proof follows the same course as the integer valued case. We must define the appropriate averages, and the transform. The average of s is still zero, and the mean square displacement is defined similarly:

$$<s> = \int ds\, sf(s) = 0 \tag{1.2.46'}$$

$$<s^2> = \int ds\, s^2 f(s) = \sigma_0^2 \tag{1.2.47'}$$

To avoid problems of notation we substitute the variable x for the state variable d:

$$<x(t)> = <\sum_{t'=1}^{t} s(t')> = t<s> = 0 \tag{1.2.48'}$$

Skipping steps that are the same we find:

$$<x(t)^2> = <\left(\sum_{t'=1}^{t} s(t')\right)^2> = \sum_{t'=1}^{t} <s(t')^2> = t\sigma_0^2 \tag{1.2.51'}$$

since $s(t')$ and $s(t'')$ are still independent for $t' \neq t''$. Eq. (1.2.53) is also essentially unchanged:

$$P(x;t) = \int dx'\, f(x-x') P(x';t-1) \tag{1.2.53'}$$

The transform and inverse transform must now be defined using

$$\tilde{P}(k;t) \equiv \int dx\, e^{-ikx} P(x;t)$$
$$\tilde{f}(k) \equiv \int ds\, e^{-iks} f(s) \tag{1.2.54'}$$

$$P(d;t) = \frac{1}{2\pi} \int dk e^{ikd} \tilde{P}(k;t) \tag{1.2.55'}$$

The latter is proved using the properties of the Dirac (continuum) delta function:

$$\delta(x - x') = \frac{1}{2\pi} \int dk e^{ik(x-x')}$$

$$\int dx' \delta(x - x')g(x') = g(x) \tag{1.2.56'}$$

where the latter equation holds for an arbitrary function $g(x)$.

The remainder of the derivation carries forward unchanged. ∎

1.2.3 Biased random walk

We now return to the simple random walk with binary steps of ± 1. The model we consider is a random walk that is biased in one direction. Each time a step is taken there is a probability P_+ for a step of $+1$, that is different from the probability P_- for a step of -1, or:

$$P(s;t) = P_+ \delta_{s,1} + P_- \delta_{s,-1} \tag{1.2.68}$$

$$P(d'|d) = P_+ \delta_{d',d+1} + P_- \delta_{d',d} \tag{1.2.69}$$

where

$$P_+ + P_- = 1 \tag{1.2.70}$$

What is the average distance traveled in time t?

$$<d(t)> = \sum_{t'=1}^{t} <s(t')> = \sum_{t'=1}^{t} (P_+ - P_-) = t(P_+ - P_-) \tag{1.2.71}$$

This equation justifies defining the mean velocity as

$$v = P_+ - P_- \tag{1.2.72}$$

Since we already have an average displacement it doesn't make sense to also ask for a typical displacement, as we did with the random walk—the typical displacement is the average one. However, we can ask about the spread of the displacements around the average displacement

$$\sigma(t)^2 = <(d(t) - <d(t)>)^2> = <d(t)^2> - 2<d(t)>^2 + <d(t)>^2$$
$$= <d(t)^2> - <d(t)>^2 \tag{1.2.73}$$

This is called the standard deviation and it reduces to the RMS distance in the unbiased case. For many purposes $\sigma(t)$ plays the same role in the biased random walk as in the unbiased random walk. From Eq. (1.2.71) and Eq. (1.2.72) the second term is $(vt)^2$. The first term is:

$$< d(t)^2 > = < \left(\sum_{t'=1}^{t} s(t') \right)^2 > = \sum_{t',t''=1}^{t} < s(t')s(t'') >$$

$$= \sum_{t',t''=1}^{t} \left(\delta_{t',t''} + (1-\delta_{t',t''})(P_+^2 + P_-^2 - 2P_+P_-) \right)$$

$$= t + t(t-1)v^2 = t^2v^2 + t(1-v^2)$$

(1.2.74)

Substituting in Eq. (1.2.73):

$$\sigma^2 = t(1-v^2)$$

(1.2.75)

It is interesting to consider this expression in the two limits $v = 1$ and $v = 0$. For $v = 1$ the walk is deterministic, $P_+ = 1$ and $P_- = 0$, and there is no element of chance; the walker always walks to the right. This is equivalent to the iterative map Eq. (1.1.4). Our result Eq. (1.2.66) is that $\sigma = 0$, as it must be for a deterministic system. However, for smaller velocities, the spreading of the systems σ increases until at $v = 0$ we recover the case of the unbiased random walk.

The complete probability distribution is given by:

$$P(d;t) = P_+^{(d+t)/2} P_-^{(d-t)/2} \binom{t}{(d+t)/2} \delta_{t,d}^{oddeven}$$

(1.2.76)

For large t the distribution can be found as we did for the unbiased random walk. The work is left to Question 1.2.7.

Question 1.2.7 Find the long time (continuum) distribution for the biased random walk.

Solution 1.2.7 We use the Sterling approximation as before and take the logarithm of the probability. In addition to the expression from the first line of Eq. (1.2.38) we have an additional factor due to the coefficient of Eq. (1.2.76) which appears in place of the factor of $1/2^t$. We again define $x = d/t$, and divide by 2 to allow both odd and even integers. We obtain the expression:

$$\ln(P(d,t)) = (t/2)[(1+x)\ln 2P_+ + (1-x)\ln 2P_-]$$
$$- (t/2)[(1+x)\ln(1+x) + (1-x)\ln(1-x)] - (1/2)\ln(2\pi t(1-x^2))$$

(1.2.77)

It makes the most sense to expand this around the mean of x, $<x> = v$. To simplify the notation we can use Eq. (1.2.70) and Eq. (1.2.72) to write:

$$P_+ = (1+v)/2$$
$$P_- = (1-v)/2$$

(1.2.78)

With these substitutions we have:

$$\ln(P(d,t)) = (t/2)[(1+x)\ln(1+v) + (1-x)\ln(1-v)]$$
$$- (t/2)[(1+x)\ln(1+x) + (1-x)\ln(1-x)] - (1/2)\ln(2\pi t(1-x^2))$$

(1.2.79)

We expand the first two terms in a Taylor expansion around the mean of x and expand the third term inside the logarithm. The first term of Eq. (1.2.79) has only a constant and linear term in a Taylor expansion. These cancel the constant and the first derivative of the Taylor expansion of the second term of Eq. (1.2.79) at $x = v$. Higher derivatives arise only from the second term:

$$\ln(P(d,t)) = -(t/2)[\frac{1}{(1-v^2)}(x-v)^2 + \frac{2}{3(1-v^2)^2}(x-v)^3 + K\ ]$$
$$- (1/2)\ln(2\pi t[(1-v^2) - 2v(x-v) + K\ ]) \qquad (1.2.80)$$
$$= -[\frac{(d-vt)^2}{2\sigma(t)^2} + \frac{(d-vt)^3}{3\sigma(t)^4} + K\ ] - (1/2)\ln(2\pi(\sigma(t)^2 - 2v(d-vt) + K\ ))$$

In the last line we have restored d and used Eq. (1.2.75). Keeping only the first terms in both expansions gives us:

$$P(d;t) = \frac{1}{\sqrt{2\pi\sigma(t)^2}} e^{-(d-vt)^2/2\sigma(t)^2} \qquad (1.2.81)$$

which is a Gaussian distribution around the mean we obtained before. This implies that aside from the constant velocity, and a slightly modified standard deviation, the distribution remains unchanged.

The second term in both expansions in Eq. (1.2.80) become small in the limit of large t, as long as we are not interested in the tail of the distribution. Values of $(d - vt)$ relevant to the main part of the distribution are given by the standard deviation, $\sigma(t)$. The second terms in Eq. (1.2.80) are thus reduced by a factor of $\sigma(t)$ compared to the first terms in the series. Since $\sigma(t)$ grows as the square root of the time, they become insignificant for long times. The convergence is slower, however, than in the unbiased random walk (Questions 1.2.2–1.2.5). ∎

Question 1.2.8 You are a manager of a casino and are told by the owner that you have a cash flow problem. In order to survive, you have to make sure that nine out of ten working days you have a profit. Assume that the only game in your casino is a roulette wheel. Bets are limited to only red or black with a 2:1 payoff. The roulette wheel has an equal number of red numbers and black numbers and one green number (the house always wins on green). Assume that people make a fixed number of 10^6 total $1 bets on the roulette wheel in each day.

a. What is the maximum number of red numbers on the roulette wheel that will still allow you to achieve your objective?

b. With this number of red numbers, how much money do you make on average in each day?

Solution 1.2.8 The casino wins \$1 for every wrong bet and loses \$1 for every right bet. The results of bets at the casino are equivalent to a random walk with a bias given by:

$$P_+ = (N_{red} + 1)/(N_{red} + N_{black} + 1) \tag{1.2.82}$$

$$P_- = N_{black}/(N_{red} + N_{black} + 1) \tag{1.2.83}$$

where, as the manager, we consider positive the wins of the casino. The color subscripts can be used interchangeably, since the number of red and black is equal. The velocity of the random walk is given by:

$$v = 1/(2N_{red} + 1) \tag{1.2.84}$$

To calculate the probability that the casino will lose on a particular day we must sum the probability that the random walk after 10^6 steps will result in a negative number. We approximate the sum by an integral over the distribution of Eq. (1.2.81). To avoid problems of notation we replace d with y:

$$
\begin{aligned}
P_{loss} &= \int_{-\infty}^{0} dy P(y; t = 10^6) = \frac{1}{\sqrt{2\pi\sigma(t)^2}} \int_{-\infty}^{0} dy\, e^{-(y-vt)^2/2\sigma(t)^2} \\
&= \frac{1}{\sqrt{2\pi\sigma(t)^2}} \int_{-\infty}^{-vt} dy'\, e^{-(y')^2/2\sigma(t)^2} \\
&= \frac{1}{\sqrt{\pi}} \int_{-\infty}^{z_0} dz\, e^{-z^2} = \frac{1}{2}(1 - \mathrm{erf}(z_0))
\end{aligned}
\tag{1.2.85}
$$

$$z = y'/\sqrt{2}\sigma(t)$$
$$z_0 = -vt/\sqrt{2\sigma(t)^2} = -vt/\sqrt{2t(1-v^2)} \tag{1.2.86}$$

We have written the probability of loss in a day in terms of the error function erf(x)—the integral of a Gaussian defined by

$$\mathrm{erf}(z_0) \equiv \frac{2}{\sqrt{\pi}} \int_0^{z_0} dz\, e^{-z^2} \tag{1.2.87}$$

Since

$$\mathrm{erf}(\infty) = 1 \tag{1.2.88}$$

we have the expression

$$(1 - \mathrm{erf}(z_0)) \equiv \frac{2}{\sqrt{\pi}} \int_{z_0}^{\infty} dz\, e^{-z^2} \tag{1.2.89}$$

which is also known as the complementary error function erfc(x).

To obtain the desired constraint on the number of red numbers, or equivalently on the velocity, we invert Eq. (1.2.85) to find a value of v that gives the desired $P_{loss} = 0.1$, or erf(z_0) = 0.8. Looking up the error function or using iterative guessing on an appropriate computer gives $z_0 = 0.9062$. Inverting Eq. (1.2.86) gives:

$$v = \frac{1}{\sqrt{t/2z_0 - 1}} \approx \sqrt{2z_0/t} \qquad (1.2.90)$$

The approximation holds because t is large. The numerical result is $v = 0.0013$. This gives us the desired number of each color (inverting Eq. (1.2.84)) of $N_{red} = 371$. Of course the result is a very large number and the problem of winning nine out of ten days is a very conservative problem for a casino. Even if we insist on winning ninety-nine out of one hundred days we would have erf(z_0) = 0.98, $z_0 = 1.645$, $v = 0.0018$ and $N_{red} = 275$. The profits per day in each case are given by vt, which is approximately \$1,300 and \$1,800 respectively. Of course this is much less than for bets on a more realistic roulette wheel. Eventually as we reduce the chance of the casino losing and z_0 becomes larger, we might become concerned that we are describing the properties of the tail of the distribution when we calculate the fraction of days the casino might lose, and Eq. (1.2.85) will not be very accurate. However, it is not difficult to see that casinos do not have cash flow problems. ∎

In order to generalize the proof of the central limit theorem to the case of a biased random walk, we can treat the continuum case most simply by considering the system variable $\hat{x}$, where (using $d \to x$ for the continuum case):

$$\hat{x} = x - <x>_t = x - t<s> = x - vt \qquad (1.2.91)$$

Only x is a stochastic variable on the right side, v and t are numbers. Since iterations of this variable would satisfy the conditions for the generalized random walk, the generalization of the Gaussian distribution to Eq. (1.2.81) is proved. The discrete case is more difficult to prove because we cannot shift the variable d by arbitrary amounts and continue to consider it as discrete. We can argue the discrete case to be valid on the basis of the result for the continuum case, but a separate proof can be constructed as well.

1.2.4 *Master equation approach*

The Master equation is an alternative approach to stochastic systems, an alternative to Eq. (1.2.5), that is usually applied when time is continuous. We develop it starting from the discrete time case. We can rewrite Eq. (1.2.5) in the form of a difference equation for a particular probability $P(s)$. Beginning from:

$$P(s;t) = P(s;t-1) + \left(\sum_{s'} P(s|s')P(s';t-1) - P(s;t-1) \right) \qquad (1.2.92)$$

we extract the term where the system remains in the same state:

$$P(s;t) = P(s;t-1) + \left(\sum_{s' \neq s} P(s|s')P(s';t-1) + P(s|s)P(s;t-1) - P(s;t-1) \right) \quad (1.2.93)$$

We use the normalization of probability to write it in terms of the transitions away from this site:

$$P(s;t) = P(s;t-1) + \left(\sum_{s' \neq s} P(s|s')P(s';t-1) + \left(1 - \sum_{s' \neq s} P(s'|s) \right) P(s;t-1) - P(s;t-1) \right)$$

$$(1.2.94)$$

Canceling the terms in the bracket that refer only to the probability $P(s;t-1)$ we write this as a difference equation. On the right appear only the probabilities at different values of the state variable ($s' \neq s$):

$$P(s,t) - P(s;t-1) = \sum_{s' \neq s} \left(P(s|s')P(s';t-1) - P(s'|s)P(s;t-1) \right) \quad (1.2.95)$$

To write the continuum form we reintroduce the time difference between steps Δt.

$$\frac{P(s,t) - P(s;t-\Delta t)}{\Delta t} = \sum_{s' \neq s} \left(\frac{P(s|s')}{\Delta t} P(s';t-\Delta t) - \frac{P(s'|s)}{\Delta t} P(s;t-\Delta t) \right) \quad (1.2.96)$$

When the limit of $\Delta t \to 0$ is meaningful, it is possible to make the change to the equation

$$\dot{P}(s,t) = \sum_{s' \neq s} \left(R(s|s')P(s';t) - R(s'|s)P(s;t) \right) \quad (1.2.97)$$

Where the ratio $P(s|s')/\Delta t$ has been replaced by the rate of transition $R(s|s')$. Eq. (1.2.97) is called the Master equation and we can consider Eq. (1.2.95) as the discrete time analog.

The Master equation has a simple interpretation: The rate of change of the probability of a particular state is the total rate at which probability is being added into that state from all other states, minus the total rate at which probability is leaving the state. Probability is acting like a fluid that is flowing to or from a particular state and is being conserved, as it must be. Eq. (1.2.97) is very much like the continuity equation of fluid flow, where the density of the fluid at a particular place changes according to how much is flowing to that location or from it. We will construct and use the Master equation approach to discuss the problem of relaxation in activated processes in Section 1.4.

1.3   Thermodynamics and Statistical Mechanics

The field of thermodynamics is easiest to understand in the context of Newtonian mechanics. Newtonian mechanics describes the effect of forces on objects. Thermodynamics describes the effect of heat transfer on objects. When heat is transferred, the temperature of an object changes. Temperature and heat are also intimately related to energy. A hot gas in a piston has a high pressure and it can do mechanical work by applying a force to a piston. By Newtonian mechanics the work is directly related to a transfer of energy. The laws of Newtonian mechanics are simplest to describe using the abstract concept of a point object with mass but no internal structure. The analogous abstraction for thermodynamic laws are materials that are in equilibrium and (even better) are homogeneous. It turns out that even the description of the equilibrium properties of materials is so rich and varied that this is still a primary focus of active research today.

Statistical mechanics begins as an effort to explain the laws of thermodynamics by considering the microscopic application of Newton's laws. Microscopically, the temperature of a gas is found to be related to the kinetic motion of the gas molecules. Heat transfer is the transfer of Newtonian energy from one object to another. The statistical treatment of the many particles of a material, with a key set of assumptions, reveals that thermodynamic laws are a natural consequence of many microscopic particles interacting with each other. Our studies of complex systems will lead us to discuss the properties of systems composed of many interacting parts. The concepts and tools of statistical mechanics will play an important role in these studies, as will the laws of thermodynamics that emerge from them. Thermodynamics also begins to teach us how to think about systems interacting with each other.

1.3.1 *Thermodynamics*

Thermodynamics describes macroscopic pieces of material in equilibrium in terms of macroscopic parameters. Thermodynamics was developed as a result of experience/experiment and, like Newton's laws, is to be understood as a set of self-consistent definitions and equations. As with Newtonian mechanics, where in its simplest form objects are point particles and friction is ignored, the discussion assumes an idealization that is directly experienced only in special circumstances. However, the fundamental laws, once understood, can be widely applied. The central quantities that are to be defined and related are the energy U, temperature T, entropy S, pressure P, the mass (which we write as the number of particles) N, and volume V. For magnets, the quantities should include the magnetization M, and the magnetic field H. Other macroscopic quantities that are relevant may be added as necessary within the framework developed by thermodynamics. Like Newtonian mechanics, a key aspect of thermodynamics is to understand how systems can be acted upon or can act upon each other. In addition to the quantities that describe the state of a system, there are two quantities that describe actions that may be made on a system to change its state: work and heat transfer.

The equations that relate the macroscopic quantities are known as the zeroth, first and second laws of thermodynamics. Much of the difficulty in understanding thermodynamics arises from the way the entropy appears as an essential but counter-intuitive quantity. It is more easily understood in the context of a statistical treatment included below. A second source of difficulty is that even a seemingly simple material system, such as a piece of metal in a room, is actually quite complicated thermodynamically. Under usual circumstances the metal is not in equilibrium but is emitting a vapor of its own atoms. A thermodynamic treatment of the metal requires consideration not only of the metal but also the vapor and even the air that applies a pressure upon the metal. It is therefore generally simplest to consider the thermodynamics of a gas confined in a closed (and inert) chamber as a model thermodynamic system. We will discuss this example in detail in Question 1.3.1. The translational motion of the whole system, treated by Newtonian mechanics, is ignored.

We begin by defining the concept of equilibrium. A system left in isolation for a long enough time achieves a macroscopic state that does not vary in time. The system in an unchanging state is said to be in equilibrium. Thermodynamics also relies upon a particular type of equilibrium known as thermal equilibrium. Two systems can be brought together in such a way that they interact only by transferring heat from one to the other. The systems are said to be in thermal contact. An example would be two gases separated by a fixed but thermally conducting wall. After a long enough time the system composed of the combination of the two original systems will be in equilibrium. We say that the two systems are in thermal equilibrium with each other. We can generalize the definition of thermal equilibrium to include systems that are not in contact. We say that any two systems are in thermal equilibrium with each other if they do not change their (macroscopic) state when they are brought into thermal contact. Thermal equilibrium does not imply that the system is homogeneous, for example, the two gases may be at different pressures.

The zeroth law of thermodynamics states that if two systems are in thermal equilibrium with a third they are in thermal equilibrium with each other. This is not obvious without experience with macroscopic objects. The zeroth law implies that the interaction that occurs during thermal contact is not specific to the materials, it is in some sense weak, and it matters not how many or how big are the systems that are in contact. It enables us to define the temperature T as a quantity which is the same for all systems in thermal equilibrium. A more specific definition of the temperature must wait till the second law of thermodynamics. We also define the concept of a thermal reservoir as a very large system such that any system that we are interested in, when brought into contact with the thermal reservoir, will change its state by transferring heat to or from the reservoir until it is in equilibrium with the reservoir, but the transfer of heat will not affect the temperature of the reservoir.

Quite basic to the formulation and assumptions of thermodynamics is that the macroscopic state of an isolated system in equilibrium is completely defined by a specification of three parameters: energy, mass and volume (U, N, V). For magnets we must add the magnetization M; we will leave this case for later. The confinement of the system to a volume V is understood to result from some form of containment. The state of a system can be characterized by the force per unit area—the pressure P—exerted by the system on the container or by the container on the system, which are the same. Since in equilibrium a system is uniquely described by the three quantities (U,N,V), these determine all the other quantities, such as the pressure P and temperature T. Strictly speaking, temperature and pressure are only defined for a system in equilibrium, while the quantities (U,N,V) have meaning both in and out of equilibrium.

It is assumed that for a homogeneous material, changing the size of the system by adding more material in equilibrium at the same pressure and temperature changes the mass, number of particles N, volume V and energy U, in direct proportion to each other. Equivalently, it is assumed that cutting the system into smaller parts results in each subpart retaining the same properties in proportion to each other (see Figs. 1.3.1 and 1.3.2). This means that these quantities are additive for different parts of a system whether isolated or in thermal contact or full equilibrium:

$$N = \sum_{\alpha} N^{\alpha}$$

$$V = \sum_{\alpha} V^{\alpha} \qquad (1.3.1)$$

$$U = \sum_{\alpha} U^{\alpha}$$

where α indexes the parts of the system. This would not be true if the parts of the system were strongly interacting in such a way that the energy depended on the relative location of the parts. Properties such as (U,N,V) that are proportional to the size of the system are called extensive quantities. Intensive quantities are properties that do not change with the size of the system at a given pressure and temperature. The ratio of two extensive quantities is an intensive quantity. Examples are the particle density N/V and the energy density U/V. The assumption of the existence of extensive and intensive quantities is also far from trivial, and corresponds to the intuition that for a macroscopic object, the local properties of the system do not depend on the size of the system. Thus a material may be cut into two parts, or a small part may be separated from a large part, without affecting its local properties.

The simplest thermodynamic systems are homogeneous ones, like a gas in an inert container. However we can also use Eq. (1.3.1) for an inhomogeneous system. For example, a sealed container with water inside will reach a state where both water and vapor are in equilibrium with each other. The use of intensive quantities and the proportionality of extensive quantities to each other applies only within a single phase—a single homogeneous part of the system, either water or vapor. However, the additivity of extensive quantities in Eq. (1.3.1) still applies to the whole system. A homogeneous as well as a heterogeneous system may contain different chemical species. In this case the quantity N is replaced by the number of each chemical species N_i and the first line of Eq. (1.3.1) may be replaced by a similar equation for each species.

Figure 1.3.1 Thermodynamics considers macroscopic materials. A basic assumption is that cutting a system into two parts will not affect the local properties of the material and that the energy U, mass (or number of particles) N and the volume V will be divided in the same proportion. The process of separation is assumed to leave the materials under the same conditions of pressure and temperature. ∎

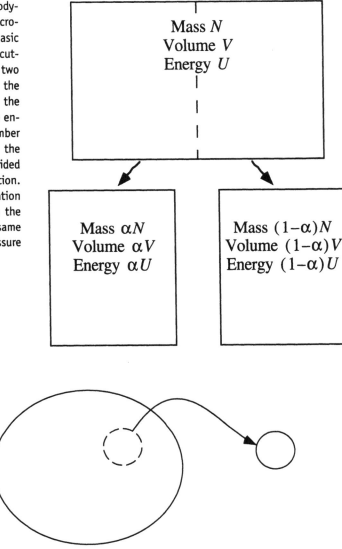

Mass N
Volume V
Energy U

Mass αN
Volume αV
Energy αU

Mass $(1-\alpha)N$
Volume $(1-\alpha)V$
Energy $(1-\alpha)U$

Figure 1.3.2 The assumption that the local properties of a system are unaffected by subdivision applies also to the case where a small part of a much larger system is removed. The local properties, both of the small system and of the large system are assumed to remain unchanged. Even though the small system is much smaller than the original system, the small system is understood to be a macroscopic piece of material. Thus it retains the same local properties it had as part of the larger system. ∎

The first law of thermodynamics describes how the energy of a system may change. The energy of an isolated system is conserved. There are two macroscopic processes that can change the energy of a system when the number of particles is fixed.

The first is work, in the sense of applying a force over a distance, such as driving a piston that compresses a gas. The second is heat transfer. This may be written as:

$$dU = q + w \tag{1.3.2}$$

where q is the heat transfer into the system, w is the work done on the system and U is the internal energy of the system. The differential d signifies the incremental change in the quantity U as a result of the incremental process of heat transfer and work. The work performed on a gas (or other system) is the force times the distance applied Fdx, where we write F as the magnitude of the force and dx as an incremental distance. Since the force is the pressure times the area $F = PA$, the work is equal to the pressure times the volume change or:

$$w = -PAdx = -PdV \tag{1.3.3}$$

The negative sign arises because positive work on the system, increasing the system's energy, occurs when the volume change is negative. Pressure is defined to be positive.

If two systems act upon each other, then the energy transferred consists of both the work and heat transfer. Each of these are separately equal in magnitude and opposite in sign:

$$\begin{aligned} dU_1 &= q_{21} + w_{21} \\ dU_2 &= q_{12} + w_{12} \\ q_{12} &= -q_{21} \\ w_{12} &= -w_{21} \end{aligned} \tag{1.3.4}$$

where q_{21} is the heat transfer from system 2 to system 1, and w_{21} is the work performed by system 2 on system 1. q_{12} and w_{12} are similarly defined. The last line of Eq. (1.3.4) follows from Newton's third law. The other equations follow from setting $dU = 0$ (Eq. (1.3.2)) for the total system, composed of both of the systems acting upon each other.

The second law of thermodynamics given in the following few paragraphs describes a few key aspects of the relationship of the equilibrium state with nonequilibrium states. The statement of the second law is essentially a definition and description of properties of the entropy. Entropy enables us to describe the process of approach to equilibrium. In the natural course of events, any system in isolation will change its state toward equilibrium. A system which is not in equilibrium must therefore undergo an irreversible process leading to equilibrium. The process is irreversible because the reverse process would take us away from equilibrium, which is impossible for a macroscopic system. Reversible change can occur if the state of a system in equilibrium is changed by transfer of heat or by work in such a way (slowly) that it always remains in equilibrium.

For every macroscopic state of a system (not necessarily in equilibrium) there exists a quantity S called the entropy of the system. The change in S is positive for any natural process (change toward equilibrium) of an isolated system

$$dS \geq 0 \tag{1.3.5}$$

For an isolated system, equality holds only in equilibrium when no change occurs. The converse is also true—any possible change that increases S is a natural process. Therefore, for an isolated system S achieves its maximum value for the equilibrium state.

The second property of the entropy describes how it is affected by the processes of work and heat transfer during reversible processes. The entropy is affected only by heat transfer and not by work. If we only perform work and do not transfer heat the entropy is constant. Such processes where $q = 0$ are called adiabatic processes. For adiabatic processes $dS = 0$.

The third property of the entropy is that it is extensive:

$$S = \sum_{\alpha} S^{\alpha} \tag{1.3.6}$$

Since in equilibrium the state of the system is defined by the macroscopic quantities (U,N,V), S is a function of them—$S = S(U,N,V)$—in equilibrium. The fourth property of the entropy is that if we keep the size of the system constant by fixing both the number of particles N and the volume V, then the change in entropy S with increasing energy U is always positive:

$$\left(\frac{\partial S}{\partial U} \right)_{N,V} > 0 \tag{1.3.7}$$

where the subscripts denote the (values of the) constant quantities. Because of this we can also invert the function $S = S(U,N,V)$ to obtain the energy U in terms of S, N and V: $U = U(S,N,V)$.

Finally, we mention that the zero of the entropy is arbitrary in classical treatments. The zero of entropy does attain significance in statistical treatments that include quantum effects.

Having described the properties of the entropy for a single system, we can now reconsider the problem of two interacting systems. Since the entropy describes the process of equilibration, we consider the process by which two systems equilibrate thermally. According to the zeroth law, when the two systems are in equilibrium they are at the same temperature. The two systems are assumed to be isolated from any other influence, so that together they form an isolated system with energy U_t and entropy S_t. Each of the subsystems is itself in equilibrium, but they are at different temperatures initially, and therefore heat is transferred to achieve equilibrium. The heat transfer is assumed to be performed in a reversible fashion—slowly. The two subsystems are also assumed to have a fixed number of particles N_1, N_2 and volume V_1, V_2. No work is done, only heat is transferred. The energies of the two systems U_1 and U_2 and entropies S_1 and S_2 are not fixed.

The transfer of heat results in a transfer of energy between the two systems according to Eq. (1.3.4), since the total energy

$$U_t = U_1 + U_2 \tag{1.3.8}$$

is conserved, we have

$$dU_t = dU_1 + dU_2 = 0 \tag{1.3.9}$$

We will consider the processes of equilibration twice. The first time we will identify the equilibrium condition and the second time we will describe the equilibration. At equilibrium the entropy of the whole system is maximized. Variation of the entropy with respect to any internal parameter will give zero at equilibrium. We can consider the change in the entropy of the system as a function of how much of the energy is allocated to the first system:

$$\frac{dS_t}{dU_1} = \frac{dS_1}{dU_1} + \frac{dS_2}{dU_1} = 0 \tag{1.3.10}$$

in equilibrium. Since the total energy is fixed, using Eq. (1.3.9) we have:

$$\frac{dS_t}{dU_1} = \frac{dS_1}{dU_1} - \frac{dS_2}{dU_2} = 0 \tag{1.3.11}$$

or

$$\frac{dS_1}{dU_1} = \frac{dS_2}{dU_2} \tag{1.3.12}$$

in equilibrium. By the definition of the temperature, any function of the derivative of the entropy with respect to energy could be used as the temperature. It is conventional to define the temperature T using:

$$\frac{1}{T} = \left(\frac{dS}{dU} \right)_{N,V} \tag{1.3.13}$$

This definition corresponds to the Kelvin temperature scale. The units of temperature also define the units of the entropy. This definition has the advantage that heat always flows from the system at higher temperature to the system at lower temperature.

To prove this last statement, consider a natural small transfer of heat from one system to the other. The transfer must result in the two systems raising their collective entropy:

$$dS_t = dS_1 + dS_2 \geq 0 \tag{1.3.14}$$

We rewrite the change in entropy of each system in terms of the change in energy. We recall that N and V are fixed for each of the two systems and the entropy is a function only of the three macroscopic parameters (U,N,V). The change in S for each system may be written as:

$$dS_1 = \left(\frac{\partial S}{\partial U} \right)_{N_1,V_1} dU_1$$

$$dS_2 = \left(\frac{\partial S}{\partial U} \right)_{N_2,V_2} dU_2 \tag{1.3.15}$$

to arrive at:

$$\left(\frac{\partial S}{\partial U}\right)_{N_1,V_1} dU_1 + \left(\frac{\partial S}{\partial U}\right)_{N_2,V_2} dU_2 \geq 0 \qquad (1.3.16)$$

or using Eq. (1.3.9) and the definition of the temperature (Eq. (1.3.13)) we have:

$$\left[\left(\frac{1}{T_1}\right) - \left(\frac{1}{T_2}\right)\right] dU_1 \geq 0 \qquad (1.3.17)$$

or:

$$(T_2 - T_1)\, dU_1 \geq 0 \qquad (1.3.18)$$

This implies that a natural process of heat transfer results in the energy of the first system increasing ($dU_1 > 0$) if the temperature of the second system is greater than the first ($(T_2 - T_1) > 0$), or conversely, if the temperature of the second system is less than the temperature of the first.

Using the definition of temperature, we can also rewrite the expression for the change in the energy of a system due to heat transfer or work, Eq. (1.3.2). The new expression is restricted to reversible processes. As in Eq. (1.3.2), N is still fixed. Considering only reversible processes means we consider only equilibrium states of the system, so we can write the energy as a function of the entropy $U = U(S,N,V)$. Since a reversible process changes the entropy and volume while keeping this function valid, we can write the change in energy for a reversible process as

$$\begin{aligned} dU &= \left(\frac{\partial U}{\partial S}\right)_{N,V} dS + \left(\frac{\partial U}{\partial V}\right)_{N,S} dV \\ &= TdS + \left(\frac{\partial U}{\partial V}\right)_{N,S} dV \end{aligned} \qquad (1.3.19)$$

The first term reflects the effect of a change in entropy and the second reflects the change in volume. The change in entropy is related to heat transfer but not to work. If work is done and no heat is transferred, then the first term is zero. Comparing the second term to Eq. (1.3.2) we find

$$P = -\left(\frac{\partial U}{\partial V}\right)_{N,S} \qquad (1.3.20)$$

and the incremental change in energy for a reversible process can be written:

$$dU = TdS - PdV \qquad (1.3.21)$$

This relationship enables us to make direct experimental measurements of entropy changes. The work done on a system, in a reversible or irreversible process, changes the energy of the system by a known amount. This energy can then be extracted in a reversible process in the form of heat. When the system returns to its original state, we can quantify the amount of heat transferred as a form of energy. Measured heat transfer can then be related to entropy changes using $q = TdS$.

Our treatment of the fundamentals of thermodynamics was brief and does not contain the many applications necessary for a detailed understanding. The properties of S that we have described are sufficient to provide a systematic treatment of the thermodynamics of macroscopic bodies. However, the entropy is more understandable from a microscopic (statistical) description of matter. In the next section we introduce the statistical treatment that enables contact between a microscopic picture and the macroscopic thermodynamic treatment of matter. We will use it to give microscopic meaning to the entropy and temperature. Once we have developed the microscopic picture we will discuss two applications. The first application, the ideal gas, is discussed in Section 1.3.3. The discussion of the second application, the Ising model of magnetic systems, is postponed to Section 1.6.

1.3.2 *The macroscopic state from microscopic statistics*

In order to develop a microscopic understanding of the macroscopic properties of matter we must begin by restating the nature of the systems that thermodynamics describes. Even when developing a microscopic picture, the thermodynamic assumptions are relied upon as guides. Macroscopic systems are assumed to have an extremely large number N of individual particles (e.g., at a scale of 10^{23}) in a volume V. Because the size of these systems is so large, they are typically investigated by considering the limit of $N \rightarrow \infty$ and $V \rightarrow \infty$, while the density $n = N/V$ remains constant. This is called the thermodynamic limit. Various properties of the system are separated into extensive and intensive quantities. Extensive quantities are proportional to the size of the system. Intensive quantities are independent of the size of the system. This reflects the intuition that local properties of a macroscopic object do not depend on the size of the system. As in Figs. 1.3.1 and 1.3.2, the system may be cut into two parts, or a small part may be separated from a large part without affecting its local properties.

The total energy U of an isolated system in equilibrium, along with the number of particles N and volume V, defines the macroscopic state (macrostate) of an isolated system in equilibrium. Microscopically, the energy of the system E is given in classical mechanics in terms of the complete specification of the individual particle positions, momenta and interaction potentials. Together these define the microscopic state (microstate) of the system. The microstate is defined differently in quantum mechanics but similar considerations apply. When we describe the system microscopically we use the notation E rather than U to describe the energy. The reason for this difference is that macroscopically the energy U has some degree of fuzziness in its definition, though the degree of fuzziness will not enter into our considerations. Moreover, U may also be used to describe the energy of a system that is in thermal equilibrium with another system. However, thinking microscopically, the energy of such a system is not well defined, since thermal contact allows the exchange of energy between the two systems. We should also distinguish between the microscopic and macroscopic concepts of the number of particles and the volume, but since we will not make use of this distinction, we will not do so.

There are many possible microstates that correspond to a particular macrostate of the system specified only by U, N, V. We now make a key assumption of statistical mechanics—that all of the possible microstates of the system occur with equal probability. The number of these microstates $\Omega(U, N, V)$, which by definition depends on the macroscopic parameters, turns out to be central to statistical mechanics and is directly related to the entropy. Thus it determines many of the thermodynamic properties of the system, and can be discussed even though we are not always able to obtain it explicitly.

We consider again the problem of interacting systems. As before, we consider two systems (Fig. 1.3.3) that are in equilibrium separately, with state variables (U_1, N_1, V_1) and (U_2, N_2, V_2). The systems have a number of microstates $\Omega_1(U_1, N_1, V_1)$ and $\Omega_2(U_2, N_2, V_2)$ respectively. It is not necessary that the two systems be formed of the same material or have the same functional form of $\Omega(U, N, V)$, so the function Ω is also labeled by the system index. The two systems interact in a limited way, so that they can exchange only energy. The number of particles and volume of each system remains fixed. Conservation of energy requires that the total energy $U_t = U_1 + U_2$ remains fixed, but energy may be transferred from one system to the other. As before, our objective is to identify when energy transfer stops and equilibrium is reached.

Consider the number of microstates of the whole system Ω_t. This number is a function not only of the total energy of the system but also of how the energy is allocated between the systems. So, we write $\Omega_t(U_1, U_2)$, and we assume that at any time the energy of each of the two systems is well defined. Moreover, the interaction between the two systems is sufficiently weak so that the number of states of each system

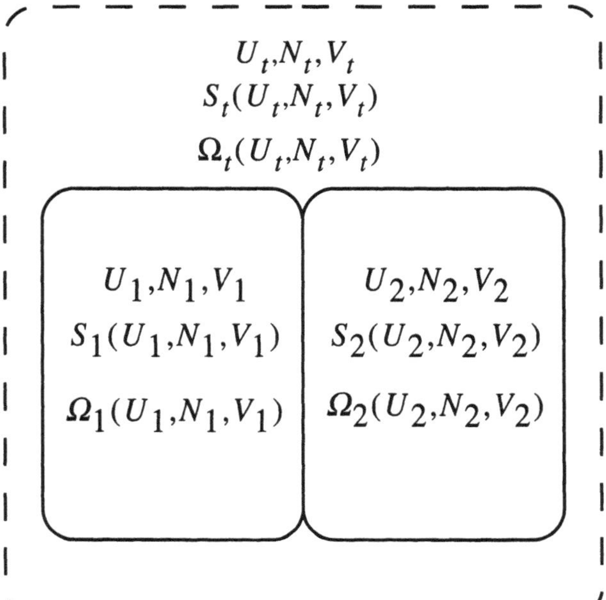

Figure 1.3.3 Illustration of a system formed out of two parts. The text discusses this system when energy is transferred from one part to the other. The transfer of energy on a microscopic scale is equivalent to the transfer of heat on a macroscopic scale, since the two systems are not allowed to change their number of particles or their volume. ∎

U_t, N_t, V_t
$S_t(U_t, N_t, V_t)$
$\Omega_t(U_t, N_t, V_t)$

U_1, N_1, V_1
$S_1(U_1, N_1, V_1)$
$\Omega_1(U_1, N_1, V_1)$

U_2, N_2, V_2
$S_2(U_2, N_2, V_2)$
$\Omega_2(U_2, N_2, V_2)$

may be counted independently. Then the total number of microstates is the product of the number of microstates of each of the two systems separately.

$$\Omega_t(U_1,U_2) = \Omega_1(U_1)\Omega_2(U_2) \tag{1.3.22}$$

where we have dropped the arguments N and V, since they are fixed throughout this discussion. When energy is transferred, the number of microstates of each of the two systems is changed. When will the transfer of energy stop? Left on its own, the system will evolve until it reaches the most probable separation of energy. Since any particular state is equally likely, the most probable separation of energy is the separation that gives rise to the greatest possible number of states. When the number of particles is large, the greatest number of states corresponding to a particular energy separation is much larger than the number of states corresponding to any other possible separation. Thus any other possibility is completely negligible. No matter when we look at the system, it will be in a state with the most likely separation of the energy. For a macroscopic system, it is impossible for a spontaneous transfer of energy to occur that moves the system away from equilibrium.

The last paragraph implies that the transfer of energy from one system to the other stops when Ω_t reaches its maximum value. Since $U_t = U_1 + U_2$ we can find the maximum value of the number of microstates using:

$$\frac{\partial \Omega_t(U_1,U_t-U_1)}{\partial U_1} = 0 = \frac{\partial \Omega_1(U_1)}{\partial U_1}\Omega_2(U_t-U_1) + \Omega_1(U_1)\frac{\partial \Omega_2(U_t-U_1)}{\partial U_1}$$

$$0 = \frac{\partial \Omega_1(U_1)}{\partial U_1}\Omega_2(U_2) - \Omega_1(U_1)\frac{\partial \Omega_2(U_2)}{\partial U_2} \tag{1.3.23}$$

or

$$\frac{1}{\Omega_1(U_1)}\frac{\partial \Omega_1(U_1)}{\partial U_1} = \frac{1}{\Omega_2(U_2)}\frac{\partial \Omega_2(U_2)}{\partial U_2}$$

$$\frac{\partial \ln\Omega_1(U_1)}{\partial U_1} = \frac{\partial \ln\Omega_2(U_2)}{\partial U_2} \tag{1.3.24}$$

The equivalence of these quantities is analogous to the equivalence of the temperature of the two systems in equilibrium. Since the derivatives in the last equation are performed at constant N and V, it appears, by analogy to Eq. (1.3.12), that we can identify the entropy as:

$$S = k\ln(\Omega(E,N,V)). \tag{1.3.25}$$

The constant k, known as the Boltzmann constant, is needed to ensure correspondence of the microscopic counting of states with the macroscopic units of the entropy, as defined by the relationship of Eq. (1.3.13), once the units of temperature and energy are defined.

The entropy as defined by Eq. (1.3.25) can be shown to satisfy all of the properties of the thermodynamic entropy in the last section. We have argued that an isolated system evolves its macrostate in such a way that it maximizes the number of microstates that correspond to the macrostate. By Eq. (1.3.25), this is the same as the first property of the entropy in Eq. (1.3.5), the maximization of the entropy in equilibrium.

Interestingly, demonstrating the second property of the entropy, that it does not change during an adiabatic process, requires further formal developments relating entropy to information that will be discussed in Sections 1.7 and 1.8. We will connect the two discussions and thus be able to demonstrate the second property of the entropy in Chapter 8 (Section 8.3.2).

The extensive property of the entropy follows from Eq. (1.3.22). This also means that the number of states at a particular energy grows exponentially with the size of the system. More properly, we can say that experimental observation that the entropy is extensive suggests that the interaction between macroscopic materials, or parts of a single macroscopic material, is such that the microstates of each part of the system may be enumerated independently.

The number of microstates can be shown by simple examples to increase with the energy of the system. This corresponds to Eq. (1.3.7). There are also examples where this can be violated, though this will not enter into our discussions.

We consider next a second example of interacting systems that enables us to evaluate the meaning of a system in equilibrium with a reservoir at a temperature T. We consider a small part of a much larger system (Fig. 1.3.4). No assumption is necessary regarding the size of the small system; it may be either microscopic or macroscopic. Because of the contact of the small system with the large system, its energy is not

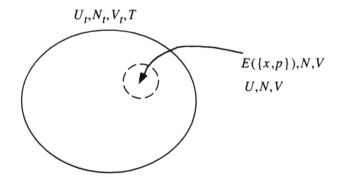

Figure 1.3.4 In order to understand temperature we consider a closed system composed of a large and small system, or equivalently a small system which is part of a much larger system. The larger system serves as a thermal reservoir transferring energy to and from the small system without affecting its own temperature. A microscopic description of this process in terms of a single microscopic state of the small system leads to the Boltzmann probability. An analysis in terms of the macroscopic state of the small system leads to the principle of minimization of the free energy to obtain the equilibrium state of a system at a fixed temperature. This principle replaces the principle of maximization of the entropy, which only applies for a closed system. ∎

always the same. Energy will be transferred back and forth between the small and large systems. The essential assumption is that the contact between the large and small system does not affect any other aspect of the description of the small system. This means that the small system is in some sense independent of the large system, despite the energy transfer. This is true if the small system is itself macroscopic, but it may also be valid for certain microscopic systems. We also assume that the small system and the large system have fixed numbers of particles and volumes.

Our objective is to consider the probability that a particular microstate of the small system will be realized. A microstate is identified by all of the microscopic parameters necessary to completely define this state. We use the notation $\{x,p\}$ to denote these coordinates. The probability that this particular state will be realized is given by the fraction of states of the whole system for which the small system attains this state. Because there is only one such state for the small system, the probability that this state will be realized is given by (proportional to) a count of the number of states of the rest of the system. Since the large system is macroscopic, we can count this number by using the macroscopic expression for the number of states of the large system:

$$P(\{x, p\}) \propto \Omega_R(U_t - E(\{x, p\}), N_t - N, V_t - V) \qquad (1.3.26)$$

where $E(\{x,p\}), N, V$ are the energy, number of particles and volume of the microscopic system respectively. $E(\{x,p\})$ is a function of the microscopic parameters $\{x,p\}$. U_t, N_t, V_t are the energy, number of particles and volume of the whole system, including both the small and large systems. Ω_R is the entropy of the large subsystem (reservoir). Since the number of states generally grows faster than linearly as a function of the energy, we use a Taylor expansion of its logarithm (or equivalently a Taylor expansion of the entropy) to find

$$\ln \Omega_R(U_t - E(\{x, p\}), N_t - N, V_t - V)$$

$$= \ln \Omega_R(U_t, N_t - N, V_t - V) + \left(\frac{\partial \ln \Omega_R(U_t, N_t - N, V_t - V)}{\partial E_t} \right)_{N_t, V_t} (-E(\{x, p\}))$$

$$= \ln \Omega_R(U_t, N_t - N, V_t - V) + \frac{1}{kT}(-E(\{x, p\})) \qquad (1.3.27)$$

where we have not expanded in the number of particles and the volume because they are unchanging. We take only the first term in the expansion, because the size of the small system is assumed to be much smaller than the size of the whole system. Exponentiating gives the relative probability of this particular microscopic state.

$$\Omega_R(U_t - E(\{x,p\}), N_t - N, V_t - V) = \Omega_R(U_t, N_t - N, V_t - V)e^{-E(\{x,p\})/kT} \qquad (1.3.28)$$

The probability of this particular state must be normalized so that the sum over all states is one. Since we are normalizing the probability anyway, the constant coefficient does not affect the result. This gives us the Boltzmann probability distribution:

$$P(\{x,p\}) = \frac{1}{Z} e^{-E(\{x,p\})/kT}$$
$$Z = \sum_{\{x,p\}} e^{-E(\{x,p\})/kT}$$

(1.3.29)

Eq. (1.3.29) is independent of the states of the large system and depends only on the microscopic description of the states of the small system. It is this expression which generally provides the most convenient starting point for a connection between the microscopic description of a system and macroscopic thermodynamics. It identifies the probability that a particular microscopic state will be realized when the system has a well-defined temperature T. In this way it also provides a microscopic meaning to the macroscopic temperature T. It is emphasized that Eq. (1.3.29) describes both microscopic and macroscopic systems in equilibrium at a temperature T.

The probability of occurrence of a particular state should be related to the description of a system in terms of an ensemble. We have found by Eq. (1.3.29) that a system in thermal equilibrium at a temperature T is represented by an ensemble that is formed by taking each of the states in proportion to its Boltzmann probability. This ensemble is known as the canonical ensemble. The canonical ensemble should be contrasted with the assumption that each state has equal probability for isolated systems at a particular energy. The ensemble of fixed energy and equal a priori probability is known as the microcanonical ensemble. The canonical ensemble is both easier to discuss analytically and easier to connect with the physical world. It will be generally assumed in what follows.

We can use the Boltzmann probability and the definition of the canonical ensemble to obtain all of the thermodynamic quantities. The macroscopic energy is given by the average over the microscopic energy using:

$$U = \frac{1}{Z} \sum_{\{x,p\}} E(\{x,p\}) e^{-E(\{x,p\})/kT}$$

(1.3.30)

For a macroscopic system, the average value of the energy will always be observed in any specific measurement, despite the Boltzmann probability that allows all energies. This is because the number of states of the system rises rapidly with the energy. This rapid growth and the exponential decrease of the probability with the energy results in a sharp peak in the probability distribution as a function of energy. The sharp peak in the probability distribution means that the probability of any other energy is negligible. This is discussed below in Question 1.3.1.

For an isolated macroscopic system, we were able to identify the equilibrium state from among other states of the system using the principle of the maximization of the entropy. There is a similar procedure for a macroscopic system in contact with a thermal reservoir at a fixed temperature T. The important point to recognize is that when we had a closed system, the energy was fixed. Now, however, the objective becomes to identify the energy at equilibrium. Of course, the energy is given by the average in

Eq. (1.3.30). However, to generalize the concept of maximizing the entropy, it is simplest to reconsider the problem of the system in contact with the reservoir when the small system is also macroscopic.

Instead of considering the probability of a particular microstate of well-defined energy E, we consider the probability of a macroscopic state of the system with an energy U. In this case, we find the equilibrium state of the system by maximizing the number of states of the whole system, or alternatively of the entropy:

$$\ln \Omega(U,N,V) + \ln \Omega_R(U_t - U, N_t - N, V_t - V)$$
$$= S(U,N,V)/k + S_R(U_t - U, N_t - N, V_t - V)/k$$
$$= S(U,N,V)/k + S_R(U_t, N_t - N, V_t - V)/k + \frac{1}{kT}(-U)$$
(1.3.31)

To find the equilibrium state, we must maximize this expression for the entropy of the whole system. We can again ignore the constant second term. This leaves us with quantities that are only characterizing the small system we are interested in, and the temperature of the reservoir. Thus we can find the equilibrium state by maximizing the quantity

$$S - U/T \tag{1.3.32}$$

It is conventional to rewrite this and, rather than maximizing the function in Eq. (1.3.32), to minimize the function known as the free energy:

$$F = U - TS \tag{1.3.33}$$

This suggests a simple physical significance of the process of change toward equilibrium. At a fixed temperature, the system seeks to minimize its energy and maximize its entropy at the same time. The relative importance of the entropy compared to the energy is set by the temperature. For high temperature, the entropy becomes more dominant, and the energy rises in order to increase the entropy. At low temperature, the energy becomes more dominant, and the energy is lowered at the expense of the entropy. This is the precise statement of the observation that "everything flows downhill." The energy entropy competition is a balance that is rightly considered as one of the most basic of physical phenomena.

We can obtain a microscopic expression for the free energy by an exercise that begins from a microscopic expression for the entropy:

$$S = k \ln(\Omega) = k \ln \left(\sum_{\{x,p\}} \delta_{E(\{x,p\}),U} \right) \tag{1.3.34}$$

The summation is over all microscopic states. The delta function is 1 only when $E(\{x,p\}) = U$. Thus the sum counts all of the microscopic states with energy U. Strictly speaking, the δ function is assumed to be slightly "fuzzy," so that it gives 1 when $E(\{x,p\})$ differs from U by a small amount on a macroscopic scale, but by a large amount in terms of the differences between energies of microstates. We can then write

$$S = k\ln(\Omega) = k\ln\left(\sum_{\{x,p\}} \delta_{E(\{x,p\}),U} e^{-E(\{x,p\})/kT} e^{U/kT}\right)$$

$$= \frac{U}{T} + k\ln\left(\sum_{\{x,p\}} \delta_{E(\{x,p\}),U} e^{-E(\{x,p\})/kT}\right) \tag{1.3.35}$$

Let us compare the sum in the logarithm with the expression for Z in Eq. (1.3.29). We will argue that they are the same. This discussion hinges on the rapid increase in the number of states as the energy increases. Because of this rapid growth, the value of Z in Eq. (1.3.29) actually comes from only a narrow region of energy. We know from the expression for the energy average, Eq. (1.3.30), that this narrow region of energy must be at the energy U. This implies that for all intents and purposes the quantity in the brackets of Eq. (1.3.35) is equivalent to Z. This argument leads to the expression:

$$S = \frac{U}{T} + k\ln Z \tag{1.3.36}$$

Comparing with Eq. (1.3.33) we have

$$F = -kT\ln Z \tag{1.3.37}$$

Since the Boltzmann probability is a convenient starting point, this expression for the free energy is often simpler to evaluate than the expression for the entropy, Eq. (1.3.34). A calculation of the free energy using Eq. (1.3.37) provides contact between microscopic models and the macroscopic behavior of thermodynamic systems. The Boltzmann normalization Z, which is directly related to the free energy is also known as the partition function. We can obtain other thermodynamic quantities directly from the free energy. For example, we rewrite the expression for the energy Eq. (1.3.30) as:

$$U = \frac{1}{Z} \sum_{\{x,p\}} E(\{x,p\}) e^{-\beta E(\{x,p\})} = -\frac{\partial \ln(Z)}{\partial \beta} = \frac{\partial \beta F}{\partial \beta} \tag{1.3.38}$$

where we use the notation $\beta = 1/kT$. The entropy can be obtained using this expression for the energy and Eq. (1.3.33) or (1.3.36).

Question 1.3.1 Consider the possibility that the macroscopic energy of a system in contact with a thermal reservoir will deviate from its typical value U. To do this expand the probability distribution of macroscopic energies of a system in contact with a reservoir around this value. How large are the deviations that occur?

Solution 1.3.1 We considered Eq. (1.3.31) in order to optimize the entropy and find the typical value of the energy U. We now consider it again to find the distribution of probabilities of values of the energy around the value U similar to the way we discussed the distribution of microscopic states $\{x, p\}$ in Eq. (1.3.27). To do this we distinguish between the observed value of the energy U' and U. Note that we consider U' to be a macroscopic energy, though the same derivation could be used to obtain the distribution of microscopic energies. The probability of U' is given by:

$$P(U') \propto \Omega(U', N, V)\Omega_R(U_t - U', N_t - N, V_t - V) = e^{S(U')/k + S_R(U_t - U')/k} \quad (1.3.39)$$

In the latter form we ignore the fixed arguments N and V. We expand the logarithm of this expression around the expected value of energy U:

$$S(U') + S_R(U_t - U')$$

$$= S(U)/k + S_R(U_t - U)/k + \frac{1}{2k}\frac{d^2S(U)}{dU^2}(U - U')^2 + \frac{1}{2k}\frac{d^2S(U_t - U)}{dU_t^2}(U - U')^2$$

$$(1.3.40)$$

where we have kept terms to second order. The first-order terms, which are of the form $(1/kT)(U' - U)$, have opposite signs and therefore cancel. This implies that the probability is a maximum at the expected energy U. The second derivative of the entropy can be evaluated using:

$$\frac{d^2S(U)}{dU^2} = \frac{d}{dU}\frac{1}{T} = -\frac{1}{T^2}\frac{1}{dU/dT} = -\frac{1}{T^2 C_V} \quad (1.3.41)$$

where C_V is known as the specific heat at constant volume. For our purposes, its only relevant property is that it is an extensive quantity. We can obtain a similar expression for the reservoir and define the reservoir specific heat C_{VR}. Thus the probability is:

$$P(U') \propto e^{-(1/2kT^2)(1/C_V + 1/C_{VR})(U - U')^2} \approx e^{-(1/2kT^2)(1/C_V)(U - U')^2} \quad (1.3.42)$$

where we have left out the (constant) terms that do not depend on U'. Because C_V and C_{VR} are extensive quantities and the reservoir is much bigger than the small system, we can neglect $1/C_{VR}$ compared to $1/C_V$. The result is a Gaussian distribution (Eq. (1.2.39)) with a standard deviation

$$\sigma = T\sqrt{kC_V} \quad (1.3.43)$$

This describes the characteristic deviation of the energy U' from the average or typical energy U. However, since C_V is extensive, the square root means that the deviation is proportional to $\sqrt{N}$. Note that the result is consistent with a random walk of N steps. So for a large system of $N \sim 10^{23}$ particles, the possible deviation in the energy is smaller than the energy by a factor of (we are neglecting everything but the N dependence) 10^{12}—i.e., it is undetectable. Thus the energy of a thermodynamic system is very well defined. ∎

1.3.3 *Kinetic theory of gases and pressure*

In the previous section, we described the microscopic analog of temperature and entropy. We assumed that the microscopic analog of energy was understood, and we developed the concept of free energy and its microscopic analog. One quantity that we have not discussed microscopically is the pressure. Pressure is a Newtonian concept—the force per unit area. For various reasons, it is helpful for us to consider the microscopic origin of pressure for the example of a simplified model of a gas called an ideal gas. In Question 1.3.2 we use the ideal gas as an example of the thermodynamic and statistical analysis of materials.

An ideal gas is composed of indistinguishable point particles with a mass m but with no internal structure or size. The interaction between the particles is neglected, so that the energy is just their kinetic energy. The particles do interact with the walls of the container in which the gas is confined. This interaction is simply that of reflection—when the particle is incident on a wall, the component of its velocity perpendicular to the wall is reversed. Energy is conserved. This is in accordance with the expectation from Newton's laws for the collision of a small mass with a much larger mass object.

To obtain an expression for the pressure, we must suffer with some notational hazards, as the pressure P, probability of a particular velocity $P(v)$ and momentum of a particular particle $\mathbf{p}_i$ are all designated by the letter P but with different case, arguments or subscripts. A bold letter $\mathbf{F}$ is used briefly for the force, and otherwise F is used for the free energy. We rely largely upon context to distinguish them. Since the objective of using an established notation is to make contact with known concepts, this situation is sometimes preferable to introducing a new notation.

Because of the absence of collisions between different particles of the gas, there is no communication between them, and each of the particles bounces around the container on its own course. The pressure on the container walls is given by the force per unit area exerted on the walls, as illustrated in Fig. 1.3.5. The force is given by the action of the wall on the gas that is needed to reverse the momenta of the incident particles between t and $t + \Delta t$:

$$P = \frac{|\mathbf{F}|}{A} = \frac{1}{A\Delta t}\left|\sum_i m\Delta\mathbf{v}_i\right| \tag{1.3.44}$$

where $|\mathbf{F}|$ is the magnitude of the force on the wall. The latter expression relates the pressure to the change in the momenta of incident particles per unit area of the wall. A is a small but still macroscopic area, so that this part of the wall is flat. Microscopic roughness of the surface is neglected. The change in velocity Δv_i of the particles during the time Δt is zero for particles that are not incident on the wall. Particles that hit the wall between t and $t + \Delta t$ are moving in the direction of the wall at time t and are near enough to the wall to reach it during Δt. Faster particles can reach the wall from farther away, but only the velocity perpendicular to the wall matters. Denoting this velocity component as $v_\perp$, the maximum distance is $v_\perp \Delta t$ (see Fig. 1.3.5).

If the particles have velocity only perpendicular to the wall and no velocity parallel to the wall, then we could count the incident particles as those in a volume $Av_\perp \Delta t$. We can use the same expression even when particles have a velocity parallel to the surface, because the parallel velocity takes particles out of and into this volume equally.

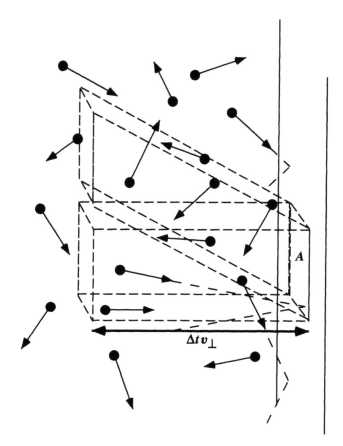

Figure 1.3.5 Illustration of a gas of ideal particles in a container near one of the walls. Particles incident on the wall are reflected, reversing their velocity perpendicular to the wall, and not affecting the other components of their velocity. The wall experiences a pressure due to the collisions and applies the same pressure to the gas. To calculate the pressure we must count the number of particles in a unit of time Δt with a particular perpendicular velocity $v_\perp$ that hit an area A. This is equivalent to counting the number of particles with the velocity $v_\perp$ in the box shown with one of its sides of length $\Delta t v_\perp$. Particles with velocity $v_\perp$ will hit the wall if and only if they are in the box. The same volume of particles applies if the particles also have a velocity parallel to the surface, since this just skews the box, as shown, leaving its height and base area the same. ∎

Another way to say this is that for a particular parallel velocity we count the particles in a sheared box with the same height and base and therefore the same volume. The total number of particles in the volume, $(N/V)Av_\perp\Delta t$, is the volume times the density (N/V).

Within the volume $Av_\perp\Delta t$, the number of particles that have the velocity $v_\perp$ is given by the number of particles in this volume times the probability $P(v_\perp)$ that a particle has its perpendicular velocity component equal to $v_\perp$. Thus the number of particles incident on the wall with a particular velocity perpendicular to the wall $v_\perp$ is given by

$$\frac{N}{V} A P(v_\perp) v_\perp \Delta t \tag{1.3.45}$$

The total change in momentum is found by multiplying this by the change in momentum of a single particle reflected by the collision, $2mv_\perp$, and integrating over all velocities.

$$\left| \sum_i m\Delta v_i \right| = \frac{1}{V} N A \Delta t \int_0^\infty dv_\perp P(v_\perp) v_\perp (2mv_\perp) \tag{1.3.46}$$

Divide this by $A\Delta t$ to obtain the change in momentum per unit time per unit area, which is the pressure (Eq. (1.3.44)),

$$P = \frac{1}{V} N \int_0^\infty dv_\perp P(v_\perp) v_\perp (2mv_\perp) \tag{1.3.47}$$

We rewrite this in terms of the average squared velocity perpendicular to the surface

$$P = \frac{N}{V} m2 \int_0^\infty dv_\perp P(v_\perp) v_\perp^2 = \frac{N}{V} m \int_{-\infty}^\infty dv_\perp P(v_\perp) v_\perp^2 = \frac{N}{V} m <v_\perp^2> \tag{1.3.48}$$

where the equal probability of having positive and negative velocities enables us to extend the integral to $-\infty$ while eliminating the factor of two. We can rewrite Eq. (1.3.48) in terms of the average square magnitude of the total velocity. There are three components of the velocity (two parallel to the surface). The squares of the velocity components add to give the total velocity squared and the averages are equal:

$$<v^2> = <v_1^2 + v_2^2 + v_3^2> = 3 <v_\perp^2> \tag{1.3.49}$$

where v is the magnitude of the particle velocity. The pressure is:

$$P = \frac{N}{V} m \frac{1}{3} <v^2> \tag{1.3.50}$$

Note that the wall does not influence the probability of having a particular velocity nearby. Eq. (1.3.50) is a microscopic expression for the pressure, which we can calculate using the Boltzmann probability from Eq. (1.3.29). We do this as part of Question 1.3.2.

Question 1.3.2 Develop the statistical description of the ideal gas by obtaining expressions for the thermodynamic quantities Z, F, U, S and P, in terms of N, V, and T. For hints read the first three paragraphs of the solution.

Solution 1.3.2 The primary task of statistics is counting. To treat the ideal gas we must count the number of microscopic states to obtain the entropy, or sum over the Boltzmann probability to obtain Z and the free energy. The ideal gas presents us with two difficulties. The first is that each particle has a continuum of possible locations. The second is that we must treat the particles as microscopically indistinguishable. To solve the first problem, we have to set some interval of position at which we will call a particle here different from a particle there. Moreover, since a particle at any location may have many different velocities, we must also choose a difference of velocities that will be considered as distinct. We define the interval of position to be Δx and the interval of momentum to be Δp. In each spatial dimension, the positions between x and $x + \Delta x$ correspond to a single state, and the momenta between p and $p + \Delta p$ correspond to a single state. Thus we consider as one state of the system a particle which has position and momenta in a six-dimensional box of a size $\Delta x^3 \Delta p^3$. The size of this box enters only as a constant in classical statistical mechanics, and we will not be concerned with its value. Quantum mechanics identifies it with $\Delta x^3 \Delta p^3 = h^3$, where h is Planck's constant, and for convenience we adopt this notation for the unit volume for counting.

There is a subtle but important choice that we have made. We have chosen to make the counting intervals have a fixed width Δp in the momentum. From classical mechanics, it is not entirely clear that we should make the intervals of fixed width in the momentum or, for example, make them fixed in the energy ΔE. In the latter case we would count a single state between E and $E + \Delta E$. Since the energy is proportional to the square of the momentum, this would give a different counting. Quantum mechanics provides an unambiguous answer that the momentum intervals are fixed.

To solve the problem of the indistinguishability of the particles, we must remember every time we count the number of states of the system to divide by the number of possible ways there are to interchange the particles, which is $N!$.

The energy of the ideal gas is given by the kinetic energy of all of the particles:

$$E(\{x, p\}) = \sum_{i=1}^{N} \frac{1}{2} m v_i^2 = \sum_{i=1}^{N} \frac{p_i^2}{2m} \qquad (1.3.51)$$

where the velocity and momentum of a particle are three-dimensional vectors with magnitude v_i and p_i respectively. We start by calculating the partition function (Boltzmann normalization) Z from Eq. (1.3.29)

$$Z = \frac{1}{N!} \sum_{\{x,p\}} e^{-\sum_{i=1}^{N} \frac{p_i^2}{2mkT}} = \frac{1}{N!} \int e^{-\sum_{i=1}^{N} \frac{p_i^2}{2mkT}} \prod_{i=1}^{N} \frac{d^3 x_i \, d^3 p_i}{h^3} \qquad (1.3.52)$$

where the integral is to be evaluated over all possible locations of each of the N particles of the system. We have also included the correction to over-counting, $N!$. Since the particles do not see each other, the energy is a sum over each particle energy. The integrals separate and we have:

$$Z = \frac{1}{N!}\left(\frac{1}{h^3}\int e^{-\frac{p^2}{2mkT}} d^3 x d^3 p\right)^N \qquad (1.3.53)$$

The position integral gives the volume V, immediately giving the dependence of Z on this macroscopic quantity. The integral over momentum can be evaluated giving:

$$\int e^{-\frac{p^2}{2mkT}} d^3 p = 4\pi \int_0^\infty p^2 dp\, e^{-\frac{p^2}{2mkT}} = 4\pi(2mkT)^{3/2} \int_0^\infty y^2 dy e^{-y^2}$$

$$= 4\pi(2mkT)^{3/2}\left(-\frac{\partial}{\partial a}\bigg|_{a=1}\right)\int_0^\infty dy e^{-ay^2} = 4\pi(2mkT)^{3/2}\left(-\frac{\partial}{\partial a}\bigg|_{a=1}\right)\frac{1}{2}\sqrt{\frac{\pi}{a}}$$

$$= (2\pi mkT)^{3/2} \qquad (1.3.54)$$

and we have that

$$Z(V,T,N) = \frac{V^N}{N!}\left(2\pi mkT/h^2\right)^{3N/2} \qquad (1.3.55)$$

We could have simplified the integration by recognizing that each component of the momentum p_x, p_y and p_z can be integrated separately, giving $3N$ independent one-dimensional integrals and leading more succinctly to the result. The result can also be written in terms of a natural length $\lambda(T)$ that depends on temperature (and mass):

$$\lambda(T) = (h^2/2\pi mkT)^{1/2} \qquad (1.3.56)$$

$$Z(V,T,N) = \frac{V^N}{N!\lambda(T)^{3N}} \qquad (1.3.57)$$

From the partition function we obtain the free energy, making use of Sterling's approximation (Eq. (1.2.36)):

$$F = kTN(\ln N - 1) - kTN\ln(V/\lambda(T)^3) \qquad (1.3.58)$$

where we have neglected terms that grow less than linearly with N. Terms that vary as $\ln(N)$ vanish on a macroscopic scale. In this form it might appear that we have a problem, since the $N\ln(N)$ term from Sterling's approximation to the factorial does not scale proportional to the size of the system, and F is an extensive quantity. However, we must also note the $N\ln(V)$ term, which we can combine with the $N\ln(N)$ term so that the extensive nature is apparent:

$$F = kTN[\ln N\lambda(T)^3/V) - 1] \qquad (1.3.59)$$

It is interesting that the factor of $N!$, and thus the indistinguishability of particles, is necessary for the free energy to be extensive. If the particles were distinguishable, then cutting the system in two would result in a different counting, since we would lose the states corresponding to particles switching from one part to the other. If we combined the two systems back together, there would be an effect due to the mixing of the distinguishable particles (Question 1.3.3).

The energy may be obtained from Eq. (1.3.38) (any of the forms) as:

$$U = \frac{3}{2} NkT \tag{1.3.60}$$

which provides an example of the equipartition theorem, which says that each degree of freedom (position-momentum pair) of the system carries $kT/2$ of energy in equilibrium. Each of the three spatial coordinates of each particle is one degree of freedom.

The expression for the entropy $(S = (U - F)/T)$

$$S = kN[\ln(V/N\lambda(T)^3) + 5/2] \tag{1.3.61}$$

shows that the entropy per particle S/N grows logarithmically with the volume per particle V/N. Using the expression for U, it may be written in a form $S(U,N,V)$.

Finally, the pressure may be obtained from Eq. (1.3.20), but we must be careful to keep N and S constant rather than T. We have

$$P = -\frac{\partial U}{\partial V}\bigg|_{N,S} = -\frac{3}{2} Nk \frac{\partial T}{\partial V}\bigg|_{N,S} \tag{1.3.62}$$

Taking the same derivative of the entropy Eq. (1.3.61) gives us (the derivative of S with S fixed is zero):

$$0 = -\frac{1}{V} - \frac{3}{2} \frac{\partial T}{\partial V}\bigg|_{N,S} \tag{1.3.63}$$

Substituting, we obtain the ideal gas equation of state:

$$PV = NkT \tag{1.3.64}$$

which we can also obtain from the microscopic expression for the pressure—Eq. (1.3.50). We describe two ways to do this. One way to obtain the pressure from the microscopic expression is to evaluate first the average of the energy

$$U = <E(\{x,p\})> = \sum_{i=1}^{N} \frac{1}{2}m <v_i^2> = N\frac{1}{2}m<v^2> \tag{1.3.65}$$

This may be substituted in to Eq. (1.3.60) to obtain

$$\frac{1}{2}m<v^2>=\frac{3}{2}kT \qquad (1.3.66)$$

which may be substituted directly in to Eq. (1.3.50). Another way is to obtain the average squared velocity directly. In averaging the velocity, it doesn't matter which particle we choose. We choose the first particle:

$$<v_1^2>=3<v_{1\perp}^2>=\frac{\dfrac{1}{N!}\displaystyle\int 3v_{1\perp}^2 e^{-\sum\limits_{i=1}^{N}\frac{p_i^2}{2mkT}}\prod\limits_{i=1}^{N}\dfrac{d^3x_i d^3p_i}{h^3}}{\dfrac{1}{N!}\displaystyle\int e^{-\sum\limits_{i=1}^{N}\frac{p_i^2}{2mkT}}\prod\limits_{i=1}^{N}\dfrac{d^3x_i d^3p_i}{h^3}} \qquad (1.3.67)$$

where we have further chosen to average over only one of the components of the velocity of this particle and multiply by three. The denominator is the normalization constant Z. Note that the factor $1/N!$, due to the indistinguishability of particles, appears in the numerator in any ensemble average as well as in the denominator, and cancels. It does not affect the Boltzmann probability when issues of distinguishability are not involved.

There are $6N$ integrals in the numerator and in the denominator of Eq. (1.3.67). All integrals factor into one-dimensional integrals. Each integral in the numerator is the same as the corresponding one in the denominator, except for the one that involves the particular component of the velocity we are interested in. We cancel all other integrals and obtain:

$$<v_1^2>=3<v_{1\perp}^2>=3\frac{\displaystyle\int v_{1\perp}^2 e^{-\frac{p_{1\perp}^2}{2mkT}}dp_{1\perp}}{\displaystyle\int e^{-\frac{p_{1\perp}^2}{2mkT}}dp_{1\perp}}=3(\frac{2kT}{m})\frac{\displaystyle\int y^2 e^{-y^2}dy}{\displaystyle\int e^{-y^2}dy}=3(\frac{2kT}{m})(\frac{1}{2})$$

$$(1.3.68)$$

The integral is performed by the same technique as used in Eq. (1.3.54). The result is the same as by the other methods. ∎

Question 1.3.3 An insulated box is divided into two compartments by a partition. The two compartments contain two different ideal gases at the same pressure P and temperature T. The first gas has N_1 particles and the second has N_2 particles. The partition is punctured. Calculate the resulting change in thermodynamic parameters (N, V, U, P, S, T, F). What changes in the analysis if the two gases are the same, i.e., if they are composed of the same type of molecules?

Solution 1.3.3 By additivity the extrinsic properties of the whole system before the puncture are (Eq. (1.3.59)–Eq. (1.3.61)):

$$U_0 = U_1 + U_2 = \frac{3}{2}(N_1 + N_2)kT$$

$$V_0 = V_1 + V_2 \tag{1.3.69}$$

$$S_0 = kN_1[\ln(V_1/N_1\lambda(T)^3) + 5/2] + kN_2[\ln(V_2/N_2\lambda(T)^3) + 5/2]$$

$$F_0 = kTN_1[\ln(N_1\lambda(T)^3/V_1) - 1] + kTN_2[\ln(N_2\lambda(T)^3/V_2) - 1]$$

The pressure is intrinsic, so before the puncture it is (Eq. (1.3.64)):

$$P_0 = N_1 kT/V_1 = N_2 kT/V_2 \tag{1.3.70}$$

After the puncture, the total energy remains the same, because the whole system is isolated. Because the two gases do not interact with each other even when they are mixed, their properties continue to add after the puncture. However, each gas now occupies the whole volume, $V_1 + V_2$. The expression for the energy as a function of temperature remains the same, so the temperature is also unchanged. The pressure in the container is now additive: it is the sum of the pressure of each of the gases:

$$P = N_1 kT/(V_1 + V_2) + N_2 kT/(V_1 + V_2) = P_0 \tag{1.3.71}$$

i.e., the pressure is unchanged as well.

The only changes are in the entropy and the free energy. Because the two gases do not interact with each other, as with other quantities, we can write the total entropy as a sum over the entropy of each gas separately:

$$S = kN_1[\ln((V_1 + V_2)/N_1\lambda(T)^3) + 5/2]$$
$$\quad + kN_2[\ln((V_1 + V_2)/N_2\lambda(T)^3) + 5/2] \tag{1.3.72}$$
$$\quad = S_0 + (N_1 + N_2)k\ln(V_1 + V_2) - N_1 k\ln(V_1) - N_2 k\ln(V_2)$$

If we simplify to the case $V_1 = V_2$, we have $S = S_0 + (N_1 + N_2)k \ln(2)$. Since the energy is unchanged, by the relationship of free energy and entropy (Eq. (1.3.33)) we have:

$$F = F_0 - T(S - S_0) \tag{1.3.73}$$

If the two gases are composed of the same molecule, there is no change in thermodynamic parameters as a result of a puncture. Mathematically, the difference is that we replace Eq. (1.3.72) with:

$$S = k(N_1 + N_2)[\ln((V_1 + V_2)/(N_1 + N_2)\lambda(T)^3) + 5/2] = S_0 \tag{1.3.74}$$

where this is equal to the original entropy because of the relationship $N_1/V_1 = N_2/V_2$ from Eq. (1.3.70). This example illustrates the effect of indistinguishability. The entropy increases after the puncture when the gases are different, but not when they are the same. ∎

Question 1.3.4 An ideal gas is in one compartment of a two-compartment sealed and thermally insulated box. The compartment it is in has a volume V_1. It has an energy U_0 and a number of particles N_0. The second compartment has volume V_2 and is empty. Write expressions for the changes in all thermodynamic parameters (N, V, U, P, S, T, F) if

a. the barrier between the two compartments is punctured and the gas expands to fill the box.
b. the barrier is moved slowly, like a piston, expanding the gas to fill the box.

Solution 1.3.4 Recognizing what is conserved simplifies the solution of this type of problem.

a. The energy U and the number of particles N are conserved. Since the volume change is given to us explicitly, the expressions for T (Eq. (1.3.60)), F (Eq. (1.3.59)), S (Eq. (1.3.61)), and P (Eq. (1.3.64)) in terms of these quantities can be used.

$$N = N_0$$
$$U = U_0$$
$$V = V_1 + V_2 \tag{1.3.75}$$
$$T = T_0$$
$$F = kTN[\ln(N\lambda(T)^3/(V_1 + V_2)) - 1] = F_0 + kTN \ln(V_1 + V_2))$$
$$S = kN[\ln((V_1 + V_2)/N\lambda T)^3) + 5/2] = S_0 + kN \ln((V_1 + V_2)/V_1)$$
$$P = NkT/V = NkT/(V_1 + V_2) = P_0 V_1/(V_1 + V_2)$$

b. The process is reversible and no heat is transferred, thus it is adiabatic—the entropy is conserved. The number of particles is also conserved:

$$N = N_0$$
$$S = S_0 \tag{1.3.76}$$

Our main task is to calculate the effect of the work done by the gas pressure on the piston. This causes the energy of the gas to decrease, and the temperature decreases as well. One way to find the change in temperature is to use the conservation of entropy, and Eq. (1.3.61), to obtain that $V/\lambda(T)^3$ is a constant and therefore:

$$T \propto V^{2/3} \tag{1.3.77}$$

Thus the temperature is given by:

$$T = T_0 \left(\frac{V_1 + V_2}{V_1} \right)^{-2/3} \tag{1.3.78}$$

Since the temperature and energy are proportional to each other (Eq. (1.3.60)), similarly:

$$U = U_0 \left(\frac{V_1 + V_2}{V_1} \right)^{-2/3} \tag{1.3.79}$$

The free-energy expression in Eq. (1.3.59) changes only through the temperature prefactor:

$$F = kTN[\ln(N\lambda(T)^3/V) - 1] = F_0 \frac{T}{T_0} = F_0 \left(\frac{V_1 + V_2}{V_1}\right)^{-2/3} \quad (1.3.80)$$

Finally, the pressure (Eq. (1.6.64)):

$$P = NkT/V = P_0 \frac{TV_0}{T_0 V} = P_0 \left(\frac{V_1 + V_2}{V_1}\right)^{-5/3} \quad (1.3.81) \blacksquare$$

The ideal gas illustrates the significance of the Boltzmann distribution. Consider a single particle. We can treat it either as part of the large system or as a subsystem in its own right. In the ideal gas, without any interactions, its energy would not change. Thus the particle would not be described by the Boltzmann probability in Eq. (1.3.29). However, we can allow the ideal gas model to include a weak or infrequent interaction (collision) between particles which changes the particle's energy. Over a long time compared to the time between collisions, the particle will explore all possible positions in space and all possible momenta. The probability of its being at a particular position and momentum (in a region $d^3x\,d^3p$) is given by the Boltzmann distribution:

$$\frac{e^{-\frac{p^2}{2mkT}} d^3p\,d^3x/h^3}{\int e^{-\frac{p^2}{2mkT}} d^3p\,d^3x/h^3} \quad (1.3.82)$$

Instead of considering the trajectory of this particular particle and the effects of the (unspecified) collisions, we can think of an ensemble that represents this particular particle in contact with a thermal reservoir. The ensemble would be composed of many different particles in different boxes. There is no need to have more than one particle in the system. We do need to have some mechanism for energy to be transferred to and from the particle instead of collisions with other particles. This could happen as a result of the collisions with the walls of the box if the vibrations of the walls give energy to the particle or absorb energy from the particle. If the wall is at the temperature T, this would also give rise to the same Boltzmann distribution for the particle. The probability of a particular particle in a particular box being in a particular location with a particular momentum would be given by the same Boltzmann probability.

Using the Boltzmann probability distribution for the velocity, we could calculate the average velocity of the particle as:

$$<v^2>=3<v_\perp^2>=3\frac{\int v_\perp^2 e^{-\frac{p^2}{2mkT}}d^3pd^3x/h^3}{\int e^{-\frac{p^2}{2mkT}}d^3pd^3x/h^3}=\frac{\int v_\perp^2 e^{-\frac{p_\perp^2}{2mkT}}dp_\perp}{\int e^{-\frac{p_\perp^2}{2mkT}}dp_\perp}=\frac{3kT}{m} \quad (1.3.83)$$

which is the same result as we obtained for the ideal gas in the last part of Question 1.3.2. We could even consider one coordinate of one particle as a separate system and arrive at the same conclusion. Our description of systems is actually a description of coordinates.

There are differences when we consider the particle to be a member of an ensemble and as one particle of a gas. In the ensemble, we do not need to consider the distinguishability of particles. This does not affect any of the properties of a single particle.

This discussion shows that the ideal gas model may be viewed as quite close to the basic concept of an ensemble. Generalize the physical particle in three dimensions to a point with coordinates that describe a complete system. These coordinates change in time as the system evolves according to the rules of its dynamics. The ensemble represents this system in the same way as the ideal gas is the ensemble of the particle. The lack of interaction between the different members of the ensemble, and the existence of a transfer of energy to and from each of the systems to generate the Boltzmann probability, is the same in each of the cases. This analogy is helpful when thinking about the nature of the ensemble.

1.3.4 *Phase transitions—first and second order*

In the previous section we constructed some of the underpinnings of thermodynamics and their connection with microscopic descriptions of materials using statistical mechanics. One of the central conclusions was that by minimizing the free energy we can find the equilibrium state of a material that has a fixed number of particles, volume and temperature. Once the free energy is minimized to obtain the equilibrium state of the material, the energy, entropy and pressure are uniquely determined. The free energy is also a function of the temperature, the volume and the number of particles.

One of the important properties of materials is that they can change their properties suddenly when the temperature is changed by a small amount. Examples of this are the transition of a solid to a liquid, or a liquid to a gas. Such a change is known as a phase transition. Each well-defined state of the material is considered a particular phase. Let us consider the process of minimizing the free energy as we vary the temperature. Each of the properties of the material will, in general, change smoothly as the temperature is varied. However, special circumstances might occur when the minimization of the free energy at one temperature results in a very different set of properties of the material from this minimization at a slightly different temperature. This is illustrated in a series of frames in Fig. 1.3.6, where a schematic of the free energy as a function of some macroscopic parameter is illustrated.

The temperature at which the jump in properties of the material occurs is called the critical or transition temperature, T_c. In general, all of the properties of the material except for the free energy jump discontinuously at T_c. This kind of phase transition is known as a first-order phase transition. Some of the properties of a first-order phase transition are that the two phases can coexist at the transition temperature so that part of the material is in one phase and part in the other. An example is ice floating in water. If we start from a temperature below the transition temperature—with ice—and add heat to the system gradually, the temperature will rise until we reach the transition temperature. Then the temperature will stay fixed as the material converts from one phase to the other—from ice to water. Once the whole system is converted to the higher temperature phase, the temperature will start to increase again.

Figure 1.3.6 Each of the curves represents the variation of the free energy of a system as a function of macroscopic parameters. The different curves are for different temperatures. As the temperature is varied the minimum of the free energy all of a sudden switches from one set of macroscopic parameters to another. This is a first-order phase transition like the melting of ice to form water, or the boiling of water to form steam. Below the ice-to-water phase transition the macroscopic parameters that describe ice are the minimum of the free energy, while above the phase transition the macroscopic parameters that describe water are the minimum of the free energy. ∎

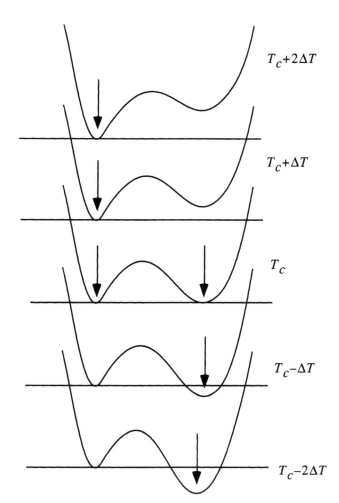

$T_c+2\Delta T$

$T_c+\Delta T$

T_c

$T_c-\Delta T$

$T_c-2\Delta T$

The temperature T_c at which a transition occurs depends on the number of particles and the volume of the system. Alternatively, it may be considered a function of the pressure. We can draw a phase-transition diagram (Fig. 1.3.7) that shows the transition temperature as a function of pressure. Each region of such a diagram corresponds to a particular phase.

There is another kind of phase transition, known as a second-order phase transition, where the energy and the pressure do not change discontinuously at the phase-transition point. Instead, they change continuously, but they are nonanalytic at the transition temperature. A common way that this can occur is illustrated in Fig. 1.3.8. In this case the single minimum of the free energy breaks into two minima as a function of temperature. The temperature at which the two minima appear is the transition temperature. Such a second-order transition is often coupled to the existence of first-order transitions. Below the second-order transition temperature, when the two minima exist, the variation of the pressure can change the relative energy of the two minima and cause a first-order transition to occur. The first-order transition occurs at a particular pressure $P_c(T)$ for each temperature below the second-order transition temperature. This gives rise to a line of first-order phase transitions. Above the second-order transition temperature, there is only one minimum, so that there are

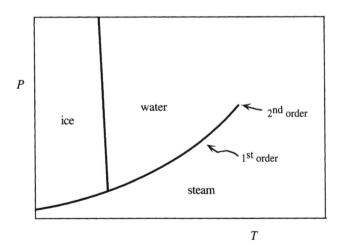

Figure 1.3.7 Schematic phase diagram of H_2O showing three phases — ice, water and steam. Each of the regions shows the domain of pressures and temperatures at which a pure phase is in equilibrium. The lines show phase transition temperatures, $T_c(P)$, or phase transition pressures, $P_c(T)$. The different ways of crossing lines have different names. Ice to water: melting; ice to steam: sublimation; water to steam: boiling; water to ice: freezing; steam to water: condensation; steam to ice: condensation to frost. The transition line from water to steam ends at a point of high pressure and temperature where the two become indistinguishable. At this high pressure steam is compressed till it has a density approaching that of water, and at this high temperature water molecules are energetic like a vapor. This special point is a second-order phase transition point (see Fig. 1.3.8). ∎

Figure 1.3.8 Similar to Fig. 1.3.6, each of the curves represents the variation of the free energy of a system as a function of macroscopic parameters. In this case, however, the phase transition occurs when two minima emerge from one. This is a second-order phase transition. Below the temperature at which the second-order phase transition occurs, varying the pressure can give rise to a first-order phase transition by changing the relative energies of the two minima (see Figs. 1.3.6 and 1.3.7). ∎

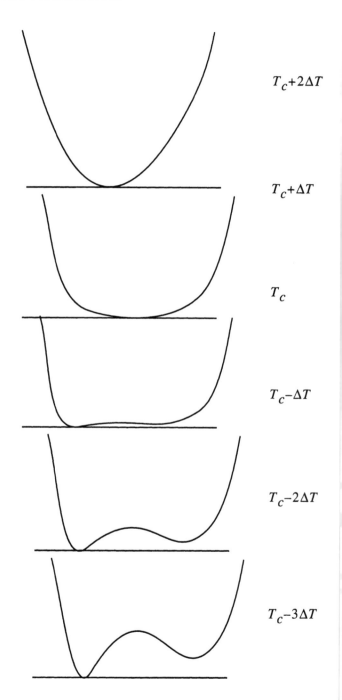

$T_c+2\Delta T$

$T_c+\Delta T$

T_c

$T_c-\Delta T$

$T_c-2\Delta T$

$T_c-3\Delta T$

also no first-order transitions. Thus, the second-order transition point occurs as the end of a line of first-order transitions. A second-order transition is found at the end of the liquid-to-vapor phase line of water in Fig. 1.3.7.

The properties of second-order phase transitions have been extensively studied because of interesting phenomena that are associated with them. Unlike a first-order phase transition, there is no coexistence of two phases at the phase transition, because there is only one phase at that point. Instead, there exist large fluctuations in the local properties of the material at the phase transition. A suggestion of why this occurs can be seen from Fig. 1.3.8, where the free energy is seen to be very flat at the phase transition. This results in large excursions (fluctuations) of all the properties of the system except the free energy. These excursions, however, are not coherent over the whole material. Instead, they occur at every length scale from the microscopic on up. The closer a material is to the phase transition, the longer are the length scales that are affected. As the temperature is varied so that the system moves away from the transition temperature, the fluctuations disappear, first on the longest length scales and then on shorter and shorter length scales. Because at the phase transition itself even the macroscopic length scales are affected, thermodynamics itself had to be carefully rethought in the vicinity of second-order phase transitions. The methodology that has been developed, the renormalization group, is an important tool in the investigation of phase transitions. We will discuss it in Section 1.10. We note that, to be consistent with Question 1.3.1, the specific heat C_V must diverge at a second-order phase transition, where energy fluctuations can be large.

1.3.5 *Use of thermodynamics and statistical mechanics in describing the real world*

How do we generalize the notions of thermodynamics that we have just described to apply to more realistic situations? The assumptions of thermodynamics—that systems are in equilibrium and that dividing them into parts leads to unchanged local properties—do not generally apply. The breakdown of the assumptions of thermodynamics occurs for even simple materials, but are more radically violated when we consider biological organisms like trees or people. We still are able to measure their temperature. How do we extend thermodynamics to apply to these systems?

We can start by considering a system quite close to the thermodynamic ideal—a pure piece of material that is not in equilibrium. For example, a glass of water in a room. We generally have no trouble placing a thermometer in the glass and measuring the temperature of the water. We know it is not in equilibrium, because if we wait it will evaporate to become a vapor spread out throughout the room (even if we simplify by considering the room closed). Moreover, if we wait longer (a few hundred years to a few tens of thousands of years), the glass itself will flow and cover the table or flow down to the floor, and at least part of it will also sublime to a vapor. The table will undergo its own processes of deterioration. These effects will occur even in an idealized closed room without considerations of various external influences or traffic through the room. There is one essential concept that allows us to continue to apply thermodynamic principles to these materials, and measure the temperature of the water, glass or table, and generally to discover that they are at the same (or close to the same) temperature. The concept is the separation of time scales. This concept is as basic as the other principles of thermodynamics. It plays an essential role in discussions of the dynamics of physical systems and in particular of the dynamics of complex systems. The separation of time scales assumes that our observations of systems have a limited time resolution and are performed over a limited time. The processes that occur in a material are then separated into fast processes that are much faster than the time resolution of our observation, slow processes that occur on longer time scales than the duration of observation, and dynamic processes that occur on the time scale of our observation. Macroscopic averages are assumed to be averages over the fast processes. Thermodynamics allows us to deal with the slow and the fast processes but only in very limited ways with the dynamic processes. The dynamic processes are dealt with separately by Newtonian mechanics.

Slow processes establish the framework in which thermodynamics can be applied. In formal terms, the ensemble that we use in thermodynamics assumes that all the parameters of the system described by slow processes are fixed. To describe a system using statistical mechanics, we consider all of the slowly varying parameters of the system to be fixed and assume that equilibrium applies to all of the fast processes. Specifically, we assume that all possible arrangements of the fast coordinates exist in the ensemble with a probability given by the Boltzmann probability. Generally, though not always, it is the microscopic processes that are fast. To justify this we can consider that an atom in a solid vibrates at a rate of 10^{10}–10^{12} times per second, a gas molecule at room temperature travels five hundred meters per second. These are, however, only a couple of select examples.

Sometimes we may still choose to perform our analysis by averaging over many possible values of the slow coordinates. When we do this we have two kinds of ensembles—the ensemble of the fast coordinates and the ensemble of the different values of the slow coordinates. These ensembles are called the annealed and quenched ensembles. For example, say we have a glass of water in which there is an ice cube. There are fast processes that correspond to the motion of the water molecules and the vibrations of the ice molecules, and there are also slow processes corresponding to the movement of the ice in the water. Let's say we want to determine the average amount of ice. If we perform several measurements that determine the coordinates and size of the ice, we may want to average the size we find over all the measurements even though they are measurements corresponding to different locations of the ice. In contrast, if we wanted to measure the motion of the ice, averaging the measurements of location would be absurd.

Closely related to the discussion of fast coordinates is the ergodic theorem. The ergodic theorem states that a measurement performed on a system by averaging a property over a long time is the same as taking the average over the ensemble of the fast coordinates. This theorem is used to relate experimental measurements that are assumed to occur over long times to theoretically obtained averages over ensembles. The ergodic theorem is not a theorem in the sense that it has been proven in general, but rather a statement of a property that applies to some macroscopic systems and is known not to apply to others. The objective is to identify when it applies. When it does not apply, the solution is to identify which quantities may be averaged and which may not, often by separating fast and slow coordinates or equivalently by identifying quantities conserved by the fast dynamics of the system.

Experimental measurements also generally average properties over large regions of space compared to microscopic lengths. It is this spatial averaging rather than time averaging that often enables the ensemble average to stand for experimental measurements when the microscopic processes are not fast compared to the measurement time. For example, materials are often formed of microscopic grains and have n any dislocations. The grain boundaries and dislocations do move, but they often chai ge very slowly over time. When experiments are sensitive to their properties, they often average over the effects of many grains and dislocations because they do not have sufficient resolution to see a single grain boundary or dislocation.

In order to determine what is the relevant ensemble for a particular experiment, both the effect of time and space averaging must be considered. Technically, this requires an understanding of the correlation in space and time of the properties of an individual system. More conceptually, measurements that are made for particular quantities are in effect made over many independent systems both in space and in time, and therefore correspond to an ensemble average. The existence of correlation is the opposite of independence. The key question (like in the case of the ideal gas) becomes what is the interval of space and time that corresponds to an independent system. These quantities are known as the correlation length and the correlation time. If we are able to describe theoretically the ensemble over a correlation length and correlation time, then by appropriate averaging we can describe the measurement.

In summary, the program of use of thermodynamics in the real world is to use the separation of the different time scales to apply equilibrium concepts to the fast degrees of freedom and discuss their influence on the dynamic degrees of freedom while keeping fixed the slow degrees of freedom. The use of ensembles simplifies consideration of these systems by systematizing the use of equilibrium concepts to the fast degrees of freedom.

1.3.6 *From thermodynamics to complex systems*

Our objective in this book is to consider the dynamics of complex systems. While, as discussed in the previous section, we will use the principles of thermodynamics to help us in this analysis, another important reason to review thermodynamics is to recognize what complex systems are not. Thermodynamics describes macroscopic systems without structure or dynamics. The task of thermodynamics is to relate the very few macroscopic parameters to each other. It suggests that these are the only relevant parameters in the description of these systems. Materials and complex systems are both formed out of many interacting parts. The ideal gas example described a material where the interaction between the particles was weak. However, thermodynamics also describes solids, where the interaction is strong. Having decided that complex systems are not described fully by thermodynamics, we must ask, Where do the assumptions of thermodynamics break down? There are several ways the assumptions may break down, and each one is significant and plays a role in our investigation of complex systems. Since we have not yet examined particular examples of complex systems, this discussion must be quite abstract. However, it will be useful as we study complex systems to refer back to this discussion. The abstract statements will have concrete realizations when we construct models of complex systems.

The assumptions of thermodynamics separate into space-related and time-related assumptions. The first we discuss is the divisibility of a macroscopic material. Fig. 1.3.2 (page 61) illustrates the property of divisibility. In this process, a small part of a system is separated from a large part of the system without affecting the *local* properties of the material. This is inherent in the use of extensive and intensive quantities. Such divisibility is not true of systems typically considered to be complex systems. Consider, for example, a person as a complex system that cannot be separated and continue to have the same properties. In words, we would say that complex systems are formed out of not only interacting, but also interdependent parts. Since both thermodynamic and complex systems are formed out of interacting parts, it is the concept of interdependency that must distinguish them. We will dedicate a few paragraphs to defining a sense in which "interdependent" can have a more precise meaning.

We must first address a simple way in which a system may have a nonextensive energy and still not be a complex system. If we look closely at the properties of a material, say a piece of metal or a cup of water, we discover that its surface is different from the bulk. By separating the material into pieces, the surface area of the material is changed. For macroscopic materials, this generally does not affect the bulk properties of the material. A characteristic way to identify surface properties, such as the surface energy, is through their dependence on particle number. The surface energy scales as $N^{2/3}$, in contrast to the extensive bulk energy that is linear in N. This kind of correction can be incorporated directly in a slightly more detailed treatment of thermodynamics, where every macroscopic parameter has a surface term. The presence of such surface terms is not sufficient to identify a material as a complex system. For this reason, we are careful to identify complex systems by requiring that the scenario of Fig. 1.3.2 is violated by changes in the local (i.e., everywhere including the bulk) properties of the system, rather than just the surface.

It may be asked whether the notion of "local properties" is sufficiently well defined as we are using it. In principle, it is not. For now, we adopt this notion from thermodynamics. When only a few properties, like the energy and entropy, are relevant, "affect locally" is a precise concept. Later we would like to replace the use of local thermodynamic properties with a more general concept—the behavior of the system.

How is the scenario of Fig. 1.3.2 violated for a complex system? We can find that the local properties of the small part are affected without affecting the local properties of the large part. Or we can find that the local properties of the large part are affected as well. The distinction between these two ways of affecting the system is important, because it can enable us to distinguish between different kinds of complex systems. It will be helpful to name them for later reference. We call the first category of systems complex materials, the second category we call complex organisms.

Why don't we also include the possibility that the large part is affected but not the small part? At this point it makes sense to consider generic subdivision rather than special subdivision. By generic subdivision, we mean the ensemble of possible subdivisions rather than a particular one. Once we are considering complex systems, the effect of removal of part of a system may depend on which part is removed. However, when we are trying to understand whether or not we have a complex system, we can limit ourselves to considering the generic effects of removing a part of the system. For this reason we do not consider the possibility that subdivision affects the large system and not the small. This might be possible for the removal of a particular small part, but it would be surprising to discover a system where this is generically true.

Two examples may help to illustrate the different classes of complex systems. At least superficially, plants are complex materials, while animals are complex organisms. The reason that plants are complex materials is that the cutting of parts of a plant, such as leaves, a branch, or a root, typically does not affect the local properties of the rest of the plant, but does affect the excised part. For animals this is not generically the case. However, it would be better to argue that plants are in an intermediate category, where some divisions, such as cutting out a lateral section of a tree trunk, affect both small and large parts, while others affect only the smaller part. For animals, essentially all divisions affect both small and large parts. We believe that complex organisms play a special role in the study of complex system behavior. The essential quality of a complex organism is that its properties are tied to the existence of all of its parts.

How large is the small part we are talking about? Loss of a few cells from the skin of an animal will not generally affect it. As the size of the removed portion is decreased, it may be expected that the influence on the local properties of the larger system will be reduced. This leads to the concept of a robust complex system. Qualitatively, the larger the part that can be removed from a complex system without affecting its local properties, the more robust the system is. We see that a complex material is the limiting case of a highly robust complex system.

The flip side of subdivision of a system is aggregation. For thermodynamic systems, subdivision and aggregation are the same, but for complex systems they are quite different. One of the questions that will concern us is what happens when we place a few or many complex systems together. Generally we expect that the individual complex systems will interact with each other. However, one of the points we can make at this time is that just placing together many complex systems, trees or people, does not make a larger complex system by the criteria of subdivision. Thus, a collection of complex systems may result in a system that behaves as a thermodynamic system under subdivision—separating it into parts does not affect the behavior of the parts.

The topic of bringing together many pieces or subdividing into many parts is also quite distinct from the topic of subdivision by removal of a single part. This brings us to a second assumption we will discuss. Thermodynamic systems are assumed to be composed of a very large number of particles. What about complex systems? We know that the number of molecules in a cup of water is not greater than the number of molecules in a human being. And yet, we understand that this is not quite the right point. We should not be counting the number of water molecules in the person, instead we might count the number of cells, which is much smaller. Thus appears the problem of counting the number of components of a system. In the context of correlations in materials, this was briefly discussed at the end of the last section. Let us assume for the moment that we know how to count the number of components. It seems clear that systems with only a few components should not be treated by thermodynamics. One of the interesting questions we will discuss is whether in the limit of a very large number of components we will always have a thermodynamic system. Stated in a simpler way from the point of view of the study of complex systems, the question becomes how large is too large or how many is too many. From the thermodynamic perspective the question is, Under what circumstances do we end up with the thermodynamic limit?

We now switch to a discussion of time-related assumptions. One of the basic assumptions of thermodynamics is the ergodic theorem that enables the description of a single system using an ensemble. When the ergodic theorem breaks down, as discussed in the previous section, additional fixed or quenched variables become important. This is the same as saying that there are significant differences between different examples of the macroscopic system we are interested in. This is a necessary condition for the existence of a complex system. The alternative would be that all realizations of the system would be the same, which does not coincide with intuitive notions of complexity. We will discuss several examples of the breaking of the ergodic theorem later. The simplest example is a magnet. The orientation of the magnet is an additional parameter that must be specified, and therefore the ergodic theorem is violated for this system. Any system that breaks symmetry violates the ergodic theorem. However, we do not accept a magnet as a complex system. Therefore we can assume that the breaking of ergodicity is a necessary but not sufficient condition for complexity. All of the systems we will discuss break ergodicity, and therefore it is always necessary to specify which coordinates of the complex system are fixed and which are to be assumed to be so rapidly varying that they can be assigned equilibrium Boltzmann probabilities.

A special case of the breaking of the ergodic theorem, but one that strikes even more deeply at the assumptions of thermodynamics, is a violation of the separation of time scales. If there are dynamical processes that occur on every time scale, then it becomes impossible to treat the system using the conventional separation of scales into fast, slow and dynamic processes. As we will discuss in Section 1.10, the techniques of renormalization that are used in phase transitions to deal with the existence of many spatial scales may also be used to describe systems changing on many time scales.

Finally, inherent in thermodynamics, the concept of equilibrium and the ergodic theorem is the assumption that the initial condition of the system does not matter. For a complex system, the initial condition of the system does matter over the time scales relevant to our observation. This brings us back to the concept of correlation time. The correlation time describes the length of time over which the initial conditions are relevant to the dynamics. This means that our observation of a complex system must be shorter than a correlation time. The spatial analog, the correlation length, describes the effects of surfaces on the system. The discussion of the effects of subdivision also implies that the system must be smaller than a correlation length. This means that complex systems change their internal structure—adapt—to conditions at their boundaries. Thus, a suggestive though incomplete summary of our discussion of complexity in the context of thermodynamics is that a complex system is contained within a single correlation distance and correlation time.

1.4   Activated Processes (and Glasses)

In the last section we saw figures (Fig. 1.3.7) showing the free energy as a function of a macroscopic parameter with two minima. In this section we analyze a single particle system that has a potential energy with a similar shape (Fig. 1.4.1). The particle is in equilibrium with a thermal reservoir. If the average energy is lower than the energy of the barrier between the two wells, then the particle generally resides for a time in one well and then switches to the other. At very low temperatures, in equilibrium, it will be more and more likely to be in the lower well and less likely to be in the higher well. We use this model to think about a system with two possible states, where one state is higher in energy than the other. If we start the system in the higher energy state, the system will relax to the lower energy state. Because the process of relaxation is enabled or accelerated by energy from the thermal reservoir, we say that it is activated.

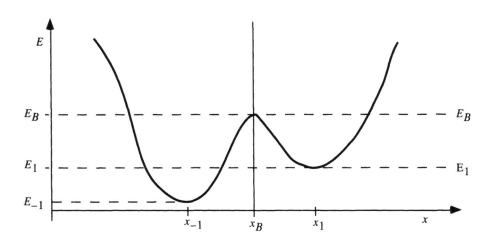

Figure 1.4.1 Illustration of the potential energy of a system that has two local minimum energy configurations x_1 and x_{-1}. When the temperature is lower than the energy barriers $E_B - E_{-1}$ and $E_B - E_1$, the system may be considered as a two-state system with transitions between them. The relative probability of the two states varies with temperature and the relative energy of the bottom of the two wells. The rate of transition also varies with temperature. When the system is cooled systematically the two-state system is a simple model of a glass (Fig. 1.4.2). At low temperatures the system can not move from one well to the other, but is in equilibrium within a single well. ∎

1.4.1 *Two-state systems*

It might seem that a system with only two different states would be easy to analyze. Eventually we will reach a simple problem. However, building the simple model will require us to identify some questions and approximations relevant to our understanding of the application of this model to physical systems (e. g. the problem of protein folding found in Chapter 4). Rather than jumping to the simple two-state problem (Eq. (1.4.40) below), we begin from a particle in a double-well potential. The kinetics and thermodynamics in this system give some additional content to the thermodynamic discussion of the previous section and introduce new concepts.

We consider Fig. 1.4.1 as describing the potential energy $V(x)$ experienced by a classical particle in one dimension. The region to the right of x_B is called the right well and to the left is called the left well. A classical trajectory of the particle with conserved energy would consist of the particle bouncing back and forth within the potential well between two points that are the solution of the equation $V(x) = E$, where E is the total energy of the particle. The kinetic energy at any time is given by

$$E(x, p) - V(x) = \tfrac{1}{2} m v^2 \qquad (1.4.1)$$

which determines the magnitude of the velocity at any position but not the direction. The velocity switches direction every bounce. When the energy is larger than E_B, there is only one distinct trajectory at each energy. For energies larger than E_1 but smaller than E_B, there are two possible trajectories, one in the right well—to the right of x_B— and one in the left well. Below E_1, which is the minimum energy of the right well, there is again only one trajectory possible, in the left well. Below E_{-1} there are no possible locations for the particle.

If we consider this system in isolation, there is no possibility that the particle will change from one trajectory to another. Our first objective is to enable the particle to be in contact with some other system (or coordinate) with which it can transfer energy and momentum. For example, we could imagine that the particle is one of many moving in the double well—like the ideal gas. Sometimes there are collisions that change the energy and direction of the motion. The same effect would be found for many other ways we could imagine the particle interacting with other systems. The main approximation, however, is that the interaction of the particle with the rest of the universe occurs only over short times. Most of the time it acts as if it were by itself in the potential well. The particle follows a trajectory and has an energy that is the sum of its kinetic and potential energies (Eq. (1.4.1)). There is no need to describe the energy associated with the interaction with the other systems. All of the other particles of the gas (or whatever picture we imagine) form the thermal reservoir, which has a well-defined temperature T.

We can increase the rate of collisions between the system and the reservoir without changing our description. Then the particle does not go very far before it forgets the direction it was traveling in and the energy that it had. But as long as the collisions themselves occur over a short time compared to the time between collisions, any time we look at the particle, it has a well-defined energy and momentum. From moment to moment, the kinetic energy and momentum changes unpredictably. Still, the position of the particle must change continuously in time. This scenario is known as diffusive motion. The different times are related by:

collision (interaction) time << time between collisions << transit time

where the transit time is the time between bounces from the walls of the potential well if there were no collisions—the period of oscillation of a particle in the well. The particle undergoes a kind of random walk, with its direction and velocity changing randomly from moment to moment. We will assume this scenario in our treatment of this system.

When the particle is in contact with a thermal reservoir, the laws of thermodynamics apply. The Boltzmann probability gives the probability that the particle is found at position x with momentum p:

$$P(x, p) = e^{-E(x,p)/kT} / Z$$

$$Z = \sum_{x,p} e^{-E(x,p)/kT} = \frac{1}{h} \int dx dp \, e^{-E(x,p)/kT} \tag{1.4.2}$$

Formally, this expression describes a large number of independent systems that make up a canonical ensemble. The ensemble of systems provides a formally precise way of describing probabilities as the number of systems in the ensemble with a particular value of the position and momentum. As in the previous section, Z guarantees that the sum over all probabilities is 1. The factor of h is not relevant in what follows, but for completeness we keep it and associate it with the momentum integral, so that $\Sigma_p \to \int dp/h$.

If we are interested in the position of the particle, and are not interested in its momentum, we can simplify this expression by integrating over all values of the momentum. Since the energy separates into kinetic and potential energy:

$$P(x) = \frac{e^{-V(x)/kT} \int (dp/h) \, e^{-p^2/2mkT}}{\int dx \, e^{-V(x)/kT} \int (dp/h) \, e^{-p^2/2mkT}} = \frac{e^{-V(x)/kT}}{\int dx \, e^{-V(x)/kT}} \tag{1.4.3}$$

The resulting expression looks similar to our original expression. Its meaning is somewhat different, however, because $V(x)$ is only the potential energy of the system. Since the kinetic energy contributes equivalently to the probability at every location, $V(x)$ determines the probability at every x. An expression of the form $e^{-E/kT}$ is known as the Boltzmann factor of E. Thus Eq. (1.4.3) says that the probability $P(x)$ is proportional to the Boltzmann factor of $V(x)$. We will use this same trick to describe the probability of being to the right or being to the left of x_B in terms of the minimum energy of each well.

To simplify to a two-state system, we must define a variable that specifies only which of the two wells the particle is in. So we label the system by $s = \pm 1$, where $s = +1$ if $x > x_B$ and $s = -1$ if $x < x_B$ for a particular realization of the system at a particular time, or:

$$s = \text{sign}(x - x_B) \tag{1.4.4}$$

Probabilistically, the case $x = x_B$ never happens and therefore does not have to be accounted for.

We can calculate the probability $P(s)$ of the system having a value of $s = +1$ using:

$$P(1) = \frac{\int\limits_{x_B}^{\infty} dx\, e^{-V(x)/kT}}{\int dx\, e^{-V(x)/kT}} \tag{1.4.5}$$

The largest contribution to this probability occurs when $V(x)$ is smallest. We assume that kT is small compared to E_B, then the value of the integral is dominated by the region immediately in the vicinity of the minimum energy. Describing this as a two-state system is only meaningful when this is true. We simplify the integral by expanding it in the vicinity of the minimum energy and keeping only the quadratic term:

$$V(x) = E_1 + \tfrac{1}{2}m\omega_1^2(x - x_1)^2 + \ldots = E_1 + \tfrac{1}{2}k_1(x - x_1)^2 + \ldots \tag{1.4.6}$$

where

$$k_1 = m\omega_1^2 = \left. \frac{d^2V(x)}{dx^2} \right|_{x_1} \tag{1.4.7}$$

is the effective spring constant and ω_1 is the frequency of small oscillations in the right well. We can now write Eq. (1.4.5) in the form

$$P(1) = \frac{e^{-E_1/kT} \int\limits_{x_B}^{\infty} dx\, e^{-k_1(x-x_1)^2/2kT}}{\int dx\, e^{-V(x)/kT}} \tag{1.4.8}$$

Because the integrand in the numerator falls rapidly away from the point $x = x_1$, we could extend the lower limit to $-\infty$. Similarly, the probability of being in the left well is:

$$P(-1) = \frac{e^{-E_{-1}/kT} \int\limits_{-\infty}^{x_B} dx\, e^{-k_{-1}(x-x_{-1})^2/2kT}}{\int dx\, e^{-V(x)/kT}} \tag{1.4.9}$$

Here the upper limit of the integral could be extended to ∞. It is simplest to assume that $k_1 = k_{-1}$. This assumption, that the shape of the wells are the same, does not significantly affect most of the discussion (Question 1.4.1–1.4.2). The two probabilities are proportional to a new constant times the Boltzmann factor $e^{-E/kT}$ of the energy at the bottom of the well. This can be seen even without performing the integrals in Eq. (1.4.8) and Eq. (1.4.9). We redefine Z for the two-state representation:

$$P(-1) = \frac{e^{-E_{-1}/kT}}{Z_s} \qquad (1.4.10)$$

$$P(1) = \frac{e^{-E_1/kT}}{Z_s} \qquad (1.4.11)$$

The new normalization Z_s can be obtained from:

$$P(1) + P(-1) = 1 \qquad (1.4.12)$$

giving

$$Z_s = e^{-E_1/kT} + e^{-E_{-1}/kT} \qquad (1.4.13)$$

which is different from the value in Eq. (1.4.2). We arrive at the desired two-state result:

$$P(1) = \frac{e^{-E_1/kT}}{e^{-E_1/kT} + e^{-E_{-1}/kT}} = \frac{1}{1 + e^{(E_1 - E_{-1})/kT}} = f(E_1 - E_{-1}) \qquad (1.4.14)$$

where f is the Fermi probability or Fermi function:

$$f(x) = \frac{1}{1 + e^{x/kT}} \qquad (1.4.15)$$

For readers who were introduced to the Fermi function in quantum statistics, it is not unique to that field, it occurs anytime there are exactly two different possibilities. Similarly,

$$P(-1) = \frac{e^{-E_{-1}/kT}}{e^{-E_1/kT} + e^{-E_{-1}/kT}} = \frac{1}{1 + e^{(E_{-1} - E_1)/kT}} = f(E_{-1} - E_1) \qquad (1.4.16)$$

which is consistent with Eq. (1.4.12) above since

$$f(x) + f(-x) = 1 \qquad (1.4.17)$$

Question 1.4.1 Discuss how $k_1 \neq k_{-1}$ would affect the results for the two-state system in equilibrium. Obtain expressions for the probabilities in each of the wells.

Solution 1.4.1 Extending the integrals to $\pm\infty$, as described in the text after Eq. (1.4.8) and Eq. (1.4.9), we obtain:

$$P(1) = \frac{e^{-E_1/kT}\sqrt{2\pi kT/k_1}}{\int dx\, e^{-V(x)/kT}} \qquad (1.4.18)$$

$$P(-1) = \frac{e^{-E_1/kT}\sqrt{2\pi kT/k_{-1}}}{\int dx\, e^{-V(x)/kT}} \qquad (1.4.19)$$

Because of the approximate extension of the integrals, we are no longer guaranteed that the sum of these probabilities is 1. However, within the accuracy of the approximation, we can reimpose the normalization condition. Before we do so, we choose to rewrite $k_1 = m\omega_1^2 = m(2\pi v_1)^2$, where v_1 is the natural frequency of the well. We then ignore all common factors in the two probabilities and write

$$P(1) = \frac{v_1^{-1} e^{-E_1/kT}}{Z_s'} \tag{1.4.20}$$

$$P(-1) = \frac{v_{-1}^{-1} e^{-E_{-1}/kT}}{Z_s'} \tag{1.4.21}$$

$$Z_s' = v_{-1}^{-1} e^{-E_1/kT} + v_{-1}^{-1} e^{-E_{-1}/kT} \tag{1.4.22}$$

Or we can write, as in Eq. (1.4.14)

$$P(1) = \frac{1}{1 + (v_1/v_{-1}) e^{(E_1 - E_{-1})/kT}} \tag{1.4.23}$$

and similarly for $P(-1)$. ∎

Question 1.4.2 Redefine the energies E_1 and E_{-1} to include the effect of the difference between k_1 and k_{-1} so that the probability $P(1)$ (Eq. (1.4.23)) can be written like Eq. (1.4.14) with the new energies. How is the result related to the concept of free energy and entropy?

Solution 1.4.2 We define the new energy of the right well as

$$F_1 = E_1 + kT \ln(v_1) \tag{1.4.24}$$

This definition can be seen to recover Eq. (1.4.23) from the form of Eq. (1.4.14) as

$$P(1) = f(F_1 - F_{-1}) \tag{1.4.25}$$

Eq. (1.4.24) is very reminiscent of the definition of the free energy Eq. (1.3.33) if we use the expression for the entropy:

$$S_1 = -k \ln(v_1) \tag{1.4.26}$$

Note that if we consider the temperature dependence, Eq. (1.4.25) is not identical in its behavior with Eq. (1.4.14). The free energy, F_1, depends on T, while the energy at the bottom of the well, E_1, does not. ∎

In Question 1.4.2, Eq. (1.4.24), we have defined what might be interpreted as a free energy of the right well. In Section 1.3 we defined only the free energy of the system as a whole. The new free energy is for part of the ensemble rather than the whole ensemble. We can do this quite generally. Start by identifying a certain subset of all possible states of a system. For example, $s = 1$ in Eq. (1.4.4). Then we define the free energy using the expression:

$$F_s(1) = -kT\ln(\sum_{\{x,p\}} \delta_{s,1}\, e^{-E(\{x,p\})/kT}) = -kT\ln(Z_1) \tag{1.4.27}$$

This is similar to the usual expression for the free energy in terms of the partition function Z, but the sum is only over the subset of states. When there is no ambiguity, we often drop the subscript and write this as $F(1)$. From this definition we see that the probability of being in the subset of states is proportional to the Boltzmann factor of the free energy

$$P(1) \propto e^{-F_s(1)/kT} \tag{1.4.28}$$

If we have several different subsets that account for all possibilities, then we can normalize Eq. (1.4.28) to find the probability itself. If we do this for the left and right wells, we immediately arrive at the expression for the probabilities in Eq. (1.4.14) and Eq. (1.4.16), with E_1 and E_{-1} replaced by $F_s(1)$ and $F_s(-1)$ respectively. From Eq. (1.4.28) we see that for a collection of states, the free energy plays the same role as the energy in the Boltzmann probability.

We note that Eq. (1.4.24) is not the same as Eq. (1.4.27). However, as long as the relative energy is the same, $F_1 - F_{-1} = F_s(1) - F_s(-1)$, the normalized probability is unchanged. When $k_1 = k_{-1}$, the entropic part of the free energy is the same for both wells. Then direct use of the energy instead of the free energy is valid, as in Eq. (1.4.14). We can evaluate the free energy of Eq. (1.4.27), including the momentum integral:

$$Z_1 = \int_{x_B}^{\infty} dx \int (dp/h)\, e^{-E(x,p)/kT} = \int_{x_B}^{\infty} dx e^{-V(x)/kT} \int (dp/h)\, e^{-p^2/2mkT}$$

$$\approx e^{-E_1/kT} \int_{x_B}^{\infty} dx\, e^{-k_1(x-x_1)^2/2kT} \sqrt{2\pi mkT}\,/h \approx e^{-E_1/kT}\sqrt{m/k_1}\,2\pi kT/h \tag{1.4.29}$$

$$= e^{-E_1/kT} kT/h\nu_1$$

$$F_s(1) = E_1 + kT\ln(h\nu_1/kT) \tag{1.4.30}$$

where we have used the definition of the well oscillation frequency above Eq. (1.4.20) to simplify the expression. A similar expression holds for Z_{-1}. The result would be exact for a pure harmonic well.

The new definition of the free energy of a set of states can also be used to understand the treatment of macroscopic systems, specifically to explain why the energy is determined by minimizing the free energy. Partition the possible microstates by the value of the energy, as in Eq. (1.3.35). Define the free energy as a function of the energy analogous to Eq. (1.4.27)

$$F(U) = -kT\ln\left(\sum_{\{x,p\}} \delta_{E(\{x,p\}),U}\, e^{-E(\{x,p\})/kT}\right) \tag{1.4.31}$$

Since the relative probability of each value of the energy is given by

$$P(U) \propto e^{-F(U)/kT} \tag{1.4.32}$$

the most likely energy is given by the lowest free energy. For a macroscopic system, the most likely value is so much more likely than any other value that it is observed in any measurement. This can immediately be generalized. The minimization of the free energy gives not only the value of the energy but the value of any macroscopic parameter.

1.4.2 Relaxation of a two-state system

To investigate the kinetics of the two-state system, we assume an ensemble of systems that is not an equilibrium ensemble. Instead, the ensemble is characterized by a time-dependent probability of occupying the two wells:

$$P(1) \rightarrow P(1;t)$$
$$P(-1) \rightarrow P(-1;t) \tag{1.4.33}$$

Normalization continues to hold at every time:

$$P(1;t) + P(-1;t) = 1 \tag{1.4.34}$$

For example, we might consider starting a system in the upper well and see how the system evolves in time. Or we might consider starting a system in the lower well and see how the system evolves in time. We answer the question using the time-evolving probabilities that describe an ensemble of systems with the same starting condition. To achieve this objective, we construct a differential equation describing the rate of change of the probability of being in a particular well in terms of the rate at which systems move from one well to the other. This is just the Master equation approach from Section 1.2.4.

The systems that make transitions from the left to the right well are the ones that cross the point $x = x_B$. More precisely, the rate at which transitions occur is the probability current per unit time of systems at x_B, moving toward the right. Similar to Eq. (1.3.47) used to obtain the pressure of an ideal gas on a wall, the number of particles crossing x_B is the probability of systems at x_B with velocity v, times their velocity:

$$J(1|-1) = \int_0^\infty (dp/h)vP(x_B,p;t) \tag{1.4.35}$$

where $J(1|-1)$ is the number of systems per unit time moving from the left to the right. There is a hidden assumption in Eq. (1.4.35). We have adopted a notation that treats all systems on the left together. When we are considering transitions, this is only valid if a system that crosses $x = x_B$ from right to left makes it down into the well on the left, and thus does not immediately cross back over to the side it came from.

We further assume that in each well the systems are in equilibrium, even when the two wells are not in equilibrium with each other. This means that the probability of being in a particular location in the right well is given by:

$$P(x, p; t) = P(1; t)e^{-E(x,p)/kT} / Z_1$$

$$Z_1 = \int_{x_B}^{\infty} dx dp \, e^{-E(x,p)/kT} \tag{1.4.36}$$

In equilibrium, this statement is true because then $P(1) = Z_1/Z$. Eq. (1.4.36) presumes that the rate of collisions between the particle and the thermal reservoir is faster than both the rate at which the system goes from one well to the other and the frequency of oscillation in a well.

In order to evaluate the transition rate Eq. (1.4.35), we need the probability at x_B. We assume that the systems that cross x_B moving from the left well to the right well (i.e., moving to the right) are in equilibrium with systems in the left well from where they came. Systems that are moving from the right well to the left have the equilibrium distribution characteristic of the right well. With these assumptions, the rate at which systems hop from the left to the right is given by:

$$J(1|-1) = \int_0^{\infty} (dp/h)(p/m) \left(P(-1; t)e^{-(E_B + p^2/2m)/kT} / Z_{-1} \right)$$
$$= P(-1; t)e^{-E_B/kT}(kT/h)/Z_{-1} \tag{1.4.37}$$

We find using Eq. (1.4.29) that the current of systems can be written in terms of a transition rate per system:

$$J(1|-1) = R(1|-1)P(-1; t)$$
$$R(1|-1) = v_{-1}e^{-(E_B - E_{-1})/kT} \tag{1.4.38}$$

Similarly, the current and rate at which systems hop from the right to the left are given by:

$$J(-1|1) = R(-1|1)P(1; t)$$
$$R(-1|1) = v_1 e^{-(E_B - E_1)/kT} \tag{1.4.39}$$

When $k_1 = k_{-1}$ then $v_1 = v_{-1}$. We continue to deal with this case for simplicity and define $v = v_1 = v_{-1}$. The expressions for the rate of transition suggest the interpretation that the frequency v is the rate of attempt to cross the barrier. The probability of crossing in each attempt is given by the Boltzmann factor, which gives the likelihood that the energy exceeds the barrier. While this interpretation is appealing, and is often given, it is misleading. It is better to consider the frequency as describing the width of the well in which the particle wanders. The wider the well is, the less likely is a barrier crossing. This interpretation survives better when more general cases are considered.

The transition rates enable us to construct the time variation of the probability of occupying each of the wells. This gives us the coupled equations for the two probabilities:

$$\dot{P}(1; t) = R(1|-1)P(-1; t) - R(-1|1)P(1; t) \tag{1.4.40}$$
$$\dot{P}(-1; t) = R(-1|1)P(1; t) - R(1|-1)P(-1; t)$$

These are the Master equations (Eq. (1.2.86)) for the two-state system. We have arrived at these equations by introducing a set of assumptions for treating the kinetics of a single particle. The equations are much more general, since they say only that there is a rate of transition between one state of the system and the other. It is the correspondence between the two-state system and the moving particle that we have established in Eqs. (1.4.38) and (1.4.39). This correspondence is approximate. Eq. (1.4.40) does not rely upon the relationship between E_B and the rate at which systems move from one well to the other. However, it does rely upon the assumption that we need to know only which well the system is in to specify its rate of transition to the other well. On average this is always true, but it would not be a good description of the system, for example, if energy is conserved and the key question determining the kinetics is whether the particle has more or less energy than the barrier E_B.

We can solve the coupled equations in Eq. (1.4.40) directly. Both equations are not necessary, given the normalization constraint Eq. (1.4.34). Substituting $P(-1;t) = 1 - P(1;t)$ we have the equation

$$\dot{P}(1;t) = R(-1|1) - P(1;t)(R(1|-1) + R(-1|1)) \tag{1.4.41}$$

We can rewrite this in terms of the equilibrium value of the probability. By definition this is the value at which the time derivative vanishes.

$$P(1;\infty) = R(-1|1)/(R(1|-1) + R(-1|1)) = f(E_1 - E_{-1}) \tag{1.4.42}$$

where the right-hand side follows from Eq. (1.4.38) and Eq. (1.4.39) and is consistent with Eq. (1.4.13), as it must be. Using this expression, Eq. (1.4.24) becomes

$$\dot{P}(1;t) = (P(1;\infty) - P(1;t))/\tau \tag{1.4.43}$$

where we have defined an additional quantity

$$1/\tau = (R(1|-1) + R(-1|1)) = v(e^{-(E_B - E_1)/kT} + e^{-(E_B - E_{-1})/kT}) \tag{1.4.44}$$

The solution of Eq. (1.4.43) is

$$P(1;t) = (P(1;0) - P(1;\infty))e^{-t/\tau} + P(1;\infty) \tag{1.4.45}$$

This solution describes a decaying exponential that changes the probability from the starting value to the equilibrium value. This explains the definition of τ, called the relaxation time. Since it is inversely related to the sum of the rates of transition between the wells, it is a typical time taken by a system to hop between the wells. The relaxation time does not depend on the starting probability. We note that the solution of Eq. (1.4.41) does not depend on the explicit form of $P(1; \infty)$ or τ. The definitions implied by the first equal signs in Eq. (1.4.42) and Eq. (1.4.44) are sufficient. Also, as can be quickly checked, we can replace the index 1 with the index -1 without changing anything else in Eq (1.4.45). The other equations are valid (by symmetry) after the substitution $1 \leftrightarrow -1$.

There are several intuitive relationships between the equilibrium probabilities and the transition rates that may be written down. The first is that the ratio of the equilibrium probabilities is the ratio of the transition rates:

$$P_1(\infty)/P_{-1}(\infty) = R(-1|1)/R(1|-1) \tag{1.4.46}$$

The second is that the equilibrium probability divided by the relaxation time is the rate of transition:

$$P_1(\infty)/\tau = R(-1|1) \tag{1.4.47}$$

Question 1.4.3 Eq. (1.4.45) implies that the relaxation time of the system depends largely on the smaller of the two energy barriers $E_B - E_1$ and $E_B - E_{-1}$. For Fig. 1.4.1 the smaller barrier is $E_B - E_1$. Since the relaxation time is independent of the starting probability, this barrier controls the rate of relaxation whether we start the system from the lower well or the upper well. Why does the barrier $E_B - E_1$ control the relaxation rate when we start from the lower well?

Solution 1.4.3 Even though the rate of transition from the lower well to the upper well is controlled by $E_B - E_{-1}$, the fraction of the ensemble that must make the transition in order to reach equilibrium depends on E_1. The higher it is, the fewer systems must make the transition from $s = -1$ to $s = 1$. Taking this into consideration implies that the time to reach equilibrium depends on $E_B - E_1$ rather than $E_B - E_{-1}$. ∎

1.4.3 *Glass transition*

Glasses are materials that when cooled from the liquid do not undergo a conventional transition to a solid. Instead their viscosity increases, and in the vicinity of a particular temperature it becomes so large that on a reasonable time scale they can be treated as solids. However, on long enough time scales, they flow as liquids. We will model the glass transition using a two-state system by considering what happens as we cool down the two-state system. At high enough temperatures, the system hops back and forth between the two minima with rates given by Eqs. (1.4.38) and (1.4.39). v is a microscopic quantity; it might be a vibration rate in the material. Even if the barriers are higher than the temperature, $E_B - E_{\pm 1} \gg kT$, the system will still be able to hop back and forth quite rapidly from a macroscopic perspective.

As the system is cooled down, the hopping back and forth slows down. At some point the rate of hopping will become longer than the time we are observing the system. Systems in the higher well will stay there. Systems in the lower well will stay there. This means that the population in each well becomes fixed. Even when we continue to cool the system down, there will be no change, and the ensemble will no longer be in equilibrium. Within each well the system will continue to have a probability distribution for its energy given by the Boltzmann probability, but the relative populations of the two wells will no longer be described by the equilibrium Boltzmann probability.

To gain a feeling for the numbers, a typical atomic vibration rate is 10^{12}/sec. For a barrier of 1eV, at twice room temperature, $kT \approx 0.05$eV (600°K), the transition rate would be of order 10^3/sec. This is quite slow from a microscopic perspective, but at room temperature it would be only 10^{-6}/sec, or one transition per year.

The rate at which we cool the system down plays an essential role. If we cool faster, then the temperature at which transitions stop is higher. If we cool at a slower rate, then the temperature at which the transitions stop is lower. This is found to be the case for glass transitions, where the cooling rate determines the departure point from the equilibrium trajectory of the system, and the eventual properties of the glass are also determined by the cooling rate. Rapid cooling is called quenching. If we raise the temperature and lower it slowly, the procedure is called annealing.

Using the model two-state system we can simulate what would happen if we perform an experiment of cooling a system that becomes a glass. Fig. 1.4.2 shows the probability of being in the upper well as a function of the temperature as the system is cooled down. The curves depart from the equilibrium curve in the vicinity of a transition temperature we might call a freezing transition, because the kinetics become frozen. The glass transition is not a transition like a first- or second-order transition (Section 1.3.4) because it is a transition of the kinetics rather than of the equilibrium structure of the system. Below the freezing transition, the relative probability of the system being in the upper well is given approximately by the equilibrium probability at the transition.

The freezing transition of the relative population of the upper state and the lower state is only a simple model of the glass transition; however, it is also more widely applicable. The freezing does not depend on cooperative effects of many particles. To find examples, a natural place to look is the dynamics of individual atoms in solids. Potential energies with two wells occur for impurities, defects and even bulk atoms in a solid. Impurities may have two different local configurations that differ in energy and are separated by a barrier. This is a direct analog of our model two-state system. When the temperature is lowered, the relative population of the two configurations becomes frozen. If we raise the temperature, the system can equilibrate again.

It is also possible to artificially cause impurity configurations to have unequal energies. One way is to apply uniaxial stress to a crystal—squeezing it along one axis. If an impurity resides in a bond between two bulk atoms, applying stress will raise the energy of impurities in bonds oriented with the stress axis compared to bonds perpendicular to the stress axis. If we start at a relatively high temperature, apply stress and then cool down the material, we can freeze unequal populations of the impurity. If we have a way of measuring relaxation, then by raising the temperature gradually and observing when the defects begin to equilibrate we can discover the barrier to relaxation. This is one of the few methods available to study the kinetics of impurity reorientation in solids.

The two-state system provides us with an example of how a simple system may not be able to equilibrate over experimental time scales. It also shows how an equi-

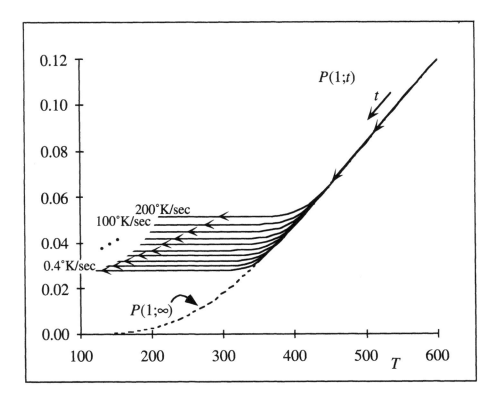

Figure 1.4.2 Plot of the fraction of the systems in the higher energy well as a function of temperature. The equilibrium value is shown with the dashed line. The solid lines show what happens when the system is cooled from a high temperature at a particular cooling rate. The example given uses $E_1 - E_{-1} = 0.1eV$ and $E_B - E_{-1} = 1.0eV$. Both wells have oscillation frequencies of $v = 10^{12}$/sec. The fastest cooling rate is 200°K/sec and each successive curve is cooled at a rate that is half as fast, with the slowest rate being 0.4°K/sec. For every cooling rate the system stops making transitions between the wells at a particular temperature that is analogous to a glass transition in this system. Below this temperature the probability becomes essentially fixed. ∎

librium ensemble can be used to treat relative probabilities within a subset of states. Because the motion within a particular well is fast, the relative probabilities of different positions or momenta within a well may be described using the Boltzmann probability. At the same time, the relative probability of finding a system in each of the two wells depends on the initial conditions and the history of the system—what temperature the system experienced and for how long. At sufficiently low temperatures, this relative probability may be treated as fixed. Systems that are in the higher well may be assumed to stay there. At intermediate temperatures, a treatment of the dynamics of the transition between the two wells can (and must) be included. This manifests a violation of the ergodic theorem due to the divergence of the time scale for equilibration between the two wells. Thus we have identified many of the features that are necessary in describing nonequilibrium systems: divergent time scales, violation of the ergodic theorem, frozen and dynamic coordinates. We have illustrated a method for treating systems where there is a separation of long time scales and short time scales.

> **Question 1.4.4** Write a program that can generate the time dependence of the two-state system for a specified time history. Reproduce Fig. 1.4.2. For an additional "experiment," try the following quenching and annealing sequence:
>
> a. Starting from a high enough temperature to be in equilibrium, cool the system at a rate of 10°K/sec down to $T = 0$.
>
> b. Heat the system up to temperature T_a and keep it there for one second.
>
> c. Cool the system back down to $T = 0$ at rate of 100°K/sec.
>
> Plot the results as a function of T_a. Describe and explain them in words. ∎

1.4.4 Diffusion

In this section we briefly consider a multiwell system. An example is illustrated in Fig. 1.4.3, where the potential well depths and barriers vary from site to site. A simpler case is found in Fig. 1.4.4, where all the well depths and barriers are the same. A concrete example would be an interstitial impurity in an ideal crystal. The impurity lives in a periodic energy that repeats every integral multiple of an elementary length a.

We can apply the same analysis from the previous section to describe what happens to a system that begins from a particular well at $x = 0$. Over time, the system makes transitions left and right at random, in a manner that is reminiscent of a random walk. We will see in a moment that the connection with the random walk is valid but requires some additional discussion.

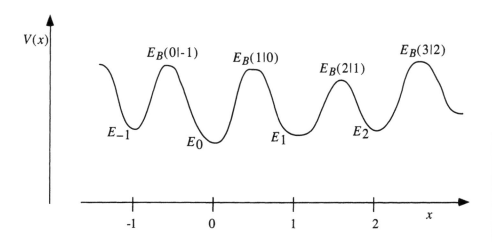

Figure 1.4.3 Illustration of a multiple-well system with barrier heights and well depths that vary from site to site. We focus on the uniform system in Fig. 1.4.4. ∎

The probability of the system being in a particular well is changed by probability currents into the well and out from the well. Systems can move to or from the well immediately to their right and immediately to their left. The Master equation for the ith well in Fig. 1.4.3 is:

$$\dot{P}(i;t) = R(i|i-1)P(i-1;t) + R(i|i+1)P(i+1;t) - (R(i+1|i) + R(i-1|i))P(i;t) \quad (1.4.48)$$

$$R(i+1|i) = v_i e^{-(E_B(i+1|i)-E_i)/kT}$$
$$R(i-1|i) = v_i e^{-(E_B(i|i-1)-E_i)/kT} \quad (1.4.49)$$

where E_i is the depth of the ith well and $E_B(i+1|i)$ is the barrier to its right. For the periodic system of Fig. 1.4.4 ($v_i \to v$, $E_B(i+1|i) \to E_B$) this simplifies to:

$$\dot{P}(i;t) = R(P(i-1;t) + P(i+1;t) - 2P(i;t)) \quad (1.4.50)$$

$$R = v e^{-(E_B-E_0)/kT} \quad (1.4.51)$$

Since we are already describing a continuum differential equation in time, it is convenient to consider long times and write a continuum equation in space as well. Allowing a change in notation we write

$$P(i;t) \to P(x_i;t) \quad (1.4.52)$$

Introducing the elementary distance between wells a we can rewrite Eq. (1.4.50) using:

$$\frac{(P(i-1;t) + P(i+1;t) - 2P(i;t))}{a^2}$$
$$\to \frac{(P(x_i-a;t) + P(x_i+a;t) - 2P(x_i;t))}{a^2} \to \frac{\partial^2}{\partial x^2}P(x;t) \quad (1.4.53)$$

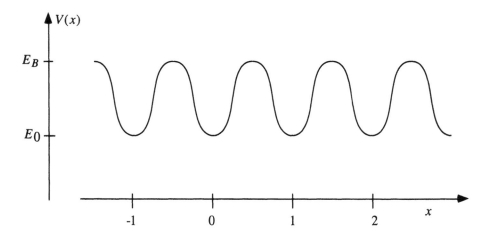

Figure 1.4.4 When the barrier heights and well depths are the same, as illustrated, the long time behavior of this system is described by the diffusion equation. The evolution of the system is controlled by hopping events from one well to the other. The net effect over long times is the same as for the random walk discussed in Section 1.2. ∎

where the last expression assumes a is small on the scale of interest. Thus the continuum version of Eq. (1.4.50) is the conventional diffusion equation:

$$\dot{P}(x;t) = D\frac{\partial^2}{\partial x^2}P(x;t) \tag{1.4.54}$$

The diffusion constant D is given by:

$$D = a^2 R = a^2 v e^{-(E_B - E_0)/kT} \tag{1.4.55}$$

The solution of the diffusion equation, Eq. (1.4.54), depends on the initial conditions that are chosen. If we consider an ensemble of a system that starts in one well and spreads out over time, the solution can be checked by substitution to be the Gaussian distribution found for the random walk in Section 1.2:

$$P(x,t) = \frac{1}{\sqrt{4\pi Dt}}e^{-x^2/4Dt} = \frac{1}{\sqrt{2\pi}\sigma}e^{-x^2/2\sigma^2} \tag{1.4.56}$$

$$\sigma = \sqrt{2Dt}$$

We see that motion in a set of uniform wells after a long time reduces to that of a random walk.

How does the similarity to the random walk arise? This might appear to be a natural result, since we showed that the Gaussian distribution is quite general using the central limit theorem. The scenario here, however, is quite different. The central limit theorem was proven in Section 1.2.2 for the case of a distribution of probabilities of steps taken at specific time intervals. Here we have a time continuum. Hopping events may happen at any time. Consider the case where we start from a particular well. Our differential equation describes a system that might hop to the next well at any time. A hop is an event, and we might concern ourselves with the distribution of such events in time. We have assumed that these events are uncorrelated. There are unphysical consequences of this assumption. For example, no matter how small an interval of time we choose, the particle has some probability of traveling arbitrarily far away. This is not necessarily a correct microscopic picture, but it is the continuum model we have developed.

There is a procedure to convert the event-controlled hopping motion between wells into a random walk that takes steps with a certain probability at specific time intervals. We must select a time interval. For this time interval, we evaluate the total probability that hops move a system from its original position to all possible positions of the system. This would give us the function $f(s)$ in Eq. (1.2.34). As long as the mean square displacement is finite, the central limit theorem continues to apply to the probability distribution after a long enough time. The generality of the conclusion also implies that the result is more widely applicable than the assumptions indicate. However, there is a counter example in Question 1.4.5.

Question 1.4.5 Discuss the case of a particle that is not in contact with a thermal resevoir moving in the multiple well system (energy is conserved).

Solution 1.4.5 If the energy of the system is lower than E_B, the system stays in a single well bouncing back and forth. A model that describes how transitions occur between wells would just say there are none.

For the case where the energy is larger than E_B, the system will move with a periodically varying velocity in one direction. There is a problem in selecting an ensemble to describe it. If we choose the ensemble with only one system moving in one direction, then it is described as a deterministic walk. This description is consistent with the motion of the system. However, we might also think to describe the system using an ensemble consisting of particles with the same energy. In this case it would be one particle moving to the right and one moving to the left. Taking an interval of time to be the time needed to move to the next well, we would find a transition probability of 1/2 to move to the right and the same to the left. This would lead to a conventional random walk and will give us an incorrect result for all later times.

This example illustrates the need for an assumption that has not yet been explicitly mentioned. The ensemble must describe systems that can make transitions to each other. Since the energy-conserving systems cannot switch directions, the ensemble cannot include both directions. It is enough, however, for there to be a small nonzero probability for the system to switch directions for the central limit theorem to apply. This means that over long enough times, the distribution will be Gaussian. Over short times, however, the probability distribution from the random walk model and an almost ballistic system would not be very similar. ∎

We can generalize the multiple well picture to describe a biased random walk. The potential we would use is a "washboard potential," illustrated in Fig. 1.4.5. The Master equation is:

$$\dot{P}(i;t) = R_+ P(i-1;t) + R_- P(i+1;t) - (R_+ + R_-)P(i;t) \tag{1.4.57}$$

$$\begin{aligned} R_+ &= v_i e^{-\Delta E_+/kT} \\ R_- &= v_i e^{-\Delta E_-/kT} \end{aligned} \tag{1.4.58}$$

To obtain the continuum limit, replace $i \rightarrow x$: $P(i+1;t) \rightarrow P(x+a,t)$, and $P(i-1;t) \rightarrow P(x-a,t)$, and expand in a Taylor series to second order in a to obtain:

$$\dot{P}(x;t) = -v\frac{\partial}{\partial x}P(x;t) + D\frac{\partial^2}{\partial x^2}P(x;t) \tag{1.4.59}$$

$$\begin{aligned} v &= a(R_+ - R_-) \\ D &= a^2(R_+ + R_-)/2 \end{aligned} \tag{1.4.60}$$

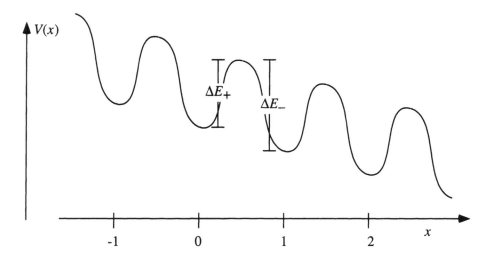

Figure 1.4.5 The biased random walk is also found in a multiple-well system when the illustrated washboard potential is used. The velocity of the system is given by the difference in hopping rates to the right and to the left. ∎

The solution is a moving Gaussian:

$$P(x,t) = \frac{1}{\sqrt{4\pi Dt}} e^{-(x-vt)^2/4Dt} = \frac{1}{\sqrt{2\pi}\sigma} e^{-(x-vt)^2/2\sigma^2}$$

$$\sigma = \sqrt{2Dt}$$

(1.4.61)

Since the description of diffusive motion always allows the system to stay where it is, there is a limit to the degree of bias that can occur in the random walk. For this limit set $R_- = 0$. Then $D = av/2$ and the spreading of the probability is given by $\sigma = \sqrt{avt}$. This shows that unlike the biased random walk in Section 1.2, diffusive motion on a washboard with a given spacing a cannot describe ballistic or deterministic motion in a single direction.

1.5 Cellular Automata

The first four sections of this chapter were dedicated to systems in which the existence of many parameters (degrees of freedom) describing the system is hidden in one way or another. In this section we begin to describe systems where many degrees of freedom are explicitly represented. Cellular automata (CA) form a general class of models of dynamical systems which are appealingly simple and yet capture a rich variety of behavior. This has made them a favorite tool for studying the generic behavior of and modeling complex dynamical systems. Historically CA are also intimately related to the development of concepts of computers and computation. This connection continues to be a theme often found in discussions of CA. Moreover, despite the wide differences between CA and conventional computer architectures, CA are convenient for computer simulations in general and parallel computer simulations in particular. Thus CA have gained importance with the increasing use of simulations in the development of our understanding of complex systems and their behavior.

1.5.1 Deterministic cellular automata

The concept of cellular automata begins from the concept of space and the locality of influence. We assume that the system we would like to represent is distributed in space, and that nearby regions of space have more to do with each other than regions far apart. The idea that regions nearby have greater influence upon each other is often associated with a limit (such as the speed of light) to how fast information about what is happening in one place can move to another place.*

Once we have a system spread out in space, we mark off the space into cells. We then use a set of variables to describe what is happening at a given instant of time in a particular cell.

$$s(i, j, k;t) = s(x_i, y_j, z_k;t) \tag{1.5.1}$$

where i, j, k are integers (i, j, $k \in Z$), and this notation is for a three-dimensional space (3-d). We can also describe automata in one or two dimensions (1-d or 2-d) or higher than three dimensions. The time dependence of the cell variables is given by an iterative rule:

$$s(i, j, k;t) = R(\{s(i' - i, j' - j, k' - k;t-1)\}_{i',j',k' \in z}) \tag{1.5.2}$$

where the rule R is shown as a function of the values of all the variables at the previous time, at positions relative to that of the cell $s(i, j, k;t-1)$. The rule is assumed to be the same everywhere in the space—there is no space index on the rule. Differences between what is happening at different locations in the space are due only to the values of the variables, not the update rule. The rule is also homogeneous in time; i.e., the rule is the same at different times.

The locality of the rule shows up in the form of the rule. It is assumed to give the value of a particular cell variable at the next time only in terms of the values of cells in the vicinity of the cell at the previous time. The set of these cells is known as its neighborhood. For example, the rule might depend only on the values of twenty-seven cells in a cube centered on the location of the cell itself. The indices of these cells are obtained by independently incrementing or decrementing once, or leaving the same, each of the indices:

$$s(i, j, k;t) = R(s(i \pm 1,0, j \pm 1, 0, k \pm 1, 0;t-1)) \tag{1.5.3}$$

---

*These assumptions are both reasonable and valid for many systems. However, there are systems where this is not the most natural set of assumptions. For example, when there are widely divergent speeds of propagation of different quantities (e.g., light and sound) it may be convenient to represent one as instantaneous (light) and the other as propagating (sound). On a fundamental level, Einstein, Podalsky and Rosen carefully formulated the simple assumptions of local influence and found that quantum mechanics violates these simple assumptions. A complete understanding of the nature of their paradox has yet to be reached.

where the informal notation $i \pm 1,0$ is the set $\{i-1, i, i+1\}$. In this case there are a total of twenty-seven cells upon which the update rule $R(s)$ depends. The neighborhood could be smaller or larger than this example.

CA can be usefully simplified to the point where each cell is a single binary variable. As usual, the binary variable may use the notation $\{0,1\}$, $\{-1,1\}$, $\{\text{ON}, \text{OFF}\}$ or $\{\uparrow, \downarrow\}$. The terminology is often suggested by the system to be described. Two 1-d examples are given in Question 1.5.1 and Fig. 1.5.1. For these 1-d cases we can show the time evolution of a CA in a single figure, where the time axis runs vertically down the page and the horizontal axis is the space axis. Each figure is a CA space-time diagram that illustrates a particular history.

In these examples, a finite space is used rather than an infinite space. We can define various boundary conditions at the edges. The most common is to use a periodic boundary condition where the space wraps around to itself. The one-dimensional examples can be described as circles. A two-dimensional example would be a torus and a three-dimensional example would be a generalized torus. Periodic boundary conditions are convenient, because there is no special position in the space. Some care must be taken in considering the boundary conditions even in this case, because there are rules where the behavior depends on the size of the space. Another standard kind of boundary condition arises from setting all of the values of the variables outside the finite space of interest to a particular value such as 0.

> **Question 1.5.1** Fill in the evolution of the two rules of Fig. 1.5.1. The first CA (Fig. 1.5.1(a)) is the majority rule that sets a cell to the majority of the three cells consisting of itself and its two neighbors in the previous time. This can be written using $s(i; t) = \pm 1$ as:
>
> $$s(i; t+1) = \text{sign}(s(i-1; t) + s(i; t) + s(i+1; t)) \qquad (1.5.4)$$
>
> In the figure $\{-1, +1\}$ are represented by $\{\uparrow, \downarrow\}$ respectively.
>
> The second CA (Fig. 1.5.1(b)), called the mod2 rule, is obtained by setting the ith cell to be OFF if the number of ON squares in the neighborhood is even, and ON if this number is odd. To write this in a simple form use $s(i; t) = \{0, 1\}$. Then:
>
> $$s(i; t+1) = \text{mod}_2 (s(i-1; t) + s(i; t) + s(i+1; t)) \qquad (1.5.5)$$

Solution 1.5.1 Notes:

1. The first rule (a) becomes trivial almost immediately, since it achieves a fixed state after only two updates. Many CA, as well as many physical systems on a macroscopic scale, behave this way.

2. Be careful about the boundary conditions when updating the rules, particularly for rule (b).

3. The second rule (b) goes through a sequence of states very different from each other. Surprisingly, it will recover the initial configuration after eight updates. ∎

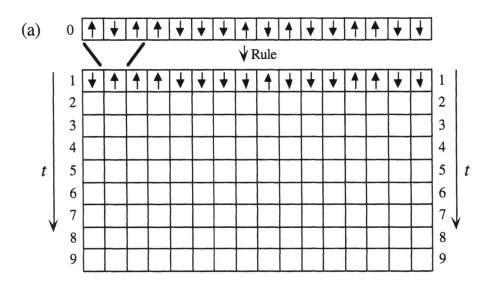

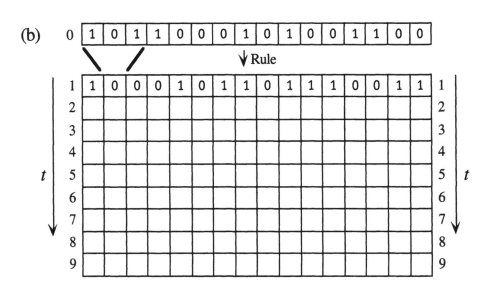

Figure 1.5.1 Two examples of one dimensional (1-d) cellular automata. The top row in each case gives the initial conditions. The value of a cell at a particular time is given by a rule that depends on the values of the cells in its neighborhood at the previous time. For these rules the neighborhood consists of three cells: the cell itself and the two cells on either side. The first time step is shown below the initial conditions for (a) the majority rule, where each cell is equal to the value of the majority of the cells in its neighborhood at the previous time and (b) the mod2 rule which sums the value of the cells in the neighborhood modulo two to obtain the value of the cell in the next time. The rules are written in Question 1.5.1. The rest of the time steps are to be filled in as part of this question. ∎

Question 1.5.2 The evolution of the mod2 rule is periodic in time. After eight updates, the initial state of the system is recovered in Fig. 1.5.1(b). Because the state of the system at a particular time determines uniquely the state at every succeeding time, this is an 8-cycle that will repeat itself. There are sixteen cells in the space shown in Fig. 1.5.1(b). Is the number of cells connected with the length of the cycle? Try a space that has eight cells (Fig. 1.5.2(a)).

Solution 1.5.2 For a space with eight cells, the maximum length of a cycle is four. We could also use an initial condition that has a space periodicity of four in a space with eight cells (Fig. 1.5.2(b)). Then the cycle length would only be two. From these examples we see that the mod2 rule returns to the initial value after a time that depends upon the size of the space. More precisely, it depends on the periodicity of the initial conditions. The time periodicity (cycle length) for these examples is simply related to the space periodicity. ∎

Question 1.5.3 Look at the mod2 rule in a space with six cells (Fig. 1.5.2(c)) and in a space with five cells (Fig. 1.5.2(d)). What can you conclude from these trials?

Solution 1.5.3 The mod2 rule can behave quite differently depending on the periodicity of the space it is in. The examples in Question 1.5.1 and 1.5.2 considered only spaces with a periodicity given by 2^k for some k. The new examples in this question show that the evolution of the rule may lead to a fixed point much like the majority rule. More than one initial condition leads to the same fixed point. Both the example shown and the fixed point itself does. Systematic analyses of the cycles and fixed points (cycles of period one) for this and other rules of this type, and various boundary conditions have been performed. ∎

The choice of initial conditions is an important aspect of the operation of many CA. Computer investigations of CA often begin by assuming a "seed" consisting of a single cell with the value +1 (a single ON cell) and all the rest −1 (OFF). Alternatively, the initial conditions may be chosen to be random: $s(i, j, k;0) = \pm 1$ with equal probability. The behavior of the system with a particular initial condition may be assumed to be generic, or some quantity may be averaged over different choices of initial conditions.

Like the iterative maps we considered in Section 1.1, the CA dynamics may be described in terms of cycles and attractors. As long as we consider only binary variables and a finite space, the dynamics must repeat itself after no more than a number of steps equal to the number of possible states of the system. This number grows exponentially with the size of the space. There are 2^N states of the system when there are a total of N cells. For 100 cells the length of the longest possible cycle would be of order 10^{30}. To consider such a long time for a small space may seem an unusual model of space-time. For most analogies of CA with physical systems, this model of space-time is not the most appropriate. We might restrict the notion of cycles to apply only when their length does not grow exponentially with the size of the system.

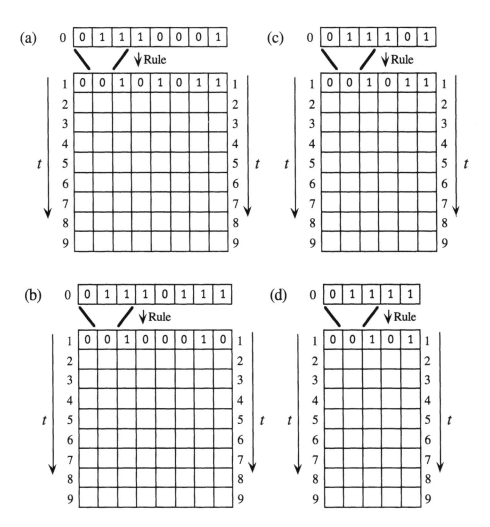

Figure 1.5.2 Four additional examples for the mod2 rule that have different initial conditions with specific periodicity: (a) is periodic in 8 cells, (b) is periodic in 4 cells, though it is shown embedded in a space of periodicity 8, (c) is periodic in 6 cells, (d) is periodic in 5 cells. By filling in the spaces it is possible to learn about the effect of different periodicities on the iterative properties of the mod2 rule. In particular, the length of the repeat time (cycle length) depends on the spatial periodicity. The cycle length may also depend on the specific initial conditions. ∎

Rules can be distinguished from each other and classified according to a variety of features they may possess. For example, some rules are reversible and others are not. Any reversible rule takes each state onto a unique successor. Otherwise it would be impossible to construct a single valued inverse mapping. Even when a rule is reversible, it is not guaranteed that the inverse rule is itself a CA, since it may not depend only on the local values of the variables. An example is given in question 1.5.5.

Question 1.5.4 Which if any of the two rules in Fig 1.5.1 is reversible?

Solution 1.5.4 The majority rule is not reversible, because locally we cannot identify in the next time step the difference between sequences that contain (11111) and (11011), since both result in a middle three of (111).

A discussion of the mod2 rule is more involved, since we must take into consideration the size of the space. In the examples of Questions 1.5.1–1.5.3 we see that in the space of six cells the rule is not reversible. In this case several initial conditions lead to the same result. The other examples all appear to be reversible, since each initial condition is part of a cycle that can be run backward to invert the rule. It turns out to be possible to construct explicitly the inverse of the mod2 rule. This is done in Question 1.5.5. ∎

Extra Credit Question 1.5.5 Find the inverse of the mod2 rule, when this is possible. This question involves some careful algebraic manipulation and may be skipped.

Solution 1.5.5 To find the inverse of the mod2 rule, it is useful to recall that equality modulo 2 satisfies simple addition properties including:

$$s_1 = s_2 \implies s_1 + s = s_2 + s \qquad \text{mod}_2 \qquad (1.5.6)$$

as well as the special property:

$$2s = 0 \qquad \text{mod}_2 \qquad (1.5.7)$$

Together these imply that variables may be moved from one side of the equality to the other:

$$s_1 + s = s_2 \implies s_1 = s_2 + s \qquad \text{mod}_2 \qquad (1.5.8)$$

Our task is to find the value of all $s(i;t)$ from the values of $s(j;t+1)$ that are assumed known. Using Eq. (1.5.8), the mod2 update rule (Eq. (1.5.5))

$$s(i;t+1) = (s(i-1;t) + s(i;t) + s(i+1;t)) \qquad \text{mod}_2 \qquad (1.5.9)$$

can be rewritten to give us the value of a cell in a layer in terms of the next layer and its own neighbors:

$$s(i-1;t) = s(i;t+1) + s(i;t) + s(i+1;t) \qquad \text{mod}_2 \qquad (1.5.10)$$

Substitute the same equation for the second term on the right (using one higher index) to obtain

$$s(i-1;t) = s(i;t+1) + [s(i+1;t+1) + s(i+1;t) + s(i+2;t)] + s(i+1;t) \qquad \text{mod}_2 \qquad (1.5.11)$$

the last term cancels against the middle term of the parenthesis and we have:

$$s(i-1;t) = s(i;t+1) + s(i+1;t+1) + s(i+2;t) \qquad \text{mod}_2 \qquad (1.5.12)$$

It is convenient to rewrite this with one higher index:

$$s(i;t) = s(i+1;t+1) + s(i+2;t+1) + s(i+3;t) \qquad \text{mod}_2 \qquad (1.5.13)$$

Interestingly, this is actually the solution we have been looking for, though some discussion is necessary to show this. On the right side of the equation appear three cell values. Two of them are from the time $t+1$, and one from the time t that we are trying to reconstruct. Since the two cell values from $t+1$ are assumed known, we must know only $s(i+3;t)$ in order to obtain $s(i;t)$. We can iterate this expression and see that instead we need to know $s(i+6;t)$ as follows:

$$s(i;t) = s(i+1;t+1) + s(i+2;t+1)$$
$$+ s(i+4;t+1) + s(i+5;t+1) + s(i+6;t) \quad \mathrm{mod}_2 \quad (1.5.14)$$

There are two possible cases that we must deal with at this point. The first is that the number of cells is divisible by three, and the second is that it is not. If the number of cells N is divisible by three, then after iterating Eq. (1.5.13) a total of $N/3$ times we will have an expression that looks like

$$s(i;t) = s(i+1;t+1) + s(i+2;t+1)$$
$$+ s(i+4;t+1) + s(i+5;t+1) + s(i+6;t)$$
$$+\ldots \quad \mathrm{mod}_2 \quad (1.5.15)$$
$$+ s(i+N-2;t+1) + s(i+N-1;t+1) + s(i;t)$$

where we have used the property of the periodic boundary conditions to set $s(i+n;t) = s(i;t)$. We can cancel this value from both sides of the equation. What is left is an equation that states that the sum over particular values of the cell variables at time $t+1$ must be zero.

$$0 = s(i+1;t+1) + s(i+2;t+1)$$
$$+ s(i+4;t+1) + s(i+5;t+1) + s(i+6;t)$$
$$+\ldots \quad \mathrm{mod}_2 \quad (1.5.16)$$
$$+ s(i+N-2;t+1) + s(i+N-1;t+1)$$

This means that any set of cell values that is the result of the mod2 rule update must satisfy this condition. Consequently, not all possible sets of cell values can be a result of mod2 updates. Thus the rule is not one-to-one and is not invertible when N is divisible by 3.

When N is not divisible by three, this problem does not arise, because we must go around the cell ring three times before we get back to $s(i;t)$. In this case, the analogous equation to Eq. (1.5.16) would have every cell value appearing exactly twice on the right of the equation. This is because each cell appears in two out of the three travels around the ring. Since the cell values all appear twice, they cancel, and the equation is the tautology $0 = 0$. Thus in this case there is no restriction on the result of the mod2 rule.

We almost have a full procedure for reconstructing $s(i; t)$. Choose the value of one particular cell variable, say $s(1;t) = 0$. From Eq. (1.5.13), obtain in sequence each of the cell variables $s(N-2;t)$, $s(N-5;t)$, . . . By going around the ring three times we can find uniquely all of the values. We now have to decide whether our original choice was correct. This can be done by directly applying the mod2 rule to find the value of say, $s(1; t+1)$. If we obtain the right value, then we have the right choice; if the wrong value, then all we have to do is switch all of the cell values to their opposites. How do we know this is correct?

There was only one other possible choice for the value of $s(1; t) = 1$. If we were to choose this case we would find that each cell value was the opposite, or one's complement, $1 - s(i; t)$ of the value we found. This can be seen from Eq. (1.5.13). Moreover, the mod2 rule preserves complementation. Which means that if we complement all of the values of $s(i; t)$ we will find the complements of the values of $s(1; t+1)$. The proof is direct:

$$1 - s(i;t+1) = 1 - (s(i-1;t) + s(i;t) + s(i+1;t))$$
$$= (1 - s(i-1;t)) + (1 - s(i;t)) + (1 - s(i+1;t))) - 2 \qquad \mathrm{mod}_2 \quad (1.5.17)$$
$$= (1 - s(i-1;t)) + (1 - s(i;t)) + (1 - s(i+1;t)))$$

Thus we can find the unique predecessor for the cell values $s(i;t+1)$. With some care it is possible to write down a fully algebraic expression for the value of $s(i;t)$ by implementing this procedure algebraically. The result for $N = 3k + 1$ is:

$$s(i;t) = s(i;t+1) + \sum_{j=1}^{(N-1)/3} (s(i+3j-2;t+1) + s(i+3j;t+1)) \quad \mathrm{mod}_2 \quad (1.5.18)$$

A similar result for $N = 3k + 2$ can also be found.

Note that the inverse of the mod2 rule is not a CA because it is not a local rule. ∎

One of the interesting ways to classify CA—introduced by Wolfram—separates them into four classes depending on the nature of their limiting behavior. This scheme is particularly interesting for us, since it begins to identify the concept of complex behavior, which we will address more fully in a later chapter. The notion of complex behavior in a spatially distributed system is at least in part distinct from the concept of chaotic behavior that we have discussed previously. Specifically, the classification scheme is:

Class-one CA: evolve to a fixed homogeneous state

Class-two CA: evolve to fixed inhomogeneous states or cycles

Class-three CA: evolve to chaotic or aperiodic behavior

Class-four CA: evolve to complex localized structures

One example of each class is given in Fig. 1.5.3. It is assumed that the length of the cycles in class-two automata does not grow as the size of the space increases. This classification scheme has not yet found a firm foundation in analytical work and is supported largely by observation of simulations of various CA.

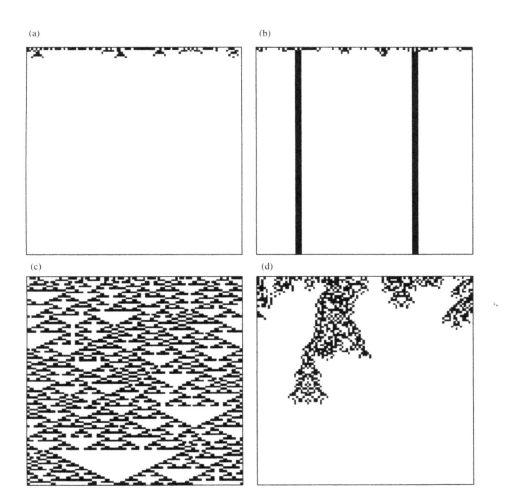

Figure 1.5.3 Illustration of four CA update rules with random initial conditions that are in a periodic space with a period of 100 cells. The initial conditions are shown at the top and time proceeds downward. Each is updated for 100 steps. ON cells are indicated as filled squares. OFF cells are not shown. Each of the rules gives the value of a cell in terms of a neighborhood of five cells at the previous time. The neighborhood consists of the cell itself and the two cells to the left and to the right. The rules are known as "totalistic" rules since they depend only on the sum of the variables in the neighborhood. Using the notation $s_i = 0,1$, the rules may be represented using $\Sigma_i(t) = s_{i-2}(t-1) + s_{i-1}(t-1) + s_i(t-1) + s_{i+1}(t-1) + s_{i+2}(t-1)$ by specifying the values of $\Sigma_i(t)$ for which $s_i(t)$ is ON. These are (a) only $\Sigma_i(t) = 2$, (b) only $\Sigma_i(t) = 3$, (c) $\Sigma_i(t) = 1$ and 2, and (d) $\Sigma_i(t) = 2$ and 4. See paper 1.3 in Wolfram's collection of articles on CA. ∎

It has been suggested that class-four automata have properties that enable them to be used as computers. Or, more precisely, to simulate a computer by setting the initial conditions to a set of data representing both the program and the input to the program. The result of the computation is to be obtained by looking some time later at the state of the system. A criteria that is clearly necessary for an automaton to be able to act as a computer is that the result of the dynamics is sensitive to the initial conditions. We will discuss the topic of computation further in Section 1.8.

The flip side of the use of a CA as a model of computation is to design a computer that will simulate CA with high efficiency. Such machines have been built, and are called cellular automaton machines (CAMs).

1.5.2 2-d cellular automata

Two- and three-dimensional CA provide more opportunities for contact with physical systems. We illustrate by describing an example of a 2-d CA that might serve as a simple model of droplet growth during condensation. The rule, illustrated in part pictorially in Fig. 1.5.4, may be described by saying that a particular cell with four or

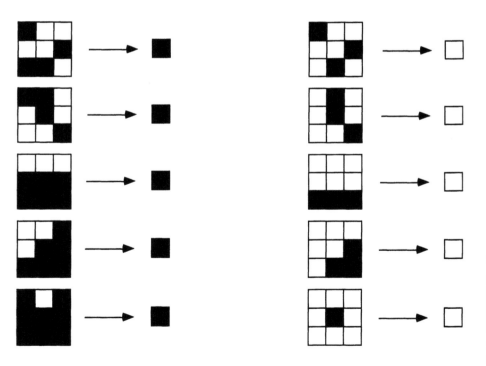

Figure 1.5.4 Illustration of a 2-d CA that may be thought of as a simple model of droplet condensation. The rule sets a cell to be ON (condensed) if four or more of its neighbors are condensed in the previous time, and OFF (uncondensed) otherwise. There are a total of $2^9=512$ possible initial configurations; of these only 10 are shown. The ones on the left have 4 or more cells condensed and the ones on the right have less than 4 condensed. This rule is explained further by Fig. 1.5.5 and simulated in Fig. 1.5.6. ∎

more "condensed" neighbors at time t is condensed at time $t + 1$. Neighbors are counted from the 3×3 square region surrounding the cell, including the cell itself.

Fig. 1.5.5 shows a simulation of this rule starting from a random initial starting point of approximately 25% condensed (ON) and 75% uncondensed (OFF) cells. Over the first few updates, the random arrangement of dots resolves into droplets, where isolated condensed cells disappear and regions of higher density become the droplets. Then over a longer time, the droplets grow and reach a stable configuration.

The characteristics of this rule may be understood by considering the properties of boundaries between condensed and uncondensed regions, as shown in Fig. 1.5.6. Boundaries that are vertical, horizontal or at a 45° diagonal are stable. Other boundaries will move, increasing the size of the condensed region. Moreover, a concave corner of stable edges is not stable. It will grow to increase the condensed region. On the other hand, a convex corner is stable. This means that convex droplets are stable when they are formed of the stable edges.

It can be shown that for this size space, the 25% initial filling is a transition density, where sometimes the result will fill the space and sometimes it will not. For higher densities, the system almost always reaches an end point where the whole space is condensed. For lower densities, the system almost always reaches a stable set of droplets.

This example illustrates an important point about the dynamics of many systems, which is the existence of phase transitions in the kinetics of the system. Such phase transitions are similar in some ways to the thermodynamic phase transitions that describe the equilibrium state of a system changing from, for example, a solid to a liquid. The kinetic phase transitions may arise from the choice of initial conditions, as they did in this example. Alternatively, the phase transition may occur when we consider the behavior of a class of CA as a function of a parameter. The parameter gradually changes the local kinetics of the system; however, measures of its behavior may change abruptly at a particular value. Such transitions are also common in CA when the outcome of a particular update is not deterministic but stochastic, as discussed in Section 1.5.4.

1.5.3 *Conway's Game of Life*

One of the most popular CA is known as Conway's Game of Life. Conceptually, it is designed to capture in a simple way the reproduction and death of biological organisms. It is based on a model where, locally, if there are too few organisms or too many organisms the organisms will disappear. On the other hand, if the number of organisms is just right, they will multiply. Quite surprisingly, the model takes on a life of its own with a rich dynamical behavior that is best understood by direct observation.

The specific rule is defined in terms of the 3×3 neighborhood that was used in the last section. The rule, illustrated in Fig. 1.5.7, specifies that when there are less than three or more than four ON (populated) cells in the neighborhood, the central cell will be OFF (unpopulated) at the next time. If there are three ON cells, the central cell will be ON at the next time. If there are four ON cells, then the central cell will keep its previous state—ON if it was ON and OFF if it was OFF.

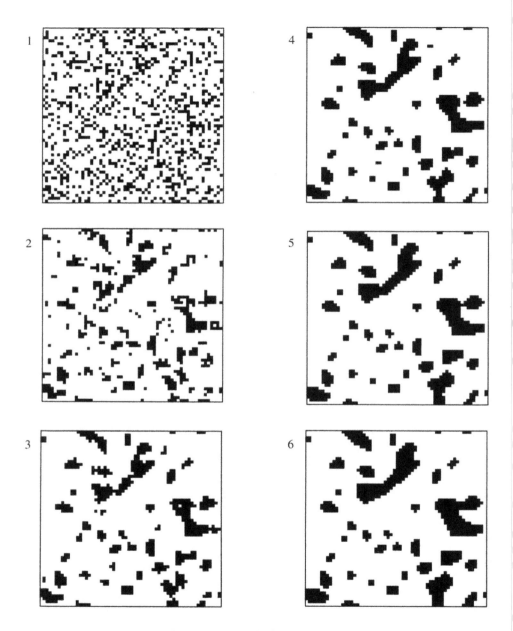

Figure 1.5.5 Simulation of the condensation CA described in Fig. 1.5.4. The initial conditions are chosen by setting randomly each site ON with a probability of 1 in 4. The initial few steps result in isolated ON sites disappearing and small ragged droplets of ON sites forming in higher-density regions. The droplets grow and smoothen their boundaries until at the sixtieth frame a static arrangement of convex droplets is reached. The first few steps are shown on the first page. Every tenth step is shown on the second page up to the sixtieth.

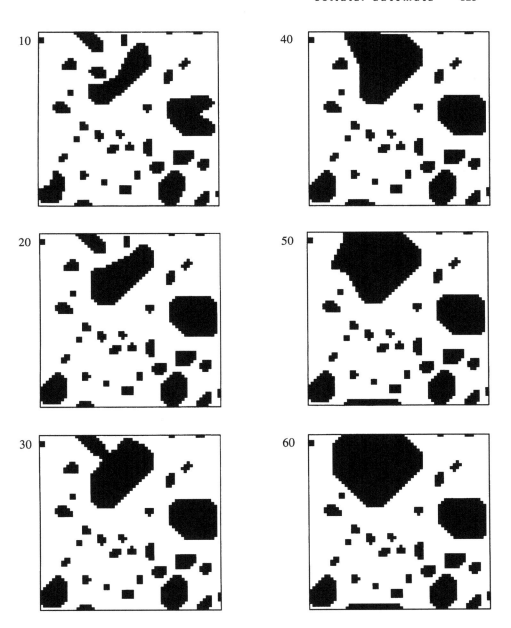

Figure 1.5.5 *Continued.* The initial occupation probability of 1 in 4 is near a phase transition in the kinetics of this model for a space of this size. For slightly higher densities the final configuration consists of a droplet covering the whole space. For slightly lower densities the final configuration is of isolated droplets. At a probability of 1 in 4 either may occur depending on the specific initial state. ∎

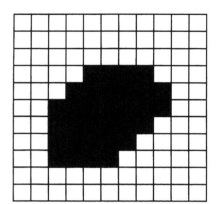

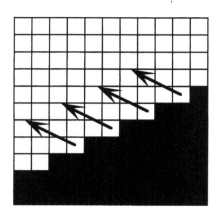

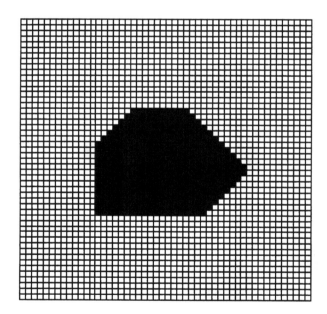

Figure 1.5.6 The droplet condensation model of Fig. 1.5.4 may be understood by noting that certain boundaries between condensed and uncondensed regions are stable. A completely stable shape is illustrated in the upper left. It is composed of boundaries that are horizontal, vertical or diagonal at 45°. A boundary that is at a different angle, such as shown on the upper right, will move, causing the droplet to grow. On a longer length scale a stable shape (droplet) is illustrated in the bottom figure. A simulation of this rule starting from a random initial condition is shown in Fig. 1.5.5. ∎

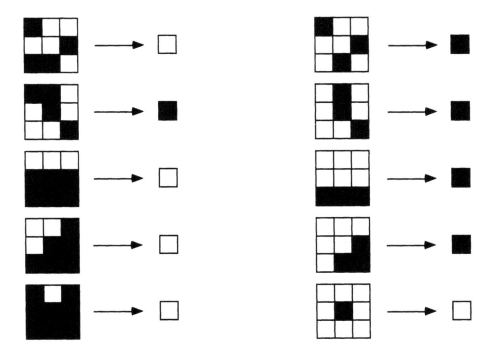

Figure 1.5.7 The CA rule Conway's Game of Life is illustrated for a few cases. When there are fewer than three or more than four neighbors in the 3×3 region the central cell is OFF in the next step. When there are three neighbors the central cell is ON in the next step. When there are four neighbors the central cell retains its current value in the next step. This rule was designed to capture some ideas about biological organism reproduction and death where too few organisms would lead to disappearance because of lack of reproduction and too many would lead to overpopulation and death due to exhaustion of resources. The rule is simulated in Fig. 1.5.8 and 1.5.9. ∎

Fig. 1.5.8 shows a simulation of the rule starting from the same initial conditions used for the condensation rule in the last section. Three sequential frames are shown, then after 100 steps an additional three frames are shown. Frames are also shown after 200 and 300 steps. After this amount of time the rule still has dynamic activity from frame to frame in some regions of the system, while others are apparently static or undergo simple cyclic behavior. An example of cyclic behavior may be seen in several places where there are horizontal bars of three ON cells that switch every time step between horizontal and vertical. There are many more complex local structures that repeat cyclically with much longer repeat cycles. Moreover, there are special structures called gliders that translate in space as they cycle through a set of configurations. The simplest glider is shown in Fig. 1.5.9, along with a structure called a glider gun, which creates them periodically.

We can make a connection between Conway's Game of Life and the quadratic iterative map considered in Section 1.1. The rich behavior of the iterative map was found because, for low values of the variable the iteration would increase its value, while for

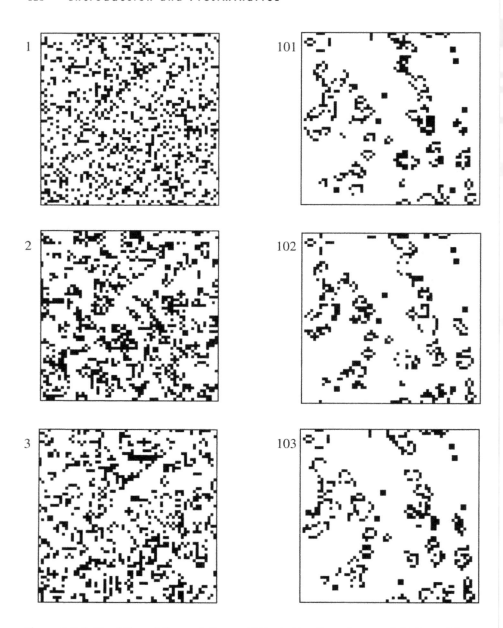

Figure 1.5.8 Simulation of Conway's Game of Life starting from the same initial conditions as used in Fig. 1.5.6 for the condensation rule where 1 in 4 cells are ON. Unlike the condensation rule there remains an active step-by-step evolution of the population of ON cells for many cycles. Illustrated are the three initial steps, and three successive steps each starting at steps 100, 200 and 300.

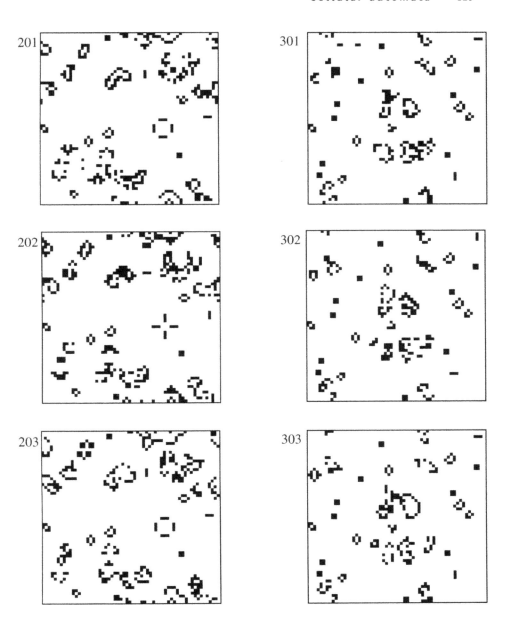

Figure 1.5.8 *Continued.* After the initial activity that occurs everywhere, the pattern of activity consists of regions that are active and regions that are static or have short cyclical activity. However, the active regions move over time around the whole space leading to changes everywhere. Eventually, after a longer time than illustrated here, the whole space becomes either static or has short cyclical activity. The time taken to relax to this state increases with the size of the space. ∎

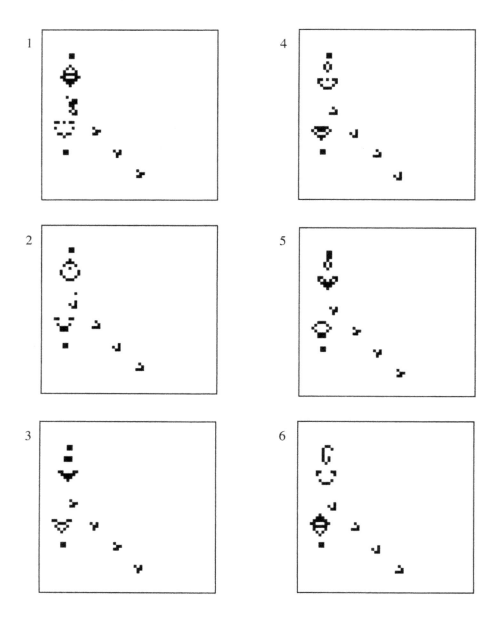

Figure 1.5.9 Special initial conditions simulated using Conway's Game of Life result in structures of oₙ cells called gliders that travel in space while progressing cyclically through a set of configurations. Several of the simplest type of gliders are shown moving toward the lower right. The more complex set of oₙ cells on the left, bounded by a 2 × 2 square of oₙ cells on top and bottom, is a glider gun. The glider gun cycles through 30 configurations during which a single glider is emitted. The stream of gliders moving to the lower right resulted from the activity of the glider gun. ∎

high values the iteration would decrease its value. Conway's Game of Life and other CA that exhibit interesting behavior also contain similar nonlinear feedback. Moreover, the spatial arrangement and coupling of the cells gives rise to a variety of new behaviors.

1.5.4 *Stochastic cellular automata*

In addition to the deterministic automaton of Eq. (1.5.3), we can define a stochastic automaton by the probabilities of transition from one state of the system to another:

$$P(\{s(i, j, k; t)\}|\{s(i, j, k; t-1)\}) \tag{1.5.19}$$

This general stochastic rule for the 2^N states of the system may be simplified. We have assumed for the deterministic rule that the rule for updating one cell may be performed independently of others. The analog for the stochastic rule is that the update probabilities for each of the cells is independent. If this is the case, then the total probability may be written as the product of probabilities of each cell value. Moreover, if the rule is local, the probability for the update of a particular cell will depend only on the values of the cell variables in the neighborhood of the cell we are considering.

$$P(\{s(i, j, k; t)\}|\{s(i, j, k; t-1)\}) = \prod_{i,j,k} P_0(s(i, j, k; t)|N(i, j, k; t-1)) \tag{1.5.20}$$

where we have used the notation $N(i, j, k; t)$ to indicate the values of the cell variables in the neighborhood of (i, j, k). For example, we might consider modifying the droplet condensation model so that a cell value is set to be ON with a certain probability (depending on the number of ON neighbors) and OFF otherwise.

Stochastic automata can be thought of as modeling the effects of noise and more specifically the ensemble of a dynamic system that is subject to thermal noise. There is another way to make the analogy between the dynamics of a CA and a thermodynamic system that is exact—if we consider not the space of the automaton but the $d + 1$ dimensional space-time. Consider the ensemble of all possible histories of the CA. If we have a three-dimensional space, then the histories are a set of variables with four indices $\{s(i, j, k, t)\}$. The probability of a particular set of these variables occurring (the probability of this history) is given by

$$P(\{s(i, j, k, t)\}) = \prod_{t} \prod_{i,j,k} P_0(s(i, j, k; t)|N(i, j, k; t-1))P(\{s(i, j, k; 0)\}) \tag{1.5.21}$$

This expression is the product of the probabilities of each update occurring in the history. The first factor on the right is the probability of a particular initial state in the ensemble we are considering. If we consider only one starting configuration, its probability would be one and the others zero.

We can relate the probability in Eq. (1.5.21) to thermodynamics using Boltzmann probability. We simply set it to the expression for the Boltzmann probability at a particular temperature T.

$$P(\{s(i, j, k, t)\}) = e^{-E(\{s(i, j, k, t)\})/kT} \tag{1.5.22}$$

There is no need to include the normalization constant Z because the probabilities are automatically normalized. What we have done is to define the energy of the particular state as:

$$E(\{s(i, j, k, t)\}) = kT\ln\left(P(\{s(i, j, k, t)\})\right) \tag{1.5.23}$$

This expression shows that any d dimensional automaton can be related to a $d+1$ dimensional system described by equilibrium Boltzmann probabilities. The ensemble of the $d+1$ dimensional system is the set of time histories of the automaton.

There is an important cautionary note about the conclusion reached in the last paragraph. While it is true that time histories are directly related to the ensemble of a thermodynamic system, there is a hidden danger in this analogy. These are not typical thermodynamic systems, and therefore our intuition about how they should behave is not trustworthy. For example, the time direction may be very different from any of the space directions. For the $d+1$ dimensional thermodynamic system, this means that one of the directions must be singled out. This kind of asymmetry does occur in thermodynamic systems, but it is not standard. Another example of the difference between thermodynamic systems and CA is in their sensitivity to boundary conditions. We have seen that many CA are quite sensitive to their initial conditions. While we have shown this for deterministic automata, it continues to be true for many stochastic automata as well. The analog of the initial conditions in a $d+1$ dimensional thermodynamic system is the surface or boundary conditions. Thermodynamic systems are typically insensitive to their boundary conditions. However, the relationship in Eq. (1.5.23) suggests that at least some thermodynamic systems are quite sensitive to their boundary conditions. An interesting use of this analogy is to attempt to discover special thermodynamic systems whose behavior mimics the interesting behavior of CA.

1.5.5 CA generalizations

There are a variety of generalizations of the simplest version of CA which are useful in developing models of particular systems. In this section we briefly describe a few of them as illustrated in Fig. 1.5.10.

It is often convenient to consider more than one variable at a particular site. One way to think about this is as multiple spaces (planes in 2-d, lines in 1-d) that are coupled to each other. We could think about each space as a different physical quantity. For example, one might represent a magnetic field and the other an electric field. Another possibility is that we might use one space as a thermal reservoir. The system we are actually interested in might be simulated in one space and the thermal reservoir in another. By considering various combinations of multiple spaces representing a physical system, the nature of the physical system can become quite rich in its structure.

We can also consider the update rule to be a compound rule formed of a sequence of steps. Each of the steps updates the cells. The whole rule consists of cycling through the set of individual step rules. For example, our update rule might consist of two different steps. The first one is performed on every odd step and the second is performed on every even step. We could reduce this to the previous single update step case by looking at the composite of the first and second steps. This is the same as looking at only every even state of the system. We could also reduce this to a multiple space rule, where both the odd and even states are combined together to be a single step.

However, it may be more convenient at times to think about the system as performing a cycle of update steps.

Finally, we can allow the state of the system at a particular time to depend on the state of the system at several previous times, not just on the state of the system at the previous time. A rule might depend on the most recent state of the system and the previous one as well. Such a rule is also equivalent to a rule with multiple spaces, by considering both the present state of the system and its predecessor as two spaces. One use of considering rules that depend on more than one time is to enable systematic construction of reversible deterministic rules from nonreversible rules. Let the original (not necessarily invertible) rule be $R(N(i, j, k; t))$. A new invertible rule can be written using the form

$$s(i, j, k; t) = \text{mod}_2(R(N(i, j, k; t-1)) + s(i, j, k; t-2)) \tag{1.5.24}$$

The inverse of the update rule is immediately constructed using the properties of addition modulo 2 (Eq. (1.5.8)) as:

$$s(i, j, k; t-2) = \text{mod}_2(R(N(i, j, k; t-1)) + s(i, j, k; t)) \tag{1.5.25}$$

1.5.6 Conserved quantities and Margolus dynamics

Standard CA are not well suited to the description of systems with constraints or conservation laws. For example, if we want to conserve the number of ON cells we must establish a rule where turning OFF one cell (switching it from ON to OFF) is tied to turning ON another cell. The standard rule considers each cell separately when an update is performed. This makes it difficult to guarantee that when this particular cell is turned OFF then another one will be turned ON. There are many examples of physical systems where the conservation of quantities such as number of particles, energy and momentum are central to their behavior.

A systematic way to construct CA that describe systems with conserved quantities has been developed. Rules of this kind are known as partitioned CA or Margolus rules (Fig. 1.5.11). These rules separate the space into nonoverlapping partitions (also known as neighborhoods). The new value of each cell in a partition is given in terms of the previous values of the cells in the same partition. This is different from the conventional automaton, since the local rule has more than one output as well as more than one input. Such a rule is not sufficient in itself to describe the system update, since there is no communication in a single update between different partitions. The complete rule must specify how the partitions are shifted after each update with respect to the underlying space. This shifting is an essential part of the dynamical rule that restores the cellular symmetry of the space.

The convenience of this kind of CA is that specification of the rule gives us direct control of the dynamics within each partition, and therefore we can impose conservation rules within the partition. Once the conservation rule is imposed inside the partition, it will be maintained globally—throughout the whole space and through every time step. Fig. 1.5.12 illustrates a rule that conserves the number of ON cells inside a 2 × 2 neighborhood. The ON cells may be thought of as particles whose num-

(a)

(b)

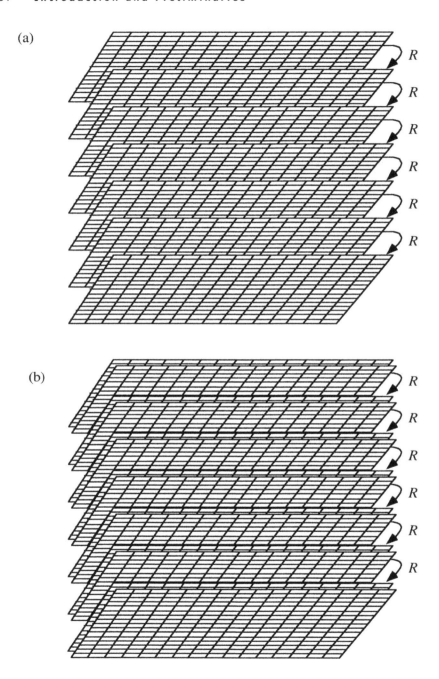

Figure 1.5.10 Schematic illustrations of several modifications of the simplest CA rule. The basic CA rule updates a set of spatially arrayed cell variables shown in (a). The first modification uses more than one variable in each cell. Conceptually this may be thought of as describing a set of coupled spaces, where the case of two spaces is shown in (b). The second modification makes use of a compound rule that combines several different rules, where the

(c)

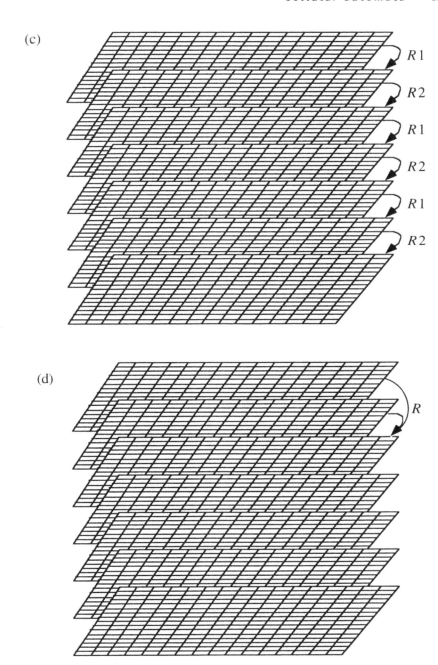

(d)

case of two rules is shown in (c). The third modification shown in (d) makes use of a rule that depends on not just the most recent value of the cell variables but also the previous one. Both (c) and (d) may be described as special cases of (b) where two successive values of the cell variables are considered instead as occurring at the same time in different spaces. ∎

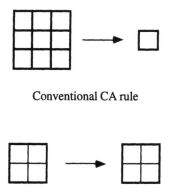

Conventional CA rule

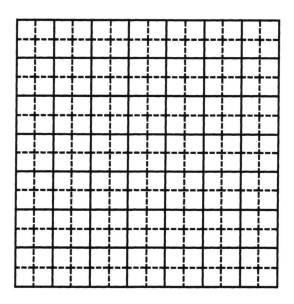

Partitioned (Margolus) CA rule

Partition Alternation

Figure 1.5.11 Partitioned CA (Margolus rules) enable the imposition of conservation laws in a direct way. A conventional CA gives the value of an individual cell in terms of the previous values of cells in its neighborhood (top). A partitioned CA gives the value of several cells in a particular partition in terms of the previous values of the same cells (center). This enables conservation rules to be imposed directly within a particular partition. An example is given in Fig. 1.5.12. In addition to the rule for updating the partition, the dynamics must specify how the partitions are to be shifted from step to step. For example (bottom), the use of a 2 × 2 partition may be implemented by alternating the partitions from the solid lines to the dashed lines. Every even update the dashed lines are used and every odd update the solid lines are used to partition the space. This restores the cellular periodicity of the space and enables the cells to communicate with each other, which is not possible without the shifting of partitions. ∎

ber is conserved. The only requirement is that each of the possible arrangement of particles on the left results in an arrangement on the right with the same number of particles. This rule is augmented by specifying that the 2×2 partitions are shifted by a single cell to the right and down after every update. The motion of these particles is that of an unusual gas of particles.

The rule shown is only one of many possible that use this 2×2 neighborhood and conserve the number of particles. Some of these rules have additional properties or symmetries. A rule that is constructed to conserve particles may or may not be reversible. The one illustrated in Fig. 1.5.12 is not reversible. There exist more than one predecessor for particular values of the cell variables. This can be seen from the two mappings on the lower left that have the same output but different input. A rule that conserves particles also may or may not have a particular symmetry, such as a symmetry of reflection. A symmetry of reflection means that reflection of a configuration across a particular axis before application of the rule results in the same effect as reflection after application of the rule.

The existence of a well-defined set of rules that conserves the number of particles enables us to choose to study one of them for a specific reason. Alternatively, by randomly constructing a rule which conserves the number of particles, we can learn what particle conservation does in a dynamical system independent of other regularities of the system such as reversibility and reflection or rotation symmetries. More systematically, it is possible to consider the class of automata that conserve particle number and investigate their properties.

Question 1.5.6 Design a 2-d Margolus CA that represents a particle or chemical reaction: $A + B \leftrightarrow C$. Discuss some of the parameters that must be set and how you could use symmetries and conservation laws to set them.

Solution 1.5.6 We could use a 2×2 partition just like that in Fig. 1.5.12. On each of the four squares there can appear any one of the four possibilities (O, A, B, C). There are $4^4 = 256$ different initial conditions of the partition. Each of these must be paired with one final condition, if the rule is deterministic. If the rule is probabilistic, then probabilities must be assigned for each possible transition.

To represent a chemical reaction, we choose cases where A and B are adjacent (horizontally or vertically) and replace them with a C and a 0. If we prefer to be consistent, we can always place the C where A was before. To go the other direction, we take cases where C is next to a 0 and replace them with an A and a B. One question we might ask is, Do we want to have a reaction whenever it is possible, or do we want to assign some probability for the reaction? The latter case is more interesting and we would have to use a probabilistic CA to represent it. In addition to the reaction, the rule would include particle motion similar to that in Fig. 1.5.12.

To apply symmetries, we could assume that reflection along horizontal or vertical axes, or rotations of the partition by 90° before the update, will have the same effect as a reflection or rotation of the partition after the

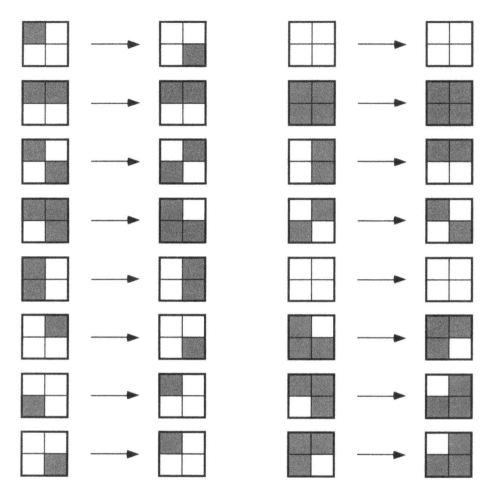

Figure 1.5.12 Illustration of a particular 2-d Margolus rule that preserves the number of ON cells which may be thought of as particles in a gas. The requirement for conservation of number of particles is that every initial configuration is matched with a final configuration having the same number of ON cells. This particular rule does not observe conventional symmetries such as reflection or rotation symmetries that might be expected in a typical gas. Many rules that conserve particles may be constructed in this framework by changing around the final states while preserving the number of particles in each case. ∎

update. We could also assume that A, B and C move in the same way when they are by themselves. Moreover, we might assume that the rule is symmetric under the transformation $A \leftrightarrow B$.

There is a simpler approach that requires enumerating many fewer states. We choose a 2×1 rectangular partition that has only two cells, and $4^2 = 16$ possible states. Of these, four do not change: $[A,A]$, $[B,B]$, $[C,C]$ and $[0,0]$.

Eight others are paired because the cell values can be switched to achieve particle motion (with a certain probability): $[A,0] \leftrightarrow [0,A]$, $[B,0] \leftrightarrow [0,B]$, $[C,A] \leftrightarrow [A,C]$, and $[C,B] \leftrightarrow [B,C]$. Finally, the last four, $[C,0]$, $[0,C]$, $[A,B]$ and $[B,A]$, can participate in reactions. If the rule is deterministic, they must be paired in a unique way for possible transitions. Otherwise, each possibility can be assigned a probability: $[C,0] \leftrightarrow [A,B]$, $[0,C] \leftrightarrow [B,A]$, $[C,0] \leftrightarrow [B,A]$ and $[0,C] \leftrightarrow [A,B]$. The switching of the particles without undergoing reaction for these states may also be allowed with a certain probability. Thus, each of the four states can have a nonzero transition probability to each of the others. These probabilities may be related by the symmetries mentioned before. Once we have determined the update rule for the 2x1 partition, we can choose several ways to map the partitions onto the plane. The simplest are obtained by dividing each of the 2×2 partitions in Fig. 1.5.11 horizontally or vertically. This gives a total of four ways to partition the plane. These four can alternate when we simulate this CA. ∎

1.5.7 *Differential equations and CA*

Cellular automata are an alternative to differential equations for the modeling of physical systems. Differential equations when modeled numerically on a computer are often discretized in order to perform integrals. This discretization is an approximation that might be considered essentially equivalent to setting up a locally discrete dynamical system that in the macroscopic limit reduces to the differential equation. Why not then start from a discrete system and prove its relevance to the problem of interest? This a priori approach can provide distinct computational advantages. This argument might lead us to consider CA as an approximation to differential equations. However, it is possible to adopt an even more direct approach and say that differential equations are themselves approximation to aspects of physical reality. CA are a different but equally valid approach to approximating this reality. In general, differential equations are more convenient for analytic solution while CA are more convenient for simulations. Since complex systems of differential equations are often solved numerically anyway, the alternative use of CA appears to be worth systematic consideration.

While both cellular automata and differential equations can be used to model macroscopic systems, this should not be taken to mean that the relationship between differential equations and CA is simple. Recognizing a CA analog to a standard differential equation may be a difficult problem. One of the most extensive efforts to use CA for simulation of a system more commonly known by its differential equation is the problem of hydrodynamics. Hydrodynamics is typically modeled by the Navier-Stokes equation. A type of CA called a lattice gas (Section 1.5.8) has been designed that on a length scale that is large compared to the cellular scale reproduces the behavior of the Navier-Stokes equation. The difficulties of solving the differential equation for specific boundary conditions make this CA a powerful tool for studying hydrodynamic flow.

A frequently occurring differential equation is the wave equation. The wave equation describes an elastic medium that is approximated as a continuum. The wave equation emerges as the continuum limit of a large variety of systems. It is to be expected that many CA will also display wavelike properties. Here we use a simple example to illustrate one way that wavelike properties may arise. We also show how the analogy may be quite different than intuition might suggest. The wave equation written in 1-d as

$$\frac{\partial^2 f}{\partial t^2} = c^2 \frac{\partial^2 f}{\partial x^2} \tag{1.5.26}$$

has two types of solutions that are waves traveling to the right and to the left with wave vectors k and frequencies of oscillation $\omega_k = ck$:

$$f = \sum_k \left(A_k e^{i(kx - \omega_k t)} + B_k e^{i(kx + \omega_k t)} \right) \tag{1.5.27}$$

A particular solution is obtained by choosing the coefficients A_k and B_k. These solutions may also be written in real space in the form:

$$f = \tilde{A}(x - ct) + \tilde{B}(x + ct) \tag{1.5.28}$$

where

$$\tilde{A}(x) = \sum_k A_k e^{ikx}$$
$$\tilde{B}(x) = \sum_k B_k e^{ikx} \tag{1.5.29}$$

are two arbitrary functions that specify the initial conditions of the wave in an infinite space.

We can construct a CA analog of the wave equation as illustrated in Fig. 1.5.13. It should be understood that the wave equation will arise only as a continuum or long wave limit of the CA dynamics. However, we are not restricted to considering a model that mimics a vibrating elastic medium. The rule we construct consists of a 1-d partitioned space dynamics. Each update, adjacent cells are paired into partitions of two cells each. The pairing switches from update to update, analogous to the 2-d example in Fig. 1.5.11. The dynamics consists solely of switching the contents of the two adjacent cells in a single partition. Starting from a particular initial configuration, it can be seen that the contents of the odd cells moves systematically in one direction (right in the figure), while the contents of the even cells moves in the opposite direction (left in the figure). The movement proceeds at a constant velocity of $c = 1$ cell/update. Thus we identify the contents of the odd cells as the rightward traveling wave, and the even cells as the leftward traveling wave.

The dynamics of this CA is the same as the dynamics of the wave equation of Eq. (1.5.28) in an infinite space. The only requirement is to encode appropriately the initial conditions $\tilde{A}(x)$, $\tilde{B}(x)$ in the cells. If we use variables with values in the conven-

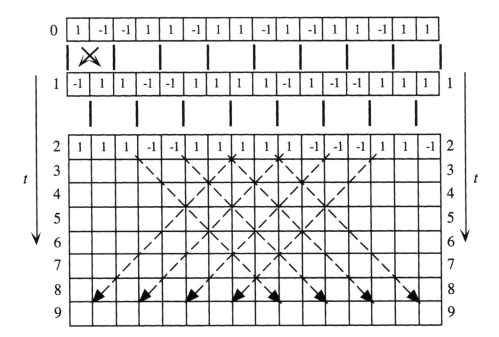

Figure 1.5.13 A simple 1-d CA using a Margolus rule, which switches the values of the two adjacent cells in the partition, can be used to model the wave equation. The partitions alternate between the two possible ways of partitioning the cells every time step. It can be seen that the initial state is propagated in time so that the odd (even) cells move at a fixed rate of one cell per update to the right (left). The solutions of the wave equation likewise consist of a right and left traveling wave. The initial conditions of the wave equation solution are the analog of the initial condition of the cells in the CA. ∎

tional real continuum $s_i \in \mathfrak{R}$, then the (discretized) waves may be encoded directly. If a binary representation $s_i = \pm 1$ is used, the local average over odd cells represents the right traveling wave $\tilde{A}(x - ct)$, and the local average over even cells represents the left traveling wave $\tilde{B}(x + ct)$.

1.5.8 *Lattice gases*

A lattice gas is a type of CA designed to model gases or liquids of colliding particles. Lattice gases are formulated in a way that enables the collisions to conserve momentum as well as number of particles. Momentum is represented by setting the velocity of each particle to a discrete set of possibilities. A simple example, the HPP gas, is illustrated in Fig. 1.5.14. Each cell contains four binary variables that represent the presence (or absence) of particles with unit velocity in the four compass directions NESW. In the figure, the presence of a particle in a cell is indicated by an arrow. There can be up to four particles at each site. Each particle present in a single cell must have a distinct velocity.

The dynamics of the HPP gas is performed in two steps that alternate: propagation and collision. In the propagation step, particles move from the cell they are in to the neighboring cell in the direction of their motion. In the collision step, each cell acts independently, changing the particles from incoming to outgoing according to prespecified collision rules. The rule for the HPP gas is illustrated in Fig. 1.5.15. Because of momentum conservation in this rule, there are only two possibilities for changes in the particle velocity as a result of a collision. A similar lattice gas, the FHP gas, which is implemented on a hexagonal lattice of cells rather than a square lattice, has been proven to give rise to the Navier-Stokes hydrodynamic equations on a macroscopic scale. Due to properties of the square lattice in two dimensions, this behavior does not occur for the HPP gas. One way to understand the limitation of the square lattice is to realize that for the HPP gas (Fig. 1.5.14), momentum is conserved in any individual horizontal or vertical stripe of cells. This type of conservation law is not satisfied by hydrodynamics.

1.5.9 *Material growth*

One of the natural physical systems to model using CA is the problem of layer-by-layer material growth such as is achieved in molecular beam epitaxy. There are many areas of study of the growth of materials. For example, in cases where the material is formed of only a single type of atom, it is the surface structure during growth that is of interest. Here, we focus on an example of an alloy formed of several different atoms, where the growth of the atoms is precisely layer by layer. In this case the surface structure is simple, but the relative abundance and location of different atoms in the material is of interest. The simplest case is when the atoms are found on a lattice that is prespecified, it is only the type of atom that may vary.

The analogy with a CA is established by considering each layer of atoms, when it is deposited, as represented by a 2-d CA at a particular time. As shown in Fig. 1.5.16 the cell values of the automaton represent the type of atom at a particular site. The values of the cells at a particular time are preserved as the atoms of the layer deposited at that time. It is the time history of the CA that is to be interpreted as representing the structure of the alloy. This picture assumes that once an atom is incorporated in a complete layer it does not move.

In order to construct the CA, we assume that the probability of a particular atom being deposited at a particular location depends on the atoms residing in the layer immediately preceding it. The stochastic CA rule in the form of Eq. (1.5.20) specifies the probability of attaching each kind of atom to every possible atomic environment in the previous layer.

We can illustrate how this might work by describing a specific example. There exist alloys formed out of a mixture of gallium, arsenic and silicon. A material formed of equal proportions of gallium and arsenic forms a GaAs crystal, which is exactly like a silicon crystal, except the Ga and As atoms alternate in positions. When we put silicon together with GaAs then the silicon can substitute for either the Ga or the As atoms. If there is more Si than GaAs, then the crystal is essentially a Si crystal with small regions of GaAs, and isolated Ga and As. If there is more GaAs than Si, then the

Figure 1.5.14 Illustration of the update of the HPP lattice gas. In a lattice gas, binary variables in each cell indicate the presence of particles with a particular velocity. Here there are four possible particles in each cell with unit velocities in the four compass directions, NESW. Pictorially the presence of a particle is indicated by an arrow in the direction of its velocity. Updating the lattice gas consists of two steps: propagating the particles according to their velocities, and allowing the particles to collide according to a collision rule. The propagation step consists of moving particles from each cell into the neighboring cells in the direction of their motion. The collision step consists of each cell independently changing the velocities of its particles. The HPP collision rule is shown in Fig. 1.5.15, and implemented here from the middle to the bottom panel. For convenience in viewing the different steps the arrows in this figure alternate between incoming and outcoming. Particles before propagation (top) are shown as outward arrows from the center of the cell. After the propagation step (middle) they are shown as incoming arrows. After collision (bottom) they are again shown as outgoing arrows. ∎

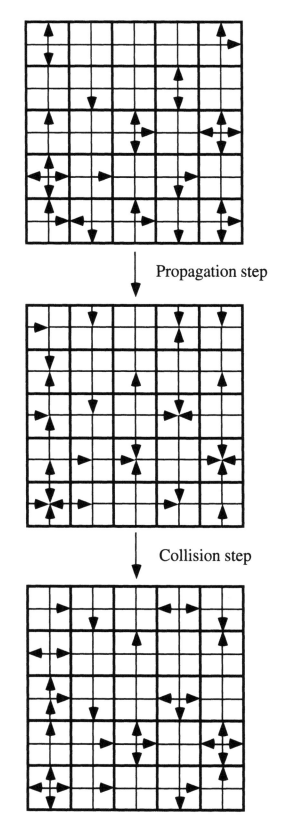

Propagation step

Collision step

Figure 1.5.15 The collision rule for the HPP lattice gas. With the exception of the case of two particles coming in from N and S and leaving from E and W, or vice versa (dashed box), there are no changes in the particle velocities as a result of collisions in this rule. Momentum conservation does not allow any other changes. ∎

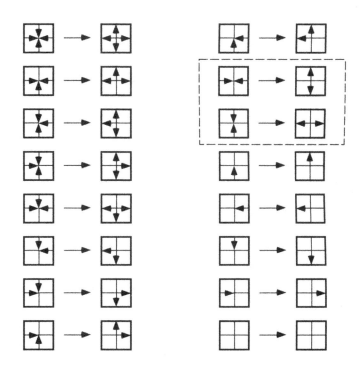

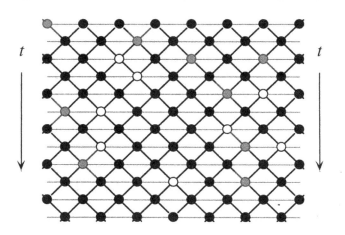

Figure 1.5.16 Illustration of the time history of a CA and its use to model the structure of a material (alloy) formed by a layer by layer growth. Each horizontal dashed line represents a layer of the material. The alloy has three types of atoms. The configuration of atoms in each layer depends only on the atoms in the layer preceding it. The type of atom, indicated in the figure by filled, empty and shaded dots, are determined by the values of the cell variables of the CA at a particular time, $s_i(t) = \pm1, 0$. The time history of the CA is the structure of the material. ∎

crystal will be essentially a GaAs crystal with isolated Si atoms. We can model the growth of the alloys formed by different relative proportions of GaAs and Si of the form $(GaAs)_{1-x}Si_x$ using a CA. Each cell of the CA has a variable with three possible values $s_i = \pm 1, 0$ that would represent the occupation of a crystal site by Ga, As and Si respectively. The CA rule (Eq. (1.5.20)) would then be constructed by assuming different probabilities for adding a Si, Ga and As atom at the surface. For example, the likelihood of finding a Ga next to a Ga atom or an As next to an As is small, so the probability of adding a Ga on top of a Ga can be set to be much smaller than other probabilities. The probability of an Si atom $s_i = 0$ could be varied to reflect different concentrations of Si in the growth. Then we would be able to observe how the structure of the material changes as the Si concentration changes.

This is one of many examples of physical, chemical and biological systems that have been modeled using CA to capture some of their dynamical properties. We will encounter others in later chapters.

1.6   Statistical Fields

In real systems as well as in kinetic models such as cellular automata (CA) discussed in the previous section, we are often interested in finding the state of a system—the time averaged (equilibrium) ensemble when cycles or randomness are present—that arises after the fast initial kinetic processes have occurred. Our objective in this section is to treat systems with many degrees of freedom using the tools of equilibrium statistical mechanics (Section 1.3). These tools describe the equilibrium ensemble directly rather than the time evolution. The simplest example is a collection of interacting binary variables, which is in many ways analogous to the simplest of the CA models. This model is known as the Ising model, and was introduced originally to describe the properties of magnets. Each of the individual variables corresponds to a microscopic magnetic region that arises due to the orbital motion of an electron or the internal degree of freedom known as the spin of the electron.

The Ising model is the simplest model of interacting degrees of freedom. Each of the variables is binary and the interactions between them are only specified by one parameter—the strength of the interaction. Remarkably, many complex systems we will be considering can be modeled by the Ising model as a first approximation. We will use several versions of the Ising model to discuss neural networks in Chapter 2 and proteins in Chapter 4. The reason for the usefulness of this model is the very existence of interactions between the elements. This interaction is not present in simpler models and results in various behaviors that can be used to understand some of the key aspects of complex systems. The concepts and tools that are used to study the Ising model also may be transferred to more complicated models. It should be understood, however, that the Ising model is a simplistic model of magnets as well as of other systems.

In Section 1.3 we considered the ideal gas with collisions. The collisions were a form of interaction. However, these interactions were incidental to the model because they were assumed to be so short that they were not present during observation. This is no longer true in the Ising model.

1.6.1 *The Ising model without interactions*

The Ising model describes the energy of a collection of elements (spins) represented by binary variables. It is so simple that there is no kinetics, only an energy $E[\{s_i\}]$. Later we will discuss how to reintroduce a dynamics for this model. The absence of a dynamics is not a problem for the study of the equilibrium properties of the system, since the Boltzmann probability (Eq. (1.3.29)) depends only upon the energy. The energy is specified as a function of the values of the binary variables $\{s_i = \pm 1\}$. Unless necessary, we will use one index for all of the spin variables regardless of dimensionality. The use of the term "spin" originates from the magnetic analogy. There is no other specific term, so we adopt this terminology. The term "spin" emphasizes that the binary variable represents the state of a physical entity such that the collection of spins is the system we are interested in. A spin can be illustrated as an arrow of fixed length (see Fig. 1.6.1). The value of the binary variable describes its orientation, where $+1$ indicates a spin oriented in the positive z direction (UP), and -1 indicates a spin oriented in the negative z direction (DOWN).

Before we consider the effects of interactions between the spins, we start by considering a system where there are no interactions. We can write the energy of such a system as:

$$E[\{s_i\}] = \sum_i e_i(s_i)$$

(1.6.1)

Where $e_i(s_i)$ is the energy of the ith spin that does not depend on the values of any of the other spins. Since s_i are binary we can write this as:

$$E[\{s_i\}] = \frac{1}{2}\sum_i (e_i(1) - e_i(-1))s_i + (e_i(1) + e_i(-1)) = E_0 - \sum_i h_i s_i \rightarrow -\sum_i h_i s_i$$

(1.6.2)

All of the terms that do not depend on the spin variables have been collected together into a constant. We set this constant to zero by redefining the energy scale. The quantities $\{h_i\}$ describe the energy due to the orientation of the spins. In the magnetic system they correspond to an external magnetic field that varies from location to location. Like small magnets, spins try to orient along the magnetic field. A spin oriented along the magnetic field (s_i and h_i have the same sign) has a lower energy than if it is antiparallel to the magnetic field. As in Eq. (1.6.2), the contribution of the magnetic field to the energy is $-|h_i|(|h_i|)$ when the spin is parallel (antiparallel) to the field direction. When convenient we will simplify to the case of a uniform magnetic field, $h_i = h$.

When the spins are noninteracting, the Ising model reduces to a collection of two-state systems that we investigated in Section 1.4. Later, when we introduce interactions between the spins, there will be differences. For the noninteracting case we can write the probability for a particular configuration of the spins using the Boltzmann probability:

$$P[\{s_i\}] = \frac{e^{-\beta E[\{s_i\}]}}{Z} = \frac{e^{\beta \sum_i h_i s_i}}{Z} = \frac{\prod_i e^{\beta h_i s_i}}{Z}$$

(1.6.3)

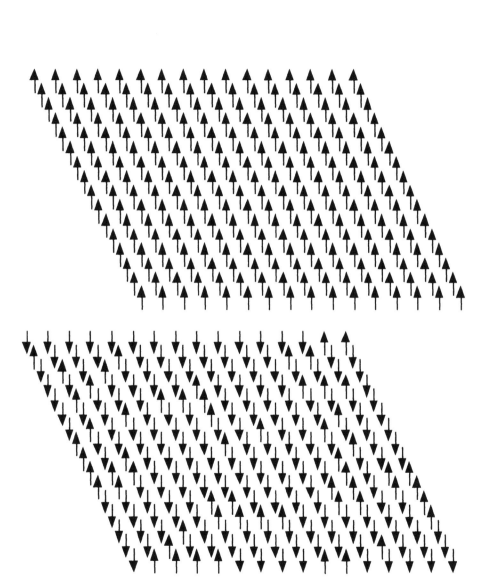

Figure 1.6.1 One way to visualize the Ising model is as a spatial array of binary variables called spins, represented as UP or DOWN arrows. A one-dimensional (1-d) example with all spins UP is shown on top. The middle and lower figures show two-dimensional (2-d) arrays which have all spins UP (middle) or have some spins UP and some spins DOWN (bottom). ∎

where $\beta = 1/kT$. The partition function Z is given by:

$$Z = \sum_{\{s_i\}} e^{-\beta E[\{s_i\}]} = \sum_{\{s_i\}} \prod_i e^{\beta h_i s_i} = \prod_i \sum_{s_i} e^{\beta h_i s_i} = \prod_i \left(e^{\beta h_i} + e^{-\beta h_i} \right) \quad (1.6.4)$$

where the second to last equality replaces the sum over all possible values of the spin variables with a sum over each spin variable $s_i = \pm 1$ within the product. Thus the probability factors as:

$$P[\{s_i\}] = \prod_i P(s_i) = \prod_i \left(\frac{e^{\beta h_i s_i}}{e^{\beta h_i} + e^{-\beta h_i}} \right) \quad (1.6.5)$$

This is a product over the result we found for probability of the two-state system (Eq. (1.4.14)) if we write the energy of a single spin using the notation $E_i(s_i) = -h_i s_i$.

Now that we have many spin variables, we can investigate the thermodynamics of this model by writing down the free energy and entropy of this model. This is discussed in Question 1.6.1.

Question 1.6.1 Evaluate the thermodynamic free energy, energy and entropy for the Ising model without interactions.

Solution 1.6.1 The free energy is given in terms of the partition function by Eq. (1.3.37):

$$F = -kT \ln(Z) = -kT \sum_i \ln\left(e^{\beta h_i} + e^{-\beta h_i} \right) = -kT \sum_i \ln\left(2\cosh\left(\beta h_i \right) \right) \quad (1.6.6)$$

The latter expression is a more common way of writing this result.

The thermodynamic energy of the system is found from Eq. (1.3.38) as

$$U = -\frac{\partial \ln(Z)}{\partial \beta} = -\sum_i \frac{h_i (e^{\beta h_i} - e^{-\beta h_i})}{(e^{\beta h_i} + e^{-\beta h_i})} = -\sum_i h_i \tanh(\beta h_i) \quad (1.6.7)$$

There is another way to obtain the same result. The thermodynamic energy is the average energy of the system (Eq. (1.3.30)), which can be evaluated directly:

$$U = \left\langle E[\{s_i\}] \right\rangle = \left\langle -\sum_i h_i s_i \right\rangle = -\sum_i h_i \left\langle s_i \right\rangle = -\sum_i h_i \sum_{s_i} s_i P(s_i)$$
$$= -\sum_i h_i \frac{(e^{\beta h_i} - e^{-\beta h_i})}{(e^{\beta h_i} + e^{-\beta h_i})} = -\sum_i h_i \tanh(\beta h_i) \quad (1.6.8)$$

which is the same as before. We have used the possibility of writing the probability of a single spin variable independent of the others in order to perform this average. It is convenient to define the local magnetization m_i as the average value of a particular spin variable:

$$m_i = \left\langle s_i \right\rangle = \sum_{s_i = \pm 1} s_i P_{s_i}(s_i) = P_{s_i}(1) - P_{s_i}(-1) \quad (1.6.9)$$

Or using Eq. (1.6.5):

$$m_i = \langle s_i \rangle = \tanh(\beta h_i) \tag{1.6.10}$$

In Fig. 1.6.2, the magnetization at a particular site is plotted as a function of the magnetic field for several different temperatures ($\beta = 1/kT$). The magnetization increases with increasing magnetic field and with decreasing temperature until it saturates asymptotically to a value of $+1$ or -1. In terms of the magnetization the energy is:

$$U = -\sum_i h_i m_i \tag{1.6.11}$$

We can calculate the entropy of the Ising model using Eq. (1.3.36)

$$S = k\beta U + k\ln Z = -k\sum_i \beta h_i \tanh(\beta h_i) + k\sum_i \ln\left(2\cosh\left(\beta h_i\right)\right) \tag{1.6.12}$$

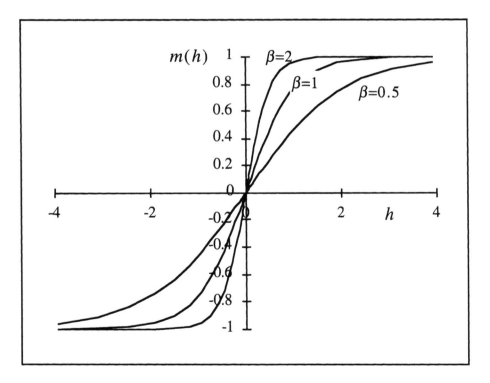

Figure 1.6.2 Plot of the magnetization at a particular site as a function of the magnetic field for independent spins in a magnetic field. The magnetization is the average of the spin value, so the magnetization shows the degree to which the spin is aligned to the magnetic field. The different curves are for several temperatures $\beta = 0.5, 1, 2$ ($\beta = 1/kT$). The magnetization has the same sign as the magnetic field. The magnitude of the spin increases with increasing magnetic field. Increasing temperature, however, decreases the alignment due to increased random motion of the spins. The maximum magnitude of the magnetization is 1, corresponding to a fully aligned spin. ∎

which is not particularly enlightening. However, we can rewrite this in terms of the magnetization using the identity:

$$\cosh(x) = \frac{1}{\sqrt{1 - \tanh^2(x)}} \tag{1.6.13}$$

and the inverse of Eq. (1.6.10):

$$\beta h_i = \frac{1}{2}\ln\left(\frac{1+m_i}{1-m_i}\right) \tag{1.6.14}$$

Substituting into Eq. (1.6.12) gives

$$S = -k\sum_i m_i \frac{1}{2}\ln\left(\frac{1+m_i}{1-m_i}\right) + kN\ln(2) - k\frac{1}{2}\sum_i \ln\left(1-m_i^2\right) \tag{1.6.15}$$

Rearranging slightly, we have:

$$S = +k\left[N\ln(2) - \frac{1}{2}\sum_i\left((1+m_i)\ln(1+m_i) + (1-m_i)\ln(1-m_i)\right)\right] \tag{1.6.16}$$

The final expression can be derived, at least for the case when all m_i are the same, by counting the number of states directly. It is worth deriving the entropy twice, because it may be used more generally than this treatment indicates. We will assume that all $h_i = h$ are the same. The energy then depends only on the total magnetization:

$$E[\{s_i\}] = -h\sum_i s_i$$
$$U = -h\sum_i m_i = -hNm \tag{1.6.17}$$

To obtain the entropy from the counting of states (Eq. (1.3.25)) we evaluate the number of states within a particular narrow energy range. Since the energy is the sum over the values of the spins, it may also be written as the difference between the number of UP spins $N(1)$ and DOWN spins $N(-1)$:

$$E[\{s_i\}] = -h(N(1) - N(-1)) \tag{1.6.18}$$

Thus, to find the entropy for a particular energy we must count how many states there are with a particular number of UP and DOWN spins. Moreover, flipping a spin from DOWN to UP causes a fixed increment in the energy. Thus there is no need to include in the counting the width of the energy interval in which we are counting states. The number of states with $N(1)$ UP spins and $N(-1)$ DOWN spins is:

$$\Omega(E,N) = \binom{N}{N(1)} = \frac{N!}{N(1)!N(-1)!} \tag{1.6.19}$$

The entropy can be written using Sterling's approximation (Eq. (1.2.27)), neglecting terms that are less than of order N, as:

$$S = k \ln(\Omega\,(E,N)) = k[N(\ln N - 1) - N(1)(\ln N(1) - 1) - N(-1)(\ln N(-1) - 1]$$

$$= k[N \ln N - N(1)\ln N(1) - N(-1)\ln N(-1)] \qquad (1.6.20)$$

the latter following from $N = N(1) + N(-1)$. To simplify this expression further, we write it in terms of the magnetization. Using $P_{s_i}(-1) + P_{s_i}(1) = 1$ and Eq. (1.6.9) for the magnetization we have the probability that a particular spin is UP and DOWN in terms of the magnetization as:

$$P_{s_i}(1) = (1 + m)\,/\,2$$
$$P_{s_i}(-1) = (1 - m)\,/\,2 \qquad (1.6.21)$$

Since there are many spins in the system, we can obtain the number of UP spins using

$$N(1) = NP_{s_i}(1) = N(1 + m)\,/\,2$$
$$N(-1) = NP_{s_i}(1) = N(1 - m)\,/\,2 \qquad (1.6.22)$$

Using these expressions, Eq. (1.6.20) becomes the same as Eq. (1.6.16), with $h_i = h$.

There is an important difference between the two derivations, in that the second assumed that all of the magnetic fields were the same. Thus, the first derivation appears more general. However, since the original system has no interactions, we could consider each of the spins with its own field h_i as a separate system. If we want to calculate the entropy of the individual spin, we would consider an ensemble of such spins. The ensemble consists of many spins with the same field $h = h_i$. The derivation of the entropy using the ensemble would be identical to the derivation we have just given, except that at the end we would divide by the number of different systems in the ensemble N. Adding together the entropies of different spins would then give exactly Eq. (1.6.16).

The entropy of a spin from Eq. (1.6.16) is maximal for a magnetization of zero when it has the value $k \ln(2)$. From the original definition of the entropy, this corresponds to the case when there are exactly two different possible states of the system. It thus corresponds to the case where the probability of each state $s = \pm 1$ is 1/2. The minimal entropy is for either $m = 1$ or $m = -1$—when there is only one possible state of the spin, so the entropy must be zero. ∎

1.6.2 *The Ising model*

We now add the essential aspect of the Ising model—interactions between the spins. The location of the spins in space was unimportant in the case of the noninteracting model. However, for the interacting model, we consider the spins to be located on a periodic lattice in space. Similar to the CA models of Section 1.5, we allow the spins to interact only with their nearest neighbors. It is conventional to interpret neighbors strictly as the spins with the shortest Euclidean distance from a particular site. This means that for a cubic lattice there are two, four and six neighbors in one, two and three dimensions respectively. We will assume that the interaction with each of the neighbors is the same and we write the energy as:

$$E[\{s_i\}] = -\sum_i h_i s_i - J\sum_{<ij>} s_i s_j \tag{1.6.23}$$

The notation $<ij>$ under the summation indicates that the sum is to be performed over all i and j that are nearest neighbors. For example, in one dimension this could be written as:

$$E[\{s_i\}] = -\sum_i h_i s_i - J\sum_i s_i s_{i+1} \tag{1.6.24}$$

If we wanted to emphasize that each spin interacts with its two neighbors, we could write this as

$$E[\{s_i\}] = -\sum_i h_i s_i - J\frac{1}{2}\sum_i (s_i s_{i+1} + s_i s_{i-1}) \tag{1.6.25}$$

where the factor of 1/2 corrects for the double counting of the interaction between every two neighboring spins. In two and three dimensions (2-d and 3-d), there is need of additional indices to represent the spatial dependence. We could write the energy in 2-d as:

$$E[\{s_{i,j}\}] = -\sum_{i,j} h_{i,j} s_{i,j} - J\sum_{i,j} (s_{i,j} s_{i+1,j} + s_{i,j} s_{i,j+1}) \tag{1.6.26}$$

and in 3-d as:

$$E[\{s_{i,j,k}\}] = -\sum_{i,j,k} h_{i,j,k} s_{i,j,k} - J\sum_{i,j,k} (s_{i,j,k} s_{i+1,j,k} + s_{i,j,k} s_{i,j+1,k} + s_{i,j,k} s_{i,j,k+1}) \tag{1.6.27}$$

In these sums, each nearest neighbor pair appears only once. We will be able to hide the additional indices in 2-d and 3-d by using the nearest neighbor notation $<ij>$ as in Eq. (1.6.23).

The interaction J between spins may arise from many different sources. Similar to the derivation of h_i in Eq. (1.6.2), this is the only form that an interaction between two spins can take (Question 1.6.2). There are two distinct possibilities for the behavior of the system depending on the sign of the interaction. Either the interaction tries to orient the spins in the same direction ($J > 0$) or in the opposite direction ($J < 0$). The former is called a ferromagnet and is the common form of a magnet. The other is called an antiferromagnet (Section 1.6.4) and has very different external properties but can be represented by the same model, with J having the opposite sign.

Question **1.6.2** Show that the form of the interaction given in Eq. (1.6.24) Jss' is the most general interaction between two spins.

Solution 1.6.2 We write as a general form of the energy of two spins:

$$e(s,s') = e(1,1)\frac{(1+s)(1+s')}{4} + e(1,-1)\frac{(1+s)(1-s')}{4}$$
$$+e(1,-1)\frac{(1-s)(1+s')}{4} + e(-1,-1)\frac{(1-s)(1-s')}{4}$$

(1.6.28)

If we expand this we will find a constant term, terms that are linear in s and s' and a term that is proportional to ss'. The linear terms give rise to the local field h_i, and the final term is the interaction. There are other possible interactions that could be written that would include three or more spins. ∎

In a magnetic system, each microscopic spin is itself the source of a small magnetic field. Magnets have the property that they can be the source of a macroscopic magnetic field. When a material is a source of a magnetic field, we say that it is magnetized. The magnetic field arises from constructive superposition of the microscopic sources of the magnetic field that we represent as spins. In effect, the small spins combine together to form a large spin. We have seen in Section 1.6.1 that when there is a magnetic field h_i, each spin will orient itself with the magnetic field. This means that in an external field—a field due to a source outside of the magnet—there will be a macroscopic orientation of the spins and they will in turn give rise to a magnetic field. Magnets, however, can be the source of a magnetic field even when there is no external field. This occurs only below a particular temperature known as the Curie temperature of the material. At higher temperatures, a magnetization exists only in an external magnetic field. The Ising model captures this behavior by showing that the interactions between the spins can cause a spontaneous orientation of the spins without any external field. The spontaneous magnetization is a collective phenomenon. It would not exist for an isolated spin or even for a small collection of interacting spins.

Ultimately, the reason that the spontaneous magnetization is a collective phenomenon has more to do with the kinetics than the thermodynamics of the system. The spontaneous magnetization must occur in a particular direction. Without an external field, there is no reason for any particular direction, but the system must choose one. In our case, it must choose between one of two possibilities—UP or DOWN. Once the magnetization occurs, it breaks a symmetry of the system, because we can now tell the difference between UP and DOWN on the macroscopic scale. At this point, the kinetics of the system must reenter. If the system were able to flip between UP and DOWN very rapidly, we would not be able to measure either case. However, we know that if all of the spins have to flip at once, the likelihood of this happening becomes vanishingly small as the number of spins grows. Thus for a large number of spins in a macroscopic material, this flipping becomes slower than our observation of the magnet. On the other hand, if we had only a few spins, they would still flip back and forth. It is this property of the system that makes the spontaneous magnetization a collective phenomenon.

Returning briefly to the discussion at the end of Section 1.3, we see that by choosing a direction for the magnetization, the magnet breaks the ergodic theorem. It is no longer possible to represent the system using an ensemble with all possible states of the system. We must exclude half of the states that have the opposite magnetization. The reason, as we described there, is because of the existence of a slow process, or a long time scale, that prevents the system from going from one choice of magnetization to the other.

The existence of a spontaneous magnetization arises because of the energy lowering of the system when neighboring spins align with each other. At sufficiently low temperatures, this causes the system to align collectively one way or another. Above the Curie temperature, T_c, the energy gain by alignment is destroyed by the temperature-induced random flipping of individual spins. We say that the higher temperature phase is a disordered phase, as compared to the ordered low temperature phase, where all spins are aligned. When we think about this thermodynamically, the disorder is an effect of optimizing the entropy, which promotes the disordered state and competes with the energy as the temperature is increased.

1.6.3 *Mean field theory*

Despite the simplicity of the Ising model, it has never been solved exactly except in one dimension, and in two dimensions for $h_i = 0$. The techniques that are useful in these cases do not generalize well. We will emphasize instead a powerful approximation technique for describing systems of many interacting parts known as the mean field approximation. The idea of this approximation is to treat a single element of the system under the average influence of the rest of the system. The key to doing this correctly is to recognize that this average must be performed self-consistently. The meaning of self-consistency will be described shortly. The mean field approximation cannot be applied to all interacting systems. However, when it can be, it enables the system to be understood in a direct way.

To use the mean field approximation we single out a particular spin s_i and find the effective field (or mean field) it experiences h_i'. This field is obtained by replacing all variables in the energy by their average values, except for s_i. This leads to an effective energy $E_{MF}(s_i)$ for s_i. To obtain it we can neglect all terms in the energy (Eq. (1.6.23)) that do not include s_i.

$$E_{MF}(s_i) = -h_i s_i - J \sum_{jnn} s_i < s_j > = -h_i' s_i$$

$$h_i' = h_i + J \sum_{jnn} < s_j >$$

(1.6.29)

The sum is over all nearest neighbors of s_i. If we are able to find what the mean field h_i' is, then we can solve this interacting Ising model using the solution of the Ising model without interactions. The problem is that in order to find the field we have to know the average value of the spins, which in turn depends on the effective fields. This is the self-consistency. We will develop a single algebraic equation for the solution. It is interesting first to consider this problem when the external fields h_i are zero. Eq. (1.6.29) shows that a mean field might still exist. When the external field is zero, each of the spin variables has the same equation. We might guess that the average value of the spin in one location will be the same as that in any other location:

$$m = m_i = < s_i > \tag{1.6.30}$$

In this case our equations become

$$E_{MF}(s_i) = - h_i' s_i$$

$$h_i' = J \sum_{jnn} m = zJm \tag{1.6.31}$$

where z is the number of nearest neighbors, known as the coordination number of the system. Eq. (1.6.10) gives us the value of the average magnetization when the spin is subject to a field. Using this same expression under the influence of the mean field we have

$$m = \tanh(\beta h_i') = \tanh(\beta zJm) \tag{1.6.32}$$

This is the self-consistent equation, which gives the value of the magnetization in terms of itself. The solution of this equation may be found graphically, as illustrated in Fig. 1.6.3, by plotting the functions $y = m$ and $y = \tanh(\beta zJm)$ and finding their intersections. There is always a solution $m = 0$. In addition, for values of $\beta zJ > 1$, there are two more solutions related by a change of sign $m = \pm m_0(\beta zJ)$, where we name the positive solution $m_0(\beta zJ)$. When $\beta zJ = 1$, the line $y = m$ is tangent to the plot of $y = \tanh(\beta zJm)$ at $m = 0$. For values $\beta zJ > 1$, the value of $y = \tanh(\beta zJm)$ must rise above the line $y = m$ for small positive m and then cross it. The crossing point is the solution $m_0(\beta zJ)$. $m_0(\beta zJ)$ approaches one asymptotically as $\beta zJ \to \infty$, e. g. as the temperature goes to zero. A plot of $m_0(\beta zJ)$ from a numerical solution of Eq. (1.6.32) is shown in Fig. 1.6.4.

We see that there are two different regimes for this model with a transition at a temperature T_c given by $\beta zJ = 1$ or

$$kT_c = zJ \tag{1.6.33}$$

To understand what is happening it is helpful to look at the energy $U(m)$ and the free energy $F(m)$ as a function of the magnetization, assuming that all spins have the same magnetization. We will treat the magnetization as a parameter that can be varied. The actual magnetization is determined by minimizing the free energy.

To determine the energy, we must average Eq. (1.6.23), which includes a product of spins on neighboring sites. The mean field approximation treats each spin as if it were independent of other spins except for their average field. This implies that we have neglected correlations between the value of one spin and the others around it. Assuming that the spins are uncorrelated means the average over the product over two spins may be approximated by the product over the averages:

$$<s_i s_j> \approx <s_i><s_j> = m^2 \tag{1.6.34}$$

The average over the energy without any external fields is then:

$$U(m) = < -J \sum_{<ij>} s_i s_j > = -\frac{1}{2} NJzm^2 \tag{1.6.35}$$

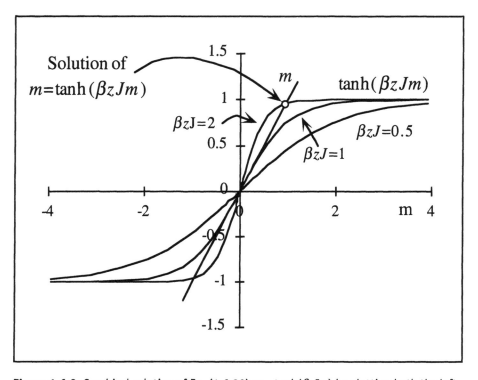

Figure 1.6.3 Graphical solution of Eq. (1.6.32) $m = \tanh(\beta z Jm)$ by plotting both the left- and right-hand sides of the equation as a function of m and looking for the intersections. $m = 0$ is always a solution. To consider other possible solutions we note that both functions are antisymmetric in m so we need only consider positive values of m. For every positive solution there is a negative solution of equal magnitude. When $\beta z J = 1$ the slope of both sides of the equation is the same at $m = 0$. For $\beta z J > 1$ the slope of the right is greater than the left side. For large positive values of m the right side of the equation is always less than the left side. Thus for $\beta z J > 1$, there must be an additional solution. The solution is plotted in Fig. 1.6.4. ∎

The factor of 1/2 arises because we count each interaction only once (see Eqs. (1.6.24)–(1.6.27)). A sum over the average of $E_{MF}(s_i)$ would give twice as much, due to counting each of the interactions twice.

Since we have fixed the magnetization of all spins to be the same, we can use the entropy we found in Question 1.6.1 to obtain the free energy as:

$$F(m) = -\frac{1}{2}NJzm^2 - NkT\left[\ln(2) - \frac{1}{2}\left((1+m)\ln(1+m) + (1-m)\ln(1-m)\right)\right] \quad (1.6.36)$$

This free energy is plotted in Fig. 1.6.5 as a function of m/Jz for various values of kT/Jz. We see that the behavior of this system is precisely the behavior of a second-order phase transition described in Section 1.3. Above the transition temperature T_c there is only one possible phase and below T_c there are two phases of equal en-

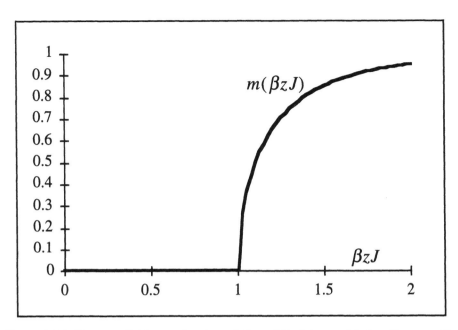

Figure 1.6.4 The mean field approximation solution of the Ising model gives the magnetization (average value of the spin) as a solution of Eq. (1.6.32). The solution is shown as a function of $\beta z J$. As discussed in Fig. 1.6.3 and the text for $\beta z J > 1$ there are three solutions. Only the positive one is shown. The solution $m = 0$ is unstable, as can be seen by analysis of the free energy shown in Fig. 1.6.5. The other solution is the negative of that shown. ∎

ergy. Question 1.6.3 clarifies a technical point in this derivation, and Question 1.6.4 generalizes the solution to include nonzero magnetic fields $h_i \neq 0$.

Question 1.6.3 Show that the minima of the free energy are the solutions of Eq. (1.6.32). This shows that our derivation is internally consistent. Specifically, that our two ways of defining the mean field approximation, first using Eq. (1.6.29) and then using Eq. (1.6.34), are compatible.

Solution 1.6.3 Taking the derivative of Eq. (1.6.35) with respect to m and setting it to zero gives:

$$0 = -Jzm - kT\left[-\frac{1}{2}\left(\ln(1+m) - \ln(1-m)\right)\right] \qquad (1.6.37)$$

Recognizing the inverse of tanh, as in Eq. (1.6.14), gives back Eq. (1.6.32) as desired. ∎

Question 1.6.4 Find the replacements for Eq. (1.6.31)–(1.6.36) for the case where there is a uniform external magnetic field $h_i = h$. Plot the free energy for a few cases.

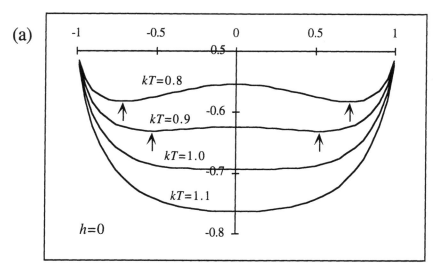

(a)

$h=0$

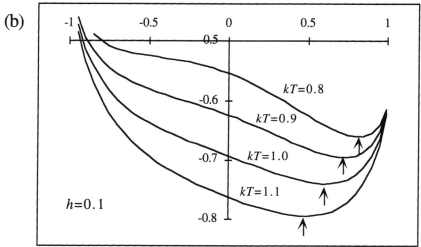

(b)

$h=0.1$

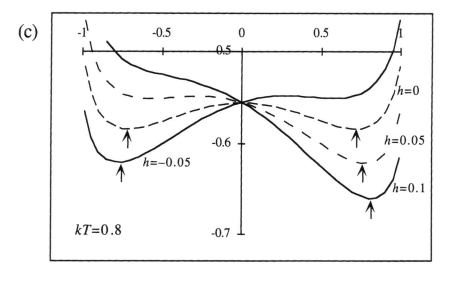

(c)

$kT=0.8$

Solution 1.6.4 Applying an external magnetic field breaks the symmetry between the two different minima in the energy that we have found. In this case we have instead of Eq. (1.6.29)

$$E_{MF}(s_i) = -h_i' s_i$$
$$h_i' = h + zJm$$

(1.6.38)

The self-consistent equation instead of Eq. (1.6.32) is:

$$m = \tanh(\beta h + \beta zJm)$$

(1.6.39)

Averaging over the energy gives:

$$U(m) = <-h\sum_i s_i - J\sum_{<ij>} s_i s_j> = -Nhm - \frac{1}{2}NJzm^2$$

(1.6.40)

The entropy is unchanged, so the free energy becomes:

$$F(m) = -Nhm - \frac{1}{2}NJzm^2 - NkT\left[\ln(2) - \frac{1}{2}\big((1+m)\ln(1+m) + (1-m)\ln(1-m)\big)\right]$$

(1.6.41)

Several plots are shown in Fig. 1.6.5. Above kT_c of Eq. (1.6.33) the application of an external magnetic field gives rise to a magnetization by shifting the location of the single minimum. Below this temperature there is a tilting of the two minima. Thus, going from a positive to a negative value of h would give an abrupt transition—a first-order transition which occurs at exactly $h=0$. ∎

In discussing the mean field equations, we have assumed that we could specify the magnetization as a parameter to be optimized. However, the prescription we have from thermodynamics is that we should take all possible states of the system with a Boltzmann probability. What is the justification for limiting ourselves to only one value of the magnetization? We can argue that in a macroscopic system, the optimal

Figure 1.6.5 Plots of the mean field approximation to the free energy. (a) shows the free energy for $h = 0$ as a function of m for various values of kT. The free energy m and kT are measured in units of Jz. As the temperature is lowered below $kT/zJ = 1$ there are two minima instead of one (shown by arrows). These minima are the solutions of Eq. (1.6.32) (see Question 1.6.3). The solutions are illustrated in Fig. 1.6.4. (b) Shows the same curves as (a) but with a magnetic field $h/zJ = 0.1$. The location of the minimum gives the value of the magnetization. The magnetic field causes a magnetization to exist at all temperatures, but it is larger at lower temperatures. At the lowest temperature shown $kT/zJ = 0.8$ the effect of the phase transition can be seen in the beginnings of a second (metastable) minimum at negative values of the magnetization. (c) shows plots at a fixed temperature of $kT/zJ = 0.8$ for different values of the magnetic field. As the value of the field goes from positive to negative, the minimum of the free energy switches from positive to negative values discontinuously. At exactly $h = 0$ there is a discontinuous jump from positive to negative magnetization—a first-order phase transition. ∎

value of the magnetization will so dominate other magnetizations that any other possibility is negligible. This is reasonable except for the case when the magnetic field is close to zero, below T_c, and we have two equally likely magnetizations. In this case, the usual justification does not hold, though it is often implicitly applied. A more complete justification requires a discussion of kinetics given in Section 1.6.6.

Using the results of Question 1.6.4, we can draw a phase diagram like that illustrated in Section 1.3 for water (Fig. 1.3.7). The phase diagram of the Ising model (Fig. 1.6.6) describes the transitions as a function of temperature (or β) and magnetic field h. It is very simple for the case of the magnetic system, since the first-order phase transition line lies along the $h = 0$ axis and ends at the second-order transition point given by Eq. (1.6.33).

1.6.4 Antiferromagnets

We found the existence of a phase transition in the last section from the self-consistent mean field result (Eq. (1.6.32)), which showed that there was a nonzero magnetization for $\beta zJ > 1$. This condition is satisfied for small enough temperature as long as $J > 0$. What about the case of $J < 0$? There are no additional solutions of Eq. (1.6.32) for this case. Does this mean there is no phase transition? Actually, it means that one of our assumptions is not a good one. When $J < 0$, each spin would like (has a lower energy if...) its neighbors to antialign rather than align their spins. However, we have assumed that all spins have the same magnetization, Eq. (1.6.30). The self-consistent equation assumes and does not guarantee that all spins have the same magnetization. This assumption is not a good one when the spins are trying to antialign.

Figure 1.6.6 The phase diagram of the Ising model found from the mean field approximation. The line of first-order phase transitions at $h = 0$ ends at the second-order phase transition point given by Eq. (1.6.32). For positive values of h there is a net positive magnetization and for negative values there is a negative magnetization. The change through $h = 0$ is continuous above the second-order transition point, and discontinuous below it. ∎

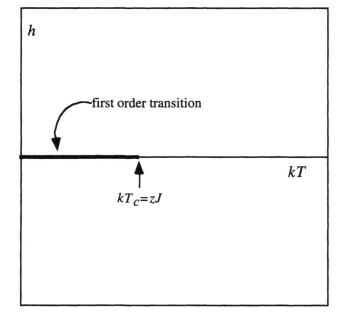

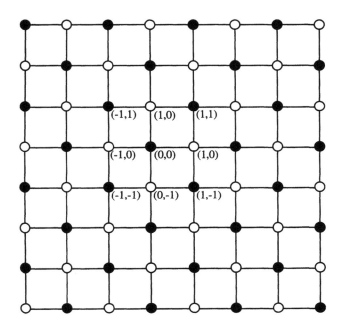

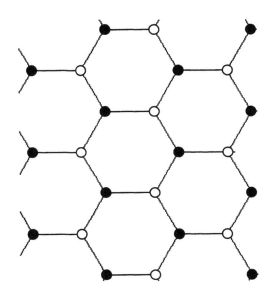

Figure 1.6.7 In order to obtain mean field equations for the anti-ferromagnetic case $J < 0$ we consider a square lattice (top) and label every site according to the sum of its rectilinear indices as odd (open circles) or even (filled circles). A few sites are shown with indices. Each site is understood to be the location of a spin. We then invert the spins (redefine them by $s \rightarrow -s$) that are on odd sites and find that the new system satisfies the same equations as the ferromagnet. The same trick works for any bipartite lattice; for example the hexagonal lattice shown (bottom). By using this trick we learn that at low temperatures the system will have a spontaneous magnetism that is positive on odd sites and negative on even sites or the opposite. ∎

We can solve the case of a system with $J < 0$ on a square or cubic lattice directly using a trick. We label every spin by indices (i,j) in 2-d, as indicated in Fig. 1.6.7, or (i,j,k) in 3-d. Then we consider separately the spins whose indices sum to an odd number ("odd spins") and those whose indices sum to an even number ("even spins"). Note that all the neighbors of an odd spin are even and all neighbors of an even spin are odd. Now we invert all of the odd spins. Explicitly we define new spin variables in 3-d as

$$s'_{ijk} = (-1)^{i+j+k} s_{ijk} \tag{1.6.42}$$

In terms of these new spins, the energy without an external magnetic field is the same as before, except that each term in the sum has a single additional factor of (-1). There is only one factor of (-1) because every nearest neighbor pair has one odd and one even spin. Thus:

$$E[\{s'_i\}] = -J \sum_{<ij>} s_i s_j = -(-J) \sum_{<ij>} s'_i s'_j = -J' \sum_{<ij>} s'_i s'_j \tag{1.6.43}$$

We have completed the transformation by defining a new interaction $J' = -J > 0$. In terms of the new variables, we are back to the ferromagnet. The solution is the same, and below the temperature given by $kT_c = zJ'$ there will be a spontaneous magnetization of the new spin variables. What happens in terms of the original variables? They become antialigned. All of the even spins have magnetization in one direction, UP, and the odd spins have magnetization in the opposite direction, DOWN, or vice versa. This lowers the energy of the system, because the negative interaction $J < 0$ means that all of the neighboring spins want to antialign. This is called an antiferromagnet.

The trick we have used to solve the antiferromagnet works for certain kinds of periodic arrangements of spins called bipartite lattices. A bipartite lattice can be divided into two lattices so that all the nearest neighbors of a member of one lattice are members of the other lattice. This is exactly what we need in order for our redefinition of the spin variables to work. Many lattices are bipartite, including the cubic lattice and the hexagonal honeycomb lattice illustrated in Fig. 1.6.7. However, the triangular lattice, illustrated in Fig. 1.6.8, is not.

The triangular lattice exemplifies an important concept in interacting systems known as frustration. Consider what happens when we try to assign magnetizations to each of the spins on a triangular lattice in an effort to create a configuration with a lower energy than a disordered system. We start at a position marked (1) on Fig. 1.6.8 and assign it a magnetization of m. Then, since it wants its neighbors to be antialigned, we assign position (2) a magnetization of $-m$. What do we do with the spin at (3)? It has interactions both with the spin at (1) and with the spin at (2). These interactions would have it be antiparallel with both—an impossible task. We say that the spin at (3) is frustrated, since it cannot simultaneously satisfy the conflicting demands upon it. It should not come as a surprise that the phenomenon of frustration becomes a commonplace occurrence in more complex systems. We might even say that frustration is a source of complexity.

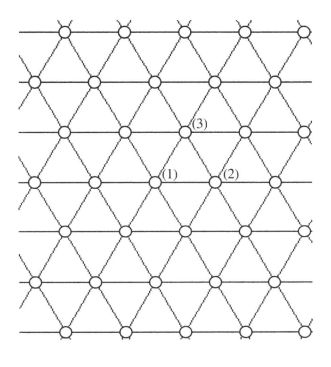

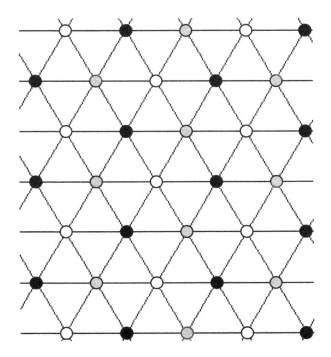

Figure 1.6.8 A triangular lattice (top) is not a bi-partite lattice. In this case we cannot solve the antiferromagnet $J < 0$ by the same method as used for the square lattice (see Fig. 1.6.7). If we try to assign magnetizations to different sites we find that assigning a magnetization to site (1) would lead site (2) to be antialigned. This combination would, however require site (3) to be antialigned to both sites (1) and (2), which is impossible. We say that site (3) is "frustrated." The bottom illustration shows what happens when we take the hexagonal lattice from Fig. 1.6.7 and superpose the magnetizations on the triangular lattice leaving the additional sites (shaded) as unmagnetized (see Questions 1.6.5–1.6.7). ∎

Question 1.6.5 Despite the existence of frustration, it is possible to construct a state with lower energy than a completely disordered state on the triangular lattice. Construct one of them and evaluate its free energy.

Solution 1.6.5 We construct the state by extending the process discussed in the text for assigning magnetizations to individual sites. We start by assigning a magnetization m to site (1) in Fig. 1.6.8 and $-m$ to site (2). Because site (3) is frustrated, we assign it no magnetization. We continue by assigning magnetizations to any site that already has two neighbors that are assigned magnetizations. We assign a magnetization of m when the neighbors are $-m$ and 0, a magnetization of $-m$ when the neighbors are m and 0 and a magnetization of 0 when the neighbors are m and $-m$. This gives the illustration at the bottom of Fig. 1.6.8. Comparing with Fig. 1.6.7, we see that the magnetized sites correspond to the honeycomb lattice. One-third of the triangular lattice sites have a magnetization of $+m$, $-m$ and 0. Each magnetized site has three neighbors of the opposite magnetization and three unmagnetized sites. The free energy of this state is given by:

$$F(m) = NJm^2 - \frac{1}{3}NkT\ln(2)$$
$$- \frac{2}{3}NkT\left[\ln(2) - \frac{1}{2}\left((1+m)\ln(1+m) + (1-m)\ln(1-m)\right)\right] \tag{1.6.44}$$

The first term is the energy. Each nearest neighbor pair of spins that are antialigned provides an energy Jm^2. Let us call this a bond between two spins. There are a total of three interactions for every spin (each spin interacts with six other spins but we can count each interaction only once). However, on average there is only one out of three interactions that is a bond in this system. To count the bonds, note that one out of three spins (with $m_i = 0$) has no bonds, while the other two out of three spins each have three bonds. This gives a total of six bonds for three sites, but each bond must be counted only once for a pair of interacting spins. We divide by two to get three bonds for three spins, or an average of one bond per site. The second term in Eq. (1.6.44) is the entropy of the $N/3$ unmagnetized sites, and the third term is the entropy of the $2N/3$ magnetized sites.

There is another way to systematically construct a state with an energy lower than a completely disordered state. Assign magnetizations $+m$ and $-m$ alternately along one straight line—a one-dimensional antiferromagnet. Then skip both neighboring lines by setting all of their magnetizations to zero. Then repeat the antiferromagnetic line on the next parallel line. This configuration of alternating antiferromagnetic lines is also lower in energy than the disordered state, but it is higher in energy than the configuration shown in Fig. 1.6.8 at low enough temperatures, as discussed in the next question. ∎

Question 1.6.6 Show that the state illustrated on the bottom of Fig. 1.6.8 has the lowest possible free energy as the temperature goes to zero, at least in the mean field approximation.

Solution 1.6.6 As the temperature goes to zero, the entropic contribution to the free energy is irrelevant. The energy of the Ising model is minimized in the mean field approximation when the magnetization is +1 if the local effective field is positive, or –1 if it is negative. The magnetization is arbitrary if the effective field is zero. If we consider three spins arranged in a triangle, the lowest possible energy of the three interactions between them is given by having one with $m = +1$, one with $m = -1$ and the other arbitrary. This is forced, because we must have at least one +1 and one –1 and then the other is arbitrary. This is the optimal energy for any triangle of interactions. The configuration of Fig. 1.6.8 achieves this optimal arrangement for all triangles and therefore must give the lowest possible energy of any state. ∎

Question 1.6.7 In the case of the ferromagnet and the antiferromagnet, we found that there were two different states of the system with the same energy at low temperatures. How many states are there of the kind shown in Fig. 1.6.8 and described in Questions 1.6.5 and 1.6.6?

Solution 1.6.7 There are two ways to count the states. The first is to count the number of distinct magnetization structures. This counting is as follows. Once we assign the values of the magnetization on a single triangle, we have determined them everywhere in the system. This follows by inspection or by induction on the size of the assigned triangle. Since we can assign arbitrarily the three different magnetizations $(m, -m, 0)$ within a triangle, there are a total of six such distinct magnetization structures.

We can also count how many distinct arrangements of spins there are. This is relevant at low temperatures when we want to know the possible states at the lowest energy. We see that there are $2^{N/3}$ arrangements of the arbitrary spins for each of the magnetizations. If we want to count all of the states, we can almost multiply this number by 6. We have to correct this slightly because of states where the arbitrary spins are all aligned UP or DOWN. There are two of these for each arrangement of the magnetizations, and these will be counted twice. Making this correction gives $6(2^{N/3} - 1)$ states. We see that frustration gives rise to a large number of lowest energy states.

We have not yet proven that these are the only states with the lowest energy. This follows from the requirement that every triangle must have its lowest possible energy, and the observation that setting the value of the magnetizations of one triangle then forces the values of all other magnetizations uniquely. ∎

Question 1.6.8 We discovered that our assumption that all spins should have the same magnetization does not always apply. How do we know that we found the lowest energy in the case of the ferromagnet? Answer this for the case of $h = 0$ and $T = 0$.

Solution 1.6.8 To minimize the energy, we can consider each term of the energy, which is just the product of spins on adjacent sites. The minimum possible value for each term of a ferromagnet occurs for aligned spins. The two states we found at $T = 0$ with $m_i = 1$ and $m_i = -1$ are the only possible states with all spins aligned. Since they give the minimum possible energy, they must be the correct states. ∎

1.6.5 Beyond mean field theory (correlations)

Mean field theory treats only the average orientation of each spin and assumes that spins are uncorrelated. This implies that when one spin changes its sign, the other spins do not respond. Since the spins are interacting, this must not be true in a more complete treatment. We expect that even above T_c, nearby spins align to each other. Below T_c, nearby spins should be more aligned than would be suggested by the average magnetization. Alignment of spins implies their values are correlated. How do we quantify the concept of correlation? When two spins are correlated they are more likely to have the same value. So we might define the correlation of two spins as the average of the product of the spins:

$$< s_i s_j > = \sum_{s_i, s_j} s_i s_j P(s_i, s_j) = P_{s_i s_j}(1,1) + P_{s_i s_j}(-1,-1) - P_{s_i s_j}(-1,1) - P_{s_i s_j}(1,-1) \quad (1.6.45)$$

According to this definition, they are correlated if they are both always $+1$, so that $P_{s_i s_j}(1,1) = 1$. Then $< s_i s_j >$ achieves its maximum possible value $+1$. The problem with this definition is that when s_i and s_j are both always $+1$ they are completely independent of each other, because each one is $+1$ independently of the other. Our concept of correlation is the opposite of independence. We know that if spins are independent, then their joint probability distribution factors (see Section 1.2)

$$P(s_i, s_j) = P(s_i) P(s_j) \quad (1.6.46)$$

Thus we define the correlation as a measure of the departure of the joint probability from the product of the individual probabilities.

$$\sum_{s_i, s_j} s_i s_j (P(s_i, s_j) - P(s_i) P(s_j)) = < s_i s_j > - < s_i >< s_j > \quad (1.6.47)$$

This definition means that when the correlation is zero, we can say that s_i and s_j are independent. However, we must be careful not to assume that they are not aligned with each other. Eq. (1.6.45) measures the spin alignment.

Question 1.6.9 One way to think about the difference between Eq. (1.6.45) and Eq. (1.6.47) is by considering a hierarchy of correlations. The first kind of correlation is of individual spins with themselves and is just the average of the spin. The second kind are correlations between pairs of spins that are not contained in the first kind. Define the next kind of correlation in the hierarchy that would describe correlations between three spins but exclude the correlations that appear in the first two.

Solution 1.6.9 The first three elements in the hierarchy of correlations are:

$$< s_i >$$

$$< s_i s_j > - < s_i > < s_j > \qquad (1.6.48)$$

$$< s_i s_j s_k > - < s_i s_j > < s_k > - < s_i s_k > < s_j > - < s_j s_k > < s_i > + 2 < s_i > < s_j > < s_k >$$

The expression for the correlation of three spins can be checked by seeing what happens if the variables are independent. When variables are independent, the average of their product is the same as the product of their averages. Then all averages become products of averages of single variables and everything cancels. Similarly, if the first two variables s_i and s_j are correlated and the last one s_k is independent of them, then the first two terms cancel and the last three terms also cancel. Thus, this expression measures the correlations of three variables that are not present in any two of them. ∎

Question 1.6.10 To see the difference between Eqs. (1.6.45) and (1.6.47), evaluate them for two cases: (a) s_i is always equal to 1 and s_j is always equal to -1, and (b) s_i is always the opposite of s_j but each of them averages to zero (i.e., is equally likely to be $+1$ or -1).

Solution 1.6.10

a. $P_{s_i s_j}(1,-1) = 1$, so $< s_i s_j > = -1$, but $< s_i s_j > - < s_i > < s_j > = 0$.

b. $< s_i s_j > = -1$, and $< s_i s_j > - < s_i > < s_j > = -1$. ∎

Comparing Eq. (1.6.34) with Eq. (1.6.47), we see that correlations measure the departure of the system from mean field theory. When there is an average magnetization, such as there is below T_c in a ferromagnet, the effect of the average magnetization is removed by our definition of the correlation. This can also be seen from rewriting the expression for correlations as:

$$< s_i s_j > - < s_i > < s_j > = < (s_i - < s_i >)(s_j - < s_j >) > \qquad (1.6.49)$$

Correlations measure the behavior of the difference between the spin and its average value. In the rest of this section we discuss qualitatively the correlations that are found in a ferromagnet and the breakdown of the mean field approximation.

The energy of a ferromagnet is determined by the alignment of neighboring spins. Positive correlations between neighboring spins reduce its energy. Positive or negative correlations diminish the possible configurations of spins and therefore reduce the entropy. At very high temperatures, the competition between the energy and the entropy is dominated by the entropy, so there should be no correlations and each spin is independent. At low temperatures, well below the transition temperature, the average value of the spins is close to one. For example, for $\beta z J = 2$, which corresponds to $T = T_c / 2$, the value of $m_0(\beta z J)$ is 0.96 (see Fig. 1.6.4). So the correlations given by Eq. (1.6.47) play almost no role. Correlations are most significant near T_c, so it is near the transition that the mean field approximation is least valid.

For all $T > T_c$ and for $h = 0$, the magnetization is zero. However, starting from high temperature, the correlation between neighboring spins increases as the temperature is lowered. Moreover, the correlation of one spin with its neighbors, and their correlation with their neighbors, induces a correlation of each spin with spins farther away. The distance over which spins are correlated increases as the temperature decreases. The correlation decays exponentially, so a correlation length $\xi(T)$ may be defined as the decay constant of the correlation:

$$< s_i s_j > - < s_i > < s_j > \propto e^{-r_{ij}/\xi(T)} \tag{1.6.50}$$

where r_{ij} is the Euclidean distance between s_i and s_j. At T_c the correlation length diverges. This is one way to think about how the phase transition occurs. The divergence of the correlation length implies that two spins anywhere in the system become correlated. As mentioned previously, in order for the instantaneous magnetization to be measured, there must also be a divergence of the relaxation time between opposite values of the magnetization. This will be discussed in Sections 1.6.6 and 1.6.7.

For temperatures just below T_c, the average magnetization is small. The correlation length of the spins is large. The average alignment (Eq. (1.6.45)) is essentially the same as the correlation (Eq. (1.6.47)). However, as T is further reduced below T_c, the average magnetization grows precipitously and the correlation measures the difference between the spin-spin alignment and the average spin value. Both the correlation and the correlation length decrease away from T_c. As the temperature goes to zero, the correlation length also goes to zero, even as the correlation itself vanishes.

At $T = T_c$ there is a special circumstance where the correlation length is infinite. This does not mean that the correlation is unchanged as a function of the distance between spins, r_{ij}. Since the magnetization is zero, the correlation is the same as the spin alignment. If the alignment did not decay with distance, the magnetization would be unity, which is not correct. The infinite correlation length corresponds to power law rather than exponential decay of the correlations. A power law decay of the correlations is more gradual than exponential and implies that there is no characteristic size for the correlations: we can find correlated regions of spins that are of any size. Since the correlated regions fluctuate, we say that there are fluctuations on every length scale.

The existence of correlations on every length scale near the phase transition and the breakdown of the mean field approximation that neglects these correlations played an important role in the development of the theory of phase transitions. The discrepancy between mean field predictions and experiment was one of the great unsolved problems of statistical physics. The development of renormalization techniques that directly consider the behavior of the system on different length scales solved this problem. This will be discussed in greater detail in Section 1.10.

In Section 1.3 we discussed the nature of ensemble averages and indicated that one of the central issues was determining the size of an independent system. For the Ising model and other systems that are spatially uniform, it is the correlation length that determines the size of an independent system. If a physical system is much larger than a correlation length then the system is self-averaging, in that experimental measurements average over many independent samples. We see that far from a phase transition, uniform systems are generally self-averaging; near a phase transition, the physical size of a system may enter in a more essential way.

The mean field approximation is sufficient to capture the collective behavior of the Ising model. However, even T_c is not given correctly by mean field theory, and indeed it is difficult to calculate. The actual transition temperature differs from the mean field value by a factor that depends on the dimensionality and structure of the lattice. In 1-d, the failure of mean field theory is most severe, since there is actually no real transition. Magnetization does not occur, except in the limit of $T \to 0$. The reason that there is no magnetization in 1-d, is that there is always a finite probability that at some point along the chain there will be a switch from having spins DOWN to having spins UP. This is true no matter how low the temperature is. The probability of such a boundary between UP and DOWN spins decreases exponentially with the temperature. It is given by $1/(1 + e^{2J/kT}) \approx e^{-2J/kT}$ at low temperature. Even one such boundary destroys the average magnetization for an arbitrarily large system. While formally there is no phase transition in one dimension, under some circumstances the exponentially growing distance between boundaries may have consequences like a phase transition. The effect is, however, much more gradual than the actual phase transitions in 2-d and 3-d.

The mean field approximation improves as the dimensionality increases. This is a consequence of the increase in the number of neighbors. As the number of neighbors increases, the averaging used for determining the mean field becomes more reliable as a measure of the environment of the spin. This is an important point that deserves some thought. As the number of different influences on a particular variable increases, they become better represented as an average influence. Thus in 3-d, the mean field approximation is better than in 2-d. Moreover, it turns out that rather than just gradually improving as the number of dimensions increases, for 4-d the mean field approximation becomes essentially exact for many of the properties of importance in phase transitions. This happens because correlations become irrelevant on long length scales in more than 4-d. The number of effective neighbors of a spin also increases if we increase the range of the interactions. Several different models with long-range interactions are discussed in the following section.

The Ising model has no built-in dynamics; however, we often discuss fluctuations in this model. The simplest fluctuation would be a single spin flipping in time. Unless the average value of a spin is $+1$ or -1, a spin must spend some time in each state. We can see that the presence of correlations implies that there must be fluctuations in time that affect more than one spin. This is easiest to see if we consider a system above the transition, where the average magnetization is zero. When one spin has the value $+1$, then the average magnetization of spins around it will be positive. On average, a region of spins will tend to flip together from one sign to the other. The amount of time that the region takes to flip depends on the length of the correlations. We have defined correlations in space between two spins. We could generalize the definition in Eq. (1.6.47) to allow the indices i and j to refer to different times as well as spatial positions. This would tell us about the fluctuations over time in the system. The analog of the correlation length Eq. (1.6.50) would be the relaxation time (Eq. (1.6.69) below).

The Ising model is useful for describing a large variety of systems; however, there are many other statistical models using more complex variables and interactions that have been used to represent various physical systems. In general, these models are treated first using the mean field approximation. For each model, there is a lower dimension (the lower critical dimension) below which the mean field results are completely invalid. There is also an upper critical dimension, where mean field is exact. These dimensions are not necessarily the same as for the Ising model.

1.6.6 *Long-range interactions and the spin glass*

Long-range interactions enable the Ising model to serve as a model of systems that are much more complex than might be expected from the magnetic analog that motivated its original introduction. If we just consider ferromagnetic interactions separately, the model with long-range interactions actually behaves more simply. If we just consider antiferromagnetic interactions, larger scale patterns of UP and DOWN spins arise. When we include both negative and positive interactions together, there will be additional features that enable a richer behavior. We will start by considering the case of ferromagnetic long-range interactions.

The primary effect of the increase in the range of ferromagnetic interactions is improvement of the mean field approximation. There are several ways to model interactions that extend beyond nearest neighbors in the Ising model. We could set a sphere of a particular radius r_0 around each spin and consider all of the spins within the sphere to be neighbors of the spin at the center.

$$E[\{s_i\}] = -\sum_i h_i s_i - \frac{1}{2} J \sum_{r_{ij} < r_0} s_i s_j \tag{1.6.51}$$

Here we do not restrict the summations over i and j in the second term, so we explicitly include a factor of $1/2$ to avoid counting interactions twice. Alternatively, we could use an interaction $J(r_{ij})$ that decays either exponentially or as a power law with distance from each spin:

$$E[\{s_i\}] = -\sum_i h_i s_i - \frac{1}{2} \sum_{i,j} J(r_{ij}) s_i s_j \tag{1.6.52}$$

In both Eqs. (1.6.51) and (1.6.52) the self-interaction terms $i = j$ are generally to be excluded. Since $s_i^2 = 1$ they only add a constant to the energy.

Quite generally and independent of the range or even the variability of interactions, when all interactions are ferromagnetic, $J > 0$, then all the spins will align at low temperatures. The mean field approximation may be used to estimate the behavior. All cases then reduce to the same free energy (Eq. (1.6.36) or Eq. (1.6.41)) with a measure of the strength of the interactions replacing zJ. The only difference from the nearest neighbor model then relates to the accuracy of the mean field approximation. It is simplest to consider the model of a fixed interaction strength with a cutoff length. The mean field is accurate when the correlation length is shorter than the interaction distance. When this occurs, a spin is interacting with other spins that are uncorrelated with it. The averaging used to obtain the mean field is then correct. Thus the approximation improves if the interaction between spins becomes longer ranged. However, the correlation length becomes arbitrarily long near the phase transition. Thus, for longer interaction lengths, the mean field approximation holds closer to T_c but eventually becomes inaccurate in a narrow temperature range around T_c. There is one model for which the mean field approximation is exact independent of temperature or dimension. This is a model of infinite range interactions discussed in Question 1.6.11. The distance-dependent interaction model of Eq. (1.6.52) can be shown to behave like a finite range interaction model for interactions that decay more rapidly than $1/r$ in 3-d. For weaker decay than $1/r$ this model is essentially the same as the long-range interaction model of Question 1.6.11. Interactions that decay as $1/r$ are a borderline case.

Question 1.6.11 Solve the Ising model with infinite ranged interactions in a uniform magnetic field. The infinite range means that all spins interact with the same interaction strength. In order to keep the energy extrinsic (proportional to the volume) we must make the interactions between pairs of spins weaker as the system becomes larger, so replace $J \to J/N$. The energy is given by:

$$E[\{s_i\}] = -h\sum_i s_i - \frac{1}{2N}J\sum_{i,j} s_i s_j \qquad (1.6.53)$$

For simplicity, keep the $i = j$ terms in the second sum even though they add only a constant.

Solution 1.6.11 We can solve this problem exactly by rewriting the energy in terms of a collective coordinate which is the average over the spin variables

$$m = \frac{1}{N}\sum_i s_i \qquad (1.6.54)$$

in terms of which the energy becomes:

$$E(\{s_i\}) = hNm - \frac{1}{2}JNm^2 \qquad (1.6.55)$$

This is the same as the mean field Eq. (1.6.39) with the substitution $Jz \to J$. Here the equation is exact. The result for the entropy is the same as before, since we have fixed the average value of the spin by Eq. (1.6.54). The solution for the value of m for $h = 0$ is given by Eq. (1.6.32) and Fig. 1.6.4. For $h \neq 0$ the discussion in Question 1.6.4 applies. ∎

The case of antiferromagnetic interactions will be considered in greater detail in Chapter 7. If all interactions are antiferromagnetic $J < 0$, then extending the range of the interactions tends to reduce their effect, because it is impossible for neighboring spins to be antialigned and lower the energy. To be antialigned with a neighbor is to be aligned with a second neighbor. However, by forming patches of UP and DOWN spins it is possible to lower the energy. In an infinite-ranged antiferromagnetic system, all possible states with zero magnetization have the same lowest energy at $h = 0$.

This can be seen from the energy expression in Eq. (1.6.55). In this sense, frustration from many sources is almost the same as no interaction.

In addition to the ferromagnet and antiferromagnet, there is a third possibility where there are both positive and negative interactions. The physical systems that have motivated the study of such models are known as spin glasses. These are materials where magnetic atoms are found or placed in a nonmagnetic host. The randomly placed magnetic sites interact via long-range interactions that oscillate in sign with distance. Because of the randomness in the location of the spins, there is a randomness in the interactions between them. Experimentally, it is found that such systems also undergo a transition that has been compared to a glass transition, and therefore these systems have become known as spin glasses.

A model for these materials, known as the Sherrington-Kirkpatrick spin glass, makes use of the Ising model with infinite-range random interactions:

$$E[\{s_i\}] = -\frac{1}{2N}\sum_{ij} J_{ij} s_i s_j$$

$$J_{ij} = \pm J$$

(1.6.56)

The interactions J_{ij} are fixed uncorrelated random variables—quenched variables. The properties of this system are to be averaged over the random variables J_{ij} but only after it is solved.

Similar to the ferromagnetic or antiferromagnetic Ising model, at high temperatures $kT \gg J$ the spin glass model has a disordered phase where spins do not feel the effect of the interactions beyond the existence of correlations. As the temperature is lowered, the system undergoes a transition that is easiest to describe as a breaking of ergodicity. Because of the random interactions, some arrangements of spins are much lower in energy than others. As with the case of the antiferromagnet on a triangular lattice, there are many of these low-energy states. The difference between any two of these states is large, so that changing from one state to the other would involve the flipping of a finite fraction of the spins of the system. Such a flipping would have to be cooperative, so that overcoming the barrier between low-energy states becomes impossible below the transition temperature during any reasonable time. The low-energy states have been shown to be organized into a hierarchy determined by the size of the overlaps between them.

Question 1.6.12 Solve a model that includes a special set of correlated random interactions of the type of the Sherrington-Kirkpatrick model, where the interactions can be written in the *separable* form

$$J_{ij} = \xi_i \xi_j$$

$$\xi_i = \pm 1$$

(1.6.57)

This is the Mattis model. For simplicity, keep the terms where $i = j$.

Solution 1.6.12 We can solve this problem by defining a new set of variables

$$s'_i = \xi_i s_i$$

(1.6.58)

In terms of these variables the energy becomes:

$$E[\{s_i\}] = -\frac{1}{2N}\sum_{ij}\xi_i\xi_j s_i s_j = -\frac{1}{2N}\sum_{ij}s_i' s_j' \qquad (1.6.59)$$

which is the same as the ferromagnetic Ising model. The phase transition of this model would lead to a spontaneous magnetization of the new variables. This corresponds to a net orientation of the spins toward (or opposite) the state $s_i = \xi_i$. This can be seen from

$$m = <s_i'> = \xi_i <s_i> \qquad (1.6.60)$$

This model shows that a set of mixed interactions can cause the system to choose a particular low-energy state that behaves like the ordered state found in the ferromagnet. By extension, this makes it plausible that fully random interactions lead to a variety of low-energy states. ∎

The existence of a large number of randomly located energy minima in the spin glass might suggest that by engineering such a system we could control where the minima occur. Then we might use the spin glass as a memory. The Mattis model provides a clue to how this might be accomplished. The use of an outer product representation for the matrix of interactions turns out to be closely related to the model developed by Hebb for biological imprinting of memories on the brain. The engineering of minima in a long-range-interaction Ising model is precisely the model developed by Hopfield for the behavior of neural networks that we will discuss in Chapter 2.

In the ferromagnet and antiferromagnet, there were intuitive ways to deal with the breaking of ergodicity, because we could easily define a macroscopic parameter (the magnetization) that differentiated between different macroscopic states of the system. More general ways to do this have been developed for the spin glass and applied to the study of neural networks.

1.6.7 *Kinetics of the Ising model*

We have introduced the Ising model without the benefit of a dynamics. There are many choices of dynamics that would lead to the equilibrium ensemble given by the Ising model. One of the most natural would arise from considering each spin to have the two-state system dynamics of Section 1.4. In this dynamics, transitions between UP and DOWN occur across an intermediate barrier that sets the transition rate. We call this the activated dynamics and will use it to discuss protein folding in Chapter 4 because it can be motivated microscopically. The activated dynamics describes a continuous rate of transition for each of the spins. It is often convenient to consider transitions as occurring at discrete times. A particularly simple dynamics of this kind was introduced by Glauber for the Ising model. It also corresponds to the dynamics popular in studies of neural networks that we will discuss in Chapter 2. In this section we will show that the two different dynamics are quite closely related. In Section 1.7 we will consider several other forms of dynamics when we discuss Monte Carlo simulations.

If there are many different possible ways to assign a dynamics to the Ising model, how do we know which one is correct? As for the model itself, it is necessary to consider the system that is being modeled in order to determine which kinetics is appropriate. However, we expect that there are many different choices for the kinetics that will provide essentially the same results as long as we consider its long time behavior. The central limit theorem in Section 1.2 shows that in a stochastic process, many independent steps lead to the same Gaussian distribution of probabilities, independent of the specific steps that are taken. Similarly, if we choose a dynamics for the Ising model that allows individual spin flips, the behavior of processes that involve many spin flips should not depend on the specific dynamics chosen. Having said this, we emphasize that the conditions under which different dynamic rules provide the same long time behavior are not fully established. This problem is essentially the same as the problem of classifying dynamic systems in general. We will discuss it in more detail in Section 1.7.

Both the activated dynamics and the Glauber dynamics assume that each spin relaxes from its present state toward its equilibrium distribution. Relaxation of each spin is independent of other spins. The equilibrium distribution is determined by the relative energy of its UP and DOWN state at a particular time. The energy difference between having the ith spin s_i UP and DOWN is:

$$E_{+i}(\{s_j\}_{j \neq i}) = E(s_i = +1, \{s_j\}_{j \neq i}) - E(s_i = -1, \{s_j\}_{j \neq i}) \qquad (1.6.61)$$

The probability of the spin being UP or DOWN is given by Eq. (1.4.14) as:

$$P_{s_i}(1) = \frac{1}{1 + e^{E_{+i}/kT}} = f(E_{+i}) \qquad (1.6.62)$$

$$P_{s_i}(-1) = 1 - f(E_{+i}) = f(-E_{+i}) \qquad (1.6.63)$$

In the activated dynamics, all spins perform transitions at all times with rates $R(1|-1)$ and $R(-1|1)$ given by Eqs. (1.4.38) and (1.4.39) with a site-dependent energy barrier E_{Bi} that sets the relaxation time for the dynamics τ_i. As with the two-state system, it is assumed that each transition occurs essentially instantaneously. The choice of the barrier E_{Bi} is quite important for the kinetics, particularly since it may also depend on the state of other spins with which the ith spin interacts. As soon as one of the spins makes a transition, all of the spins with which it interacts must change their rate of relaxation accordingly. Instead of considering directly the rate of transition, we can consider the evolution of the probability using the Master equation, Eq. (1.4.40) or (1.4.43). This would be convenient for Master equation treatments of the whole system. However, the necessity of keeping track of all of the probabilities makes this impractical for all but simple considerations.

Glauber dynamics is simpler in that it considers only one spin at a time. The system is updated in equal time intervals. Each time interval is divided into N small time increments. During each time increment, we select a particular spin and only consider its dynamics. The selected spin then relaxes completely in the sense that its state is set to be UP or DOWN according to its equilibrium probability, Eq. (1.6.62). The transitions of different spins occur sequentially and are not otherwise coupled. The way we select which spin to update is an essential part of the Glauber dynamics. The simplest and most commonly used approach is to select a spin at random in each time increment. This means that we do not guarantee that every spin is selected during a time interval consisting of N spin updates. Likewise, some spins will be updated more than once in a time interval. On average, however, every spin is updated once per time interval.

In order to show that the Glauber dynamics are intimately related to the activated dynamics, we begin by considering how we would implement the activated dynamics on an ensemble of independent two-state systems whose dynamics are completely determined by the relaxation time $\tau = (R(1|-1) + R(1|-1))^{-1}$ (Eq. (1.4.44)). We can think about this ensemble as representing the dynamics of a single two-state system, or, in a sense that will become clear, as representing a noninteracting Ising model. The total number of spins in our ensemble is N. At time t the ensemble is described by the number of UP spins given by $NP(1;t)$ and the number of DOWN spins $NP(-1;t)$.

We describe the activated dynamics of the ensemble using a small time interval Δt, which eventually we would like to make as small as possible. During the interval of time Δt, which is much smaller than the relaxation time τ, a certain number of spins make transitions. The probability that a particular spin will make a transition from UP to DOWN is given by $R(-1|1)\Delta t$. The total number of spins making a transition from DOWN to UP, and from UP to DOWN, is:

$$NP(-1;t)R(1|-1)\Delta t$$
$$NP(1;t)R(-1|1)\Delta t$$

(1.6.64)

respectively. To implement the dynamics, we must randomly pick out of the whole ensemble this number of UP spins and DOWN spins and flip them. The result would be a new number of UP and DOWN spins $NP(1;t + \Delta t)$ and $NP(-1;t + \Delta t)$. The process would then be repeated.

It might seem that there is no reason to randomly pick the ensemble elements to flip, because the result is the same if we rearrange the spins arbitrarily. However, if each spin represents an identifiable physical system (e.g., one spin out of a noninteracting Ising model) that is performing an internal dynamics we are representing, then we must randomly pick the spins to flip.

It is somewhat inconvenient to have to worry about selecting a particular number of UP and DOWN spins separately. We can modify our prescription so that we select a subset of the spins regardless of orientation. To achieve this, we must allow that some of the selected spins will be flipped and some will not. We select a fraction η of the spins of the ensemble. The number of these that are DOWN is $\eta NP(-1;t)$. In order to flip the same number of spins from DOWN to UP, as in Eq. (1.6.64), we must flip UP a fraction $R(1|-1)\Delta t/\eta$ of the $\eta NP(-1;t)$ spins. Consequently, the fraction of spins we do not flip is $(1 - R(1|-1)\Delta t/\eta)$. Similarly, the number of selected UP spins is $\eta NP(1;t)$ the fraction of these to be flipped is $R(-1|1)\Delta t/\eta$, and the fraction we do not flip is $(1 - R(-1|1) \Delta t/\eta)$. In order for these expressions to make sense (to be positive) η must be large enough so that at least one spin will be flipped. This implies $\eta > \max (R(1|-1)\Delta t, R(-1|1)\Delta t)$. Moreover, we do not want η to be larger than it must be because this will just force us to select additional spins we will not be flipping. A convenient choice would be to take

$$\eta = (R(1|-1) + R(-1|1))\Delta t = \Delta t / \tau \qquad (1.6.65)$$

The consequences of this choice are quite interesting, since we find that the fraction of selected DOWN spins to be flipped UP is $R(1|-1) / (R(1|-1) + R(-1|1)) = P(1)$, the equilibrium fraction of UP spins. The fraction not to be flipped is the equilibrium fraction of DOWN spins. Similarly, the fraction of selected UP spins that are to be flipped DOWN is the equilibrium fraction of DOWN spins, and the fraction to be left UP is the equilibrium fraction of UP spins. Consequently, the outcome of the dynamics of the selected spin does not depend at all on the initial state of the spin. The revised prescription for the dynamics is to select a fraction η of spins from the ensemble and set them according to their equilibrium probability.

We still must choose the time interval Δt. The smallest time interval that makes sense is the interval for which the number of selected spins would be just one. A smaller number would mean that sometimes we would not choose any spins. Setting the number of selected spins $\eta N = 1$ using Eq. (1.6.65) gives:

$$\Delta t = \frac{1}{N(R(1|-1)+R(-1|1))} = \frac{\tau}{N} \qquad (1.6.66)$$

which also implies the condition $\Delta t \ll \tau$, and means that the approximation of a finite time increment Δt is directly coupled to the size of the ensemble. Our new prescription is that we select a single spin and set it UP or DOWN according to its equilibrium probability. This would be the prescription of Glauber dynamics if the ensemble were considered to be the Ising model without interactions. Thus for a noninteracting Ising model, the Glauber dynamics and the activated dynamics are the same. So far we have made no approximation except the finite size of the ensemble. We still have one more step to go to apply this to the interacting Ising model.

The activated dynamics is a stochastic dynamics, so it does not make sense to discuss only the dynamics of a particular system but the dynamics of an ensemble of Ising models. At any moment, the activated dynamics treats the Ising model as a collection of several kinds of spins. Each kind of spin is identified by a particular value of E_+ and E_B. These parameters are controlled by the local environment of the spin. The dynamics is not concerned with the source of these quantities, only their values. The dynamics are that of an ensemble consisting of several kinds of spins with a different number N_k of each kind of spin, where k indexes the kind of spin. According to the result of the previous paragraph, and specifically Eq. (1.6.65), we can perform this dynamics over a time interval Δt by selecting $N_k \Delta t / \tau_k$ spins of each kind and updating them according to the Glauber method. This is strictly applicable only for an ensemble of Ising systems. If the Ising system that we are considering contains many correlation lengths, Eq. (1.6.50), then it represents the ensemble by itself. Thus for a large enough Ising model, we can apply this to a single system.

If we want to select spins arbitrarily, rather than of a particular kind, we must make the assumption that all of the relaxation times are the same, $\tau_k \to \tau$. This assumption means that we would select a total number of spins:

$$\sum_k \frac{N_k \Delta t}{\tau_k} \to N \frac{\Delta t}{\tau} \tag{1.6.67}$$

As before, Δt may also be chosen so that in each time interval only one spin is selected.

Using two assumptions, we have been able to derive the Glauber dynamics directly from the activated dynamics. One of the assumptions is that the dynamics must be considered to apply only as the dynamics of an ensemble. Even though both dynamics are stochastic dynamics, applying the Glauber dynamics directly to a single system is only the same as the activated dynamics for a large enough system. The second assumption is the equivalence of the relaxation times τ_k. When is this assumption valid? The expression for the relaxation time in terms of the two-state system is given by Eq. (1.4.44) as

$$1/\tau = (R(1|-1) + R(-1|1)) = \nu(e^{-(E_B - E_1)/kT} + e^{-(E_B - E_{-1})/kT}) \tag{1.6.68}$$

When the relative energy of the two states E_1 and E_{-1} varies between different spins, this will in general vary. The size of the relaxation time is largely controlled by the smaller of the two energy differences $E_B - E_1$ and $E_B - E_{-1}$. Thus, maintaining the same relaxation time would require that the smaller energy difference is nearly constant. This is essential, because the relaxation time changes exponentially with the energy difference.

We have shown that the Glauber dynamics and the activated dynamics are closely related despite appearing to be quite different. We have also found how to generalize the Glauber dynamics if we must allow different relaxation times for different spins. Finally, we have found that the time increment for a single spin update corresponds to τ/N. This means that a single Glauber time step consisting of N spin updates corresponds to a physical time τ—the microscopic relaxation time of the individual spins.

At this point we have introduced a dynamics for the Ising model, and it should be possible for us to investigate questions about its kinetics. Often questions about the kinetics may be described in terms of time correlations. Like the correlation length, we can introduce a correlation time τ_s that is given by the decay of the spin-spin correlation

$$< s_i(t')s_i(t) > - < s_i >^2 \propto e^{-|t-t'|/\tau_s} \tag{1.6.69}$$

For the case of a relaxing two-state system, the correlation time is the relaxation time τ. This follows from Eq. (1.4.45), with some attention to notation as described in Question 1.6.13.

Question 1.6.13 Show that for a two-state system, the correlation time is the relaxation time τ.

Solution 1.6.13 The difficulty in this question is restoring some of the no-tational details that we have been leaving out for convenience. From Eq. (1.6.45) we have for the average:

$$
\begin{aligned}
<s_i(t')s_i(t)> = &\ P_{s_i(t'),s_i(t)}(1,1) + P_{s_i(t'),s_i(t)}(-1,-1) \\
&- P_{s_i(t'),s_i(t)}(1,-1) - P_{s_i(t'),s_i(t)}(-1,1)
\end{aligned}
\tag{1.6.70}
$$

Let's assume that $t' > t$, then each of these joint probabilities of the form $P_{s_i(t'),s_i(t)}(s_2,s_1)$ is given by the probability that the two-state system starts in the state s_1 at time t, multiplied by the probability that it will evolve from s_1 into s_2 at time t'.

$$
P_{s_i(t'),s_i(t)}(s_2,s_1) = P_{s_i(t'),s_i(t)}(s_2 \mid s_1)P_{s_i(t)}(s_1)
\tag{1.6.71}
$$

The first factor on the right is called the conditional probability. The probability for a particular state of the spin is the equilibrium probability that we wrote as $P(1)$ and $P(-1)$. The conditional probabilities satisfy $P_{s_i(t'),s_i(t)}(1 \mid s_1) + P_{s_i(t'),s_i(t)}(-1 \mid s_1) = 1$, so we can simplify Eq. (1.6.70) to:

$$
<s_i(t')s_i(t)> = (2P_{s_i(t'),s_i(t)}(1 \mid 1)-1)P(1) + (2P_{s_i(t'),s_i(t)}(-1 \mid -1)-1)P(-1)
\tag{1.6.72}
$$

The evolution of the probabilities are described by Eq. (1.4.45), repeated here:

$$
P(1;t) = (P(1;0) - P(1;\infty))e^{-t/\tau} + P(1;\infty)
\tag{1.6.73}
$$

Since the conditional probability assumes a definite value for the initial state (e.g., $P(1;0) = 1$ for $P_{s(t'),s(t)}(1 \mid 1)$), we have:

$$
\begin{aligned}
P_{s(t'),s(t)}(1 \mid 1) &= (1 - P(1))e^{-(t'-t)/\tau} + P(1) \\
P_{s(t),s(t)}(-1 \mid -1) &= (1 - P(-1))e^{-(t'-t)/\tau} + P(-1)
\end{aligned}
\tag{1.6.74}
$$

Inserting these into Eq. (1.6.72) gives:

$$
\begin{aligned}
<s_i(t')s_i(t)> = &\ (2\left[(1-P(1))e^{-(t'-t)/\tau} + P(1)\right]-1)P(1) \\
&+ (2\left[(1-P(-1))e^{-(t'-t)/\tau} + P(-1)\right]-1)P(-1) \\
= &\ 4P(1)P(-1)e^{-(t'-t)/\tau} + (P(1)-P(-1))^2
\end{aligned}
\tag{1.6.75}
$$

The constant term on the right is the same as the square of the average of the spin:

$$
<s_i(t)>^2 = (P(1) - P(-1))^2
\tag{1.6.76}
$$

Inserting into Eq. (1.6.69) leads to the desired result (we have assumed that $t' > t$):

$$
<s_i(t')s_i(t)> - <s_i(t)>^2 = 4P(1)P(-1)e^{-(t'-t)/\tau} \propto e^{-(t'-t)/\tau}
\tag{1.6.77} \ \blacksquare
$$

From the beginning of our discussion of the Ising model, a central issue has been the breaking of the ergodic theorem associated with the spontaneous magnetization. Now that we have introduced a kinetic model, we will tackle this problem directly. First we describe the problem fully. The ergodic theorem states that a time average may be replaced by an ensemble average. In the ensemble, all possible states of the system are included with their Boltzmann probability. Without formal justification, we have treated the spontaneous magnetization of the Ising model at $h = 0$ as a macroscopically observable quantity. According to our prescription, this is not the case. Let us perform the average $< s_i >$ over the ensemble at $T = 0$ and $h = 0$. There are two possible states of the system with the same energy, one with $\{s_i = 1\}$ and one with $\{s_i = -1\}$. Since they must occur with equal probability by our assumption, we have that the average $< s_i >$ is zero.

This argument breaks down because of the kinetics of the system that prevents a transition from one state to the other during the course of a measurement. Thus we measure only one of the two possible states and find a magnetization of 1 or −1. How can we prove that this system breaks the ergodic theorem? The most direct test is to start from a system with a slightly positive magnetic field near $T = 0$ where the magnetization is +1, and reverse the sign of the magnetic field. In this case the equilibrium state of the system should have a magnetization of −1. Instead the system will maintain its magnetization as +1 for a long time before eventually switching from one to the other. The process of switching corresponds to the kinetics of a first-order transition.

1.6.8 *Kinetics of a first-order phase transition*

In this section we discuss the first-order transition kinetics in the Ising model. Similar arguments apply to other first-order transitions like the freezing or boiling of water. If we start with an Ising model in equilibrium at a temperature $T < T_c$ and a small positive magnetic field $h << zJ$, the magnetization of the system is essentially $m_0(\beta zJ)$. If we change the magnetic field suddenly to a small negative value, the equilibrium state of the system is $-m_0(\beta zJ)$; however, the system will require some time to change its magnetization. The change in the magnetic field has very little effect on the energy of an individual spin s_i. This energy is mostly due to the interaction with its neighbors, with a relatively small contribution due to the external field. Most of the time the neighbors are oriented UP, and this makes the spin have a lower energy when it is UP. This gives rise to the magnetization $m_0(\beta zJ)$. Until s_i's neighbors change their average magnetization, s_i has no reason to change its magnetization. But then neither do the neighbors. Thus, because each spin is in its own local equilibrium, the process that eventually equilibrates the system requires a cooperative effect including more than one spin. The process by which such a first-order transition occurs is not the simultaneous switching of all of the spins from one value to the other. This would require an impossibly long time. Instead the transition occurs by nucleation and growth of the equilibrium phase.

It is easiest to describe the nucleation process when T is sufficiently less than T_c, so that the spins are almost always +1. In mean field, already for $T < 0.737 T_c$ the probability of a spin being UP is greater than 90% ($P(1) = (1 + m)/2 > 0.9$), and for $T < 0.61T_c$ the probability of a spin being UP is greater than 95%. As long as T is greater than zero, individual spins will flip from time to time. However, even though the magnetic field would like them to be DOWN, their local environment consisting of UP spins does not. Since the interaction with their neighbors is stronger than the interaction with the external field, the spin will generally flip back UP after a short time. There is a smaller probability that a second spin, a neighbor of the first spin, will also flip DOWN. Because one of the neighbors of the second spin is already DOWN, there is a lower energy cost than for the first one. However, the energy of the second spin is still higher when it is DOWN, and the spins will generally flip back, first one then the other. There is an even smaller probability that three interacting spins will flip DOWN. The existence of two DOWN spins makes it more likely for the third to do so. If the first two spins were neighbors, than the third spin can have only one of them as its neighbor. So it still costs some energy to flip DOWN the third spin. If there are three spins flipped DOWN in an L shape, the spin that completes a 2×2 square has two neighbors that are $+1$ and two neighbors that are -1, so the interactions with its neighbors cancel. The external field then gives a preference for it to be DOWN. There is still a high probability that several of the spins that are DOWN will flip UP and the little cluster will then disappear. Fig. 1.6.9 shows various clusters and their energies compared to a uniform region of $+1$ spins. As more spins are added, the internal region of the cluster becomes composed of spins that have four neighbors that are all DOWN. Beyond a certain size (see Question 1.6.14) the cluster of DOWN spins will grow, because adding spins lowers the energy of the system. At some point the growing region of DOWN spins encounters another region of DOWN spins and the whole system reaches its new equilibrium state, where most spins are DOWN.

Question 1.6.14 Using an estimate of how the energy of large clusters of DOWN spins grows, show that large enough clusters must have a lower energy than the same region if it were composed of UP spins.

Solution 1.6.14 The energy of a cluster of DOWN spins is given by its interaction with the external magnetic field and the number of antialigned bonds that form its boundary. The change in energy due to the external magnetic field is exactly $2hN_c$, which is proportional to the number of spins in the

Figure 1.6.9 Illustration of small clusters of DOWN spins shown as filled dark squares residing in a background of UP spins on a square lattice. The energies for creating the clusters are shown. The magnetic field, h, is negative. The formation of such clusters is the first step towards nucleation of a DOWN region when the system undergoes a first-order transition from UP to DOWN. The energy is counted by the number of spins that are DOWN times the magnetic field strength, plus the interaction strength times the number of antialigned neighboring spins, which is the length of the boundary of the cluster. In a first-order transition, as the size of the clusters grows the gain from orienting toward the magnetic field eventually becomes greater than the loss from the boundary energy. Then the cluster becomes more likely to grow than shrink. See Question 1.6.14 and Fig. 1.6.10. ∎

$2h+8J$

$4h+12J$

$6h+16J$

$8h+16J$ $8h+20J$

$10h+20J$ $10h+24J$

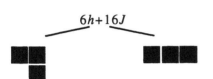
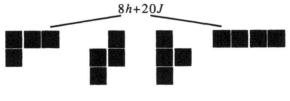

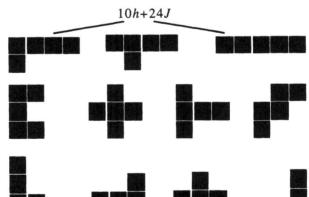

$12h+22J$ $12h+24J$ $12h+26J$

cluster N_c. This is negative since h is negative. The energy of the boundary is proportional to the number of antialigned bonds, and it is always positive. Because every additional antialigned bond raises the cluster energy, the boundary of the cluster tends to be smooth at low temperatures. Therefore, we can estimate the boundary energy using a simple shape like a square or circular cluster in 2-d (a cube or ball in 3-d). Either way the energy will increase as $fJN_c^{(d-1)/d}$, where d is the dimensionality and f is a constant accounting for the shape. Since the negative contribution to the energy increases, in proportion to the area (volume) of the cluster, and the positive contribution to the energy increases in proportion to the perimeter (surface area) of the cluster, the negative term eventually wins. Once a cluster is large enough so that its energy is dominated by the interaction with the magnetic field, then, on-average, adding an additional spin to the cluster will lower the system energy. ∎

Question **1.6.15** Without looking at Fig. 1.6.9, construct all of the different possible clusters of as many as five DOWN spins. Label them with their energy.

Solution 1.6.15 See Fig. 1.6.9. ∎

The scenario just described, known as nucleation and growth, is generally responsible for the kinetics of first-order transitions. We can illustrate the process schematically (Fig. 1.6.10) using a one dimensional plot indicating the energy per spin of a cluster as a function of the number of atoms in the cluster. The energy of the cluster increases at first when there are very few spins in the cluster, and then decreases once it is large enough. Eventually the energy decreases linearly with the number of spins in the cluster. The decrease per spin is the energy difference per spin between the two phases. The first cluster size that is "over the hump" is known as the critical cluster. The process of reaching this cluster is known as nucleation. A first estimate of the time to nucleate a critical cluster at a particular place in space is given by the inverse of the Boltzmann factor of the highest energy barrier in Fig. 1.6.10. This corresponds to the rate of transition over the barrier given by a two-state system with this same barrier (see Eq. (1.4.38) and Eq. (1.4.44)). The size of the critical cluster depends on the magnitude of the magnetic field. A larger magnetic field implies a smaller critical cluster. Once the critical cluster is reached, the kinetics corresponds to the biased diffusion described at the end of Section 1.4. The primary difficulty with an illustration such as Fig. 1.6.10 is that it is one-dimensional. We would need to show the energy of each type of cluster and all of the ways one cluster can transform into another. Moreover, the clusters themselves may move in space and merge or separate. In Fig. 1.6.11 we show frames from a simulation of nucleation in the Ising model using Glauber dynamics. The frames illustrate the process of nucleation and growth.

Experimental studies of nucleation kinetics are sometimes quite difficult. In physical systems, impurities often lower the barrier to nucleation and therefore control the rate at which the first-order transition occurs. This can be a problem for the investigation of the inherent nucleation because of the need to study highly purified

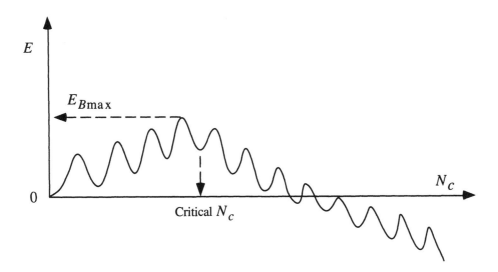

Figure 1.6.10 Schematic illustration of the energies that control the kinetics of a first-order phase transition. The horizontal axis is the size of a cluster of DOWN spins N_c that are the equilibrium phase. The cluster is in a background of UP spins that are the metastable phase. The vertical axis is the energy of the cluster. Initially the energy increases with cluster size until the cluster reaches the critical cluster size. Then the energy decreases. Each spin flip has its own barrier to overcome, leading to a washboard potential. The highest barrier E_{Bmax} that the system must overcome to create a critical nucleus controls the rate of nucleation. This is similar to the relaxation of a two-level system discussed in Section 1.4. However, this simple picture neglects the many different possible clusters and the many ways they can convert into each other by the flipping of spins. A few different types of clusters are shown in Fig. 1.6.9. ∎

systems. However, this sensitivity should be understood as an opportunity for control over the kinetics. It is similar to the sensitivity of electrical properties to dopant impurities in a semiconductor, which enables the construction of semiconductor devices. There is at least one direct example of the control of the kinetics of a first-order transition. Before describing the example, we review a few properties of the water-to-ice transition. The temperature of the water-to-ice transition can be lowered significantly by the addition of impurities. The freezing temperature of salty ocean water is lower than that of pure water. This suppression is thermodynamic in origin, which means that the T_c is actually lower. There exist fish that live in sub-zero-degrees ocean water whose blood has less salt than the surrounding ocean. These fish use a family of so-called antifreeze proteins that are believed to kinetically suppress the freezing of their blood. Instead of lowering the freezing temperature, these proteins suppress ice nucleation.

The existence of a long nucleation time implies that it is often possible to create metastable materials. For example, supercooled water is water whose temperature has been lowered below its freezing point. For many years, particle physicists used a superheated fluid to detect elementary particles. Ultrapure liquids in large tanks were

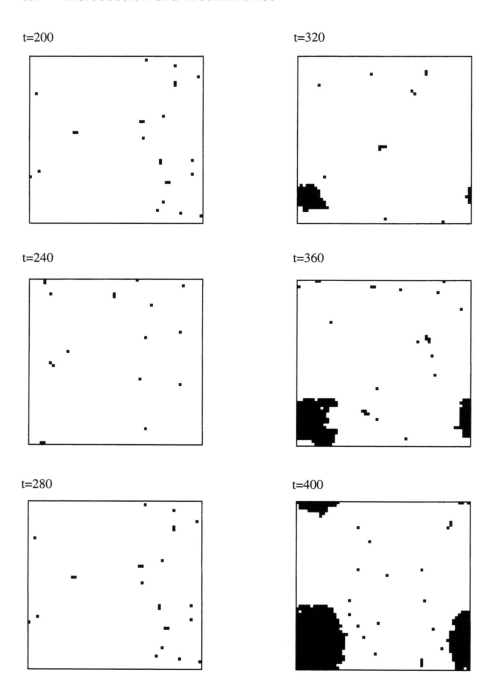

Figure 1.6.11 Frames from a simulation illustrating nucleation and growth in an Ising model in 2-d. The temperature is $T = zJ/3$ and the magnetic field is $h = -0.25$. Glauber dynamics was used. Each time step consists of N updates where the space size is $N = 60 \times 60$. Frames shown are in intervals of 40 time steps. The first frame shown is at $t = 200$ steps after the beginning of the simulation. Black squares are DOWN spins and white areas are UP spins. The

t=440

t=560

t=480

t=600

t=520

t=640

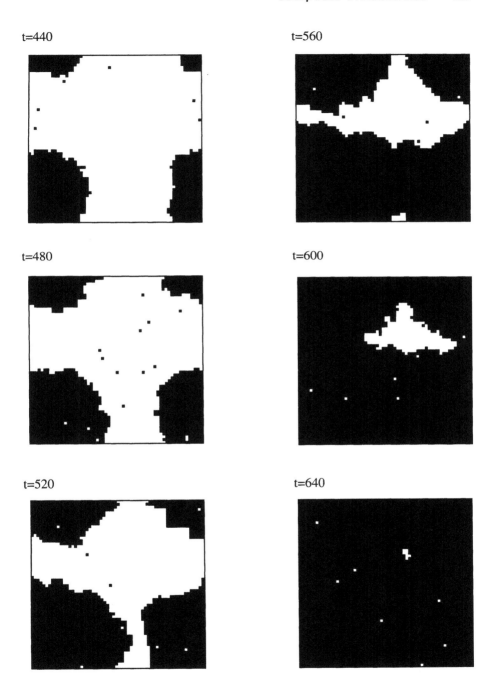

metastability of the UP phase is seen in the existence of only a few DOWN spins until the frame at $t = 320$. All earlier frames are qualitatively the same as the frames at $t = 200, 240$ and 280. A critical nucleus forms between $t = 280$ and $t = 320$. This nucleus grows systematically until the final frame when the whole system is in the equilibrium DOWN phase. ∎

suddenly shifted above their boiling temperature. Small bubbles would then nucleate along the ionization trail left by charged particles moving through the tank. The bubbles could be photographed and the tracks of the particles identified. Such detectors were called bubble chambers. This methodology has been largely abandoned in favor of electronic detectors. There is a limit to how far a system can be supercooled or superheated. The limit is easy to understand in the Ising model. If a system with a positive magnetization m is subject to a negative magnetic field of magnitude greater than zJm, then each individual spin will flip DOWN independent of its neighbors. This is the ultimate limit for nucleation kinetics.

1.6.9 *Connections between CA and the Ising model*

Our primary objective throughout this section is the investigation of the equilibrium properties of interacting systems. It is useful, once again, to consider the relationship between the equilibrium ensemble and the kinetic CA we considered in Section 1.5. When a deterministic CA evolves to a unique steady state independent of the initial conditions, we can identify the final state as the $T = 0$ equilibrium ensemble. This is, however, not the way we usually consider the relationship between a dynamic system and its equilibrium condition. Instead, the equilibrium state of a system is generally regarded as the time average over microscopic dynamics. Thus when we use the CA to represent a microscopic dynamics, we could also identify a long time average of a CA as the equilibrium ensemble. Alternatively, we can consider a stochastic CA that evolves to a unique steady-state distribution where the steady state is the equilibrium ensemble of a suitably defined energy function.

1.7   Computer Simulations (Monte Carlo, Simulated Annealing)

Computer simulations enable us to investigate the properties of dynamical systems by directly studying the properties of particular models. Originally, the introduction of computer simulation was viewed by many researchers as an undesirable adjunct to analytic theory. Currently, simulations play such an important role in scientific studies that many analytic results are not believed unless they are tested by computer simulation. In part, this reflects the understanding that analytic investigations often require approximations that are not necessary in computer simulations. When a series of approximations has been made as part of an analytic study, a computer simulation of the original problem can directly test the approximations. If the approximations are validated, the analytic results often generalize the simulation results. In many other cases, simulations can be used to investigate systems where analytic results are unknown.

1.7.1 *Molecular dynamics and deterministic simulations*

The simulation of systems composed of microscopic Newtonian particles that experience forces due to interparticle interactions and external fields is called molecular dynamics. The techniques of molecular dynamics simulations, which integrate

Newton's laws for individual particles, have been developed to optimize the efficiency of computer simulation and to take advantage of parallel computer architectures. Typically, these methods implement a discrete iterative map (Section 1.1) for the particle positions. The most common (Verlet) form is:

$$r(t) = 2r(t - \Delta t) - r(t - 2\Delta t) + \Delta t^2 a(t - \Delta t) \qquad (1.7.1)$$

where $a(t) = F(t)/m$ is the force on the particle calculated from models for interparticle and external forces. As in Section 1.1, time would be measured in units of the time interval Δt for convenience and efficiency of implementation. Eq. (1.7.1) is algebraically equivalent to the iterative map in Question 1.1.4, which is written as an update of both position and velocity:

$$r(t) = r(t - \Delta t) + \Delta t v(t - \Delta t/2)$$
$$v(t + \Delta t/2) = v(t - \Delta t/2) + \Delta t a(t) \qquad (1.7.2)$$

As indicated, the velocity is interpreted to be at half integral times, though this does not affect the result of the iterative map.

For most such simulations of physical systems, the accuracy is limited by the use of models for interatomic interactions. Modern efforts attempt to improve upon this approach by calculating forces from quantum mechanics. However, such simulations are very limited in the number of particles and the duration of a simulation. A useful measure of the extent of a simulation is the product Nt_{max} of the amount of physical time t_{max}, and the number of particles that are simulated N. Even without quantum mechanical forces, molecular dynamics simulations are still far from being able to describe systems on a space and time scale comparable to human senses. However, there are many questions that can be addressed regarding microscopic properties of molecules and materials.

The development of appropriate simplified macroscopic descriptions of physical systems is an essential aspect of our understanding of these systems. These models may be based directly upon macroscopic phenomenology obtained from experiment. We may also make use of the microscopic information obtained from various sources, including both theory and experiment, to inform our choice of macroscopic models. It is more difficult, but important as a strategy for the description of both simple and complex systems, to develop systematic methods that enable macroscopic models to be obtained directly from microscopic models. The development of such methods is still in its infancy, and it is intimately related to the issues of emergent simplicity and complexity discussed in Chapter 8.

Abstract mathematical models that describe the deterministic dynamics for various systems, whether represented in the form of differential equations or deterministic cellular automata (CA, Section 1.5), enable computer simulation and study through integration of the differential equations or through simulation of the CA. The effects of external influences, not incorporated in the parameters of the model, may be modeled using stochastic variables (Section 1.2). Such models, whether of fluids or of galaxies, describe the macroscopic behavior of physical systems by assuming that the microscopic (e.g., molecular) motion is irrelevant to the macroscopic phenomena being described. The microscopic behavior is summarized by parameters such as density, elasticity or viscosity. Such model simulations enable us to describe macroscopic phenomena on a large range of spatial and temporal scales.

1.7.2 Monte Carlo simulations

In our investigations of various systems, we are often interested in average quantities rather than a complete description of the dynamics. This was particularly apparent in Section 1.3, when equilibrium thermodynamic properties of systems were discussed. The ergodic theorem (Section 1.3.5) suggested that we can use an ensemble average instead of the space-time average of an experiment. The ensemble average enables us to treat problems analytically, when we cannot integrate the dynamics explicitly. For example, we studied equilibrium properties of the Ising model in Section 1.6 without reference to its dynamics. We were able to obtain estimates of its free energy, energy and magnetization by averaging various quantities using ensemble probabilities.

However, we also found that there were quite severe limits to our analytic capabilities even for the simplest Ising model. It was necessary to use the mean field approximation to obtain results analytically. The essential difficulty that we face in performing ensemble averages for complex systems, and even for the simple Ising model, is that the averages have to be performed over the many possible states of the system. For as few as one hundred spins, the number of possible states of the system—2^{100}— is so large that we cannot average over all of the possible states. This suggests that we consider approximate numerical techniques for studying the ensemble averages. In order to perform the averages without summing over all the states, we must find some way to select a representative sample of the possible states.

Monte Carlo simulations were developed to enable numerical averages to be performed efficiently. They play a central role in the use of computers in science. Monte Carlo can be thought of as a general way of estimating averages by selecting a limited sample of states of the system over which the averages are performed. In order to optimize convergence of the average, we take advantage of information that is known about the system to select the limited sample. As we will see, under some circumstances, the sequence of states selected in a Monte Carlo simulation may itself be used as a model of the dynamics of a system. Then, if we are careful about designing the Monte Carlo, we can separate the time scales of a system by treating the fast degrees of freedom using an ensemble average and still treat explicitly the dynamic degrees of freedom.

To introduce the concept of Monte Carlo simulation, we consider finding the average of a function $f(s)$, where the system variable s has the probability $P(s)$. For simplicity, we take s to be a single real variable in the range $[-1,+1]$. The average can be approximated by a sum over equally spaced values s_i:

$$<f(s)> = \int_{-1}^{1} f(s)P(s)ds \approx \sum_{s_i} f(s_i)P(s_i)\Delta s = \frac{1}{M}\sum_{n=-M}^{M} f(n/M)P(n/M) \quad (1.7.3)$$

This formula works well if the functions $f(s)$ and $P(s)$ are reasonably smooth and uniform in magnitude. However, when they are not smooth, this sum can be a very inefficient way to perform the integral. Consider this integral when $P(s)$ is a Gaussian, and $f(s)$ is a constant:

$$< f(s) > \propto \int_{-1}^{1} e^{-s^2/2\sigma^2} ds \approx \frac{1}{M} \sum_{n=-M}^{M} e^{-(n/M)^2/2\sigma^2} \qquad (1.7.4)$$

A plot of the integrand in Fig. 1.7.1 shows that for $\sigma \ll 1$ we are performing the integral by summing many values that are essentially zero. These values contribute nothing to the result and require as much computational effort as the comparatively few points that do contribute to the integral near $s = 0$, where the function is large. The few points near $s = 0$ will not give a very accurate estimate of the integral. Thus, most of the computational work is being wasted and the integral is not accurately evaluated. If we want to improve the accuracy of the sum, we have to increase the value of M. This means we will be summing many more points that are almost zero.

To avoid this problem, we would like to focus our attention on the region in Eq. (1.7.4) where the integrand is large. This can be done by changing how we select the points where we perform the average. Instead of picking the points at equal intervals along the line, we pick them with a probability given by $P(s)$. This is the same as saying that we have an ensemble representing the system with the state variable s. Then we perform the ensemble average:

$$< f(s) > = \int f(s)P(s)ds = \frac{1}{N} \sum_{s:P(s)}^{N} f(s) \qquad (1.7.5)$$

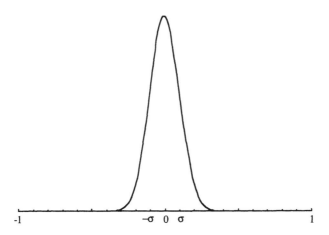

Figure 1.7.1 Plot of the Gaussian distribution illustrating that an integral that is performed by uniform sampling will use a lot of points to represent regions where the Gaussian is vanishingly small. The problem gets worse as σ becomes smaller compared to the region over which the integral must be performed. It is much worse in typical multidimensional averages where the Boltzmann probability is used. Monte Carlo simulations make such integrals computationally feasible by sampling the integrand in regions of high probability. ∎

The latter expression represents the sum over N values of s, where these values have the probability distribution $P(s)$. We have implicitly assumed that the function $f(s)$ is relatively smooth compared to $P(s)$. In Eq. (1.7.5) we have replaced the integral with a sum over an ensemble. The problem we now face is to obtain the members of the ensemble with probability $P(s)$. To do this we will invert the ergodic theorem of Section 1.3.5.

Since Section 1.3 we have described an ensemble as representing a system, if the dynamics of the system satisfied the ergodic theorem. We now turn this around and say that the ensemble sum in Eq. (1.7.5) can be represented by any dynamics that satisfies the ergodic theorem, and which has as its equilibrium probability $P(s)$. To do this we introduce a time variable t that, for our current purposes, just indicates the order of terms in the sum we are performing. The value of s appearing in the tth term would be $s(t)$. We then rewrite the ergodic theorem by considering the time average as an approximation to the ensemble average (rather than the opposite):

$$<f(s)>=\frac{1}{T}\sum_{t=1}^{T}f(s(t))\tag{1.7.6}$$

The problem remains to sequentially generate the states $s(t)$, or, in other words, to specify the dynamics of the system. If we know the probability $P(s)$, and s is a few binary or real variables, this may be done directly with the assistance of a random number generator (Question 1.7.1). However, often the system coordinate s represents a large number of variables. A more serious problem is that for models of physical systems, we generally don't know the probability distribution explicitly.

Thermodynamic systems are described by the Boltzmann probability (Section 1.3):

$$P(\{x,p\})=\frac{1}{Z}e^{-E(\{x,p\})/kT}$$
$$Z=\sum_{\{x,p\}}e^{-E(\{x,p\})/kT}\tag{1.7.7}$$

where $\{x,p\}$ are the microscopic coordinates of the system, and $E(\{x,p\})$ is the microscopic energy. An example of a quantity we might want to calculate would be the average energy:

$$U=\frac{1}{Z}\sum_{\{x,p\}}E(\{x,p\})e^{-E(\{x,p\})/kT}\tag{1.7.8}$$

In many cases, as discussed in Section 1.4, the quantity that we would like to find the average of depends only on the position of particles and not on their momenta. We then write more generally

$$P(s)=\frac{1}{Z_s}e^{-F(s)/kT}$$
$$Z_s=\sum_{s}e^{-F(s)/kT}\tag{1.7.9}$$

where we use the system state variable s to represent the relevant coordinates of the system. We make no assumption about the dimensionality of the coordinate s which may, for example, be the coordinates $\{x\}$ of all of the particles. $F(s)$ is the free energy of the set of states associated with the coordinate s. A precise definition, which indicates both the variable s and its value s', is given in Eq. (1.4.27):

$$F_s(s') = -kT \ln(\sum_{\{x,p\}} \delta_{s,s'} \, e^{-E(\{x,p\})/kT})$$ (1.7.10)

We note that Eq. (1.7.9) is often written using the notation $E(s)$ (the energy of s) instead of $F(s)$ (the free energy of s), though $F(s)$ is more correct. An average we might calculate, of a quantity $Q(s)$, would be:

$$U = \frac{1}{Z} \sum_s Q(s) e^{-F(s)/kT}$$ (1.7.11)

where $Q(s)$ is assumed to depend only on the variable s and not directly on $\{x,p\}$.

The problem with the evaluation of either Eq. (1.7.8) or Eq. (1.7.11) is that the Boltzmann probability does not explicitly give us the probability of a particular state. In order to find the actual probability, we need to find the partition function Z. To calculate Z we need to perform a sum over all states of the system, which is computationally impossible. Indeed, if we were able to calculate Z, then, as discussed in Section 1.3, we would know the free energy and all the other thermodynamic properties of the system. So a prescription that relies upon knowing the actual value of the probability doesn't help us. However, it turns out that we don't need to know the actual probability in order to construct a dynamics for the system, only the relative probabilities of particular states. The relative probability of two states, $P(s) / P(s')$, is directly given by the Boltzmann probability in terms of their relative energy:

$$P(s) / P(s') = e^{-(F(s)-F(s'))/kT}$$ (1.7.12)

This is the key to Monte Carlo simulations. It is also a natural result, since a system that is evolving in time does not know global properties that relate to all of its possible states. It only knows properties that are related to the energy it has, and how this energy changes with its configuration. In classical mechanics, the change of energy with configuration would be the force experienced by a particle.

Our task is to describe a dynamics that generates a sequence of states of a system $s(t)$ with the proper probability distribution, $P(s)$. The classical (Newtonian) approach to dynamics implies that a deterministic dynamics exists which is responsible for generating the sequence of states of a physical system. In order to generate the equilibrium ensemble, however, there must be contact with a thermal reservoir. Energy transfer between the system and the reservoir introduces an external interaction that disrupts the system's deterministic dynamics.

We will make our task simpler by allowing ourselves to consider a stochastic Markov chain (Section 1.2) as the dynamics of the system. The Markov chain is described by the probability $P_s(s'|s'')$ of the system in a state $s = s''$ making a transition to the state $s = s'$. A particular sequence $s(t)$ is generated by starting from one configuration and choosing its successors using the transition probabilities.

The general formulation of a Markov chain includes the classical Newtonian dynamics and can also incorporate the effects of a thermal reservoir. However, it is generally convenient and useful to use a Monte Carlo simulation to evaluate averages that do not depend on the momenta, as in Eq. (1.7.11). There are some drawbacks to this approach. It limits the properties of the system whose averages can be evaluated. Systems where interactions between particles depend on their momenta cannot be easily included. Moreover, averages of quantities that depend on both the momentum and the position of particles cannot be performed. However, if the energy separates into potential and kinetic energies as follows:

$$E(\{x, p\}) = V(\{x\}) + \sum_i p_i^2 / 2m \qquad (1.7.13)$$

then averages over all quantities that just depend on momenta (such as the kinetic energy) can be evaluated directly without need for numerical computation. These averages are the same as those of an ideal gas. Monte Carlo simulations can then be used to perform the average over quantities that depend only upon position $\{x\}$, or more generally, on position-related variables s. Thus, in the remainder of this section we focus on describing Markov chains for systems described only by position-related variables s.

As described in Section 1.2 we can think about the Markov dynamics as a dynamics of the probability rather than the dynamics of a system. Then the dynamics are specified by

$$P_s(s'; t) = \sum_{s''} P_s(s' | s'') P_s(s''; t-1) \qquad (1.7.14)$$

In order for the stochastic dynamics to represent the ensemble, we must have the time average over the probability distribution $P_s(s', t)$ equal to the ensemble probability. This is true for a long enough time average if the probability converges to the ensemble probability distribution, which is a steady-state distribution of the Markov chain:

$$P_s(s') = P_s(s'; \infty) = \sum_{s''} P_s(s' | s'') P_s(s''; \infty) \qquad (1.7.15)$$

Thermodynamics and stochastic Markov chains meet when we construct the Markov chain so that the Boltzmann probability, Eq. (1.7.9), is the limiting distribution.

We now make use of the Perron-Frobenius theorem (see Section 1.7.4 below), which says that a Markov chain governed by a set of transition probabilities $P_s(s'|s'')$ converges to a unique limiting probability distribution as long as it is irreducible and acyclic. Irreducible means that there exist possible paths between each state and all other possible states of the system. This does not mean that all states of the system are connected by nonzero transition probabilities. There can be transition probabilities that are zero. However, it must be impossible to separate the states into two sets for which there are no transitions from one set to the other. Acyclic means that the system is not ballistic—the states are not organized by the transition matrix into a ring with a deterministic flow around it. There may be currents, but they must not be deterministic. It is sufficient for there to be a single state which has a nonzero probability of making a transition to itself for this condition to be satisfied, thus it is often assumed and unstated.

We can now summarize the problem of identifying the desired Markov chain. We must construct a matrix $P_s(s'|s'')$ that satisfies three properties. First, it must be an allowable transition matrix. This means that it must be nonnegative, $P_s(s''|s')\geq0$, and satisfy the normalization condition (Eq (1.2.4)):

$$\sum_{s''} P_s(s''\,|\,s')=1 \tag{1.7.16}$$

Second, it must have the desired probability distribution, Eq. (1.7.9), as a fixed point. Third, it must not be reducible—it is possible to construct a path between any two states of the system.

These conditions are sufficient to guarantee that a long enough Markov chain will be a good approximation to the desired ensemble. There is no guarantee that the convergence will be rapid. As we have seen in Section 1.4, in the case of the glass transition, the ergodic theorem may be violated on all practical time scales for systems that are following a particular dynamics. This applies to realistic or artificial dynamics. In general such violations of the ergodic theorem, or even just slow convergence of averages, are due to energy barriers or entropy "bottlenecks" that prevent the system from reaching all possible configurations of the system in any reasonable time. Such obstacles must be determined for each system that is studied, and are sometimes but not always apparent. It should be understood that different dynamics will satisfy the conditions of the ergodic theorem over very different time scales. The equivalence of results of an average performed using two distinct dynamics is only guaranteed if they are both simulated for long enough so that each satisfies the ergodic theorem.

Our discussion here also gives some additional insights into the conditions under which the ergodic theorem applies to the actual dynamics of physical systems. We note that any proof of the applicability of the ergodic theorem to a real system requires considering the actual dynamics rather than a model stochastic process. When the ergodic theorem does not apply to the actual dynamics, then the use of a Monte Carlo simulation for performing an average must be considered carefully. It will not give the same results if it satisfies the ergodic theorem while the real system does not.

We are still faced with the task of selecting values for the transition probabilities $P_s(s'|s'')$ that satisfy the three requirements given above. We can simplify our search for transition probabilities $P_s(s'|s'')$ for use in Monte Carlo simulations by imposing the additional constraint of microscopic reversibility, also known as detailed balance:

$$P_s(s''|s')P_s(s';\infty) = P_s(s'|s'')\,P_s(s'';\infty) \tag{1.7.17}$$

This equation implies that the transition currents between two states of the system are equal and therefore cancel in the steady state, Eq. (1.7.15). It corresponds to true equilibrium, as would be present in a physical system. Detailed balance implies the steady-state condition, but is not required by it. Steady state can also include currents that do not change in time. We can prove that Eq. (1.7.17) implies Eq. (1.7.15) by summing over s':

$$\sum_{s'} P_s(s''|s')P_s(s';\infty) = \sum_{s'} P_s(s'|s'')P_s(s'';\infty) = P_s(s'';\infty) \quad (1.7.18)$$

We do not yet have an explicit prescription for $P_s(s'|s'')$. There is still a tremendous flexibility in determining the transition probabilities. One prescription that enables direct implementation, called Metropolis Monte Carlo, is:

$$P_s(s'|s'') = \lambda(s'|s'') \qquad\qquad P_s(s')/P_s(s'') \geq 1 \quad s'' \neq s'$$
$$P_s(s'|s'') = \lambda(s'|s'')P_s(s')/P_s(s'') \quad P_s(s')/P_s(s'') < 1 \quad s'' \neq s'$$
$$P_s(s''|s'') = 1 - \sum_{s' \neq s''} P_s(s'|s'') \quad (1.7.19)$$

These expressions specify the transition probability $P_s(s'|s'')$ in terms of a symmetric stochastic matrix $\lambda(s'|s'')$. $\lambda(s'|s'')$ is independent of the limiting equilibrium distribution. The constraint associated with the limiting distribution has been incorporated explicitly into Eq. (1.7.19). It satisfies detailed balance by direct substitution in Eq. (1.7.17), since for $P_s(s') \geq P_s(s'')$ (similarly for the opposite) we have

$$P_s(s''|s')P_s(s') = \lambda(s''|s')P_s(s') = \lambda(s'|s'')P_s(s')$$
$$= (\lambda(s'|s'')P_s(s')/P_s(s''))P_s(s'') = P_s(s'|s'')P_s(s'') \quad (1.7.20)$$

The symmetry of the matrix $\lambda(s'|s'')$ is essential to the proof of detailed balance. One must often be careful in the design of specific algorithms to ensure this property. It is also important to note that the limiting probability appears in Eq. (1.7.19) only in the form of a ratio $P_s(s')/P_s(s'')$ which can be given directly by the Boltzmann distribution.

To understand Metropolis Monte Carlo, it is helpful to describe a few examples. We first describe the movement of the system in terms of the underlying stochastic process specified by $\lambda(s'|s'')$, which is independent of the limiting distribution. This means that the limiting distribution of the underlying process is uniform over the whole space of possible states.

A standard way to choose the matrix $\lambda(s'|s'')$ is to set it to be constant for a few states s' that are near s''. For example, the simplest random walk is such a case, since it allows a probability of $1/2$ for the system to move to the right and to the left. If s is a continuous variable, we could choose a distance r_0 and allow the walker to take a step anywhere within the distance r_0 with equal probability. Both the discrete and continuous random walk have d-dimensional analogs or, for a system of interacting particles, N-dimensional analogs. When there is more than one dimension, we can choose to move in all dimensions simultaneously. Alternatively, we can choose to move in only one of the dimensions in each step. For an Ising model (Section 1.6), we could allow equal probability for any one of the spins to flip.

Once we have specified the underlying stochastic process, we generate the sequence of Monte Carlo steps by applying it. However, we must modify the probabilities according to Eq. (1.7.19). This takes the form of choosing a step, but sometimes rejecting it rather than taking it. When a step is rejected, the system does not change its state. This gives rise to the third equation in Eq. (1.7.19) where the system does not move. Specifically, we can implement the Monte Carlo process according to the following prescription:

1. Pick one of the possible moves allowed by the underlying process. The selection is random from all of the possible moves. This guarantees that we are selecting it with the underlying probability $\lambda(s'|s'')$.

2. Calculate the ratio of probabilities between the location we are going to, compared to the location we are coming from

$$P_s(s'') \, / \, P_s(s') = e^{-(E(s')-E(s''))/kT} \tag{1.7.21}$$

If this ratio of probabilities is greater than one, which means the energy is lower where we are going, the step is accepted. This gives the probability for the process to occur as $\lambda(s'|s'')$, which agrees with the first line of Eq. (1.7.19). However, if this ratio is less than one, we accept it with a probability given by the ratio. For example, if the ratio is 0.6, we accept the move 60% of the time. If the move is rejected, the system stays in its original location. Thus, if the energy where we are trying to go is higher, we do not accept it all the time, only some of the time. The likelihood that we accept it decreases the higher the energy is.

The Metropolis Monte Carlo prescription makes logical sense. It tends to move the system to regions of lower energy. This must be the case in order for the final distribution to satisfy the Boltzmann probability. However, it also allows the system to climb up in energy so that it can reach, with a lower probability, states of higher energy. The ability to climb in energy also enables the system to get over barriers such as the one in the two-state system in Section 1.4.

For the Ising model, we can see that the Monte Carlo dynamics that uses all single spin flips as its underlying stochastic process is not the same as the Glauber dynamics (Section 1.6.7), but is similar. Both begin by selecting a particular spin. After selection of the spin, the Monte Carlo will set the spin to be the opposite with a probability:

$$\min(1, e^{-(E(1)-E(-1))\,/\,kT}) \tag{1.7.22}$$

This means that if the energy is lower for the spin to flip, it is flipped. If it is higher, it may still flip with the indicated probability. This is different from the Glauber prescription, which sets the selected spin to UP or DOWN according to its equilibrium probability (Eq. (1.6.61)–Eq. (1.6.63)). The difference between the two schemes can be shown by plotting the probability of a selected spin being UP as a function of the energy difference between UP and DOWN, $E_+ = E(1) - E(-1)$ (Fig. 1.7.2). The Glauber dynamics prescription is independent of the starting value of the spin. The Metropolis Monte Carlo prescription is not. The latter causes more changes, since the spin is more likely to flip. Unlike the Monte Carlo prescription, the Glauber dynamics explicitly requires knowledge of the probabilities themselves. For a single spin flip in an Ising system this is fine, because there are only two possible states and the probabilities depend only on E_+. However, this is difficult to generalize when a system has many more possible states.

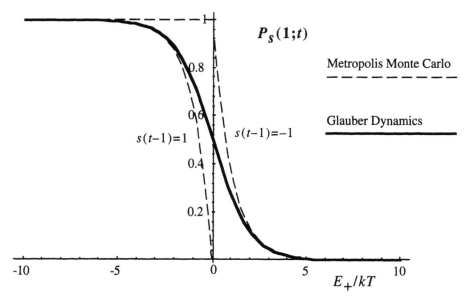

Figure 1.7.2 Illustration of the difference between Metropolis Monte Carlo and Glauber dynamics for the update of a spin in an Ising model. The plots show the probability $P_s(1;t)$ of a spin being UP at time t. The Glauber dynamics probability does not depend on the starting value of the spin. There are two curves for the Monte Carlo probability, for $s(t-1) = 1$ and $s(t-1) = -1$. ∎

There is a way to generalize further the use of Monte Carlo by recognizing that we do not even have to use the correct equilibrium probability distribution when generating the time series. The generalized expression for an arbitrary probability distribution $P'(s)$ is:

$$<f(s)>_{P(s)} = \int \frac{f(s)P(s)}{P'(s)} P'(s)ds = \frac{1}{N} \sum_{s:P'(s)}^{N} \frac{f(s)P(s)}{P'(s)} \qquad (1.7.23)$$

The subscript $P(s)$ indicates that the average assumes that s has the probability distribution $P(s)$ rather than $P'(s)$. This equation generalizes Eq. (1.7.5). The problem with this expression is that it requires that we know explicitly the probabilities $P(s)$ and $P'(s)$. This can be remedied. We illustrate for a specific case, where we use the Boltzmann distribution at one temperature to evaluate the average at another temperature:

$$<f(s)>_{P(s)} = \frac{1}{N} \sum_{s:P'(s)}^{N} \frac{f(s)P(s)}{P'(s)} = \left(\frac{Z'}{Z}\right) \frac{1}{N} \sum_{s:P'(s)}^{N} f(s)e^{-E(s)(1/kT - 1/kT')} \qquad (1.7.24)$$

The ratio of partition functions can be directly evaluated as an average:

$$\left(\frac{Z}{Z'}\right) = \frac{\sum_s e^{-E(s)/kT}}{\sum_s e^{-E(s)/kT'}} = \frac{\sum_s e^{-E(s)(1/kT-1/kT')}e^{-E(s)/kT'}}{\sum_s e^{-E(s)/kT'}}$$

$$= \left\langle e^{-E(s)(1/kT-1/kT')}\right\rangle_{P'(s)} = \frac{1}{N}\sum_{s:P'(s)}^{N} e^{-E(s)(1/kT-1/kT')} \quad (1.7.25)$$

Thus we have the expression:

$$<f(s)>_{P(s)} = \sum_{s:P'(s)}^{N} f(s)e^{-E(s)(1/kT-1/kT')} \Bigg/ \sum_{s:P'(s)}^{N} e^{-E(s)(1/kT-1/kT')} \quad (1.7.26)$$

This means that we can obtain the average at various temperatures using only a single Monte Carlo simulation. However, the whole point of using the ensemble average is to ensure that the average converges rapidly. This may not happen if the ensemble temperature T' is much different from the temperature T. On the other hand, there are circumstances where the function $f(s)$ may have an energy dependence that makes it better to perform the average using an ensemble that is not the equilibrium ensemble.

The approach of Monte Carlo simulations to the study of statistical averages ensures that we do not have to be concerned that the dynamics we are using for the system is a real dynamics. The result is the same for a broad class of artificial dynamics. The generality provides a great flexibility; however, this is also a limitation. We cannot use the Monte Carlo dynamics to study dynamics. We can only use it to perform statistical averages. Must we be resigned to this limitation? The answer, at least in part, is no. The reason is rooted in the central limit theorem. For example, the implementations of Metropolis Monte Carlo and the Glauber dynamics are quite different. We know that in the limit of long enough times, the distribution of configurations generated by both is the same. We expect that since each of them flips only one spin, if we are interested in changes in many spins, the two should give comparable results in the sense of the central limit theorem. This means that aside from an overall scale factor, the time evolution of the distribution of probabilities for long times is the same. Since we already know that the limiting distribution is the same in both cases, we are asserting that the approach to this limiting distribution, which is the long time dynamics, is the same.

The claim that for a large number of steps all dynamics is the same is not true about all possible Monte Carlo dynamics. If we allow all of the spins in an Ising model to change their values in one step of the underlying dynamics $\lambda(s'|s'')$, then this step would be equivalent to many steps in a dynamics that allows only one spin to flip at a time. In order for two different dynamics to give the same results, there are two types of constraints that are necessary. First, both must have similar kinds of allowed steps. Specifically, we define steps to the naturally proximate configurations as local moves. As long as the Monte Carlo allows only local moves, the long time dynamics should be the same. Such dynamics correspond to a local diffusion in the space of possible configurations of the system. More generally, two different dynamics should be the same if configuration changes that require many steps in one also require many steps in the other. The second type of constraint is related to symmetries of the problem. A lack of bias in the random walk was necessary to guarantee that the Gaussian distribution resulted from a generalized random walk in Section 1.2. For systems with more than one dimension, we must also ensure that there is no relative bias between motion in different directions.

We can think about Monte Carlo dynamics as diffusive dynamics of a system that interacts frequently with a reservoir. There are properties of more realistic dynamics that are not reproduced by such configuration Monte Carlo simulations. Correlations between steps are not incorporated because of the assumptions underlying Markov chains. This rules out ballistic motion, and exact or approximate momentum conservation. Momentum conservation can be included if both position and momentum are included as system coordinates. The method called Brownian dynamics incorporates both ballistic and diffusive dynamics in the same simulation. However, if correlations in the dynamics of a system have a shorter range than the motion we are interested in, momentum conservation may not matter to results that are of interest, and conventional Monte Carlo simulations can be used directly.

In summary, Monte Carlo simulations are designed to reproduce an ensemble rather than the dynamics of a particular system. As such, they are ideally suited to investigating the equilibrium properties of thermodynamic systems. However, Monte Carlo dynamics with local moves often mimic the dynamics of real systems. Thus, Monte Carlo simulations may be used to investigate the dynamics of systems when they are appropriately designed. This property will be used in Chapter 5 to simulate the dynamics of long polymers.

There is a flip side to the design of Monte Carlo dynamics to simulate actual dynamics. If our objective is the traditional objective of a Monte Carlo simulation, of obtaining an ensemble average, then the ability to simulate dynamics may not be an advantage. In some systems, the real dynamics is slow and we would prefer to speed up the process. This can often be done by knowingly introducing nonlocal moves that displace the state of the system by large distances in the space of conformations. Such nonlocal Monte Carlo dynamics have been designed for various systems. In particular, both local and nonlocal Monte Carlo dynamics for the problem of polymer dynamics will be described in Chapter 5.

Question 1.7.1 In order to perform Monte Carlo simulations, we must be able to choose steps at random and accept or reject steps with a certain probability. These operations require the availability of random numbers. We might think of the source of these random numbers as a thermal reservoir. Computers are specifically deigned to be completely deterministic. This means that inherently there is no randomness in their operation. To obtain random numbers in a computer simulation requires a deterministic algorithm that generates a sequence of numbers that look random but are not random. Such sequences are called pseudo-random numbers. Random numbers should not be correlated to each other. However, using pseudo-random numbers, if we start a program over again we must get exactly the same sequence of numbers. The difficulties associated with the generation of random numbers are central to performing Monte Carlo computer simulations. If we assume that we have random numbers, and they are not really uncorrelated, then our results may very well be incorrect. Nevertheless, pseudo-random numbers often give results that are consistent with those expected from random numbers.

There are a variety of techniques to generate pseudo-random numbers. Many of these pseudo-random number generators are designed to provide, with equal "probability," an integer between 0 and the maximal integer possible. The maximum integer used by a particular routine on a particular machine should be checked before using it in a simulation. Some use a standard short integer which is represented by 16 bits (2 bytes). One bit represents the unused sign of the integer. This leaves 15 bits for the magnitude of the number. The pseudo-random number thus ranges up to $2^{15} - 1 = 32767$. An example of a routine that provides pseudo-random integers is the subroutine rand() in the ANSI C library, which is executed using a line such as:

$$k = \text{rand}();   \hspace{2cm} (1.7.27)$$

The following three questions discuss how to use such a pseudo-random number generator. Assume that it provides a standard short integer.

1. Explain how to use a pseudo-random number generator to choose a move in a Metropolis Monte Carlo simulation, Eq. (1.7.19).

2. Explain how to use a pseudo-random number generator to accept or reject a move in a Metropolis Monte Carlo simulation, Eq. (1.7.19).

3. Explain how to use a pseudo-random number generator to provide values of x with a probability $P(x)$ for x in the interval $[0,1]$. Hint: Use two pseudo-random numbers every step.

Solution 1.7.1

1. Given the necessity of choosing one out of M possible moves, we create a one-to-one mapping between the M moves and the integers $\{0, \ldots, M-1\}$ If M is smaller than 2^{15} we can use the value of $k = \text{rand}()$ to determine which move is taken next. If k is larger than $M-1$, we don't make any move. If M is much smaller than 2^{15} then we can use only some of the bits of k. This avoids making many unused calls to rand(). Fewer bits can be obtained using a modulo operation. For example, if $M = 10$ we might use k modulo 16. We could also ignore values above 32759, and use k modulo 10. This also causes each move to occur with equal frequency. However, a standard word of caution about using only a few bits is that we shouldn't use the lowest order bits (i.e., the units, twos and fours bits), because they tend to be more correlated than the

higher order bits. Thus it may be best first to divide k by a small number, like 8 (or equivalently to shift the bits to the right), if it is desired to use fewer bits. If M is larger than 2^{15} it is necessary to use more than one call to rand() (or a random number generator that provides a 4-byte integer) so that all possible moves are accounted for.

2. Given the necessity of determining whether to accept a move with the probability P, we compare $2^{15} P$ with a number given by $k =$ rand(). If the former is bigger we accept the move, and if it is smaller we reject the move.

3. One way to do this is to generate two random numbers r_1 and r_2. Dividing both by 32767 (or 2^{15}), we use the first random number to be the location in the interval $x = r_1/32767$. However, we use this location only if the second random number $r_2/32767$ is smaller than $P(x)$. If the random number is not used, we generate two more and proceed. This means that we will use the position x with a probability $P(x)$ as desired. Because it is necessary to generate many random numbers that are rejected, this method for generating numbers for use in performing the integral Eq. (1.7.3) is only useful if evaluations of the function $f(x)$ are much more costly than random number generation. ∎

Question 1.7.2 To compare the errors that arise from conventional numerical integration and Monte Carlo sampling, we return to Eq. (1.7.4) and Eq. (1.7.5) in this and the following question. We choose two integrals that can be evaluated analytically and for which the errors can also be evaluated analytically.

Evaluate two examples of the integral $\int P(x)f(x)dx$ over the interval $x \in [1,1]$. For the first example (1) take $f(x) = 1$, and for the second (2) $f(x) = x$. In both cases assume the probability distribution is an exponential

$$P(x) = Ae^{-\lambda x} = \left[\frac{\lambda}{e^\lambda - e^{-\lambda}}\right]e^{-\lambda x} \qquad (1.7.28)$$

where the normalization constant A is given by the expression in square brackets.

Calculate the two integrals exactly (analytically). Then evaluate approximations to the integrals using sums over N equally spaced points, Eq. (1.7.4). These sums can also be evaluated analytically. To improve the result of the sum, you can use Simpson's rule. This modifies Eq. (1.7.4) only by subtracting 1/2 of the value of the integrand at the first and last points. The errors in evaluation of the same integral by Monte Carlo simulation are to be calculated in Question 1.7.3.

Solution 1.7.2

1. The value of the integral of $P(x)$ is unity as required by normalization. If we use a sum over equally spaced points we would have:

$$A\int_{-1}^{1}dxe^{-\lambda x} \approx \frac{A}{M}\sum_{n=-M}^{M}e^{-\lambda(n/M)} = \frac{A}{M}\sum_{n=-M}^{M}a^{n} \qquad (1.7.29)$$

where we used the temporary definition $a = e^{-\lambda/M}$ to obtain

$$A\int_{-1}^{1}dxe^{-\lambda x} \approx \frac{A}{M}\frac{(a^{M+1}-a^{-M})}{a-1} = \frac{A}{M}\frac{(e^{\lambda}e^{\lambda/M}-e^{-\lambda})}{e^{\lambda/M}-1} \qquad (1.7.30)$$

Expanding the answer in powers of λ/M gives:

$$A\int_{-1}^{1}dxe^{-\lambda x} \approx A\frac{(e^{\lambda}-e^{-\lambda})}{\lambda} + A\frac{(e^{\lambda}+e^{-\lambda})}{2M} + A\frac{\lambda(e^{\lambda}-e^{-\lambda})}{2M^{2}} + ...$$

$$(1.7.31)$$

$$= 1 + \frac{\lambda}{2M}\tanh(\lambda) + \frac{\lambda^{2}}{2M^{2}} + ...$$

The second term can be eliminated by noting that the sum could be evaluated using Simpson's rule by subtracting 1/2 of the contribution of the end points. Then the third term gives an error of $\lambda^{2}/2M^{2}$. This is the error in the numerical approximation to the average of $f(x) = 1$.

2. For $f(x) = x$ the exact integral is:

$$A\int_{-1}^{1}dx\,xe^{-\lambda x} = -A\frac{d}{d\lambda}\int_{-1}^{1}dxe^{-\lambda x} = -A\frac{d}{d\lambda}\left[\frac{e^{\lambda}-e^{-\lambda}}{\lambda}\right] \qquad (1.7.32)$$

$$= -\coth(\lambda) + (1/\lambda)$$

while the sum is:

$$A\int_{-1}^{1}dx\,xe^{-\lambda x} \approx \frac{A}{M^{2}}\sum_{n=-M}^{M}ne^{-\lambda(n/M)} = \frac{A}{M^{2}}\sum_{n=-M}^{M}na^{n}$$

$$= \frac{A}{M^{2}}a\frac{d}{da}\sum_{n=-M}^{M}a^{n} = \frac{A}{M^{2}}a\frac{d}{da}\frac{(a^{M+1}-a^{-M})}{a-1}$$

$$(1.7.33)$$

$$= \frac{A}{M^{2}}\frac{((M+1)a^{M+1}+Ma^{-M})}{a-1} + \frac{A}{M^{2}}\frac{a(a^{M+1}-a^{-M})}{(a-1)^{2}}$$

$$= \frac{A}{M}\frac{(e^{\lambda}e^{\lambda/M}+e^{-\lambda})}{e^{\lambda/M}-1} + \frac{A}{M^{2}}\frac{(e^{\lambda}e^{\lambda/M})}{e^{\lambda/M}-1} + \frac{A}{M^{2}}\frac{e^{\lambda/M}(e^{\lambda}e^{\lambda/M}-e^{-\lambda})}{(e^{\lambda/M}-1)^{2}}$$

With some assistance from Mathematica, the expansion to second order in λ/M is:

$$= -A\frac{(e^{\lambda}-e^{-\lambda})}{\lambda^{2}} + A\frac{(e^{\lambda}+e^{-\lambda})}{\lambda} + \frac{A}{M}\frac{(e^{\lambda}-e^{-\lambda})}{2} + \frac{A}{M^{2}}\frac{11}{12}(e^{\lambda}-e^{-\lambda}) + ...$$

$$= -1/\lambda + \coth(\lambda) + \frac{\lambda}{2M} + \frac{11}{12}\frac{\lambda}{M^{2}} + ... \qquad (1.7.34)$$

The first two terms are the correct result. The third term can be seen to be eliminated using Simpson's rule. The fourth term is the error. ∎

Question 1.7.3 Estimate the errors in performing the same integrals as in Question 1.7.2 using a Monte Carlo ensemble sampling with N terms as in Eq. (1.7.5). It is not necessary to evaluate the integrals to evaluate the errors.

Solution 1.7.3

1. The errors in performing the integral for $f(x) = 1$ are zero, since the Monte Carlo sampling would be given by the expression:

$$<1>_{P(s)} = \frac{1}{N} \sum_{s:P(s)}^{N} 1 = 1 \qquad (1.7.35)$$

One way to think about this result is that Monte Carlo takes advantage of the normalization of the probability, which the technique of summing the integrand over equally spaced points cannot do. This knowledge makes this integral trivial, but it is also of use in performing other integrals.

2. To evaluate the error for the integral over $f(x) = x$ we use an argument based on the sampling error of different regions of the integral. We break up the domain $[-1,1]$ into q regions of size $\Delta x = 2/q$. Each region is assumed to have a significant number of samples. The number of these samples is approximately given by:

$$NP(x)\Delta x \qquad (1.7.36)$$

If this were the exact number of samples as q increased, then the integral would be exact. However, since we are picking the points at random, there will be a deviation in the number of these from this ideal value. The typical deviation, according to the discussion in Section 1.2 of random walks, is the square root of this number. Thus the error in the sum

$$\sum_{s:P(s)}^{N} f(x) \qquad (1.7.37)$$

from a particular interval Δx is

$$(NP(x)\Delta x)^{1/2} f(x) \qquad (1.7.38)$$

Since this error could have either a positive or negative sign, we must take the square root of the sum of the squares of the error in each region to give us the total error:

$$\left| \int P(x)f(x) - \frac{1}{N} \sum_{s:P(s)}^{N} f(x) \right| \approx \frac{1}{N} \sqrt{\sum NP(x)\Delta x \, f(x)^2} \approx \frac{1}{\sqrt{N}} \sqrt{\int P(x)f(x)^2}$$

$$(1.7.39)$$

For $f(x) = x$ the integral in the square root is:

$$\int Ae^{-\lambda x} f(x)^2 \, dx = \int Ae^{-\lambda x} x^2 \, dx = A \frac{d^2}{d\lambda^2} \frac{(e^{\lambda} - e^{-\lambda})}{\lambda} = \frac{2}{\lambda^2} - \frac{2\coth(\lambda)}{\lambda} + 1$$

(1.7.40)

The approach of Monte Carlo is useful when the exponential is rapidly decaying. In this case, $\lambda \gg 1$, and we keep only the third term and have an error that is just of magnitude $1/\sqrt{N}$. Comparing with the sum over equally spaced points from Question 1.7.2, we see that the error in Monte Carlo is independent of λ for large λ, while it grows for the sum over equally spaced points. This is the crucial advantage of the Monte Carlo method. However, for a fixed value of λ we also see that the error is more slowly decreasing with N than the sum over equally spaced points. So when a large number of samples is possible, the sum over equally spaced points is more rapidly convergent. ∎

Question 1.7.4 How would the discrete nature of the integer random numbers described in Question 1.7.1 affect the ensemble sampling? Answer qualitatively. Is there a limit to the accuracy of the integral in this case?

Solution 1.7.4 The integer random numbers introduce two additional sources of error, one due to the sampling interval along the x axis and the other due to the imperfect approximation of $P(x)$. In the limit of a large number of samples, each of the possible values along the x axis would be sampled equally. Thus, the ensemble sum would reduce to a sum of the integrand over equally spaced points. The number of points is given by the largest integer used (e.g., 2^{15}). This limits the accuracy accordingly. ∎

1.7.3 *Perron-Frobenius theorem*

The Perron-Frobenius theorem is tied to our understanding of the ergodic theorem and the use of Monte Carlo simulations for the representation of ensemble averages. The theorem only applies to a system with a finite space of possible states. It says that a transition matrix that is irreducible must ultimately lead to a stable limiting probability distribution. This distribution is unique, and thus depends only on the transition matrix and not on the initial conditions. The Perron-Frobenius theorem assumes an irreducible matrix, so that starting from any state, there is some path by which it is possible to reach every other state of the system. If this is not the case, then the theorem can be applied to each subset of states whose transition matrix is irreducible.

In a more general form than we will discuss, the Perron-Frobenius theorem deals with the effect of matrix multiplication when all of the elements of a matrix are positive. We will consider it only for the case of a transition matrix in a Markov chain, which also satisfies the normalization condition, Eq. (1.7.16). In this case, the proof of the Perron-Frobenius theorem follows from the statement that there cannot be any eigenvalues of the transition matrix that are larger than one. Otherwise there would be a vector that would increase everywhere upon matrix multiplication. This is not possible, because probability is conserved. Thus if the probability increases in one place it must decrease someplace else, and tend toward the limiting distribution.

A difficulty in the proof of the theorem arises from dealing with the case in which there are deterministic currents through the system: e.g., ballistic motion in a circular path. An example for a two-state system would be

$$P(1|1) = 0 \quad P(1|-1) = 1$$
$$P(-1|1) = 1 \quad P(-1|-1) = 0 \tag{1.7.41}$$

In this case, a system in the state $s = +1$, goes into $s = -1$, and a system in the state $s = -1$ goes into $s = +1$. The limiting behavior of this Markov chain is of two probabilities that alternate in position without ever settling down into a limiting distribution. An example with three states would be

$$P(1|1) = 0 \quad P(1|2) = 1 \quad P(1|3) = 1$$
$$P(2|1) = .5 \quad P(2|2) = 0 \quad P(2|3) = 0 \tag{1.7.42}$$
$$P(3|1) = .5 \quad P(3|2) = 0 \quad P(3|3) = 0$$

Half of the systems with $s = 1$ make transitions to $s = 2$ and half to $s = 3$. All systems with $s = 2$ and $s = 3$ make transitions to $s = 1$. In this case there is also a cyclical behavior that does not disappear over time. These examples are special cases, and the proof shows that they are special. It is sufficient, for example, for there to be a single state where there is some possibility of staying in the same state. Once this is true, these examples of cyclic currents do not apply and the system will settle down into a limiting distribution.

We will prove the Perron-Frobenius theorem in a few steps enumerated below. The proof is provided for completeness and reference, and can be skipped without significant loss for the purposes of this book. The proof relies upon properties of the eigenvectors and eigenvalues of the transition matrix. The eigenvectors need not always be positive, real or satisfy the normalization condition that is usually applied to probability distributions, $P(s)$. Thus we use $v(s)$ to indicate complex vectors that have a value at every possible state of the system.

Given an irreducible real nonnegative matrix $(P(s'|s) \geq 0)$ satisfying

$$\sum_{s'} P(s'|s) = 1 \tag{1.7.43}$$

we have:

1. Applying $P(s'|s)$ cannot increase the value of all elements of a nonnegative vector, $v(s') \geq 0$:

$$\min_{s'} \left(\frac{1}{v(s')} \sum_{s} P(s'|s)v(s) \right) \leq 1 \tag{1.7.44}$$

To avoid infinities, we can assume that the minimization only includes s' such that $v(s') \neq 0$.

Proof: Assume that Eq. (1.7.44) is not true. In this case

$$\sum_s P(s'|s)v(s) > v(s')$$ (1.7.45)

for all $v(s') \neq 0$, which implies

$$\sum_{s'} \sum_s P(s'|s)v(s) > \sum_{s'} v(s')$$ (1.7.46)

Using Eq. (1.7.43), the left is the same as the right and the inequality is impossible.

2. The magnitude of eigenvalues of $P(s'|s)$ is not greater than one.

Proof: Let $v(s)$ be an eigenvector of $P(s'|s)$ with eigenvalue λ:

$$\sum_s P(s'|s)v(s) = \lambda v(s')$$ (1.7.47)

Then:

$$\sum_s P(s'|s)|v(s)| \geq |\lambda||v(s')|$$ (1.7.48)

This inequality follows because each term in the sum on the left has been made positive. If all terms started with the same phase, then equality holds. Otherwise, inequality holds. Comparing Eq. (1.7.48) with Eq. (1.7.44), we see that $|\lambda| \leq 1$.

If $|\lambda| = 1$, then equality must hold in Eq. (1.7.48), and this implies that $|v(s)|$, the vector whose elements are the magnitudes of $v(s)$, is an eigenvector with eigenvalue 1. Steps 3–5 show that there is one such vector which is strictly positive (greater than zero) everywhere.

3. $P(s'|s)$ has an eigenvector with eigenvalue $\lambda = 1$. We use the notation $v_1(s)$ for this vector.

Proof: The existence of such an eigenvector follows from the existence of an eigenvector of the transpose matrix with eigenvalue $\lambda = 1$. Eq. (1.7.43) implies that the vector $v(s) = 1$ (one everywhere) is an eigenvector of the transpose matrix with eigenvalue $\lambda = 1$. Thus $v_1(s)$ exists, and by step 2 we can take it to be real and nonnegative, $v_1(s) \geq 0$. We can, however, assume more, as the following shows.

4. An eigenvector of $P(s'|s)$ with eigenvalue 1 must be strictly positive, $v_1(s) > 0$.

Proof: Define a new Markov chain given by the transition matrix

$$Q(s'|s) = (P(s'|s) + \delta_{s,s'}) / 2$$ (1.7.49)

Applying $Q(s'|s)$ $N-1$ times to any vector $v_1(s) \geq 0$ must yield a vector that is strictly positive. This follows because $P(s'|s)$ is irreducible. Starting with unit probability at any one value of s, after $N-1$ steps we will move some probability everywhere. Also, by the construction of $Q(s'|s)$, any s which has a nonzero probability at one time will continue to have a nonzero probability at all later times. By linear superposition, this applies to any initial probability distribution. It also applies to any unnormalized vector $v_1(s) \geq 0$. Moreover, if $v_1(s)$ is an eigenvector of $P(s'|s)$ with eigenvalue one, then it is also an eigenvector of $Q(s'|s)$ with the same eigenvalue. Since applying $Q(s'|s)$ to $v_1(s)$ changes nothing, applying it $N - 1$ times also changes nothing. We have just proven that $v_1(s)$ must be strictly positive.

5. There is only one linearly independent eigenvector of $P(s'|s)$ with eigenvalue $\lambda = 1$.

Proof: Assume there are two such eigenvectors: $v_1(s)$ and $v_2(s)$. Then we can make a linear combination $c_1 v_1(s) + c_2 v_2(s)$, so that at least one of the elements is zero and others are positive. This linear combination is also an eigenvector of $P(s'|s)$ with eigenvalue $\lambda = 1$, which violates step 4. Thus there is exactly one eigenvector of $P(s'|s)$ with eigenvalue $\lambda = 1$, $v_1(s)$:

$$\sum_s P(s'|s)v_1(s) = v_1(s') \tag{1.7.50}$$

6. Either $P(s'|s)$ has only one eigenvalue with $|\lambda| = 1$ (in which case $\lambda = 1$), or it can be written as a cyclical flow.

Proof: Steps 2 and 5 imply that all eigenvectors of $P(s'|s)$ with eigenvalues λ satisfying $|\lambda| = 1$ can be written as:

$$v_i(s) = D_i(s)v_1(s) = e^{i\phi_i(s)}v_1(s) \tag{1.7.51}$$

As indicated, $D_i(s)$ is a vector with elements of magnitude one, $|D_i(s)| = 1$. We can write

$$\sum_s P(s'|s)D_i(s)v_1(s) = \lambda_i D_i(s')v_1(s') \tag{1.7.52}$$

There cannot be any terms in the sum on the left of Eq. (1.7.52) that add terms of different phase. If there were, then we would have a smaller magnitude than adding the absolute values, which would not agree with Eq. (1.7.50). Thus we can assign all of the elements of $D_i(s)$ into groups that have the same phase. $P(s'|s)$ cannot allow transitions to occur from any two of these groups into the same group. Since $P(s'|s)$ is irreducible, the only remaining possibility is that the different groups are connected in a ring with the first mapped onto the second, and the second mapped onto the third, and so on until we return to the first group. In particular, if there are any transitions between a site and itself this would violate the requirements and we could have no complex eigenvalues.

7. A Markov chain governed by an irreducible transition matrix, which has only one eigenvector, $v_1(s)$ with $|\lambda| = 1$, has a limiting distribution over long enough times which is proportional to this eigenvector. Using $P^t(s'|s)$ to represent the effect of applying $P(s'|s)$ t times, we must prove that:

$$\lim_{t \to \infty} v(s;t) = \lim_{t \to \infty} \sum_s P^t(s'|s)v(s) = cv_1(s') \tag{1.7.53}$$

for $v(s) \geq 0$. The coefficient c depends on the normalization of $v(s)$ and $v_1(s)$. If both are normalized so that the total probability is one, then conservation of probability implies that $c = 1$.

Proof: We write the matrix $P(s'|s)$ in the Jordan normal form using a similarity transformation. In matrix notation:

$$\mathbf{P} = \mathbf{S}^{-1}\mathbf{J}\mathbf{S} \tag{1.7.54}$$

J consists of a block diagonal matrix. Each of the block matrices along the diagonal is of the form

$$\mathbf{N} = \begin{pmatrix} \lambda & 1 & 0 & 0 \\ 0 & \lambda & \ddots & 0 \\ 0 & 0 & \ddots & 1 \\ 0 & 0 & 0 & \lambda \end{pmatrix} \tag{1.7.55}$$

where λ is an eigenvalue of **P**. In this block the only nonzero elements are λs on the diagonal, and 1s just above the diagonal.

Since $\mathbf{P}^t = \mathbf{S}^{-1}\mathbf{J}^t\mathbf{S}$, we consider $\mathbf{J}^t$, which consists of diagonal blocks $\mathbf{N}^t$. We prove that $\mathbf{N}^t \to 0$ as $t \to \infty$ for $|\lambda| < 1$. This can be shown by evaluating explicitly the matrix elements. The qth element above the diagonal of $\mathbf{N}^t$ is:

$$\lambda^{t-q} \binom{t}{q} \tag{1.7.56}$$

which vanishes as $t \to \infty$.

Since 1 is an eigenvalue with only one eigenvector, there must be one 1×1 block along the diagonal of **J** for the eigenvalue 1. Then $\mathbf{J}^t$ as $t \to \infty$ has only one nonzero element which is a 1 on the diagonal. Eq. (1.7.53) follows, because applying the matrix $\mathbf{P}^t$ always results in the unique column of $\mathbf{S}^{-1}$ that corresponds to the nonzero diagonal element of $\mathbf{J}^t$. By our assumptions, this column must be proportional to $v_1(s)$. This completes our proof and discussion of the Perron-Frobenius theorem.

1.7.4 Minimization

At low temperatures, a thermodynamic system in equilibrium will be found in its minimum energy configuration. For this and other reasons, it is often useful to identify the minimum energy configuration of a system without describing the full ensemble. There are also many other problems that can be formulated as minimization or optimization problems.

Minimization problems are often described in a d-dimensional space of continuous variables. When there is only a single valley in the parameter space of the problem, there are a variety of techniques that can be used to obtain this minimum. They may be classified into direct search and gradient-based techniques. In this section we focus on the single-valley problem. In Section 1.7.5 we will discuss what happens when there is more than one valley.

Direct search techniques involve evaluating the energy at various locations and closing in on the minimum energy. In one dimension, search techniques can be very effective. The key to a search is bracketing the minimum energy. Then each energy evaluation is used to geometrically shrink the possible domain of the minimum.

We start in one dimension by looking at the energy at two positions s_1 and s_2 that are near each other. If the left of the two positions s_1 is higher in energy $E(s_1) > E(s_2)$, then the minimum must be to its right. This follows from our assumption that there is only a single valley—the energy rises monotonically away from the minimum and therefore cannot be lower than $E(s_2)$, anywhere to the left of s_1. Evaluating the energy at a third location s_3 to the right of s_2 further restricts the possible locations of the minimum. If $E(s_3)$ is also greater than the middle energy location $E(s_3) > E(s_2)$, then the minimum must lie between s_1 and s_3. Thus, we have successfully bracketed the minimum. Otherwise, we have that $E(s_3) < E(s_2)$, and the minimum must lie to the right of s_2. In this case we look at the energy at a location s_4 to the right of s_3. This process is continued until the energy minimum is bracketed. To avoid taking many steps to the right, the size of the steps to the right can be taken to be an increasing geometric series, or may be based on an extrapolation of the function using the values that are available.

Once the energy minimum is bracketed, the segment is bisected again and again to locate the energy minimum. This is an iterative process. We describe a simple version of this process that can be easily implemented. An iteration begins with three locations $s_1 < s_2 < s_3$. The values of the energy at these locations satisfy $E(s_1), E(s_3) > E(s_2)$. Thus the minimum is between s_1 and s_3. We choose a new location s_4, which in even steps is $s_4 = (s_1 + s_2) / 2$ and in odd steps is $s_4 = (s_2 + s_3)/2$. Then we eliminate either s_1 or s_3. The one that is eliminated is the one next to s_2 if $E(s_2) > E(s_4)$, or the one next to s_4 if $E(s_2) < E(s_4)$. The remaining three locations are relabled to be s_1, s_2, s_3 for the next step. Iterations stop when the distance between s_1 and s_3 is smaller than an error tolerance which is set in advance. More sophisticated versions of this algorithm use improved methods for selecting s_4 that accelerate the convergence.

In higher-dimension spaces, direct search can be used. However, mapping a multidimensional energy surface is much more difficult. Moreover, the exact logic that enables an energy minimum to be bracketed within a particular domain in one dimension is not possible in higher-dimension spaces. Thus, techniques that make use of a gradient of the function are typically used even if the gradient must be numerically evaluated. The most common gradient-based minimization techniques include steepest descent, second order and conjugate gradient.

Steepest descent techniques involve taking steps in the direction of the most rapid descent direction as determined by the gradient of the energy. This is the same as using a first-order expansion of the energy to determine the direction of motion toward lower energy. Illustrating first in one dimension, we start from a position s_1 and write the expansion as:

$$E(s) = E(s_1) + (s - s_1) \frac{dE(s)}{ds}\bigg|_{s_1} + O((s - s_1)^2) \tag{1.7.57}$$

We now take a step in the direction of the minimum by setting:

$$s_2 = s_1 - c\frac{dE(s)}{ds}\bigg|_{s_1} \qquad (1.7.58)$$

From the expansion we see that for small enough c, $E(s_2)$ must be smaller than $E(s_1)$. The problem is to carefully select c so that we do not go too far. If we go too far we may reach beyond the energy minimum and increase the energy. We also do not want to make such a small step that many steps will be needed to reach the minimum. We can think of the sequence of configurations we pick as a time sequence, and the process we use to pick the next location as an iterative map. Then the minimum energy configuration is a fixed point of the iterative map given by Eq. (1.7.58). From a point near to the minimum we can have all of the behaviors described in Section 1.1—stable (converging) and unstable (diverging), both of these with or without alternation from side to side of the minimum. Of particular relevance is the discussion in Question 1.1.12 that suggests how c may be chosen to stabilize the iterative map and obtain rapid convergence.

When s is a multidimensional variable, Eq. (1.7.57) and Eq. (1.7.58) both continue to apply as long as the derivative is replaced by the gradient:

$$E(s) = E(s_1) + (s - s_1)\cdot \nabla_s E(s)\big|_{s_1} + O((s - s_1)^2) \qquad (1.7.59)$$

$$s_2 = s_1 - c\nabla_s E(s)\big|_{s_1} \qquad (1.7.60)$$

Since the direction opposite to the gradient is the direction in which the energy decreases most rapidly, this is known as a steepest descent technique. For the multidimensional case it is more difficult to choose a consistent value of c, since the behavior of the function may not be the same in different directions. The value of c may be chosen "on the fly" by making sure that the new energy is smaller than the old. If the current value of c gives a value $E(s_2)$ which is larger than $E(s_1)$ then c is reduced. We can improve upon this by looking along the direction of the gradient and considering the energy to be a function of c:

$$E(s_1 - c\nabla_s E(s)\big|_{s_1}) \qquad (1.7.61)$$

Then c can be chosen by finding the actual minimum in this direction using the search technique that works well in one dimension.

Gradient techniques work well when different directions in the energy have the same behavior in the vicinity of the minimum energy. This means that the second derivative in different directions is approximately the same. If the second derivatives are very different in different directions, then the gradient technique tends to bounce back and forth perpendicular to the direction in which the second derivative is very small, without making much progress toward the minimum (Fig. 1.7.3). Improvements of the gradient technique fall into two classes. One class of techniques makes direct use of the second derivatives, the other does not. If we expand the energy to second order at the present best guess for the minimum energy location s_1 we have

$$E(s) = E(s_1) + (s - s_1)\cdot \nabla_s E(s)\big|_{s_1} + (s - s_1)\cdot \vec{\nabla}_s\vec{\nabla}_s E(s)\big|_{s_1} \cdot (s - s_1) + O((s - s_1)^3) \qquad (1.7.62)$$

Figure 1.7.3 Illustration of the difficulties in finding a minimum energy by steepest descent when the second derivative is very different in different directions. The steps tend to oscillate and do not make progress toward the minimum along the flat direction. ∎

Setting the gradient of this expression to zero gives the next approximation for the minimum energy location s_2 as:

$$s_2 = s_1 - \frac{1}{2}\left(\vec{\nabla}_s\vec{\nabla}_s E(s)\Big|_{s_1}\right)^{-1}\cdot\nabla_s E(s)\Big|_{s_1} \tag{1.7.63}$$

This, in effect, gives a better description of the value of c for Eq. 1.7.60, which turns out to be a matrix inversely related to the second-order derivatives. Steps are large in directions in which the second derivative is small. If the second derivatives are not easily available, approximate second derivatives are used that may be improved upon as the minimization is being performed. Because of the need to evaluate the matrix of second-order derivatives and invert the matrix, this approach is not often convenient. In addition, the use of second derivatives assumes that the expansion is valid all the way to the minimum energy. For many minimization problems, this is not valid enough to be a useful approximation. Fortunately, there is a second approach called the conjugate gradient technique that often works as well and sometimes better.

Conjugate gradient techniques make use of the gradient but are designed to avoid the difficulties associated with long narrow wells where the steepest descent techniques result in oscillations. This is done by starting from a steepest descent in the first step of the minimization. In the second step, the displacement is taken to be along a direction that does not include the direction taken in the first step. Explicitly, let v_i be the direction taken in the ith step, then the first two directions would be:

$$v_1 = -\nabla_s E(s)\Big|_{s_1}$$

$$v_2 = -\nabla_s E(s)\Big|_{s_2} + v_1 \frac{(v_1 \cdot \nabla_s E(s)\Big|_{s_2})}{v_1 \cdot v_1} \tag{1.7.64}$$

This ensures that v_2 is orthogonal to v_1. Subsequent directions are made orthogonal to some number of previous steps. The use of orthogonal directions avoids much of the problem of bouncing back and forth in the energy well.

Monte Carlo simulation can also be used to find minimum energy configurations if the simulations are done at zero temperature. A zero temperature Monte Carlo means that the steps taken always reduce the energy of the system. This approach works not only for continuous variables, but also for the discrete variables like in the Ising model. For the Ising model, the zero temperature Monte Carlo described above and the zero temperature Glauber dynamics are the same. Every selected spin is placed in its low energy orientation—aligned with the local effective field.

None of these techniques are suited to finding the minimum energy configuration if there are multiple energy minima, and we do not know if we are located near the correct minimum energy location. One way to address this problem is to start from various initial configurations and to look for the local minimum nearby. By doing this many times it might be possible to identify the global minimum energy. This works when there are only a few different energy minima. There are no techniques that guarantee finding the global minimum energy for an arbitrary energy function $E(s)$. However, by using Monte Carlo simulations that are not at $T = 0$, a systematic approach called simulated annealing has been developed to try to identify the global minimum.

1.7.5 *Simulated annealing*

Simulated annealing was introduced relatively recently as an approach to finding the global minimum when the energy or other optimization function contains many local minima. The approach is based on the physical process of heating a system and cooling it down slowly. The minimum energy for many simple materials is a crystal. If a material is heated to a liquid or vapor phase and cooled rapidly, the material does not crystallize. It solidifies as a glass or amorphous solid. On the other hand, if it is cooled slowly, crystals may form. If the material is formed out of several different kinds of atoms, the cooling may also result in phase separation into particular compounds or atomic solids. The separated compounds are lower in energy than a rapidly cooled mixture.

Simulated annealing works in much the same way. A Monte Carlo simulation is started at a high temperature. Then the temperature is lowered according to a cooling schedule until the temperature is so low that no additional movements are likely. If the procedure is effective, the final energy should be the lowest energy of the simulation. We could also keep track of the energy during the simulation and take the lowest value, and the configuration at which the lowest value was reached.

In general, simulated annealing improves upon methods that find only a local minimum energy, such as steepest descent, discussed in the previous section. For some problems, the improvement is substantial. Even if the minimum energy that is found is not the absolute minimum in energy of the system, it may be close. For example, in problems where there are many configurations that have roughly the same low energy, simulated annealing may find one of the low-energy configurations.

However, simulated annealing does not work well for all problems, and for some problems it fails completely. It is also true that annealing of physical materials does not always result in the lowest energy conformation. Many materials, even when cooled slowly, result in polycrystalline materials, disordered solids and mixtures. When it is important for technological reasons to reach the lowest energy state, special techniques are often used. For example, the best crystal we know how to make is silicon. In order to form a good silicon crystal, it is grown using careful nonuniform cooling. A single crystal can be gradually pulled from a liquid that solidifies only on the surfaces of the existing crystal. Another technique for forming crystals is growth from the vapor phase, where atoms are deposited on a previously formed crystal that serves as a template for the continuing growth. The difficulties inherent in obtaining materials in their lowest energy state are also apparent in simulations.

In Section 1.4 we considered the cooling of a two-state system as a model of a glass transition. We can think about this simulation to give us clues about why both physical and simulated annealing sometimes fail to find low energy states of the system. We saw that using a constant cooling rate leaves some systems stuck in the higher energy well. When there are many such high energy wells then the system will not be successful in finding a low energy state. The problem becomes more difficult if the height of the energy barrier between the two wells is much larger than the energy difference between the upper and lower wells. In this case, at higher temperatures the system does not care which well it is in. At low temperatures when it would like to be in the lower energy well, it cannot overcome the barrier. How well the annealing works in finding a low energy state depends on whether we care about the energy scale characteristic of the barrier, or characteristic of the energy difference between the two minima.

There is another characteristic of the energy that can help or hurt the effectiveness of simulated annealing. Consider a system where there are many local minimum energy states (Fig. 1.7.4). We can think about the effect of high temperatures as placing the system in one of the many wells of the energy minima. These wells are called basins of attraction. A system in a particular basin of attraction will go into the minimum energy configuration of the basin if we suddenly cool to zero temperature. We

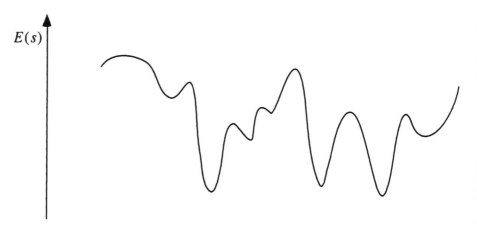

$E(s)$

Figure 1.7.4 Schematic plot of a system energy $E(s)$ as a function of a system coordinate s. In simulated annealing, the location of a minimum energy is sought by starting from a high temperature Monte Carlo and cooling the system to a low temperature. At the high temperature the system has a high kinetic energy and explores all of the possible configurations. As the temperature is cooled it descends into one of the wells, called basins of attraction, and cannot escape. Finally, when the temperature is very low it loses all kinetic energy and sits in the bottom of the well. Minima with larger basins of attraction are more likely to capture the system. Simulated annealing works best when the lowest-energy minima have the largest basins of attraction. ∎

also can see that the gradual cooling in simulated annealing will result in low energy states if the size of the basin of attraction increases with the depth of the well. This means that at high temperatures the system is more likely to be in the basin of attraction of a lower energy minimum. Thus, simulated annealing works best when energy varies in the space in such a way that deep energy minima also have large basins of attraction. This is sometimes but not always true both in physical systems and in mathematical optimization problems.

Another way to improve the performance of simulated annealing is to introduce nonlocal Monte Carlo steps. If we understand the characteristics of the energy, we can design steps that take us through energy barriers. The problem with this approach is that if we don't know the energy surface well enough, then moving around in the space by arbitrary nonlocal steps will result in attempts to move to locations where the energy is high. These steps will be rejected by the Monte Carlo and the nonlocal moves will not help. An example where nonlocal Monte Carlo moves can help is treatments of low-energy atomic configurations in solids. Nonlocal steps can allow atoms to move through each other, switching their relative positions, instead of trying to move gradually around each other.

Finally, for the success of simulated annealing, it is often necessary to design carefully the cooling schedule. Generally, the slower the cooling the more likely the simulation will end up in a low energy state. However, given a finite amount of computer and human time, it is impossible to allow an arbitrarily slow cooling. Often there are particular temperatures where the cooling rate is crucial. This happens at phase transitions, such as at the liquid-to-solid boundary. If we know of such a transition, then we can cool rapidly down to the transition, cool very slowly in its vicinity and then speed up thereafter. The most difficult problems are those where there are barriers of varying heights leading to a need to cool slowly at all temperatures.

For some problems the cooling rate should be slowed as the temperature becomes lower. One way to achieve this is to use a logarithmic cooling schedule. For example, we set the temperature $T(t)$ at time step t of the Monte Carlo, to be:

$$T(t) = T_0 / \ln(t / t_0 + 1) \qquad (1.7.65)$$

where t_0 and T_0 are parameters that must be chosen for the particular problem. In Question 1.7.5 we show that for the two-state system, if $kT_0 > (E_B - E_1)$, then the system will always relax into its ground state.

Question 1.7.5: Show that by using a logarithmic cooling schedule, Eq. (1.7.65), where $kT_0 > (E_B - E_1)$, the two-state system of Section 1.4 always relaxes into the ground state. To simplify the problem, consider an incremental time Δt during which the temperature is fixed. Show that the system will still relax to the equilibrium probability over this incremental time, even at low temperatures.

Solution 1.7.5: We write the solution of the time evolution during the incremental time Δt from Eq. (1.4.45) as:

$$P(1; t + \Delta t) - P(1; \infty) = (P(1; t) - P(1; \infty))e^{-t/\tau(t)} \qquad (1.7.66)$$

where $P(1;\infty)$ is the equilibrium value of the probability for the temperature $T(t)$. $\tau(t)$ is the relaxation time for the temperature $T(t)$. In order for relaxation to occur we must have that $e^{-t/\tau(t)} \ll 1$, equivalently:

$$t / \tau(t) \gg 1 \qquad (1.7.67)$$

We calculate $\tau(t)$ from Eq. (1.4.44):

$$1/\tau(t) = v(e^{-(E_B-E_1)/kT(t)} + e^{-(E_B-E_{-1})/kT(t)})$$
$$> ve^{-(E_B-E_1)/kT(t)} = v(t/t_0 + 1)^{-\gamma} \qquad (1.7.68)$$

where we have substituted Eq. (1.7.65) and defined $\gamma = (E_B - E_1)/kT_0$. We make the reasonable assumption that we start our annealing at a high temperature where relaxation is not a problem. Then by the time we get to the low temperatures that are of interest, $t \gg t_0$, so:

$$1/\tau(t) > 2v(t/t_0)^{-\gamma} \qquad (1.7.69)$$

and

$$t/\tau(t) > vt_0^\gamma t^{1-\gamma} \qquad (1.7.70)$$

For $\gamma < 1$ the right-hand side increases with time and thus the relaxation improves with time according to Eq. (1.7.67). If relaxation occurs at higher temperatures, it will continue to occur at all lower temperatures despite the increasing relaxation time. ∎

1.8   Information

Ultimately, our ability to quantify complexity (How complex is it?) requires a quantification of information (How much information does it take to describe it?). In this section, we discuss information. We will also need computation theory described in Section 1.9 to discuss complexity in Chapter 8. A quantitative theory of information was developed by Shannon to describe the problem of communication. Specifically, how much information can be communicated through a transmission channel (e.g., a telephone line) with a specified alphabet of letters and a rate at which letters can be transmitted. The simplest example is a binary alphabet consisting of two characters (digits) with a fixed rate of binary digits (bits) per second. However, the theory is general enough to describe quite arbitrary alphabets, letters of variable duration such as are involved in Morse code, or even continuous sound with a specified band-width. We will not consider many of the additional applications, our objective is to establish the basic concepts.

1.8.1 *The amount of information in a message*

We start by considering the information content of a string of digits $s = (s_1 s_2 ... s_N)$. One might naively expect that information is contained in the state of each digit. However, when we receive a digit, we not only receive information about what the digit is, but also what the digit is not. Let us assume that a digit in the string of digits we receive is the number 1. How much information does this provide? We can contrast two different scenarios—binary and hexadecimal digits:

1. There were two possibilities for the number, either 0 or 1.

2. There were sixteen possibilities for the number {0, 1, 2, 3, 4, 5, 6, 7, 8, 9, A, B, C, D, E, F}.

In which of these did the "1" communicate more information? Since the first case provides us with the information that it is "not 0," while the second provides us with the information that it is "not 0," "not 2," "not 3," etc., the second provides more information. Thus there is more information in a digit that can have sixteen states than a digit that can have only two states. We can quantify this difference if we consider a binary representation of hexadecimal digits {0000,0001,0010,0011,...,1111}. It takes four binary digits to represent one hexadecimal digit. The hexadecimal number 1 is represented as 0001 in binary form and uses four binary digits. Thus a hexadecimal 1 contains four times as much information as a binary 1.

We note that the amount of information does not depend on the particular value that is taken by the digit. For hexadecimal digits, consider the case of a digit that has the value 5. Is there any difference in the amount of information given by the 5 than if it were 1? No, either number contains the same amount of information.

This illustrates that information is actually contained in the distinction between the state of a digit compared to the other possible states the digit may have. In order to quantify the concept of information, we must specify the number of possible states. Counting states is precisely what we did when we defined the entropy of a system in Section 1.3. We will see that it makes sense to define the information content of a string in the same way as the entropy—the logarithm of the number of possible states of the string:

$$I(s) = \log_2(\Omega) \tag{1.8.1}$$

By convention, the information is defined using the logarithm base two. Thus, the information contained in a single binary digit which has two possible states is $\log_2(2) = 1$. More generally, the number of possible states in a string of N bits, with each bit taking one of two values (0 or 1) is 2^N. Thus the information in a string of N bits is (in what follows the function log( ) will be assumed to be base two):

$$I(s) = \log(2^N) = N \tag{1.8.2}$$

Eq. (1.8.2) says that each bit provides one unit of information. This is consistent with the intuition that the amount of information grows linearly with the length of the string. The logarithm is essential, because the number of possible states grows exponentially with the length of the string, while the information grows linearly.

It is important to recognize that the definition of information we have given assumes that each of the possible realizations of the string has equal a priori probability. We use the phrase a priori to emphasize that this refers to the probability prior to receipt of the string—once the string has arrived there is only one possibility.

To think about the role of probability we must discuss further the nature of the message that is being communicated. We construct a scenario involving a sender and a receiver of a message. In order to make sure that the recipient of the message could not have known the message in advance (so there is information to communicate), we assume that the sender of the information is sending the result of a random occurrence, like the flipping of a coin or the throwing of a die. To enable some additional flexibility, we assume that the random occurrence is the drawing of a ball from a bag. This enables us to construct messages that have different probabilities. To be specific, we assume there are ten balls in the bag numbered from 0 to 9. All of them are red except the ball marked 0, which is green. The person communicating the message only reports if the ball drawn from the bag is red (using the digit 1) or green (using the digit 0). The recipient of the message is assumed to know about the setup. If the recipient receives the number 0, he then knows exactly which ball was selected, and all that were not selected. However, if he receives a 1, this provides less information, because he only knows that one of nine was selected, not which one. We notice that the digit 1 is nine times as likely to occur as the digit 0. This suggests that a higher probability digit contains less information than a lower probability digit.

We generalize the definition of the information content of a string of digits to allow for the possibility that different strings have different probabilities. We assume that the string is one of an ensemble of possible messages, and we define the information as:

$$I(s) = -\log(P(s)) \qquad (1.8.3)$$

where $P(s)$ is the probability of the occurrence of the message s in the ensemble. Note that in the case of equal a priori probability $P(s) = 1/\Omega$, Eq. (1.8.3) reduces to Eq. (1.8.1). The use of probabilities in the definition of information makes sense in one of two cases: (1) The recipient knows the probabilities that represent the conventions of the transmission, or (2) A large number of independent messages are sent, and we are considering the information communicated by one of them. Then we can approximate the probability of a message by its proportion of appearance among the messages sent. We will discuss these points in greater detail later.

Question 1.8.1 Calculate the information, according to Eq. (1.8.3), that is provided by a single digit in the example given in the text of drawing red and green balls from a bag.

Solution 1.8.1 For the case of a 0, the information is the same as that of a decimal digit:

$$I(0) = -\log(1/10) \approx 3.32 \qquad (1.8.4)$$

For the case of a 1 the information is

$$I(0) = -\log(9/10) \approx 0.152 \qquad (1.8.5) \quad \blacksquare$$

We can specialize the definition of information in Eq. (1.8.3) to a message $s = (s_1 s_2 ... s_N)$ composed of individual characters (bits, hexadecimal characters, ASCII characters, decimals, etc.) that are completely independent of each other (for example, each corresponding to the result of a separate coin toss). This means that the total probability of the message is the product of the probability of each character, $P(s) = \prod_i P(s_i)$. Then the information content of the message is given by:

$$I(s) = -\sum_i \log(P(s_i)) \tag{1.8.6}$$

If all of the characters have equal probability and there are k possible characters in the alphabet, then $P(s_i) = 1/k$, and the information content is:

$$I(s) = N \log(k) \tag{1.8.7}$$

For the case of binary digits, this reduces to Eq. (1.8.2). For other cases like the hexadecimal case, $k = 16$, this continues to make sense: the information $I = 4N$ corresponds to the requirement of representing each hexadecimal digit with four bits. Note that the previous assumption of equal a priori probability for the whole string is stronger than the independence of the digits and implies it.

Question 1.8.2 Apply the definition of information content in Eq. (1.8.3) to each of the following cases. Assume messages consist of a total of N bits subject to the following constraints (aside for the constraints assume equal probabilities):

1. Every even bit is 1.
2. Every (odd, even) pair of bits is either 11 or 00.
3. Every eighth bit is a parity bit (the sum modulo 2 of the previous seven bits).

Solution 1.8.2: In each case, we first give an intuitive argument, and then we show that Eq. (1.8.3) or Eq. (1.8.6) give the same result.

1. The only information that is transferred is the state of the odd bits. This means that only half of the bits contain information. The total information is $N / 2$. To apply Eq. (1.8.6), we see that the even bits, which always have the value 1, have a probability $P(1) = 1$ which contributes no information. Note that we never have to consider the case $P(0) = 0$ for these bits, which is good, because by the formula it would give infinite information. The odd bits with equal probabilities, $P(1) = P(0) = 1/2$, give an information of one for either value received.

2. Every pair of bits contains only two possibilities, giving us the equivalent of one bit of information rather than two. This means that total information is $N/2$. To apply Eq. (1.8.6), we have to consider every (odd, even) pair of bits as a single character. These characters can never have the value 01 or 10, and they have the value 11 or 00 with probability $P(11) = P(00) = 1/2$, which gives the expected result. We will see later that there is another way to think about this example by using conditional probabilities.

3. The number of independent pieces of information is $7N / 8$. To see this from Eq. (1.8.6), we group each set of eight bits together and consider

them as a single character (a byte). There are only 2^7 different possibilities for each byte, and each one has equal probability according to our constraints and assumptions. This gives the desired result.

Note: Such representations are used to check for noise in transmission. If there is noise, the redundancy of the eighth bit provides additional information. The noise-dependent amount of additional information can also be quantified; however, we will not discuss it here. ∎

Question 1.8.3 Consider a transmission of English characters using an ASCII representation. ASCII characters are the conventional method for computer representation of English text including small and capital letters, numerals and punctuation. Discuss (do not evaluate for this question) how you would determine the information content of a message. We will evaluate the information content of English in a later question.

Solution 1.8.3 In ASCII, characters are represented using eight bits. Some of the possible combinations of bits are not used at all. Some are used very infrequently. One way to determine the information content of a message is to assume a model where each of the characters is independent. To calculate the information content using this assumption, we must find the probability of occurrence of each character in a sample text. Using these probabilities, the formula Eq. (1.8.6) could be applied. However, this assumes that the likelihood of occurrence of a character is independent of the preceding characters, which is not correct. ∎

Question 1.8.4: Assume that you know in advance that the number of ones in a long binary message is M. The total number of bits is N. What is the information content of the message? Is it similar to the information content of a message of N independent binary characters where the probability that any character is one is $P(1) = M/N$?

Solution 1.8.4: We count the number of possible messages with M ones and take the logarithm to obtain the information as

$$I = \log\left(\binom{N}{M}\right) = \log\left(\frac{N!}{M!(N-M)!}\right) \qquad (1.8.8)$$

We can show that this is almost the same as the information of a message of the same length with a particular probability of ones, $P(1) = M/N$, by use of the first two terms of Sterling's approximation Eq. (1.2.36). Assuming $1 \ll M \ll N$ (A correction to this would grow logarithmically with N and can be found using the additional terms in (Eq. (1.2.36)):

$$I \sim N(\log(N) - 1) - M(\log(M) - 1) - (N - M)(\log(N - M) - 1)$$
$$= -N[P(1)\log(P(1)) + (1 - P(1))\log(1 - P(1))] \qquad (1.8.9)$$

This is the information from a string of independent characters where $P(1) = M / N$. For such a string, the number of ones is approximately $NP(1)$ and the number of zeros $N(1 - P(1))$ (see also Question 1.8.7). ∎

1.8.2 *Characterizing sources of information*

The information content of a particular message is defined in terms of the probability that it, out of all possible messages, will be received. This means that we are characterizing not just a message but the source of the message. A direct characterization of the source is not the information of a particular message, but the average information over the ensemble of possible messages. For a set of possible messages with a given probability distribution $P(s)$ this is:

$$<I>=-\sum_{s} P(s)\log(P(s)) \tag{1.8.10}$$

If the messages are composed out of characters $s = (s_1 s_2 ... s_N)$, and each character is determined independently with a probability $P(s_i)$, then we can write the information content as:

$$<I>=-\sum_{s}\left(\prod_{i}P(s_i)\right)\log\left(\prod_{i'}P(s_{i'})\right)=-\sum_{s}\left(\prod_{i}P(s_i)\right)\sum_{i'}\log(P(s_{i'})) \tag{1.8.11}$$

We can move the factor in parenthesis inside the inner sum and interchange the order of the summations.

$$<I>=-\sum_{i'}\sum_{s}\left(\prod_{i}P(s_i)\right)\log(P(s_{i'}))=-\sum_{i'}\left(\sum_{\{s_i\}_{i\neq i'}}\prod_{i\neq i'}P(s_i)\right)\sum_{s_{i'}}P(s_{i'})\log(P(s_{i'})) \tag{1.8.12}$$

The latter expression results from recognizing that the sum over all possible states is a sum over all possible values of each of the letters. The sum and product can be interchanged:

$$\sum_{\{s_i\}_{i\neq i'}}\prod_{i\neq i'}P(s_i)=\prod_{i\neq i'}\sum_{s_i}P(s_i)=1 \tag{1.8.13}$$

giving the result:

$$<I>=-\sum_{i'}\sum_{s_{i'}}P(s_{i'})\log(P(s_{i'})) \tag{1.8.14}$$

This shows that the average information content of the whole message is the average information content of each character summed over the whole character string. If the characters have the same probability, this is just the average information content of an individual character times the number of characters. If all letters of the alphabet have the same probability, this reduces to Eq. (1.8.7).

The average information content of a binary variable is given by:

$$<I> = -(P(1)\log(P(1)) + P(0)\log(P(0))) \tag{1.8.15}$$

Aside from the use of a logarithm base two, this is the same as the entropy of a spin (Section 1.6) with two possible states $s = \pm 1$ (see Question 1.8.5). The maximum information content occurs when the probabilities are equal, and the information goes to zero when one of the two becomes one, and the other zero (see Fig. 1.8.1). The information reflects the uncertainty in, or the lack of advance knowledge about, the value received.

Question 1.8.5 Show that the expression for the entropy S given in Eq. (1.6.16) of a set of noninteracting binary spins is the same as the information content defined in Eq. (1.8.15) aside from a normalization constant $k \ln(2)$. Consider the binary notation $s_i = 0$ to be the same as $s_i = -1$ for the spins.

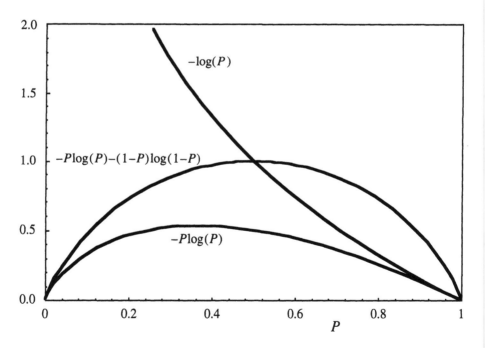

Figure 1.8.1 Plots of functions related to the information content of a message with probability P. $-\log(P)$ is the information content of a single message of probability P. $-P\log(P)$ is the contribution of this message to the average information given by the source. While the information content of a message diverges as P goes to zero, it appears less frequently so its contribution to the average information goes to zero. If there are only two possible messages, or two possible (binary) characters with probability P and $1 - P$ then the average information given by the source per message or per character is given by $-P\log(P) - (1 - P)\log(1 - P)$. ∎

Solution 1.8.5 The local magnetization m_i is the average value of a particular spin variable:

$$m_i = P_{s_i}(1) - P_{s_i}(-1) \qquad (1.8.16)$$

Using $P_{s_i}(1) + P_{s_i}(-1) = 1$ we have:

$$P_{s_i}(1) = (1 + m_i) / 2$$
$$P_{s_i}(-1) = (1 - m_i) / 2 \qquad (1.8.17)$$

Inserting these expressions into Eq. (1.8.15) and summing over a set of binary variables leads to the expression:

$$I = \left[N - \frac{1}{2} \sum_i \Big((1+m_i)\log(1+m_i) + (1-m_i)\log(1-m_i) \Big) \right] = S/k\ln(2)$$

$$(1.8.18)$$

The result is more general than this derivation suggests and will be discussed further in Chapter 8. ∎

Question 1.8.6 For a given set of possible messages, prove that the ensemble where all messages have equal probability provides the highest average information.

Solution 1.8.6 Since the sum over all probabilities is a fixed number (1), we consider what happens when we transfer some probability from one message to another. We start with the information given by

$$<I> = -P(s')\ln(P(s')) - P(s'')\ln(P(s'')) - \sum_{s \neq s', s''} P(s)\ln(P(s)) \qquad (1.8.19)$$

and after shifting a probability of δ from one to the other we have:

$$<I'> = -(P(s')-\delta)\ln(P(s')-\delta) - (P(s'')+\delta)\ln(P(s'')+\delta) - \sum_{s \neq s', s''} P(s)\ln(P(s))$$

$$(1.8.20)$$

We need to expand the change in information to first nonzero order in δ. We simplify the task by using the expression:

$$<I'> - <I> = f(P(s') + \delta) - f(P(s')) + f(P(s'') - \delta) - f(P(s'')) \qquad (1.8.21)$$

where

$$f(x) = -x\log(x) \qquad (1.8.22)$$

Taking a derivative, we have

$$\frac{d}{dx} f(x) = -(\log(x)+1) \qquad (1.8.23)$$

This gives the result:

$$<I'> - <I> = -(\log(P(s')) - \log(P(s'')))\delta \qquad (1.8.24)$$

Since $\log(x)$ is a monotonic increasing function, we see that the average information increases $((<I'>-<I>)>0)$ when probability $\delta>0$ is transferred from a higher-probability character to a lower-probability character $(P(s'')>P(s')\Rightarrow-(\log(P(s'))-\log(P(s'')))>0)$. Thus, any change of the probability toward a more uniform probability distribution increases the average information. ∎

Question 1.8.7 A source produces strings of characters of length N. Each character that appears in the string is independently selected from an alphabet of characters with probabilities $P(s_i)$. Write an expression for the probability $P(s)$ of a typical string of characters. Show that this expression implies that the string gives N times the average information content of an individual character. Does this mean that every string must give this amount of information?

Solution 1.8.7 For a long string, each character will appear $NP(s_i)$ times. The probability of such a string is:

$$P(s)=\prod_{s_i}P(s_i)^{NP(s_i)} \tag{1.8.25}$$

The information content is:

$$I(s)=-\log(P(s))=-N\sum_{s_i}P(s_i)\log(P(s_i)) \tag{1.8.26}$$

which is N times the average information of a single character. This is the information of a typical string. A particular string might have information significantly different from this. However, as the number of characters in the string increases, by the central limit theorem (Section 1.2), the fraction of times a particular character appears (i.e., the distance traveled in a random walk divided by the total number of steps) becomes more narrowly distributed around the expected probability $P(s_i)$. This means the proportion of messages whose information content differs from the typical value decreases with increasing message length. ∎

1.8.3 Correlations between characters

Thus far we have considered characters that are independent of each other. We can also consider characters whose values are correlated. We describe the case of two correlated characters. Because there are two characters, the notation must be more complete. As discussed in Section 1.2, we use the notation $P_{s_1,s_2}(s'_1,s'_2)$ to denote the probability that in the same string the character s_1 takes the value s'_1 and the variable s_2 takes the value s'_2. The average information contained in the two characters is given by:

$$<I_{s_1,s_2}>=-\sum_{s'_1,s'_2}P_{s_1,s_2}(s'_1,s'_2)\log(P_{s_1,s_2}(s'_1,s'_2)) \tag{1.8.27}$$

Note that the notation $I(s_1,s_2)$ is often used for this expression. We use $<I_{s_1,s_2}>$ because it is not a function of the values of the characters—it is the average information carried by the characters labeled by s_1 and s_2. We can compare the information content of the two characters with the information content of each character separately:

$$P_{s_1}(s_1') = \sum_{s_2'} P_{s_1,s_2}(s_1',s_2')$$

$$P_{s_2}(s_2') = \sum_{s_1'} P_{s_1,s_2}(s_1',s_2') \tag{1.8.28}$$

$$<I_{s_1}> = -\sum_{s_1',s_2'} P_{s_1,s_2}(s_1',s_2')\log(\sum_{s_2''} P_{s_1,s_2}(s_1',s_2''))$$

$$<I_{s_2}> = -\sum_{s_1',s_2'} P_{s_1,s_2}(s_1',s_2')\log(\sum_{s_1''} P_{s_1,s_2}(s_1'',s_2')) \tag{1.8.29}$$

It is possible to show (see Question 1.8.8) the inequalities:

$$<I_{s_2}> + <I_{s_1}> \ \geq \ <I_{s_1,s_2}> \ \geq \ <I_{s_2}>,<I_{s_1}> \tag{1.8.30}$$

The right inequality means that we receive more information from both characters than from either one separately. The left inequality means that information we receive from both characters together cannot exceed the sum of the information from each separately. It can be less if the characters are dependent on each other. In this case, receiving one character reduces the information given by the second.

The relationship between the information from a character s_1 and the information from the same character after we know another character s_2 can be investigated by defining a contingent or conditional probability:

$$P_{s_1,s_2}(s_1' \mid s_2') = \frac{P_{s_1,s_2}(s_1',s_2')}{\sum_{s_1''} P_{s_1,s_2}(s_1'',s_2')} \tag{1.8.31}$$

This is the probability that s_1 takes the value s_1' assuming that s_2 takes the value s_2'. We used this notation in Section 1.2 to describe the transitions from one value to the next in a chain of events (random walk). Here we are using it more generally. We could recover the previous meaning by writing the transition probability as $P_s(s_1' \mid s_2') = P_{s(t),s(t-1)}(s_1' \mid s_2')$. In this section we will be concerned with the more general definition, Eq. (1.8.31).

We can find the information content of the character s_1 when s_2 takes the value s_2'

$$<I_{s_1}>\big|_{s_2=s_2'} = -\sum_{s_1'} P_{s_1,s_2}(s_1' \mid s_2')\log(P_{s_1,s_2}(s_1' \mid s_2'))$$

$$= \frac{-\sum_{s_1'} P_{s_1,s_2}(s_1',s_2')\left(\log(P_{s_1,s_2}(s_1',s_2')) - \log(\sum_{s_1'''} P_{s_1,s_2}(s_1''',s_2'))\right)}{\sum_{s_1''} P_{s_1,s_2}(s_1'',s_2')} \tag{1.8.32}$$

This can be averaged over possible values of s_2, giving us the average information content of the character s_1 when the character s_2 is known.

$$\ll I_{s_1|s_2} \gg \equiv \ll I_{s_1} \gg |_{s_2=s_2'} >$$

$$= -\sum_{s_2'} P_{s_2}(s_2') \sum_{s_1'} P_{s_1,s_2}(s_1'|s_2') \log(P_{s_1,s_2}(s_1'|s_2'))$$

$$= -\sum_{s_2'} \sum_{s_1'''} P_{s_1,s_2}(s_1''',s_2') \sum_{s_1'} \frac{P_{s_1,s_2}(s_1',s_2')}{\sum_{s_1''} P_{s_1,s_2}(s_1'',s_2')} \log(P_{s_1,s_2}(s_1'|s_2')) \quad (1.8.33)$$

$$= -\sum_{s_1',s_2'} P_{s_1,s_2}(s_1',s_2') \log(P_{s_1,s_2}(s_1'|s_2'))$$

The average we have taken should be carefully understood. The unconventional double average notation is used to indicate that the two averages are of a different nature. One way to think about it is as treating the information content of a dynamic variable s_1 when s_2 is a quenched (frozen) random variable. We can rewrite this in terms of the information content of the two characters, and the information content of the character s_2 by itself as follows:

$$\ll I_{s_1|s_2} \gg = -\sum_{s_1',s_2'} P_{s_1,s_2}(s_1',s_2') \left(\log(P_{s_1,s_2}(s_1',s_2')) - \log(\sum_{s_1''} P_{s_1,s_2}(s_1'',s_2')) \right) \quad (1.8.34)$$

$$= <I_{s_1,s_2}> - <I_{s_2}>$$

Thus we have:

$$<I_{s_1,s_2}> = <I_{s_1}> + \ll I_{s_2|s_1} \gg = <I_{s_2}> + \ll I_{s_1|s_2} \gg \quad (1.8.35)$$

This is the intuitive result that the information content given by both characters is the same as the information content gained by sequentially obtaining the information from the characters. Once the first character is known, the second character provides only the information given by the conditional probabilities. There is no reason to restrict the use of Eq. (1.8.27) – Eq. (1.8.35) to the case where s_1 is a single character and s_2 is a single character. It applies equally well if s_1 is one set of characters, and s_2 is another set of characters.

Question 1.8.8 Prove the inequalities in Eq. (1.8.30).

Hints for the left inequality:

1. It is helpful to use Eq. (1.8.35).
2. Use convexity ($f(\langle x \rangle) > \langle f(x) \rangle$) of the function $f(x) = -x\log(x)$.

Solution 1.8.8 The right inequality in Eq. (1.8.30) follows from the inequality:

$$P_{s_1}(s_1') = \sum_{s_1''} P_{s_1,s_2}(s_1'',s_2') > P_{s_1,s_2}(s_1',s_2') \quad (1.8.36)$$

The logarithm is a monotonic increasing function, so we can take the logarithm:

$$\log(\sum_{s_1''} P_{s_1,s_2}(s_1'',s_2')) > \log(P_{s_1,s_2}(s_1',s_2')) \tag{1.8.37}$$

Changing sign and averaging leads to the desired result:

$$<I_{s_2}> = -\sum_{s_1',s_2'} P_{s_1,s_2}(s_1',s_2')\log(\sum_{s_1''} P_{s_1,s_2}(s_1'',s_2'))$$
$$< -\sum_{s_1',s_2'} P_{s_1,s_2}(s_1',s_2')\log(P_{s_1,s_2}(s_1',s_2')) = <I_{s_1,s_2}> \tag{1.8.38}$$

The left inequality in Eq. (1.8.30) may be proven from Eq. (1.8.35) and the intuitive inequality

$$(<I_{s_1}>) > (<<I_{s_1|s_2}>>) \tag{1.8.39}$$

To prove this inequality we make use of the convexity of the function $f(x) = -x\log(x)$. Convexity of a function means that its value always lies above line segments (secants) that begin and end at points along its graph. Algebraically:

$$f((ax+by)/(a+b)) > (af(x)+bf(y))/(a+b) \tag{1.8.40}$$

More generally, taking a set of values of x and averaging over them gives:

$$f(\langle x \rangle) > \langle f(x) \rangle \tag{1.8.41}$$

Convexity of $f(x)$ follows from the observation that

$$\frac{d^2f}{dx^2} = -\frac{1}{x\ln(2)} < 0 \tag{1.8.42}$$

for all $x > 0$, which is where the function $f(x)$ is defined.

We then note the relationship:

$$P_{s_1}(s_1') = \sum_{s_2'} P_{s_2}(s_2')P_{s_1,s_2}(s_1'|s_2') = <P_{s_1,s_2}(s_1'|s_2')>_{s_2} \tag{1.8.43}$$

where, to simplify the following equations, we use a subscript to indicate the average with respect to s_2. The desired result follows from applying convexity as follows:

$$<I_{s_1}> = -\sum_{s_1'} P_{s_1}(s_1')\log(P_{s_1}(s_1')) = \sum_{s_1'} f(P_{s_1}(s_1')) = \sum_{s_1'} f(<P_{s_1,s_2}(s_1'|s_2')>_{s_2})$$
$$> \sum_{s_1'} <f(P_{s_1,s_2}(s_1'|s_2'))>_{s_2}$$
$$= -\sum_{s_2'} P_{s_2}(s_2')\sum_{s_1'} P_{s_1,s_2}(s_1'|s_2')\log(P_{s_1,s_2}(s_1'|s_2')) = <<I_{s_1|s_2}>>$$

$$\tag{1.8.44}$$

the final equality following from the definition in Eq. (1.8.33). We can now make use of Eq. (1.8.35) to obtain the desired result. ∎

1.8.4 *Ergodic sources*

We consider a source that provides arbitrarily long messages, or simply continues to give characters at a particular rate. Even though the messages are infinitely long, they are still considered elements of an ensemble. It is then convenient to measure the average information per character. The characterization of such an information source is simplified if each (long) message contains within it a complete sampling of the possibilities. This means that if we wait long enough, the entire ensemble of possible character sequences will be represented in any single message. This is the same kind of property as an ergodic system discussed in Section 1.3. By analogy, such sources are known as ergodic sources. For an ergodic source, not only the characters appear with their ensemble probabilities, but also the pairs of characters, the triples of characters, and so on.

For ergodic sources, the information from an ensemble average over all possible messages is the same as the information for a particular long string. To write this down we need a notation that allows variable length messages. We write $\underline{s}_N = (s_1 s_2 ... s_N)$, where N is the length of the string. The average information content per character may be written as:

$$<i_s> = \lim_{N \to \infty} \frac{<I_{\underline{s}_N}>}{N} = -\lim_{N \to \infty} \frac{1}{N} \sum_{\underline{s}_N} P(\underline{s}_N) \log(P(\underline{s}_N)) = -\lim_{N \to \infty} \frac{1}{N} \log(P(\underline{s}_N))$$

$$(1.8.45)$$

The rightmost equality is valid for an ergodic source. An example of an ergodic source is a source that provides independent characters—i.e., selects each character from an ensemble. For this case, Eq. (1.8.45) was shown in Question 1.8.7. More generally, for a source to be ergodic, long enough strings must break up into independent substrings, or substrings that are more and more independent as their length increases.

Assuming that N is large enough, we can use the limit in Eq. (1.8.45) and write:

$$P(\underline{s}_N) \approx 2^{-N<i_s>} \qquad (1.8.46)$$

Thus, for large enough N, there are a set of strings that are equally likely to be generated by the source. The number of these strings is

$$2^{N<i_s>} \qquad (1.8.47)$$

Since any string of characters is possible, in principle, this statement must be formally understood as saying that the total probability of all other strings becomes arbitrarily small.

If the string of characters is a Markov chain (Section 1.2), so that the probability of each character depends only on the previous character, then there are general conditions that can ensure that the source is ergodic. Similar to the discussion of Monte Carlo simulations in Section 1.7, for the source to be ergodic, the transition probabilities between characters must be irreducible and acyclic. Irreducibility guarantees that all characters are accessible from any starting character. The acyclic property guarantees that starting from one substring, all other substrings are accessible. Thus, if we can reach any particular substring, it will appear with the same frequency in all long strings.

We can generalize the usual Markov chain by allowing the probability of a character to depend on several (n) previous characters. A Markov chain may be constructed to represent such a chain by defining new characters, where each new character is formed out of a substring of n characters. Then each new character depends only on the previous one. The essential behavior of a Markov chain that is important here is that correlations measured along the chain of characters disappear exponentially. Thus, the statistical behavior of the chain in one place is independent of what it was in the sufficiently far past. The number of characters over which the correlations disappear is the correlation length. By allowing sufficiently many correlation lengths along the string—segments that are statistically independent—the average properties of one string will be the same as any other such string.

Question 1.8.9 Consider ergodic sources that are Markov chains with two characters $s_i = \pm 1$ with transition probabilities:

a. $P(1|1) = .999, P(-1|1) = .001, P(-1|-1) = 0.5, P(1|-1) = 0.5$

b. $P(1|1) = .999, P(-1|1) = .001, P(-1|-1) = 0.999, P(1|-1) = 0.001$

c. $P(1|1) = .999, P(-1|1) = .001, P(-1|-1) = 0.001, P(1|-1) = 0.999$

d. $P(1|1) = .001, P(-1|1) = .999, P(-1|-1) = 0.5, P(1|-1) = 0.5$

e. $P(1|1) = .001, P(-1|1) = .999, P(-1|-1) = 0.999, P(1|-1) = 0.001$

f. $P(1|1) = .001, \ P(-1|1) = .999, P(-1|-1) = 0.001, P(1|-1) = 0.999$

Describe the appearance of the strings generated by each source, and (roughly) its correlation length.

Solution 1.8.9 (a) has long regions of 1s of typical length 1000. In between there are short strings of -1s of average length $2 = 1 + 1/2 + 1/4 + \ldots$ (there is a probability of $1/2$ that a second character will be -1 and a probability of $1/4$ that both the second and third will be -1, etc.). (b) has long regions of 1s and long regions of -1s, both of typical length 1000. (c) is like (a) except the regions of -1s are of length 1. (d) has no extended regions of 1 or -1 but has slightly longer regions of -1s. (e) inverts (c). (f) has regions of alternating 1 and -1 of length 1000 before switching to the other possibility (odd and even indices are switched). We see that the characteristic correlation length is of order 1000 in (a),(b),(c),(e) and (f) and of order 2 in (d). ∎

We have considered in detail the problem of determining the information content of a message, or the average information generated by a source, when the characteristics of the source are well defined. The source was characterized by the ensemble of possible messages and their probabilities. However, we do not usually have a well-defined characterization of a source of messages, so a more practical question is to determine the information content from the message itself. The definitions that we have provided do not guide us in determining the information of an arbitrary message. We must have a model for the source. The model must be constructed out of the information we have—the string of characters it produces. One possibility is to model the source as ergodic. An ergodic source can be modeled in two ways, as a source of independent substrings or as a generalized Markov chain where characters depend on a certain number of previous characters. In each case we construct not one, but an infinite sequence of models. The models are designed so that if the source is ergodic then the information estimates given by the models converge to give the correct information content.

There is a natural sequence of independent substring models indexed by the number of characters in the substrings n. The first model is that of a source producing independent characters with a probability specified by their frequency of occurrence in the message. The second model would be a source producing pairs of correlated characters so that every pair of characters is described by the probability given by their occurrence (we allow character pairs to overlap in the message). The third model would be that of a source producing triples of correlated characters, and so on. We use each of these models to estimate the information. The nth model estimate of the information per character given by the source is:

$$<i_s>_{1,n}= \lim_{N\to\infty} \frac{1}{n}\sum_{\underline{s}_n} \tilde{P}_N(\underline{s}_n)\log(\tilde{P}_N(\underline{s}_n)) \tag{1.8.48}$$

where we indicate using the subscript $1,n$ that this is an estimate obtained using the first type of model (independent substring model) using substrings of length n. We also make use of an approximate probability for the substring defined as

$$\tilde{P}_N(\underline{s}_n)= N(\underline{s}_n)/(N-n+1) \tag{1.8.49}$$

where $N(\underline{s}_n)$ is the number of times s_n appears in the string of length N. The information of the source might then be estimated as the limit $n \to \infty$ of Eq. (1.8.48):

$$<i_s>= \lim_{n\to\infty}\lim_{N\to\infty} \frac{1}{n}\sum_{\underline{s}_n} \tilde{P}_N(\underline{s}_n)\log(\tilde{P}_N(\underline{s}_n)) \tag{1.8.50}$$

For an ergodic source, we can see that this converges to the information of the message. The n limit converges monotonically from above. This is because the additional information in $\underline{s}_{n+1}$ given by s_{n+1} is less than the information added by each previous character (see Eq. 1.8.59 below). Thus, the estimate of information per character based on $\underline{s}_n$ is higher than the estimate based on $\underline{s}_{n+1}$. Therefore, for each value of n the estimate $<i_s>_{1,n}$ is an upper bound on the information given by the source.

How large does N have to be? Since we must have a reasonable sample of the occurrence of substrings in order to estimate their probability, we can only estimate probabilities of substrings that are much shorter than the length of the string. The number of possible substrings grows exponentially with n as k^n, where k is the number of possible characters. If substrings occur with roughly similar probabilities, then to estimate the probability of a substring of length n would require at least a string of length k^n characters. Thus, taking the large N limit should be understood to correspond to N greater than k^n. This is a very severe requirement. This means that to study a model of English character strings of length $n = 10$ (ignoring upper and lower case, numbers and punctuation) would require $26^{10} \sim 10^{14}$ characters. This is roughly the number of characters in all of the books in the Library of Congress (see Question 1.8.15).

The generalized Markov chain model assumes a particular character is dependent only on n previous characters. Since the first n characters do not provide a significant amount of information for a very long chain ($N \gg n$), we can obtain the average information per character from the incremental information given by a character. Thus, for the nth generalized Markov chain model we have the estimate:

$$< i_s >_{2,n} = << I_{s_n | \underline{s}_{n-1}} >> = \lim_{N \to \infty} \sum_{\underline{s}_{n-1}} \tilde{P}_N(\underline{s}_{n-1}) \sum_{s_n} \tilde{P}(s_n | \underline{s}_{n-1}) \log(\tilde{P}(s_n | \underline{s}_{n-1})) \quad (1.8.51)$$

where we define the approximate conditional probability using:

$$\tilde{P}_N(s_n | \underline{s}_{n-1}) = N(\underline{s}_{n-1} s_n) / N(\underline{s}_{n-1}) \quad (1.8.52)$$

Taking the limit $n \to \infty$ we have an estimate of the information of the source per character:

$$< i_s > = \lim_{n \to \infty} \lim_{N \to \infty} \sum_{\underline{s}_{n-1}} \tilde{P}_N(\underline{s}_{n-1}) \sum_{s_n} \tilde{P}(s_n | \underline{s}_{n-1}) \log(\tilde{P}(s_n | \underline{s}_{n-1})) \quad (1.8.53)$$

This also converges from above as a function of n for large enough N. For a given n, a Markov chain model takes into account more correlations than the previous independent substring model and thus gives a better estimate of the information (Question 1.8.10).

Question 1.8.10 Prove that the Markov chain model gives a better estimate of the information for ergodic sources than the independent substring model for a particular n. Assume the limit $N \to \infty$ so that the estimated probabilities become actual and we can substitute $\tilde{P}_N \to P$ in Eq. (1.8.48) and Eq. (1.8.51).

Solution 1.8.10 The information in a substring of length n is given by the sum of the information provided incrementally by each character, where the previous characters are known. We derive this statement algebraically (Eq. (1.8.59)) and use it to prove the desired result. Taking the N limit in Eq. (1.8.48), we define the nth approximation using the independent substring model as:

$$< i_s >_{1,n} = \frac{1}{n} \sum_{\underline{s}_n} P(\underline{s}_n) \log(P(\underline{s}_n)) \quad (1.8.54)$$

and for the nth generalized Markov chain model we take the same limit in Eq. (1.8.51):

$$< i_s >_{2,n} = \sum_{\underline{s}_{n-1}} P(\underline{s}_{n-1}) \sum_{s_n} P(s_n | \underline{s}_{n-1}) \log(P(s_n | \underline{s}_{n-1})) \qquad (1.8.55)$$

To relate these expressions to each other, follow the derivation of Eq. (1.8.34), or use it with the substitutions $s_1 \to \underline{s}_{n-1}$ and $s_2 \to s_n$, to obtain

$$< i_s >_{2,n} = - \sum_{\underline{s}_{n-1}, s_n} P(\underline{s}_{n-1} s_n) \left(\log(P(\underline{s}_{n-1} s_n)) - \log(\sum_{s_n} P(\underline{s}_{n-1} s_n)) \right) \qquad (1.8.56)$$

Using the identities

$$P(\underline{s}_{n-1} s_n) = P(\underline{s}_n)$$
$$P(\underline{s}_{n-1}) = \sum_{s_n} P(\underline{s}_{n-1} s_n) \qquad (1.8.57)$$

this can be rewritten as:

$$< i_s >_{2, n} = n < i_s >_{1, n} - (n - 1) < i_s >_{1, n-1} \qquad (1.8.58)$$

This result can be summed over n from 1 to n (the $n = 1$ case is $<i_s>_{2,1} = <i_s>_{1,1}$) to obtain:

$$\sum_{n'=1}^{n} < i_s >_{2,n'} = n < i_s >_{1,n} \qquad (1.8.59)$$

since $< i_s >_{2,n}$ is monotonic decreasing and $< i_s >_{1,n}$ is seen from this expression to be an average over $< i_s >_{2,n}$ with lower values of n, we must have that

$$< i_s >_{2,n} \leq < i_s >_{1,n} \qquad (1.8.60)$$

as desired. ∎

Question 1.8.11 We have shown that the two models—the independent substring models and the generalized Markov chain model—are upper bounds to the information in a string. How good is the upper bound? Think up an example that shows that it can be terrible for both, but better for the Markov chain.

Solution 1.8.11 Consider the example of a long string formed out of a repeating substring, for example (000000010000000100000001...). The average information content per character of this string is zero. This is because once the repeat structure has become established, there is no more information. Any model that gives a nonzero estimate of the information content per character will make a great error in its estimate of the information content of the string, which is N times as much as the information per character.

For the independent substring model, the estimate is never zero. For the Markov chain model it is nonzero until n reaches the repeat distance. A Markov model with n the same size or larger than the repeat length will give the correct answer of zero information per character. This means that even for the Markov chain model, the information estimate does not work very well for n less than the repeat distance. ∎

Question 1.8.12 Write a computer program to estimate the information in English and find the estimate. For simple, easy-to-compute estimates, use single-character probabilities, two-character probabilities, and a Markov chain model for individual characters. These correspond to the above definitions of $< i_s >_{2,1} = < i_s >_{1,1}$, $< i_s >_{1,2}$, and $< i_s >_{2,2}$ respectively.

Solution 1.8.12 A program that evaluates the information content using single-character probabilities applied to the text (excluding equations) of Section 1.8 of this book gives an estimate of information content of 4.4 bits/character. Two-character probabilities gives 3.8 bits/character, and the one-character Markov chain model gives 3.3 bits/character. A chapter of a book by Mark Twain gives similar results. These estimates are decreasing in magnitude, consistent with the discussion in the text. They are also still quite high as estimates of the information in English per character.

The best estimates are based upon human guessing of the next character in a written text. Such experiments with human subjects give estimates of the lower and upper bounds of information content per character of English text. These are 0.6 and 1.2 bits/character. This range is significantly below the estimates we obtained using simple models. Remarkably, these estimates suggest that it is enough to give only one in four to one in eight characters of English in order for text to be decipherable. ∎

Question 1.8.13 Construct an example illustrating how correlations can arise between characters over longer than, say, ten characters. These correlations would not be represented by any reasonable character-based Markov chain model. Is there an example of this type relevant to the English language?

Solution 1.8.13 Example 1: If we have information that is read from a matrix row by row, where the matrix entries have correlations between rows, then there will be correlations that are longer than the length of the matrix rows.

Example 2: We can think about successive English sentences as rows of a matrix. We would expect to find correlations between rows (i.e., between words found in adjacent sentences) rather than just between letters. ∎

Question 1.8.14 Estimate the amount of information in a typical book (order of magnitude is sufficient). Use the best estimate of information content per character of English text of about 1 bit per character.

Solution 1.8.14 A rough estimate can be made using as follows: A 200 page novel with 60 characters per line and 30 lines per page has 4×10^5 characters. Textbooks can have several times this many characters. A dictionary, which is significantly longer than a typical book, might have 2×10^7 characters. Thus we might use an order of magnitude value of 10^6 bits per book. ∎

Question 1.8.15 Obtain an estimate of the number of characters (and thus the number of bits of information) in the Library of Congress. Assume an average of 10^6 characters per book.

Solution 1.8.15 According to information provided by the Library of Congress, there are presently (in 1996) 16 million books classified according to the Library of Congress classification system, 13 million other books at the Library of Congress, and approximately 80 million other items such as newspapers, maps and films. Thus with 10^7–10^8 book equivalents, we estimate the number of characters as 10^{13}–10^{14}. ∎

Inherent in the notion of quantifying information content is the understanding that the same information can be communicated in different ways, as long as the amount of information that can be transmitted is sufficient. Thus we can use binary, decimal, hexadecimal or typed letters to communicate both numbers and letters. Information can be communicated using any set of (two or more) characters. The presumption is that there is a way of translating from one to another. Translation operations are called codes; the act of translation is encoding or decoding. Among possible codes are those that are invertible. Encoding a message cannot add information, it might, however, lose information (Question 1.8.16). Invertible codes must preserve the amount of information.

Once we have determined the information content, we can compare different ways of writing the same information. Assume that one source generates a message of length N characters with information I. Then a different source may transmit the same information using fewer characters. Even if characters are generated at the same rate, the information may be more rapidly transmitted by one source than another. In particular, regardless of the value of N, by definition of information content, we could have communicated the same information using a binary string of length I. It is, however, impossible to use fewer than I bits because the maximum information a binary message can contain is equal to its length. This amount of information occurs for a source with equal a priori probability.

Encoding the information in a shorter form is equivalent to data compression. Thus a completely compressed binary data string would have an amount of information given by its length. The source of such a message would be characterized as a source of messages with equal a priori probability—a random source. We see that randomness and information are related. Without a translation (decoding) function it would be impossible to distinguish the completely compressed information from random numbers. Moreover, a random string could not be compressed.

> **Question 1.8.16** Prove that an encoding operation that takes a message as input and converts it into another well-defined message (i.e., for a particular input message, the same output message is always given) cannot add information but may reduce it. Describe the necessary conditions for it to keep the same amount of information.

Solution 1.8.16 Our definition of information relies upon the specification of the ensemble of possible messages. Consider this ensemble and assume that each message appears in the ensemble a number of times in proportion to its probability, like the bag with red and green balls. The effect of a coding operation is to label each ball with the new message (code) that will be delivered after the coding operation. The amount of information depends not on the nature of the label, but rather on the number of balls with the same label. The requirement that a particular message is encoded in a well-defined way means that two balls that start with the same message cannot be labeled with different codes. However, it is possible for balls with different original messages to be labeled the same. The average information is not changed if and only if all distinct messages are labeled with distinct codes. If any distinct messages become identified by the same label, the information is reduced.

We can prove this conclusion algebraically using the result of Question 1.8.8, which showed that transferring probability from a less likely to a more likely case reduced the information content. Here we are, in effect, transferring all of the probability from the less likely to the more likely case. The change in information upon labeling two distinct messages with the same code is given by ($f(x) = -x\log(x)$, as in Question 1.8.8):

$$\Delta I = f(P(s_1) + P(s_2)) - (f(P(s_1)) + f(P(s_2)))$$
$$= (f(P(s_1) + P(s_2)) + f(0)) - (f(P(s_1)) + f(P(s_2))) < 0$$

(1.8.61)

where the inequality follows because $f(x)$ is convex in the range $0 < x < 1$. ∎

1.8.5 *Human communication*

The theory of information, like other theories, relies upon idealized constructs that are useful in establishing the essential concepts, but do not capture all features of real systems. In particular, the definition and discussion of information relies upon sources that transmit the result of random occurrences, which, by definition, cannot be known by the recipient. The sources are also completely described by specifying the nature of the random process. This model for the nature of the source and the recipient does not adequately capture the attributes of communication between human beings. The theory of information can be applied directly to address questions about information channels and the characterization of communication in general. It can also be used to develop an understanding of the complexity of systems. In this section, however, we will consider some additional issues that should be kept in mind when applying the theory to the communication between human beings. These issues will arise again in Chapter 8.

The definition of information content relies heavily on the concepts of probability, ensembles, and processes that generate arbitrarily many characters. These concepts are fraught with practical and philosophical difficulties—when there is only one transmitted message, how can we say there were many that were possible? A book may be considered as a single communication. A book has finite length and, for a particular author and a particular reader, is a unique communication. In order to understand both the strengths and the limitations of applying the theory of information, it is necessary to recognize that the information content of a message depends on the information that the recipient of the message already has. In particular, information that the recipient has about the source. In the discussion above, a clear distinction has been made. The only information that characterizes the source is in the ensemble probabilities $P(s)$. The information transmitted by a single message is distinct from the ensemble probabilities and is quantified by $I(s)$. It is assumed that the characterization of the source is completely known to the recipient. The content of the message is completely unknown (and unknowable in advance) to the recipient.

A slightly more difficult example to consider is that of a recipient who does not know the characterization of the source. However, such a characterization in terms of an ensemble $P(s)$ does exist. Under these circumstances, the amount of information transferred by a message would be more than the amount of information given by $I(s)$. However, the maximum amount of information that could be transferred would be the sum of the information in the message, and the information necessary to characterize the source by specifying the probabilities $P(s)$. This upper bound on the information that can be transferred is only useful if the amount of information necessary to characterize the source is small compared to the information in the message.

The difficulty with discussing human communication is that the amount of information necessary to fully characterize the source (one human being) is generally much larger than the information transmitted by a particular message. Similarly, the amount of information possessed by the recipient (another human being) is much larger than the information contained in a particular message. Thus it is reasonable to assume that the recipient does not have a full characterization of the source. It is also reasonable to assume that the model that the recipient has about the source is more sophisticated than a typical Markov chain model, even though it is a simplified model of a human being. The information contained in a message is, in a sense, the additional information not contained in the original model possessed by the recipient. This is consistent with the above discussion, but it also recognizes that specifying the probabilities of the ensemble may require a significant amount of information. It may also be convenient to summarize this information by a different type of model than a Markov chain model.

Once the specific model and information that the recipient has about the source enters into an evaluation of the information transfer, there is a certain and quite reasonable degree of relativity in the amount of information transferred. An extreme example would be if the recipient has already received a long message and knows the same message is being repeated, then no new information is being transmitted. A person who has memorized the Gettysburg Address will receive very little new information upon hearing or reading it again. The prior knowledge is part of the model possessed by the recipient about the source.

Can we incorporate this in our definition of information? In every case where we have measured the information of a message, we have made use of a model of the source of the information. The underlying assumption is that this model is possessed by the recipient. It should now be recognized that there is a certain amount of information necessary to describe this model. As long as the amount of information in the model is small compared to the amount of information in the message, we can say that we have an absolute estimate of the information content of the message. As soon as the information content of the model approaches that of the message itself, then the amount of information transferred is sensitive to exactly what information is known. It might be possible to develop a theory of information that incorporates the information in the model, and thus to arrive at a more absolute measure of information. Alternatively, it might be necessary to develop a theory that considers the recipient and source more completely, since in actual communication between human beings, both are nonergodic systems possessed of a large amount of information. There is significant overlap of the information possessed by the recipient and the source. Moreover, this common information is essential to the communication itself.

One effort to arrive at a universal definition of information content of a message has been made by formally quantifying the information contained in models. The resulting information measure, Kolmogorov complexity, is based on computation theory discussed in the next section. While there is some success with this approach, two difficulties remain. In order for a universal definition of information to be agreed upon, models must still have an information content which is less than the message—knowledge possessed must be smaller than that received. Also, to calculate the information contained in a particular message is essentially impossible, since it requires computational effort that grows exponentially with the length of the message. In any practical case, the amount of information contained in a message must be estimated using a limited set of models of the source. The utilization of a limited set of models means that any estimate of the information in a message is an upper bound.

1.9 Computation

The theory of computation describes the operations that we perform on numbers, including addition, subtraction, multiplication and division. More generally, a computation is a sequence of operations each of which has a definite/unique/well-defined result. The fundamental study of such operations is the theory of logic. Logical operations do not necessarily act upon numbers, but rather upon abstract objects called statements. Statements can be combined together using operators such as AND and OR, and acted upon by the negation operation NOT. The theory of logic and the theory of computation are at root the same. All computations that have been conceived of can be constructed out of logical operations. We will discuss this equivalence in some detail.

We also discuss a further equivalence, generally less well appreciated, between computation and deterministic time evolution. The theory of computation strives to describe the class of all possible discrete deterministic or causal systems. Computations are essentially causal relationships. Computation theory is designed to capture all such possible relationships. It is thus essential to our understanding not just of the behavior of computers, or of human logic, but also to the understanding of causal relationships in all physical systems. A counterpoint to this association of computation and causality is the recognition that certain classes of deterministic dynamical systems are capable of the property known as universal computation.

One of the central findings of the theory of computation is that many apparently different formulations of computation turn out to be equivalent. The sense in which they are equivalent is that each one can simulate the other. In the early years of computation theory, there was an effort to describe sets of operations that would be more powerful than others. When all of them were shown to be equivalent it became generally accepted (the Church-Turing hypothesis) that there is a well-defined set of possible computations realized by any of several conceptual formulations. This has become known as the theory of universal computation.

1.9.1 Propositional logic

Logic is the study of reasoning, inference and deduction. Propositional logic describes the manipulation of statements that are either true or false. It assumes that there exists a set of statements that are either true or false at a particular time, but not both. Logic then provides the possibility of using an assumed set of relationships between the statements to determine the truth or falsehood of other statements.

For example, the statements $Q_1 = $ "I am standing" and $Q_2 = $ "I am sitting" may be related by the assumption: Q_1 is true implies that Q_2 is not true. Using this assumption, it is understood that a statement "Q_1 AND Q_2" must be false. The falsehood depends only on the relationship between the two sentences and not on the particular meaning of the sentences. This suggests that an abstract construction that describes mechanisms of inference can be developed. This abstract construction is propositional logic.

Propositional logic is formed out of statements (propositions) that may be true (T) or false (F), and operations. The operations are described by their actions upon statements. Since the only concern of logic is the truth or falsehood of statements, we can describe the operations through tables of truth values (truth tables) as follows. NOT ($\wedge$) is an operator that acts on a single statement (a unary operator) to form a new statement. If Q is a statement then $\wedge$Q (read "not Q") is the symbolic representation of "It is not true that Q." The truth of $\wedge$Q is directly (causally) related to the truth of Q by the relationship in the table:

Q	$\wedge$Q
T	F
F	T

(1.9.1)

The value of the truth or falsehood of Q is shown in the left column and the corresponding value of the truth or falsehood of $\wedge$Q is given in the right column.

Similarly, we can write the truth tables for the operations AND (&) and OR (|):

Q_1	Q_2	$Q_1 \& Q_2$
T	T	T
T	F	F
F	T	F
F	F	F

(1.9.2)

Q_1	Q_2	$Q_1	Q_2$
T	T	T
T	F	T
F	T	T
F	F	F

(1.9.3)

As the tables show, $Q_1 \& Q_2$ is only true if both Q_1 is true and Q_2 is true. $Q_1 | Q_2$ is only false if both Q_1 is false and Q_2 is false.

Propositional logic includes logical theorems as statements. For example, the statement Q_1 is true if and only if Q_2 is true can also be written as a binary operation $Q_1 \equiv Q_2$ with the truth table:

Q_1	Q_2	$Q_1 \equiv Q_2$
T	T	T
T	F	F
F	T	F
F	F	T

(1.9.4)

Another binary operation is the statement Q_1 implies Q_2, $Q_1 \Rightarrow Q_2$. When this statement is translated into propositional logic, there is a difficulty that is usually bypassed by the following convention:

Q_1	Q_2	$Q_1 \Rightarrow Q_2$
T	T	T
T	F	F
F	T	T
F	F	T

(1.9.5)

The difficulty is that the last two lines suggest that when the antecedent Q_1 is false, the implication is true, whether or not the consequent Q_2 is true. For example, the statement "If I had wings then I could fly" is as true a statement as "If I had wings then I couldn't fly," or the statement "If I had wings then potatoes would be flat." The problem originates in the necessity of assuming that the result is true or false in a unique way based upon the truth values of Q_1 and Q_2. Other information is not admissible, and a third choice of "nonsense" or "incomplete information provided" is not allowed within propositional logic. Another way to think about this problem is to say that there are many operators that can be formed with definite outcomes. Regardless of how we relate these operators to our own logical processes, we can study the system of operators that can be formed in this way. This is a model, but not a complete one, for human logic. Or, if we choose to define logic as described by this system, then human thought (as reflected by the meaning of the word "implies") is not fully characterized by logic (as reflected by the meaning of the operation "$\Rightarrow$").

In addition to unary and binary operations that can act upon statements to form other statements, it is necessary to have parentheses that differentiate the order of operations to be performed. For example a statement $((Q_1 \equiv Q_2)\&(^\wedge Q_3)|Q_1)$ is a series of operations on primitive statements that starts from the innermost parenthesis and progresses outward. As in this example, there may be more than one innermost parenthesis. To be definite, we could insist that the order of performing these operations is from left to right. However, this order does not affect any result.

Within the context of propositional logic, it is possible to describe a systematic mechanism for proving statements that are composed of primitive statements. There are several conclusions that can be arrived at regarding a particular statement. A tautology is a statement that is always true regardless of the truth or falsehood of its component statements. Tautologies are also called theorems. A contradiction is a statement that is always false. Examples are given in Question 1.9.1.

Question 1.9.1 Evaluate the truth table of:

a. $(Q_1 \Rightarrow Q_2)|((^\wedge Q_2)\&Q_1)$

b. $(^\wedge(Q_1 \Rightarrow Q_2))\equiv((^\wedge Q_1)|Q_2)$

Identify which is a tautology and which is a contradiction.

Solution 1.9.1 Build up the truth table piece by piece:

a. Tautology:

Q_1	Q_2	$Q_1 \Rightarrow Q_2$	$(^\wedge Q_2)\&Q_1$	$(Q_1 \Rightarrow Q_2)|((^\wedge Q_2)\&Q_1)$
T	T	T	F	T
T	F	F	T	T
F	T	T	F	T
F	F	T	F	T

$$(1.9.6)$$

b. Contradiction:

Q_1	Q_2	$^\wedge(Q_1 \Rightarrow Q_2)$	$(^\wedge Q_1)|Q_2$	$(^\wedge(Q_1 \Rightarrow Q_2)) \equiv ((^\wedge Q_1)|Q_2)$
T	T	F	T	F
T	F	T	F	F
F	T	F	T	F
F	F	F	T	F

$$(1.9.7) \quad \blacksquare$$

Question 1.9.2: Construct a theorem (tautology) from a contradiction.

Solution 1.9.2: By negation. $\blacksquare$

1.9.2 *Boolean algebra*

Propositional logic is a particular example of a more general symbolic system known as a Boolean algebra. Set theory, with the operators complement, union and intersection, is another example of a Boolean algebra. The formulation of a Boolean algebra is convenient because within this more general framework a number of important theorems can be proven. They then hold for propositional logic, set theory and other Boolean algebras.

A Boolean algebra is a set of elements $B=\{Q_1, Q_2, ...\}$, a unary operator ($^\wedge$), and two binary operators, for which we adopt the notation $(+,\cdot)$, that satisfy the following properties for all Q_1, Q_2, Q_3 in B:

1. Closure: $^\wedge Q_1$, Q_1+Q_2, and $Q_1 \cdot Q_2$ are in B
2. Commutative law: $Q_1+Q_2=Q_2+Q_1$, and $Q_1 \cdot Q_2=Q_2 \cdot Q_1$
3. Distributive law: $Q_1 \cdot (Q_2+Q_3)=(Q_1 \cdot Q_2)+(Q_1 \cdot Q_3)$ and
   $Q_1+(Q_2 \cdot Q_3)=(Q_1+Q_2) \cdot (Q_1+Q_3)$
4. Existence of identity elements, 0 and 1: $Q_1+0=Q_1$, and $Q_1 \cdot 1=Q_1$
5. Complementarity law: $Q_1+(^\wedge Q_1)=1$ and $Q_1 \cdot (^\wedge Q_1)=0$

The statements of properties 2 through 5 consist of equalities. These equalities indicate that the element of the set that results from operations on the left is the same as the element resulting from operations on the right. Note particularly the second part of the distributive law and the complementarity law that would not be valid if we interpreted + as addition and • as multiplication.

Assumptions 1 to 5 allow the proof of additional properties as follows:

6. Associative property: $Q_1+(Q_2+Q_3)=(Q_1+Q_2)+Q_3$ and $Q_1 \cdot (Q_2 \cdot Q_3)=(Q_1 \cdot Q_2) \cdot Q_3$
7. Idempotent property: $Q_1+Q_1=Q_1$ and $Q_1 \cdot Q_1=Q_1$
8. Identity elements are nulls: $Q_1+1=1$ and $Q_1 \cdot 0=0$
9. Involution property: $^\wedge(^\wedge Q_1)=Q_1$

10. Absorption property: $Q_1+(Q_1 \cdot Q_2)=Q_1$ and $Q_1 \cdot (Q_1+Q_2)=Q_1$
11. DeMorgan's Laws: $^\wedge(Q_1+Q_2)=(^\wedge Q_1) \cdot (^\wedge Q_2)$ and $^\wedge(Q_1 \cdot Q_2)=(^\wedge Q_1)+(^\wedge Q_2)$

To identify propositional logic as a Boolean algebra we use the set B={T,F} and map the operations of propositional logic to Boolean operations as follows: ($^\wedge$ to $^\wedge$), (I to +) and (& to •). The identity elements are mapped: (1 to T) and (0 to F). The proof of the Boolean properties for propositional logic is given as Question 1.9.3.

Question 1.9.3: Prove that the identification of propositional logic as a Boolean algebra is correct.

Solution 1.9.3: (1) is trivial; (2) is the invariance of the truth tables of $Q_1 \& Q_2$, $Q_1 I Q_2$ to interchange of values of Q_1 and Q_2; (3) requires comparison of the truth tables of $Q_1 I (Q_2 \& Q_3)$ and $(Q_1 I Q_2) \& (Q_1 I Q_3)$ (see below). Comparison of the truth tables of $Q_1 \& (Q_2 I Q_3)$ and $(Q_1 \& Q_2) I (Q_1 \& Q_3)$ is done similarly.

Q_1	Q_2	Q_3	$Q_2\&Q_3$	$Q_1I(Q_2\&Q_3)$	Q_1IQ_2	Q_1IQ_3	$(Q_1IQ_2)\&(Q_1IQ_3)$
T	T	T	T	T	T	T	T
T	T	F	F	T	T	T	T
T	F	T	F	T	T	T	T
T	F	F	F	T	T	T	T
F	T	T	T	T	T	T	T
F	T	F	F	F	T	F	F
F	F	T	F	F	F	T	F
F	F	F	F	F	F	F	F

(1.9.8)

(4) requires verifying $Q_1\&T=T$, and $Q_1 I F=F$ (see the truth tables for & and I above); (5) requires constructing a truth table for $QI^\wedge Q$ and verifying that it is always T (see below). Similarly, the truth table for $Q\&^\wedge Q$ shows that it is always F.

Q	$^\wedge Q$	$QI^\wedge Q$
T	F	T
F	T	T

(1.9.9) ∎

1.9.3 *Completeness*

Our objective is to show that an arbitrary truth table, an arbitrary logical statement, can be constructed out of only a few logical operations. Truth tables are also equivalent to numerical functions—specifically, functions of binary variables that have binary results (binary functions of binary variables). This can be seen using the Boolean representation of T and F as {1,0} that is more familiar as a binary notation for numerical functions. For example, we can write the AND and OR operations (functions) also as:

Q_1	Q_2	$Q_1 \cdot Q_2$	$Q_1 + Q_2$
1	1	1	1
1	0	0	1
0	1	0	1
0	0	0	0

$$(1.9.10)$$

Similarly for all truth tables, a logical operation is a binary function of a set of binary variables. Thus, the ability to form an arbitrary truth table from a few logical operators is the same as the ability to form an arbitrary binary function of binary variables from these same logical operators.

To prove this ability, we use the properties of the Boolean algebra to systematically discuss truth tables. We first construct an alternative Boolean expression for $Q_1 + Q_2$ by a procedure that can be generalized to arbitrary truth tables. The procedure is to look at each line in the truth table that contains an outcome of 1 and write an expression that provides unity for that line only. Then we combine the lines to achieve the desired table. $Q_1 \cdot Q_2$ is only unity for the first line, as can be seen from its column. Similarly, $Q_1 \cdot (^\wedge Q_2)$ is unity for the second line and $(^\wedge Q_1) \cdot Q_2$ is unity for the third line. Using the properties of $+$ we can then combine the terms together in the form:

$$Q_1 \cdot Q_2 + Q_1 \cdot (^\wedge Q_2) + (^\wedge Q_1) \cdot Q_2 \qquad (1.9.11)$$

Using associative and identity properties, this gives the same result as $Q_1 + Q_2$.

We have replaced a simple expression with a much more complicated expression in Eq. (1.9.11). The motivation for doing this is that the same procedure can be used to represent any truth table. The general form we have constructed is called the disjunctive normal form. We can construct a disjunctive normal representation for an arbitrary binary function of binary variables. For example, given a specific binary function of binary variables, $f(Q_1, Q_2, Q_3)$, we construct its truth table, e.g.,

Q_1	Q_2	Q_3	$f(Q_1, Q_2, Q_3)$
1	1	1	1
1	0	1	0
0	1	1	1
0	0	1	0
1	1	0	0
1	0	0	1
0	1	0	0
0	0	0	0

$$(1.9.12)$$

The disjunctive normal form is given by:

$$f(Q_1, Q_2, Q_3) = Q_1 \cdot Q_2 \cdot Q_3 + (^\wedge Q_1) \cdot Q_2 \cdot Q_3 + Q_1 \cdot (^\wedge Q_2) \cdot (^\wedge Q_3) \qquad (1.9.13)$$

as can be verified by inspection. An analogous construction can represent any binary function.

We have demonstrated that an arbitrary truth table can be constructed out of the three operations $(^\wedge, +, \cdot)$. We say that these form a complete set of operations. Since there are 2^n lines in a truth table formed out of n binary variables, there are 2^{2^n} possible functions of these n binary variables. Each is specified by a particular choice of the 2^n possible outcomes. We have achieved a dramatic simplification by recognizing that all of them can be written in terms of only three operators. We also know that at most $(1/2)n2^n$ ($\wedge$) operations, $(n-1)\,2^n$ ($\cdot$) operations and 2^n-1 (+) operations are necessary. This is the number of operations needed to represent the identity function 1 in disjunctive normal form.

It is possible to further simplify the set of operations required. We can eliminate either the + or the $\cdot$ operations and still have a complete set. To prove this we need only display an expression for either of them in terms of the remaining operations:

$$Q_1 \cdot Q_2 = {\wedge}(({\wedge}Q_1)+({\wedge}Q_2))$$
$$Q_1 + Q_2 = {\wedge}(({\wedge}Q_1)\cdot({\wedge}Q_2)) \tag{1.9.14}$$

Question 1.9.4: Verify Eq. (1.9.14).

Solution 1.9.4: They may be verified using DeMorgan's Laws and the involution property, or by construction of the truth tables, e.g.:

Q_1	Q_2	${\wedge}Q_1$	${\wedge}Q_2$	$Q_1 \cdot Q_2$	$({\wedge}Q_1)+({\wedge}Q_2)$
1	1	0	0	1	0
1	0	0	1	0	1
0	1	1	0	0	1
0	0	1	1	0	1

$$\tag{1.9.15}$$ ∎

It is possible to go one step further and identify binary operations that can represent all possible functions of binary variables. Two possibilities are the NAND ($\&$) and NOR ($\hat{\mathrm{I}}$) operations defined by:

$$Q_1 \,\&\, Q_2 = {\wedge}(Q_1 \& Q_2) \rightarrow {\wedge}(Q_1 \cdot Q_2)$$
$$Q_1 \,\hat{\mathrm{I}}\, Q_2 = {\wedge}(Q_1 | Q_2) \rightarrow {\wedge}(Q_1 + Q_2) \tag{1.9.16}$$

Both the logical and Boolean forms are written above. The truth tables of these operators are:

Q_1	Q_2	${\wedge}(Q_1 \cdot Q_2)$	${\wedge}(Q_1 + Q_2)$
1	1	0	0
1	0	1	0
0	1	1	0
0	0	1	1

$$\tag{1.9.17}$$

We can prove that each is complete by itself (capable of representing all binary functions of binary variables) by showing that they are capable of representing one of the earlier complete sets. We prove the case for the NAND operation and leave the NOR operation to Question 1.9.5.

$$^\wedge Q_1=^\wedge(Q_1\cdot Q_1)=Q_1 \;\hat{\&}\; Q_1$$

$$(Q_1\cdot Q_2)=^\wedge(^\wedge(Q_1\cdot Q_2))=^\wedge(Q_1 \;\hat{\&}\; Q_2)=(Q_1 \;\hat{\&}\; Q_2) \;\hat{\&}\; (Q_1 \;\hat{\&}\; Q_2)$$

(1.9.18)

Question 1.9.5: Verify completeness of the NOR operation.

Solution 1.9.5: We can use the same formulas as in the proof of the completeness of NAND by replacing • with + and $\hat{\&}$ with $\hat{|}$ everywhere. ∎

1.9.4 Turing machines

We have found that logical operators can represent any binary function of binary variables. This means that all well-defined mathematical operations on integers can be represented in this way. One of the implications is that we might make machines out of physical elements, each of which is capable of performing a Boolean operation. Such a machine would calculate a mathematical function and spare us a tedious task. We can graphically display the operations of a machine performing a series of Boolean operations as shown in Fig. 1.9.1. This is a simplified symbolic form similar to forms used in the design of computer logic circuits.

By looking carefully at Fig. 1.9.1 we see that there are several additional kinds of actions that are necessary in addition to the elementary Boolean operation. These actions are indicated by the lines that might be thought of as wires. One action is to transfer information from the location where it is input into the system, to the place where it is used. The second is to duplicate the information. Duplication is represented in the figure by a branching of the lines. The branching enables the same

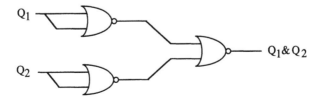

Figure 1.9.1 Graphical representation of Boolean operations. The top figure shows a graphical element representing the NOR operation $Q_1\hat{|}Q_2 = {}^\wedge(Q_1|Q_2)$. In the bottom figure we combine several operations together with lines (wires) indicating input, output, data duplication and transfer to form the AND operation, $(Q_1\hat{|}Q_1)\hat{|}(Q_2\hat{|}Q_2) = (^\wedge Q_1)\hat{|}(^\wedge Q_2) = Q_1\&Q_2$. This equation may be used to prove completeness of the NOR operation. ∎

information to be used in more than one place. Additional implicit actions involve timing, because the representation makes an assumption that time causes the information to be moved and acted upon in a sequence from left to right. It is also necessary to have mechanisms for input and output.

The kind of mathematical machine we just described is limited to performing one prespecified function of its inputs. The process of making machines is time consuming. To physically rearrange components to make a new function would be inconvenient. Thus it is useful to ask whether we might design a machine such that part of its input could include a specification of the mathematical operation to be performed. Both information describing the mathematical function, and the numbers on which it is to be performed, would be encoded in the input which could be described as a string of binary characters.

This discussion suggests that we should systematically consider the properties/ qualities of machines able to perform computations. The theory of computation is a self-consistent discussion of abstract machines that perform a sequence of prespecified well-defined operations. It extends the concept of universality that was discussed for logical operations. While the theory of logic determined that all Boolean functions could be represented using elementary logic operations, the theory of computation endeavors to establish what is possible to compute using a sequence of more general elementary operations. For this discussion many of the practical matters of computer design are not essential. The key question is to establish a relationship between machines that might be constructed and mathematical functions that may be computed. Part of the problem is to define what a computation is.

There are several alternative models of computation that have been shown to be equivalent in a formal sense since each one of them can simulate any other. Turing introduced a class of machines that represent a particular model of computation. Rather than maintaining information in wires, Turing machines (Fig. 1.9.2) use a storage device that can be read and written to. The storage is represented as an infinite one-dimensional tape marked into squares. On the tape can be written characters, one to a square. The total number of possible characters, the alphabet, is finite. These characters are often taken to be digits plus a set of markers (delimiters). In addition to the characters, the tape squares can also be blank. All of the tape is blank except for a finite number of nonblank places. Operations on the tape are performed by a roving read-write head that has a specified (finite) number of internal storage elements and a simple kind of program encoded in it. We can treat the program as a table similar to the tables discussed in the context of logic. The table operation acts upon the value of the tape at the current location of the head, and the value of storage elements within the read head. The result of an operation is not just a single binary value. Instead it corresponds to a change in the state of the tape at the current location (write), a change in the internal memory of the head, and a shift of the location of the head by one square either to the left or to the right.

We can also think about a Turing machine (TM) as a dynamic system. The internal table does not change in time. The internal state $s(t)$, the current location $l(t)$, the current character $a(t)$ and the tape $c(t)$ are all functions of time. The table consists of

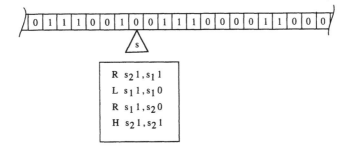

Figure 1.9.2 Turing's model of computation — the Turing machine (TM) — consists of a tape divided into squares with characters of a finite alphabet written on it. A roving "head" indicated by the triangle has a finite number of internal states and acts by reading and writing the tape according to a prespecified table of rules. Each rule consists of a command to read the tape, write the tape, change the internal state of the TM head and move either to the left or right. A simplified table is shown consisting of several rules of the form $\{\mu, s', a', s, a\}$ where a and a' are possible tape characters, s and s' are possible states of the head and μ is a movement of the head right (R), left (L) or halt (H). Each update the TM starts by finding the rule $\{\mu, s', a', s, a\}$ in the table such that a is the character on the tape at the current location of the head, and s is its current state. The tape is written with the corresponding a' and the state of the TM head is changed to s'. Then the TM head moves according to the corresponding μ right or left. The illustration simplifies the characters to binary digits 0 and 1 and the states of the TM head to s_1 and s_2. ∎

a set of instructions or rules of the form $\{\mu, s', a', s, a\}$ corresponding to a deterministic transition matrix. s and a are the current internal state and current tape character respectively. s' and a' are the new internal state and character. μ is the move to be made, either right or left (R or L).

Using either conceptual model, the TM starts from an initial state and location and a specified tape. In each time interval the TM head performs the following operations:

1. Read the current tape character

2. Find the instruction that corresponds to the existing combination of (s,a)

3. Change the internal memory to the corresponding s'

4. Write the tape with the corresponding character a'

5. Move the head to the left or right as specified by μ

When the TM head reaches a special internal state known as the halt state, then the outcome of the computation may be read from the tape. For simplicity, in what follows we will indicate entering the halt state by a move $\mu = H$ which is to halt.

The best way to understand the operation of a TM is to construct particular tables that perform particular actions (Question 1.9.6). In addition to logical operations, the possible actions include moving and copying characters. Constructing particular actions using a TM is tedious, in large part because the movements of the head are limited to a single displacement right or left. Actual computers use direct addressing that enables access to a particular storage location in its memory using a number (address) specifying its location. TMs do not generally use this because the tape is arbitrarily long, so that an address is an arbitrarily large number, requiring an arbitrarily large storage in the internal state of the head. Infinite storage in the head is not part of the computational model.

Question 1.9.6 The following TM table is designed to move a string of binary characters (0 and 1) that are located to the left of a special marker M to blank squares on the tape to the right of the M and then to stop on the M. Blank squares are indicated by B. The internal states of the head are indicated by $s_1, s_2 \ldots$ These are not italicized, since they are values rather than variables. The movements of the head right and left are indicated by R and L. As mentioned above, we indicate entering the halt state by a movement H. Each line has the form $\{\mu, s', a', s, a\}$.

Read over the program and convince yourself that it does what it is supposed to. Describe how it works. The TM must start from state s_1 and must be located at the leftmost nonblank character. The line numbering is only for convenience in describing the TM, and has no role in its operation.

1.	R	s_2	B	s_1	0	
2.	R	s_3	B	s_1	1	
3.	R	s_2	0	s_2	0	
4.	R	s_2	1	s_2	1	
5.	R	s_2	M	s_2	M	
6.	R	s_3	0	s_3	0	
7.	R	s_3	1	s_3	1	
8.	R	s_3	M	s_3	M	(1.9.19)
9.	L	s_4	0	s_2	B	
10.	L	s_4	1	s_3	B	
11.	L	s_4	0	s_4	0	
12.	L	s_4	1	s_4	1	
13.	L	s_4	M	s_4	M	
14.	R	s_1	B	s_4	B	
15.	H	s_1	M	s_1	M	

Solution 1.9.6 This TM works by (lines 1 or 2) reading a nonblank character (0 or 1) into the internal state of the head; 0 is represented by s_2 and 1 is represented by s_3. The character that is read is set to a blank B. Then the TM moves to the right, ignoring all of the tape characters 0, 1 or M (lines 3 through 8) until it reaches a blank B. It writes the stored character (lines 9 or 10), changing its state to s_4. This state specifies moving to the left, ignoring all characters 0,1 or M (lines 11 through 13) until it reaches a blank B. Then

(line 14) it moves one step right and resets its state to s_1. This starts the procedure from the beginning. If it encounters the marker M in the state s_1 instead of a character to be copied, then it halts (line 15). ∎

Since each character can also be represented by a set of other characters (i.e., 2 in binary is 10), we can allow the TM head to read and write not one but a finite prespecified number of characters without making a fundamental change. The following TM, which acts upon pairs of characters and moves on the tape by two characters at a time, is the same as the one given in Question 1.9.6.

1.	01	01	00	00	01
2.	01	11	00	00	11
3.	01	01	01	01	01
4.	01	01	11	01	11
5.	01	01	10	01	10
6.	01	11	01	11	01
7.	01	11	11	11	11
8.	01	11	10	11	10
9.	10	10	01	01	00
10.	10	10	11	11	00
11.	10	10	01	10	01
12.	10	10	11	10	11
13.	10	10	10	10	10
14.	01	00	00	10	00
15.	00	00	10	00	10

$$(1.9.20)$$

The particular choice of the mapping from characters and internal states onto the binary representation is not unique. This choice is characterized by using the left and right bits to represent different aspects. In columns 3 or 5, which represent the tape characters, the right bit represents the type of element (marker or digit), and the left represents which element or marker it is: 00 represents the blank B, 10 represents M, 01 represents the digit 0, and 11 represents the digit 1. In columns 2 or 4, which represent the state of the head, the states s_1 and s_4 are represented by 00 and 10, s_2 and s_3 are represented by 01 and 11 respectively. In column 1, moving right is 01, left is 10, and halt is 00.

The architecture of a TM is very general and allows for a large variety of actions using complex tables. However, all TMs can be simulated by transferring all of the responsibility for the table and data to the tape. A TM that can simulate all TMs is called a universal Turing machine (UTM). As with other TMs, the responsibility of arranging the information lies with the "programmer." The UTM works by representing the table, current state, and current letter on the UTM tape. We will describe the essential concepts in building a UTM but will not explicitly build one.

The UTM acts on its own set of characters with its own set of internal states. In order to use it to simulate an arbitrary TM, we have to represent the TM on the tape of the UTM in the characters that the UTM can operate on. On the UTM tape, we

Figure 1.9.3 The universal Turing machine (UTM) is a special TM that can simulate the computation performed by any other TM. The UTM does this by executing the rules of the TM that are encoded on the tape of the UTM. There are three parts to the UTM tape, the part where the TM table is encoded (on the left), the part where the tape of the TM is encoded (on the right) and a workspace (in the middle) where information representing the current state of the TM head, the current character of the TM tape, and the movement command, are encoded. See the text for a description of the operation of the UTM based on its own rule table. ∎

must be able to represent four types of entities: a TM character, the state of the TM head, the movement to be taken by the TM head, and markers that indicate to the UTM what is where on the tape. The markers are special to the UTM and must be carefully distinguished from the other three. For later reference, we will build a particular type of UTM where the tape can be completely represented in binary.

The UTM tape has three parts, the part that represents the table of the TM, a work area, and the part that represents the tape of the TM (Fig. 1.9.3). To represent the tape and table of a particular but arbitrary TM, we start with a binary representation of its alphabet and of its internal states

$$a_1 \to 00000, a_2 \to 00001, a_3 \to 00010, \dots$$

$$s_1 \to 000, s_2 \to 001, \dots \tag{1.9.21}$$

where we keep the left zeros, as needed for the number of bits in the longest binary number. We then make a doubled binary representation like that used in the previous example, where each bit becomes two bits with the low order bit a 1. The doubled binary notation will enable us to distinguish between UTM markers and all other entities on the tape. Thus we have:

$$a_1 \to 01\ 01\ 01\ 01\ 01, a_2 \to 01\ 01\ 01\ 01\ 11, a_3 \to 01\ 01\ 01\ 11\ 01, \dots$$

$$s_1 \to 01\ 01\ 01, s_2 \to 01\ 01\ 11, \dots \tag{1.9.22}$$

These labels of characters and states are in a sense arbitrary, since the transition table is what gives them meaning.

We also encode the movement commands. The movement commands are not arbitrary, since the UTM must know how to interpret them. We have allowed the TM to displace more than one character, so we must encode a set of movements such as R_1, L_1, R_2, L_2, and H. These correspond respectively to moving one character right, one character left, two characters right, two characters left, and entering the halt state. Because the UTM must understand the move that is to be made, we must agree once and for all on a coding of these movements. We use the lowest order bit as a direction bit (1 Right, 0 Left) and the rest of the bits as the number of displacements in binary

$$R_1 \to 011, R_2 \to 101, ...,$$

$$L_1 \to 010, L_2 \to 100, ..., \qquad (1.9.23)$$

$$H \to 000 \text{ or } 001$$

The doubled binary representation is as before: each bit becomes two bits with the low order bit a 1,

$$R_1 \to 01\ 11\ 11\ , R_2 \to 11\ 01\ 11\ , ...,$$

$$L_1 \to 01\ 11\ 01\ , L_2 \to 11\ 01\ 01\ , ..., \qquad (1.9.24)$$

$$H \to 01\ 01\ 01 \text{ or } 01\ 01\ 11$$

Care is necessary in the UTM design because we do not know in advance how many types of TM moves are possible. We also don't know how many characters or internal states the TM has. This means that we don't know the length of their binary representations.

We need a number of markers that indicate to the UTM the beginning and end of encoded characters, states and movements described above. We also need markers to distinguish different regions of the tape. A sufficient set of markers are:

M_1—the beginning of a TM character,

M_2—the beginning of a TM internal state,

M_3—the beginning of a TM table entry, which is also the beginning of a movement command,

M_4—a separator between the TM table and the workspace,

M_5—a separator between the workspace and the TM tape,

M_6—the beginning of the current TM character (the location of the TM head),

M_7—the identified TM table entry to be used in the current step, and

B—the blank, which we include among the markers.

Depending on the design of the UTM, these markers need not all be distinct. In any case, we encode them also in binary

$$B \to 000, M_1 \to 001, M_2 \to 010, ... \qquad (1.9.25)$$

and then doubled binary form where the second character is now zero:

$$B \to 00\ 00\ 00, M_1 \to 00\ 00\ 10, M_2 \to 00\ 10\ 00, ... \qquad (1.9.26)$$

We are now in a position to encode both the tape and table of the TM on the tape of the UTM. The representation of the table consists of a sequence of representations of the lines of the table, $L_1 L_2 ...$, where each line is represented by the doubled binary representation of

$$M_3\, \mu\, M_2\, s'\, M_1\, a'\, M_2\, s\, M_1 a \qquad (1.9.27)$$

The markers are definite but the characters and states and movements correspond to those in a particular line in the table. The UTM representation of the tape of the TM, $a_1 a_2 \ldots$, is a doubled binary representation of

$$M_1 \, a_1 \, M_1 \, a_2 \, M_1 \, a_3 \ldots \qquad (1.9.28)$$

The workspace starts with the character M_4 and ends with the character M_5. There is room enough for the representation of the current TM machine state, the current tape character and the movement command to be executed. At a particular time in execution it appears as:

$$M_4 \, \mu \, M_2 \, s \, M_1 \, a \, M_5 \qquad (1.9.29)$$

We describe in general terms the operation of the UTM using this representation of a TM. Before execution we must indicate the starting location of the TM head and its initial state. This is done by changing the corresponding marker M_1 to M_6 (at the UTM tape location to the left of the character corresponding to the initial location of the TM), and the initial state of the TM is encoded in the workspace after M_2.

The UTM starts from the leftmost nonblank character of its tape. It moves to the right until it encounters M_6. It then copies the character after M_6 into the work area after M_1. It compares the values of (s,a) in the work area with all of the possible (s,a) pairs in the transition table pairs until it finds the same pair. It marks this table entry with M_7. The corresponding s' from the table is copied into the work area after M_2. The corresponding a' is copied to the tape after M_6. The corresponding movement command μ is copied to the work area after M_4. If the movement command is H the TM halts. Otherwise, the marker M_6 is moved according to the value of μ. It is moved one step at a time (i.e., the marker M_6 is switched with the adjacent M_1) while decrementing the value of the digits of μ (except the rightmost bit) and in the direction specified by the rightmost bit. When the movement command is decremented to zero, the TM begins the cycle again by copying the character after M_6 into the work area.

There is one detail we have overlooked: the TM can write to the left of its nonblank characters. This would cause problems for the UTM we have designed, since to the left of the TM tape representation is the workspace and TM table. There are various ways to overcome this difficulty. One is to represent the TM tape by folding it upon itself and interleaving the characters. Starting from an arbitrary location on the TM tape we write all characters on the UTM tape to the right of M_5, so that odd characters are the TM tape to the right, and even ones are the TM tape to the left. Movements of the M_6 marker are doubled, and it is reflected (bounces) when it encounters M_5.

A TM is a dynamic system. We can reformulate Turing's model of computation in the form of a cellular automaton (Section 1.5) in a way that will shed some light on the dynamics that are being discussed. The most direct way to do this is to make an automaton with two adjacent tapes. The only information in the second strip is a single nonblank character at the location of the head that represents its internal state. The TM update is entirely contained within the update rule of the automaton. This update rule may be constructed so that it acts at every point in the space, but is enabled by the nonblank character in the adjacent square on the second tape. When the dynamics reaches a steady state (it is enough that two successive states of the automaton are the same), the computation is completed. If desired we could reduce this CA to one tape by placing each pair of squares in the two tapes adjacent to each other, interleaving the two tapes. While a TM can be represented as a CA, any CA with only a finite number of active cells can be updated by a Turing machine program (it is computable). There are many other CA that can be programmed by their initial state to perform computations. These can be much simpler than using the TM model as a starting point. One example is Conway's Game of Life, discussed in Section 1.5. Like a UTM, this CA is a universal computer—any computation can be performed by starting from some initial state and looking at the final steady state for the result.

When we consider the relationship of computation theory to dynamic systems, there are some intentional restrictions in the theory that should be recognized. The conventional theory of computation describes a single computational unit operating on a character string formed from a finite alphabet of characters. Thus, computation theory does not describe a continuum in space, an infinite array of processors, or real numbers. Computer operations only mimic approximately the formal definition of real numbers. Since an arbitrary real number requires infinitely many digits to specify, computations upon them in finite time are impossible. The rejection by computation theory of operations upon real numbers is not a trivial one. It is rooted in fundamental results of computation theory regarding limits to what is inherently possible in any computation.

This model of computation as dynamics can be summarized by saying that a computation is the steady-state result of a deterministic CA with a finite alphabet (finite number of characters at each site) and finite domain update rule. One of the characters (the blank or vacuum) must be such that it is unchanged when the system is filled with these characters. The space is infinite but the conditions are such that all space except for a finite region must be filled with the blank character.

1.9.5 *Computability and the halting problem*

The construction of a UTM guarantees that if we know how to perform a particular operation on numbers, we can program a UTM to perform this computation. However, if someone gives you such a program—can you determine what it will compute? This seemingly simple question turns out to be at the core of a central problem of logic theory. It turns out that it is not only difficult to determine what it will compute, it is, in a formal sense that will be described below, impossible to figure out if it will compute anything at all. The requirement that it will compute something is that eventually it will halt. By halting, it declares its computation completed and the answer given. Instead of halting, it might loop forever or it might continue to write on ever larger regions of tape. To say that we can determine whether it will compute something is equivalent to saying that it will eventually halt. This is called the halting problem. How could we determine if it would halt? We have seen above how to represent an arbitrary TM on the tape of a particular TM. Consistent with computation theory, the halting problem is to construct a special TM, T_H, whose input is a description of a TM and whose output is a single bit that specifies whether or not the

TM will halt. In order for this to make sense, the program T_H must itself halt. We can prove by contradiction that this is not possible in general, and therefore we say that the halting problem is not computable. The proof is based on constructing a paradoxical logical statement of the form "This statement is false."

A proof starts by assuming we have a TM called T_H that accepts as input a tape representing a TM Y and its tape y. The output, which can be represented in functional form as $T_H(Y, y)$, is always well-defined and is either 1 or 0 representing the statement that the TM Y halts on y or doesn't halt on y respectively. We now construct a logical contradiction by constructing an additional TM based on T_H. First we consider $T_H(Y, Y)$, which asks whether Y halts when acting on a tape representing itself. We design a new TM T_{H1} that takes only Y as input, copies it and then acts in the same way as T_H. So we have

$$T_{H1}(Y) = T_H(Y, Y) \tag{1.9.30}$$

We now define a TM T_{H2} that is based on T_{H1} but whenever T_{H1} gives the answer 0 it gives the answer 1, and whenever T_{H1} gives the answer 1 it enters a loop and computes forever. A moment's meditation shows that this is possible if we have T_{H1}. Applying T_{H2} to itself then gives us the contradiction, since $T_{H2}(T_{H2})$ gives 1 if

$$T_{H1}(T_{H2}) = T_H(T_{H2}, T_{H2}) = 0 \tag{1.9.31}$$

By definition of T_H this means that $T_{H2}(T_{H2})$ does not halt, which is a contradiction. Alternatively, $T_{H2}(T_{H2})$ computes forever if

$$T_{H1}(T_{H2}) = T_H(T_{H2}, T_{H2}) = 1$$

by definition of T_H this means that $T_{H2}(T_{H2})$ halts, which is a contradiction.

The noncomputability of the halting problem is similar to Gödel's theorem and other results denying the completeness of logic, in the sense that we can ask a question about a logical construction that cannot be answered by it. Gödel's theorem may be paraphrased as: In any axiomatic formulation of number theory (i.e., integers), it is possible to write a statement that cannot be proven T or F. There has been a lot of discussion about the philosophical significance of these theorems. A basic conclusion that may be reached is that they describe something about the relationship of the finite and infinite. Turing machines can be represented, as we have seen, by a finite set of characters. This means that we can enumerate them, and they correspond one-to-one to the integers. Like the integers, there are (countably) infinitely many of them. Gödel's theorem is part of our understanding of how an infinite set of numbers must be described. It tells us that we cannot describe their properties using a finite set of statements. This is appealing from the point of view of information theory since an arbitrary integer contains an arbitrarily large amount of information. The noncomputability of the halting problem tells us more specifically that we can ask a question about a system that is described by a finite amount of information whose answer (in the sense of computation) is not contained within it. We have thus made a vague connection between computation and information theory. We take this connection one step further in the following section.

1.9.6 *Computation and information in brief*

One of our objectives will be to relate computation and information. We therefore ask, Can a calculation produce information? Let us think about the results of a TM calculation which is a string of characters—the nonblank characters on the output tape. How much information is necessary to describe it? We could describe it directly, or use a Markov model as in Section 1.8. However, we could also give the input of the TM and the TM description, and this would be enough information to enable us to obtain the output by computation. This description might contain more or fewer characters than the direct description of the output. We now return to the problem of defining the information content of a string of characters. Utilizing the full power of computation, we can define this as the length of the shortest possible input tape for a UTM that gives the desired character string as its output. This is called the algorithmic (or Kolmogorov) complexity of a character string. We have to be careful with the definition, since there are many different possible UTM. We will discuss this in greater detail in Chapter 8. However, this discussion does imply that a calculation cannot produce information. The information present at the beginning is sufficient to obtain the result of the computation. It should be understood, however, that the information that seems to us to be present in a result may be larger than the original information unless we are able to reconstruct the starting point and the TM used for the computation.

1.9.7 *Logic, computation and human thought*

Both logic and computation theory are designed to capture aspects of human thought. A fundamental question is whether they capture enough of this process— are human beings equivalent to glorified Turing machines? We will ask this question in several ways throughout the text and arrive at various conclusions, some of which support this identification and some that oppose it. One way to understand the question is as one of progressive approximation. Logic was originally designed to model human thought. Computation theory, which generalizes logic, includes additional features not represented in logic. Computers as we have defined them are instruments of computation. They are given input (information) specifying both program and data and provide well-defined output an indefinite time later. One of the features that is missing from this kind of machine is the continuous input-output interaction with the world characteristic of a sensory-motor system. An appropriate generalization of the Turing machine would be a robot. As it is conceived and sometimes realized, a robot has both sensory and motor capabilities and an embedded computer. Thus it has more of the features characteristic of a human being. Is this sufficient, or have we missed additional features?

Logic and computation are often contrasted with the concept of creativity. One of the central questions about computers is whether they are able to simulate creativity. In Chapter 3 we will produce a model of creativity that appears to be possible to simulate on a computer. Hidden in this model, however, is a need to use random numbers. This might seem to be a minor problem, since we often use computers to generate random numbers. However, computers do not actually generate randomness, they generate pseudo-random numbers. If we recall that randomness is the same as information, by the discussion in the previous section, a computer cannot generate true randomness. A Turing machine cannot generate a result that has more information than it is given in its initial data. Thus creativity appears to be tied at least in part to randomness, which has often been suggested, and this may be a problem for conventional computers. Conceptually, this problem can be readily resolved by adding to the description of the Turing machine an infinite random tape in addition to the infinite blank tape. This new system appears quite similar to the original TM specification. A reasonable question would ask whether it is really inherently different. The main difference that we can ascertain at this time is that the new system would be capable of generating results with arbitrarily large information content, while the original TM could not. This is not an unreasonable distinction to make between a creative and a logical system. There are still key problems with understanding the practical implications of this distinction.

The subtlety of this discussion increases when we consider that one branch of theoretical computer science is based on the commonly believed assumption that there exist functions that are inherently difficult to invert—they can only be inverted in a time that grows exponentially with the length of the nonblank part of the tape. For all practical purposes, they cannot be inverted, because the estimated lifetime of the universe is insufficient to invert such functions. While their existence is not proven, it has been proven that if they do exist, then such a function can be used to generate a string of characters that, while not random, cannot be distinguished from a random string in less than exponential time. This would suggest that there can be no practical difference between a TM with a random tape, and one without. Thus, the possibility of the existence of noninvertible functions is intimately tied to questions about the relationship between TM, randomness and human thought.

1.9.8 *Using computation and information to describe the real world*

In this section we review the fundamental relevance of the theories of computation and information in the real world. This relevance ultimately arises from the properties of observations and measurements.

In our observations of the world, we find that quantities we measure vary. Indeed, without variation there would be no such thing as an observation. There are variations over time as well as over space. Our intellectual effort is dedicated to classifying or understanding this variation. To concretize the discussion, we consider observations of a variable s which could be as a function of time $s(t)$ or of space $s(x)$. Even though x or t may appear continuous, our observations may often be described as a finite discrete set $\{s_i\}$. One of the central (meta)observations about the variation in value of $\{s_i\}$ is that sometimes the value of the variable s_i can be inferred from, is correlated with, or is not independent from its value or values at some other time or position s_j.

These concepts have to do with the relatedness of s_i to s_j. Why is this important? The reason is that we would like to know the value of s_i without having to observe it.

We can understand this as a problem in prediction—to anticipate events that will occur. We would also like to know what is located at unobserved positions in space; e.g., around the corner. And even if we have observed something, we do not want to have to remember all observations we make. We could argue more fundamentally that knowledge/information is important only if prediction is possible. There would be no reason to remember past observations if they were uncorrelated with anything in the future. If correlations enable prediction, then it is helpful to store information about the past. We want to store as little as possible in order to make the prediction. Why? Because storage is limited, or because accessing the right information requires a search that takes time. If a search takes more time than we have till the event we want to predict, then the information is not useful. As a corollary (from a simplified utilitarian point of view), we would like to retain only information that gives us the best, most rapid prediction, under the most circumstances, for the least storage.

Inference is the process of logic or computation. To be able to infer the state of a variable s_i means that we have a definite formula $f(s_j)$ that will give us the value of s_i with complete certainty from a knowledge of s_j. The theory of computation describes what functions f are possible. If the index i corresponds to a later time than j we say that we can predict its value. In addition to the value of s_j we need to know the function f in order to predict the value of s_i. This relationship need not be from a single value s_j to a single value s_i. We might need to know a collection of values $\{s_j\}$ in order to obtain the value of s_i from $f(\{s_j\})$.

As part of our experience of the world, we have learned that observations at a particular time are more closely related to observations at a previous time than observations at different nearby locations. This has been summarized by the principle of causality. Causality is the ability to determine what happens at one time from what happened at a previous time. This is more explicitly stated as microcausality—what happens at a particular time and place is related to what happened at a previous time in its immediate vicinity. Causality is the principle behind the notion of determinism, which suggests that what occurs is determined by prior conditions. One of the ways that we express the relationship between system observations over time is by conservation laws. Conservation laws are the simplest form of a causal relationship.

Correlation is a looser relationship than inference. The statement that values s_i and s_j are correlated implies that even if we cannot tell exactly what the value s_i is from a knowledge of s_j, we can describe it at least partially. This partial knowledge may also be inherently statistical in the context of an ensemble of values as discussed below. Correlation often describes a condition where the values s_i and s_j are similar. If they are opposite, we might say they are anticorrelated. However, we sometimes use the term "correlated" more generally. In this case, to say that s_i and s_j are correlated would mean that we can construct a function $f(s_j)$ which is close to the value of s_i but not exactly the same. The degree of correlation would tell us how close we expect them to be. While correlations in time appear to be more central than correlations in space, systems with interactions have correlations in both space and time.

Concepts of relatedness are inherently of an ensemble nature. This means that they do not refer to a particular value s_i or a pair of values (s_i, s_j) but rather to a collection of such values or pairs. The ensemble nature of relationships is often more explicit for correlations, but it also applies to inference. This ensemble nature is hidden by functional terminology that describes a relationship between particular values. For example, when we say that the temperature at 1:00 P.M. is correlated with the temperature at 12:00 P.M., we are describing a relationship between two temperature values. Implicitly, we are describing the collection of all pairs of temperatures on different days or at different locations. The set of such pairs are analogs. The concept of inference also generally makes sense only in reference to an ensemble. Let us assume for the moment that we are discussing only a single value s_i. The statement of inference would imply that we can obtain s_i as the value $f(s_j)$. For a single value, the easiest way (requiring the smallest amount of information) to specify $f(s_j)$ would be to specify s_i. We do not gain by using inference for this single case. However, we can gain if we know that, for example, the velocity of an object will remain the same if there are no forces upon it. This describes the velocity $v(t)$ in terms of $v(t')$ of any one object out of an ensemble of objects. We can also gain from inference if the function $f(s_j)$ gives a string of more than one s_i.

The notion of independence is the opposite of inference or correlation. Two values s_i and s_j are independent if there is no way that we can infer the value of one from the other, and if they are not correlated. Randomness is similar to independence. The word "independent" is used when there is no correlation between two observations. The word "random" is stronger, since it means that there is no correlation between an observed value and anything else. A random process, like a sequence of coin tosses, is a sequence where each value is independent of the others. We have seen in Section 1.8 that randomness is intimately related with information. Random processes are unpredictable, therefore it makes no sense for us to try to accumulate information that will help predict it. In this sense, a random process is simple to describe. However, once a random process has occurred, other events may depend upon it. For example, someone who wins a lottery will be significantly affected by an event presumed to be random. Thus we may want to remember the results of the random process after it occurs. In this case we must remember each value. We might ask, Once the random process has occurred, can we summarize it in some way? The answer is that we cannot. Indeed, this property has been used to define randomness.

We can abstract the problem of prediction and description of observations to the problem of data compression. Assume there are a set of observations $\{s_i\}$ for which we would like to obtain the shortest possible description from which we can reconstruct the complete set of observations. If we can infer one value from another, then the set might be compressed by eliminating the inferable values. However, we must make sure that the added information necessary to describe how the inference is to be done is less than the information in the eliminated values. Correlations also enable compression. For example, let us assume that the values are biased ON with a probability $P(1) = .999$ and OFF with a probability $P(-1) = 0.001$. This means that one in a thousand values is OFF and the others are ON. In this case we can remember which ones are OFF rather than keeping a list of all of the values. We would say they are ON except for numbers 3, 2000, 2403, 5428, etc. This is one way of coding the information. This method of encoding has a problem in that the numbers representing the locations of the OFF values may become large. They will be correlated because the first few digits of successive locations will be the same (..., 431236, 432112, 434329,...). We can further reduce the list if we are willing to do some more processing, by giving the intervals between successive OFF values rather than the absolute numbers of their location.

Ultimately, when we have reached the limits of our ability to infer one observation from another, the rest is information that we need. For example, differential equations are based on the presumption that boundary conditions (initial conditions in time, and boundary conditions in space) are sufficient to predict the behavior of a system. The values of the initial conditions and the boundary conditions are the information we need. This simple model of a system, where information is clearly and simply separated from the problem of computation, is not always applicable.

Let us assume that we have made extensive observations and have separated from these observations a minimal set that then can be used to infer all the rest. A minimal set of information would have the property that no one piece of information in it could be obtained from other pieces of information. Thus, as far as the set itself is concerned, the information appears to be random. Of course we would not be satisfied with any random set; it would have to be this one in particular, because we want to use this information to tell us about all of the actual observations.

One of the difficulties with random numbers is that it is inherently difficult to prove that numbers are random. We may simply not have thought of the right function f that can predict the value of the next number in a sequence from the previous numbers. We could argue that this is one of the reasons that gambling is so attractive to people because of the use of "lucky numbers" that are expected by the individual to have a better-than-random chance of success. Indeed, it is the success of science to have shown that apparently uncorrelated events may be related. For example, the falling of a ball and the motion of the planets. At the same time, science provides a framework in which noncausal correlations, otherwise called superstitions, are rejected.

We have argued that the purpose of knowledge is to succinctly summarize information that can be used for prediction. Thus, in its most abstract form, the problem of deduction or prediction is a problem in data compression. It can thus be argued that science is an exercise in data compression. This is the essence of the principle of Occam's razor and the importance of simplicity and universality in science. The more universal and the more general a law is, and the simpler it is, then the more data compression has been achieved. Often this is considered to relate to how valuable is the contribution of the law to science. Of course, even if the equations are general and simple, if we cannot solve them then they are not particularly useful from a practical point of view. The concept of simplicity has always been poorly defined. While science seeks to discover correlations and simplifications in observations of the universe around us, ultimately the minimum description of a system (i.e., the universe) is given by the number of independent pieces of information required to describe it.

Our understanding of information and computation enters also into a discussion of our models of systems discussed in previous sections. In many of these models, we assumed the existence of random variables, or random processes. This randomness represents either unknown or complex phenomena. It is important to recognize that this represents an assumption about the nature of correlations between different aspects of the problem that we are modeling. It assumes that the random process is independent of (uncorrelated with) the aspects of the system we are explicitly studying. When we model the random process on a computer by a pseudo-random number generator, we are assuming that the computations in the pseudo-random number generator are also uncorrelated with the system we are studying. These assumptions may or may not be valid, and tests of them are not generally easy to perform.

1.10   Fractals, Scaling and Renormalization

The physics of Newton and the related concepts of calculus, which have dominated scientific thinking for three hundred years, are based upon the understanding that at smaller and smaller scales—both in space and in time—physical systems become simple, smooth and without detail. A more careful articulation of these ideas would note that the fine scale structure of planets, materials and atoms is not without detail. However, for many problems, such detail becomes irrelevant at the larger scale. Since the details are irrelevant, formulating theories in a way that assumes that the detail does not exist yields the same results as a more exact description.

In the treatment of complex systems, including various physical and biological systems, there has been a recognition that the concept of progressive smoothness on finer scales is not always a useful mathematical starting point. This recognition is an important fundamental change in perspective whose consequences are still being explored.

We have already discussed in Section 1.1 the subject of chaos in iterative maps. In chaotic maps, the smoothness of dynamic behavior is violated. It is violated because fine scale details matter. In this section we describe fractals, mathematical models of the spatial structure of systems that have increasing detail on finer scales. Geometric fractals have a self-similar structure, so that the structure on the coarsest scale is repeated on finer length scales. A more general framework in which we can articulate questions about systems with behavior on all scales is that of scaling theory introduced in Section 1.10.3. One of the most powerful analytic tools for studying systems that have scaling properties is the renormalization group. We apply it to the Ising model in Section 1.10.4, and then return full cycle by applying the renormalization group to chaos in Section 1.10.5. A computational technique, the multigrid method, that enables the description of problems on multiple scales is discussed in Section 1.10.6. Finally, we discuss briefly the relevance of these concepts to the study of complex systems in Section 1.10.7.

1.10.1 *Fractals*

Traditional geometry is the study of the properties of spaces or objects that have integral dimensions. This can be generalized to allow effective fractional dimensions of objects, called fractals, that are embedded in an integral dimension space. In recent years the recognition that fractals can play an important role in modeling natural phenomena has fueled a whole area of research investigating the occurrence and properties of fractal objects in physical and biological systems.

Fractals are often defined as geometric objects whose spatial structure is self-similar. This means that by magnifying one part of the object, we find the same structure as of the original object. The object is characteristically formed out of a collection of elements: points, line segments, planar sections or volume elements. These elements exist in a space of the same or higher dimension to the elements themselves. For example, line segments are one-dimensional objects that can be found on a line, plane, volume or higher dimensional space. We might begin to describe a fractal by the objects of which it is formed. However, geometric fractals are often described by a procedure (algorithm) that creates them in an explicitly self-similar manner.

One of the simplest examples of a fractal object is the Cantor set (Fig. 1.10.1). This set is formed by a procedure that starts from a single line segment. We remove the middle third from the segment. There are then two line segments left. We then remove the middle third from both of these segments, leaving four line segments. Continuing iteratively, at the kth iteration there are 2^k segments. The Cantor set, which is the limiting set of points obtained from this process, has no line segments in it. It is self-similar by direct construction, since the left and right third of the original line segment can be expanded by a factor of three to appear as the original set.

An analog of the Cantor set in two dimensions is the Sierpinski gasket (Fig. 1.10.2). It is constructed from an equilateral triangle by removing an internal triangle which is half of the size of the original triangle. This procedure is then iterated for all of the smaller triangles that result. We can see that there are no areas that are left in this shape. It is self-similar, since each of the three corner triangles can be expanded by a factor of two to appear as the original set.

For self-similar objects, we can obtain the effective fractal dimension directly by considering their composition from parts. We do this by analogy with conventional

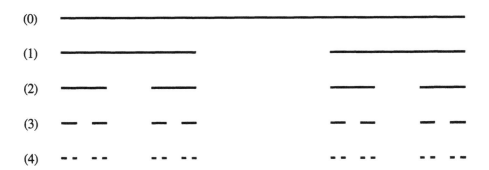

Figure 1.10.1 Illustration of the construction of the Cantor set, one of the best-known fractals. The Cantor set is formed by iteratively removing the middle third from a line segment, then the middle third from the two remaining line segments, and so on. Four iterations of the procedure are shown starting from the complete line segment at the top. ∎

Figure 1.10.2 The Sierpinski gasket is formed in a similar manner to the Cantor set. Starting from an equilateral triangle, a similar triangle one half the size is removed from the middle leaving three triangles at the corners. The procedure is then iteratively applied to the remaining triangles. The figure shows the set that results after four iterations of the procedure. ∎

geometric objects which are also self-similar. For example, a line segment, a square, or a cube can be formed from smaller objects of the same type. In general, for a d-dimensional cube, we can form the cube out of smaller cubes. If the size of the smaller cubes is reduced from that of the large cube by a factor of η, where η is inversely proportional to their diameter, $\eta \propto 1/R$, then the number of smaller cubes necessary to form the original is $N = \eta^d$. Thus we could obtain the dimension as:

$$d = \ln(N) \,/\, \ln(\eta) \qquad (1.10.1)$$

For self-similar fractals we can do the same, where N is the number of parts that make up the whole. Each of the parts is assumed to have the same shape, but reduced in size by a factor of η from the original object.

We can generalize the definition of fractal dimension so that we can use it to characterize geometric objects that are not strictly self-similar. There is more than one way to generalize the definition. We will adopt an intuitive definition of fractal dimension which is closely related to Eq. (1.10.1). If the object is embedded in d-dimensions, we cover the object with d-dimensional disks. This is illustrated in Fig. 1.10.3 for a line segment and a rectangle in a two-dimensional space. If we cover the object with two-dimensional disks of a fixed radius, R, using the minimal number of disks possible, the number of these disks changes with the radius of the disks according to the power law:

$$N(R) \propto R^{-d} \qquad (1.10.2)$$

where d is defined as the fractal dimension. We note that the use of disks is only illustrative. We could use squares and the result can be proven to be equivalent.

(a)

(b)

(c)

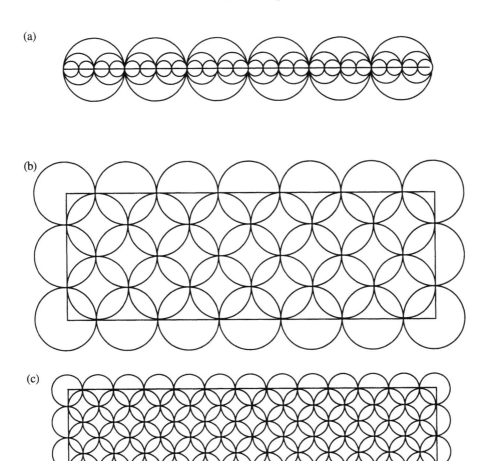

Figure 1.10.3 In order to define the dimension of a fractal object, we consider the problem of covering a set with a minimal number of disks of radius R. (a) shows a line segment with three different coverings superimposed. (b) and (c) show a rectangle with two different coverings respectively. As the size of the disks decreases the number of disks necessary to cover the shape grows as R^{-d}. This behavior becomes exact only in the limit $R \to 0$. The fractal dimension defined in this way is sometimes called the box-counting dimension, because d-dimensional boxes are often used rather than disks. ∎

We can use either Eq. (1.10.1) or Eq. (1.10.2) to calculate the dimension of the Cantor set and the Sierpinski gasket. We illustrate the use of Eq. (1.10.2). For the Cantor set, by construction, 2^k disks (or line segments) of radius $1/3^k$ will cover the set. Thus we can write:

$$N(R/3^k) = 2^k N(R) \tag{1.10.3}$$

Using Eq. (1.10.2) this is:

$$(R/3^k)^{-d} = 2^k R^{-d} \tag{1.10.4}$$

or:

$$3^d = 2 \tag{1.10.5}$$

which is:

$$d = \ln(2)/\ln(3) \cong 0.631 \tag{1.10.6}$$

We would arrive at the same result more directly from Eq. (1.10.1).

For the Sierpinski gasket, we similarly recognize that the set can be covered by three disks of radius $1/2$, nine disks of radius $1/4$, and more generally 3^k disks of radius $1/2^k$. This gives a dimension of:

$$d = \ln(3)/\ln(2) \cong 1.585 \tag{1.10.7}$$

For these fractals there is a deterministic algorithm that is used to generate them. We can also consider a kind of stochastic fractal generated in a similar way, however, at each level the algorithm involves choices made from a probability distribution. The simplest modification of the sets is to assume that at each level a choice is made with equal probability from several possibilities. For example, in the Cantor set, rather than removing the middle third from each of the line segments, we could choose at random which of the three thirds to remove. Similarly for the Sierpinski gasket, we could choose which of the four triangles to remove at each stage. These would be stochastic fractals, since they are not described by a deterministic self-similarity but by a statistical self-similarity. Nevertheless, they would have the same fractal dimension as the deterministic fractals.

Question 1.10.1 How does the dimension of a fractal, as defined by Eq. (1.10.2), depend on the dimension of the space in which it is embedded?

Solution 1.10.1 The dimension of a fractal is independent of the dimension of the space in which it is embedded. For example, we might start with a d-dimensional space and increase the dimension of the space to $d+1$ dimensions. To show that Eq. (1.10.2) is not changed, we form a covering of the fractal by $d+1$ dimensional spheres whose intersection with the d-dimensional space is the same as the covering we used for the analysis in d dimensions. ∎

Question 1.10.2 Prove that the fractal dimension does not change if we use squares or circles for covering an object.

Solution 1.10.2 Assume that we have minimal coverings of a shape using $N_1(R) = c_1 R^{-d_1}$ squares, and minimal coverings by $N_2(R) = c_2 R^{-d_2}$ circles, with $d_1 \neq d_2$. The squares are characterized using R as the length of their side, while the circles are characterized using R as their radius. If d_1 is less than d_2, then for smaller and smaller R the number of disks becomes arbitrarily smaller than the number of squares. However, we can cover the same shape using squares that circumscribe the disks. The number of these squares is $N_1'(R) = c_1 (R / 2)^{-d_1}$. This is impossible, because for small enough R, $N_1'(R)$ will be smaller than $N_1(R)$, which violates the assumption that the latter is a minimal covering. Similarly, if d is greater than d', we use disks circumscribed around the squares to arrive at a contradiction. ∎

Question 1.10.3 Calculate the fractal dimension of the Koch curve given in Fig. 1.10.4.

Solution 1.10.3 The Koch curve is composed out of four Koch curves reduced in size from the original by a factor of 3. Thus, the fractal dimension is $d = \ln(4) / \ln(3) \approx 1.2619$. ∎

Question 1.10.4 Show that the length of the Koch curve is infinite.

Solution 1.10.4 The Koch curve can be constructed by taking out the middle third of a line segment and inserting two segments equivalent to the one that was removed. They are inserted so as to make an equilateral triangle with the removed segment. Thus, at every iteration of the construction procedure, the length of the perimeter is multiplied by 4/3, which means that it diverges to infinity. It can be proven more generally that any fractal of dimension 2 > $d > 1$ must have an infinite length and zero area, since these measures of size are for one-dimensional and two-dimensional objects respectively. ∎

Eq. (1.10.2) neglects the jumps in $N(R)$ that arise as we vary the radius R. Since $N(R)$ can only have integral values, as we lower R and add additional disks there are discrete jumps in its value. It is conventional to define the fractal dimension by taking the limit of Eq. (1.10.2) as $R \rightarrow 0$, where this problem disappears. This approach, however, is linked philosophically to the assumption that systems simplify in the limit of small length scales. The assumption here is not that the system becomes smooth and featureless, but rather that the fractal properties will continue to all finer scales and remain ideal. In a physical system, the fractal dimension cannot be taken in this limit. Thus, we should allow the definition to be applied over a limited domain of length scales as is appropriate for the problem. As long as the domain of length scales is large, we can use this definition. We then solve the problem of discrete jumps by treating the leading behavior of the function $N(R)$ over this domain.

The problem of treating distinct dimensions at different length scales is only one of the difficulties that we face in discussing fractal systems. Another problem is inhomogeneity. In the following section we discuss objects that are inherently inhomogeneous but for which an alternate natural definition of dimension can be devised to describe their structure on all scales.

1.10.2 *Trees*

Iterative procedures like those used to make fractals can also be used to make geometric objects called trees. An example of a geometric tree, which bears vague

(0)

(1)

(2)

(3)

(4)

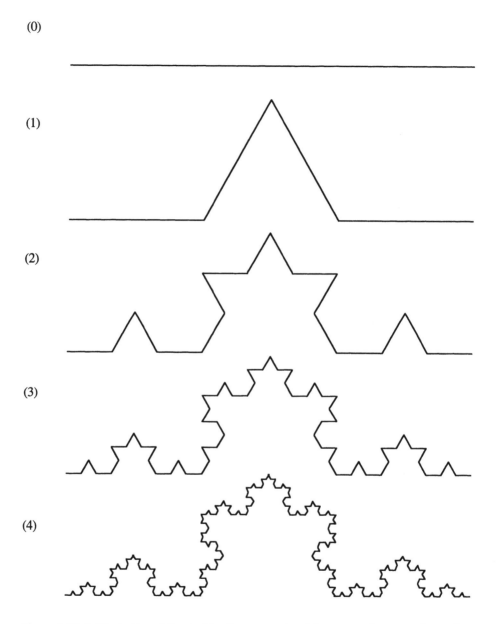

Figure 1.10.4 Illustration of the starting line segment and four successive stages in the formation of the Koch curve. For further discussion see Questions 1.10.3 and 1.10.4. ∎

resemblance to physical trees, is shown in Fig. 1.10.5. The tree is formed by starting with a single object (a line segment), scaling it by a factor of 1/2, duplicating it two times and attaching the parts to the original object at its boundary. This process is then iterated for each of the resulting parts. The iterations create structure on finer and finer scales.

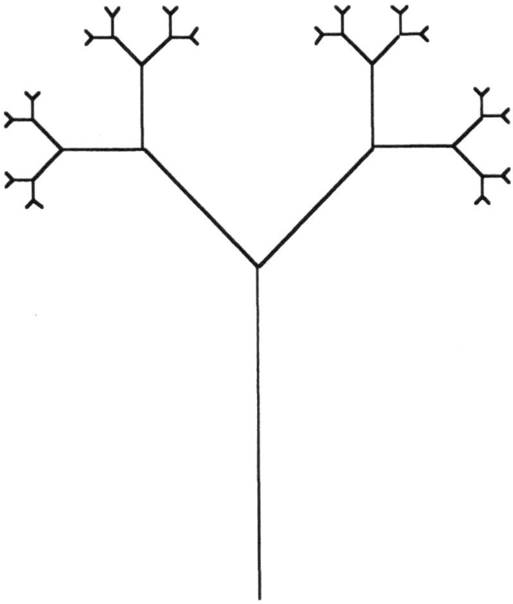

Figure 1.10.5 A geometric tree formed by an iterative algorithm similar to those used in forming fractals. This tree can be formed starting from a single line segment. Two copies of it are then reduced by a factor of 2, rotated by 45° left and right and attached at one end. The procedure is repeated for each of the resulting line segments. Unlike a fractal, a tree is not solely composed out of parts that are self-similar. It is formed out of self-similar parts, along with the original shape — its trunk. ∎

We can generalize the definition of a tree to be a set formed by iteratively adding to an object copies of itself. At iteration t, the added objects are reduced in size by a factor η^t and duplicated N^t times, the duplicated versions being rotated and then shifted by vectors whose lengths converge to zero as a function of t. A tree is different from a fractal because the smaller versions of the original object, are not contained within the original object.

The fractal dimension of trees is not as straightforward as it is for self-similar fractals. The effective fractal dimension can be calculated; however, it gives results that are not intuitively related to the tree structure. We can see why this is a problem in Fig. 1.10.6. The dimension of the region of the tree which is above the size R is that of the embedded entity (line segments), while the fractal dimension of the region which is less than the size R is determined by the spatial structure of the tree. Because of the changing value of R in the scaling relation, an intermediate value for the fractal dimension would typically be found by a direct calculation (Question 1.10.5).

It is reasonable to avoid this problem by classifying trees in a different category than fractals. We can define the tree dimension by considering the self-similarity of the tree structure using the same formula as Eq. (1.10.1), but now applying the definition to the number N and scaling η of the displaced parts of the generating structure, rather than the embedded parts as in the fractal. In Section 1.10.7 we will encounter a treelike structure; however, it will be more useful to describe it rather than to give a dimension that might characterize it.

Q **uestion 1.10.5** A simple version of a tree can be constructed as a set of points $\{1/k\}$ where k takes all positive integer values. The tree dimension

Figure 1.10.6 Illustration of the covering of a geometric tree by disks. The covering shows that the larger-scale structures of the tree (the trunk and first branches in this case) have an effective dimension given by the dimension of their components. The smaller scale structures have a dimension that is determined by the algorithm used to make the tree. This inhomogeneity implies that the fractal dimension is not always the natural way to describe the tree. ∎

of this set is zero because it can be formed from a point which is duplicated and then displaced by progressively smaller vectors. Calculate the fractal dimension of this set.

Solution 1.10.5 We construct a covering of scale R from line segments of this length. The covering that we construct will be formed out of two parts. One part is constructed from segments placed side by side. This part starts from zero and covers infinitely many points of the set. The other part is constructed from segments that are placed on individual points. The crossing point between the two sets can be calculated as the value of k where the difference between successive points is R. For k below this value, it is not possible to include more than one point in one line segment. For k above this value, there are two or more points per line segment. The critical value of k is found by setting:

$$\frac{1}{k_c} - \frac{1}{k_c+1} = \frac{1}{k_c(k_c+1)} \approx \frac{1}{k_c^2} = R \qquad (1.10.8)$$

or $k_c = R^{-1/2}$. This means that the number of segments needed to cover individual points is given by this value. Also, the number of segments that are placed side by side must be enough to go up to this point, which has the value $1/k_c$. This number of segments is given by

$$\frac{1/k_c}{R} = R^{-1/2} \approx k_c \qquad (1.10.9)$$

Thus we must cover the line segment up to the point $R^{1/2}$ with $R^{-1/2}$ line segments, and use an additional $R^{-1/2}$ line segments to cover the rest of the points. This gives a total number of line segments in a covering of $2R^{-1/2}$. The fractal dimension is thus $d = 1/2$.

We could have used fewer line segments in the covering by covering pairs of points and triples of points rather than covering the whole line segment below $1/k_c$. However, each partial covering of the set that is concerned with pairs, triples and so on consists of a number of segments that grows as $R^{-1/2}$. Thus our conclusion remains unchanged by this correction. ∎

Trees illustrate only one example of how system properties may exist on many scales, but are not readily described as fractals in the conventional sense. In order to generalize our concepts to enable the discussion of such properties, we will introduce the concept of scaling.

1.10.3 *Scaling*

Geometric fractals suggest that systems may have a self-similar structure on all length scales. This is in contrast with the more typical approach of science, where there is a specific scale at which a phenomenon appears. We can think about the problem of describing the behavior of a system on multiple length scales in an abstract manner. A phenomenon (e.g., a measurable quantity) may be described by some function of scale, $f(x)$. Here x represents the characteristic scale rather than the position. When there is a well-defined length scale at which a particular effect occurs, for longer length scales the function would typically decay exponentially:

$$f(x) \sim e^{-x/\lambda} \tag{1.10.10}$$

This functional dependence implies that the characteristic scale at which this property disappears is given by λ.

In order for a system property to be relevant over a large range of length scales, it must vary more gradually than exponentially. In such cases, typically, the leading behavior is a power law:

$$f(x) \sim x^\alpha \tag{1.10.11}$$

A function that follows such power-law behavior can also be characterized by the scaling rule:

$$f(ax) = a^\alpha f(x) \tag{1.10.12}$$

This means that if we characterize the system on one scale, then on a scale that is larger by the factor a it has a similar appearance, but scaled by the factor a^α. α is called the scaling exponent. In contrast to the behavior of an exponential, for a power law there is no particular length at which the property disappears. Thus, it may extend over a wide range of length scales. When the scaling exponent is not an integer, the function $f(x)$ is nonanalytic. Non-analyticity is often indicative of a property that cannot be treated by assuming that it becomes smooth on small or large scales. However, fractional scaling exponents are not necessary in order for power-law scaling to be applicable.

Even when a system property follows power-law scaling, the same behavior cannot continue over arbitrarily many length scales. The disappearance of a certain power law may occur because of the appearance of a new behavior on a longer scale. This change is characterized by a crossover in the scaling properties of $f(x)$. An example of crossover occurs when we have a quantity whose scaling behavior is

$$f(x) \sim A_1 x^{\alpha_1} + A_2 x^{\alpha_2} \qquad (1.10.13)$$

If $A_1 > A_2$ and $\alpha_1 < \alpha_2$ then the first term will dominate at smaller length scales, and the second at larger length scales. Alternatively, the power-law behavior may eventually succumb to exponential decay at some length scale.

There are three related approaches to applying the concept of scaling in model or physical systems. The first approach is to consider the scale x to be the physical size of the system, or the amount of matter it contains. The quantity $f(x)$ is then a property of the system measured as the size of the system changes. The second approach is to keep the system the same, but vary the scale of our observation. We assume that our ability to observe the system has a limited degree of discernment of fine details—a finest scale of observation. Finer details are to be averaged over or disregarded. By moving toward or away from the system, we change the physical scale at which our observation can no longer discern details. x then represents the smallest scale at which we can observe variation in the system structure. Finally, in the third approach we consider the relationship between a property measured at one location in the system and the same property measured at another location separated by the distance x. The function $f(x)$ is a correlation of the system measurements as a function of the distance between regions that are being considered.

Examples of quantities that follow scaling relations as a function of system size are the extensive properties of thermodynamic systems (Section 1.3) such as the energy, entropy, free energy, volume, number of particles and magnetization:

$$U(ax) = a^d U(x) \qquad (1.10.14)$$

These properties measure quantities of the whole system as a function of system size. All have the same scaling exponent—the dimension of space. Intrinsic thermodynamic quantities are independent of system size and therefore also follow a scaling behavior where the scaling exponent is zero.

Another example of scaling can be found in the random walk (Section 1.2). We can generalize the discussion in Section 1.2 to allow a walk in d dimensions by choosing steps which are ± 1 in each dimension independently. A random walk of N steps in three dimensions can be thought of as a simple model of a molecule formed as a chain of molecular units—a polymer. If we measure the average distance between the ends of the chain as a function of the number of steps $R(N)$, we have the scaling relation:

$$R(aN) = a^{1/2} R(N) \qquad (1.10.15)$$

This scaling of distance traveled in a random walk with the number of steps taken is independent of dimension. We will consider random walks and other models of polymers in Chapter 5.

Often our interest is in knowing how different parts of the system affect each other. Direct interactions do not always reflect the degree of influence. In complex systems, in which many elements are interacting with each other, there are indirect means of interacting that transfer influence between one part of a system and another. The simplest example is the Ising model, where even short-range interactions can lead to longer-range correlations in the magnetization. The correlation function introduced in Section 1.6.5 measures the correlations between different locations. These correlations show the degree to which the interactions couple the behavior of different parts of the system. Correlations of behavior occur in both space and time. As we mentioned in Section 1.3.4, near a second-order phase transition, there are correlations between different places and times on every length and time scale, because they follow a power-law behavior. This example will be discussed in greater detail in the following section.

Our discussion of scaling also finds application in the theory of computation (Section 1.9) and the practical aspects of simulation (Section 1.7). In addition to the question of computability discussed in Section 1.9, we can also ask how hard it is to compute something. Such questions are generally formulated by describing a class of problems that can be ordered by a parameter N that describes the size of the problem. The objective of the theory of computational complexity is to determine how the number of operations necessary to solve a problem grows with N. A scaling analysis can also be used to compare different algorithms that may solve the same problem. We are often primarily concerned with the scaling behavior (exponential, power law and the value of the scaling exponent) rather than the coefficients of the scaling behavior, because in the comparison of the difficulty of solving different problems or different methodologies this is often, though not always, the most important issue.

1.10.4 *Renormalization group*

General method The renormalization group is a formalism for studying the scaling properties of a system. It starts by assuming a set of equations that describe the behavior of a system. We then change the length scale at which we are describing the system. In effect, we assume that we have a finite ability to see details. By moving away from a system, we lose some of the detail. At the new scale we assume that the same set of equations can be applied, but possibly with different coefficients. The objective is to relate the set of equations at one scale to the set of equations at the other scale. Once this is achieved, the scale-dependent properties of the system can be inferred.

Applications of the renormalization group method have been largely to the study of equilibrium systems, particularly near second-order phase transitions where mean field approaches break down (Section 1.6). The premise of the renormalization group is that exactly at a second-order phase transition, the equations describing the system are independent of scale. In recent years, dynamic renormalization theory has been developed to describe systems that evolve in time. In this section we will describe the more conventional renormalization group for thermodynamic systems.

We illustrate the concepts of renormalization using the Ising model. The Ising model, discussed in Section 1.6, describes the interactions of spins on a lattice. It is a first model of any system that exhibits simple cooperative behavior, such as a magnet.

In order to appreciate the concept of renormalization, it is useful to recognize that the Ising model is not a true microscopic theory of the behavior of a magnet. It might seem that there is a well-defined way to identify an individual spin with a single electron at the atomic level. However, this is far from apparent when equations that describe quantum mechanics at the atomic level are considered. Since the relationship between the microscopic system and the spin model is not manifest, it is clear that our description of the magnet using the Ising model relies upon the macroscopic properties of the model rather than its microscopic nature. Statistical mechanics does not generally attempt to derive macroscopic properties directly from microscopic reality. Instead, it attempts to describe the macroscopic phenomena from simple models. We might not give up hope of identifying a specific microscopic relationship between a particular material and the Ising model, however, the use of the model does not rely upon this identification.

Essential to this approach is that many of the details of the atomic regime are somehow irrelevant at longer length scales. We will return later to discuss the relevance or irrelevance of microscopic details. However, our first question is: What is a single spin variable? A spin variable represents the effective magnetic behavior of a region of the material. There is no particular reason that we should imagine an individual spin variable as representing a small or a large region of the material. Sometimes it might be possible to consider the whole magnet as a single spin in an external field. Identifying the spin with a region of the material of a particular size is an assignment of the length scale at which the model is being applied.

What is the difference between an Ising model describing the system at one length scale and the Ising model describing it on another? The essential point is that the interactions between spins will be different depending on the length scale at which we choose to model the system. The renormalization group takes this discussion one step further by explicitly relating the models at different scales.

In Fig. 1.10.7 we illustrate an Ising model in two dimensions. There is a second Ising model that is used to describe this same system but on a length scale that is twice as big. The first Ising model is described by the energy function (Hamiltonian):

$$E[\{s_i\}] = -c\sum_i 1 - h\sum_i s_i - J\sum_{<ij>} s_i s_j \qquad (1.10.16)$$

For convenience, in what follows we have included a constant energy term $-cN = -c\Sigma 1$. This term does not affect the behavior of the system, however, its variation from scale to scale should be included. The second Ising model is described by the Hamiltonian

$$E'[\{s_i'\}] = -c'\sum_i 1 - h'\sum_i s_i' - J'\sum_{<ij>} s_i' s_j' \qquad (1.10.17)$$

where both the variables and the coefficients have primes. While the first model has N spins, the second model has N' spins. Our objective is to relate these two models. The general process is called renormalization. When we go from the fine scale to the coarse scale by eliminating spins, the process is called decimation.

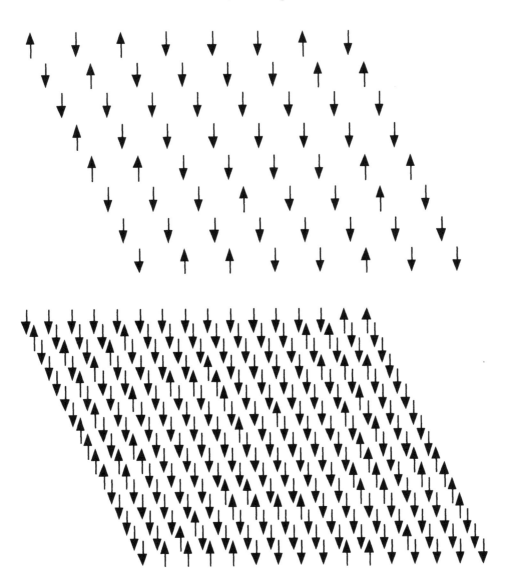

Figure 1.10.7 Schematic illustration of two Ising models in two dimensions. The spins are indicated by arrows that can be UP or DOWN. These Ising models illustrate the modeling of a system with different levels of detail. In the upper model there are one-fourth as many spins as in the lower model. In a renormalization group treatment the parameters of the lower model are related to the parameters of the upper model so that the same system can be described by both. Each of the spins in the upper model, in effect, represents four spins in the lower model. The interactions between adjacent spins in the upper model represent the net effect of the interactions between groups of four spins in the lower model. ∎

There are a variety of methods used for relating models at different scales. Each of them provides a distinct conceptual and practical approach. While in principle they should provide the same answer, they are typically approximated at some stage of the calculation and therefore the answers need not be the same. All the approaches we describe rely upon the partition function to enable direct connection from the microscopic statistical treatment to the macroscopic thermodynamic quantities. For a particular system, the partition function can be written so that it has the same value, independent of which representation is used:

$$Z = \sum_{\{s_i\}} e^{-\beta E[\{s_i\}]} = \sum_{\{s_i'\}} e^{-\beta E'[\{s_i'\}]} \tag{1.10.18}$$

It is conventional and convenient when performing renormalization transformations to set $\beta = 1/kT = 1$. Since β multiplies each of the parameters of the energy function, it is a redundant parameter. It can be reinserted at the end of the calculations.

The different approaches to renormalization are useful for various models that can be studied. We will describe three of them in the following paragraphs because of the importance of the different conceptual treatments. The three approaches are (1) summing over values of a subset of the spins, (2) averaging over a local combination of the spins, and (3) summing over the short wavelength degrees of freedom in a Fourier space representation.

1. Summing over values of a subset of the spins. In the first approach we consider the spins on the larger scale to be a subset of the spins on the finer scale. To find the energy of interaction between the spins on the larger scale we need to eliminate (decimate) some of the spins and replace them by new interactions between the spins that are left. Specifically, we identify the larger scale spins as corresponding to a subset $\{s_i\}_A$ of the smaller scale spins. The rest of the spins $\{s_i\}_B$ must be eliminated from the fine scale model to obtain the coarse scale model. We can implement this directly by using the partition function:

$$e^{-E'[\{s_i'\}]} = \sum_{\{s_i\}_B} e^{-E[\{s_i'\}_A,\{s_i\}_B]} = \sum_{\{s_i\}} e^{-E[\{s_i\}]} \prod_{i\in A} \delta_{s_i',s_i} \tag{1.10.19}$$

In this equation we have identified the spins on the larger scale as a subset of the finer scale spins and have summed over the finer scale spins to obtain the effective energy for the larger scale spins.

2. Averaging over a local combination of the spins. We need not identify a particular spin of the finer scale with a particular spin of the coarser scale. We can choose to identify some function of the finer scale spins with the coarse scale spin. For example, we can identify the majority rule of a certain number of fine scale spins with the coarse scale spins:

$$e^{-E[\{s_i'\}]} = \sum_{\{s_i\}} \prod_{i\in A} \delta_{s_i',\mathrm{sign}(\sum s_i)} e^{-E[\{s_i\}]} \tag{1.10.20}$$

This is easier to think about when an odd number of spins are being renormalized to become a single spin. Note that this is quite similar to the concept of defining a collective coordinate that we used in Section 1.4 in discussing the two-state system. The difference here is that we are defining a collective coordinate out of only a few original coordinates, so that the reduction in the number of degrees of freedom is comparatively small. Note also that by convention we continue to use the term "energy," rather than "free energy," for the collective coordinates.

3. Summing over the short wavelength degrees of freedom in a Fourier space representation. Rather than performing the elimination of spins directly, we may recognize that our procedure is having the effect of removing the fine scale variation in the problem. It is natural then to consider a Fourier space representation where we can remove the rapid changes in the spin values by eliminating the higher Fourier components. To do this we need to represent the energy function in terms of the Fourier transform of the spin variables:

$$s_k = \sum_i e^{ikx_i} s_i \tag{1.10.21}$$

Writing the Hamiltonian in terms of the Fourier transformed variables, we then sum over the values of the high frequency terms:

$$e^{-E[\{s_k\}]} = \sum_{\{s_k\}_{k<k_0}} e^{-E[\{s_k\}]} \tag{1.10.22}$$

The remaining coordinates s_k have $k > k_0$.

All of the approaches described above typically require some approximation in order to perform the analysis. In general there is a conservation of effort in that the same difficulties tend to arise in each approach, but with different manifestation. Part of the reason for the difficulties is that the Hamiltonian we use for the Ising model is not really complete. This means that there can be other parameters that should be included to describe the behavior of the system. We will see this by direct application in the following examples.

Ising model in one dimension We illustrate the basic concepts by applying the renormalization group to a one-dimensional Ising model where the procedure can be done exactly. It is convenient to use the first approach (number 1 above) of identifying a subset of the fine scale spins with the larger scale model. We start with the case where there is an interaction between neighboring spins, but no magnetic field:

$$E[\{s_i\}] = -c\sum_i 1 - J\sum_{<ij>} s_i s_j \tag{1.10.23}$$

We sum the partition function over the odd spins to obtain

$$Z = \sum_{\{s_i\}_{even}} \sum_{\{s_i\}_{odd}} e^{c\sum_i 1 + J\sum_i s_i s_{i+1}} = \sum_{\{s_i\}_{even}} \prod_{i\,even} 2\cosh(J(s_i + s_{i+2}))e^{2c} \tag{1.10.24}$$

We equate this to the energy for the even spins by themselves, but with primed quantities:

$$Z = \sum_{\{s_i\}_{even}} e^{\sum_i c' + J' \sum_i s_i s_{i+2}} = \sum_{\{s_i\}_{even}} \prod_{i even} 2\cosh(J(s_i + s_{i+2}))e^{2c} \qquad (1.10.25)$$

This gives:

$$e^{c' + J' s_i s_{i+2}} = 2\cosh(J(s_i + s_{i+2}))e^{2c} \qquad (1.10.26)$$

or

$$c' + J' s_i s_{i+2} = \ln(2\cosh(J(s_i + s_{i+2}))) + 2c \qquad (1.10.27)$$

Inserting the two distinct combinations of values of s_i and s_{i+2} ($s_i = s_{i+2}$ and $s_i = -s_{i+2}$), we have:

$$c' + J' = \ln(2\cosh(2J)) + 2c$$
$$c' - J' = \ln(2\cosh(0)) + 2c = \ln(2) + 2c \qquad (1.10.28)$$

Solving these equations gives the primed quantities for the larger scale model as:

$$J' = (1/2)\ln(\cosh(2,J))$$
$$c' = 2c + (1/2)\ln(4\cosh(2J)) \qquad (1.10.29)$$

This is the renormalization group relationship that we have been looking for. It relates the values of the parameters in the two different energy functions at the different scales.

While it may not be obvious by inspection, this iterative map always causes J to decrease. We can see this more easily if we transform the relationship of J to J' to the equivalent form:

$$\tanh(J') = \tanh(J)^2 \qquad (1.10.30)$$

This means that on longer and longer scales the effective interaction between neighboring spins becomes smaller and smaller. Eventually the system on long scales behaves as a string of decoupled spins.

The analysis of the one-dimensional Ising model can be extended to include a magnetic field. The decimation step becomes:

$$Z = \sum_{\{s_i\}_{even}} \sum_{\{s_i\}_{odd}} e^{c\sum_i 1 + h\sum_i s_i + J\sum_i s_i s_{i+1}} = \sum_{\{s_i\}_{even}} \prod_{i odd} 2\cosh(h + J(s_i + s_{i+2}))e^{2c} \qquad (1.10.31)$$

We equate this to the coarse scale partition function:

$$Z = \sum_{\{s_i\}_{odd}} e^{c' + h'\sum_i s_i' + J'\sum_i s_i' s_{i+1}'} = \sum_{\{s_i\}_{odd}} \prod_{i odd} 2\cosh(h + J(s_i + s_{i+2}))e^{2c} \qquad (1.10.32)$$

which requires that:

$$c' + h' + J' = h + \ln(2\cosh(h + 2J)) + 2c$$
$$c' - J' = \ln(2\cosh(h)) + 2c \qquad (1.10.33)$$
$$c' - h' + J' = -h + \ln(2\cosh(h - 2J)) + 2c$$

We solve these equations to obtain:

$$c' = 2c + (1/4)\ln(16\cosh(h + 2J)\cosh(h - 2J)\cosh(h)^2)$$
$$J' = (1/4)\ln(\cosh(h + 2J)\cosh(h - 2J)/\cosh(h)^2) \qquad (1.10.34)$$
$$h' = h + (1/2)\ln(\cosh(h + 2J)/\cosh(h - 2J))$$

which is the desired renormalization group transformation. The renormalization transformation is an iterative map in the parameter space (c, h, J).

We can show what happens in this iterative map using a plot of changes in the values of J and h at a particular value of these parameters. Such a diagram of flows in the parameter space is illustrated in Fig. 1.10.8. We can see from the figure or from Eq. (1.10.34) that there is a line of fixed points of the iterative map at $J = 0$ with arbitrary

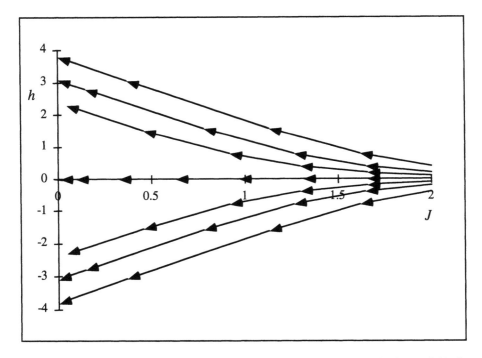

Figure 1.10.8 The renormalization transformation for the one-dimensional Ising model is illustrated as an iterative flow diagram in the two-dimensional (h,J) parameter space. Each of the arrows represents the effect of decimating half of the spins. We see that after a few iterations the value of J becomes very small. This indicates that the spins become decoupled from each other on a larger scale. The absence of any interaction on this scale means that there is no phase transition in the one-dimensional Ising model. ∎

value of h. This simply means that the spins are decoupled. For $J = 0$ on any scale, the behavior of the spins is determined by the value of the external field.

The line of fixed points at $J = 0$ is a stable (attracting) set of fixed points. The flow lines of the iterative map take us to these fixed points on the attractor line. In addition, there is an unstable fixed point at $J = \infty$. This would correspond to a strongly coupled line of spins, but since this fixed point is unstable it does not describe the large scale behavior of the model. For any finite value of J, changing the scale rapidly causes the value of J to become small. This means that the large scale behavior is always that of a system with $J = 0$.

Ising model in two dimensions In the one-dimensional case treated in the previous section, the renormalization group works perfectly and is also, from the point of view of studying phase transitions, uninteresting. We will now look at two dimensions, where the renormalization group must be approximated and where there is also a phase transition.

We can simplify our task in two dimensions by eliminating half of the spins (Fig. 1.10.9) instead of three out of four spins as illustrated previously in Fig. 1.10.7. Eliminating half of the spins causes the square cell to be rotated by 45°, but this should not cause any problems. Labeling the spins as in Fig. 1.10.9 we write the decimation step for a Hamiltonian with $h = 0$:

$$Z = \sum_{\{s_i\}_A} \sum_{\{s_i\}_B} e^{c\sum_i 1 + J\sum_i s_0(s_1 + s_2 + s_3 + s_4)}$$

$$= \sum_{\{s_i\}_A} \prod_{i \in B} 2\cosh(J(s_1 + s_2 + s_3 + s_4))e^c \qquad (1.10.35)$$

$$= \sum_{\{s_i\}_A} \prod_{i \in B} e^{c' + (J'/2)(s_1 s_2 + s_2 s_3 + s_3 s_4 + s_4 s_1)}$$

In the last expression we take into consideration that each bond of the form $s_1 s_2$ appears in two squares and each spin appears in four squares.

In order to solve Eq. (1.10.35) for the values of c' and J' we must insert all possible values of the spins (s_1, s_2, s_3, s_4). However, this leads to a serious problem. There are four distinct equations that arise from the different values of the spins. This is reduced from $2^4 = 8$ because, by symmetry, inverting all of the spins gives the same answer. The problem is that while there are four equations, there are only two unknowns to solve for, c' and J'. The problem can be illustrated by recognizing that there are two distinct ways to have two spins UP and two spins DOWN. One way is to have the spins that are the same be adjacent to each other, and the other way is to have them be opposite each other across a diagonal. The two ways give the same result for the value of $(s_1 + s_2 + s_3 + s_4)$ but different results for $(s_1 s_2 + s_2 s_3 + s_3 s_4 + s_4 s_1)$.

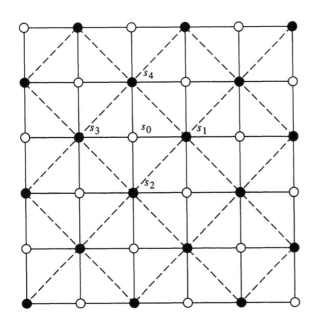

Figure 1.10.9 In a renormalization treatment of the two-dimensional Ising model it is possible to decimate one out of two spins as illustrated in this figure. The black dots represent spins that remain in the larger-scale model, and the white dots represent spins that are decimated. The nearest-neighbor interactions in the larger-scale model are shown by dashed lines. As discussed in the text, the process of decimation introduces new interactions between spins across the diagonal, and four spin interactions between spins around a square. ∎

In order to solve this problem, we must introduce additional parameters which correspond to other interactions in the Hamiltonian. To be explicit, we would make a table of symmetry-related combinations of the four spins as follows:

(s_1,s_2,s_3,s_4)	(1,1,1,1)	(1,1,1,−1)	(1,1,−1,−1)	(1,−1,1,−1)	
1	1	1	1	1	
$(s_1 + s_2 + s_3 + s_4)$	4	2	0	0	
$(s_1 s_2 + s_2 s_3 + s_3 s_4 + s_4 s_1)$	4	0	0	−4	(1.10.36)
$(s_1 s_3 + s_2 s_4)$	2	0	−2	2	
$s_1 s_2 s_3 s_4$	1	−1	1	1	

In order to make use of these to resolve the problems with Eq. (1.10.35), we must introduce new interactions in the Hamiltonian and new parameters that multiply them. This leads to second-neighbor interactions (across a cell diagonal), and four spin interactions around a square:

$$E[\{s_i\}] = -c'\sum_i 1 - J'\sum_{<ij>} s_i s_j - K' \sum_{<<ij>>} s_i s_j - L' \sum_{<ijkl>} s_i s_j s_k s_l \qquad (1.10.37)$$

where the notation $<< ij >>$ indicates second-neighbor spins across a square diagonal, and $< ijkl >$ indicates spins around a square. This might seem to solve our problem. However, we started out from a Hamiltonian with only two parameters, and now we are switching to a Hamiltonian with four parameters. To be self-consistent, we should start from the same set of parameters we end up with. When we start with the additional parameters K and L this will, however, lead to still further terms that should be included.

Relevant and irrelevant parameters In general, as we eliminate spins by renormalization, we introduce interactions between spins that might not have been included in the original model. We will have interactions between second or third neighbors or between more than two spins at a time. In principle, by using a complete set of parameters that describe the system we can perform the renormalization transformation and obtain the flows in the parameter space. These flows tell us about the scale-dependent properties of the system.

We can characterize the flows by focusing on the fixed points of the iterative map. These fixed points may be stable or unstable. When a fixed point is unstable, renormalization takes us away from the fixed point so that on a larger scale the properties of the system are found to be different from the values at the unstable fixed point. Significantly, it is the unstable fixed points that represent the second-order phase transitions. This is because deviating from the fixed point in one direction causes the parameters to flow in one direction, while deviating from the fixed point in another direction causes the parameters to flow in a different direction. Thus, the macroscopic properties of the system depend on the direction microscopic parameters deviate from the fixed point—a succinct characterization of the nature of a phase transition.

Using this characterization of fixed points, we can now distinguish between different types of parameters in the model. This includes all of the additional parameters that might be introduced in order to achieve a self-consistent renormalization transformation. There are two major categories of parameters: relevant or irrelevant. Starting near a particular fixed point, changes in a relevant parameter grow under renormalization. Changes in an irrelevant parameter shrink. Because renormalization indicates the values of system parameters on a larger scale, this tells us which microscopic parameters are important to the macroscopic scale. When observed on the macroscopic scale, relevant parameters change at the phase transition, while irrelevant parameters do not. A relevant parameter should be included in the Hamiltonian because its value affects the macroscopic behavior. An irrelevant parameter may often be included in the model in a more approximate way. Marginal parameters are the borderline cases that neither grow nor shrink at the fixed point.

Even when we are not solely interested in the behavior of a system at a phase transition, but rather are concerned with its macroscopic properties in general, the definition of "relevant" and "irrelevant" continues to make sense. If we start from a particular microscopic description of the system, we can ask which parameters are relevant for the macroscopic behavior. The relevant parameters are the ones that can affect the macroscopic behavior of the system. Thus, a change in a relevant microscopic parameter changes the macroscopic behavior. In terms of renormalization, changes in relevant parameters do not disappear as a result of renormalization.

We see that the use of any model, such as the Ising model, to model a physical system assumes that all of the parameters that are essential in describing the system have been included. When this is true, the results are universal in the sense that all microscopic Hamiltonians will give rise to the same behavior. Additional terms in the Hamiltonian cannot affect the macroscopic behavior. We know that the microscopic behavior of the physical system is not really described by the Ising model or any other simple model. Thus, in creating models we always rely upon the concept, if not the process, of renormalization to make many of the microscopic details disappear, enabling our simple models to describe the macroscopic behavior of the physical system.

In the Ising model, in addition to longer range and multiple spin interactions, there is another set of parameters that may be relevant. These parameters are related to the use of binary variables to describe the magnetization of a region of the material. It makes sense that the process of renormalization should cause the model to have additional spin values that are intermediate between fully magnetized UP and fully magnetized DOWN. In order to accommodate this, we might introduce a continuum of possible magnetizations. Once we do this, the amplitude of the magnetization has a probability distribution that will be controlled by additional parameters in the Hamiltonian. These parameters may also be relevant or irrelevant. When they are irrelevant, the Ising model can be used without them. However, when they are relevant, a more complete model should be used.

The parameters that are relevant generally depend on the dimensionality of space. From our analysis of the behavior of the one-dimensional Ising model, the parameter J is irrelevant. It is clearly irrelevant because not only variations in J but J itself disappears as the scale increases. However, in two dimensions this is not true.

For our purposes we will be satisfied by simplifying the renormalization treatment of the two-dimensional Ising model so that no additional parameters are introduced. This can be done by a fourth renormalization group technique which has some conceptual as well as practical advantages over the others. However, it does hide the importance of determining the relevant parameters.

Bond shifting We simplify our analysis of the two-dimensional Ising model by making use of the Migdal-Kadanoff transformation. This renormalization group technique is based on the recognition that the correlation between adjacent spins can enable us to, in effect, substitute the role of one spin for another. As far as the coarser scale model is concerned, the function of the finer scale spins is to mediate the interaction between the coarser scale spins. Because one spin is correlated to the behavior of its neighbor, we can shift the responsibility for this interaction to a neighbor, and use this shift to simplify elimination of the spins.

To apply these ideas to the two-dimensional Ising model, we move some of the interactions (bonds) between spins, as shown in Fig. 1.10.10. We note that the distance over which the bonds act is preserved. The net result of the bond shifting is that we form short linear chains that can be renormalized just like a one-dimensional chain. The renormalization group transformation is thus done in two steps. First we shift the bonds, then we decimate. Once the bonds are moved, we write the renormalization of the partition function as:

$$
\begin{aligned}
Z &= \sum_{\{s_i\}_A} \sum_{\{s_i\}_B} \sum_{\{s_i\}_C} e^{c\sum_i 1 + 2J\sum_i s_0(s_1+s_2)} \\
&= \sum_{\{s_i\}} \prod_{i\in A} 2\cosh(2J(s_1+s_2))e^{4c} \\
&= \sum_{\{s_i\}} \prod_{i\in A} e^{c'+J'(s_1 s_2)}
\end{aligned}
\tag{1.10.38}
$$

(a)

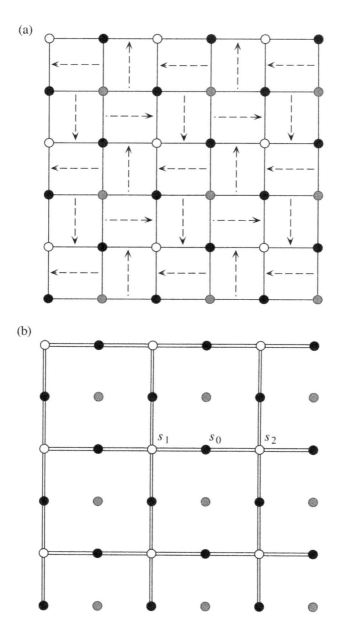

(b)

Figure **1.10.10** Illustration of the Migdal-Kadanoff renormalization transformation that enables us to bypass the formation of additional interactions. In this approach some of the interactions between spins are moved to other spins. If all the spins are aligned (at low temperature or high J), then shifting bonds doesn't affect the spin alignment. At high temperature, when the spins are uncorrelated, the interactions are not important anyway. Near the phase transition, when the spins are highly correlated, shifting bonds still makes sense. A pattern of bond movement is illustrated in (a) that gives rise to the pattern of doubled bonds in (b). Note that we are illustrating only part of a periodic lattice, so that bonds are moved into and out of the region illustrated. Using the exact renormalization of one-dimensional chains, the gray spins and the black spins can be decimated to leave only the white spins. ∎

The spin labels s_0, s_1, s_2 are assigned along each doubled bond, as indicated in Fig. 1.10.10. The three types of spin A, B and C correspond to the white, black and gray dots in the figure. The resulting equation is the same as the one we found when performing the one-dimensional renormalization group transformation with the exception of factors of two. It gives the result:

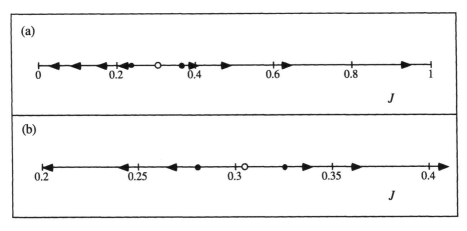

Figure 1.10.11 The two-dimensional Ising model renormalization group transformation obtained from the Migdal-Kadanoff transformation is illustrated as a flow diagram in the one-dimensional parameter space (J). The arrows show the effect of successive iterations starting from the black dots. The white dot indicates the position of the unstable fixed point, J_c, which is the phase transition in this model. Starting from values of J slightly below J_c, iteration results in the model on a large scale becoming decoupled with no interactions between spins ($J \to 0$). This is the high-temperature phase of the material. However, starting from values of J slightly above J_c iteration results in the model on the large scale becoming strongly coupled ($J \to \infty$) and spins are aligned. (a) shows only the range of values from 0 to 1, though the value of J can be arbitrarily large. (b) shows an enlargement of the region around the fixed point. ∎

$$J' = (1/2)\ln(\cosh(4J))$$
$$c' = 4c + (1/2)\ln(4\cosh(4J)) \tag{1.10.39}$$

The renormalization of J in the two-dimensional Ising model turns out to behave qualitatively different from the one-dimensional case. Its behavior is plotted in Fig. 1.10.11 using a flow diagram. There is an unstable fixed point of the iterative map at $J \approx .305$. This nonzero and noninfinite fixed point indicates that we have a phase transition. Reinserting the temperature, we see that the phase transition occurs at $\beta J = .305$ which is significantly larger than the mean field result $\beta zJ = 1$ or $\beta J = .25$ found in Section 1.6. The exact value for the phase transition for this lattice, $\beta J \approx .441$, which can be obtained analytically by other techniques, is even larger.

It turns out that there is a trick that can give us the exact transition point using a similar renormalization transformation. This trick begins by recognizing that we could have moved bonds in a larger square. For a square with b cells on a side, we would end up with each bond on the perimeter being replaced by a bond of strength b. Using Eq. (1.10.30) we can infer that a chain of b bonds of strength bJ gives rise to an effective interaction whose strength is

$$J'(b) = \tanh^{-1}(\tanh(bJ)^b) \tag{1.10.40}$$

The trick is to take the limit $b \to 1$, because in this limit we are left with the original Ising model. Extending b to nonintegral values by analytic continuation may seem a little strange, but it does make a kind of sense. We want to look at the incremental change in J as a result of renormalization, with b incrementally different from 1. This can be most easily found by taking the hyperbolic tangent of both sides of Eq. (1.10.40), and then taking the derivative with respect to b. The result is:

$$\left.\frac{dJ'(b)}{db}\right|_{b=1} = J + \sinh(J)\cosh(J)\ln(\tanh(J)) \tag{1.10.41}$$

Setting this equal to zero to find the fixed point of the transformation actually gives the exact result for the phase transition.

The renormalization group gives us more information than just the location of the phase transition. Fig. 1.10.11 shows changes that occur in the parameters as the length scale is varied. We can use this picture to understand the behavior of the Ising model in some detail. It shows what happens on longer length scales by the direction of the arrows. If the flow goes toward a particular point, then we can tell that on the longest (thermodynamic) length scale the behavior will be characterized by the behavior of the model at that point. By knowing how close we are to the original phase transition, we can also learn from the renormalization group what is the length scale at which the behavior characteristic of the phase transition will disappear. This is the length scale at which the iterative map leaves the region of the repelling fixed point and moves to the attracting one.

We can also characterize the relationship between systems at different values of the parameters: temperatures or magnetic fields. Renormalization takes us from a system at one value of βJ to another. Thus, we can relate the behavior of a system at one temperature to another by performing the renormalization for both systems and stopping both at a particular value of βJ. At this point we can directly relate properties of the two systems, such as their free energies. Different numbers of renormalization steps in the two cases mean that we are relating the two systems at different scales. Such descriptions of relationships of the properties of one system at one scale with another system at a different scale are known as scaling functions because they describe how the properties of the system change with scale.

The renormalization group was developed as an analytic tool for studying the scaling properties of systems with spatially arrayed interacting parts. We will study another use of renormalization in Section 1.10.5. Then in Section 1.10.6 we will introduce a computational approach—the multigrid method.

Question 1.10.6 In this section we displayed our iterative maps graphically as flow diagrams, because in renormalization group transformations we are often interested in maps that involve more than one variable. Make a diagram like Fig. 1.1.1 for the single variable J showing the iterative renormalization group transformation for the two-dimensional Ising model as given in Eq. (1.10.39).

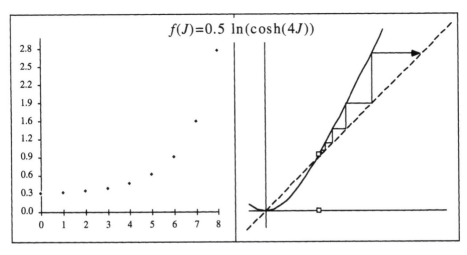

Figure 1.10.12 The iterative map shown as a flow diagram in Fig. 1.10.11 is shown here in the same manner as the iterative maps in Section 1.1. On the left are shown the successive values of J as iteration proceeds. Each iteration should be understood as a loss of detail in the model and hence an observation of the system on a larger scale. Since in general our observations of the system are macroscopic, we typically observe the limiting behavior as the iterations go to ∞. This is similar to considering the limiting behavior of a standard iterative map. On the right is the graphical method of determining the iterations as discussed in Section 1.1. The fixed points are visible as intersections of the iterating function with the diagonal line. ∎

Solution 1.10.6 See Fig. 1.10.12. The fixed point and the iterative behavior are readily apparent. ∎

1.10.5 *Renormalization and chaos*

Our final example of renormalization brings us back to Section 1.1, where we studied the properties of iterative maps and the bifurcation route to chaos. According to our discussion, cycles of length 2^k, $k = 0,1,2,...$, appeared as the parameter a was varied from 0 to $a_c = 3.56994567$, at which point chaotic behavior appeared. Fig. 1.1.3 summarizes the bifurcation route to chaos. A schematic of the bifurcation part of this diagram is reproduced in Fig. 1.10.13. A brief review of Section 1.1 may be useful for the following discussion.

The process of bifurcation appears to be a self-similar process in the sense that the appearance of a 2-cycle for $f(s)$ is repeated in the appearance of a 2-cycle for $f^2(s)$, but over a smaller range of a. The idea of self-similarity seems manifest in Fig. 1.10.13, where we would only have to change the scale of magnification in the s and a directions in order to map one bifurcation point onto the next one. While this mapping might not work perfectly for smaller cycles, it becomes a better and better

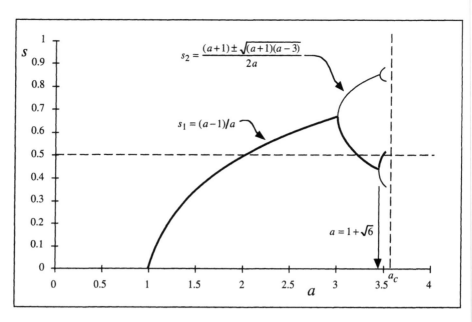

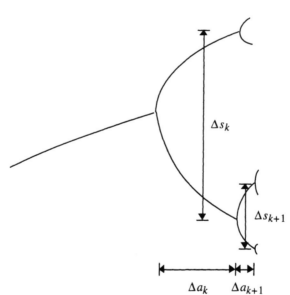

Figure 1.10.13 Schematic reproduction of Fig. 1.1.4, which shows the bifurcation route to chaos. Successive branchings are approximately self-similar. The bottom figure shows the definition of the scaling factors that relate the successive branchings. The horizontal rescaling of the branches, δ, is given by the ratio of Δa_k to Δa_{k+1}. The vertical rescaling of the branches, α, is given by the ratio of Δs_k to Δs_{k+1}. The top figure shows the values from which we can obtain a first approximation to the values of α and δ, by taking the ratios from the zeroth, first and second bifurcations. The zeroth bifurcation point is actually the point $a = 1$. The first bifurcation point occurs at $a = 3$. the second occurs at $a = 1 + \sqrt{6}$. The values of s at the bifurcation points were obtained in Section 1.1, and formulas are indicated on the figure. When the scaling behavior of the tree is analyzed using a renormalization group treatment, we focus on the tree branches that cross $s = 1/2$. These are indicated by bold lines in the top figure. ∎

approximation as the number of cycles increases. The bifurcation diagram is thus a treelike object. This means that the sequence of bifurcation points forms a geometrically converging sequence, and the width of the branches is also geometrically converging. However, the distances in the s and a directions are scaled by different factors. The factors that govern the tree rescaling at each level are δ and α, as shown in Fig. 1.10.13 (b):

$$\delta = \lim_{k \to \infty} \frac{\Delta a_k}{\Delta a_{k+1}}$$

$$\alpha = \lim_{k \to \infty} \frac{\Delta s_k}{\Delta s_{k+1}}$$

(1.10.42)

By this convention, the magnitude of both α and δ is greater than one. α is defined to be negative because the longer branch flips up to down at every branching. The values are to be obtained by taking the limit as $k \to \infty$ where these scale factors have well-defined limits.

We can find a first approximation to these scaling factors by using the values at the first and second bifurcations that we calculated in Section 1.1. These values, given in Fig. 1.10.13, yield:

$$\delta \approx (3-1)/(1+\sqrt{6}-3) = 4.449$$

(1.10.43)

$$\alpha \approx \frac{2s_1 \big|_{a=3}}{\left(s_2^+ - s_2^-\right)\big|_{a=1+\sqrt{6}}} = \frac{4}{3} \frac{a}{\sqrt{(a+1)(a-3)}}\bigg|_{a=1+\sqrt{6}} = 3.252$$

(1.10.44)

Numerically, the asymptotic value of δ for large k is found to be 4.6692016. This differs from our first estimate by only 5%. The numerical value for α is 2.50290787, which differs from our first estimate by a larger margin of 30%.

We can determine these constants with greater accuracy by studying directly the properties of the functions $f, f^2, \ldots f^{2^k} \ldots$ that are involved in the formation of 2^k cycles. In order to do this we modify our notation to explicitly include the dependence of the function on the parameter a. $f(s,a), f^2(s,a)$, etc. Note that iteration of the function f only applies to the first argument.

The tree is formed out of curves $s_{2^k}(a)$ that are obtained by solving the fixed point equation:

$$s_{2^k}(a) = f^{2^k}(s_{2^k}(a), a)$$

(1.10.45)

We are interested in mapping a segment of this curve between the values of s where

$$\frac{df^{2^k}(s,a)}{ds} = 1$$

(1.10.46)

and

$$\frac{df^{2^k}(s,a)}{ds} = -1$$

(1.10.47)

onto the next function, where k is replaced everywhere by $k + 1$. This mapping is a kind of renormalization process similar to that we discussed in the previous section. In order to do this it makes sense to expand this function in a power series around an intermediate point, which is the point where these derivatives are zero. This is known as the superstable point of the iterative map. The superstable point is very convenient for study, because for any value of k there is a superstable point at $s = 1/2$. This follows because $f(s,a)$ has its maximum at $s = 1/2$, and so its derivative is zero. By the chain rule, the derivative of $f^{2^k}(s,a)$, is also zero. As illustrated in Fig. 1.10.13, the line at $s = 1/2$ intersects the bifurcation tree at every level of the hierarchy at an intermediate point between bifurcation points. These intersection points must be superstable.

It is convenient to displace the origin of s to be at $s = 1/2$, and the origin of a to be at the convergence point of the bifurcations. We thus define a function g which represents the structure of the tree. It is approximately given by:

$$g(s,a) \approx f(s + 1/2, a + a_c) - 1/2 \qquad (1.10.48)$$

However, we would like to represent the idealized tree rather than the real tree. The idealized tree would satisfy the scaling relation exactly at all values of a. Thus g should be the analog of the function f which would give us an ideal tree. To find this function we need to expand the region near $a = a_c$ by the appropriate scaling factors. Specifically we define:

$$g(s,a) = \lim_{k \to \infty} \alpha^k f^{2^k}(s/\alpha^k + 1/2, a/\delta^k + a_c) - 1/2 \qquad (1.10.49)$$

The easiest way to think about the function $g(s,a)$ is that it is quite similar to the quadratic function $f(s,a)$ but it has the form necessary to cause the bifurcation tree to have the ideal scaling behavior at every branching. We note that $g(s,a)$ depends on the behavior of $f(s,a)$ only very near to the point $s = 1/2$. This is apparent in Eq. (1.10.49) since the region near $s = 1/2$ is expanded by a factor of α^k.

We note that $g(s,a)$ has its maximum at $s = 0$. This is a consequence of the shift in origin that we chose to make in defining it.

Our objective is to find the form of $g(s,a)$ and, with this form, the values of α and δ. The trick is to recognize that what we need to know can be obtained directly from its scaling properties. To write the scaling properties we look at the relationship between successive iterations of the map and write:

$$g(s,a) = \alpha g^2(s/\alpha, a/\delta) \qquad (1.10.50)$$

This follows either from our discussion and definition of the scaling parameters α and δ or directly from Eq. (1.10.49).

For convenience, we analyze Eq. (1.10.50) first in the limit $a \to 0$. This corresponds to looking at the function $g(s,a)$ as a function of s at the limit of the bifurcation sequence. This function (Fig. 1.10.14) still looks quite similar to our original function $f(s)$, but its specific form is different. It satisfies the relationship:

$$g(s,0) = g(s) = \alpha g^2(s/\alpha) \qquad (1.10.51)$$

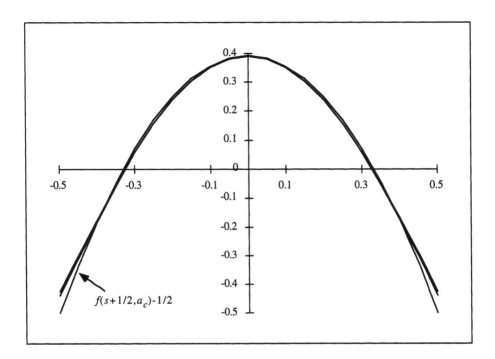

Figure 1.10.14 Three functions are plotted that are successive approximations to $g(s) = g(s, 0)$. This function is essentially the limiting behavior of the quadratic iterative map $f(s)$ at the end of the bifurcation tree a_c. The functions plotted are the first three k values inserted in Eq. (1.10.49): $f(s + 1/2, a + a_c) - 1/2$, $af^2(s/\alpha + 1/2, a_c) - 1/2$ and $\alpha^2 f^4(s/\alpha^2 + 1/2, a_c) - 1/2$. The latter two are almost indistinguishable, indicating that the sequence of functions converges rapidly. ∎

We approximate this function by a quadratic with no linear term because $g(s)$ has its maximum at $s = 0$:

$$g(s) \approx g_0 + cs^2 \qquad (1.10.52)$$

Inserting into Eq. (1.10.51) we obtain:

$$g_0 + cs^2 \approx \alpha(g_0 + c(g_0 + c(s/\alpha)^2)^2) \qquad (1.10.53)$$

Equating separately the coefficients of the first and second terms in the expansion gives the solution:

$$\alpha = 1 / (1 + cg_0)$$
$$\alpha = 2cg_0 \qquad (1.10.54)$$

We see that c and g_0 only appear in the combination cg_0, which means that there is one parameter that is not determined by the scaling relationship. However, this does not prevent us from determining α. Eq. (1.10.54) can be solved to obtain:

$$cg_0 = (-1 \pm \sqrt{3})/2 = -1.3660254$$
$$\alpha = (-1 \pm \sqrt{3}) = -2.73205081 \tag{1.10.55}$$

We have chosen the negative solutions because the value of α and the value of cg_0 must be negative.

We return to consider the dependence of $g(s,a_,)$ on a to obtain a new estimate for δ. Using a first-order linear dependence on a we have:

$$g(s,a_,) \approx g_0 + cs^2 + ba \tag{1.10.56}$$

Inserting into Eq. (1.10.50) we have:

$$g_0 + cs^2 + ba \approx \alpha(g_0 + c(g_0 + c(s/\alpha)^2 + ba/\delta)^2 + ba/\delta) \tag{1.10.57}$$

Taking only the first order term from this equation in a we have:

$$\delta = 2\alpha cg_0 + \alpha = 4.73205 \tag{1.10.58}$$

Eq. (1.10.55) and Eq. (1.10.58) are a significant improvement over Eq. (1.10.44) and Eq. (1.10.43). The new value of α is less than 10% from the exact value. The new value of δ is less than 1.5% from the exact value. To improve the accuracy of the results, we need only add additional terms to the expansion of $g(s,a)$ in s. The first-order term in a is always sufficient to obtain the corresponding value of δ.

It is important, and actually central to the argument in this section, that the explicit form of $f(s,a)$ never entered into our discussion. The only assumption was that the functional behavior near the maximum is quadratic. The rest of the argument follows independent of the form of $f(s,a)$ because we are looking at its properties after many iterations. These properties depend only on the region right in the vicinity of the maximum of the function. Thus only the first-order term in the expansion of the original function $f(s,a)$ matters. This illustrates the notion of universality so essential to the concept of renormalization—the behavior is controlled by very few parameters. All other parameters are irrelevant—changing their values in the original iterative map is irrelevant to the behavior after many iterations (many renormalizations) of the iterative map. This is similar to the study of renormalization in models like the Ising model, where most of the details of the behavior at small scales no longer matter on the largest scales.

1.10.6 Multigrid

The multigrid technique is designed for the solution of computational problems that benefit from a description on multiple scales. Unlike renormalization, which is largely an analytic tool, the multigrid method is designed specifically as a computational tool. It works well when a problem can be approximated using a description on a coarse lattice, but becomes more and more accurate as the finer-scale details on finer-scale lattices are included. The idea of the method is not just to solve an equation on finer and finer levels of description, but also to correct the coarser-scale equations based on the finer-scale results. In this way the methodology creates an improved description of the problem on the coarser-scale.

The multigrid approach relies upon iterative refinement of the solution. Solutions on coarser scales are used to approximate the solutions on finer scales. The finer-scale solutions are then iteratively refined. However, by correcting the coarser-scale equations, it is possible to perform most of the iterative refinement of the fine-scale solution on the coarser scales. Thus the iterative refinement of the solution is based both upon correction of the solution and correction of the equation. The idea of correcting the equation is similar in many ways to the renormalization group approach. However, in this case it is a particular solution, which may be spatially dependent, rather than an ensemble averaging process, which provides the correction.

We explain the multigrid approach using a conventional problem, which is the solution of a differential equation. To solve the differential equation we will find an approximate solution on a grid of points. Our ultimate objective is to find a solution on a fine enough grid so that the solution is within a prespecified accuracy of the exact answer. However, we will start with a much coarser grid solution and progressively refine it to obtain more accurate results. Typically the multigrid method is applied in two or three dimensions, where it has greater advantages than in one dimension. However, we will describe the concepts in one dimension and leave out many of the subtleties.

For concreteness we will assume a differential equation which is:

$$\frac{d^2 f(x)}{dx^2} = g(x) \tag{1.10.59}$$

where $g(x)$ is specified. The domain of the equation is specified, and boundary conditions are provided for $f(x)$ and its derivative. On a grid of equally spaced points we might represent this equation as:

$$\frac{1}{d^2}(f(i+1) + f(i-1) - 2f(i)) = g(i) \tag{1.10.60}$$

This can be written as a matrix equation:

$$\sum_j A(i, j) f(j) = g(i) \tag{1.10.61}$$

The matrix equation can be solved for the values of $f(i)$ by matrix inversion (using matrix diagonalization). However, diagonalization is very costly when the matrix is large, i.e., when there are many points in the grid.

A multigrid approach to solving this equation starts by defining a set of lattices (grids), $L_j, j \in \{0, \ldots, q\}$, where each successive lattice has twice as many points as the previous one (Fig. 1.10.15). To explain the procedure it is simplest to assume that we start with a good approximation for the solution on grid L_{j-1} and we are looking for a solution on the grid L_j. The steps taken are then:

1. Interpolate to find $f_0^j(i)$, an approximate value of the function on the finer grid L_j.

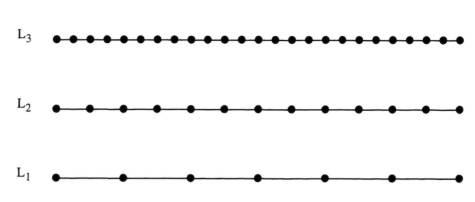

Figure 1.10.15 Illustration of four grids for a one-dimensional application of the multigrid technique to a differential equation by the procedure illustrated in Fig. 1.10.16. ∎

2. Perform an iterative improvement (relaxation) of the solution on the finer grid. This iteration involves calculating the error

$$\sum_{i'} A(i, i')\, f_0^j(i') - g(i) = r^j(i) \tag{1.10.62}$$

where all indices refer to the grid L_j. This error is used to improve the solution on the finer grid, as in the minimization procedures discussed in Section 1.7.5:

$$f_1^j(i) = f_0^j(i) - c r^j(i) \tag{1.10.63}$$

The scalar c is generally replaced by an approximate inverse of the matrix $A(i, j)$ as discussed in Section 1.7.5. This iteration captures much of the correction of the solution at the fine-scale level; however, there are resulting corrections at coarser levels that are not captured. Rather than continuing to iteratively improve the solution at this fine-scale level, we move the iteration to the next coarser level.

3. Recalculate the value of the function on the coarse grid L_{j-1} to obtain $f_1^{j-1}(i)$. This might be just a restriction from the fine-grid points to the coarse-grid points. However, it often involves some more sophisticated smoothing. Ideally, it should be such that interpolation will invert this process to obtain the values that were found on the finer grid. The correction for the difference between the interpolated and exact fine-scale results are retained.

4. Correct the value of $g(i)$ on the coarse grid using the values of $r^j(i)$ restricted to the coarser grid. We do this so that the coarse-grid equation has an exact solution that is consistent with the fine-grid equation. From Eq. (1.10.62) this essentially means adding $r^j(i)$ to $g(i)$. The resulting corrected values we call $g_1^{j-1}(i)$.

5. Relax the solution $f_1^{j-1}(i)$ on the coarse grid to obtain a new approximation to the function on the coarse grid $f_2^{j-1}(i)$. This is done using the same procedure for relaxation described in step 3; however $g(i)$ is replaced by $g_1^{j-1}(i)$.

    The procedure of going to coarser grids in steps 3 through 5 is repeated for all grids $L_{j-2}, L_{j-3}, \ldots$ till the coarsest grid, L_0. The values of the function $g(i)$ are progressively corrected by the finer-scale errors. Step 5 on the coarsest grid is replaced by exact solution using matrix diagonalization. The subsequent steps are designed to bring all of the iterative refinements to the finest-scale solution.

6. Interpolate the coarse-grid solution L_0 to the finer-grid L_1.

7. Add the correction that was previously saved when going from the fine to the coarse grid.

    Steps 6–7 are then repeated to take us to progressively finer-scale grids all the way back to L_j.

    This procedure is called a V-cycle since it appears as a V in a schematic that shows the progressive movement between levels. A V-cycle starts from a relaxed solution on grid L_{j-1} and results in a relaxed solution on the grid L_j. A full multigrid procedure involves starting with an exact solution at the coarsest scale L_0 and then performing V-cycles for progressively finer grids. Such a multigrid procedure is graphically illustrated in Fig. 1.10.16.

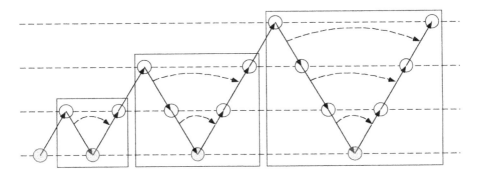

Figure 1.10.16 The multigrid algorithm used to obtain the solution to a differential equation on the finest grid is described schematically by this sequence of operations. The operation sequence is to be read from left to right. The different grids that are being used are indicated by successive horizontal lines with the coarsest grid on the bottom and the finest grid on the top. The sequence of operations starts by solving a differential equation on the coarsest grid by exact matrix diagonalization (shaded circle). Then iterative refinement of the equations is performed on finer grids. When the finer-grid solutions are calculated, the coarse-grid equations are corrected so that the iterative refinement of the fine-scale solution can be performed on the coarse grids. This involves a V-cycle as indicated in the figure by the boxes. The horizontal curved arrows indicate the retention of the difference between coarse- and fine-scale solutions so that subsequent refinements can be performed. ∎

There are several advantages of the multigrid methodology for the solution of differential equations over more traditional integration methods that use a single-grid representation. With careful implementation, the increasing cost of finer-scale grids grows slowly with the number of grid points, scaling as $N \ln(N)$. The solution of multiple problems of similar type can be even more efficient, since the corrections of the coarse-scale equations can often be carried over to similar problems, accelerating the iterative refinement. This is in the spirit of developing universal coarse-scale representations as discussed earlier. Finally, it is natural in this method to obtain estimates of the remaining error due to limited grid density, which is important to achieving a controlled error in the solution.

1.10.7 *Levels of description, emergence of simplicity and complexity*

In our explorations of the world we have often discovered that the natural world may be described in terms of underlying simple objects, concepts, and laws of behavior (mechanics) and interactions. When we look still closer we see that these simple objects are composite objects whose internal structure may be complex and have a wealth of possible behavior. Somehow, the wealth of behavior is not relevant at the larger scale. Similarly, when we look at longer length scales than our senses normally are attuned to, we discover that the behavior at these length scales is not affected by objects and events that appear important to us.

Examples are found from the behavior of galaxies to elementary particles: galaxies are composed of suns and interstellar gases, suns are formed of complex plasmas and are orbited by planets, planets are formed from a diversity of materials and even life, materials and living organisms are formed of atoms, atoms are composed of nuclei and electrons, nuclei are composed of protons and neutrons (nucleons), and nucleons appear to be composed of quarks.

Each of these represents what we may call a level of description of the world. A level is an internally consistent picture of the behavior of interacting elements that are simple. When taken together, many such elements may or may not have a simple behavior, but the rules that give rise to their collective behavior are simple. We note that the interplay between levels is not always just a self-contained description of one level by the level immediately below. At times we have to look at more than one level in order to describe the behavior we are interested in.

The existence of these levels of description has led science to develop the notion of fundamental law and unified theories of matter and nature. Such theories are the self-consistent descriptions of the simple laws governing the behavior and interplay of the entities on a particular level. The laws at a particular level then give rise to the larger-scale behavior.

The existence of simplicity in the description of underlying fundamental laws is not the only way that simplicity arises in science. The existence of multiple levels implies that simplicity can also be an emergent property. This means that the collective behavior of many elementary parts can behave simply on a much larger scale.

The study of complex systems focuses on understanding the relationship between simplicity and complexity. This requires both an understanding of the emergence of complex behavior from simple elements and laws, as well as the emergence of simplicity from simple or complex elements that allow a simple larger-scale description to exist.

Much of our discussion in this section was based upon the understanding that macroscopic behavior of physical systems can be described or determined by only a few relevant parameters. These parameters arise from the underlying microscopic description. However, many of the aspects of the microscopic description are irrelevant. Different microscopic models can be used to describe the same macroscopic phenomenon. The approach of scaling and renormalization does not assume that all the details of the microscopic description become irrelevant, however, it tries to determine self-consistently which of the microscopic parameters are relevant to the macroscopic behavior in order to enable us to simplify our analysis and come to a better understanding.

Whenever we are describing a simple macroscopic behavior, it is natural that the number of microscopic parameters relevant to model this behavior must be small. This follows directly from the simplicity of the macroscopic behavior. On the other hand, if we describe a complex macroscopic behavior, the number of microscopic parameters that are relevant must be large.

Nevertheless, we know that the renormalization group approach has some validity even for complex systems. At long length scales, all of the details that occur on the smallest length scale are not relevant. The vibrations of an individual atom are not generally relevant to the behavior of a complex biological organism. Indeed, there is a pattern of levels of description in the structure of complex systems. For biological organisms, composed out of atoms, there are additional levels of description that are intermediate between atoms and the organism: molecules, cells, tissues, organs and systems. The existence of these levels implies that many of the details of the atomic behavior are not relevant at the macroscopic level. This should also be understood from the perspective of the multi-grid approach. In this picture, when we are describing the behavior of a complex system, we have the possibility of describing it at a very coarse level or a finer and yet finer level. The number of levels that are necessary depends on the level of precision or level of detail we wish to achieve in our description. It is not always necessary to describe the behavior in terms of the finest scale. It is essential, however, to identify properly a model that can capture the essential underlying parameters in order to discuss the behavior of any system.

Like biological organisms, man-made constructs are also built from levels of structure. This method of organization is used to simplify the design and enable us to understand and work with our own creations. For example, we can consider the construction of a factory from machines and computers, machines constructed from individual moving parts, computers constructed from various components including computer chips, chips constructed from semiconductor devices, semiconductor devices composed out of regions of semiconductor and metal. Both biology and engineering face problems of design for function or purpose. They both make use of interacting building blocks to engineer desired behavior and therefore construct the complex out of the simple. The existence of these building blocks is related to the existence of levels of description for both natural and artificial systems.

Our discussion thus brings us to recognize the importance of studying the properties of substructure and its relationship to function in complex systems. This relationship will be considered in Chapter 2 in the context of our study of neural networks.

2

Neural Networks I:
Subdivision and Hierarchy

Conceptual Outline

▋ 2.1 ▋ Motivated by the properties of biological neural networks, we introduce simple mathematical models whose properties may be explored and related to aspects of human information processing.

▋ 2.2 ▋ The attractor network embodies the properties of an associative content-addressable memory. Memories are imprinted and are accessed by presenting the network with part of their content. Properties of the network can be studied using a signal-to-noise analysis and simulations. The capacity of the attractor network for storage of memories is proportional to the number of neurons.

▋ 2.3 ▋ The feedforward network acts as an input-output system formed out of several layers of neurons. Using prototypes that indicate the desired outputs for a set of possible inputs, the feedforward network is trained by minimizing a cost function which measures the output error. The resulting training algorithm is called back-propagation of error.

▋ 2.4 ▋ In order to study the overall function of the brain, an understanding of substructure and the interactions between parts of the brain is necessary. Feedforward networks illustrate one way to build a network out of parts. A second model of interacting subnetworks is a subdivided attractor network. A subdivided attractor network stores more than just the imprinted patterns—it stores composite patterns formed out of parts of the imprinted patterns. If these are patterns that an organism might encounter, then this is an advantage. Features of human visual processing, language and motor control illustrate the relevance of composite patterns.

▋ 2.5 ▋ Analysis and simulations of subdivided attractor networks reveal that partial subdivision can balance a decline in the storage capacity of imprinted patterns with the potential advantages of composite patterns. However, this balance only allows direct control over composite pattern stability when the number of subdivisions is no more than approximately seven, suggesting a connection to the 7 ± 2 rule of short-term memory.

■ **2.6** ■   The limitation in the number of subdivisions in an effective architecture suggests that a hierarchy of functional subdivisions is best for complex pattern-recognition tasks, consistent with the observed hierarchical brain structure.

■ **2.7** ■   More general arguments suggest the necessity of substructure, and applicability of the 7 ± 2 rule, in complex systems.

2.1   Neural Networks: Brain and Mind

The functioning of the brain as part of the nervous system is generally believed to account for the complexity of human (or animal) interaction with its environment. The brain is considered responsible for sensory processing, motor control, language, common sense, logic, creativity, planning, self-awareness and most other aspects of what might be called higher information processing. The elements believed responsible for brain function are the nerve cells—neurons—and the interactions between them. The interactions are mediated by a variety of chemicals transferred through synapses. The brain is also affected by diverse substances (e.g., adrenaline) produced by other parts of the body and transported through the bloodstream. Neurons are cells that should not be described in only one form, as they have diverse forms that vary between different parts of the brain and within particular brain sections (Fig. 2.1.1). Specifying the complete behavior of an individual neuron is a detailed and complex problem. However, it is reasonable to assume that many of the general principles upon which the nervous system is designed may be described through a much-simplified model that takes into account only a few features of each neuron and the interactions between them. This is expected, in part, because of the large number, on the order of 10^{11}, neurons in the brain.

A variety of mathematical models have been described that attempt to capture particular features of the neurons and their interactions. All such models are incomplete. Some models are particularly well suited for theoretical investigations, others for pattern-recognition tasks. Much of the modern effort in the modeling of the nervous system is of commercial nature in seeking to implement pattern-recognition strategies for artificial intelligence tasks. Our approach will be to introduce two of the simpler models of neural networks, one of which has been used for extensive theoretical studies, the other for commercial applications. We will then take advantage of the simple analysis of the former to develop an understanding of subdivision in neural networks. Subdivision and substructure is a key theme that appears in many forms in the study of complex systems.

There have been many efforts to demonstrate the connection between mathematical models of neural networks and the biological brain. These are important in order to bridge the gap between the biological and mathematical models. The additional readings located at the end of this text may be consulted for detailed discussions. We do not review these efforts here; instead we motivate more loosely the arti-

Figure 2.1.1 Several different types of neurons adapted from illustrations obtained by various staining techniques. ∎

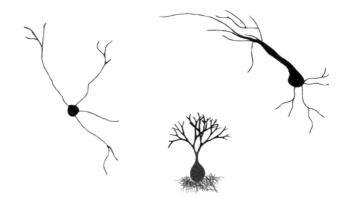

Figure 2.1.2 Schematic illustration of a biological neural network showing several nerve cells with branching axons. The axons end at synapses connecting to the dendrites of the next neuron that lead to its cell body. This schematic illustration is further simplified to obtain the artificial network models shown in Fig. 2.1.3. ∎

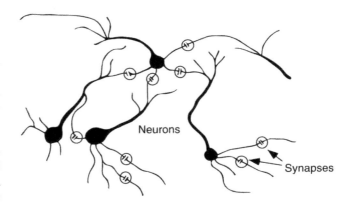

ficial models and rely upon investigations of the properties of these models to establish the connection, or to suggest investigations of the biological system.

To motivate the artificial models of neural networks, we show in Fig. 2.1.2 a schematic of a biological neural network that consists of a few neurons. Each neuron has a cell body with multiple projections called dendrites, and a longer projection called an axon which branches into terminal fibers. The terminal fibers of the axon of one neuron generally end proximate to the dendrites of a different cell body. The cell walls of a neuron support the transmission of electrochemical pulses that travel along the axon from the cell body to the terminal fibers. A single electrochemical pulse is not usually considered to be the quantum of information. Instead it is the "activity"—the rate of pulsing—that is considered to be the relevant parameter describing the state of the neuron. Pulses that arrive at the end of a terminal fiber release various chemicals into the narrow intracellular region separating them from the dendrites of the adjacent cell. This region, known as a synapse, provides the medium of influence of one neuron on the next. The chemicals released across the gap may either stimulate (an excitatory synapse) or depress (an inhibitory synapse) the activity of the next neuron.

It is generally assumed, though not universally accepted, that the "state of the mind" at a particular time is described by the activities of all the neurons—the pattern of neural activity. This activity pattern evolves in time, because the activity of each neuron is determined by the activity of neurons at an earlier time and the excitatory or inhibitory synapses between them. The influence of the external world on the neurons occurs through the activity of sensory neurons that are affected by sensory receptors. Actions are effected by the influence of motor-neuron activity on the muscle cells. Synaptic connections are in part "hardwired" performing functions that are prespecified by genetic programming. However, memory and experience are also believed to be encoded into the strength (or even the existence) of the synapses between neurons. It has been demonstrated that synaptic strengths are affected by the state of neuronal excitation. This influence, called imprinting, is considered to be the principle mechanism for adaptive learning. The most established and well-studied form of imprinting was originally proposed by Hebb in 1949. The plasticity of synapses should not be underestimated, because the development of even basic functions of vision is known to be influenced by sensory stimulation.

Hebbian imprinting suggests that when two neurons are both firing at a particular time, an excitatory synapse between them is strengthened and an inhibitory synapse is weakened. Conversely, when one is firing and the other is not, the inhibitory synapse is strengthened and the excitatory synapse is weakened. Intuitively, this results in the possibility of reconstructing the neural activity pattern from a part of it, because the synapses have been modified so as to reinforce the pattern. Thus, the imprinted pattern of neural activity becomes a memory. This will be demonstrated explicitly and explained more fully in the context of artificial networks that successfully reproduce this process and help explain its function.

The two types of artificial neural networks we will consider are illustrated in Fig. 2.1.3. The first kind is called an attractor network, and consists of mathematical neurons identified as variables s_i that represent the neuron activity. i is the neuron index. Neurons are connected by synapses consisting of variables J_{ij} that represent the strength of the synapse between two neurons i and j. The synapses are taken to be symmetric, so that $J_{ij} = J_{ji}$. A positive value of J_{ij} indicates an excitatory synapse. A negative value indicates an inhibitory synapse. A more precise definition follows in Section 2.2. The second kind of network, discussed in Section 2.3, is called a feedforward network. It consists of a set of two or more layers of mathematical neurons consisting of variables s_i^l that represent the neuron activity. For convenience, l is added as a layer index. Synapses represented by variables J_{ij}^l act only in one direction and sequentially from one layer to the next.

Our knowledge of biological neural networks indicates that it would be more realistic to represent synapses as unidirectional, as the feedforward network does, but to allow neurons to be connected in loops. Some of the effects of feedback in loops are represented in the attractor network by the symmetric synapses.

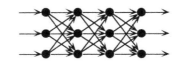

Figure 2.1.3 Schematic illustration of two types of artificial neural networks that are used in modeling biological networks either for formal studies or for application to pattern recognition. On the left is a schematic of an attractor network. The dots represent the neurons and the lines represent the synapses that mediate the influence between them. The synapses are symmetric carrying equal influence in both directions. On the right is a feedforward network consisting of several layers (here four) of neurons that influence each other in a unidirectional fashion. The input arriving from the left sets the values of the first layer of neurons. These neurons influence the second layer of neurons through the synapses between layer one and two. After several stages, the output is read from the final layer of neurons. ∎

A second distinction between the two types of networks is in their choice of representation of the neural activity. The attractor network typically uses binary variables, while the feedforward network uses a real number in a limited range. These choices are related to the nonlinear response of neurons. The activity of a neuron at a particular time is thought to be a sigmoidal function of the influence of other neurons. This means that at moderate levels of excitation, the activity of the neuron is proportional to the excitation. However, for high levels of excitation, the activity saturates. The question arises whether the brain uses the linear regime or generally drives the neurons to saturation. The most reasonable answer is that it depends on the function of the neuron. This is quite analogous to the use of silicon transistors, which are used both for linear response and switching tasks. The neurons that are used in signal-processing functions in the early stages of the auditory or visual systems are likely to make use of the linear regime. However, a linear operation is greatly limited in its possible effects. For example, any number of linear operations are equivalent to a single linear operation. If only the linear regimes of neurons were utilized, the whole operation of the network would be reducible to application of a linear operator to the input information—multiplication by a matrix. Thus, while for initial signal processing the linear regime should play an important role, in other parts of the brain the saturation regime should be expected to be important. The feedforward network uses a model of nonlinear response that includes both linear and saturation regimes, while the attractor network typically represents only the saturation regime. Generalizing the attractor network to include a linear regime adds analytic difficulty, but does not significantly change the results. In contrast, both the linear and nonlinear regimes are necessary for the feedforward network to be a meaningful model.

Each of the two artificial network models represents drastic simplifications over more realistic network models. These simplifications enable intuitive mathematical

treatments and capture behaviors that are likely to be an important part of more realistic models. The attractor network with symmetric synapses is the most convenient
for analytic treatments because it can be described using the stochastic field formalism discussed in Section 1.6. The feedforward network is more easily used as an input-
output system and has found more use in applications.

2.2   Attractor Networks

2.2.1 Defining attractor networks

Attractor networks, also known as Hopfield networks, in their simplest form, have
three features:

a. Symmetric synapses:

$$J_{ij} = J_{ji} \tag{2.2.1}$$

b. No self-action by a neuron:

$$J_{ii} = 0 \tag{2.2.2}$$

c. Binary variables for the neuron activity values:

$$s_i = \pm 1 \tag{2.2.3}$$

There are N neurons, so the neuron indices i, j take values in the range $\{1,...,N\}$. By
Eq. (2.2.1) and Eq. (2.2.2), the synapses J_{ij} form a symmetric $N \times N$ matrix with all diagonal elements equal to zero.

The binary representation of neuron activity suggests that the activity has only
two values which are active or "firing," $s_i = +1$ (ON), and inactive or "quiescent," $s_i = -1$
(OFF). The activity of a particular neuron, updated at time t, is given by:

$$s_i(t) = \text{sign}(\sum_j J_{ij} s_j(t-1)) \tag{2.2.4}$$

where the values of all the other neurons at time $t-1$ are polled through the synapses
to determine the ith neuron activity at time t. Specifically, this expression states that
a particular neuron fires or does not fire depending on the result of performing a sum
of all of the messages it is receiving through synapses. This sum is formed from the
activity of every neuron multiplied by the strength of the synapse between the two
neurons. Thus, for example, a firing neuron j, $s_j = +1$, which has a positive (excitatory)
synapse to the neuron i, $J_{ij} > 0$, will increase the likelihood of neuron i firing. If neuron j is not firing, $s_j = -1$, then the likelihood of neuron i firing is reduced. On the
other hand, if the synapse is inhibitory, $J_{ij} < 0$, the opposite occurs — a firing neuron
j, $s_j = +1$, will decrease the likelihood of neuron i firing, and a quiescent neuron j, $s_j =$
-1, will increase the likelihood of neuron i firing. When necessary, it is understood
that sign(0) takes the value ± 1 with equal probability.

The activity of the whole network of neurons may be determined either synchronously (all neurons at once) or asynchronously (selecting one neuron at a time).
Asynchronous updating is probably more realistic in models of the brain. However,

for many purposes the difference is not significant, and in such cases we can assume a synchronous update.

2.2.2 *Operating and training attractor networks*

Conceptually, the operation of an attractor network proceeds in the following steps. First a pattern of neural activities, the "input", is imposed on the network. Then the network is evolved by updating several times the neurons according to the neuron update rule, Eq. (2.2.4). The evolution continues until either a steady state is reached or a prespecified number of updates have been performed. Then the state of the network is read as the "output." The next pattern is then imposed on the network.

At the same time as the network is performing this process, the synapses themselves are modified by the state of the neurons according to a mathematical formulation of the Hebbian rule:

$$J_{ij}(t) = J_{ij}(t-1) + cs_i(t-1)s_j(t-1) \qquad\qquad i \neq j \qquad (2.2.5)$$

where the rate of change of the synapses is controlled by the parameter c. This is a mathematical description of Hebbian imprinting, because the synapse between two neurons is changed in the direction of being excitatory if both neurons are either ON or OFF, and the synapse is changed in the direction of being inhibitory if one neuron is ON and the other is OFF.

The update of a neuron is considered to be a much faster process than the Hebbian changes in the synapses—the synaptic dynamics. Thus we assume that c is small compared to the magnitude of the synapse values, so that each imprint causes only an incremental change. Because the change in synapses occurs much more slowly than the neuron update, for modeling purposes it is convenient to separate it completely from the process of neuron update. We then describe the operation of the network in terms of a training period and an operating period.

The training of the network consists of imprinting a set of selected neuron firing patterns $\{\xi_i^\mu\}$ where i is the neuron index $i \in \{1,...,N\}$, μ is the pattern index $\mu \in \{1,...,p\}$, and ξ_i^μ is the value of a particular neuron s_i in the μth pattern. It is assumed that there are a fixed number p of patterns that are to be trained. The synapses are then set to:

$$J_{ij} = \begin{cases} \dfrac{1}{N}\sum_{\mu=1}^{p} \xi_i^\mu \xi_j^\mu & i \neq j \\ 0 & i = j \end{cases} \qquad (2.2.6)$$

The prefactor $1/N$ is a choice of normalization of the synapses that is often convenient, but it does not affect in an essential way any results described here.

2.2.3 *Energy analog*

The formulation of the attractor network can be recognized as a generalization of the Ising model discussed in Section 1.6. Neurons are analogous to spins, and the interaction between two spins s_i and s_j is the synapse J_{ij}.

We can thus identify the effective energy of the system as:

$$E[\{s_i\}] = -\frac{1}{2}\sum_{i,j}J_{ij}s_is_j \tag{2.2.7}$$

The update of a particular neuron, Eq. (2.2.4), consists of "aligning" it with the effective local field (known as the postsynaptic potential):

$$h_i(t) = \sum_{j\neq i}J_{ij}s_j(t-1) \tag{2.2.8}$$

This is the same dynamics as the Glauber or Monte Carlo dynamics of an Ising model at zero temperature. At zero temperature the system evolves to a local minimum energy state. In this state each spin is aligned with the effective local field.

The analogy between a neural network and a model with a well-defined energy enables us to consider the operation of the network in a natural way. The pattern of neural activities evolves in time to decrease the energy of the pattern until it reaches a local energy minimum, where each neuron activity is consistent with the influences upon it as measured by the postsynaptic potential. Imprinting a pattern of neural activity lowers the energy of this pattern and, to a lesser degree, the energy of patterns that are similar. In lowering the energy of these patterns, imprinting creates a basin of attraction. The basin of attraction is the region of patterns near the imprinted pattern that will evolve under the neural updating back to the imprinted pattern (Fig. 2.2.1).

The network operation can now be understood schematically as follows (Fig. 2.2.2). We imprint a particular pattern. If we then impose a different pattern on the network, the evolution of the neurons will recover the original pattern if the imposed pattern is within its basin of attraction. The more similar are the two patterns, the more likely the imposed pattern will be in the basin of attraction of the imprinted one. Since part of the imprinted pattern was retrieved, the network acts as a kind of memory.

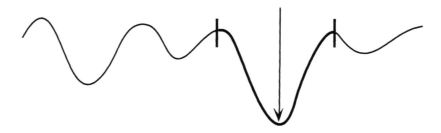

Figure 2.2.1 Schematic illustration of the energy analog of imprinting on an attractor network. Imprinting a pattern lowers its energy and the energy of all patterns in its vicinity. This creates a basin of attraction. If we initialize the network to any pattern within the basin of attraction, the network will relax to the imprinted pattern by its own neural evolution. The network acts as a memory that is "content-addressable." When a pattern is imprinted we can recover it by starting from partial information about it (see Fig. 2.2.2). This is also a form of associative memory. ∎

The operation of the network may be described as an associative memory. The restoration of the complete imprinted pattern, in effect, associates the reconstructed part of the pattern with the part of the pattern that was imposed. We can also say that the network has the ability to perform a kind of generalization (Fig. 2.2.3). The network has generalized the imprinted pattern to the set of patterns that are in its basin of attraction. Moreover, the retrieval process is also a form of categorization, since it assigns the imprinted pattern as a category label to the set of patterns in the basin of attraction. All of these properties of the neural network are suggestive of some of the basic features that are thought to apply to human memory, and thus to the biological neural network. Their natural relationship to each other and the simplicity by which they are achieved in this model is one of the main reasons for the interest in this neural network representation.

We can contrast the properties of the network memory with a computer memory. In a computer, the memory is accessed by an address that specifies the location of a particular piece of information. In order to retrieve information, it is necessary to have the address, or to search systematically through the possibilities. On the other hand, for a human being, retrieving the rest of the sentence "To be or not to be ..." is generally much easier than retrieving line 64 from act 3, scene 1, of *Hamlet*, by William Shakespeare. To emphasize the difference in the nature of addressing between

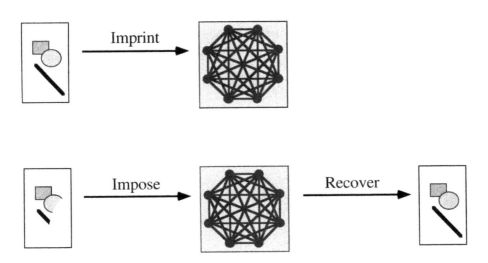

Figure 2.2.2 Schematic operation of the attractor network as a content-addressable memory. Imprinting a pattern on the network in the training stage (top) enables us to use the network as a content-addressable memory (bottom). By imposing a pattern that has a significant overlap with the imprinted pattern the original pattern can be recovered. This is analogous to being able to complete the sentence "To be or not to be ..." ∎

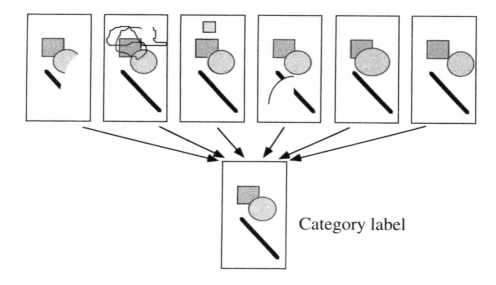

Figure 2.2.3 The neural dynamics of an attractor network maps a variety of patterns onto an imprinted pattern. This is equivalent to a classification of patterns by a category label, which is the imprinted pattern. The category of patterns labeled by the imprinted pattern is its basin of attraction. Classification is also a form of pattern recognition. Moreover, we can say that the basin of attraction is a generalization of the imprinted pattern. Thus the attractor network has properties very unlike those of a conventional computer memory, which is accessed using a numerical address that is distinct from the memory itself. It works much more like human memories that are accessed through information related to the information that is sought after.

The behavior of the attractor network may be summarized as follows:

Attractor network training and operation:
Training — Imprint a neural state.
Operation — Recover original state from part of it.

Analogies for operation:
• Content-addressable memory
• Limited form of classification
• Limited form of pattern recognition
• Limited form of generalization

The relationship between human information-processing and various network models will be discussed further in Chapter 3. ∎

a computer memory and a network memory, we say that the network memory is content addressable.

The associative nature of the attractor network thus captures some of the properties of human memory that are quite distinct from those of a computer memory. It should be understood, however, that this is not expected to be the last word in development of such models or in the understanding of these processes.

One of the important properties of a memory is its capacity. If we try to imprint more than one pattern on the network, the basin of attraction of each pattern will take up some of the space of all possible patterns. It is natural to expect that there will be a limit to how many patterns we can imprint before the basins of attraction will interfere destructively with each other. When the memory is full, the basins of attraction are small and the memory is not usable because it can only recover a pattern if we already know it. When the destructive interference is complete, the basins of attraction disappear. At that point, a typical imprinted pattern will no longer even be stable, because stability is a basin of attraction of one. Studying the number of imprints that are possible before the network reaches this condition gives us an understanding of the network capacity and how this capacity depends on network size. Thus we can determine the storage capacity by measuring the stability of patterns that are imprinted on the network.

Our mathematical study of attractor networks begins in Section 2.2.4 with an analysis of the network behavior when there are a few imprints. This analysis shows the retrieval of patterns and the relevance of their basin of attraction. In Section 2.2.5 we use a signal-to-noise analysis to determine the stability of an imprinted pattern, and thus the storage capacity of the network. Simulations of an attractor network are discussed in Section 2.2.6. Finally, some aspects of the overload catastrophe that occurs when network capacity is exceeded are discussed in Section 2.2.7.

2.2.4 One or two imprinted patterns

We first consider the case of imprinting a single pattern $\{\xi_i\}$. The synapses are constructed as the "outer product" of the neural activities and we have:

$$J_{ij} = \begin{cases} \dfrac{1}{N}\xi_i\xi_j & i \neq j \\[2mm] 0 & i = j \end{cases} \tag{2.2.9}$$

Using these synapses, we start the network at a set of neural activities $\{s_i(0)\}$ and evolve the network using the definition of the dynamics (Eq. (2.2.4)):

$$s_i(1) = \text{sign}(\sum_j J_{ij}s_j(0)) = \text{sign}(\sum_{j\neq i}\frac{1}{N}\xi_i\xi_j s_j(0)) = \text{sign}(\xi_i)\text{sign}(\sum_{j\neq i}\xi_j s_j(0))$$
$$= \xi_i\,\text{sign}(\sum_{j\neq i}\xi_j s_j(0)) \tag{2.2.10}$$

The second line is a consequence of the normalization $|\xi_i| = |\pm 1| = 1$.

If we didn't have the restriction of $j \neq i$ in the sum in Eq. (2.2.10), the factor multiplying ξ_i would be independent of i. We would then have

$$s_i(1) \stackrel{\ast}{=} \xi_i\text{sign}(\sum_j \xi_j s_j(0)) = \pm\xi_i \tag{2.2.11}$$

where the $\pm$ sign in front is independent of i. This means that in one iteration the network neurons reached either the imprinted pattern or its inverse. This implies that recall of the pattern has been achieved. Why either the pattern or its inverse? It shouldn't be too surprising that we can arrive at either the imprinted pattern or its inverse because the form of the energy function (Eq. (2.2.7)) and the neural update (Eq. (2.2.4)) is invariant under the transformation $s_i \rightarrow -s_i$ for all i simultaneously. Thus, in an attractor network we automatically store both the pattern and its inverse.

How do we treat the actual case with the $i = j$ term missing? We write

$$s_i(1) = \xi_i \text{sign}(\sum_j \xi_j s_j(0) - \xi_i s_i(0)) = \xi_i \text{sign}(\xi \cdot s(0) - \xi_i s_i(0)) \tag{2.2.12}$$

where $\xi \cdot s(0) = \sum \xi_j s_j(0)$ is the inner product of the imprinted pattern with the initial state of the network. As long as this inner product is greater than 1 or less than -1, the extra term doesn't affect the result. This means that for $|\xi \cdot s(0)| > 1$, recall is achieved and $s(0)$ is within the basin of attraction of the imprinted pattern or its inverse. This is nearly all of the possible choices for $s(0)$. Note that the imprinted pattern is a stable fixed point of the network dynamics—once the imprinted pattern is reached it will be repeated.

The case of a single imprinted pattern is somewhat unusual in that even if the initial pattern $\{s_i(0)\}$ is not correlated with the imprinted pattern, we will still recover the imprinted pattern. If we take random numbers for $\{s_i(0)\}$, then the sum over j in Eq. (2.2.12) is a random walk—the sum over N uncorrelated values of ± 1 (Section 1.2). The typical size of this number is $\sqrt{N}$, which places the pattern solidly within the basin of attraction of the imprinted pattern or its inverse. It has been suggested that the case of a single dominant imprinted pattern has properties analogous to human obsession, compulsion or fixation—because the imprinted pattern is the output regardless of the input—and is a natural mode of failure of the network that can arise in Hebbian imprinting.

We can also ask what will happen if the magnitude of $\xi \cdot s(0)$ is equal to $-1, 0$, or 1. In these cases the first iteration should not lead to the imprinted pattern. However, it is still most likely that after two updates the network will be in either the imprinted pattern or its inverse. When the inner product $\xi \cdot s(0)$ is $+1$, the result of the first update is a new pattern $s(1)$ that is likely to have a larger than unit overlap with ξ, $\xi \cdot s(1) > 1$. When $\xi \cdot s(0)$ is -1, it is likely that $\xi \cdot s(1) < -1$. The second iteration would be analogous to Eq. (2.2.12):

$$s_i(2) = \xi_i \text{ sign}(\sum_j \xi_j s_j(1) - \xi_i s_i(1))$$
$$= \xi_i \text{ sign}(\xi \cdot s(1) - \xi_i s_i(1)) \tag{2.2.13}$$

resulting in retrieval of the imprinted pattern.

The case of $\xi \cdot s(0) = 0$ is special, and a synchronous update in this case simply leads to oscillation of the pattern—a 2-cycle:

$$s_i(1) = \xi_i \text{ sign}(\sum_j \xi_j s_j(0) - \xi_i s_i(0))$$
$$= \xi_i \text{ sign}(-\xi_i s_i(0)) = -\xi_i^2 s_i(0) = -s_i(0) \tag{2.2.14}$$

More generally:

$$s_i(t+1) = -s_i(t) \tag{2.2.15}$$

This is one of the few cases where asynchronous updating would lead to a different result, since the randomness inherent in the asynchronous updating would lead to the evolution of the network to the imprinted pattern or its inverse.

We ran into some additional trouble in the preceding discussion because of the omission of the $i = j$ term in the synapses. Why don't we just include this term? The answer, from the point of view of the formal analysis, is that this corresponds to self-action by a neuron on itself. Such self-action is inconsistent with an energy function. In a real network, self-action is not impossible. It might correspond, for example, to an inhibition of neural activity during a period of time after activity has happened. This does occur in biological neurons where the period of self-inhibition is known as a refractory period. The implications of such terms are, however, outside the present discussion.

Our conclusion from the analysis of the single imprint case is that the basin of attraction of the single imprint is large. To measure the size of the basin of attraction, we define the Hamming distance $d(\mathbf{s},\mathbf{s}')$ between two patterns as the number of neurons that differ between the two patterns. The Hamming distance is related to the inner product by

$$d(\mathbf{s},\mathbf{s}') = \frac{N}{2} - \frac{1}{2}\sum_i s_i s_i' \tag{2.2.16}$$

as can be verified using a few examples. The Hamming distance of a pattern from itself is zero, from an orthogonal state is $N/2$, and from its opposite is N. For a single imprint in a neural network the initial pattern $\mathbf{s}(0)$ is in the basin of attraction of the imprinted pattern if the inner product between them is positive. This implies that the Hamming distance must be less than $N/2$. This is the effective size (radius) of the basin of attraction.

We can now ask what happens if two patterns are imprinted instead of just one. The synapses are given by Eq. (2.2.6) with $p = 2$. Following through the same steps we have the expression:

$$s_i(1) = \text{sign}(\sum_j J_{ij} s_j(0)) = \text{sign}(\xi_i^1 \sum_{j \neq i} \xi_j^1 s_j(0) + \xi_i^2 \sum_{j \neq i} \xi_j^2 s_j(0)) \tag{2.2.17}$$

Qualitatively, we can understand this result by considering an initial pattern $\mathbf{s}(0)$ that is close to one of the imprinted patterns, say ξ^1. Let us assume that there is no particular relationship between the two patterns that were imprinted and, quite reasonably, that there is also no relationship between the initial pattern $\mathbf{s}(0)$ and the second imprinted pattern ξ^2. The first sum in Eq. (2.2.17) will give us a number that has a magnitude $N - 2d(\mathbf{s},\xi^1)$. The assumption that the initial configuration is close to the first imprinted pattern means that $d(\mathbf{s},\xi^1)$ is small. The magnitude of the second sum is given by the inner product of the initial pattern with the second imprinted pattern. If there is no relationship between them, then each term in the inner product is independent.

Since each term has the value ±1 with equal probability, it is a random walk with $(N-1)$ steps. The typical magnitude of the second term is thus $\sqrt{N-1}$. For large enough N there is essentially no chance that the second term is as large as the first term. Neglecting the second term, the first term gives us the same result we had before, which is recovery of the imprinted pattern. We see from this argument that retrieval depends on the proximity of the initial state with the pattern that will be retrieved. If the initial pattern is close to the second imprinted pattern, then the second imprinted pattern will be retrieved. Successful retrieval also depends on the number of neurons in the network.

The retrieval of two patterns can be extended to more patterns. For a large enough number of neurons, retrieval will still occur. We can make this argument more rigorous by considering a "signal-to-noise" analysis that pits the term that is trying to retrieve the pattern—the signal—against the rest of the terms—the noise. To do this formally we will assume that all the imprinted patterns are truly uncorrelated with each other. The neural activities are randomly selected values ±1. These values are fixed over the duration of the discussion. In the language of Section 1.3 they are quenched variables. Including correlations between the patterns would be important in understanding how the real brain works. We will consider correlations between patterns later in this chapter in Section 2.4.

2.2.5 Signal-to-noise analysis of memory stability

In this section, we formulate what is called a signal-to-noise analysis that enables us to determine statistically the stability of an imprinted pattern. This in turn enables us to determine the storage capacity of the network. Question 2.2.1 generalizes the analysis to give an estimate of the basin of attraction of an imprinted pattern.

We start from a network imprinted with p uncorrelated patterns. From Eq. (2.2.4) and Eq. (2.2.8), an imprinted pattern, imposed as the neural state $\{s_i | s_i = \xi_i^\mu\}$, is stable when $s_i = \text{sign}(h_i)$ or equivalently:

$$s_i h_i > 0 \qquad i \in \{1,...,N\} \qquad (2.2.18)$$

which implies that the local field at each neuron has the same sign as the value of the imprinted pattern.

Without loss of generality, we consider the stability (retrieval) of the first pattern $\{s_i | s_i = \xi_i^1\}$, since by symmetry the choice of pattern is arbitrary. To simplify slightly the notation, we consider the stability of the first neuron s_1, from the results we will be able to infer the stability of the others. The stability of s_1 depends on the sign of

$$s_1 h_1 = s_1 \sum_{j=2}^{N} J_{1j} s_j = \xi_1^1 \sum_{j=2}^{N} J_{1j} \xi_j^1 \qquad (2.2.19)$$

Inserting the Hebbian form for the synapses after p imprints, Eq. (2.2.6), we have:

$$s_1 h_1 = \frac{1}{N} \sum_{j=2}^{N} \sum_{\mu=1}^{p} \xi_1^1 \xi_1^\mu \xi_j^\mu \xi_j^1 \qquad (2.2.20)$$

We separate this expression into two parts. The first part is due to the imprint of the pattern we are trying to retrieve. The other part is due to all of the other patterns:

$$
s_1 h_1 = \frac{1}{N} \sum_{j=2}^{N} \xi_1^1 \xi_1^1 \xi_j^1 \xi_j^1 + \frac{1}{N} \sum_{j=2}^{N} \sum_{\mu=2}^{p} \xi_1^1 \xi_1^\mu \xi_j^\mu \xi_j^1
$$
$$
= \frac{N-1}{N} + \frac{1}{N} \sum_{j=2}^{N} \sum_{\mu=2}^{p} \xi_1^1 \xi_1^\mu \xi_j^\mu \xi_j^1
$$

(2.2.21)

The first sum was explicitly evaluated because the square of either ±1 is 1. The second sum we can also evaluate, at least statistically. Since ξ_1^μ is not correlated with ξ_j^μ for $j \neq 1$, and ξ_j^1 is not correlated with ξ_j^μ for $\mu \neq 1$, the four factors in each term of the sum are independent of each other. Moreover, each term in the sum has a factor that is independent of the factors in every other term. Therefore, each term in this sum is ±1 with equal probability, and each term is uncorrelated with the others. Thus it is just a random walk with $(N-1)(p-1)$ terms.

We can see that the two parts of $s_1 h_1$ play distinct roles. The first part, called the signal, is positive, and therefore tries to satisfy the stability condition of the imprinted pattern. This is consistent with the idea that a Hebbian imprint contributes to stability of the imprinted pattern. For $N \gg 1$, the size of the signal is 1.

The second part, called the noise, can be either positive or negative. The average value of the noise is zero, but the typical value (root mean square value) is given by:

$$
\sigma = \frac{1}{N} \sqrt{(N-1)(p-1)} \approx \sqrt{\frac{p}{N}}
$$

(2.2.22)

where the latter expression is valid for $N, p \gg 1$. When the typical value of the noise is much less than the size of the signal, then most of the time the neurons will be stable and we can say that we have a stable imprinted pattern. When the noise is the same size as the signal, then each neuron may either stay the same or switch after a single update, and the pattern will not be stable.

From the expression for the noise term, we see that the stability of the pattern depends on p, the number of patterns that are imprinted on the network. For low storage, $p \ll N$, the noise term becomes negligible and the imprinted patterns are stable.

If we want to understand how large p can be before the storage will deteriorate, we need to estimate the probability that a single neuron will be unstable—the probability that $s_i h_i$ is negative (see Fig. 2.2.4). The probability that a particular neuron will be unstable is given by the probability that the noise is less than −1. This depends on the distribution of the values of the noise, not just its typical value. We can find the distribution of the noise using the central limit theorem (Section 1.2). When the number of steps in the random walk is large, the distribution of values of the noise can be approximated by a Gaussian (Eq. (1.2.39)). Then we can find the probability that a neuron is unstable using:

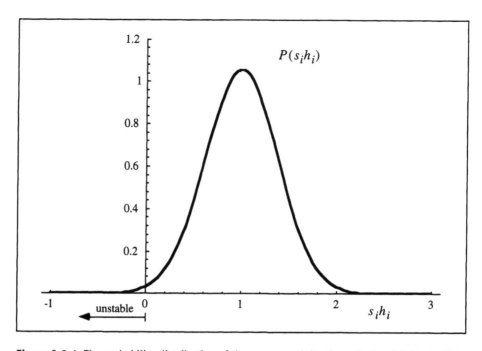

Figure 2.2.4 The probability distribution of the neuron activity times the local field $s_i h_i$. This figure illustrates the signal-to-noise analysis of the stability of an imprinted pattern. The average value of $s_i h_i$ (the signal) is 1. The standard deviation σ of the distribution $P(s_i h_i)$ (the noise) is given by Eq. (2.2.22). Neurons that are unstable have a negative value of $s_i h_i$. The figure is drawn for $\sigma = .379$, when less than 1% are unstable. If σ is larger than this critical value there are more unstable neurons, and when they switch after one update of the network they destabilize the whole pattern. When σ is smaller than this critical value and there are fewer unstable neurons, the rest of the pattern remains stable. The critical value of σ corresponds to a maximum number of patterns that can be stored in the network. ∎

$$P(s_i h_i < 0) = \frac{1}{\sqrt{2\pi\sigma^2}} \int_{-\infty}^{-1} dx\, e^{-x^2/2\sigma^2}$$

$$= \frac{1}{2}\left(1 - \mathrm{erf}(\frac{1}{\sigma\sqrt{2}})\right)$$

$$(2.2.23)$$

The latter expression follows from the definition of the error function $\mathrm{erf}(x)$. For $\sigma < 0.430$ this probability, and therefore the fraction of unstable neurons in the imprinted pattern, is less than 1%. The unstable neurons will switch their values when the network updates itself. We now make the assumption that a few unstable neurons will not, when they flip their values, destabilize many other neurons. This makes sense for few enough unstable neurons. If we are satisfied with a small fraction of error of about 1%, we can store a number of patterns that is given by

$$\alpha_c = \frac{p}{N} = \sigma^2 \qquad (2.2.24)$$

or approximately 0.185 for $\sigma = 0.430$. A more formal analysis, which we do not reproduce here, shows that there is a critical value of p/N at which the error fraction in a pattern jumps from about 1% to essentially no useful retrieval. The critical value $\alpha_c = 0.144$ ($\sigma = 0.379$) can be obtained from techniques developed in the study of spin glasses.

Question 2.2.1 Generalize the signal-to-noise analysis to describe the behavior of an initial pattern which is a Hamming distance B away from one of the imprinted patterns. Assume that the initial pattern is not correlated with any of the other imprinted patterns. Use the analysis to obtain an estimate of the basin of attraction of the imprinted patterns.

Solution 2.2.1 We initialize the network with a pattern that is a Hamming distance B from the first pattern. For convenience, we choose the pattern so that the first $N - B$ neurons are given by the first pattern, and the last B neurons are inverted:

$$s_i(0) = \begin{cases} \xi_i^1 & i \in \{1, ..., N - B\} \\ -\xi_i^1 & i \in \{N - B + 1, ..., N\} \end{cases} \tag{2.2.25}$$

Our objective is to see whether the neural update will recover the imprinted pattern. To simplify the analysis, we assume that the recovery of the pattern must occur in the first update of the network. This will occur if the first $N - B$ neurons are stable and the last B neurons are unstable.

We check the stability of the first $N - B$ neurons by studying the stability of the first neuron. It is stable if

$$s_1(0)h_1 = \xi_1^1 h_1 = \xi_1^1 \sum_{j=2}^{N-B} J_{1j}\xi_j^1 + \xi_1^1 \sum_{j=N-B+1}^{N} J_{1j}\left(-\xi_j^1\right) \tag{2.2.26}$$

is positive. Similarly, we check the stability of the last B neurons by studying the stability of the last neuron. It is unstable if:

$$s_N(0)h_N = \left(-\xi_N^1\right)h_N = \left(-\xi_N^1\right)\sum_{j=2}^{N-B} J_{Nj}\xi_j^1 + \left(-\xi_N^1\right)\sum_{j=N-B+1}^{N} J_{Nj}\left(-\xi_j^1\right) \tag{2.2.27}$$

is negative. Multiplying Eq. (2.2.27) by -1, we see that the condition that Eq. (2.2.27) is negative is actually the same as the condition that Eq. (2.2.26) is positive. Thus we can study the stability of the first neuron in order to verify whether the imprinted pattern will be recovered after one update.

Inserting the Hebbian form for the synapses after p imprints into Eq. (2.2.26), we have:

$$\xi_1^1 h_1 = \frac{1}{N}\sum_{j=2}^{N-B}\sum_{\mu=1}^{p}\xi_1^1\xi_1^\mu\xi_j^\mu\xi_j^1 - \frac{1}{N}\sum_{j=(N-B+1)}^{N}\sum_{\mu=1}^{p}\xi_1^1\xi_1^\mu\xi_j^\mu\xi_j^1 \tag{2.2.28}$$

We separate each sum in this expression into two parts. The first part is due to the imprint of the pattern we are trying to retrieve. The other part is due to all of the other patterns

$$
\xi_1^1 h_1 = \frac{1}{N}\sum_{j=2}^{N-B}\xi_1^1\xi_1^1\xi_j^1\xi_j^1 + \frac{1}{N}\sum_{j=2}^{N-B}\sum_{\mu=2}^{p}\xi_1^1\xi_1^\mu\xi_j^\mu\xi_j^1
$$
$$
- \frac{1}{N}\sum_{j=(N-B+1)}^{N}\xi_1^1\xi_1^1\xi_j^1\xi_j^1 - \frac{1}{N}\sum_{j=(N-B+1)}^{N}\sum_{\mu=2}^{p}\xi_1^1\xi_1^\mu\xi_j^\mu\xi_j^1 \tag{2.2.29}
$$
$$
= \frac{N-2B-1}{N} + \frac{1}{N}\sum_{j=2}^{N-B}\sum_{\mu=2}^{p}\xi_1^1\xi_1^\mu\xi_j^\mu\xi_j^1 - \frac{1}{N}\sum_{j=(N-B+1)}^{N}\sum_{\mu=2}^{p}\xi_1^1\xi_1^\mu\xi_j^\mu\xi_j^1
$$

where the parts due to the first imprinted pattern are explicitly evaluated because the square of either ±1 is 1. As before, the remaining sums constitute a random walk. It doesn't matter that there is a minus sign, since all of the terms are either ±1 and are uncorrelated. The total number of terms is $(N-1)(p-1)$. The typical magnitude (root mean square) of the noise is the same as before, but the signal term is different. The ratio of signal to noise is:

$$
\frac{(N-2B)/N}{\sqrt{p/N}} = \frac{(N-2B)}{\sqrt{pN}} \tag{2.2.30}
$$

where we have neglected 1 compared to both N and p.

To obtain an approximation to the size of the basin of attraction, we set the signal-to-noise ratio equal to the critical value for pattern stability $1/\sqrt{\alpha_c}$ obtained before. This gives a basin of attraction of size:

$$
B = \frac{N}{2}\left(1 - \sqrt{\frac{p}{\alpha_c N}}\right) \tag{2.2.31}
$$

The result is consistent with two limiting results that we already know. The basin of attraction for a small number of imprints is just $N/2$, which is consistent with our discussion of a single imprint in Section 2.2.4. The basin of attraction vanishes when p reaches $\alpha_c N$. ∎

2.2.6 *Simulations*

The attractor neural network is well suited for simulation. In Fig. 2.2.5 we show the probability that an imprinted pattern is unstable and in Fig. 2.2.6 we show the number of stable imprinted patterns. Both are plotted as a function of the number of imprints p. The network used in the simulations has $N = 100$ neurons. The results are obtained by following the procedure:

1. Generate p random neural states $\{\xi_i^\mu\}$:

$$
\xi_i^\mu = \pm 1 \qquad \mu = \{1,...,p\}, i = \{1,...,N\} \tag{2.2.32}
$$

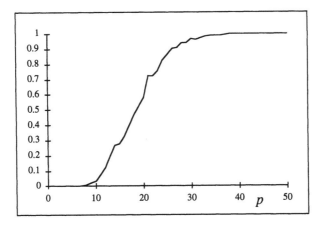

Figure 2.2.5 Fraction of unstable imprints as a function of the number of imprints p on a neural network of 100 neurons using Hebbian imprinting. For p less than 10 the stability of all of the stored patterns is perfect. Above this value the percentage of unstable patterns increases until all patterns are unstable. ∎

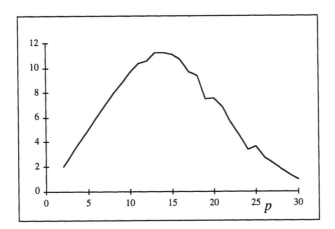

Figure 2.2.6 Number of stable imprints as a function of the number of imprints p on a neural network of 100 neurons using Hebbian imprinting. For p less than 10 all patterns are stable. The maximum number of stable imprinted patterns is less than 12. Above 15 imprints the number of stable patterns decreases gradually to zero. However, throughout this regime the basins of attraction of the patterns are very small and the system is not usable as a memory. ∎

2. Imprint them on the synapses of the neural network:

$$J_{ij} = \begin{cases} \dfrac{1}{N}\displaystyle\sum_{\mu=1}^{p} \xi_i^{\mu}\xi_j^{\mu} & i \neq j \\[2ex] 0 & i = j \end{cases} \qquad (2.2.33)$$

3. Find the number of imprinted neural states that are stable:

$$P_{stable} = \sum_{\mu=1}^{p} \prod_i \delta\left(\xi_i^\mu, \text{sign}\left(\sum_j J_{ij} \xi_j^\mu \right) \right) \tag{2.2.34}$$

where $\delta(i,j)$ is the Kronecker delta function defined by

$$\delta(i,j) = \begin{cases} 1 & i=j \\ 0 & i \ne j \end{cases} \tag{2.2.35}$$

4. Find the probability that a pattern is stable:

$$P_{stable} = p_{stable}/p \tag{2.2.36}$$

5. Average p_{stable} and P_{stable} over a number of trials (steps 1–4) with fixed N and p.

We can also investigate the basin of attraction. Consider a particular stable state of the neural network $\{s_i\}$ which may be an imprinted state. The basin of attraction of $\{s_i\}$ measures the size of the region of possible network states which is "attracted" to $\{s_i\}$. A neural state $\{s_i'\}$ is attracted to $\{s_i\}$ if $\{s_i'\}$ evolves to $\{s_i\}$ upon multiple application of the neuron update rule. Measuring the size of the basin of attraction is important, because the functioning of the neural network as an associative memory depends upon it.

In Fig. 2.2.7(a), the distribution of sizes of the basins of attraction B is shown for different numbers of imprints p on a network with 100 neurons. Each curve is normalized to 1 so that it gives the probability of finding a particular imprinted state with the specified basin of attraction. Fig. 2.2.7(b) shows the corresponding histograms of sizes of the basins of attraction for p imprinted states (the normalization of each curve is p). The maximum possible size of the basin of attraction is 50, half of the number of neurons. We can see from these figures that as the number of imprints increases, the average size of the basins of attraction decreases, and the width of their distribution increases. In addition, the number of imprinted states that are unstable increases. The algorithm used to obtain these figures includes the unstable states as having a basin of attraction of zero.

Fig. 2.2.7 was obtained using the following procedure:

1. Generate p random neural states $\{\xi_i^\mu\}$:

$$\xi_i^\mu = \pm 1 \qquad \mu = \{1,...,p\}, i = \{1,...,N\} \tag{2.2.37}$$

2. Imprint them on the synapses of the neural network:

$$J_{ij} = \begin{cases} \dfrac{1}{N}\displaystyle\sum_{\mu=1}^{p} \xi_i^\mu \xi_j^\mu & i \ne j \\ \\ 0 & i = j \end{cases} \tag{2.2.38}$$

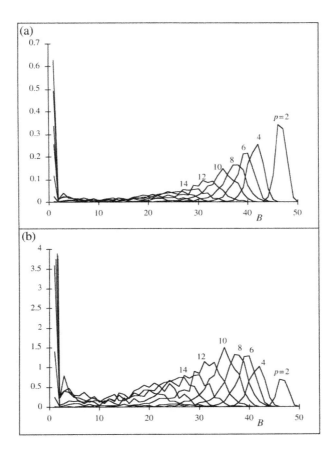

Figure 2.2.7 (a) Probability distribution of the size of the basin of attraction of imprinted patterns for a neural network of 100 neurons, and (b) histograms of the number of imprinted patterns with a particular basin of attraction. The horizontal axis is the Hamming distance, which measures the size of the basin of attraction. The probability distributions are normalized to 1, while the histograms are normalized to p. Each curve is for a different number of imprinted patterns as shown. The size of the basin of attraction decreases as the number of imprints increases. The probability distribution also broadens. When the number of imprints becomes greater than 10, the number of imprints with basins of attraction of zero begins to increase. This is the probability that a pattern is unstable as shown in Fig. 2.2.4. ∎

3. Find the basin of attraction of each of the imprinted patterns ξ^μ. The following steps measure the size of the basin of attraction by finding the average Hamming distance to a state which is not in the basin of attraction of the pattern ξ^μ:

a. Set the neural state to ξ_i^μ.

b. Pick (at random) an ordering of neurons $l(i)$.

c. Switch the state of each neuron in sequence according to the ordering $l(i)$ and find the minimum number of switches for which: the neural state resulting after at most a prespecified number n (taken to be 10) neural updates is *not* equal to the neural state ξ^μ:

$$w_j^{b,0} = \prod_{j'=1}^{b}(1-2\delta(j,l(j')))\xi_i^\mu \qquad b = \{0,...,N/2\}$$

$$w_i^{b,r} = \mathrm{sign}\left(\sum_j J_{ij}w_j^{b,r-1}\right) \qquad r = \{1,...,n\} \qquad (2.2.39)$$

$$B(\xi^\mu) = \min_{\prod_i \delta\left(\xi_i^\mu,w_i^{b,n}\right)=0} b$$

The last expression specifies that we take the minimal value of b subject to the constraint that the state $w^{b,n}$ is not equal to ξ^μ. A straightforward procedure would increment b, evaluate $w^{b,n}$, and stop incrementing b when the condition of the last equation is satisfied.

d. Average over choices of neuron orderings $l(i)$.

4. Make a histogram of the basins of attraction for different ξ^μ, with fixed N and p.

Question 2.2.2 Use Glauber dynamics (Section 1.6.7) to introduce noise into the neural dynamics. Show that the noise actually can improve the retrieval of imprinted states. Specifically, use a network with 100 neurons and imprint 8 (random) neural states. Starting from a random neural state that was not imprinted, evolve the network a number of times with a measured amount of noise. Find the fraction of times that the network recovers one of the imprinted states. Vary the amount of noise to see its effect.

Why would noise increase the probability of retrieving the imprinted states? The imprinting not only creates basins of attraction for the imprinted states, it also causes the existence of many small shallow local energy minima that are called spurious memories. The noise enables the network to escape these shallow local energy minima and fall into the deeper energy minima that are the imprinted states. Spurious memories are discussed in Section 2.2.7.

Solution 2.2.2 Glauber dynamics is a standard implementation of a noisy update rule for neural networks. It uses a statistical rule for the state of each neuron at the next time step. At each time step the value of the neuron is set according to a probability given by the local field. The probability is written as:

$$P_{s_i}(+1;t) = \left(1+\tanh\left(\beta\sum_i J_{ij}s_j(t-1)\right)\right)/2$$

$$P_{s_i}(-1;t) = 1 - P_{s_i}(+1;t)$$

$$(2.2.40)$$

where $\beta = 1/kT$ and T is the effective temperature associated with the noise. While Glauber dynamics uses asynchronous updating by selecting neurons to update at random, synchronous updating does not give significantly different results in many cases.

We can write the Glauber dynamics in an implicit way by introducing a temperature-dependent sign function which implies the probabilistic rule:

$$s_i(t) = \text{sign}_T\left(\sum_j J_{ij}s_j(t-1)\right) \tag{2.2.41}$$

where $\text{sign}_T(x)$ is suggestive of a finite temperature version of the sign function. However, this notation means nothing else than the previous probabilistic expression. In the limit as T approaches zero, the original $T = 0$ update rule is recovered. It should be noted that the temperature as used here does not necessarily correspond to the physical temperature. In the brain there is noisiness in the neural firing that may depend on the physical temperature, but may also be controlled by other factors.

With the introduction of the temperature-dependent update rule, care must be taken in defining the normalization of the synapse matrix J_{ij}. This is one of the reasons for the introduction of the normalization $1/N$ in the definition of the Hebbian imprinting rule (Eq. (2.2.6)).

Fig. 2.2.8 shows the probability of retrieving an imprinted state as a function of the amount of noise. This is the fraction of evolved-random states that result in imprinted states of the network (memories) for different values of β. No noise ($T = 0$) corresponds to $\beta = \infty$. The optimal noise for retrieval in these simulations is around $\beta = 4$. One problem in the simulations is how to identify when we have arrived at an imprinted state. Since we are evolving the network with noise, we should not arrive precisely at the imprinted state. One way to solve this problem is to assume that all states within a small Hamming distance are accepted as the imprinted state. For these simulations we avoid this problem by evolving the network at zero temperature, after it is evolved with noise.

Fig. 2.2.8 was generated using the following procedure:

1. Generate $p = 8$ random neural states $\{\xi_i^\mu\}$:

$$\xi_i^\mu = \pm 1 \qquad\qquad \mu = \{1,...,p\}, i = \{1,...,N\} \tag{2.2.42}$$

2. Imprint them on the synapses of the neural network:

$$J_{ij} = \begin{cases} \dfrac{1}{N}\displaystyle\sum_{\mu=1}^{p}\xi_i^\mu\xi_j^\mu & i \neq j \\ 0 & i = j \end{cases} \tag{2.2.43}$$

3. Generate a random neural state $\{w_i\}$:

$$w_i = \pm 1 \qquad\qquad i = \{1,...,N\} \tag{2.2.44}$$

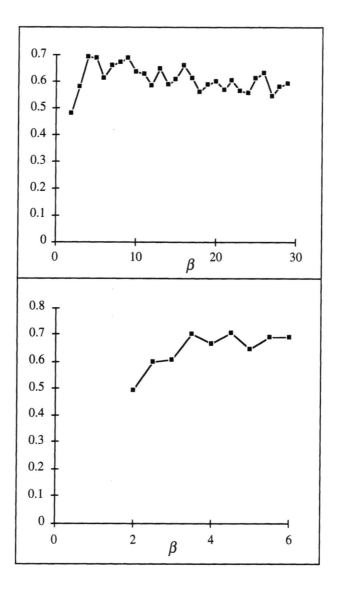

Figure 2.2.8 Tests of the influence of noise on the recovery of patterns imprinted on a network. The curves show the fraction of times neural evolution from a random initial state results in an imprinted state. The simulations use a network of $N = 100$ neurons and $p = 8$ imprinted states. The horizontal axis is the inverse of the effective temperature that describes
the noise. The smallest amount of noise corresponds to the highest value of β. For low levels
of noise the probability of recovering an imprinted state is less than 0.6. When noise is included the recovery rate can reach almost 0.7. The recovery rate improves gradually with increasing noise until about $\beta \approx 0.04$ when the recovery rate decreases dramatically. (a) shows
a broader range of β, and (b) shows a narrower range near the optimal value of β for these
simulations. The variability in the result, despite averaging in the simulations, reflects the
importance of the particular (random) choice of imprinted patterns, and the use of only a limited number of updates of the network. ∎

4. Update the neural state $\{w_i\}$ according to the Glauber dynamics update rule $r_1 = 20$ times at temperature T:

$$w_i^r = \text{sign}_T \left(\sum_j J_{ij} w_j^{r-1} \right) \qquad r = \{1,...,r_1\}, i = \{1,...,N\} \qquad (2.2.45)$$

5. Update the neural state $\{w_i^{r_1}\}$ according to the update rule $r_2 = 5$ times at $T = 0$:

$$w_i^r = \text{sign} \left(\sum_j J_{ij} w_j^{r-1} \right) \qquad r = \{r_1 + 1,...,r_1 + r_2\}, i = \{1,...,N\} \qquad (2.2.46)$$

6. Find if the evolved neural state is equal to one of the originally imprinted states:

$$P_{stable} = \sum_\mu \prod_i \delta \left(\xi_i^\mu, w_i^{r_1 + r_2} \right) \qquad (2.2.47)$$

7. Average P_{stable} over different trials to find the proportion of evolved random states equal to one of the imprinted states. ∎

2.2.7 Overload and spurious states

We have discussed the storage capacity of attractor networks with Hebbian imprinting. As part of this discussion we showed that the basins of attraction of the imprinted states go to zero when the memory becomes overloaded. This implies that there is a catastrophic failure of the network—when we exceed capacity, all of the memories are forgotten. The reason that this occurs is that all of the memories are treated the same by the imprinting process. When the capacity is exceeded, there is no mechanism for the network to select which of them to remember.

There are modifications of the imprinting rule that enable the memory to retain some of the imprinted patterns as memories, at the expense of losing the others. The simplest way to determine which imprints to remember is by the order of the imprint. Rather than keeping the first few imprints, it makes sense to retain the most recent (last few) imprints. A memory that retains the most recent imprints is known as a palimpsest memory, after the name of parchments that were erased and reused in medieval times. Historians benefited from the residuals of earlier writings that remained visible. For our neural network implementation, we could modify the Hebbian imprinting by progressively increasing the strength of the imprint of patterns by a factor $e^{\varepsilon / N}$:

$$J_{ij}(t) = J_{ij}(t-1) + e^{+\varepsilon t / N} \xi_i^t \xi_j^t \qquad (i \neq j) \qquad (2.2.48)$$

where we assume that each imprint is performed in a unit time interval, and the patterns are indexed by time. A value of $\varepsilon = 8.44$ has been calculated as optimal for storing random neural states. In general this and other palimpsest memories reduce the effective capacity of the network. Because of the difference in treatment of recent versus older memories, there is a significant degradation in total number of memories retained. This sacrifice occurs for the benefit of ensuring that some memories are retained after overload would otherwise occur.

Note that multiplying all the synapses J_{ij} by a constant does not affect any results of retrieval or stability. Thus, only the relative strength of different imprints is important. If a bound on the magnitude of synapses is desired, we can adopt the expression

$$J_{ij}(t) = e^{-\varepsilon/N} J_{ij}(t-1) + \xi_i^t \xi_j^t \quad (i \neq j) \tag{2.2.49}$$

instead of Eq. (2.2.48). This is more like the erasure of previous writing, because all the synapses are reduced by the factor $e^{-\varepsilon/N}$ before the next imprint.

While there are methods, like these palimpsest memories, to ensure that overload does not occur, it is important to understand how overload occurs. Overload is a natural mode of failure of the attractor neural network. Therefore, it is likely to occur for biological networks under some circumstances. Detailed studies of attractor networks at high capacity indicate that the behavior of the network near overload becomes dominated by what are called spurious memories. We have spoken about the imprinted patterns as if they are the only stable states of a network. This is not the case. Spurious memories are stable states of the network that were not imprinted. Without an independent way of telling whether they were imprinted or not, these states masquerade as memories, but they are not. Spurious memories are not completely unrelated to the imprinted states. Instead they are generally a mixture of states. One example is a state formed by a majority rule from three imprinted states:

$$s_i = \text{sign}(\xi_i^1 + \xi_i^2 + \xi_i^3) \tag{2.2.50}$$

In Question 2.2.3 the stability of this state is shown using a signal-to-noise analysis. The problem that arises as the number of imprints increase is that the number of such spurious states increases combinatorially (greater than exponentially) with the number of imprints. The growth in the number of spurious states occurs because they are formed from all possible combinations of the imprinted states. When overload occurs, it is actually the basins of attraction of these states that swamp the basins of attraction of the imprinted states.

Once the spurious states swamp the imprinted states, the network becomes essentially equivalent to a spin glass (Section 1.6) that has random weights for each of the synapses. We can understand this qualitatively because the noise becomes larger than the signal in the signal-to-noise analysis. Thus the energy of any state is given by a sum over random variables. Beyond overload, the characteristics of the neural network become similar to those of the spin glass, where there are a hierarchically structured set of minimum energy configurations with large barriers between them. The lowest energy states are not the imprinted ones.

Question 2.2.3 Evaluate the stability of the symmetric mixture of three states given by Eq. (2.2.50) using a signal-to-noise analysis. Hint: convince yourself that the noise is essentially the same as that for an imprinted state and evaluate only the signal.

Solution 2.2.3 A difficulty in studying the stability of the spurious pattern given by Eq. (2.2.50) is that there are two distinct types of neurons—those where all three patterns in the mixture have the same activity, and those where only two out of the three have the same activity. It is important to distinguish these two cases. We evaluate first the signal and then discuss the noise.

The stability of s_1 for the state given by Eq. (2.2.50) is determined by

$$s_1 h_1 = \text{sign}(\xi_1^1 + \xi_1^2 + \xi_1^3) \sum_{j=2}^{N} J_{1j} \text{sign}(\xi_j^1 + \xi_j^2 + \xi_j^3) \qquad (2.2.51)$$

Inserting the Hebbian form for the synapses after p imprints gives

$$s_1 h_1 = \frac{1}{N} \sum_{j=2}^{N} \sum_{\mu=1}^{p} \text{sign}(\xi_1^1 + \xi_1^2 + \xi_1^3) \xi_1^\mu \xi_j^\mu \text{sign}(\xi_j^1 + \xi_j^2 + \xi_j^3) \quad (2.2.52)$$

Our objective is to determine the probability that this is negative, in which case s_1 is unstable. As in the treatment of the imprinted patterns in the text, we do this by evaluating the average (the signal) and the standard deviation (the noise) of the distribution, and approximate the distribution as a Gaussian.

The signal arises from the first three terms in the sum over μ, which we separate to obtain:

$$s_1 h_1 = \frac{1}{N} \sum_{j=2}^{N} \text{sign}(\xi_1^1 + \xi_1^2 + \xi_1^3)\left(\xi_1^1 \xi_j^1 + \xi_1^2 \xi_j^2 + \xi_1^3 \xi_j^3 \right) \text{sign}(\xi_j^1 + \xi_j^2 + \xi_j^3)$$

$$+ \frac{1}{N} \sum_{j=2}^{N} \sum_{\mu=4}^{p} \text{sign}(\xi_1^1 + \xi_1^2 + \xi_1^3) \xi_1^\mu \xi_j^\mu \text{sign}(\xi_j^1 + \xi_j^2 + \xi_j^3) \qquad (2.2.53)$$

The average value of the first sum depends on whether ξ_1^1, ξ_1^2 and ξ_1^3 have the same sign. If they do we have a signal given by:

$$<s_1 h_1> = \frac{1}{N} \sum_{j=2}^{N} \left(\xi_j^1 + \xi_j^2 + \xi_j^3 \right) \text{sign}(\xi_j^1 + \xi_j^2 + \xi_j^3) = 3 \times \frac{1}{4} + 1 \times \frac{3}{4} = \frac{3}{2} \;\; (2.2.54)$$

where the intermediate equation indicates the value of the term multiplied by the probability of its occurrence. Thus, terms that have a magnitude of 3 occur 1/4 of the time, while terms that have a magnitude of 1 occur 3/4 of the time. Similarly, if ξ_1^1, ξ_1^2 and ξ_1^3 do not have the same sign, then two out of three of them have the same sign, and they have a signal given by:

$$<s_1 h_1> = \frac{1}{N} \sum_{j=2}^{N} \left(\xi_j^1 + \xi_j^2 - \xi_j^3 \right) \text{sign}(\xi_j^1 + \xi_j^2 - \xi_j^3) = 1 \times \frac{1}{4} + 3 \times \frac{1}{4} - 1 \times \frac{1}{2} = \frac{1}{2}$$

$$(2.2.55)$$

We see that for the first type of neuron (1/4 of the neurons are of this type) the signal is higher than the signal of an imprinted pattern. On the other hand for the second type of neuron (the remaining 3/4 of the neurons) the signal is lower than that of an imprinted pattern.

The noise can be determined by direct evaluation of the standard deviation of Eq. (2.2.53). However, we can convince ourselves that it is not much different than the noise found for an imprinted pattern in Eq. (2.2.22). The last sum in Eq. (2.2.53) is a sum over $(N-1)(p-3)$ uncorrelated random values of ± 1. Its root mean square magnitude is $\sigma_2 = \sqrt{(p-3)/N}$. This is most of the noise for $p \gg 1$ because the first sum, after we subtract the mean, contains no more than $3N$ uncorrelated terms with magnitude one. Thus the standard deviation of the first sum is no more than roughly $\sigma_1 \approx \sqrt{3/N}$, and the total standard deviation satisfies

$$\sigma^2 = \sigma_1^2 + \sigma_2^2 \approx \sigma_2^2 \approx \sqrt{p/N} \qquad (2.2.56)$$

for $p \gg 1$. This is the same as the noise term found for imprinted patterns.

The main conclusion that we reach from this analysis is that for low storage, $p \ll N$, the neurons have a signal that is much greater than the noise, so the pattern will be stable. The observation above that 3/4 of the neurons have a signal that is half of the signal in an imprinted pattern implies that the basin of attraction of the spurious patterns is shallower and smaller than that of the imprinted patterns. ∎

2.3 Feedforward Networks*

2.3.1 Defining feedforward networks

Feedforward networks (Fig. 2.3.1) are convenient for visualizing input-output systems. They have also been more extensively used in the construction of commercial applications than other neural network models. A feedforward network is composed of several layers. The number of these layers is not large, in part because of difficulties in training these networks. The synapses of a feedforward network are unidirectional. The neuron activity is represented by a continuous variable over a limited range of possible values. We take the range of values to be $(-1,+1)$:

$$s_i^l \in (-1,+1) \qquad i \in \{1,...,N_l\}, l \in \{1,...,L\} \qquad (2.3.1)$$

where l is the layer index, and the number of neurons in a layer N_l may vary from layer to layer. For the synapses we adopt the notation:

$$J_{ij}^l \qquad i \in \{1,...,N_{l+1}\}, j \in \{1,...,N_l\}, l \in \{1,...,L-1\} \qquad (2.3.2)$$

For L layers of neurons there are $L-1$ sets of synapses. By our indexing conventions, the last set of synapses is J_{ij}^{L-1}.

---

*This section may be omitted without significant loss of continuity.

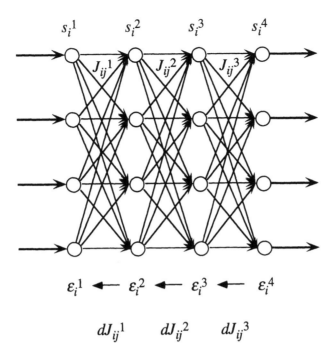

Figure 2.3.1 Schematic of a feedforward network showing the notation for the neurons s_i^l and synapses J_{ij}^l running in only one direction between the layers of neurons. The input to the network enters from the left and the output is read from the right. The most commonly used training algorithm for the feedforward network is the back-propagation algorithm. Starting from an initial set of values of the synapses, the output is calculated from a preselected input for which a desired output is known. The desired output is compared with the output that is calculated. The difference is the error on the last layer of neurons (here the fourth layer) ε_i^4. The error is used to change incrementally the synapses J_{ij}^3 so as to reduce the error. The error is also used to obtain the corresponding error on the previous layer ε_i^3 so that the previous layers of synapses J_{ij}^2 can also be corrected accordingly. In this way the error is propagated backward through the network to correct all of the synapses. ∎

The propagation of input through the network proceeds in a layer by layer fashion. We can picture this as a signal that is propagating through the network, so that the layer index l in the neuron variable s_i^l becomes the analog of a time index. However, it is important to recognize that the propagation through the network is both the space and time coordinate when the processing of an individual input pattern is considered. There is no other time coordinate in the network operation.

The update rule that determines the activity of a neuron at a particular time is a function of the influence of the neurons of the previous layer, usually taken to be sigmoidal:

$$s_i^{l+1} = \text{sgm}\left(\sum_j J_{ij}^l s_j^l\right)$$

(2.3.3)

The sigmoidal function may be the function $\tanh(x)$, which may also be generalized to

$$\text{sgm}(x) = \tanh(\beta(x - h)) \tag{2.3.4}$$

The parameter β is an overall multiplier of the synaptic weights, and therefore is redundant. It may nevertheless be convenient to use it under some circumstances. h is an additional parameter that could vary from neuron to neuron (with the notation h_i^l), and may also be adjusted as part of the training procedure described in the following section. Other forms of sigmoidal function may be used as well.

2.3.2 *Operating and training feedforward networks*

The operation of a feedforward network begins with the imposition of a pattern of activity on the first layer of the neurons. This pattern is assumed to represent the operation of sensory neurons. The activity of each successive layer is then determined according to Eq. (2.3.3). The action of the network on this input ends with the extraction of the neuron activities from the final layer of neurons. This extraction may be considered to be an effect—an action caused by motor neurons. Alternatively, we could consider the activities of the final layer of neurons to be a new representation of the input sensory information—the recognition of a pattern.

To train the network synapses, we begin from a set of examples of input and output pairs that the network should emulate. The objective is to produce the specified outputs from the specified inputs. Once the network is trained, as far as possible, to produce the desired output from each input, it automatically generalizes from these training examples. The generalization is obtained by inputting other patterns to the network and obtaining the resulting output. In effect, the network interpolates between the training examples.

We designate the input and output training pairs (of which there are p) as:

$$(\xi_i^v, \eta_j^v) \qquad i \in \{1,...,N_1\}, j \in \{1,...,N_L\}, v = \{1,...,p\} \tag{2.3.5}$$

The training of the feedforward network can be performed in many ways. The most common method begins from the recognition that it is only the values of the neurons in the final layer that explicitly matter to the operation of the system. The layers between the input and output layer are "hidden." The objective is to optimize the agreement between the action of the network and the desired output. To achieve this we write a cost function (energy), which measures the error—the difference between the value of the output neurons after action on the input trial state and the desired output. The cost function is:

$$E[\{J_{ij}^l\}] = \sum_{v=1}^{p} \sum_{i=1}^{N_L} (s_i^L(v) - \eta_i^v)^2 \tag{2.3.6}$$

where we have introduced the notation $s_i^L(v)$ to indicate the activities of the Lth layer of neurons that result from application of the network to the vth input. For simplicity, the different errors are weighted equally. Implicitly, $s_i^L(v)$ and the cost function depend on the values of all of the synaptic variables J_{ij}^l. The cost function should be minimized with respect to them. The cost function may be minimized in a variety of ways (Section 1.7.4), however, as usual for a problem with a high-dimensional minimization space, there may be problems with many local minima. Nevertheless, the simplest approach is a steepest-descent minimization algorithm.

Before proceeding with a mathematical derivation of the most common approach to minimizing the cost function, we briefly summarize the results. At each step of the procedure, we present to the network a particular one of the input examples. Once the network has operated on the input, we compare the activities of the last neuron layer to the desired output. Our objective is to decrease the error. The easiest way to improve the agreement is to change the last layer of synapses. We calculate the direction to change these synapses to improve the agreement. In general, making this change in the synapses will not be sufficient. To improve the agreement further, we take the error at each of the output neurons and determine what changes would be needed in the activity of the previous layer of neurons in order to correct the output-neuron values. This is done using the existing synapses between the two layers. This step, taking the error in the final layer of neurons and identifying the corresponding error in the previous layer of neurons, is called back-propagation of error. Once we know the error in the second to last layer of neurons, we can find the direction to change the second to last layer of synapses. We repeat the procedure for earlier layers, and correct incrementally all of the synapses of the network.

The following derivation is difficult only because of the number of subscripts and superscripts. We adopt standard practice and assume that we will minimize the cost function by modifying J_{ij}^l in steps that reduce the cost function successively for each of the patterns separately. Convergence is not guaranteed, but will work if the cost function is well behaved. We thus adopt the partial cost function

$$E^v[\{J_{ij}^l\}] = \sum_{i=1}^{N_L} (s_i^L(v) - \eta_i^v)^2 \tag{2.3.7}$$

To minimize this function we change J_{ij}^l in the direction of steepest descent:

$$J_{ij}^l(t+1) = J_{ij}^l(t) + \delta J_{ij}^l(t)$$
$$\delta J_{ij}^l(t) = -c \frac{\partial E^v[\{J_{ij}^l\}]}{\partial J_{ij}^l(t)} \tag{2.3.8}$$

We use the time variable to indicate repetitive cycling over the different patterns v. It keeps track of the steps in the minimization, not propagation of the signal through the network. c is chosen small enough, and possibly time-dependent, to provide for convergence. In principle it doesn't matter in which order we consider the J_{ij}^l, but it is convenient to start from the synapses leading to the final (Lth) layer J_{ij}^{L-1}.

$$\delta J_{ij}^{L-1}(t) = -c \frac{\partial E^v[\{J_{ij}^l\}]}{\partial J_{ij}^{L-1}(t)} = -c \frac{\partial \sum_{k=1}^{N_L} (s_k^L(v) - \eta_k^v)^2}{\partial J_{ij}^{L-1}(t)} \tag{2.3.9}$$

where we have taken care to use a new index for the sum in the numerator. Taking the derivative inside the sum we have:

$$\delta J_{ij}^{L-1}(t) = -2c \sum_{k=1}^{N_L} (s_k^L(v) - \eta_k^v) \frac{\partial s_k^L(v)}{\partial J_{ij}^{L-1}(t)} \tag{2.3.10}$$

The value of the kth neuron in the Lth layer can only depend on synapses leading directly to it, and not on other synapses. Thus the derivative has to be zero unless $k = i$, and the other terms in the k sum can be neglected. We show this explicitly using the expression for $s_k^L(v)$ in terms of the previous layer of neurons:

$$\frac{\partial s_k^L(v)}{\partial J_{ij}^{L-1}(t)} = \frac{\partial \mathrm{sgm}(\sum_m J_{km}^{L-1}(t)s_m^{L-1}(v))}{\partial J_{ij}^{L-1}(t)}$$

$$= \mathrm{sgm}'(\sum_m J_{km}^{L-1}(t)s_m^{L-1}(v)) \sum_n s_n^{L-1}(v) \frac{\partial J_{kn}^{L-1}(t)}{\partial J_{ij}^{L-1}(t)} \tag{2.3.11}$$

$$= \mathrm{sgm}'(\sum_m J_{km}^{L-1}(t)s_m^{L-1}(v)) \sum_n s_n^{L-1}(v)\delta_{k,i}\delta_{n,j}$$

$$= \mathrm{sgm}'(\sum_m J_{km}^{L-1}(t)s_m^{L-1}(v))s_j^{L-1}(v)\delta_{k,i}$$

The prime on the sigmoidal function has the conventional meaning of a derivative with respect to its argument. In the third line we made use of the knowledge that $s_m^{L-1}(v)$ is independent of J_{ij}^{L-1} because the $(L-1)$st layer of neurons precede the $(L-1)$st layer of synapses.

Returning to the evaluation of the change in J_{ij}^{L-1} we find

$$\delta J_{ij}^{L-1}(t) = -2c \sum_{k=1}^{N_L} (s_k^L(v) - \eta_k^v)\mathrm{sgm}'(\sum_m J_{km}^{L-1}(t)s_m^{L-1}(v))s_j^{L-1}(v)\delta_{k,i}$$

$$= -2c(s_i^L(v) - \eta_i^v)\mathrm{sgm}'(\sum_m J_{im}^{L-1}(t)s_m^{L-1}(v))s_j^{L-1}(v) \tag{2.3.12}$$

We can simplify the notation by defining two auxiliary quantities. We define the error at the Lth layer as:

$$\varepsilon_i^L(v) = (s_i^L(v) - \eta_i^v) \tag{2.3.13}$$

The derivative of the sigmoidal function in Eq. (2.3.12) could be written as a function of the neuron $s_i^L(v)$, since it applies the sigmoidal derivative to the same argument (the postsynaptic potential) that determines the neuron value. We call this function $w(s)$:

$$w(s_i^L(v)) = \mathrm{sgm}'(\sum_m J_{im}^{L-1}(t)s_m^{L-1}(v)) = \mathrm{sgm}'(\mathrm{sgm}^{-1}(s_i^L(v))) \tag{2.3.14}$$

This leads to the simplified form of Eq. (2.3.12)

$$\delta J_{ij}^{L-1}(t) = -2c\,\varepsilon_i^L(v)w(s_i^L(v))s_j^{L-1}(v) \tag{2.3.15}$$

which is the desired result.

Before we continue to find the incremental changes in the other J_{ij}^l for $l < L-1$, we discuss the result for J_{ij}^{L-1} in a simple case. Consider Eq. (2.3.15) if all of the neurons have a level of activity that is within the linear range of the sigmoidal function, taken to be $\tanh(x)$. This would be equivalent to neglecting the nonlinear response of the neurons. Then $\mathrm{sgm}'(x) = 1$ and $w(s_i^L(v)) = 1$. Inserting Eq. (2.3.15) into Eq. (2.3.8) we have:

$$J_{ij}^{L-1}(t+1) \cong J_{ij}^{L-1}(t) - 2c(s_i^L(v,t) - \eta_i^v)s_j^{L-1}(v,t) \quad \text{(linear regime)} \tag{2.3.16}$$

Note that all of the neuron activities, indexed by v, also have a t index, since they depend on the synapses, and thus their values change during the minimization. We can act with the new synapse values on the neurons of the previous layer to obtain the new neuron values in the final layer. We assume that the values of the neurons at the previous layer have not been changed. This would be true if we chose not to modify the previous layers of synapses, or if there were only two layers of neurons (one layer of synapses), then:

$$s_i^L(v,t+1) \cong \sum_j J_{ij}^{L-1}(t+1)s_j^{L-1}(v,t) \tag{2.3.17}$$

Inserting Eq. (2.3.16)

$$
\begin{aligned}
s_i^L(v,t+1) &\cong \sum_j J_{ij}^{L-1}(t)s_j^{L-1}(v,t) - 2c(s_i^L(v,t) - \eta_i^v)\sum_j s_j^{L-1}(v,t)s_j^{L-1}(v,t) \\
&\cong s_i^L(v,t) - 2c(s_i^L(v,t) - \eta_i^v)\sum_j s_j^{L-1}(v,t)s_j^{L-1}(v,t)
\end{aligned}
\tag{2.3.18}
$$

From this we see that if the neuron values of layer $L-1$ are normalized to 1 ($\sum_j s_j^{L-1}s_j^{L-1} = 1$) and $c = 1/2$ (or if c is chosen to be 1/2 of the inverse of the normalization) then convergence will be perfect for the pattern v, since then

$$s_i^L(v,t+1) \cong \eta_i^v \tag{2.3.19}$$

More generally, a smaller value of c will bring the neuron values closer to the desired result, as should be expected from a steepest descent. This shows that for a single input-output training pair, the cost function may be readily minimized using only the linear regime of one layer of synapses. Constructing a network that will perform a desired pattern-recognition task can be much more difficult when there are many input-output training patterns representing the task.

We return to the main line of our discussion and consider the second to last layer of synapses J_{ij}^{L-2}

$$\delta J_{ij}^{L-2}(t) = -c\frac{\partial E^v[\{J_{ij}^l\}]}{\partial J_{ij}^{L-2}(t)} = -c\sum_k \frac{\partial E^v[\{J_{ij}^l\}]}{\partial s_k^L(v)} \frac{\partial s_k^L(v)}{\partial s_i^{L-1}(v)} \frac{\partial s_i^{L-1}(v)}{\partial J_{ij}^{L-2}(t)} \tag{2.3.20}$$

The latter expression uses the sequentiality of the determination of the neuron values. We have also taken into account that in the $(L-1)$st layer, it is only the ith neuron that depends on the synapse J_{ij}^{L-2}. Each of the factors is readily evaluated:

$$\frac{\partial E^{\nu}[\{J_{ij}^{l}\}]}{\partial s_k^L(v)} = 2(s_k^L(v) - \eta_k^{\nu}) = 2\varepsilon_k^L(v)$$

$$\frac{\partial s_k^L(v)}{\partial s_i^{L-1}(v)} = \frac{\partial \, sgm(\sum_m J_{km}^{L-1}(t)s_m^{L-1}(v))}{\partial s_i^{L-1}(v)} = sgm'(\sum_m J_{km}^{L-1}(t)s_m^{L-1}(v))J_{ki}^{L-1}(t)$$

$$= w(s_k^L(v))J_{ki}^{L-1}(t)$$

$$(2.3.21)$$

$$\frac{\partial s_i^{L-1}(v)}{\partial J_{ij}^{L-2}(t)} = \frac{\partial \, sgm(\sum_n J_{in}^{L-2}(t)s_n^{L-2}(v))}{\partial J_{ij}^{L-2}(t)} = sgm'(\sum_n J_{in}^{L-2}(t)s_n^{L-2}(v))s_j^{L-2}(v)$$

$$= w(s_i^{L-1}(v))s_j^{L-2}(v)$$

leading to:

$$\delta J_{ij}^{L-2}(t) = -2c\sum_k \varepsilon_k^L(v)w(s_k^L(v))J_{ki}^{L-1}(t)w(s_i^{L-1}(v))s_j^{L-2}(v) \qquad (2.3.22)$$

We can recast this expression into the same form as for the synapses J_{ij}^{L-1}

$$\delta J_{ij}^{L-2}(t) = -2c\varepsilon_i^{L-1}(v)w(s_i^{L-1}(v))s_j^{L-2}(v) \qquad (2.3.23)$$

by defining the error on the $(L-1)$st layer of neurons as:

$$\varepsilon_i^{L-1}(v) = \sum_k \varepsilon_k^L(v)w(s_k^L(v))J_{ki}^{L-1}(t) \qquad (2.3.24)$$

This expression, in effect, takes the error that was known on the Lth layer and obtains the error on the previous layer using the existing set of synapse values $J_{ki}^{L-1}(t)$. This procedure is known as back-propagation of error and gives the name back-propagation algorithm to this method of training feedforward networks.

The modification of earlier layers is obtained similarly by extending this analysis layer by layer. In each case, an expression of the form of Eq. (2.3.24) can be written that takes the error of one layer and sends it back to the previous layer. The correction to the synapses is then written as in Eq. (2.3.23).

2.4   Subdivided Neural Networks

Among the objectives of the study of neural networks is the development of a basis for an understanding of sensory processing, motor control, memory and higher information-processing functions of the brain. In previous sections, we have seen that it is possible to describe an associative content-addressable memory using an attractor neural network. The associative memory captures an important generic property that we would like to build upon to understand additional aspects of brain function. We also touched upon some aspects of the processing by feedforward networks that are suggestive of sensory-motor systems. However, most of the higher information-processing tasks, of which the brain is readily capable, appear remote from these considerations.

In order to make progress in understanding the higher information-processing functions of the brain we must construct an additional level of organization between neurons and the brain—that of brain subdivisions or subnetworks. The substructure of the brain is well known to students of neurophysiology (see Fig. 2.4.1). Experimentally, the mapping of brain function has identified sensory- or motor-related aspects of the brain—visual processing centers, auditory processing centers, the motor cortex, as well as aspects of language processing that may be counted among the higher information-processing functions. There is a long-standing debate regarding the degree of localization of function in the brain. In a simplified form, the debate is between two camps, one suggesting that specific functions are localized on individual neurons, the other suggesting complete delocalization of function throughout the brain. At present, experimental evidence has led to general agreement that at least an intermediate degree of regional specialization exists.

As we discussed in Section 1.10.7, most fields of inquiry are built upon levels of description each of which is constructed on finer scales. To neglect the description of the subdivisions of the brain and try to explain brain function directly from the behavior of individual neurons would be to skip an important and simplifying level of description. One of our primary tasks, therefore, in studying neural networks, is to investigate and identify the function and interaction of subnetworks. We hope then to build models of human information-processing using subnetworks as the analog of brain subdivisions. It is almost a separate endeavor to construct such a theory of higher information-processing. The historical efforts in this area are the theories of the mind such as those of Freud, which separated the mind in two ways. The first separation was between the conscious and subconscious and the second between the id, ego and superego. This and other theoretical models should be considered within the domain of our inquiry. However, until we have a better understanding of the function of brain subdivisions, we will not be able to evaluate the validity of the many psychological theories or propose more complete ones.

There are two forms of subdivision that can be readily identified—longitudinal and lateral. We have already considered the longitudinal form of subdivision in the example of feedforward networks. A multilayer feedforward network describes a set of neural subdivisions each of which is a single layer of the network. In this model there are no synapses within a neural subdivision, all of the synapses run between subdivisions, specifically in a feedforward direction. The input layer (or first few layers) represents sensory processing, and the output layer (or last few layers) represents motor control. Intermediate layers are less clearly identified. This longitudinal subdivision in feedforward networks is directly related to sequential stages in processing. Longitudinal subdivision in feedforward networks is necessary because of limitations on what a single layer of synapses can be trained to accomplish.

In the remainder of this chapter we consider the second type of subdivision—lateral subdivision—formed when the synaptic connections within each subdivision are of greater number or of greater strength than between the subdivisions. In contrast to

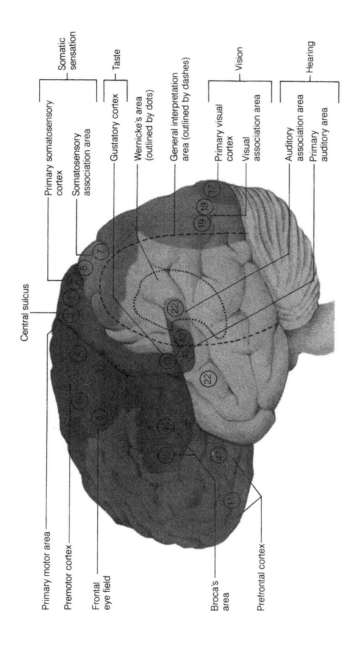

Figure 2.4.1 Illustration of the functional anatomy of the brain as determined by neurophysiological experiments (from E. N. Marieb, *Human Anatomy and Physiology*, [Benjamin Cummings, 1992]) ∎

longitudinal subdivision, lateral subdivision separates the processing of sensory or other information into parallel channels. This kind of subdivision can be treated and understood within the attractor network model (Fig. 2.4.2).

In developing an understanding of the role of subdivisions in the brain, we must begin from basic questions. The most basic is the question, Why should the brain be subdivided at all? This may seem a simple question, since it might seem obvious that, for example, language should be separate from visual processing and from auditory processing and from motor control—doesn't this make sense? But we know that all of these are also connected to each other. Why then should we not process them all together? In the attractor network, we simplify the consideration of network function to that of an associative memory. If we compare a subdivided attractor network with the fully connected attractor network, we immediately run into a fundamental problem—a lower storage capacity.

The storage capacity of a neural network increases with the degree of interconnectedness. In Section 2.2 we determined the storage capacity for the fully connected network with Hebbian imprinting. The network could store $\alpha_c N$ patterns. We can count, instead, the total number of independent bits in the stored patterns. Since each pattern has N bits, this gives a total of $\alpha_c N^2$ bits. This is somewhat deceptive, since in the limit of maximum storage, just below overload, we must present almost all of the pattern in order for it to be "retrieved." However, because any part of the pattern could be retrieved, we might still consider $\alpha_c N^2$ to be the maximum number of bits of information stored in the network. The expression $\alpha_c N^2$ should not be surprising, since the information is stored in the synapses and the number of synapses is N^2. While it is difficult to guess the value of the prefactor α_c, the maximum number of stored bits must be proportional to the number of synapses. Many efforts have been made to improve upon this storage capacity, however, for a fully connected network it is possible to prove that the maximum number of stored independent bits cannot be greater than $2N^2$, or $2N$ uncorrelated patterns. More generally, if all neurons are not connected to each other, then the maximum number of patterns that can be stored is limited by the average number of synapses per neuron.

The loss of memory on reducing the number of synapses occurs when the synapses are set to zero a priori, independent of the information to be stored. This is a clue to the motivation for subdivision. The storage capacity would not be reduced if the synapses are set to zero because a zero value is appropriate to the information that is to be stored.

There may be reasons that are quite independent of storage considerations that the brain does not make use of a fully connected network. The most well known of these is the "connection problem." Three-dimensional space does not allow us to connect all neurons because of the difficulties in packing all of the connections into a volume in the presence of communication delays and heat-dissipation constraints. This problem is familiar to those who study the problems of designing massively parallel computers. While the connection problem might explain why the brain is not fully connected, it does not reveal the reason for nonuniformity of function in the brain.

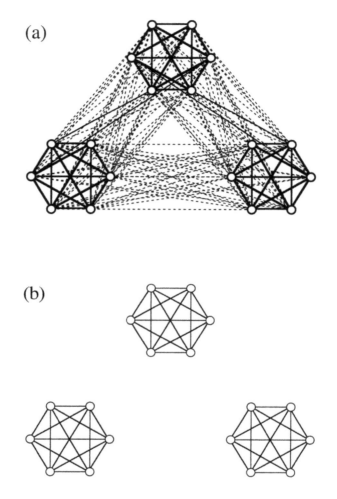

Figure 2.4.2 Subdivided neural network formed out of three subnetworks. The subdivision may be achieved either by setting the synapses between the subnetworks to be systematically weaker than those within a subnetwork or by systematically reducing the number of synapses between subnetworks. The former is indicated by the relative thickness of the lines in (a). In (b) the extreme case of a completely subdivided network is shown where the synapses between the subdivisions have been removed. ∎

We summarize our fundamental question as follows. For a network where each neuron is connected to every other neuron, the number of imprints that can be recalled, αN, is proportional to the number of neurons N with a constant of proportionality $\alpha < \alpha_c$ somewhat dependent on the particular imprinting rule and the desired properties of retrieval. When additional imprints are added, an overload catastrophe causes erasure of all information. Removing synapses or systematically weakening synapses between subdivisions *inherently* results in a decreased storage capacity. Why then subdivide the brain?

2.4.1 *The left-right universe*

To begin to answer this question, consider first an artificial world composed of pictures with independent (uncorrelated) left and right halves (Fig. 2.4.3). This means that any left half that is seen in the universe may be found with any right half. A concrete example is the set of possible first and last name initials, where all letters of the alphabet might appear on the left, as they might on the right. Our task is to design a neural network for an organism in this artificial world. We assume that the pictures are mapped directly onto the network so that they are represented point by point as the neuron activity pattern. We will consider this example in detail, since it captures many of the essential concepts that will be relevant later.

As we have discussed, a fully connected network of N neurons is capable of recalling αN distinct and uncorrelated pictures. We can represent the stored pictures using the notation (L_i, R_i), where R_i is the right half of the ith picture and L_i is the left half of the ith picture. Then the stored images are of this form with i in the range $\{1,...,\alpha N\}$. In order for the pictures not to be correlated, the left sides of all stored pictures must be different from each other, as must be the right sides. If the pictures that the organism encountered in the universe were indeed distinct and uncorrelated, this is the best that can be expected from Hebbian training. However, in the left-right universe, the pictures that might be encountered *are* correlated.

Let us divide the network into left and right hemispheres by cutting all of the synapses running between them. The left hemisphere receives the left part of each picture and the right hemisphere receives the right part of each picture. Each of the hemispheres has $N/2$ neurons. Using Hebbian imprinting, each hemisphere can store $(\alpha N/2)$ distinct half-pictures. Because the subnetworks are half as large as the full network, the number of patterns that can be stored is half as many and each pattern is also half as large. The storage of the left hemisphere is of left halves of pictures, L_i. The storage of the right hemisphere is of right halves of pictures, R_i. In both cases i takes values in the range $\{1,...,\alpha N/2\}$. Storage in the left hemisphere is independent of the storage of the right hemisphere. When we test for recall, each of the patterns stored by the left hemisphere can be combined with each of the patterns in the right hemisphere to obtain a different stored picture. These are composites of the imprinted pictures. Thus the subdivided network stores a total of $(\alpha N/2)^2$ composite pictures of the form (L_i, R_j), where both i and j are taken independently from the range $\{1,...,\alpha N/2\}$. Each of these $(\alpha N/2)^2$ pictures may be encountered in the left-right universe. Since the number of neurons N is large, $(\alpha N/2)^2$ is much larger than αN. Cutting the synapses between the hemispheres results in a huge increase in the number of pictures that can be stored in the network. For an organism in the artificial world, this is a significant advantage.

The retrieval process is different in the fully connected network and in the subdivided network. In the fully connected network, retrieval starts by presenting to the network an image that is close to one of the stored images (L_i, R_i). Somewhat over 50%

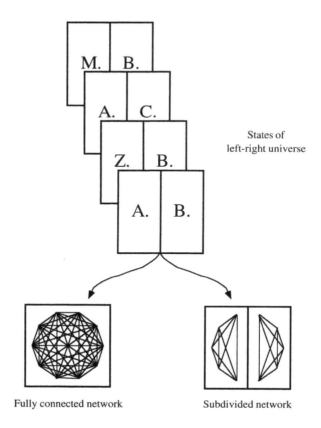

<figure>
M. B.

A. C.

Z. B.

A. B.

States of
left-right universe

Fully connected network   Subdivided network
</figure>

Figure 2.4.3 Illustration of the left-right universe, which consists of images that are composed out of independent left and right halves. The set of all initials is a simple example of such a universe. We could try to store these images on a fully connected network, or we could first subdivide the network into two hemispheres. The fully connected network would remember more completely independent images, but would fail to be able to store multiple images with the same left half or the same right half. The subdivided network would store independently the left and right half, and so would store many more images that would be composed of a selection of the stored left halves and the stored right halves. A biological analog of the left-right universe may arise in the control by the different brain hemispheres of left- and right-hand motion. ∎

of the neurons must be presented to the network in order to recover the full image. If we stored the patterns (L_1,R_1) and (L_2,R_2) and then the universe presents the pattern (L_1,R_2), the network will choose to settle into either (L_1,R_1) or (L_2,R_2), depending on which one of these is closer to (L_1,R_2). In either case we could say that the network is in error, but this error occurs for a state that the network never imprinted.

The subdivided network works on the retrieval of each half of the picture separately. It recognizes (L_1,R_2) from a pattern that is close to L_1 on the left and close to R_2

on the right. It might appear that there is a disadvantage because the subdivided network cannot use partial information from one half of the network to help in recall of the other half. However, it is important to recognize that this is actually dictated by the nature of the information in the left-right universe and only incidentally by the subdivision of the network. Using information from the left half to help on the right would only lead to errors, since it was assumed that there is no correlation between the two parts of the information.

It is significant that the training of the subdivided network was achieved with only $(\alpha N/2)$ imprints. The subdivided network recognizes many more pictures than were trained. In this way the network *generalized* from the training set to a much larger number of pictures. This works if we are able to select which pictures to train initially. We take $(\alpha N/2)$ of the pictures and make sure that all of the left halves and all of the right halves are different. We imprint these pictures. From the perspective of the storage of complete patterns, what we have done may appear quite strange. It is true that we have caused the network to store $(\alpha N/2)^2$ patterns, but aren't all of these really only a few patterns? Yes, they are all related to the imprinted $(\alpha N/2)$ patterns. The point is that in the artificial world, where the left and right parts of the image are independent, we *want* to store the $(\alpha N/2)^2$ different combinations rather than only a particular set of complete patterns.

Let us consider an alternative design of an organism in the left-right universe—a different way of subdividing the network. Instead of cutting the synapses between left and right hemispheres, we cut the synapses between the top and bottom halves of the network. In this case each half of the network acts to store $(\alpha N/2)$ half pictures. The top half stores the top part of each picture. The bottom half stores the bottom part of each picture. Now we cannot claim that the network stores $(\alpha N/2)^2$ pictures, because different combinations of top and bottom are not possible pictures in the universe. Instead the network stores at most only a total of $(\alpha N/2)$ of the possible pictures. There is an additional problem in retrieval, because information from the top cannot be used to help with retrieval of the bottom part of the image, and information from the bottom cannot be used to help with the top. In order to retrieve a particular image, we must have over 50% of the neurons correct in the top half and over 50% of the neurons correct in the bottom half. We have degraded the network storage with no compensating advantages. We have also created a whole host of undesirable memories that are not real. These undesirable memories are combinations of stored top and bottom halves of pictures.

From this discussion we learn that subdividing a network can improve dramatically the storage of patterns. However, the effectiveness of subdivision requires direct matching to the nature of the information: if we know that the organism lives in the left-right universe we can cut synapses between left and right hemispheres.

2.4.2 *Imprinting correlated patterns*

The advantage of subdivision in terms of the number of pictures that can be stored is not the whole story for the left-right universe. The fully connected network actually fails when patterns that are imprinted are significantly correlated. We have, until now, considered only uncorrelated pattern storage in the fully connected network. In the left-right universe, the independence of information between the two halves of the pictures means that the patterns themselves are correlated. Two or more patterns that are to be remembered may have the same right halves and different left halves. When we imprint such correlated information in a fully connected network using Hebbian imprinting, the patterns are not stored. Qualitatively, the problem arises because the right-hand side of the network does not know which of the left sides to reconstruct. When we try to retrieve one of the stored patterns, the result on the left is retrieval of some intermediate picture that is neither of the desired memories. The degree of failure of the network depends on the number of pictures that are imprinted with the same right halves. Memory degradation occurs for just two imprinted pictures. Failure becomes explicit when there are as few as three imprinted pictures. It is, however, simplest to consider first the case of four imprinted patterns, all of which have the same right halves.

We can see the failure of a fully connected network analytically by considering the update of neurons when starting from one of the imprinted patterns. Extending Eq. (2.2.17) to the case of four patterns we have:

$$
s_i(1) = \text{sign}(\sum_j J_{ij} s_j(0))
$$

$$
= \text{sign}(\xi_i^1 \sum_{j \neq i} \xi_j^1 s_j(0) + \xi_i^2 \sum_{j \neq i} \xi_j^2 s_j(0) + \xi_i^3 \sum_{j \neq i} \xi_j^3 s_j(0) + \xi_i^4 \sum_{j \neq i} \xi_j^4 s_j(0)) \quad (2.4.1)
$$

We assume that we are looking at the value of a neuron in the left half of the network $i \in \{1,...,N/2\}$ and the four patterns ξ_j^1, ξ_j^2, ξ_j^3, ξ_j^4, are identical in the right half of the network $j \in \{N/2,...,N\}$. We split the sums in Eq. (2.4.1) into separate sums over the left and right halves of the network so that:

$$
s_i(1) = \text{sign}(\xi_i^1 \sum_{j \neq i}^{N/2} \xi_j^1 s_j(0) + \xi_i^2 \sum_{j \neq i}^{N/2} \xi_j^2 s_j(0) + \xi_i^3 \sum_{j \neq i}^{N/2} \xi_j^3 s_j(0) + \xi_i^4 \sum_{j \neq i}^{N/2} \xi_j^4 s_j(0)
$$

$$
+ (\xi_i^1 + \xi_i^2 + \xi_i^3 + \xi_i^4) \sum_{j=N/2+1}^{N} \xi_j^1 s_j(0)) \quad (2.4.2)
$$

We have collected the sums over the right halves together since they are the same. We can test the stability of the first imprinted pattern. Setting $s_i(0) = \xi_i^1$, the first sum and the last sum are just $N/2$

$$
s_i(1) = \text{sign}(\xi_i^2 \sum_{j \neq i}^{N/2} \xi_j^2 \xi_j^1 + \xi_i^3 \sum_{j \neq i}^{N/2} \xi_j^3 \xi_j^1 + \xi_i^4 \sum_{j \neq i}^{N/2} \xi_j^4 \xi_j^1 + (2\xi_i^1 + \xi_i^2 + \xi_i^3 + \xi_i^4)N/2) \quad (2.4.3)
$$

The remaining three sums are random walks, because the left sides of the patterns are assumed to be uncorrelated. They have a typical size of $\sqrt{N}$ and are smaller than the last set of terms. So we can ignore the first three terms. As we look at different values of $i \in \{1,...,N/2\}$, one-eighth of the time we will have pattern 2,3,4 opposite the first pattern $\xi_i^1 = -\xi_i^2 = -\xi_i^3 = -\xi_i^4$. In this case the neuron will flip after the first update. If one-eighth of a pattern changes after a single update, it is not stable. Since the same argument holds for each of the imprinted patterns, they were not stored.

Let us now think about the case of three patterns imprinted with the same left halves. Instead of Eq. (2.4.3) we have:

$$s_i(1) = \text{sign}(\xi_i^2 \sum_{j\neq i}^{N/2} \zeta_j^2 \zeta_j^1 + \xi_i^3 \sum_{j\neq i}^{N/2} \zeta_j^3 \zeta_j^1 + (2\xi_i^1 + \xi_i^2 + \xi_i^3)N/2) \qquad (2.4.4)$$

This implies that whenever the other two patterns differ from the first at a particular neuron $\xi_i^1 = -\xi_i^2 = -\xi_i^3$, which happens 25% of the time, then the result is dependent upon the overlap of the three states. Specifically, in this case we have

$$s_i(1) = -\xi_i^1 \text{sign}(\sum_{j\neq i}^{N/2} \zeta_j^2 \zeta_j^1 + \sum_{j\neq i}^{N/2} \zeta_j^3 \zeta_j^1) \qquad (2.4.5)$$

The argument of the sign function would always have to be negative in order for the imprinted pattern to be recovered. Statistically, the sign function will be negative or positive with equal probability. When it is negative, the whole pattern is stable. When it is positive, 25% of the initial pattern will not be recovered after a single iteration and the pattern is unstable. Thus for three imprinted patterns, on average half of the patterns will be stable. The stability of these patterns is based upon the sign of the random-walk terms in Eq. (2.4.5). Imprinting any other pattern, correlated or uncorrelated, on the network will destabilize them.

In this section we have shown that the fully connected network fails for correlated patterns. Earlier, in Section 2.4.1, we discussed storage of uncorrelated pictures in the same network. If we have control over the order of pictures that are presented to the network, we can choose uncorrelated pictures to imprint. However, in the left-right universe, we should allow for an arbitrary order of the possible pictures. Some of the pictures will have the same left or right halves. In this case, the fully connected network with Hebbian imprinting will fail. Thus it is necessary to modify the fully connected network to work in the artificial left-right universe, and subdivision of the network is one way to do this. A network without synapses between left and right hemispheres does not suffer from this failure.

The real world is not constructed out of independent left and right pictures, at least the visual field is not. Can we make any sense of the actual subdivision of the brain (specifically the cerebrum) into left and right hemispheres from this model left-right universe? We can, at least in part, by recognizing that both tactile sensation and motor control of the arms and legs requires states that are left-right independent. Motor control requires neural activity patterns that describe (or prescribe) the motion. If we were to try to store the possible patterns of motion of the two hands in a uniform network, the actions of one hand would always be directly related to the actions of the other hand. If we want to be able to do one of several actions with the left hand for the same action of the right hand, then subdividing the network that stores the pattern of neural activity makes sense, and may even be necessary. Of course we would like there to be coordination between actions of the two hands or legs. This means that we need a balance between the independence and dependence of the patterns in the two different divisions. We will investigate partially subdivided networks as one way to achieve this balance.

2.4.3 *Separating independent information*

The example of left-right motor control is a special case in which the simple model of subdivision might help, and might even correspond to an important example of subdivision of the brain. We can generalize this example by recognizing that the information we process is highly correlated. One of the reasons for the correlation is that different aspects of the information are independent of each other in similar manner to the independence of the left-right universe. It is this independence that gives rise to correlations.

We therefore recognize that there are two tasks. First, we must in some way identify which parts of the information to be stored are independent (uncorrelated) and separate these parts of the information. Then we must store these different types of information in different parts of the network. Achieving this will enable tremendous increase in storage of the correlated patterns. Fig. 2.4.4 shows a simple model for the function of sensory processing consistent with this concept. Sensory information is separated by input processors to distinct channels. The input processors are presumed to be composed of feedforward neural networks that are illustrated only schematically. The information is then imposed on a subdivided attractor network that serves as a content-addressable memory.

If the information in the different channels were completely independent, the channels should be completely independent and the entire problem would be to identify what the channels should be. However, the information is not usually completely independent. This suggests that we adopt a model where the network is partially subdivided, with weaker or fewer synapses between the subdivisions of the attractor network. This is the model that we will adopt and investigate. Before pursuing this approach we discuss two more examples of the relevance of this architecture to human information-processing: vision and language.

2.4.4 *Sensory processing: color, shape and motion in vision*

The human visual system does not take advantage of the two hemispheres of the brain to divide the visual information right from left because the left and right parts of the visual field are not independent. There is a large interconnection area called the corpus callosum that connects the visual areas in the two hemispheres. Instead, detailed mapping of the visual cortex has revealed that visual processing separates three attributes of the information: color, shape and motion. The implication of the separate processing of these three attributes using, in effect, a preprocessing step to separate them, is that these information categories are partially independent. For example, visual fields with different shapes can have the same colors. Or, vice versa, the same shapes can have different colors. This independence has been used in the design of the genetically encoded structure of the initial visual information-processing.

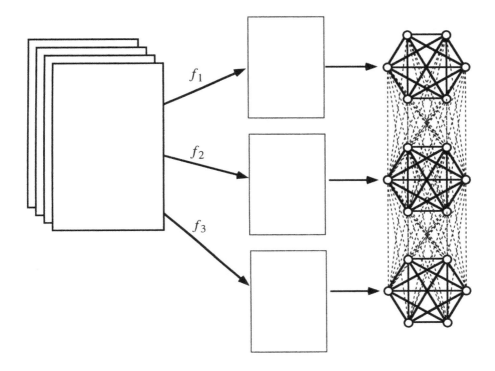

Figure 2.4.4 Schematic illustration of a model for sensory processing that first separates the information into distinct channels, each of which corresponds to a different attribute of the input. The separated attributes are then imprinted on distinct subdivisions of a neural network. This approach is effective if the different attributes of the information are independent, or at least partially independent. ∎

The use of these three channels in the visual system can be recognized in our use of attributes to identify objects. In describing objects, we generally distinguish distinct types of attributes—color, shape and action/motion. Within each of these attribute categories we can construct a list of attributes such as

color: RED, GREEN, BLUE, ORANGE, PURPLE, WHITE, BLACK, ...

shape: ROUND, OVAL, SQUARE, FLAT, TALL, ...

action/motion: STATIONARY, MOVING-LEFT, MOVING-RIGHT, RISING, FALLING, GROWING, SHRINKING, ...

The existence of three attribute categories enables a large number of descriptive categories to be constructed. A description is composed out of a selection of one attribute from each category. The number of descriptive categories is the product of the numbers of attributes of each type.

By subdividing a network into three subnetworks and separating the color information to one subnetwork, the shape information to the second, and the movement information to the third, it is possible for the network to identify categories such as: RED ROUND MOVING-LEFT, RED ROUND FALLING, BLUE SQUARE MOVING-LEFT, and BLUE ROUND FALLING. The network receiving color information identifies the color, and so on. In a fully connected network these categories would each require separate identification (and correlations would actually cause the network to fail). As with our descriptions, in the subdivided network the total number of categories is the product of the number of categories stored in each subnetwork.

We caution that the shape, color and motion attributes of the information are not completely independent, and neither is their processing in the brain. Partial subdivision implies correlations between the different attributes are also significant. Particularly in the natural world, there are important correlations between the overall shape of an object, its color, and both its direction and likelihood of motion. For example, leaves have a set of characteristic shapes and they are usually green. Tree trunks and their associated vertical or branching shapes are usually brown. The ways in which leaves are likely to move are not the same as the way tree trunks are likely to move. If we used a completely subdivided network for vision, after imprinting brown stationary trees and green rustling leaves we would also remember brown rustling tree trunks and green stationary tree trunks. In order not to lose the color-shape-motion relationships, we must be able to store the correlations between these different attributes. This may be done in a partially subdivided network using the weaker or fewer synapses that run between the different subdivisions.

2.4.5 Language and grammar: nouns, verbs and adjectives

If subdivision provides advantages for neural network function, then this should be particularly manifest in man-made constructs. These constructs are likely to reflect the architecture of the brain and therefore mirror the use of subdivision. In the context of vision, one might consider the use of color and shape in abstract art, as well as the decoration of man-made objects (e.g., package labels). Studies of these constructions might help develop an understanding of the human visual system. Another source of information is human language. Known as "natural language" in the artificial intelligence community, to contrast it with computer languages, the spoken or written language is a man-made construct that has been studied for many years by linguists as a source of information or insight into the functioning of the human brain.

Linguists differentiate between the grammatical and semantic aspects of language construction. Loosely speaking, grammar is the structure of well-formed sentences, while semantics is the content of the sentences. It is grammar, which is much more amenable to formal studies, that has been considered to reflect the architecture of the brain. A basic premise in the field of modern linguistics is that common features in the grammar of different languages exist and are the primary clue to the inherent brain architecture. The most recent widely accepted linguistic theory is the transformational grammar. It suggests that there exists an underlying representation that is transformed upon output into the usual grammatical form of sentences.

Sentences are interpreted upon input to reconstruct the underlying representation. This resembles our model of the neural architecture, with feedforward input processing that leads to the subdivided attractor networks and, we add here, output processing from the subdivided network to the motor controls. The input and output mappings that form the grammatical constructions used in a particular language are not completely universal and must be trained. This will not be the focus of our attention here.

Our objective at this stage in describing the connection between grammar and our model is quite modest: to make contact with one of the most fundamental aspects of grammatical construction that is familiar to everybody—the existence of parts of speech in sentence construction. Indeed, without the existence of parts of speech, there would be no meaning to the term "grammar." Grammar investigates the construction of sentences out of words. Words are separated into categories that are the "parts of speech," such as nouns, verbs, adjectives, and adverbs. The central role of grammar is to describe the rules by which properly formed sentences are constructed out of the parts of speech.

In order for words to be stored in the brain, some appropriate representation must exist in terms of neuron activity patterns. We do not know what this representation is, nor how universal the representation is. However, assuming some representation, we can ask how the organization of words into parts of speech can be realized in the brain. One way is to attach a label to each word that indicates what part of speech it is, and to store each word with its label as a pattern in a uniformly connected neural network. When we use a particular word, the label can serve to identify how it should be used in a sentence. This is how dictionaries are organized. After each word appears the usage—part of speech (abbreviated n, v, adj, adv, etc.)—identifying how it may be used in a sentence. There are some technical problems with storing patterns which incorporate labels in this way. Since the same label (part of speech) applies to many words, this will not work in a conventional attractor network. There are ways to overcome this problem, but we will not take this route here.

Instead we describe an alternative that makes use of network subdivisions. The architecture is similar to the model for vision that was used in the previous section, or the more general model of Fig. 2.4.4. We simplify the construction by considering only three parts of speech—nouns, verbs and adjectives. We assign each part of speech to a particular brain subdivision and assume that visual (reading) or auditory processing separates the information stream into three parts. The separation of the information is equivalent to parsing sentences using the grammatical sentence structure, a process that is reasonably well-understood. The importance of input processing provides a reason for the need for consistency in grammatical construction. After the initial processing parses the sentence, the parts (noun, verb and adjective) are transferred to distinct brain subdivisions. In order to generate sentences for writing or speaking, an output processor (presumably another set of feedforward networks) is necessary to take the content of the brain subdivisions (noun, verb and adjective) and compose a sentence. This output processor precedes the motor control of speaking, writing or typing. It reimposes the grammatical construction of sentences.

In this picture the brain as a whole does not store words, per se, but rather phrases or short sentences that consist of one each of a noun, verb and adjective (n, v, adj). If we wanted to discuss how this model is capable of continuous language we would add a time evolution of the network that causes a transition from one word triple (n, v, adj) to the next. We limit our consideration of this model to an understanding of the role of subdivision by comparing a model with completely separated subnetworks with another model that stores phrases or short sentences in a fully connected network.

The comparison is illustrated in Fig. 2.4.5. We take two networks with the same number of neurons—a uniformly connected network and a network divided into three parts. Since the independent pattern storage capacity grows linearly with the number of neurons, the number of short sentences that can be stored in the uniform network is three times the number of patterns that can be stored in each of the three pieces of the subdivided network. For this example we take quite small networks, so that each of the subdivisions can store three words coded appropriately. The uniform network would then be able to store nine sentences with three words each. We choose to imprint the nine sentences that are shown in Fig. 2.4.6 on the left. On the subdivided network we can only imprint three sentences. However, twenty-seven composite sentences would be recognized.

The difference between the set of sentences that can be remembered by the full network and the set that can be "remembered" by the subdivided network are related to the distinction between the grammatical and semantic content of a sentence. The complete network knows more full sentences, but does not have knowledge of the divisibility of the sentences into parts that can be put together in different ways. It does not even recognize the existence of word boundaries. The subdivided network knows the parts but does not know the relationship between them, thus it knows grammar and it knows the individual words, but it does not know the semantic content. For example, it does not know who it is that fell. The subdivided network generalizes from the three imprinted sentences to twenty-seven sentences. This generalization is based on the grammatical construction of the sentence. The fully connected network does not generalize in this way because it remembers the specific imprinted sentences to the exclusion of all others.

The field of linguistics as well as our intuition suggests that the actual process in the human brain lies somewhere between these extremes. Sentences make sense or are "grammatically correct" if properly put together out of largely interchangeable parts. However, a recalled event is described by a specific combination. Language, whether written or spoken, is generated by each individual out of sentences. The particular sentences that are used were not necessarily learned. Whether read or heard, language is understood by each individual by recognizing the component words. Yet much of the new meaning that is learned is contained in the interrelationship of words. A specific combination of words can be remembered by an individual and repeated. However, in general such memorization is not easy and is not as permanent as the memory of individual words.

It is possible to achieve an intermediate balance between storage of components and complete imprints by use of a partial interconnection between subnetworks. We

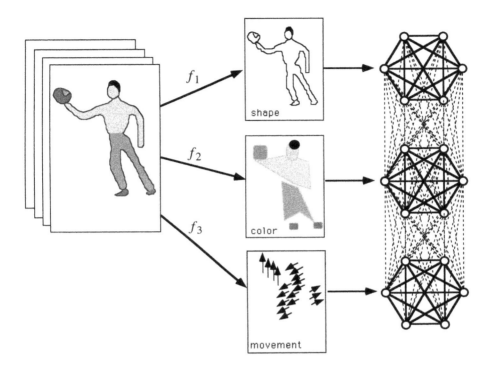

Figure 2.4.5 A practical example of the separation of processing into separate channels is the separation of visual processing into color, shape and motion. A significant body of experimental literature indicates that visual processing is separated into different channels. The channels are not fully characterized but are roughly considered to correspond to color, shape and motion. ∎

will investigate a few properties of partially subdivided networks in Section 2.5. One of the features of the partially subdivided network is that the synapses that run between subnetworks impose "compatibility" relations between the patterns stored in each subdivision. Some combinations of subpatterns are stable while others are not.

The subdivided network provides a systematic method for information organization in terms of elements (the stable states of subnetworks) which are organized in element-categories (the stable states of a particular subnetwork) and the compatibility relationships between elements as dictated by the inter-subnetwork synapses. This is indeed reminiscent of the structure of grammar, where nouns, verbs and adjectives and other parts of speech are categories that have elements, and there are compatibility relations among them. It is tempting to speculate that different subdivisions of the brain are responsible for the classification of words into parts of speech, and that the ability to combine them in different ways results from balancing the strength of inter-subnetwork synapses and the intra-subnetwork synapses which store representations

Fully connected network			Subdivided network					
Imprinting and Retreival			Imprinting			Retreival		
Big	Bob	ran.	Big	Bob	ran.	**Big**	**Bob**	**ran.**
Kind	John	ate.	Kind	John	ate.	Big	Bob	ate.
Tall	Susan	fell.	Tall	Susan	fell.	Big	Bob	fell.
Bad	Sam	sat.				Big	John	ran.
Sad	Pat	went.				Big	John	ate.
Small	Tom	jumped.				Big	John	fell.
Happy	Nate	gave.				Big	Susan	ran.
Mad	Dave	took.				Big	Susan	ate.
Shy	Cathy	slept.				Big	Susan	fell.
						Kind	Bob	ran.
						Kind	Bob	ate.
						Kind	Bob	fell.
						Kind	John	ran.
						Kind	**John**	**ate.**
						Kind	John	fell.
						Kind	Susan	ran.
						Kind	Susan	ate.
						Kind	Susan	fell.
						Tall	Bob	ran.
						Tall	Bob	ate.
						Tall	Bob	fell.
						Tall	John	ran.
						Tall	John	ate.
						Tall	John	fell.
						Tall	Susan	ran.
						Tall	Susan	ate.
						Tall	**Susan**	**fell.**

Figure 2.4.6 Illustration of the use of subdivided networks in the context of language. A fully connected network with enough neurons to store exactly nine sentences shown on the left may be imprinted with and recognize these sentences. If the network is divided into three parts it may be imprinted with only three sentences (center). However, because each subnetwork functions independently, all twenty-seven sentences (right) that are formed as composites of the imprinted sentences are recognized. Comparing left and right columns suggests the difference between semantics and grammar in sentence construction. ∎

of each word. There is even some biological evidence for the separation of nouns and verbs in different parts of the brain. We could take a step further and consider the relationship of the subdivisions of the brain that store noun and verb representations with other parts of the brain. For example, it makes sense to speculate that the subdivision that stores nouns would be more strongly connected by synapses to sensory-processing parts of the brain as compared to motor-processing parts. In contrast, verbs would be likely to be more strongly connected to motor control than most of the sensory processing (but also to the motion-detection subdivision of the visual system). This might even be part of an explanation of why words are divided into the categories of noun and verb, rather than some other categorization.

The discussion of the previous paragraph is the beginning of an approach to discussion of the architecture of the brain based on an understanding of how subdivided neural networks function. A more detailed discussion of how this approach might help in developing an understanding of neurophysiology will be given in Chapter 3. However, we mention here the implication that one might use the logic of grammar to represent more generally the function of the brain. To do this we would expand the articulated sentence to include additional "unvoiced" words in new categories representing the state of brain subdivisions other than the language-related ones.

2.5   Analysis and Simulations of Subdivided Networks

Our objective is to consider the advantages of subdivided networks in the context of sensory processing or, more generally, in the context of pattern-recognition tasks. The advantage of subdivision arises when the information is naturally subdivided so that combinations of imprinted subnetwork states also represent desirable states to be recalled by the network. We call these combinations of subnetwork states composite states. For completely subdivided networks, the analysis is immediate (Section 2.5.1). For partially subdivided networks, the analysis is discussed in Sections 2.5.2-2.5.4. Partially subdivided networks are relevant to pattern recognition when the information may be divided into partially but not completely independent parts. Thus a determination of the interdependence of recalled subnetwork states is relevant. It is assumed that for a particular pattern-recognition task, a balance is desirable between independence and correlation of subnetwork states. The central question is whether it is possible to achieve an adjustable intermediate balance between storage of complete neural patterns and storage of composite states.

We will use partially subdivided networks consisting of a conventional network of N neurons with Hebbian imprinting, where the strength of synapses between q subdivisions of $N' = N/q$ neurons are reduced by a factor g compared to the synapses between neurons within each subnetwork. We can expect that dilution of inter-subnetwork synapses, with g the fraction of remaining synapses, will lead to similar results (Question 2.5.1 on p. 364). It is important to distinguish the subdivided network from a randomly diluted network. Random dilution would sever synapses selected at random. Dilution of inter-subnetwork synapses results in storage of composite patterns. This would not occur for random dilution.

Consider a network with predefined subdivisions. The training of the network is performed by imprinting complete neural states. Since subdivision is favorable only when it is desirable to store and recognize composite patterns, we measure the stability of various composite patterns such that the state of each subnetwork corresponds to its state for one of the imprinted patterns.* Thus, two questions to be asked about the capacity of the network are: (1) How many complete neural states can be stored in the network? and (2) How many composite patterns are stored in the network?

2.5.1 *Completely subdivided networks*

For either fully connected or completely subdivided networks, the capacity is obtained from the results of Section 2.3 on fully connected networks. The maximum capacity of a large attractor network of N neurons for storage of complete neural states is $\alpha_c N$, where $\alpha_c = 0.145$ for Hebbian learning when a small fraction of errors in retrieval is allowed. For a network of N neurons completely subdivided into q subdivisions, the maximum capacity for complete neural states is $\alpha_c N/q$, which is lower than for the undivided network. However, since storing $\alpha_c N/q$ states results in all possible combinations of these substates being stored, the number of composite states stored is $(\alpha_c N/q)^q$, which is much larger than $\alpha_c N$ for large N and not too large q. This result follows directly from the linear dependence of the network capacity on the number of neurons.

As an example, for a network of 100 neurons, the storage capacity is 14.5 states for the full network. When subdivided into two halves, the network capacity is 7.25 full memories, and $(7.25)^2 - 7.25 = 46$ composite patterns in addition to the full memories. When subdivided into four subnetworks, the same 100 neurons store 3.75 complete states and 195 composite patterns. This example is described and simulated more fully below (see Fig. 2.5.1 through Fig. 2.5.4).

As an exercise we might calculate the number of subdivisions that results in storage of the largest number of composite patterns:

$$\frac{d}{dq}\left(\left[\frac{\alpha_c N}{q}\right]^q\right)_{q=q_{opt}} = q\left[\frac{\alpha_c N}{q}\right]^q\left(\ln\left(\frac{\alpha_c N}{q}\right)-1\right)_{q=q_{opt}} = 0$$

$$q_{opt} = \frac{\alpha_c}{e}N$$

(2.5.1)

For this number of subdivisions, the number of neurons in a subdivision would be only $e/\alpha_c \approx 19$. The number of memories stored (assuming that the usual formula applies, which is only approximate in this small subnetwork limit) is $e \approx 2.7$, or less than 3 on average. The total number of independent memories stored is:

$$\left[\frac{\alpha N}{q_{opt}}\right]^{q_{opt}} = e^{q_{opt}} = e^{\alpha N/e}$$

(2.5.2)

---

* In the present discussion we neglect the inversion of imprinted substates. For example, for a network subdivided into two parts, we could consider a state formed out of the right half of one imprinted pattern and the left half of the same pattern inverted. Such states may also be stable. A bias in the relative number of ON and OFF neurons (see also Section 3.2.13), as is found in the brain, would lead such states to be less relevant.

# of Neurons	# of patterns in full network	# of patterns in 2 subdivisions	# of patterns in 3 subdivisions	q_{opt}	maximal # of patterns
100	1.45×10^1	5.26×10^1	1.13×10^2	5	1.48×10^2
1000	1.45×10^2	5.26×10^3	1.13×10^5	53	1.04×10^{23}
10000	1.45×10^3	5.26×10^5	1.13×10^8	533	3.01×10^{231}

Table 2.5.1 Table of the storage capacity for composite patterns in various subdivision schemes. If the number of composite patterns stored is maximized, the number of subdivisions q_{opt} and the number of memories stored is indicated. Note that the number of imprints needed to store a large number of composite patterns is not great. In particular it is only three for all cases in the last column.

This subdivision size is based on considering the maximum number of composite states which can be stored.

The main problem with this analysis is that it only applies if the information can be divided into independent parts consisting of N/q bits. We could also consider a more extreme case where all bits are independent. In this case there would be no need for synapses (or storage) since all 2^N patterns are possible. Nevertheless, the implication that an optimal subnetwork size only stores approximately e states may be of significance when we can adjust the pattern-recognition task to suit the capabilities of the neural architecture.

Table 2.5.1 indicates the potential advantage of subdivision if the information can be appropriately subdivided into aspects that can be mapped onto subdivisions of the network. The human brain (with 10^{11} neurons) has left and right hemispheres that are further subdivided into a hierarchy of subdivisions. Small divisions are sometimes modeled as having about 10^4 neurons in number. Further subdivisions into still smaller neuron groups may also occur in the brain.

2.5.2 *Summary of results on partially subdivided networks*

In Section 2.5.3 and Section 2.5.4 we analyze networks that are partially subdivided. In Section 2.5.3 simulations are used and in Section 2.5.4 a signal-to-noise analysis is used. Before proceeding, we summarize the results. All of the results on partially subdivided networks depend on the degree of subdivision. g sets the relative strength of inter-subnetwork synapses and intra-subnetwork synapses. For $g = 1$ we have a fully connected network, and for $g = 0$ we have a completely subdivided network. The simulations and signal-to-noise analysis show that

1. For $g = 1$ the maximal number of imprinted patterns may be stored and for $g = 0$ the minimal number of imprinted patterns may be stored with a continuous interpolation between them.

2. For $g = 1$ the lowest number of composite patterns may be stored and for $g = 0$ the largest number of composite patterns may be stored with a continuous interpolation between them.

3. For particular values of g a well-defined balance between complete patterns and composite pattern storage is achieved.

4. a. We should distinguish between composite patterns that have some subdivisions with the same imprinted pattern in them. For example, for three subdivisions, there are composite patterns with two of the subdivisions having the same imprinted pattern in them and the third subdivision with a different imprinted pattern. Patterns that have more than one subdivision with the same imprinted pattern continue to be stable to higher values of g. More specifically:

   b. The simulations study a network subdivided into four subdivisions. A composite pattern formed by setting two of the subdivisions to one imprinted pattern and two subdivisions to another is stable to higher values of g.

   c. The signal-to-noise analysis considers networks with q subdivisions. Let a be the smallest number of subdivisions occupied by an imprinted pattern in a particular composite pattern. Then for low storage, this pattern will be stable for all g satisfying

$$g < \frac{1}{q - 2a + 1} \tag{2.5.3}$$

The significance of result 4 is that we can use the value of g to impose correlations between patterns stored in different subnetworks by stabilizing some composite patterns and not others.

5. When the number of subdivisions becomes large, Eq. (2.5.3) ceases to apply, and it becomes impossible to selectively impose correlations between patterns stored on different subnetworks. If we try to reduce g to allow composite states, then all composite states become possible. As a consequence, we learn that beyond a certain number of subdivisions, partial subdivision is essentially impossible. The network either behaves as a fully connected network or as a completely subdivided network. For many purposes it is thus undesirable to have more than a few subdivisions. The crossover point is calculated to occur for approximately seven subdivisions. This result has some significance for our understanding of the subdivision in the brain and brain function. For example, it is consistent with the 7 $\pm$ 2 rule of short-term memory. It is also of significance for our understanding of complex systems in general.

Another way to state result 5 is in the language of Section 1.3.6. A uniform network may be categorized as a complex material. Removing part of the network affects the smaller part but does not affect the larger part of the network. In contrast, for less than approximately seven subdivisions, at intermediate values of g, subdivided neural network function depends on each of its subdivisions. It is therefore in the category of complex organisms. For greater than seven subdivisions it can no longer be in the category of a complex organism. For large enough g it behaves as a fully connected network and is a complex material. For smaller g the network decouples and becomes a set of independent networks. In this case it is unchanged under subdivision, like a thermodynamic system.

2.5.3 *Simulations of partially subdivided networks*

To evaluate the behavior of networks that are partially subdivided rather than completely subdivided, it is natural to perform simulations. These simulations, analogous to those of fully connected networks, test the stability of imprinted and composite neural states. In the following we use Hebbian imprinting and synchronous updating of an attractor network with 100 neurons partially subdivided into either two or four subdivisions. The imprinted patterns are chosen at random with an equal probability of the two neuron activities ±1. The procedure used for performing the simulations of subdivided networks is:

1. Generate p complete random neural states:

$$\xi_i^\mu = \pm 1 \qquad\qquad \mu = \{1, ..., p\}, i = \{1, ..., N\} \qquad\qquad (2.5.4)$$

2. Imprint these neural states on the synapses of the neural network:

$$J_{ij} = \begin{cases} \dfrac{1}{N} \sum_{\mu=1}^{p} \xi_i^\mu \xi_j^\mu & i \neq j \\[2mm] 0 & i = j \end{cases} \qquad\qquad (2.5.5)$$

3. Write the matrix of synapses in block form corresponding to q equal size neuron subdivisions

$$J = \begin{pmatrix} J^1 & J^{1,2} & \cdots & J^{1,q} \\ J^{2,1} & J^2 & & J^{2,q} \\ \vdots & & \ddots & \vdots \\ J^{q,1} & J^{q,2} & \cdots & J^q \end{pmatrix} \qquad\qquad (2.5.6)$$

where each superscripted J is an $N' \times N'$ matrix, $N' = N/q$. Diminish off diagonal blocks of synapses, which connect between q different subnetworks of equal size, by a factor g.

$$\begin{aligned} J^i &\leftarrow J^i & i &= \{1, ..., q\} \\ J^{i,j} &\leftarrow gJ^{i,j} & i, j &= \{1, ..., q\}, i \neq j \end{aligned} \qquad\qquad (2.5.7)$$

4. Find the number of imprinted and composite states that are stable under updating of the neurons. A composite state is composed from imprinted states in each subdivision. In general:

$$\begin{aligned} \zeta_i^\beta &= \xi_i^{\alpha_1} & \alpha_1 &\in \{1, ..., p\}, i = \{1, ..., N'\} \\ \zeta_i^\beta &= \xi_i^{\alpha_2} & \alpha_2 &\in \{1, ..., p\}, i = \{N'+1, ..., 2N'\}: \\ &\vdots \\ \zeta_i^\beta &= \xi_i^{\alpha_q} & \alpha_q &\in \{1, ..., p\}, i = \{(q-1)N'+1, ..., N\} \end{aligned} \qquad (2.5.8)$$

The number of stable composite states is:

$$p_{stable} = \sum_{\beta} \prod_{i} \delta\left(\zeta_i^{\beta}, \theta\left(\sum_{j} J_{i,j} \zeta_j^{\beta} \right) \right) \qquad (2.5.9)$$

In the simulations, the number of stable memories for small p are counted by enumerating all combinations. For p greater than a few, the number of stable states is obtained by sampling. In both cases, averaging over many sets of imprinted states is performed. Note that no errors are allowed in recall.

Fig. 2.5.1 shows the results of simulations of a network with 100 total neurons and two subdivisions. It shows separately the number of stable composite states that were not imprinted (Fig. 2.5.1(a)), and the imprinted states (Fig. 2.5.1(b)). Fig. 2.5.2 shows the total number of stable states. Different curves show the result of diminishing the inter-subnetwork synapses by g ranging from 0 to 1 in increments of 0.1. The maximum number of stable imprinted states is for a network that has not been subdivided, $g = 1$. The maximum is obtained for 13 or 14 imprints and is about 11 memories. This is the same as the earlier Fig. 2.2.6. The maximum number of composite states recalled is for the completely subdivided network. The maximum is obtained for approximately 7 imprints, resulting in recall of 45 composite states. These numbers approximate the expected results given by the analytic treatment. Note that the analytic treatment need not give exactly the same result as the simulation, because it assumes that N, N' are very large, and it allows some error in the network recall.

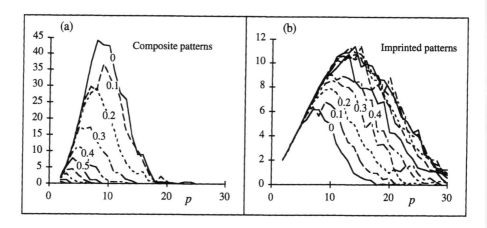

Figure 2.5.1 The number of stable memories after imprinting p patterns (horizontal axis) on a subdivided neural network with 100 total neurons and two subdivisions. (a) shows the number of stable composite patterns composed of combinations of imprinted subnetwork patterns where the complete pattern is not an imprinted one. (b) shows the number of stable imprinted patterns. Note the difference in vertical scale. The different curves are labeled by the factor g which weakens the synapses connecting different subnetworks. The curves labeled 0 are for a completely dissociated network. ∎

Of particular interest in these simulations is the possibility of partially subdividing the network and achieving an intermediate balance between storing imprinted states and composite states. For interconnection strengths reduced by $g = 0.3$, and with 9 imprints of which nearly all are recalled, the number of additional composite states recalled is about 10. This example of the network subdivided into two parts illustrates the balance between complete and composite memories. However, the nature of stability of subnetwork combinations in a subdivided network is more effectively illustrated with additional subdivisions.

Imprinting on a network with four subdivisions results in various possibilities for composite patterns. As in step 4 (p. 349), letting $\xi_i^{\alpha_1}$ be the state of the ith neuron in the α_1 imprint, we can write the composite states using the notation $\zeta_i^\beta \rightarrow (\alpha_1, \alpha_2, \alpha_3, \alpha_4)$. The distinct types of vectors whose stability can be tested are distinguished by the equality or inequality of the α_i as shown in Table 2.5.2. The number of stable memories of each type for 100 neurons and after p imprints is plotted in Fig. 2.5.3, and totaled in Fig. 2.5.4.

For a completely subdivided network ($g = 0$) the storage capacity for imprints is just over 3, compared with the full network capacity of 11. The number of composite

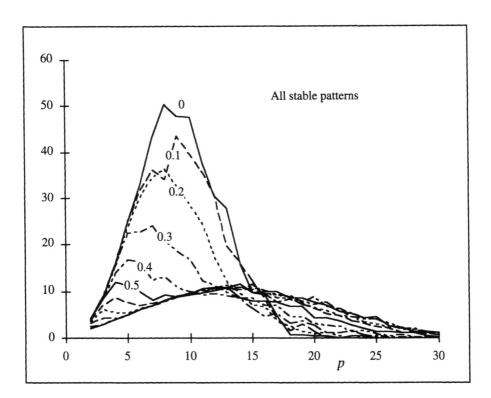

Figure 2.5.2 The total number of stable composite and imprinted states after imprinting p patterns on a subdivided neural network with 100 total neurons and two subdivisions. The value of g is indicated on each curve. ∎

Category	Number of Such States	Label	Schematic
$(\alpha_1,\alpha_1,\alpha_1,\alpha_1)$	p	Imprinted states	
$(\alpha_1,\alpha_1,\alpha_1,\alpha_2)$	$4p(p-1)$	Three equal substates	
$(\alpha_1,\alpha_1,\alpha_2,\alpha_2)$	$3p(p-1)$	Two & two equal substates	
$(\alpha_1,\alpha_1,\alpha_2,\alpha_3)$	$6p(p-1)(p-2)$	Two equal substates	
$(\alpha_1,\alpha_2,\alpha_3,\alpha_4)$	$p(p-1)(p-2)(p-3)$	Unequal neural substates	

Table 2.5.2 Different types of composite states for a network subdivided into four parts. The first type consists of substates that all originate from the same imprinted state—an imprinted state. The second type consists of substates of which three are from the same imprinted state and one originates from a different imprinted state. The other types are similarly defined. The number of states of each type is indicated (it is assumed that $p \geq 4$) in the second column. A label for each type, which is used in the figures and in the text, is given in the third column. A schematic is indicated in the last column. Note that the number of states in the last category is largest when p becomes large enough, however, for $p < 9$, the second to last category has a larger number of states. ∎

states recalled is nearly 400. When interconnection strengths are reduced by $g = 0.2$, it is possible to store between 6 and 7 complete memories while enabling the stability of 70 additional composite states at the same time. These composite states are roughly equally divided between those with two equal substates, and those with two & two equal substates. Other values for the interconnection strength can provide a distinct balance between the independence and dependence of subnetwork states.

Systematically, it is possible to see that composite patterns that have more than one subdivision containing the same imprinted pattern remain stable at higher values of g. When all substates arise from different imprinted states (Fig. 2.5.3(a)), the stability decreases very rapidly as g increases. The number of substate combinations with two equal substates (Fig. 2.5.3(b)) decreases almost as rapidly. In contrast, the stability of states with two & two equal substates (Fig. 2.5.3(c)), diminishes much more slowly. The number of states with three equal substates (Fig. 2.5.3(d)) is insignificant in these simulations. The greater stability of states with two & two equal substates at higher values of g is reasonable because the synapses between the subdivisions can contribute to the stability of each of the two parts of the composite pattern that arise from different imprints, even though the interactions between the two parts tend to destabilize each other. This will become more apparent through the analytic discussion in the following section.

2.5.4 *Signal-to-noise analysis of subdivided networks*

A signal-to-noise analysis of the stability of composite patterns in partially subdivided networks requires some care, because there are several different contributions to the signal and to the noise. Before we perform the analysis, it is helpful to discuss these

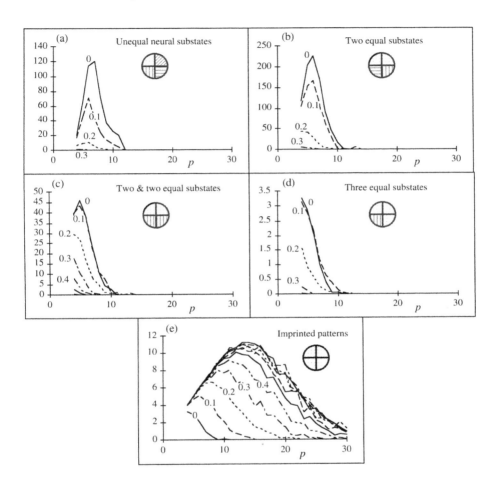

Figure 2.5.3 Same as Fig. 2.5.1 but for four subdivisions. Each panel (a)–(e) is for a different type of composite pattern (see Table 2.5.2). ∎

contributions qualitatively. Confirmation of the qualitative discussion can be found in the details of the analysis below. Fig. 2.5.5 illustrates an example of a composite pattern formed from four imprinted patterns in a subdivided network with eight total subdivisions. We will analyze the stability of a neuron in the first subdivision. The state of this neuron is initially chosen from the first imprinted pattern. We must determine if it retains this value after an update of the network. The synapse matrix contains one contribution from the imprinting of each of the imprinted patterns, and the synapses between subdivisions are reduced by the factor g.

The signal term that tries to maintain the stability of the neuron in the first subdivision arises from the imprint of the first pattern. However, only synapses to other neurons whose activity is set according to the first imprinted pattern contribute to the

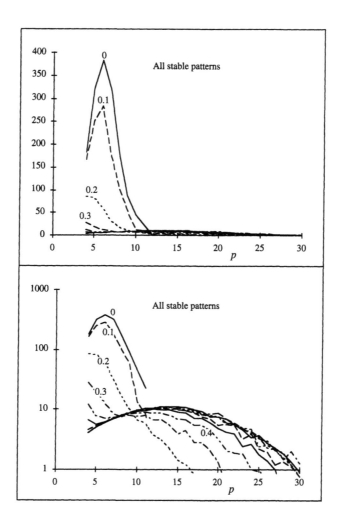

Figure 2.5.4 Total number of stable composite and imprinted patterns corresponding to the sum of Fig. 2.5.2(a)–(e). (a) shows a linear scale, and (b) shows a logarithmic scale for the same results. ∎

signal. Thus, the signal arises from synapses to other neurons within the first subdivision, and also from synapses to neurons in the second subdivision, but not to neurons in any other subdivisions. The signal from the second subdivision is reduced from what it would be in a fully connected network by the factor g.

The noise terms arise from the imprinting of all the other patterns. However, there are special problems with the subdivisions that have other patterns present in them. These subdivisions are trying to recreate their own pattern in the full network. For example, the third, fourth and fifth subdivisions in Fig. 2.5.5 all contain the sec-

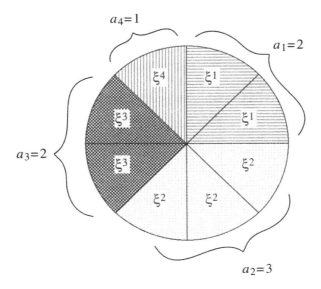

Figure 2.5.5 Schematic illustration of a composite pattern represented by a subdivided network. Each of the distinct shadings indicates the region of the network that contains a particular imprinted pattern. In the illustrated case, the network has eight subdivisions and the composite pattern is formed from parts of four imprinted patterns, ξ^1 through ξ^4. For simplicity the parts of the network that contain the same pattern are shown adjacent to each other. The values of a_1 through a_4 indicate how many subdivisions represent each of the imprinted patterns. ∎

ond imprinted pattern. The neurons in these subdivisions act coherently to try to influence the neuron in the first subdivision to take its value according to the second imprinted pattern. Its value in the first imprinted pattern, whose stability we are testing, may or may not be the same as its value in the second imprinted pattern. On average, half of the neurons in the first subdivision will receive an influence that will try to flip them. Because all the neurons in the third, fourth and fifth subdivisions act coherently to try to reconstruct their pattern in the first subdivision, we must calculate their combined influence as a contribution to the noise which may destabilize the pattern in the first subdivision. It is important that this destabilizing influence is diminished by the factor g, since in a fully connected network the composite pattern would be unstable for this reason.

An important distinction between two cases arises in our analysis when we consider the combined effect of all of the other patterns present in the composite state. There are a total of four patterns in Fig. 2.5.5. This means that there are three coherent noise terms that are trying to destabilize the first pattern. When we calculate the effect of these noise terms, we must decide whether we can average them together or whether we must add their effects. The correct answer depends on how many patterns there are. When there are only a few patterns, we cannot average them together, but when there are many, we can. By the analysis discussed below, the crossover point occurs roughly at seven patterns. A simple way to understand this result is to realize that seven equal contributions to the noise will add together (have the same sign) 1 in 2^7 times, or just under 1% of the time. As discussed in the case of a fully connected network (Section 2.2.5), this is the limiting fraction of unstable neurons that can be tolerated in a stable pattern. Thus, when there are more than seven patterns, it is not necessary to add their contributions to determine the impact of the pattern stability; it is enough to average them.

The existence of a crossover in the behavior of the subdivided network with seven parts of a composite pattern is the basis of our discussion of the 7 ± 2 rule. In essence, when we can average over the effects of other subdivisions, then each subdivision does not influence the other subdivisions directly, only the average effect is relevant. In contrast, when there are no more than seven different patterns, which is always true when there are no more than seven subdivisions, then the effect of each of the subdivisions must be considered explicitly in evaluating the stability.

We review and introduce additional notation for the signal-to-noise analysis. We assume a network comprised of q subnetworks each containing $N' = N/q$ neurons that are fully internally connected but more weakly connected to each other. The ratio of connection strengths is controlled by the parameter $g \in [0,1]$. $g = 0$ corresponds to a completely subdivided network and $g = 1$ corresponds to a fully connected network. For arbitrary g, the synaptic connection matrix is written as:

$$J_{ij} = \begin{cases} J'_{ij} & \left\lfloor \dfrac{i}{N'} \right\rfloor = \left\lfloor \dfrac{j}{N'} \right\rfloor \\ gJ'_{ij} & \text{otherwise} \end{cases} \tag{2.5.10}$$

where $\lfloor x \rfloor$ is the integer part of x. The first case corresponds to i and j in the same block along the matrix diagonal, i.e., in the same subnetwork. J' is the usual Hebbian matrix:

$$J'_{ij} = \begin{cases} \dfrac{1}{N} \displaystyle\sum_{\mu=1}^{p} \xi_i^\mu \xi_j^\mu & i \neq j \\ 0 & i = j \end{cases} \tag{2.5.11}$$

The composite pattern that we wish to test the stability of is formed out of pieces of imprinted states. As in the simulations, it is important to distinguish how many subdivisions have the same imprinted pattern. We test the stability of a trial composite pattern written in the form:

$$\left(\xi_1^1 \quad \cdots \quad \xi_{N'a_1}^1 \quad \xi_{N'a_1+1}^2 \quad \cdots \quad \xi_{N'(a_1+a_2)}^2 \quad \cdots \right) \tag{2.5.12}$$

This pattern is constructed by taking the first a_1 subdivisions from the corresponding part of the first pattern. The next a_2 subdivisions are taken from the second pattern and so on. In general there are a_μ subdivisions extracted from the pattern ξ^μ. We denote the number of $\{a_\mu\}$ that are non zero by $\hat{p}$. The sum over all a_μ is the total number of subdivisions:

$$\sum_{\mu=1}^{\hat{p}} a_\mu = q \tag{2.5.13}$$

We start by considering the stability of the first subdivision by looking at the stability of the first neuron

$$s_1 h_1 = s_1 \sum_{j=2}^{N} J_{1j} s_j \tag{2.5.14}$$

and separate J into the part due to the subdivision itself and all the rest, which is multiplied by the factor g compared to the usual expression:

$$= \frac{1}{N} \sum_{j=2}^{N'} \sum_{\mu=1}^{p} s_1 \xi_1^\mu \xi_j^\mu s_j + \frac{g}{N} \sum_{j=N'+1}^{N} \sum_{\mu=1}^{p} s_1 \xi_1^\mu \xi_j^\mu s_j \tag{2.5.15}$$

It is helpful to consider separately the part of the j sum that corresponds to each of the parts of the composite pattern. The first imprinted pattern appears in the composite pattern not only in the first subnetwork, but in the first a_1 subnetworks. We also have to consider each of the other patterns in the composite state. To simplify the notation, we will look at only the second pattern (in the subdivisions $\{a_1 + 1,...,a_1 + a_2\}$) and add the rest at the end:

$$= \frac{1}{N} \sum_{j=2}^{N'} \sum_{\mu=1}^{p} s_1 \xi_1^\mu \xi_j^\mu s_j + \frac{g}{N} \sum_{j=N'+1}^{a_1 N'} \sum_{\mu=1}^{p} s_1 \xi_1^\mu \xi_j^\mu s_j + \frac{g}{N} \left[\sum_{j=a_1 N'+1}^{(a_1+a_2)N'} \sum_{\mu=1}^{p} s_1 \xi_1^\mu \xi_j^\mu s_j + ... \right] \tag{2.5.16}$$

With this separation of the sum over j, we can now replace the values of s_j with their corresponding values in terms of the imprinted patterns. The first two sums have $s_j = \xi_j^1$ and the third term has $s_j = \xi_j^2$. We also substitute the value of $s_1 = \xi_1^1$ which appears in each of the terms:

$$= \frac{1}{N} \sum_{j=2}^{N'} \sum_{\mu=1}^{p} \xi_1^1 \xi_1^\mu \xi_j^\mu \xi_j^1 + \frac{g}{N} \sum_{j=N'+1}^{a_1 N'} \sum_{\mu=1}^{p} \xi_1^1 \xi_1^\mu \xi_j^\mu \xi_j^1 + \frac{g}{N} \left[\sum_{j=a_1 N'+1}^{(a_1+a_2)N'} \sum_{\mu=1}^{p} \xi_1^1 \xi_1^\mu \xi_j^\mu \xi_j^2 + ... \right] \tag{2.5.17}$$

The signal will arise from the imprinting of the first pattern, $\mu = 1$, but only in the first two terms above. All the rest of the terms will give rise to the noise. However, we must also be careful in the third term how we treat the contribution of the imprinting of the second pattern $\mu = 2$. So we separate these parts from each of the terms:

$$
= \frac{1}{N} \sum_{j=2}^{N'} \xi_1^1 \xi_1^1 \xi_j^1 S_j^1 + \frac{g}{N} \sum_{j=N'+1}^{a_1 N'} \xi_1^1 \xi_1^1 \xi_j^1 S_j^1
$$

$$
+ \frac{1}{N} \sum_{j=2}^{N'} \sum_{\mu=2}^{p} \xi_1^1 \xi_1^\mu \xi_j^\mu \xi_j^1 + \frac{g}{N} \sum_{j=N'+1}^{a_1 N'} \sum_{\mu=2}^{p} \xi_1^1 \xi_1^\mu \xi_j^\mu \xi_j^1
$$

$$
+ \frac{g}{N} \left[\sum_{j=a_1 N'+1}^{(a_1+a_2)N'} \xi_1^1 \xi_1^1 \xi_j^1 S_j^2 + \sum_{j=a_1 N'+1}^{(a_1+a_2)N'} \xi_1^1 \xi_1^2 \xi_j^2 S_j^2 + \sum_{j=a_1 N'+1}^{(a_1+a_2)N'} \sum_{\mu=3}^{p} \xi_1^1 \xi_1^\mu \xi_j^\mu \xi_j^2 + \ldots \right] \tag{2.5.18}
$$

We can now resolve all squares of variables to +1. When we do this, the first two sums can be directly evaluated, since they are sums over unity. Note also that the middle sum in the square brackets is a summation over terms that are independent of j, and can be evaluated by taking out the factor $\xi_1^1 = \xi_1^2$:

$$
= \frac{N'-1}{N} + \frac{g(a_1-1)N'}{N}
$$

$$
+ \frac{1}{N} \sum_{j=2}^{N'} \sum_{\mu=2}^{p} \xi_1^1 \xi_1^\mu \xi_j^\mu \xi_j^1 + \frac{g}{N} \sum_{j=N'+1}^{a_1 N'} \sum_{\mu=2}^{p} \xi_1^1 \xi_1^\mu \xi_j^\mu \xi_j^1
$$

$$
+ \frac{g}{N} \left[\sum_{j=a_1 N'+1}^{(a_1+a_2)N'} \xi_j^1 S_j^2 + \xi_1^1 \xi_1^2 a_2 N' + \sum_{j=a_1 N'+1}^{(a_1+a_2)N'} \sum_{\mu=3}^{p} \xi_1^1 \xi_1^\mu \xi_j^\mu \xi_j^2 + \ldots \right] \tag{2.5.19}
$$

The signal term is now visible. The first part of the signal arises from the first subdivision acting upon itself, and the rest is from the other subdivisions that contain the first imprinted pattern. We can take $N \gg 1$ and substitute $q = N/N'$ to obtain the expression:

$$
\text{signal} = \frac{(1-g)+ga_1}{q} \tag{2.5.20}
$$

   To evaluate the noise we must pay attention to the special terms mentioned before. We have succeeded to resolve Eq. (2.5.19) in such a way that each of the remaining terms is uncorrelated in sign. When we reached this point in the uniform network, all we had to do was to count the number of terms—the number of steps in the random walk—and use a root mean square evaluation of its magnitude. In this case, however, all of the steps do not have the same magnitude. This is a particular problem for the special term in the middle of the square bracket. There will be one such term for each of the imprinted patterns. We can rewrite the noise term by replacing the uncorrelated values with the notation ±1. This makes it easier to count how many terms there are in each sum:

$$\frac{1}{N}\sum_{j=2}^{N'}\sum_{\mu=2}^{p}(\pm1)+\frac{g}{N}\sum_{j=N'+1}^{a_1N'}\sum_{\mu=2}^{p}(\pm1)+\frac{g}{N}\left[\sum_{j=a_1N'+1}^{(a_1+a_2)N'}(\pm1)+(\pm1)a_2N'+\sum_{j=a_1N''+1}^{(a_1+a_2)N''}\sum_{\mu=3}^{p}(\pm1)+...\right]$$

$$=\frac{1}{N}\sum^{(N'-1)(p-1)}(\pm1)+\frac{g}{N}\sum^{(a_1-1)N'(p-1)}(\pm1)+\frac{g}{N}\left[\sum^{a_2N'}(\pm1)+(\pm1)a_2N'+\sum^{a_2N'(p-2)}(\pm1)+...\right]$$

$$=\frac{1}{N}\sum^{(N'-1)(p-1)}(\pm1)+\frac{g}{N}\sum^{(a_1-1)N'(p-1)}(\pm1)+\frac{g}{N}\sum_{v=2}^{\hat{p}}\left[\sum^{a_vN'}(\pm1)+(\pm1)a_vN'+\sum^{a_vN'(p-2)}(\pm1)\right]$$

$$(2.5.21)$$

In the last expression we have restored the contribution of all of the other parts of the composite pattern explicitly. We can now see that there are three kinds of steps in our random walk, those that have a coefficient of $1/N$ of which there are $(N'-1)(p-1)$, those with a coefficient of g/N of which there are a total of

$$(a_1-1)N'(p-1)+\sum_{v=2}^{\hat{p}}(a_vN'+a_vN'(p-2))=qN'(p-1)-N'(p-1)$$

$$(2.5.22)$$

$$=(N-N')(p-1)$$

and a special set of $\hat{p}-1$ terms with coefficients of the form

$$\frac{ga_vN'}{N}=\frac{ga_v}{q}$$

$$(2.5.23)$$

Because of this last set of terms—the coherent noise terms—we have to be much more careful about our analysis than in the uniform network case. The magnitude of the first two kinds of terms are small as N becomes large, and the number of terms is large. The coherent noise terms may be large even for large N, and their number can be small since it only increases with $\hat{p}$. For the uniform network we considered a root mean square estimate of the noise. This root mean square estimate only works, however, if the number of terms is large. Thus we must distinguish between the cases where $\hat{p}$ is small and when $\hat{p}$ is large.

When we studied the signal-to-noise analysis of the fully connected network and the retrieval of imprinted patterns, we found that for low storage, $p \ll N$, the noise disappeared and retrieval of the imprinted patterns would occur. Then we could consider the storage capacity as p increased. For composite patterns, the situation is different, because the noise does not disappear for low storage. Thus we first study the low-storage case.

We start by considering two estimates of the magnitude of the noise relevant for $\hat{p}$ large and small respectively. For the case of $\hat{p}$ large we can use a root mean square estimate of the noise because the number of independent terms is large. To obtain the noise we use a generalization of the random walk with different step sizes D_i (left half of Eq. 1.2.51):

$$\sigma=\sqrt{\sum D_i^2}$$

$$(2.5.24)$$

In the limit of low storage $p \ll N$ and not too many subdivisions $q \ll N$ only the coherent noise terms are important. From Eq. (2.5.23) this gives us a root mean square noise

$$\text{noise} \approx \left(\frac{g}{q}\right)\sqrt{\sum_{v \neq 1} a_v^2} \tag{2.5.25}$$

When $\hat{p}$ is small we cannot average the noise terms and consider only the typical value. Thus we consider the maximum possible effect of the coherent noise terms. This occurs when all values of ± 1 are -1. The magnitude of the maximum noise is

$$\text{maximum noise} = \frac{g}{q}\sum_{v=2} a_v = \frac{g}{q}(q - a_1) \tag{2.5.26}$$

If the signal is greater than the maximum noise, then the pattern must be stable. However, if the signal is significantly greater than the typical noise, but less than the maximum noise, then it may be stable. If the signal is about the same as the typical noise, then the pattern is almost certainly unstable.

Thus, we can guarantee retrieval of a composite pattern if the maximum possible noise is less than the signal. This places a limit on g determined by the inequality

$$\frac{(1-g) + ga_1}{q} > \frac{g}{q}(q - a_1) \tag{2.5.27}$$

If $g = 0$, this inequality is always satisfied. This is just the completely subdivided case that we know leads to stability of composite patterns. If $g = 1$, this condition becomes:

$$a_1 > q/2 \tag{2.5.28}$$

This means that the first pattern must have more than half of the subdivisions in order to be stable. How we define subnetworks is arbitrary in the case of $g = 1$, but the meaning of this statement is that a pattern is stable only if it occupies more than half of the network. However, this must apply to each of the parts of the composite pattern, and thus implies that no composite pattern except the trivial one of a single imprinted pattern can be stable in the case $g = 1$.

If we assume that $1 > g > 0$, we can simplify the inequality in Eq. (2.5.27) to obtain:

$$\frac{(1-g)}{g} > (q - 2a_1) \tag{2.5.29}$$

or

$$g < \frac{1}{q - 2a_1 + 1} \tag{2.5.30}$$

This limit on g ensures that the part of the composite pattern in the first subnetwork is stable. In order for all parts of the composite pattern to be stable, g must be smaller than the minimum value of this expression taken over all subnetworks:

$$g < \min \frac{1}{\mu} \frac{1}{q - 2a_\mu + 1} = \frac{1}{q - 2a + 1} \qquad (2.5.31)$$

where

$$a = \min a_\mu \qquad (2.5.32)$$

is the minimum number of subnetworks containing any one of the imprinted patterns. The greatest restriction on g arises from demanding the stability of the smallest part of the composite pattern (smallest a_μ). This result is the same as Eq. (2.5.3).

For $\hat{p}$ large, this limit on g is overly severe, since the maximum noise occurs infrequently. In this case we use the root mean square estimate of the noise, Eq. (2.5.26). However, stability of the pattern does not result when the signal is just greater than the noise. As in the signal-to-noise analysis of the fully connected network (Section 2.2.5), it must be sufficiently greater to ensure that only about 1% of the neurons will be unstable. The mean probability that a given neuron is unstable is given by integrating the probability of the noise being less than the signal. We thus require the signal-to-noise ratio to be greater than the number $r = 1/\sqrt{\alpha_c} \approx 2.64$:

$$\frac{ga_1 + (1-g)}{q} > r \frac{g}{q} \sqrt{\sum_{\lambda \neq 1} a_\lambda^{\,2}} \qquad (2.5.33)$$

This gives a limit on acceptable g of

$$g < \frac{1}{r \sqrt{\sum_{\lambda \neq 1} a_\lambda^{\,2}} - a_1 + 1} \qquad (2.5.34)$$

If we assume that the maximal allowable error rate at any neuron in any subdivision is given by this inequality, then we have the result:

$$g < \frac{1}{r \sqrt{\sum_{\lambda} a_\lambda^{\,2} - a^2} - a + 1} \qquad (2.5.35)$$

The limit in Eq. (2.5.35) corresponds to a certain probability, rather than a guarantee, that subnetworks of the composite pattern are stable. This is important, since requiring that all parts of the composite pattern are likely to be stable is a much stricter condition. Similarly, requiring that at least one of the parts is likely to be stable is a much weaker condition. For example, if all of the subdivisions have distinct imprinted patterns $\{a_\mu = 1\}$ and each has a probability P of being stable then the probability that all are stable is only P^q and the probability that at least one is stable is $1 - (1 - P)^q$. Thus, as a function of g, composite patterns become progressively unstable in more subdivisions in the vicinity of the limit in Eq. (2.5.35).

The analysis that we have performed reveals an interesting limitation to the degree of useful subdivision if we consider the role subdivision may play in the brain, or in an artificial pattern-recognition task. The second limit we obtained in Eq. (2.5.35) is a higher one than the first limit in Eq. (2.4.37) when there are many subdivisions. Why is this a problem? Because it indicates that once the number of subdivisions becomes large, for g values that satisfy Eq. (2.5.35), essentially all possible combinations become stable, but some parts will be stable at higher values of g and some at lower values. Moreover, individual subdivisions do not affect the stability of the state. In contrast, when there are only a few subdivisions, we can control which combinations are stable using the value of g, and the state of each subdivision matters.

We can estimate the crossover point where the number of subdivisions becomes too large by looking at a pattern where each subdivision has a different imprinted pattern $a_\mu = 1$, for $\mu = 1,\ldots,q$, and equating the two limits

$$\frac{1}{r\sqrt{q-1-1+1}} = \frac{1}{q-2+1} \tag{2.5.36}$$

or

$$q = r^2 + 1 \approx 7.94 \tag{2.5.37}$$

This suggests that q should be less than this value for effective use of partially correlated information stored by subdivisions in a network. It is possible to suggest that this limitation in the number of useful subdivisions may be related to the characteristic number of independent pieces of information a human is able to "keep in mind" at one time. This number is conventionally found to be 7 ± 2. The comparative rigidity of this number as compared with many other tests of variation between human beings suggests that it may indeed be tied to a fundamental limitation related to the architecture of neural networks as we have found.

Up to this point we have been considering the case of low storage. In the case of high storage, we must consider all three types of noise terms found in Eq. (2.5.21) rather than just the coherent terms. We should still distinguish between the cases where $\hat{p}$ is small or large. In either case, we estimate the contribution of the first two types of terms as a random walk. However, only for $\hat{p}$ large can we treat the coherent terms as a random walk.

For $\hat{p}$ large we calculate the typical noise from the three types of terms as:

$$\text{noise} \approx \sqrt{\left(\frac{1}{N}\right)^2 (N'-1)(p-1) + \left(\frac{g}{N}\right)^2 (N-N')(p-1) + \left(\frac{g}{q}\right)^2 \sum_{v \neq 1} a_v^2} \tag{2.5.38}$$

This can be simplified using $N, N', p \gg 1$ to obtain:

$$\text{noise} \approx \sqrt{\frac{(1+g^2(q-1))}{q}\frac{p}{N} + \left(\frac{g}{q}\right)^2 \sum_{v \neq 1} a_v^2} \tag{2.5.39}$$

The relationship between the storage capacity, which limits the value of p/N, and the interconnection strength g can be found by setting the signal-to-noise ratio to be less than $r = 1/\sqrt{\alpha_c}$. This gives

$$p < \frac{N\left(\alpha_c(1-g+ga)^2 - g^2\sum a_v^2 - g^2a^2\right)}{q(1+g^2(q-1))} \qquad (2.5.40)$$

We have replaced a_1 by a. Note that the numerator is zero when the maximum value of g, according to Eq. (2.5.35), is reached.

For $\hat{p}$ small we must take the maximum value of the coherent noise terms. This value should be subtracted from the signal before we compare the result with the root mean square value of the rest of the noise. Separating the two noise terms from each other we have:

$$\text{noise}_1 = \frac{g}{q}(q - a_1) \qquad (2.5.41)$$

$$\text{noise}_2 = \sqrt{\left(\frac{1}{N}\right)^2 (N'-1)(p-1) + \left(\frac{g}{N}\right)^2 (N-N')(p-1)} \qquad (2.5.42)$$

or using $N, N', p \gg 1$

$$\text{noise}_2 = \sqrt{\frac{p}{qN}}\sqrt{1+g^2(q-1)} \qquad (2.5.43)$$

Subtracting the first noise term from the signal and insisting the result is greater than r times the rest of the noise we have:

$$\frac{(1-g)+ga_1}{q} - \frac{g}{q}(q-a_1) > r\sqrt{\frac{p}{qN}}\sqrt{1+g^2(q-1)} \qquad (2.5.44)$$

This implies that the number of patterns that can be imprinted, and the corresponding composite patterns recalled, is limited by:

$$p < \frac{\alpha_c N}{q}\frac{\left(1-g-gq+2ga\right)^2}{1+g^2(q-1)} \qquad (2.5.45)$$

where we have again replaced a_1 by a. Note that the numerator is zero when the maximum value of g at low storage is reached according to Eq. (2.5.32). The storage capacity increases for lower values of g if a is small compared to q; i.e., for a composite pattern. If we ask about the retrieval of only the imprinted patterns, then we can use the same expression with $a = q$ or

$$p < \frac{\alpha_c N}{q}\frac{(1-g+gq)^2}{1+g^2(q-1)} \qquad (2.5.46)$$

In this case the maximum storage is for $g = 1$, as we would expect for the imprinted patterns.

Question 2.5.1 Compare the signal-to-noise analysis presented above with a signal-to-noise analysis of a subdivided network where the subdivision is accomplished by diluting inter-subnetwork synapses. Specifically, assume that a fraction g' of synapses between neurons in different subdivisions remain after dilution.

Solution 2.5.1 Since the signal-to-noise analysis of a diluted network follows very similar steps to the analysis of a network whose inter-subnetwork synapses are multiplied by the factor g, we mention only the differences in the two treatments.

Reducing the number of terms in a sum by the factor g' leads to the same effect on its average value as multiplying each of the terms by the same factor. However, there will be a different effect on a root mean square estimate of the magnitude of a random walk. A random walk with fewer steps, by a factor g', will be reduced in magnitude by a factor of $\sqrt{g'}$ rather than g'.

Thus, the analysis of the signal is the same for dilution as that given in the text except for the substitution of g by g'. This also applies to the low-storage analysis of the coherent noise terms, either for small $\hat{p}$ or for large $\hat{p}$. In each of these cases, the sums over individual synapses are performed directly, rather than as a random walk, and thus g can be directly replaced by g'.

The only place in the analysis where the dilution gives a different result is in the discussion of the noise terms that limit the storage capacity. These terms, in Eq. (2.5.39) and Eq. (2.5.43), can be found for the case of dilution by substituting $\sqrt{g'}$ for g. The resulting noise terms are larger for the case of dilution, resulting in a smaller storage capacity as compared to the effect of multiplying all the inter-subnetwork synapses by the same factor. ∎

2.6   From Subdivision to Hierarchy

In the last section our analysis of the properties of partially subdivided networks led to a conclusion that begs for further discussion. Our motivation for investigating the properties of subdivided networks was to discover the underlying purpose of functional subdivision. We were able to demonstrate that subdivision does provide a mechanism for storage of patterns with a particular composite structure. However, we encountered a fundamental limitation. Once there are too many subdivisions, the ability to store correlations between the subdivisions is diminished. In this section we review the argument that led to this conclusion and discuss further implications.

A fully connected network stores complete neural patterns. On the other hand, a completely subdivided network stores independent subpatterns without correlations between them. For most applications it is reasonable to assume that different aspects of the information are partially independent. This requires the ability to balance the storage of independent subpatterns and yet retain correlations between them. To achieve this balance requires an intermediate degree of interconnection between the subdivisions. This is possible, we found, when the number of subdivisions is small. However, when the number of subdivisions is large there is essentially no intermediate possibility: either the connections are strong enough to store complete states or they are weak enough to allow all combinations. What is particularly surprising is that the meaning of the term "large" is any number greater than roughly seven. While this is consistent with the well established 7±2 rule of short term memory, the implications extend yet further.

We are limited to a maximum of seven subdivisions, and yet the advantages of subdivisions for storage of independent aspects of information extends to many more subdivisions. An architecture that could provide further use of subdivision within the limitation is a hierarchically subdivided system. By keeping the branching ratio less than seven, we would construct a network formed of small networks that are strongly coupled, large networks that are weakly coupled, and still larger networks that are more weakly coupled. At each level of organization the strength of the connections within each subdivision must be strong enough compared to the connections between subdivisions to establish the integrity of the subdivision. Yet they must be weak enough to allow for the influence of the other subdivisions. Our model of the brain is no longer a model of interacting neurons, but rather of interacting units that at each level of organization attain an additional degree of complexity.

The brain has been found to be subdivided in a hierarchical fashion, and at least at the level of the major structures, this hierarchy does not have a high branching ratio. The brain is formed from the cerebrum, the cerebellum and the brain stem. The cerebrum is divided into two hemispheres. Each hemisphere is divided into four lobes. Each lobe is further divided into smaller functional regions; however, there are fewer than ten of these per lobe. The brain stem can also be subdivided into a hierarchy of functional regions. One could argue that the mapping of these structures reflects our own abilities caused by the 7±2 rule that lead us to divide the brain into only a few parts when we study it. This, however, misses the point of our observations. Our conclusions predict the relative strength of interdependence of different sections of the brain. This prediction would require additional systematic studies to confirm.

As we emphasized in the introduction to this book, our approach is primarily a statistical one. However, if the system we are investigating is composed out of only a few distinct components, then the statistical approach must have limited ability to describe it. When there are only a few components, each one should be specifically designed for the purpose to which it is assigned. A general statistical approach does not capture their specific nature. This is the reason the preface recommends the need for complementary investigation of particular aspects of individual complex systems. Here, we will continue to pursue other general principles through the statistical study of these systems. It is natural to ask whether we can generalize the conclusions from the study of neural networks to other complex systems. Several aspects of this question are discussed in the following section and others will arise in later chapters of this book.

2.7   Subdivision as a General Phenomenon

In this section we pursue the question of the necessity of subdivision and substructure in complex systems. All of the examples of complex systems that we will discuss in this text have the property that they are formed from substructures that extend to essentially the same scale as the whole system. We will review them as they are introduced more fully later in the text. Here we summarize some examples. The human body has nine physiological systems further divided into organs. Proteins have substructure formed of α-helices and β-sheets, and are often organized together with other proteins in quaternary structures. Life on earth considered as a complex system is divided among climates, ecosystems, habitats and species. Weather is formed out of large-scale air and ocean currents, storms and regions of high and low pressure. In all of these systems the largest scale of subdivision comprises fewer than 100 parts, and more typically of order 10 parts of the whole system. Why should this be the case?

Our discussion of subdivision in this chapter has been based on the function of the neural network as a memory. The importance of both combinatorial expansion of composite states and the constraints upon them due to interactions between subnetworks played a role. Similar considerations may apply in some of the other complex systems. For example, in the immune system the importance of composite states is apparent in the generation of a large variety of immune receptors by a process that combines different segments of DNA. A related discussion of substructure in the context of evolution will be included in Chapter 6.

In this section we adopt a different approach and relate substructure to the categories of complex systems that were articulated in Section 1.3. We argue qualitatively that substructure is necessary for what we generally consider to be a complex system. More specifically, we distinguish between a complex material and a complex organism. As defined in Section 1.3, a complex material is a system from which we can cut a part without significantly affecting the rest of the system. A complex organism is a system whose behavior depends on all of its parts and so is changed when a piece is removed. We propose that complex organisms require substructure.

In a system formed out of many interacting elements, each element may interact directly with a few or with many other elements. Our concern is to establish the conditions that imply that the behavior of a particular element depends on various sets of elements that comprise a significant fraction of the system. When this is true we have a complex organism. Otherwise, parts of the system may be removed without affecting the rest, and the system is a complex material or the system may even be a divisible thermodynamic system. The effective interaction between elements may be direct, or may be indirect because it is mediated by other elements. However, even if there is a direct interaction, if it does not affect the behavior of the element we are considering, then this interaction is irrelevant as far as we are concerned.

Let us start by considering generic interacting spin models such as the Ising model (Section 1.6). When the interaction between spins is local, then the system is generally a divisible thermodynamic system. When the interactions are long range, then, if there is a dominant ground state, we have a divisible thermodynamic system analogous to a magnet. If there are competing ground states, the system behaves as a spin glass or as an attractor neural network model with trained states. The neural network is the most favorable for consideration as a complex system.

We classify a fully connected neural network with Hebbian imprinting as a complex material rather than as a complex organism. This classification is based upon evaluating the impact of removing, say, 10% of the neurons. Our main concern in describing a neural network is the storage and retrieval of patterns. If we have only a few imprinted states, then separating the smaller part of the network does not affect the ability of either the large or small parts to retrieve the patterns. This is characteristic of a divisible system. If the number of stored patterns p is greater than the capacity of the smaller part, by itself, but smaller than the capacity of the larger part $(0.9\alpha_c N > p > 0.1\alpha_c N)$, then the smaller part will fail in retrieval and the larger part will be unaffected. This is the regime of operation in which the network would be expected to be used—the regime in which its storage capacity is utilized and the basins of attraction remain significant. The behavior is characteristic of a complex material. On the other hand, if we are very close to the full capacity of the network $(\alpha_c N > p > 0.9\alpha_c N)$ then both the large and small parts of the network will be affected by the removal of the small part. We could consider this to be a regime in which the network has the behavior of a complex organism. However, this is the regime in which the basins of attraction of memories are significantly degraded, and any perturbation affects the performance of the system. A better way to approach the classification problem is to consider the number of states that can be stored before and after the separation. We see that the storage capacity of the larger part of the system is weakly affected by the removal of a small part, while the small part is strongly affected. Thus the fully connected network should be classified as a complex material.

We classified the attractor network as a complex material on the basis of our investigations of its properties. However, the reason that the system is not a complex organism rests more generally on the existence of long- (infinite-) range interactions. If a particular element of the system interacts with all other elements, it is difficult for the removal of 10% of the system to affect its behavior significantly. Since there is nothing that differentiates the part of the system that was removed from any other part, the most that can be affected is 10% of the forces that act on the particular element. This is not enough, under most circumstances, to affect its behavior.

We found that short-range or long-range interactions do not give rise to complex organism behavior. Since an element cannot be affected by many other elements directly, it makes sense for us to start with a model where the element is primarily affected by only a few other elements that constitute its neighbors. This is the best that can be achieved. Then, so that it will be affected by other elements, we arrange interactions so that the neighborhood as a whole is affected by several other neighborhoods, and so on in a hierarchical fashion. This is the motivation for substructure.

What happens if we cut out one subdivision of a subdivided network which has only a few subdivisions, or a significant fraction of a subdivision? The stable states of the network are composite states. If the interactions between subdivisions are too weak, then all composite states are stable and removal will affect nothing. The system is a completely divisible system. If the interactions are too strong, then we are back to the case of a fully connected network. However, in the intermediate regime determined in Section 5.4.5, the stability of the composite states depends on the interactions between the subdivisions and the behavior of the large and small parts are both affected. Why? The reason is that only some of the composite patterns are stable. Which ones are stable depends on all of the subdivisions and their interactions. Thus, in the intermediate regime of interactions, the system behaves as a complex organism.

We could further develop the complex organism behavior of a subdivided network by recalling that the architecture is designed so that a particular aspect of the information is present in each subdivision. The loss of a subdivision would cause the loss of this aspect of the information. While this is reasonable argument, it is not a fundamental difference between a subdivided system and the fully connected network. We could choose to map different aspects of the information onto different parts of a fully connected network and arrive at the same conclusion. Even though this is more natural for a subdivided system, it is not inherent in the subdivision itself and therefore does not advance our general discussion.

An alternate way to consider the complex behavior of the subdivided network is in terms of the growth in the number of stable states of the network. For a fully connected network, the growth is linear. For a completely subdivided network, the growth is exponential, reflecting the independence of subdivisions. In the intermediate regime of interconnection, the growth in the number of composite states requires more detailed study and depends on the particular way in which the growth is performed. This suggests a level of control of properties of the system by its structure that we associate with a complex organism.

We have been considering the influence of elements of an Ising model upon each other. There is an important case that we have not included that could be represented by a feedforward network or by an Ising model sensitive to boundary conditions. In such systems, the influence of one neuron is transferred by a sequence of steps to other neurons down a chain of influence. We could consider an extreme case in which there is a long sequence of elements each affecting the subsequent one in the chain. Removing any of the elements of this sequence would break the chain of influence and all of the downstream elements would be affected. As a complex system that cannot be separated, this violates our claim that substructure is necessary or necessarily limited to only a few elements.

This counterexample has some validity; however, the argument is not as simple as it might appear. There are two general cases. Either each of the elements in the chain of influence serves only as a conduit for the information, in which case the nature of its influence is minimal, or, alternatively, each element modifies the information being transmitted, in which case generally the influence of the input dies rapidly with distance and only a few previous elements in the sequence are important for any particular element. The former case is like a pipe. All the segments of the pipe are essential for the transmission of the fluid, but they do not affect its nature. From the point of view of describing the behavior of complex systems, a conduit performs only a single function and therefore may be described using a single (or even no) variables. On the other hand, when each element affects the information, then we are back to the circumstance we have considered previously, and substructure is necessary. The reason that the influence dies with distance along the chain can be understood by considering a sequence of filters. Unless the filters are matched, then even a few filters will block all transmission.

We can revive the counterexample by considering a system whose elements represent a long sequence of logical instructions such as might be found in a computer program. Here we are faced with a problem of interpretation. If the system always represents only one particular program, then it is similar to a conduit. If the program changes, then we must consider the mechanism by which it is changed as part of the system. Nevertheless, the recognition that a narrowly construed sequence of instructions does represent an exception to the 7±2 rule about substructure can play a role in our understanding of the behavior of complex systems.

The discussion of substructure may be further generalized by considering elementary building blocks that are more complex than binary variables. Our objective is to argue that even when the building blocks are highly complex, the generalized 7±2 rule that requires substructure applies without essential modification. It applies therefore to complex systems formed from abstract or realistic neurons or to complex social systems of human beings. Our discussion has already, in a limited sense, considered elements of varied complexity, because subnetworks may contain different numbers of neurons. Our conclusion about the number of allowed subdivisions (seven) was independent of the number of neurons in the subdivision. Why should this be true?

We might imagine that a particular element with a greater complexity can be affected in one way by some elements and in another way by other elements. This would enable the whole complex system to be composed of a number of subdivisions equal to the number of aspects of the element that can be affected. Or we could even allow seven subdivisions for each aspect of the element. Then the number of subdivisions could be the product of seven times the number of aspects of an element. For example, when we think of human physiology, a cell requires both oxygen and sugars for energy production. Why couldn't we construct a system that is composed of seven subdivisions that are involved in providing oxygen and seven subdivisions that are involved in providing sugar, with a result that we have a system of fourteen subdivisions that is still a complex system. We could argue that the oxygen system and the sugar system are new subsystems that are described by the model. However, since there appears to be no limit to the number of aspects of an element, there should be no characteristic limit to the number of relevant subsystems. Make a list of all of the different chemicals required by a cell and generate a separate system for each one.

This argument, however, does not withstand detailed scrutiny. The central point is illustrated by considering a system formed out of elements each of which consists of two binary variables called TOP and BOTTOM. Each of the binary variables is part of a neural network that is composed out of seven subdivisions. How many subdivisions would there be altogether? The simplest case would be if there were fourteen subdivisions seven of which contain all of the TOP variables and seven of which contain all of the BOTTOM variables. Many other possibilities could be imagined. From the point of view of a complex system, however, as long as the two binary variables at each element behave independently, we could separately consider one set of seven subdivisions by themselves and the other seven subdivisions by themselves. They are completely decoupled. The physical proximity of the two binary variables as part of the same element does not affect the essential independence of the decoupled systems. As soon as there is a coupling between them we are back to where we started from, with interacting binary variables. Thus, increasing the complexity of the elements from which the complex system is composed does not appear to be able to change qualitatively the requirements of substructure found for the neural network model.

3

Neural Networks II:
Models of Mind

Conceptual Outline

■ 3.1 ■ The training of a model network that has subdivisions requires a process that can train synapses within subdivisions and between subdivisions without subjecting either to overload. A natural solution to this problem involves taking the network "off-line," so that a filtering of memories can occur when the network is dissociated. This is a possible model for the role of sleep in human information-processing that explains some of the unusual features of sleep and suggests new experiments that can be performed.

■ 3.2 ■ Various features of human information-processing, including the learning of associations, pattern recognition, creativity, individuality and consciousness can be discussed within the context of neural network models.

Efforts to describe and explain the higher information-processing tasks that human beings are capable of performing have always generated tension and concern. There has been a tendency to elevate these processes outside of the domain of the physical world, or to mystify them, through a characterization as infinite and incomprehensible. This tendency may arise from the desire to maintain a uniqueness of and importance to our own capabilities. We will adopt the contrary point of view that our capabilities are fundamentally comprehensible. However, it turns out that this does not diminish a quality of uniqueness and importance. If anything, it shows us how this importance arises.

We begin to tackle the task of explaining aspects of human information-processing in this chapter. However, we will not conclude our discussion until we have described complexity in the context of human civilization in Chapter 9. We start in Section 3.1 by considering the training of subdivided networks and the role of sleep in human information-processing. This section is an essential sequel to the discussion in the last chapter that introduced subdivision. The problem is to develop a systematic approach to the training of subdivided networks.

3.1   Sleep and Subdivision Training

3.1.1 *Training a partially subdivided network*

In Chapter 2 we discussed the role of functional subdivision in the brain. We showed that the storage capacity of a subdivided network was reduced; however, the ability to recall composite states may confer significant advantages on a properly designed network. The subdivided network presents us with a new set of challenges when we consider its training. The approach used in Chapter 2, where we imprinted a set of distinct patterns—each precisely what must be learned—is woefully incomplete. For more realistic modeling of neural networks we should assume that the information presented to the network is not so well-organized. When the information is presented in either a random or, even more realistically, a correlated fashion, there arise problems that relate to the storage capacity of the network and the selection of desired memories.

To illustrate the problem, we can return to the simplest example—the left-right universe (Section 2.4.1). In the first discussion it was assumed that there were no correlations between the left and right halves. The two halves were completely independent and all composite states were equally possible. Now we assume that correlations exist, and that there are some synapses that connect left and right hemispheres to capture these correlations. This means that there are many fewer possible states than $N_L N_R$ but still many more than N_L or N_R, where N_L is the number of possible left halves and N_R is the number of possible right halves. Moreover, we should expect that the number of possible imprints of each subnetwork is greater than its storage capacity ($N_L, N_R \gg \alpha N/2$).

Some selection of which states to keep in memory must be made. The point of introducing the subdivided network was to accommodate more of the possible states that can arise. However, if we try to imprint all of them, we will exceed both the capacity of the subnetworks and the capacity of the synapses between the subnetworks. When we were faced with the problem of overload in a uniformly connected network, we used a palimpsest memory to retain only the most recently imprinted memories. However, in a subdivided network this is not the best strategy for keeping memories. If the imprints are correlated, the most recent ones may all happen to have a particular right half, and this will end up being the only right half that will be remembered by the network.

Thus, without control over the number of states that are imprinted and the order in which they are imprinted, we must be concerned about the problem of overload failure. If we must stop the imprinting after only a few imprints sufficient to reach the capacity of the smallest subdivision, we will have limited the training of the network very severely. How can we overcome this problem?

To design a strategy to overcome the overload problem, we must first identify our objective in training the network. The objective should be based on achieving the best utilization of the available capacity: to enable each of the subdivisions of the brain to store a number of patterns that are commensurate with its capacity. These and no others. The stored patterns should be the most important patterns to remember. How do we identify the most important patterns? They should be the patterns that appear most frequently, and the patterns that are most orthogonal—most different. The reason for orthogonality is that it enables more patterns to be stored (Question 3.1.1). Also, if two patterns are similar, we might be able to substitute one for the other without too much loss. Thus if we cannot store two patterns as distinct, the next best thing is to store them as one and classify them together. More generally, when there are highly correlated states, we store one prototype that could be one of the states, or even a spurious state that has maximal overlap with the correlated states. This best utilization strategy enables each subdivision to retain states that are well representative, even though not necessarily exact reproductions, of the possible states.

The next objective is to train synapses that run between subdivisions to achieve correlations between the patterns imprinted in each of the subdivisions. We can think about each subdivision as itself like a neuron. The difference is that the neuron has only two states while the subdivision has approximately $\alpha N'$ possible states, where N' is the number of neurons in a subdivision. Thus we train the synapses between subdivisions to utilize the storage capacity for composite patterns, choosing those that are the most important composite patterns to remember. As before, the most important patterns are those that appear most frequently and those that are most orthogonal.

If we have a subdivision hierarchy, then at the third level of organization, the stable patterns of each of the subnetworks consists of various composite states. The overall objectives that we articulated for the storage of patterns also apply to the storage of states of the third level of the network. These objectives for training the synapses between subdivisions continue to remain the same all the way up to the complete network.

The model of a network of networks suggests a general strategy for its training. Since the training of inter-subnetwork synapses relies upon a well-defined set of subnetwork states, it seems reasonable to train first the subnetworks and then the synapses between them. To achieve this we would separate the subnetworks from each other, train each one, and then attach them and train the synapses between them. In the first stage of training, the subnetworks would be trained to their optimal capacity. Once the subnetworks are trained, the storage of patterns using the inter-subnetwork synapses would have a well-defined significance. Otherwise, if the synapses between subnetworks were trained before the synapses within a subnetwork were set, then modifying synapses within the subnetwork would change the significance of the synapses between subnetworks. Thus it seems that the brain should be trained by first training the smallest subdivisions, then connecting them into larger groupings and training the synapses between them. However, while it might appear to be convenient to train the subdivisions first, this presents us with several practical problems.

We must assume that the training requires many exposures to various environments and circumstances. We cannot wait until the training of subdivisions is complete before the brain is used. Moreover, the sensory information to which the brain is exposed does not reach the brain except through the interaction of action and sensation. In order to act, the brain must be functional, at least to some degree, and therefore we cannot train a brain when disconnected into its small parts.

There is an alternative approach that takes into account the need for separation and training of each subdivision while enabling the functioning of the brain. This approach adds an additional dynamics to the brain. In addition to the neural dynamics and the synaptic dynamics there is a dynamics of subdivision. The dynamics of subdivision is incorporated in a two-step procedure:

1. Imprint the complete network. If the number of synapses between subdivisions is smaller than within subdivisions, this already contains some predisposition to subdivision. Since the network is connected, it can also function.

2. Separate the network into its subdivisions and selectively reinforce (filter) the patterns that satisfy our objectives of optimal utilization of each subnetwork.

The cyclical repetition of steps 1 and 2 should enable the training of the subdivisions to proceed as operation continues. However, it requires the system to go "off-line" periodically for the filtering process.

The purpose of the two-stage process is to train the subdivisions separately from the training of the inter-subnetwork synapses. However, there is a need to obtain neural activity patterns for the training. These states must originate from the imprinting that occurs when the system is together. How are the imprinted patterns retrieved for the training of the subdivisions when the system is off-line? By the operation of the network itself. We call the second step reimprinting or relearning. Simulations that build an understanding of its operation are discussed in Section 3.1.2.

Having arrived at this scenario for training the subdivided network, we ask whether there is an analog of this in the actual system. The answer may be that sleep is the time during which the brain performs the subdivision training. This suggests that we consider the phenomenology of sleep and see if it can be reconciled with the possibility that the brain is separated into subdivisions and undergoes a filtering procedure. Because the training of subdivisions is central to their utilization in the brain, as well as in other complex systems, and the purpose of sleep is one of the great unsolved mysteries, we will consider their relationship in some detail in Section 3.1.3.

More generally, it is quite natural to suggest that complex systems that have identifiable function may undergo processes of dissociation and reconnection as part of their developmental process. This enables the subdivisions to develop autonomous capabilities which may then be recombined to achieve new stages of development.

Question 3.1.1 Show that the number of orthogonal states that can be stored in an attractor network is N. This is larger than the number of random states αN derived in Section 2.2.

Solution 3.1.1 The orthogonality of different states may be written as:

$$\sum_{j=1}^{N} \xi_j^\mu \xi_j^\nu = N\delta_{\mu,\nu} \tag{3.1.1}$$

Using a signal-to-noise analysis as in Section 2.2.5 we can evaluate the stability of a particular neuron $\{s_i | s_i = \xi_i^\mu\}$ when the states ξ_i^μ are orthogonal. We arrive directly at Eq. (2.2.21):

$$\xi_1^1 h_1 = \frac{1}{N}\sum_{j=2}^{N}\xi_1^1\xi_1^1\xi_j^1\xi_j^1 + \frac{1}{N}\sum_{j=2}^{N}\sum_{\mu=2}^{p}\xi_1^1\xi_1^\mu\xi_j^\mu\xi_j^1 \qquad (3.1.2)$$

The first term is the signal term and, as before, is just $(N-1)/N$. The second term, which was the noise term in the previous analysis, is essentially given by the overlap of different states, which is zero in this case by Eq. (3.1.1). However, we must take into consideration that the sum does not include $j=1$. So we have a correction to the signal

$$\xi_1^1 h_1 = \frac{N-1}{N} - \frac{1}{N}\sum_{\mu=2}^{p}\xi_1^1\xi_1^\mu\xi_1^\mu\xi_1^1 = \frac{(N-1)}{N} - \frac{(p-1)}{N} = \frac{(N-p)}{N} \qquad (3.1.3)$$

We see that the result vanishes when p reaches N. It is impossible to imprint more than N orthogonal states because there are no more than N orthogonal states of N variables. However, this analysis also shows that the basins of attraction vanish in the limit of $p=N$. ∎

Question 3.1.2 Show that after imprinting N orthogonal states, all possible neural states have the same energy in the energy analog.

Solution 3.1.2 N orthogonal states of dimension N are a complete set of states. This can be written as:

$$\frac{1}{N}\sum_{\mu=1}^{N}\xi_i^\mu\xi_j^\mu = \delta_{i,j} \qquad (3.1.4)$$

Except for the diagonal terms, this is the same as the synapses of the imprinted neural network. Since all the diagonal terms of the synapses are set to zero, and the off diagonal terms are zero by Eq. (3.1.4), we must have a completely null set of synapses. ∎

3.1.2 *Recovery and reimprinting of memories*

In this section we demonstrate the use of a neural network to recover memories and reimprint them. The reimprinting reinforces some memories at the expense of others. Effectively, the number of memories stored in the network is reduced so that further imprinting does not cause overload. The retrieval process begins from a random initial state which is evolved to a stable state. Using a random initial state means that the retrieval emphasizes the memories with the largest basins of attraction. These memories have been imprinted with the largest weight or are most different (most orthogonal) from other states that have been imprinted. In effect, the process is a filtering of memories that retains the "most important" ones and the ones that provide for best utilization of the storage capacity of the network. There are two advantages of this selection procedure over palimpsest memories discussed in Section 2.2.7. First, selection may be done after the original imprinting of the memories instead of during the imprinting. Second, the selection does not solely rely upon the specific order of imprints, it enables the persistence of particular older memories. However, when imprinting and selection are repeated many times, it is still true that recent imprints are more likely to survive the filtering process.

The reimprinting procedure is a particular way of filtering memories so that overload in the network is prevented and learning can continue. The continued learning is assumed to serve both as continued adaptation to a changing environment and as a refinement of the storage due to selection of more optimal memories—memories that are more likely to be recalled because they appear more frequently in the environment.

Reimprinting is well suited to subdivided neural architectures where each subdivision is expected to have well-defined memories that serve as the building blocks for the complete neural states of the network. In a subdivided network, reimprinting is implemented during a temporary dissociation of the network, which may correspond to sleep (Section 3.1.3). Temporary dissociation is achieved by diminishing the synaptic weights that connect between subdivisions. During the temporary dissociation, reimprinting optimizes the storage of patterns within each subdivision without regard to the associations established by inter-subnetwork synapses. When the inter-subnetwork synapses are reestablished, these associations are also reestablished.

In order to understand how the reimprinting works, we describe simulations that build the process step by step. The reimprinting of memories occurs on top of the imprinting that has already been performed. We must first understand the effect of imprinting states on top of existing memories using a palimpsest approach that weakens the previous memories before imprinting new ones. We can anticipate the results in two limits. The limit of no reduction in synapse strength corresponds to the conventional training. On the other hand, if the synapse strengths are sufficiently reduced in strength, then imprinting new patterns must be equivalent to the case of no prior imprints. Intermediate cases enable us to develop an understanding of the extent to which prior memories affect new imprints, and conversely how new imprints affect prior memories.

Assuming Hebbian imprinting, the combination of reducing the prior synapse values and imprinting a new state is described by modifying the synapses according to:

$$J_{ij}(t) = \chi J_{ij}(t-1) + \frac{1}{N} s_i(t) s_j(t) \qquad i \neq j \qquad (3.1.5)$$

Instead of diminishing the synapses with each imprint, we consider an episodic approach. We imprint a prespecified number of states and then reduce the synapses before continuing. For the simulations we use a network of $N = 100$ neurons, and imprint $p_1 = 4$ neural states with equal coefficients. Then the synapse strengths are multiplied by a factor χ. Then p_1 more neural states are imprinted and the synapse values are again multiplied by χ. This is repeated until a total of $p = 16$ neural states are imprinted. The final value of the synapses could be written using the expression:

$$J_{ij} = \chi^{p/p_1 - 1} \sum_{v=1}^{p_1} \xi_i^v \xi_j^v + \chi^{p/p_1 - 2} \sum_{v=p_1+1}^{2p_1} \xi_i^v \xi_j^v + \ldots + \sum_{v=p-p_1+1}^{p} \xi_i^v \xi_j^v \qquad (3.1.6)$$

To analyze the effect of the procedure, we calculate the basin of attraction of the imprinted states. Results averaged over many examples are plotted in Fig. 3.1.1. Each curve shows the distribution of basins of attraction for the imprinted states. The curves are labeled by the factor χ, which scales the synapses between each set of 4 imprints. The results show that for 16 imprints of equal strength ($\chi = 1$) the network is well beyond the optimal number of imprints. Only 10 imprints are stable and the basins of attraction are very small. When we diminish the synapses between successive imprints, $\chi < 1$, then the basin of attraction of the last four imprints are much larger. However, this occurs at the expense of reducing dramatically the basins of the other memories, eventually destabilizing them. Two conclusions from this analysis are: (1) diminishing the synapses is an effective mechanism for ensuring that successive imprints are learned effectively, and (2) the older memories are lost.

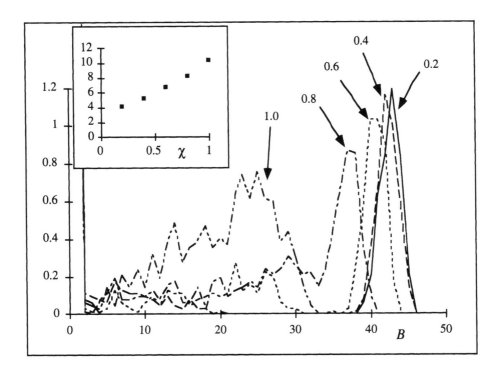

Figure 3.1.1 Simulations of an episodic palimpsest memory using a network of $N = 100$ neurons. Each episode consists of reducing the synaptic strengths by a factor χ then imprinting four new states. Plotted are the resulting histograms of basins of attraction (B). Curves are labeled by the value of χ. The results illustrated have been averaged over many examples. Unstable states are included as having basins of attraction of zero. The curves are normalized to the total number of imprints $p = 16$. By diminishing the strength of synapses ($\chi < 1$) the more recent imprints are better remembered, at the expense of forgetting the earlier sets of memories. The total number of stable states as a function of χ is shown in the inset on the upper left. ∎

The procedure used to generate Fig. 3.1.1 is:

1. Generate p random neural states $\{\xi_i^\mu\}$

$$\xi_i^\mu = \pm 1 \qquad \mu = \{1,...,p\},\ i = \{1,...,N\} \qquad (3.1.7)$$

2. Imprint the first p_1 neural states on the synapses of the neural network (Hebbian rule)

$$J_{ij} = \begin{cases} \dfrac{1}{N}\displaystyle\sum_{\mu=1}^{p_1} \xi_i^\mu \xi_j^\mu & i \neq j \\ 0 & i = j \end{cases} \qquad (3.1.8)$$

3. Rescale the network synapses by a factor χ.

$$J_{ij} \leftarrow \chi J_{ij} \qquad i,j = \{1,...,N\} \qquad (3.1.9)$$

4. Repeat steps (2) and (3) for each successive set of p_1 neural states until all p neural states are imprinted.

5. Find the basin of attraction of each of the neural states ξ_i^μ, where an unstable neural state is set to have a basin of attraction of zero (see Section 2.2.6).

6. Make a histogram of the basins of attraction for different ξ_i^μ.

Thus far we have demonstrated the performance of an episodic palimpsest memory. As mentioned above, we prefer to retain selected older memories. They can be retained if we reinforce them before the imprinting continues. We will consider several models of reimprinting.

If we know the states that were imprinted, we can select some of the older memories and reimprint them. This is not a practical approach if the earlier imprints are not known at the time of reimprinting, as when training a subdivided network. However, because it can help us understand the optimal effect of reimprinting, this will be the first case studied below.

If we do not independently know the imprinted states, then we must use the network itself to recover patterns that were imprinted. In this case selective reinforcement of memories may be achieved using the following steps: (1) initialize the network to a random state, (2) update the network a prespecified number of times, and (3) imprint the network. The neural update rule we have discussed in the text is noiseless. Because of the spurious states it is advantageous to add a small amount of noise. The noise helps the network escape from shallow minima associated with the spurious states. Glauber dynamics is just such a noisy neural update rule that was introduced in Section 1.6 and discussed in the context of neural networks in Question 2.2.2. Thus we will compare reimprinting in two cases, using a noiseless update rule and using a noisy update rule.

It is important to perform the reimprinting well before the network reaches overload. As we approach overload, the memory becomes ineffective. This is not only a problem for the network operation, it is also a problem for the filtering process that uses the network to retrieve patterns for reimprinting. If the network is too near overload, the retrieval process will find more spurious states than imprinted states. However, we would like to be able to use a significant fraction of the network capacity. In the simulations below, we balance these considerations by using an initial imprinting with 8 states on a network of 100 neurons. Eight states is below, but a significant fraction of, the maximal capacity of about twelve states.

Figs. 3.1.2 and 3.1.3 compare three different reimprinting procedures. Fig. 3.1.2 shows the distribution of basins of attraction. Fig. 3.1.3 shows the integrated number of basins of attraction higher than the value along the abscissa. The starting point, before reimprinting, consists of a network with 8 imprinted neural states. This is shown

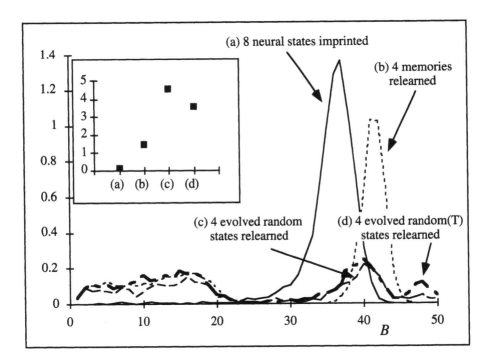

Figure 3.1.2 Plots illustrating tests of reimprinting procedures. Histograms of the basins of attraction are shown. The objective is to strengthen several memories at the expense of the others. The starting point is a network of 100 neurons with 8 imprinted neural states. This is shown as curve (a). Curves (b),(c) and (d) show the results after reimprinting. (b) is the optimal case where four of the originally imprinted states are imprinted again. In curve (c) the 4 imprinted states are evolved random states obtained by applying the deterministic neural update rule to a random initial configuration. Curve (d) is the same as (c) except the neuron evolution includes noise ($\beta = 3.5$) (see Question 2.2.2). In both cases there is an enhancement of some of the basins of attraction. For curve (c) the number of basins enhanced is about 1.5 while for curve (d) it is about 2. Note that the basins of attraction of all of the evolved random states are not necessarily included in the figure since they are not always desired memories. The insert on the upper left shows the number of unstable memories out of the eight originally imprinted states. ∎

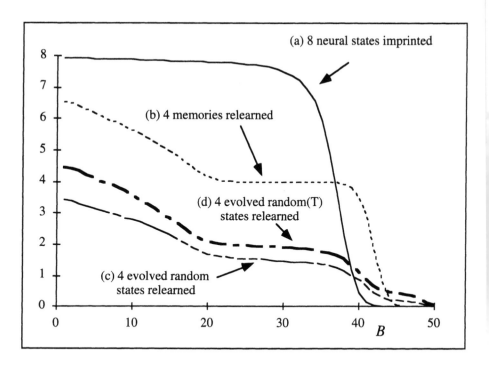

Figure 3.1.3 Plots illustrating tests of reimprinting procedures. This figure is similar to Fig. 3.1.2, except that the plots show the integrated number of imprinted states with a basin of attraction grater than a given value B. Flat regions of the curves separate states with large basins of attraction from states with small basins of attraction. This shows the ability of the reimprinting to reinforce some memories at the expense of others. ∎

as curve (a). The objective is to strengthen several memories at the expense of others. Curves (b), (c), and (d) show the results after reimprinting. (b) is the idealized case, where four of the originally imprinted states are imprinted again. This is the same as imprinting four of the states with twice the strength of the other four. As in Fig. 3.1.1, the effect is to reduce the basin of attraction of the neural states that are not reimprinted, and increase the basin of attraction of the ones that are reimprinted. In curve (c) the four states to be imprinted are obtained by applying the deterministic neural update rule to a random initial state. We call these "evolved random" states. In curve (d) the four states to be imprinted are obtained by applying the noisy, or nonzero temperature, neural update rule to a random initial state. The temperature $kT = 0.285$ or $\beta = 3.5$ was chosen based on simulations of the recovery of imprinted states at different temperatures (Question 2.2.2). Whenever the nonzero temperature update rule is used, we also evolve the network by the zero temperature rule to bring the neural state to its local minimum. In both (c) and (d) there is an enhancement of some of the basins of attraction. For curve (c) the number of memories that are enhanced is about 1.5, while for curve (d) it is about 2. The basins of attraction of memories that were reimprinted appear as a peak around the value 40 in Fig. 3.1.2. There is also a small probability that a memory will be reimprinted twice. Such memories appear in a small peak near 50.

The results of the simulations demonstrate that the imprinting of evolved random states enables selective reinforcement of prior imprints. Curve (d) is an improvement over curve (c) because more of the original states are reimprinted. This is explained by the improved retrieval of imprinted states by the noisy evolution.

The procedure for generating Fig. 3.1.2 and Fig. 3.1.3 is:

1. Generate $p = 8$ random neural states $\{\xi_i^\mu\}$

$$\xi_i^\mu = \pm 1 \qquad \mu = \{1,...,p\}, i = \{1,...,N\} \qquad (3.1.10)$$

2. Imprint the states on the synapses of the neural network:

$$J_{ij} = \begin{cases} \dfrac{1}{N}\displaystyle\sum_{\mu=1}^{p}\xi_i^\mu \xi_j^\mu & i \neq j \\ 0 & i = j \end{cases} \qquad (3.1.11)$$

3. Execute the branching instruction:

   For (a): proceed directly to step 7.

   For (b): imprint again the first $p_1 = 4$ neural states on the synapses of the neural network:

$$J_{ij} \leftarrow J_{ij} + \frac{1}{N}\sum_{\mu=1}^{p_1}\xi_i^\mu \xi_j^\mu \qquad i \neq j \qquad (3.1.12)$$

   Then proceed to step 7.

   For (c) or (d): proceed:

4. Generate $p_1 = 4$ random neural states $\{w_i^\nu\}$

$$w_i^\nu = \pm 1 \qquad \nu = \{1,...,p_2\}, i = \{1,...,N\} \qquad (3.1.13)$$

5. Execute the branching instruction:

   For (c): update the neural states $\{w_i^\nu\}$ according to the neural update rule 10 times

$$w_i^\nu \leftarrow \text{sign}\left(\sum_j J_{ij} w_j^\nu\right) \qquad (3.1.14)$$

   Proceed to step 6.

   For (d): (see Question 2.2.2) update the neural states $\{w_i^\nu\}$ according to the $T \neq 0$ neural update rule 20 times

$$w_i^\nu \leftarrow \text{sign}_T\left(\sum_j J_{ij} w_j^\nu\right) \qquad (3.1.15)$$

then update the neural states $\{w_i^v\}$ according to the $T = 0$ neural update rule 10 times.

$$w_i^v \leftarrow \text{sign}\left(\sum_j J_{ij} w_j^v\right) \tag{3.1.16}$$

6. Imprint $p_1 = 4$ evolved-random neural states $\{w_i^v\}$ on the synapses of the neural network:

$$J_{ij} \leftarrow J_{ij} + \frac{1}{N}\sum_{v=1}^{p_1} w_i^v w_j^v \qquad i \neq j \tag{3.1.17}$$

7. Find the basin of attraction of each of the originally imprinted neural states ξ^μ, where an unstable neural state is set to have a basin of attraction of zero (see Section 2.2.6).

8. Make a histogram of the basins of attraction for different ξ^μ. For Fig. 3.1.3 integrate the histogram from a specified value up to 100.

The previous simulations show the use of imprinting evolved random states to reinforce some of the memories. The next simulation takes the procedure one step further to demonstrate the effect of subsequent imprinting. The simulations consist of four main steps. The first step consists of imprinting eight neural states. The second step consists of selecting four random states, evolving them at a temperature T, then evolving them at $T = 0$ (to bring them to the local minimum) and imprinting the result on the network. The third step consists of diminishing the strength of the synapses by a factor of 2. This ensures that the effective imprinting strength of reimprinted states (which have been imprinted twice) is comparable with that of new states to be imprinted. The fourth step consists of imprinting four new states on the network. The purpose is to demonstrate the ability of the network to continue learning.

The consequences of the full procedure are shown as curve (c) of Fig. 3.1.4 and Fig. 3.1.5. It is the result of imprinting eight memories, then imprinting four evolved random states, then diminishing the strength of the synapses by a factor of 2, and then imprinting four additional memories. In Fig. 3.1.4 the distribution of basins of attraction is shown normalized to 12. Fig. 3.1.5 shows the integrated number of memories with basins of attraction greater than the value along the abscissa.

Two reference curves are included as curves (a) and (b). Curve (a) is the result of imprinting 12 memories on the network. The degree of degradation of the basins of attraction when 12 memories are imprinted is easily seen. This is essentially the maximum capacity of the network, the effectiveness of the memory of these patterns is minimal. Curve (b) is the result of imprinting 8 memories, then diminishing the strength of the synapses by a factor of 2, and then imprinting four additional memories. The total number of imprints is still 12. However, as in the simulations of Fig. 3.1.1, the recent set of 4 imprints have large basins of attraction, while the initial set of 8 imprints are effectively lost, since their basins of attraction have been completely degraded.

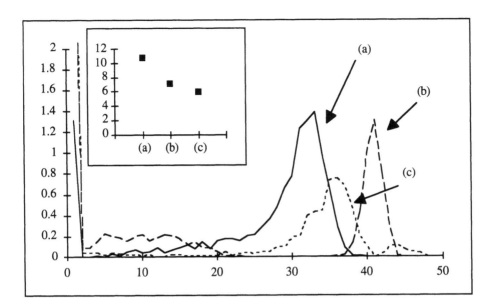

Figure 3.1.4 Continuation of the reimprinting test by imprinting some new states. Curves (a) and (b) are for reference. (a) shows the degree of degradation of the basins of attraction when 12 memories are imprinted all with the same weight. (b) shows the effect of imprinting 8 states, then reducing the strength of the synapses by a factor of 2, then imprinting 4 new states. As explained in Fig. 3.1.1 this results in effective recall of the recently imprinted neural states at the expense of the previously imprinted neural states. The difference between curves (b) and (c) is the insertion of the procedure of evolving four random states with noise, and imprinting the result. This relearning procedure results in the retention of two of the eight original memories for a total of six memories. The others are completely forgotten. ∎

The difference between the simulations leading to curve (b) and curve (c) is only the inclusion of the reimprinting procedure in curve (c). Curve (c) shows that the reimprinting was successful in isolating two of the original memories to retain. These memories survive the imprinting of new states and join them to form a total of approximately six stable memories. This is easiest to see in Fig. 3.1.5, where the relatively flat part of the curve extends down to zero and intercepts the axis at 6 states. Even though curve (b) has more stable states (insert in Fig. 3.1.4), all of these except the newly imprinted ones have very small basins of attraction.

The procedure used to generate Fig. 3.1.4 and Fig. 3.1.5 is:

1. Generate $p_1 = 8$ random neural states $\{\xi_i^\mu\}$

$$\xi_i^\mu = \pm 1 \qquad \mu = \{1,...,p\}, \, i = \{1,...,N\} \qquad (3.1.18)$$

2. Imprint the neural states on the synapses of the neural network

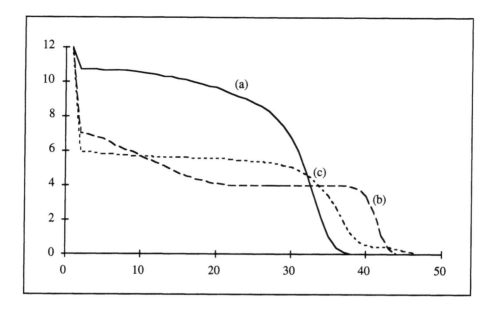

Figure 3.1.5 Similar to Fig. 3.1.4, except that the plots show the integrated number of imprinted states with a basin of attraction greater than a given value. Flat regions separate states with high basins of attraction from states with low basins of attraction. This shows the ability of the reimprinting to reinforce some memories at the expense of others. ∎

$$J_{ij} = \begin{cases} \dfrac{1}{N} \displaystyle\sum_{\mu=1}^{P_1} \xi_i^\mu \xi_j^\mu & i \neq j \\ 0 & i = j \end{cases} \tag{3.1.19}$$

3. Execute the branching instruction:

   For (a): proceed directly to step 5.

   For (b): proceed directly to step 4.

   For (c):

   c1. Generate $p_2 = 4$ random neural states $\{w_i^v\}$:

$$w_i^v = \pm 1 \qquad v = \{1,...,p_2\}, \ i = \{1,...,N\} \tag{3.1.20}$$

   c2. Update the neural states $\{w_i^v\}$ according to the $T \neq 0$ neural update rule 20 times

$$w_i^v \leftarrow \mathrm{sign}_T \left(\sum_j J_{ij} w_j^v \right) \tag{3.1.21}$$

   c3. Update the neural states $\{w_i^v\}$ according to the $T = 0$ neural update rule 5 times.

$$w_i^v \leftarrow \text{sign}\left(\sum_j J_{ij} w_j^v \right) \tag{3.1.22}$$

c4. Imprint $p_1 = 4$ evolved-random neural states $\{w_i^v\}$ on the synapses of the neural network.

$$J_{ij} \leftarrow J_{ij} + \begin{cases} \dfrac{1}{N} \displaystyle\sum_{v=1}^{p_1} w_i^v w_j^v & i \neq j \\ 0 & i = j \end{cases} \tag{3.1.23}$$

4. Rescale the network synapses by a factor $\chi = 1/2$.

$$J_{ij} \leftarrow \chi J_{ij} \qquad i,j = \{1,\dots,N\} \tag{3.1.24}$$

5. Generate p_2 additional random neural states $\{\xi_i^\mu\}$

$$\xi_i^\mu = \pm 1 \qquad \mu = \{p_1 + 1,\dots,p_1 + p_2\}, \, i = \{1,\dots,N\} \tag{3.1.25}$$

6. Imprint the neural states on the synapses of the neural network (Hebbian rule)

$$J_{ij} \leftarrow J_{ij} + \frac{1}{N} \sum_{\mu=p_1+1}^{p_1+p_2} \xi_i^\mu \xi_j^\mu \qquad i \neq j \tag{3.1.26}$$

7. Find the basin of attraction of each of the $p = p_1 + p_2$ neural states ξ^μ (see Section 2.2.6).

8. Make a histogram of the basins of attraction for different ξ^μ. For Fig. 3.1.5, integrate the histogram from a specified value up to 100.

3.1.3 *Sleep phenomenology and theory*

Sleep is one of the fundamental phenomena in biological organisms. An excellent review of the phenomenology of sleep and speculations about its nature are given in the book *Why We Sleep* by Horne. The analysis of subdivisions in complex systems in Section 3.1.1 offers an interesting but speculative theory for the role of sleep—that sleep constitutes dissociation with relearning. This theory is consistent with the suggestion advanced in recent years that dreams have a role in "memory consolidation." However, it extends this role to all of sleep. It also provides a constructive framework in which we can discuss the meaning of memory consolidation. In this section we will provide a brief overview of the phenomenology of sleep, challenge two traditional theories for its role and discuss a few modern theories for the role of dreams based upon neural networks. In Section 3.1.4 we compare the theory of sleep as dissociation with the phenomenology of sleep and suggest experiments that can directly test it. While the theory is not directly supported by current experimental evidence, it is consistent with existing results and is an example of a "good theory" because it predicts definite outcomes for novel experiments and would significantly increase our understanding if found to be correct. Finally, in Section 3.1.5, we discuss a new experimental result which provides some support for the predictions.

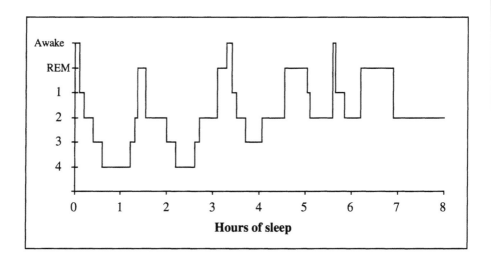

Figure 3.1.6 A sleep "hypnogram" — schematic illustration of the structure of sleep for young human adults showing the different stages as determined by EEG signals. Stages 3 and 4 are called slow wave or SWS sleep. ∎

Human beings spend almost one-third of their lifetimes asleep. Human sleep is known to have several levels identified by quite different brain electrical signals as measured by electroencephalography (EEG). There are at least five recognized levels, the two deepest are together called slow wave sleep (SWS), while the shallowest is rapid eye movement (REM) sleep. Typically, in the first part of sleep, the deepest level is attained rapidly. Then the level of sleep alternates in a pattern of shallower and deeper levels with the average level becoming shallower as sleep progresses (see Fig. 3.1.6). The sleep of animals does not have as many levels. The complexity of sleep increases with the (conventional) evolutionary complexity of the organism.

There are two conflicting popular views of sleep. One is that sleep is necessary for health and well-being. The other is that sleep is a waste of time. Modern society often pays little attention to the significance of sleep. For example, doctors, especially during training, work very extended shifts. Other professions either work long shifts or ignore the natural sleep cycle of day and night. Evidence has accumulated that such practices are counterproductive and cause errors, even fatal errors. There exist efforts to change the training of doctors, and to avoid excessive sleep disruption of airplane pilots. In addition, there are many sleep clinics that are designed to help individuals who suffer from sleep disorders, including various forms of insomnia, an inability to sleep.

Much of our understanding of the role of sleep arises from human sleep-deprivation studies. These studies reveal that sleep loss results in psychofunctional degradation. However, the precise nature of the degradation is not well-understood. Some of the effects of sleep deprivation over several nights include visual illusions or mild hallucinations, and loss of a proper time sense. There are also particular tasks that have been shown to be sensitive to the loss of sleep. However, many others do not appear to be systematically affected. An example of a test that is quite sensitive to the loss of sleep is the vigilance test. In this test a person is asked to monitor a sequence of pictures or sounds for a particular feature that should trigger a response. After sleep deprivation, individuals frequently do not respond to the special feature when it is presented.

Modern theories of sleep have suggested either that it serves a physiological restorative function or that it exists because of genetic adaptation to a survival advantage in removing primitive man from danger. Extensive experimental measurements directed at unveiling the physiological restorative function have not been successful. It appears that after exertion, physical rest rather than sleep is sufficient for reconstruction of tissue damaged during use.

The second suggestion, that sleep serves to remove primitive man from danger, does not coincide with a variety of observations about sleep and its role throughout the animal kingdom. First, sleep consists of a time of reduced awareness of environmental dangers. Even if there were an advantage in inactivity, this lack of activity could be achieved by physical rest rather than the loss of alertness that occurs in sleep. Moreover, even animals that are in constant danger sleep. An extreme example is the case of certain types of dolphins that sleep one half-brain at a time in order to monitor their environment and avoid dangers. Moreover, it is surprising that there are no animals that do not sleep. Nocturnal animals sleep during the day. Predators, whose survival does not depend upon safety from other predators, sleep. Why have no species adapted to the survival advantages of alertness and extra time to find food without sleep? Finally, sleep-deprivation studies on animals show that extended sleep deprivation is fatal. For example, rats die on average after 21 days without sleep. The direct cause of death has not been identified despite substantial efforts.

Neither of the two traditional theories explains the mechanism for psychofunctional degradation after sleep loss. They also do not explain many specific results in sleep-deprivation studies on either humans or animals.

While there has been only limited discussion of the role of sleep in human psychofunction, dreams have evoked more elaborate speculations. Many believe that dreams, or more specifically rapid eye movement (REM) sleep, are the essence of sleep even though they occupy only about one quarter of sleep. Because of their bizarre content, dreams have always invoked mystery. Various theories have suggested that dreams play an important role in human psychology. More recent theories relate dreams to aspects of human information-processing, usually memory. In particular, they suggest that dreams play a role in the conversion of short- to long-term memory—memory consolidation.

There are two specific proposals for the role of dreams that are based upon neural network models. They are precisely opposite. Crick and Mitchison suggested that dreams cause selective forgetting of undesirable or parasitic neural network states. One piece of evidence for this approach is our inability to remember most dreams. More concrete support for this proposal was gained through simulations of attractor networks. Simulations, similar to those in the previous section, were performed by Hopfield, et al. After imprinting a network, the network was initialized to a random configuration and evolved. Instead of imprinting the resulting state, the state was unimprinted, or imprinted with a small negative coefficient of 0.01. This was found to improve the retrieval of imprinted states. The improvement arises because the states with larger basins of attraction are responsible for the instability of the other imprinted states. Reducing the large basins of attraction by a small amount improves the balance between different states. In contrast, Geszti and Pázmándi suggested that dreams are a form of relearning. Their relearning procedure is the one described in the previous section. Its purpose is to filter the memories to enable continued learning.

Both of these models attributed information-processing tasks to rapid eye movement (REM) sleep, or dream sleep. The other parts of sleep, where dreams are infrequent (non-REM sleep), are still generally believed to have a physiological role. However, as described earlier, total sleep deprivation causes psychofunctional, not physiological, deterioration in humans. The primary effects occur with loss of non-REM sleep.

Based on the discussion of subdivision in neural networks and training it is reasonable to propose that the stages of sleep correspond to degrees of interconnection between subdivisions of the brain. SWS corresponds to the greatest dissociation, where small neuron groups function independently of each other. At shallower levels of sleep, larger regions of the brain are connected. Ultimately, the waking state is fully connected, including sensory and motor neurons. From EEG measurements it is known that all of the levels of sleep are neurologically active. Consistent with our discussion in the last section, it may be proposed that the activity is a filtering process that reinforces some memories at the expense of others to prevent overload and allow for additional learning. The filtering of memories occurs on all levels of organization. The ultimate purpose of this filtering process is to establish the memories within subdivisions, and the stable composite memories. It also balances the strength of synapses within subdivisions compared to the strength of synapses between them.

There are general consequences of the filtering that we can infer and use to make predictions of its effects on memory. It is to be expected that the strength of associations that are represented by synapses between subdivisions are weakened more rapidly than associations that are represented by synapses within subdivisions. Thus, memories are progressively decomposed into their aspects, stored within subdivisions. Implicit in this architecture is the assumption that the most permanent associations—stable patterns of neural activity—are stored inside the smallest subdivisions. These associations are the elements that are the building blocks for new imprints on the network, and thus are the elements for building new memories.

This theory for the role of sleep is based upon a subdivided attractor network, with no directionality to the synapses, or to the processing as a whole. The presence of directionality in the processing of sensory information should modify this picture, but may not change the essential conclusions. One modification that we can expect is that the triggering of a random neural state will also acquire directionality. The triggering should follow the usual processing path in order to be consistent with the system's natural mechanism for retrieving memories.

3.1.4 *Predictions and experimental tests*

A variety of aspects of the general phenomenology of sleep are consistent with the idea that sleep is a temporary dissociation of the brain into its components. Several of these are described in the following paragraphs.

Sleep itself consists of a dissociation of the cerebral activity from both sensory and motor neurons. This separation is accomplished by sleep substances that control particular interconnecting neurons or synapses. While the dissociation is not complete—we can still respond to sounds and lights during sleep—the degree of correlation between sensory stimuli and the activity of the brain is reduced. Similar controls could be used to further dissociate various subdivisions of the brain.

As mentioned before, sleep is a time during which there is significant neural activity. This is to be contrasted with the lack of explicit memory of this activity. The patterns of neural activity differ qualitatively between different stages of sleep. These changes can be measured using EEG signals, which are used to identify stages of sleep. Systematic differences between patterns of neural activity imply basic changes in either the activity of neurons or their synaptic efficiencies. This requires an explanation. Qualitatively, the greater simplicity of EEG signals in SWS (hence the name slow wave sleep) is consistent with a loss of complexity in the activity patterns due to a lack of correlation between different neuron groups.

The internal triggering of patterns of neural activity occurs in all stages of sleep, but is very apparent during REM sleep, where pulsed neural activity patterns extend through a significant part of the cerebral cortex.

The greater difficulty of waking during SWS is consistent with a greater degree of dissociation in deep sleep than in shallower levels of sleep. It may also be difficult to wake from REM sleep, despite its other characteristics as a shallow stage of sleep. However, in this case the internal triggering of neural activity appears to mask awareness of actual sensory stimuli.

Systematic studies of dream content indicate that specific higher-level critical faculties and a "sense of self" are absent. This includes a lack of surprise at the content of dreams, and an inability to see or perceive one's self. It has been pointed out by Hartmann that this is similar to the waking mental functioning of postlobotomy patients, where connections to the frontal lobes of the brain have been severed. Specific higher-level critical facilities related to self-awareness are often associated with these frontal lobes. This suggests that during REM sleep, specific major sections of the brain are dissociated.

Dissociation during sleep would imply that the neural activity is formed out of composite states that typically would not occur if the brain subdivisions were connected. In REM sleep, when only major sections of the brain are separated from each other, the composite states are formed out of only a few elements. The waking brain with full connections can, at least sometimes, make a kind of sense out of the juxtaposition of elements from the sleep state. This explains the possibility of recalling sleep states from REM sleep in the form of dreams. It also explains their bizarre content. Moreover, it explains why dreams are not always recalled even when experimental subjects are woken during REM sleep. For deeper levels of sleep, with smaller subdivisions, the waking brain can generally make no coherent picture of the sleeping mental state. This explains the absence of recalled dreams from deep sleep despite the ongoing neural activity.

To make further progress in our understanding of sleep and the dissociation model, we will discuss the psychofunctional effects of sleep deprivation. Our discussion will provide some understanding of the deterioration that can result from sleep deprivation. It will also explain why experimental efforts have found it difficult to identify specific psychofunctional tasks that are affected. The central point is recognizing that the deterioration of capabilities is directly linked to activities that are performed during waking. Thus the question, How does sleep deprivation affect the capabilities of an individual? is meaningless without a specification of the activities performed by the individual during the period in which he or she is awake.

The model of sleep as dissociation implies that it is basic to the functioning of the subdivided architecture of the brain. However, the manifestation of sleep deprivation would not be the complete disruption of this architecture. The shorter-term effects of sleep deprivation are related to overload failure. Overload occurs because imprinting is continued during waking hours without a periodic filtering process. Under normal circumstances, there must be a substantial buffer before overload is reached. The buffer exists because of the need to stop imprinting well before the overload threshold. However, if the buffer were very large, then the full capacity of the network would not be utilized. This explains the need for a regular sleep schedule with a consistent structure to the levels of sleep. It also explains why there are dramatic effects of only a few nights of sleep deprivation, which become catastrophic if further extended. We note that no model of the role of sleep based solely on a concept of memory consolidation would account for psychofunctional failure due to sleep deprivation.

The implications of overload failure in a fully connected network were discussed in Section 2.2.7. When overload is reached, various spurious states replace memories as the stable states of the network, and a complete loss of memory results. In order to adapt this picture to describe the effect of sleep deprivation, we must include both the correlated nature of the information that is presented to the brain over any particular period of time, and the subdivided architecture. Their implications may be understood simply. First, the correlated information implies that overload does not affect all imprinted states equally. If newly imprinted states are confined to a particular subspace of all possible neural states, then all states that are not correlated with them will not be affected. The existence of subdivisions leads to similar conclusions by emphasizing that overload should occur in particular subdivisions first, rather than uniformly throughout the network. The conclusion is that the effect of sleep deprivation is primarily confined to activities that are exercised during the waking period. This explains much of the difficulties that have arisen in the efforts to determine specific activities that are strongly affected by sleep deprivation. Many tests evaluate the degradation in an activity, such as intelligence tests, tests of ability to maintain balance, etc. However, even when significant correlations are reported between sleep loss and a particular test in one experiment, these are found not to exist under other experimental conditions. In contrast to the generally ambiguous results on specific abilities, it has been reliably shown that a degradation of ability is found for repeated or similar tests, essentially independent of the nature of the task.

Unlike other activities, the vigilance test is a self-contained test of the ability to persist in a particular activity. The vigilance test requires paying attention to a series of varying sensory images, and responding only to a particular variant. We can understand why a sleep-deprived individual finds this difficult. Imprinting various similar states up to overload would cause, in effect, the basins of attraction of the lack of action to overtake the only slightly different circumstances that require action. The inability to respond differently to a slightly different stimulus may very well be the cause of accidents that occur in early-morning hours. Consider the train conductor who is required to brake the train upon seeing a particular set of lights, after viewing many different panoramas with various sets of lights. Consider the doctor, who after seeing many patients with similar ailments is required to change the treatment based on one of many pieces of information.

The relevance of repeated tasks to the need for sleep does not require sleep-deprivation studies. It is sufficient to note the sleepiness that arises from boredom due to monotonous or repetitive activity. It is also interesting that schools (at all levels) do not schedule classes around learning a particular topic for a whole day. Instead, activities and classes vary through the day. Each subject is taught only for a limited period of time.

Sleep-deprivation studies are generally performed in a monotonous environment without many stimulating or novel activities. Stress, which is likely to increase imprinting (see Section 3.2.4), is also absent. This suggests that aside from the generally necessary activities, clinical sleep-deprivation studies do not capture the psycho-functional degradation from typical daily activities. The most commonly observed difficulties in sleep deprivation arise from visual illusions. This may be understood both from the necessity of vision even under laboratory conditions, the popularity of reading or watching TV during an experiment, as well as the monotonous laboratory environment that implies significant correlations between visual stimuli.

Modeling sleep deprivation by overload failure implies that novel, stimulating, stressful, or boring circumstances lead to an increased effect of sleep deprivation. As will be discussed in Section 3.2.4, all of these, except for boring circumstances, can be related to an increase in imprinting strength and a more rapid approach to overload. Boring circumstances, by virtue of repetition, achieve this result rapidly not because of the strength of imprinting but because of the overlap of different imprints that cause overload in a more limited domain of patterns in the network. Consistent with experience, sleepiness that results from repetitive activity can often be overcome by changing the activity.

If we wish to understand the implications of sustained sleep deprivation, we must look for individuals that inherently possess a particular form of sleep deprivation. The simplest to understand would be a loss of deep sleep, where the basic elements of neural functioning in the smallest neural networks are established. A loss of SWS would be associated with a breakdown in psychofunction well beyond a severe case of sleep deprivation. Experimentally, it has been found that there is a complete lack of SWS in about 50% of individuals diagnosed with schizophrenia. Schizophrenia includes a broad class of severe psychofunctional disorders.

We now turn to discuss new experiments using several distinct methodologies that could directly evaluate the possible role of subdivision during sleep. These tests include clinical studies, imaging and physiological experiments.

The most direct clinical tests would measure the retention of memory of associations at higher levels of the hierarchy. According to the model of temporary dissociation, associations between disparate kinds of information stored in different regions of the brain are preferentially lost during sleep. An experimental test would expose subjects to information that is composed of two or more different aspects. The subjects would be split into two groups, one would sleep and the other would not. The retention of the information would then be tested. Various experiments of this kind have been done but without specific emphasis on correlations of different aspects of information. For example, a visual image and a sound could be presented at the same time. A test would measure the ability to recognize which image-sound pairs were presented. Other combinations of information could also be tested by selecting from known subdivisions in the brain: vision, audition, somatosensory, language, and motor control. Within each category further tests could be performed. For example, tests in vision could measure the ability to retain particular combinations of shape and color. Pictures of people, each with particular color clothes, could be changed by reassigning colors. Tests would determine the ability to recall the association of color with shape.

The development of positron emission tomography (PET) and magnetic resonance imaging (MRI) has enabled more detailed mapping of brain activity in recent years. The ability to map brain activity can also enable mapping of correlations between activity in different parts of the brain. This becomes increasingly feasible as the temporal resolution of imaging is improved. Statistical studies of the correlations in neural firing could directly measure the strength of influence between different parts of the brain while a subject is awake, and during various stages of sleep.

Neurophysiological studies of animals characteristically measure the activity of a neuron under particular stimulus. Using more than one probe at a time, the correlations between neural activities in different parts of the brain could be compared in waking and in various sleep states. Such experiments can also stimulate some neurons, and measure the difference in signal transmission between neurons in different parts of the brain in animals that are awake and asleep.

The dissociation model would require a chemical mechanism for preferential inhibition of the synapses or neurons that interconnect various regions of the brain. The ability to chemically separate different regions of the brain can be directly tested by investigating the impact of sleep substances on neurons and synapses. Synapses or neurons that interconnect different regions of the brain would be expected to have a characteristically distinct sensitivity when compared to neurons and synapses within a particular region of the brain. In order to enable a difference between REM sleep and SWS it would also be necessary for there to be differences in the connections between smaller regions and larger regions. The possibility of sleep substances that preferentially isolate a particular level of the brain structure may become apparent from such tests.

3.1.5 *Recent experimental results*

In one of the first multiprobe experiments, Wilson et al. recently investigated correlations between neural activities in the hippocampus. The hippocampus is an area of the brain that is responsible for representation of information about the organism's spatial position, in particular its location with respect to large objects or boundaries. They found that new correlations in neural activity due to changes in the environment were subsequently repeated during sleep.

This experiment supports a number of aspects of neural network models of brain function. Of principal significance, it supports the attractor network model that memories are stored and can be recovered as a pattern of neural activity. It also supports the discussion in this chapter, that they are recovered during sleep. The idea that waking experiences are reflected in dreams is known. However, this is the first indication of the nature of their representation. Moreover, it is interesting (and consistent with the above discussion) that the recovery of patterns of neural activity was not particularly associated with REM sleep, but rather occurred in SWS.

3.2   Brain Function and Models of Mind

3.2.1 *The fundamental questions*

We use phenomena that are associated with neural networks to understand some of the aspects of brain function by our own recognition of their similarities. In the previous chapter, we briefly mentioned the associative memory function of the attractor network that is reminiscent of human association capabilities. We will expand upon this discussion in the following sections to cover a variety of information-processing tasks. As we do so, we will find that we have to expand our model to include additional features. We start with both the attractor network formed of symmetric synapses, and the feedforward network with unidirectional synapses. We use subdivision to clarify some of the basic issues and expand into the realm of higher information-processing tasks.

As our description of information-processing functions progresses, we must allow ourselves to expand the conventional terminology. We use our model neural network as a model of the brain. The functioning of this network is a model of the mind. We can use terminology such as the subconscious mind to describe the part of the neural network/brain whose function we identify with what is commonly understood to be the subconscious. A sentence that contains such terminology can still possess precise mathematical meaning in the context of the neural network architecture. This is similar to the use of words like "energy" and "work," which have different meanings in scientific and popular contexts.

3.2.2 *Association*

In the attractor neural network model of the human mind, the basic learning process is an imprinting of information. The information may, for example, be a visual image. This information is represented as a state of the neural network (pattern of neural activity), and the synapses between neurons are modified so as to store—remember—this information. The mechanism for retrieval is through imposing only part of the same image. The synapses force the rest of the neurons to recreate the stored pattern of activity, and thus their representation of the stored image.

In order to illustrate how this process manifests itself in behavior, we have to consider the nervous system "output" leading to action as also part of the state of the mind. Part of the pattern of neural activity specifies (controls) the muscles, and therefore behavior. Using the pattern of activity that represents both sensory information and motor control we can, in a simple way, understand how reactions to the environment are learned.

We can consider the example of a child who learns to say the name "Ma" whenever she sees her mother. Let us say that somehow (by smiling, for example) we are able to trigger imprinting. At some time, by pure coincidence, at the sight of her mother the child says something which sounds like Ma (or even quite different at first, subject to later refinement) and we encourage an imprint by smiling. Thereafter the child will say Ma whenever she sees her mother. The pattern of neural activity that arises when the mother is in the visual field has been associated with the pattern of neural activity representing motor control that manifests itself in the word "Ma." Of course this process could be enhanced by all kinds of additions, but this is one essential process for human learning and human functioning that this neural network captures.

We note that the training of a feedforward network discussed in Section 2.3 requires a comparison between the desired output and the output generated by the network. Because both the desired output and the output generated by the network must be represented at the same time, the feedforward network does not by itself provide a model of how responses can be learned. A solution to this problem will appear when we discuss consciousness in Section 3.2.12.

3.2.3 *Objects, pattern recognition and classification*

When we look at a room we do not interpret the image in the form of a mapping of the visual field as a point by point (pixel by pixel) entity. Our interpretation is based on the existence of objects and object relationships that exist in the visual field. The same is true of auditory information, where sounds, notes or auditory representations of words are the entities we differentiate. Similarly, our associations are driven not by direct overlap of sensory information but rather by objects, aspects or relationships. Why is this useful, and how is it possible to identify objects in sensory fields?

The reason objects are used rather than the visual field itself is easy to understand within the neural network model. Consider a particular visual image which is mapped pixel by pixel onto a neural network and imprinted. An attractor network relies upon the Hamming distance of a new image with the imprinted image for recall and therefore for association. Any new image mapped onto the network is characterized by the overlap (similarity measured by direct counting of the number of equivalent pixels) of the image with imprinted images.

This means, for example, that if we want the child of the last section to say "Ma" no matter how her mother appears in the visual field, then all possible ways the mother can appear in the visual field must be imprinted independently. "All possible" means essentially independent ways, ways for which the overlap of one with the other is small. This overlap is strictly a Hamming distance overlap—a count of the number of equal pixels. Since there are many ways that the mother can appear in the visual field with only a small overlap between them, this would require a large part of the neural memory. Saying that we identify objects is the same as saying that the child identifies as similar many of the possible realizations of the visual field that contain her mother. We must then ask how this is possible when the visual fields compared pixel by pixel are different.

Underlying the use of objects in describing the visual field is the assumption that objects possess attributes that are unchanged by their different possible presentations in the visual field. The existence of attributes, as discussed in Section 2.4, can be used by a subdivided network to identify the objects. We identify the attributes of a particular object as the states of each of the subnetworks when we are presented with the object. For example, in the separation of visual information into shape, color and motion, the attribute RED would be represented by a particular pattern of neural activity in the subnetwork representing color information. Extracting different aspects of the information and storing them in particular subdivisions of the network enables the object to be identified by a particular set of subnetwork states—by the pattern of common attributes. The suggestion that attributes can provide a mechanism for the identification of objects is not a complete answer to the problem of object identification. It is still necessary to examine how the characteristic attributes of objects are found in the visual field.

In recent years the field of computational vision has been dominated in large part by discussion of computational problems associated, for example, with extracting boundaries of objects. This is important because the extraction of edges provides an important clue as to the existence and nature of objects. This research has been viewed as opposed to the use of attributes for object identification. It may be better understood as providing the computational approach to extracting these attributes. Thus, the extraction of edges provides one (or several) attributes of the visual field that can be used to identify objects; other attributes can be used as well. Rather than relying upon a single algorithm to identify objects, the use of multiple attributes enables several algorithms to act together through associative links, as suggested by Fig. 2.4.4.

Once we have understood the identification of objects through their attributes, we can likewise understand pattern recognition or classification as a process of identifying common attributes. Specifically, elements of a category may be identified by the common state of a particular subnetwork or set of subnetworks. Pattern recognition, viewed in an abstract form, is equivalent to the problem of classification.

The existence of objects is often considered one of the most concrete aspects of reality. We see that the identification of objects is actually an abstraction. It is a basic abstraction central to our ability to interpret sensory information. Moreover, the same methodology of abstraction is also the key to understanding abstract concepts. Abstract concepts, like concrete objects, may be stored in the brain by combinations of aspects or attributes. The attributes are represented by patterns of neural activities of brain subdivisions. This method of representation is also related to other aspects of higher information-processing—generalization, creativity and innovation, to be discussed below.

In summary, because of the many different possible visual fields, it is impossible for the brain to be imprinted with all of the appropriate ones, and associate them with the appropriate response. Instead, the visual fields are reinterpreted as composed of combinations of attributes that reflect the existence of objects and their relationships.

3.2.4 Emotions and imprinting

One of the central properties of the neural network that we have not investigated in any detail is the strength of imprinting. Our numerical modeling of the imprinting process generally assumed that each imprint has the same coefficient. However, it is quite reasonable to include the possibility of stronger and weaker imprints, where the strength of imprinting can be controlled in various ways. Stronger imprints result in larger basins of attraction. Larger basins of attraction imply that recall is easier—triggered by a smaller set of common attributes. Even in our discussion of association in Section 3.2.2 it was necessary to invoke a mechanism for triggering stronger imprinting in order to describe the learning of a response.

There are various ways to control the strength of imprinting at a particular synapse. Our concern here is not with an individual synapse, but rather with the overall strength of a particular imprint. In the Hebbian imprinting model, this strength is controlled by the parameter c in Eq. (2.2.6). The control of the strength of imprinting must occur at every synapse in the brain. Chemicals that can be distributed throughout the brain to affect the imprinting would be most easily distributed through the bloodstream. The most natural assumption is that the relevant chemicals are associated with emotions. Emotions affect the general response by the body to external circumstances. At least some of these circumstances imply that imprinting and memory should be enhanced. One indication of this is that new circumstances, or circumstances that are important due to the existence of a threat, or circumstances that are painful, give rise to the release of such chemicals. It would make sense that the emotional reaction governs not only the immediate reaction (the traditional fight or flight response to stress) but also the recollection of such circumstances in the future. Without discussing the process in detail we may conclude that imprinting strength under these circumstances—the coefficient c—is increased by adrenaline (epinephrine/norepinephrine) and affected by other endocrine-system chemicals associated with various emotional states.

A second way to strengthen the imprinting is to repeat the same imprint more than once. In the simplest model of a constant imprinting strength, the total strength of imprinting grows linearly with the number of imprints. We will modify this assumption in the next section, because such continued imprinting is not advisable.

If we assume that selective retention and forgetting of memories is necessary, as discussed in Section 3.1, then the relative strength of the original imprinting will also affect which memories persist. We can expect that memories that persist are those associated with the greatest stress or strongest emotions, or with the largest number of repetitions. A classic example is the persistent memory of traumatic events. In a subdivided network, the persistent memories may be aspects of situations rather than the situations themselves.

We have discussed, thus far, the effect of emotional response on the chemistry of the blood and its consequent effect on imprinting. The source of emotional response must also originate in the nervous system, because the sensory stimuli that describes the environmental circumstances leading to the emotional response are received by the nervous system. We must therefore assume that neural activity affects the adrenal gland and other glands responsible for chemicals that affect the physiological response. The circle of influences between bloodstream chemicals and brain function is an important feedback loop. Part of the brain initiates the emotional state by controlling the bloodstream chemicals, which then affect the functioning and the imprinting of the brain. Physiologically, it is the diencephalon and particularly the hypothalamus, a hybrid nervous-system component and endocrine gland, that bridges between the nervous system and the endocrine system.

3.2.5 *Fixation, compulsion and obsession*

Any model of how the brain works contains within it a model for how the brain may fail. Since there are also many real occurrences of failure, we can compare and try to evaluate whether the model is properly predicting the failure. The storage of various information in the brain with different imprinting strengths enables the possibility that a single imprint will become dominant. The meaning of a single dominant imprint was discussed early in Chapter 2 when the case of a single imprint was described. Under these circumstances any initial state will evolve in time to the attractor that is the dominant imprint. This description is very reminiscent of the behavior of a person who suffers from fixation, compulsion or obsession. Such individuals repeat actions or thoughts regardless of the external circumstances and regardless of the recent history.

Examples of dysfunction include a compulsive repetitive action such as hand washing, fixation on a person or object, and obsession with an idea. In each case the persistent return to the behavior pattern or thought pattern can lead to a severe breakdown in human function. Strong imprinting of a particular thought or behavior itself arises from repetition, so that this is a self-consistent failure of the architecture. Self-consistency arises because repetition strengthens imprinting, and when imprinting is strengthened the tendency to repetition is increased. As discussed in the last section, the existence of strong emotions contributes to imprinting, and indeed strong emotions, passion or anxiety, are often associated with the development of these disorders.

When there is a natural mode of failure, one must expect that the system has built-in safeguards against it. When the failure occurs anyway, it is because of a breakdown in the safeguards. Safeguards against the creation of a dominant imprint may consist of a reduction in strength of repeated imprinting. One approach to reducing the strength of repeated imprints uses an internal feedback mechanism that changes the imprint strength depending on whether the neural activity pattern is already stable. This can be done locally at each neuron—when a neuron activity is consistent with the local field h_i, its own synapses need not be imprinted. Some such modification of the prescription of Hebbian imprinting is likely to exist in the brain to avoid excessive imprinting of existing memories. A second safeguard would require a behavior that avoids exposure to repetitions. Boredom as an emotional state may have a purpose of causing behavior that avoids continued repetition. In order for this safeguard to work, there must be an internal mechanism for recognizing repetition—for recognizing the stability of a state. The problem of recognition is discussed in Section 3.2.6.

While such safeguards may serve to help prevent this mechanism for failure, it should also be understood that the strengthening of imprinting with repetition, and the use of emotions, must not be completely curtailed by the safeguards. Otherwise, the primary function of the brain would be degraded. This limits the implementation of safeguards to an extent that enables function, but also enables failure.

Medical classification of mental disorders distinguishes between neurotic and psychotic conditions. The former are less severe than the latter. Various neurotic compulsions, fixations and obsessions may persist for years without treatment. Psychotic conditions can require severe intervention using drugs or electroshock therapy. The action of these treatments is not well-understood. However, we can speculate about electroshock therapy as a means of shaking, in an uncontrolled way, the energy landscape of the space of states of the brain as we have represented it in the neural network model. This may explain why, despite the grave concerns about its side effects, the treatment continues to be used. Interpreted simply this also suggests that for these disorders the forms of traditional psychotherapy that dwell upon the problem may actually promote it. Therapeutic strategies that emphasize other important (strongly imprinted) areas of a person's life, and change as much as possible of the circumstances of a person's life, would be expected to be more effective.

The distinction between different kinds of repetitive processes—fixation associated with the senses, obsession with abstract thoughts, and compulsion with simple actions—suggests that an overly strong imprint may be localized in different regions of the brain. However, because of the coupling between different parts of the brain these distinctions may not always be maintained.

3.2.6 *Recognition*

One of the standard tests of memory is the recognition test. In this test, for example, a person is shown a number of pictures and asked to recognize them later. The human brain is capable of successfully recognizing at least ten thousand pictures. This would seem to be a natural application of our neural network memory that imprints the images and then recalls them. But is it? The process here is different. The subject is required to identify whether or not he has seen the image before. This is different from reconstructing the image itself. In our model of the associative or content-addressable memory, the task is to provide the missing pieces. In the recognition experiment, the subject says "yes" or if he has, or "no" if he has not, recognized the image. Saying "yes" reflects a particular pattern of neural activity exercising vocal control. The recognition task requires that these neurons have the same firing pattern for all stored images, and a different firing pattern for any image that has not previously been stored. Our model does not contain a single association with all of the images that have been stored. Everything that is recovered in the attractor network is part of the original image.

One way to solve this problem is to suggest that there is a part of the brain that stores the word "yes" along with every image we see. Then we can use this part of the brain to perform the recognition task. We will not adopt this approach here. Instead we will require that the network have some way of identifying whether or not it has imprinted the picture from the behavior of the network itself.

When we impose upon the network a previously imprinted pattern of neural activity, the state of the network is a fixed point of the neural evolution—a stable state. We must find a way for the brain to know—i.e., represent the information—that it is in a fixed point, so that it can say "yes", and then when it is not in a fixed point, it can say "no". In order to act differently if the network is in a fixed point or not we need to have a particular neuron, or set of neurons, that are ON in one case and OFF in the other. Our problem is to construct a set of synapses and neurons that achieves this objective. This problem may appear superficial, but it is not. Let us try to do it in a natural way.

To distinguish between the case of a stable state and an unstable state, it is reasonable to think about comparing the value of a neuron before and after a neural update. We can do this using a synapse that includes a time delay. Biologically, the time delay would be achieved through the axon rather than the synapse, but this is irrelevant to our argument. Using the time delay in the transmission of the signal from one neuron to the next, we can arrange to have the second neuron (the recognition neuron) receive information about the time evolution of the first neuron, in the form of the neuron value at one moment and its value at a time corresponding to one update later. We might think that this is enough to enable the recognition neuron to determine whether the neuron is changing or not. Surprisingly, this is not the case.

The reason that there is still a difficulty can be explained by considering the four pictures at the bottom of Fig. 3.2.1. On the left of each picture are illustrated the two states of the first neuron, before and after the update. The four pictures are the four possibilities. On the right in each picture is shown the value that we want the recognition neuron to take. The essential point is that we would like the recognition neuron to have the same value when the two states of the original neuron are the same, and the opposite value when they are different. This function is equivalent to the logical operation *exclusive or*, XOR, which gives TRUE (ON) if either input but not both is TRUE (ON) and FALSE (OFF) otherwise. What is illustrated in Fig. 3.2.1 is the opposite or negation of XOR if we follow the usual pictorial interpretation of UP as TRUE. We

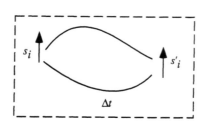

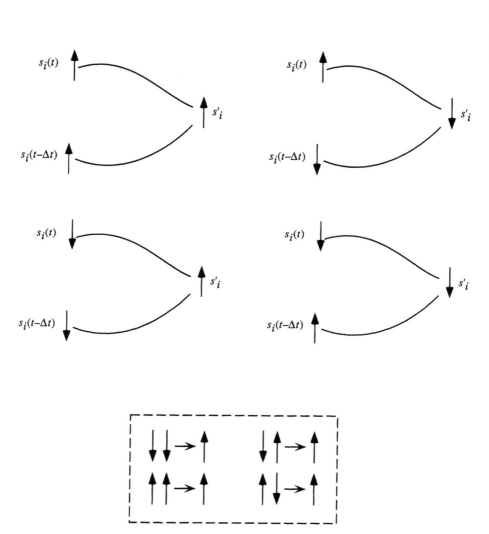

Figure 3.2.1 The problem of recognition requires the network to be able to respond the same to all patterns that have been imprinted, and differently to all patterns that have not. This requires detection of a stable state, which can be found from the time dependence of neural activity. To detect the stability of a particular pattern we use (top) two synapses, one of which is delayed by a time Δt. Both synapses run to a particular other neuron that is supposed to fire only when the two signals it receives are the same. The four possible cases of neuron firings are shown (center) where the left neuron is shown at two different times. Considering the right neuron as a function of two variables, we find (bottom) that it must represent the negation of the logical function exclusive or (XOR). This function cannot be represented by an attractor network. It can be represented by a feedforward network. ∎

can invert the definition or the picture to make the two agree. However, the problem is that the XOR logical operation cannot be built out of symmetric synapses between pairs of neurons. This is apparent when we remember the energy analog of the neural network (Section 2.2.3). Switching the activities of all neurons does not affect the energy, so the inverse of any low-energy conformation is a low-energy conformation. Inverting the XOR operation results in the opposite logical operation, not the same one. Pictorially in Fig. 3.2.1, we see that if the upper left is a minimum energy state, flipping all of the neurons would not lead to the lower left but instead would cause the recognition neuron to be inverted, giving the wrong information.

To overcome the problem of representation of the XOR operation and enable recognition requires the introduction of a new kind of synapse. There are many possible ways to do this. One is to use interactions between three neurons $s_i s_j s_k$. This breaks the inversion symmetry and enables the minimum energy states of the three neurons to correspond to the XOR operation. However, using a symmetric three-way synapse would still lead to some difficulties. A symmetric synapse, where the neurons influence each other reciprocally, does not really make sense when there is a time delay. Moreover, if a symmetric synapse is used, the recognition neuron could affect the other neurons rather than representing their state. This would not be helpful. A directed synapse such as the ones used in a feedforward network would be simpler.

There is a way to introduce an XOR operation using a feedforward network. This is discussed in Question 3.2.1. However, the feedforward network requires two stages for this operation, and it is not particularly convenient. Fortunately, we can probably do just as well with an AND logical operation. The AND operation would detect when a neuron stays ON. Ignoring the neurons that stay OFF, this would be enough to tell us when the state of a neuron is stable. The AND operation can be represented by a feedforward network using one stage (Question 3.2.2). Experimental studies of the biology of neurons also show the existence of individual directional synapses that couple three neurons. In some of these, two neurons must fire in order for the third to fire, thus directly implementing the AND operation. This solves the problem of enabling the recognition task to be performed. The recognition task is fundamental not only for the external recognition test that we have been describing but also for internal processes that lead to other capabilities. For our purposes in continuing to build models of brain function it is sufficient to note the necessity and biological plausibility for such logical operations.

Question 3.2.1 We have shown that an attractor network by itself cannot perform the XOR operation to perform a recognition task. Find a feedforward network with two layers of synapses that can perform the XOR operation. You may supplement the two neurons that you are comparing by a neuron that is always ON. Discuss the biological implementation of this feedforward network.

Solution 3.2.1 The XOR operation requires a comparison of two different binary variables. However, the feedforward network uses neurons represented by real numbers. According to the model we developed in Section 2.3, for two neurons s, s' in the first layer, we can write the value of any second-layer neuron as:

$$s_2 = \tanh(Js + J's' + h) \tag{3.2.1}$$

We can think of the constant term h as arising from a first-layer neuron that is always ON. The two independent linear combinations of the neuron activities that we have are $s + s'$ and $s - s'$. Either can be thought of as a comparator. If we construct a table from different values of s and s' we have:

s	s'	$s + s'$	$s - s'$	$\text{XOR}(s, s')$	
1	1	2	0	-1	
1	-1	0	2	1	(3.2.2)
-1	1	0	-2	1	
-1	-1	-2	0	-1	

Comparing the $s + s'$ and $s - s'$ columns with the XOR column we see that the XOR operation requires us to treat a positive sum and a negative sum the same, or a positive difference and a negative difference the same. We must therefore take an absolute value, or square the linear combinations. Two ways to write the XOR operation in terms of floating point operations are:

$$-\text{sign}(|s + s'| - 1) = -\text{sign}((s + s')^2 - 1) \tag{3.2.3}$$

The tanh function can provide us with the square of $s + s'$ by setting up a situation where we make use of its second-order expansion:

$$s_2 = \tanh(h + J(s+s')) \approx \tanh(h) + J\tanh'(h)(s+s') + \frac{1}{2}J^2\tanh''(h)(s+s')^2 + \ldots$$

$$s_2' = \tanh(h - J(s+s')) \approx \tanh(h) - J\tanh'(h)(s+s') + \frac{1}{2}J^2\tanh''(h)(s+s')^2 + \ldots$$

$$\tag{3.2.4}$$

The expansion is valid if we use a small enough value of J. Setting up two second-layer neurons with these values, we can take their sum to eliminate the first-order term and keep the second-order term that we need. We use $J = 0.1$, and $h = 0.5$ to obtain the following table:

s	s'	$s_2 = \tanh(h+J(s+s'))$	$s_2' = \tanh(h - J(s+s'))$	$s_2 + s_2'$	$\tanh(J'(s_2 + s_2) - 0.9J')$
1	1	0.604	0.291	0.896	-1.000
1	-1	0.462	0.462	0.924	1.000
-1	1	0.462	0.462	0.924	1.000
-1	-1	0.291	0.604	0.896	-1.000

$$\tag{3.2.5}$$

The final column is the value of the neuron in the third layer (after two layers of synapses) that gives the XOR operation on the first layer of neurons s, s'. $J' = 1000$ is a large number that makes the tanh function into a sign function as required to obtain ± 1. The value 0.9 that appears in the final formula is chosen to lie between the two possible values of $s_2 + s_2'$ shown in the previous column.

There are two difficulties with this representation of the XOR operation that are related to robustness and reliability in a biological context. The first is that it makes use of matched values of h and J on different synapses to ensure that s_2 and s_2' have consistent values, and a matched value of J'. The sensitivity to different values can be seen, for example, by trying to use $J = 0.11$ only for s_2' in the above table. The second is that this operation uses a second-order property of the synapse (the second derivative) that may be more variable than first-order properties. ∎

Question 3.2.2 Construct a logical AND using a feedforward network. You may supplement the neurons that you are comparing by a reference neuron that is always ON.

Solution 3.2.2 The second-layer neuron must fire if and only if both neurons of the first layer fire. If we add the activity of the two first-layer neurons, and require the result to be greater than a number greater than zero in order for the second-layer neuron to fire, then we will have the AND operation. This can be achieved in Eq. (3.2.1), for example, by setting J and J' to a large positive number, and h to its negative. Using a large number converts the tanh function into a sign function.

The existence of logical operations such as AND and XOR in the available functions of a neural network is interesting from a computer science point of view. For example, using just AND and negation (NOT), we can construct all possible logical operators (Section 1.8). This might suggest we could construct mathematical or logical operations of the neural network along the same lines as computers. One difficulty with this approach lies in the problem of representation. It is unlikely that the brain represents numbers in a conventional binary fashion. Instead, the word and number "one" are somehow represented as a state of the network involving many neurons. Thus the use of conventional logical operations on individual neurons by synapses is not likely to be the source of the brain's ability to perform addition. At the same time, we should not hesitate to make use of the logical operators at the level of individual neurons to justify the brain's ability to recognize images that have been imprinted. ∎

3.2.7 *Generalization*

One of the important properties of neural networks in pattern recognition or artificial intelligence tasks is their ability to generalize from a few training examples to a large number of cases. Generalization in a fully connected attractor network is simple to understand. The training creates a local minimum in the space of possible network states. The basin of attraction of this state becomes its generalization.

Partially subdivided networks provide an additional layer of generalization (Section 2.4). In addition to trained states, various combinations of substates (composite states) that may appear in the environment are recognized by the network. Since the network has been trained on far fewer states than it recognizes, it may be said to have generalized from the training set to the set of recognized states. This is an advantage if the architecture of subdivision of the network is in direct correspondence to the information to be presented. However, it can be a disadvantage if the new combinations are "errors"—states that do not appear in the environment. The advantage of using subdivision for language acquisition through grammatical decomposition of sentences was discussed in Section 2.4.5.

In Section 2.5.3, the simulation of a network with four subdivisions illustrated generalization in a subdivided network. The strength of the inter-subnetwork synapses determines the number and kind of composite states that appear as memories. Moreover, the combinations that are recalled are preferentially those that share some substate combinations with the originally imprinted states. For example, even though the network is divided into four parts, a state that is composed equally of two of the imprinted states is more likely to be stable than other possible combinations. This is not the same, however, as a network with two subdivisions. When there are four subdivisions, the combinations of two imprinted states can occur in three distinct ways. If we consider the substate imprints to correspond to attributes (features) of the information, this implies that novel combinations of features may be recognized if the combinations are not completely different from the imprinted states.

This description of our ability to generalize raises basic questions about the objective of brain function. We should not consider neural network models solely as a model of memory. Traditional evaluations of an individual's ability, such as in exams, relied upon direct tests of memory. However, the central purpose of the brain is not to remember experiences, but rather to obtain from them knowledge that will serve in future circumstances. The memory of prior experience can serve in future circumstances when there are correlations between them. The purpose of the subdivided network is to abstract the essential aspects of an experience and the relationships between them, enabling this information to be used in future circumstances. In order to do so it is essential both to remember relevant information and to forget information specific to the particular circumstance. As discussed in Section 3.1, a significant role of sleep may be filtering memories, keeping the more persistent associations and forgetting associations that are specific to a particular circumstance. In this model the brain architecture is constructed so that the information to be forgotten largely consists of the associations between information stored in different subnetworks.

3.2.8 *Internal dialogue*

Through most of the twentieth century, behaviorism greatly influenced the field of psychology. Behaviorism attempted to describe all of human behavior in terms of reactions to a set of stimuli. However, it has become more generally accepted in recent years that descriptions of human behavior without invoking complex internal processes (cognition) cannot provide an understanding of more than a limited number of behavioral patterns. We can contrast the behaviorist approach with the concept of an internal dialogue that describes the ongoing language-based processes that occur in the brain without specific sensory stimulation or speech. One reason for modifying the behaviorist approach is the recent ability to measure neurological activity by means other than behavior. Tools for measuring this activity include positron emission tomography (PET) and magnetic resonance imaging (MRI). Even before these techniques, the behaviorist approach was not universally adopted. However, such imaging techniques provide a scientific basis and tools for investigating the internal processes. More fundamentally, adopting a model that includes an internal process, rather than a phenomenological behaviorist approach, is justified when it is easier to describe behavior using the internal process.

In this section we discuss some features of a neural network that are necessary for the existence of processes that are, at least in part, independent of the immediate sensory information. Such independence is not found in a feedforward network, where the input is progressively transmitted through stages to the output. It is also not realized in an attractor network, where the initial state of the whole network is fixed by input and the internal dynamics evolves the state to an attractor. These models are thus incomplete, because thought, and the internal functioning of the brain, is often largely independent of the immediate sensory input. People are able to think about a problem without regard to circumstances unless the circumstances become demanding upon their attention. We rely upon this when exams are given to students, since we do not generally consider most sensory information in the room as relevant to a student's performance, unless there are significant distractions. How is this independence realized in a neural network model?

A natural solution to the problem of introducing an internal dynamics is to assume that there are internal loops whose input is their own output. These loops interact with sensory input, and with motor output, but are not dictated by them. The flow of neural influence is illustrated in Fig. 3.2.2. We can associate the internal processes, for example, with internal dialogue. The existence of such loops leads to several concerns. The first would be that the system becomes stuck in a self-consistent loop. This is the dynamical analog of the excessively strong attractor that was discussed in the context of fixation. To avoid this problem requires some protections against repetition. For example, a neuronal refractory period that is longer than the

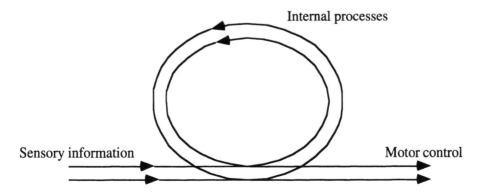

Figure 3.2.2 Internal processes such as internal dialogue can continue largely independent of sensory information. This requires internal dynamical loops that receive input both from the senses as well as from their own recursive structures. The figure illustrates a simple feedforward system with a recursive internal loop. ∎

cycle time could protect against single-cycle repetition. However, it would not protect against double-cycle repetition. While persistent loops are to be avoided, it is also possible for an internal cycle to preserve information over time. This may provide a form of short-term memory. Among other capabilities, a short-term memory enables juxtaposing, at one time in the neural state, events that are separated in time.

A second major area of concern in a discussion of internal loops is the balance that is established between the influence of the senses on the internal processes and their independence. There are dangers associated both with excessive coupling or decoupling of the internal processes from the senses. The stereotype of the absent-minded professor may be a manifestation of a particular balance between connection and independence that might be realized in this model. This balance of connection and independence is realized through the strength of particular sets of synapses. As discussed in Section 3.1, such balances may be maintained through the processes that occur during different stages of sleep.

A related question raised by the model of Fig. 3.2.2 is the relative capacity of the information paths from the senses, as compared to the information paths cycling internally. Specifically, what fraction of the information present in the brain at any one time is a direct consequence of the sensory input? This should play an important role in our understanding of the qualitative behavior of the brain. Is it largely driven by the outside or is it largely internal? When a system is completely determined by the immediate sensory information, we would identify it as a reactive system. When the sensory information determines only a part of the internal state, we can talk about the external and internal worlds as they are manifest in the state of the network. The problem then becomes to identify the relative complexity and interactions between the external and internal worlds. We will discuss the quantitative characterization of complexity in Chapter 8.

3.2.9 *Imagination, creativity and error*

There are two forms of creativity that are often discussed separately. The first is the general human ability to create new sentences, or to respond to circumstances that have not been experienced before. The second form of creativity is considered to be rare and is associated with particularly "creative" individuals. In this section we discuss the first more general creativity. The second is related to the first, but also requires an understanding of individual differences, and therefore it will be discussed in the next section.

The term "create" implies an act after which something exists that was not present before. However, acts of creation involve bringing together elements that were previously in existence, but juxtaposed in new ways. The elements may be objects, attributes or relationships. In order to create there must be an external manifestation— the act of creation. However, a precursor to the act of creation is imagination. The ability to imagine is the ability to represent internally a combination of elements that was not previously experienced. Creativity thus involves both imagination and implementation. Imagination requires a partial independence of internal representations from the external world. Otherwise, the internal state of the brain would only reflect the external reality and there would be no imagining or creativity. The previous section on internal dialogue described how such independence can be implemented. Here we focus on the problem of generating internal representations that differ from imprinted information.

The ability to imagine novel combinations of elements is implicit in the subdivided network we have been investigating. The network can generate a set of stable composite states that are, in effect, untrue memories. Assuming that the internal neural dynamics described in Section 3.2.8 explores various possible states of the network, these composite states appear from time to time as "imagined" possible combinations of partial states that were not imprinted. For example, having seen a bird in flight and a walking human, one might imagine a composite consisting of a flying human.

The extent to which composite patterns appear is controlled by the relative strength of inter-subnetwork synapses and intra-subnetwork synapses (the parameter g) discussed in Chapter 2. The progressive decomposition of memories during sleep, discussed in Section 3.1, suggests that sleep is also intimately related to the emergence of composite patterns. Composite patterns would appear first during sleep in the form of dreams, most of which would not be remembered. The partially subdivided network reflects both the concepts of divergent and convergent thinking. Divergent thinking is the ability to imagine new combinations. Convergent thinking is reflected in the inter-subnetwork synapses that limit them.

One question that might be asked is: How does the network distinguish between imagined states and real memories? A possible answer may be found in the relative strength of their basins of attraction. It can be shown that the basin of attraction of composite states is smaller than that of imprinted states. This may enable the network to distinguish them using a strategy similar to that described in the section on recognition. However, it is also apparent that some degree of confusion may arise. Isolated occurrences would result in false memories. In extreme cases, this confusion may give rise to functional disorders. This is consistent with the existence of a variety of psychological disorders involving hallucination. Thus the possibility of hallucination is rooted in the basic nature of the network architecture that enables imagination to occur.

Another consequence of the model of imagination is a trade-off between memory and creativity. In order for new composite states to be formed, the strength of associations between subdivisions must be reduced. The relationship between the elements that were originally imprinted tends to be lost. The trade-off between storage of more composite states and more imprinted states discussed in Section 2.4 appears here as a trade-off between memory and imagination, or even memory and creativity. Thus, for example, the ability to combine words into new sentences also requires a forgetting of sentences that were heard or spoken before. Memory requires maintaining the associations, while creativity requires loss of associations so that novel combinations can be imagined.

We should also make a connection between creativity and error. Even the most basic form of creativity—the application of prior experience to new circumstances—requires the possibility of error. More generally, in any act of creation there must be a possibility for error. An error can be defined as a creation that is not consistent with the external world.

The possibility of error implies the importance of limiting creativity. To manifest all possible combinations of elements, while in some sense creative, would not be effective. Creativity is only effective when the many possible combinations are limited to those that are more likely to be correct. A partially subdivided network appears to be an effective approach. It limits the number and type of composite states. Limiting creativity in this way reduces the probability of errors; however, it does not eliminate them.

The interdependence of creativity and error, two characteristics of human activity, should not be considered a limitation of our neural network model; it appears instead as a fundamental relationship. This relationship should persist despite improvements in the modeling and understanding of brain function.

Using the picture we have developed for creativity and error, we are able to begin to describe individual differences. The degree to which subdivisions of the network are isolated—the parameter g—can describe a one-parameter variability between individuals. Individuals who have a smaller value of g will be more forgetful, more creative and more prone to error. Individuals with a larger value of g will retain more memories, be less creative and less prone to error. This prediction could be tested by psychofunctional tests of a group of individuals. Here we can consider allegorical evidence from conventional stereotypes of various professions. The conventional stereotype of the most creative profession—artists—as also the least practical, can be contrasted with professions requiring few errors, such as accounting. Since the consequences of error are diminished, the arts would be expected to attract more creative individuals (lower g), with weaker memories and higher susceptibility to error. On the other hand, individuals with lower levels of creativity (higher g) and greater memory retention would be expected to be more successful in professions where consequences for error are higher.

The preceding paragraph begins to identify distinctions between individuals; however, this is only a small step toward understanding individuality or, more specifically, the second form of creativity that is attributed to specific individuals. The artistic creativity of Picasso is typically considered to be a completely different phenomenon from the commonplace ability to form new sentences. Nevertheless, it is possible to suggest they are quite similar. To do so, however, requires us to go further into an understanding of the subdivided network architecture and the source of individuality.

Question 3.2.3 Why didn't we consider spurious states introduced in Section 2.2.7 as a source of imagination/creativity?

Solution 3.2.3 Spurious states, like composite states, are formed from combinations of imprinted states. Spurious states, however, do not generally retain identifiable aspects of the original states. This is because they are formed by combining individual neuron activities from each of the imprinted patterns, rather than neurons associated with a particular attribute. The only structure imposed upon spurious patterns is by virtue of their overlap with imprinted states. Spurious states may be stable states of the network, and therefore may be imagined. However, unlike composite states, in general they will not have a coherent interpretation. ∎

3.2.10 *Individuality*

The design of modern computers relies upon a set of models that perform all computational tasks (Section 1.9). Despite various architectural differences, there is a uniformity of function. Often the objective of new models of computation, including neural network models, is to demonstrate that they have sufficient capability to be classified with computers—they are capable of universal computation. We have argued already in Section 1.3 that one of the essential characteristics of complex systems is the distinction between different realizations of the same architecture. Consistent with this, the subdivided neural network suggests an entirely different approach to computation based on a nonuniversal computation strategy. This nonuniversal strategy is the subject of this section and forms a basis for understanding human individuality.

Before proceeding, we mention that different computers, or a computer at different times operating on different information, behaves in different ways. We might suppose that this would allow us to use the universal computation approach to account for individual differences. However, one aspect of the concept of universal computation is that the basic capabilities are universal even though the particular data and the particular hardware are not. For example, certain problems that are inherently difficult for one computer running one computer program will also be inherently difficult for any other computer running any other program. There are various assumptions inherent in this statement, and it would be more correct to formulate it in terms of computational complexity classes. However, in the case of the human architecture, it appears that the capabilities are fundamentally different between different realizations of the architecture.

The reason for nonuniversality is rooted in the original motivation for subdivision—correlations and independence in information. Our environment manifests correlations that are nonuniversal. Tree leaves could be any color, or could be colored at random. Objects need not retain their shape over time. Subdivision exists because of the correlations in the information that is presented to the individual by the external world. By structuring the information internally in a way that is compatible with the structure of the external information, the subdivided architecture is designed to accommodate to it, or take advantage of it. However, from a computation theory point of view, there is no reason for the information to be structured in a particular way.

Once again our simplest example, the left-right universe (Section 2.4), is helpful. We contrasted the capabilities of a network divided right from left and the network divided top from bottom. These networks had radically different capabilities in the left-right universe. This demonstrates in a simple way how the capabilities of distinct individual realizations of the same architecture may vary drastically.

The inherent nonuniversality of the architecture of subdivision is modified by the effect of selection due to fitness, which can lead to commonality between individuals. Thus, for organisms in the left-right universe we would expect to find only left-right subdivided networks and no top-bottom subdivided networks. Similar commonalities should also be expected among people. Thus the variability in brain architecture and the resulting variation in capabilities is limited to the degree that selection imposes the architecture as a result of evolutionary processes (Chapter 6). Thus we have argued that there can be an environmental/evolutionary pressure toward commonality in brain architecture because of a commonality in the environment of different individuals. However, this commonality is limited to the actual impact of selective forces.

The nonuniversality of the subdivided network becomes clearer when we think about the hierarchical structure motivated by the 7±2 rule and the large variety of possible mappings of sensory and motor information onto this structure. Consider the many different filters of information that might be useful under different circumstances. It is possible for a single individual to have many of them and to selectively use them. In the extreme case we can ask: If there are many possible filters of information that might be useful, why doesn't each individual make use of all of them? The first answer to this is that the number of such mappings grows exponentially with the amount of information, so it would be impossible to contain them in a single realization of the architecture.

We also recall that much of the usefulness of subdivision is lost when the number of subdivisions becomes greater than seven. In a hierarchy, we can use more than seven distinct filters; however, choosing how to arrange them matters. The strongest associations are maintained between information that is connected at the lowest level of the hierarchy. Progressively weaker associations exist between subdivisions that are connected at higher levels of the hierarchy. Depending on how the filtered information is mapped onto the subdivisions, an individual will retain distinct associations leading to a wide variety of possible individual differences.

Using the individualized hierarchically subdivided architecture as a model of the brain we can return to a consideration of imagination, creativity and memory. The functional hierarchy corresponds to a nonunique selection of attributes distributed in a tree of stronger and weaker interconnection. This nonuniqueness suggests that different individuals will remember different associations and also be creative in different ways. For example, some individuals will find it easy to remember the association of names and faces while others will not. Those who remember these associations have these attributes strongly connected to each other. In this model, generic capabilities of an individual are directly related to the organization of information within the architecture of the brain.

Our conclusion is that unlike modern computation theory, the subdivided architecture of the human brain is a nonuniversal architecture whose individual realizations have widely different task-dependent capabilities. We also surmise that a universal strategy may not be effective at many human information-processing tasks. The nonuniversal architecture is consistent with the uniqueness of individuals.

3.2.11 *Nature versus nurture*

A central controversy in modern science revolves around the relative importance of the genetic code as compared to environmental influence in determining human behavior and various aspects of brain function and human information-processing. This is often called the nature versus nurture controversy. The model of the brain formed from a hierarchy of functional subdivision also provides us with a model for the relative influence of nature and nurture.

To extend the model for individuality to a first model of the influence of nature and nurture we need only suggest that the subdivided architecture itself is genetically controlled. On the other hand, the information that is imprinted upon it is a direct result of the environment. Thus, the aspects of information that are retained by a particular individual are guided by genetics. Genetics controls the type of associations that are strongly retained, and those that are readily forgotten. The environmental influence is contained in the actual associations and information that is present. This model exhibits a complementary influence of genetics and environment on an individual. It shows explicitly how genetics influences potential qualities of an individual, and how the environment influences the actual qualities.

It is important to emphasize that while this picture is appealing in its simplicity, it must be considered only as a first approximation. Aspects of the subdivided architecture are susceptible to environmental influence. For example, even the development of basic interconnections of the visual system are influenced by exposure to light. A limited amount of specific information may also be built in due to genetic programming. These include instinctive behaviors that are more prevalent in animals.

3.2.12 *Consciousness and self-awareness*

Of all the traits associated with human beings, the ability to be self-aware, and the related concept of consciousness present some of the most difficult philosophical dilemmas. The practical implications of these dilemmas are related to the concepts of free will and determinism and the related responsibility of an individual for action. We separate consciousness from the problem of selective awareness ("I am conscious of …") which is described in the following section. In this section, after discussing some conceptual obstacles, we will construct a neural network model of consciousness. We then discuss practical reasons for its existence, a test of the model's ability to recognize self, and modes of failure of the model which can be compared with psychofunctional failure.

Conceptually, a paradox in considering the problem of consciousness arises from the problem of recursive signal processing. Consider, for a moment, the problem of consciousness as that of being aware of sensory input, and consider the neural network as a form of information processor. An example is the sensory processing that is performed by the neurons that receive and process visual information. There is no indication that this provides any awareness. It is simply a mapping of sensory information into another form. This information is transferred to the input of another neural system. The second system is now handed the job of providing the awareness. However, no process that transforms the sensory information further would do anything qualitatively different. Thus we are perpetually left to the problem of deferring the consciousness to later, more internal stages, without resolution. Some have argued that the ultimate recursion must lead to something that is unphysical and therefore outside the domain of scientific inquiry.

The problem with this model of the process of consciousness is that is places the consciousness in the wrong place—as the primary recipient and interpreter of sensory information. It also uses only a limited view of the function of a neural network. We will address both from a different perspective.

We begin by considering a model that differentiates in an essential way between the conscious and subconscious mind. There is generally no disagreement that such a differentiation should be made, since casual introspection shows that the conscious mind is not aware of—does not contain—all of the internal processes taking place in the brain. Indeed, it is aware of very limited aspects of these processes. Thus we begin by constructing a subconscious part of the brain. The responsibility of the subconscious region of the brain is to receive sensory input and to act upon it. This may seem strange at first sight; isn't the conscious mind necessary? The answer is, largely, no. Habitual acts and most of the details of daily activity are performed directly without apparent input from the conscious part of the brain. The easiest way for us to represent the subconscious brain is as a conventional feedforward network that takes the sensory input and determines motor control based on this sensory input. Thus far we have done nothing at all unusual except to claim that this model does not possess consciousness, which we knew from the outset.

Now we would like to construct consciousness. We do this from a pragmatic point of view by asking, What is the information that the conscious mind possesses? Introspection suggests that the conscious mind possesses sensory information. It also possesses knowledge of motor activity. However, it does not possess information about the internal processes that lead from sensory to motor activity. This suggests that we construct a new part of the network model that represents both the sensory information and motor activity information, but not the intermediate stages. The next question to ask is, What does the conscious mind control? The first answer is that it has no primary control function. By this we mean that control is not continuous in the same way that it is for the feedforward network. We can see this from the terminology—the awareness or consciousness do not convey the meaning of action. They are rather passive terms describing the possession of information. No action is required on the basis of this information.

There is, however, a secondary control function. The awareness is capable of exercising control over the motor activity. However, this control is circumscribed. It acts as a corrective process rather than a control over moment-by-moment action. Thus the direct control over action is performed by the subconscious network, while the conscious network acts by redirecting the subconscious feedforward network.

How does the conscious network decide to exercise control over actions of the subconscious network? We answer by considering the conscious mind as an attractor network (Fig. 3.2.3). The pattern of neural activity in the attractor network represents both sensory and motor activity. It's task is to recognize their "compatibility." As discussed in Section 3.2.6, recognition can be performed by measuring the dynamics of the attractor network. If the juxtaposition of the sensory and motor activity is recog-

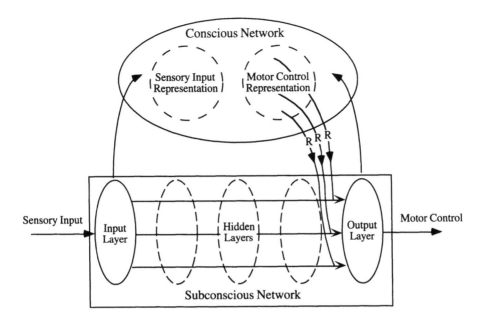

Figure 3.2.3 A model that captures some of the essential features of self-awareness and consciousness can be constructed out of two parts. The first part, representing the subconscious mind, is a feedforward network that is a sensory motor control system. The input is sensory information and the output is motor control. The second part, representing the conscious mind, is an attractor network whose state is composed of input both from the sensory information and from the motor control. It has no direct control function. However, when the sensory and motor information is not recognized as an imprinted state the network exercises control over the actions through recognition synapses similar to those discussed in Section 3.2.6 and Fig. 3.2.1. In effect, the imprinted states in the conscious network represent a model of the self. When the actions are not consistent with the model it intervenes to change the behavior. The network function is illustrated schematically in Fig. 3.2.4. ■

nized, then the conscious mind does not interfere. However, when the state of the sensory and motor activities are not recognized, then the conscious mind acts by causing the feedforward network to modify its actions. This occurs over a longer time scale than direct action by the subconscious network.

An interesting way to summarize the recognition process that the attractor network performs is as a question. The question, in this case, is: Is this me? Or specifically: Is the current situation and my actions within it consistent with my self-image? The self-image is the set of stable states of the attractor network. Summarized in this way, we see how the notion of self-awareness and consciousness are related and are represented by this model.

An additional concept that can be described is the concept of a will. It is easiest to identify the will by noting the use of the modifiers that describe an individual as having a strong or weak will. The will represents the ability of the conscious part of the mind to control the subconscious. In this model this is represented by the strength of the synapses that originate in the conscious attractor network and act to modify the state of activity of the subconscious feedforward network.

The architecture we have constructed, shown in Fig. 3.2.3, is a comparatively simple model that captures some of the features that we attribute to the function of the conscious and subconscious parts of the mind as well as the interactions between them. The interactions are described pictorially in Fig. 3.2.4. Given the conflicts that have arisen out of the concept of consciousness, the possibility of discussing a concrete model provides some new opportunities for progress in our understanding.

The practicality of consciousness can be considered by realizing that the combination of feedforward and attractor networks provides a solution to the limitations of each of these network architectures. As discussed in Sections 2.3 and 3.2.2, the training of a feedforward network requires storage of the desired input-output pairs. In our model of consciousness, this storage is performed using the attractor network. Moreover, the training of the feedforward network must be done incrementally, while that of the attractor network may be achieved by a single imprint. On the other hand, the attractor network is not capable of complex processing and will suffer overload if too many patterns are stored in it. In this model, complex processing can be left to the feedforward network, and once it is trained, the size of the basin of attraction of the attractor network pattern can decrease. Significantly, it is not necessary for the attractor network by itself to be able to generate the pattern representing a response, it must only verify and provide corrections to this response.

Consciousness is characterized by the recognition of oneself. Experimentally it is manifest in the ability to recognize oneself in a mirror. This ability is not present for animals other than apes and man. Even monkeys appear unable to recognize themselves. The neural network model indicates how self-recognition can occur. By virtue of juxtaposing sensory and motor information in an associative memory (attractor network), it enables correlations between them to be imprinted. When moving and seeing this motion in a mirror, the imprinted information is recognized by the conscious network, and thus the answer to the question, Is this me? is yes.

The physiological location of the conscious and subconscious networks can be tentatively identified. The frontal lobes that are much more developed in apes and man than in other animals have generally been identified with consciousness and planning. They also have a topographic map of the body that serves as an area of motor control. However, the motor control due to the frontal lobes has been associated with voluntary rather than habitual motion. Involuntary movement and the coordination of voluntary and involuntary movement are both centered in the cerebellum,

Figure 3.2.4 Schematic illustration of the model of consciousness described in the text. This model assumes two components of the mind—the conscious mind and the subconscious mind. The activity and interactions of the two components are illustrated in the figure as a time sequence. The subconscious mind is directly responsible for receiving sensory input and acting upon it. The conscious network corrects the subconscious based on an internal representation of the self, and triggers retraining of the subconscious network. ∎

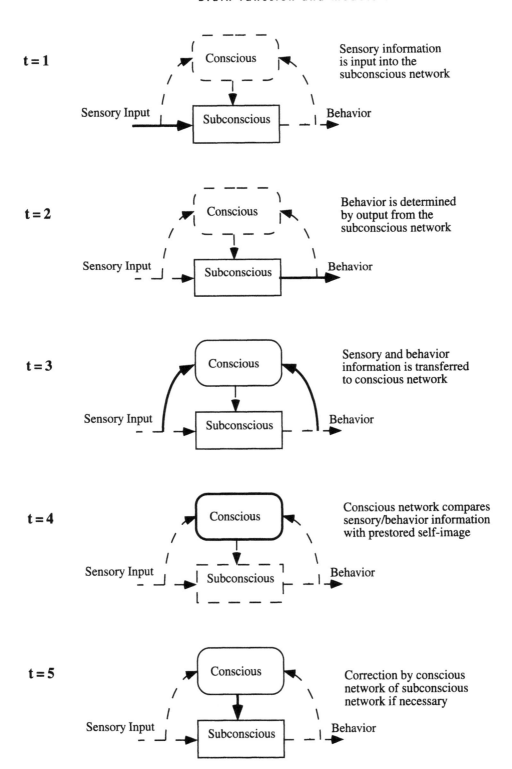

to which there are many projections from the motor areas of the frontal lobes. These observations are consistent with the neural network model.

While most animals do not recognize themselves in a mirror, they do have a sense of self associated with location/territory or smell. The part of the brain associated with representing information about spatial location is the hippocampus. This part of the brain may serve as the associative network that enables an identification of self related to location, as in: "I am here" or "This is my place." Recent experiments discussed in Section 3.1.5 identify the hippocampus as a network that stores information in correlated patterns of neural activity—an associative network. This also suggests that consciousness is not a monolithic entity; it may have various aspects related to different parts of the brain.

As with other aspects of the models we have discussed, the model of consciousness provides an understanding of some of the failure mechanisms of the network. One failure mechanism is found by considering the strength of the control by the conscious over the subconscious—the will. We see that the subconscious mind is essentially reactive. When the will is weak, the behavior would be characterized as impulsive. On the other hand, if the will is too strong, then the attractor network, which does not have the ability to process information through several layers of synapses, takes over the reactive function. This implies that actions are based upon relatively simple conscious processing. When all actions are based upon simple conscious processing, behavior is characterized as fanatic.

We can take the discussion one step further by discussing changes in the will. Similar to other synapses, the will is likely to be changed by imprinting. Thus it is strengthened by action and weakened by inaction. Since the exercise of the will is under conscious control, it may be strengthened by consciously exercising control even when the control is unnecessary, or it may be weakened by passivity when the control would otherwise be exercised.

In this section we have emphasized the limited control that the conscious mind exercises over action. However, the conscious mind appears to exercise direct control over what we are paying attention to, as discussed in the following section.

3.2.13 *Attention*

One of the phenomena associated with both internal dialogue and response to sensory stimuli is that of attention. We are able to be aware of various aspects of sensory information, or focus on a particular thought. How would we design a system that can achieve this? One approach is used by computers, where a central processing unit receives information from different parts of the memory according to its instructions. The central processing unit must label the information according to where it is taken from in the memory. This requires an addressing/labeling that distinguishes one part of the brain from another. We will discuss an alternate strategy that leaves the information in place but acts as a kind of spotlight. This approach is better suited to the neural network models we have been describing, because the nature of information is established by its location in the brain rather than by retrieval and labeling. We continue to avoid an explicit labeling scheme in this manner.

Until now we have been comfortable with the model that neurons are firing or not firing with roughly equal probability, and the pattern of activity or inactivity represents the information that is present in the mind at a particular time. We now need an additional intrinsic label that will enable us to identify which region of the brain has our attention. One way to achieve this is to give up the symmetry between activity and inactivity and assume that significantly more of the neurons are inactive; it is then the neurons with significant activity that are representing the information. This helps because we can then control the overall activity level in a particular region of the brain. If we raise the overall activity we are drawing attention to it, and if we reduce the level of activity we are reducing our attention to it. It is indeed well established that the neurons in the brain are active less than half of the time. Moreover, imaging experiments that are assumed to measure which parts of the brain are utilized at a particular time measure their average neural activity, which is higher than in other regions of the brain.

How does this change affect all of our previous analyses of the storage of patterns in attractor networks? The answer is that qualitatively very little changes. A pattern that is to be imprinted consists of a pattern of neural activity where the fraction of active ($s_i(t) = 1$) neurons is less than one-half. The imprinting rule may be modified slightly to prevent the bias itself from being imprinted. If the average activity of the neurons is consistently m, then the Hebbian imprinting (Eq. (2.2.6)) may be modified to read:

$$J_{ij}(t) = J_{ij}(t-1) + c(s_i(t-1) - m)(s_j(t-1) - m) \qquad i \neq j \qquad (3.2.6)$$

This means that imprinting results from deviations from the average activity. The network capacity as measured by the number of patterns that can be imprinted and retrieved actually goes up slightly, because, in effect, the patterns do not interfere with each other as much since they involve different sets of firing neurons. However, the overall amount of information that can be stored is diminished because each pattern does not contain as much information. For our purposes, these are minor adjustments to the results that we have already found in Chapter 2.

In order to make use of the bias in neural activity for the purpose of attention, there must be a mechanism by which the overall activity within a particular region of the brain is controlled. We will describe a mechanism for such control. The mechanism must be independent of the neural activity and synaptic transmission that we have been describing. Throughout the brain there are found cells whose function is not understood. These cells, called glial cells, may have some function in maintaining the structural integrity of the neural system. We will invest these cells with a model for how the attention system might work, recognizing that there are other possible embodiments. The essential property that we are using is that these cells are not part of the neural representation themselves. This role could also be taken by a separate set of neurons. The reason it appears natural that the cells involved would not actually be neurons is that they do not require specificity of interaction with a particular neuron. Instead they should interact more generally with a whole region of neurons. There are, however, some classes of neurons that do this as well.

In order to fulfill their function, the glial cells would need two capabilities: to control the overall activity of the cells and to measure their overall activity. Control over the activity might be achieved by control over the local blood flow supplying nutrients to a region of cells, or by control over the chemical contents of the blood. Alternatively, chemicals in the intercellular fluid might be involved. The purpose of these chemicals would be similar to that of neurotransmitters in that they inhibit or excite neuron activity; however, they are likely to be quite different in detail, since they have an effect on the overall level of activity rather than serving as one of many signals to a cell. Measuring the overall activity of a region of neurons may be achieved by sensing various by-products of their electrochemical activity. This measurement would take longer than the transmission of an individual pulse along an axon; however, the activity itself is an average over many transmission pulses.

The behavior of the glial cell then becomes similar to the behavior of a neuron, in the sense that it has either an ON or an OFF state. In the ON state it promotes the activity of the neuronal assembly it is in contact with; in the OFF state it suppresses it. It acts as a metaneuron that is related to the average neuronal activity. Control over the glial cell may then be exercised in several ways. For example, a self-consistent attention mechanism could be formed by glial cells attempting to activate the assembly of neurons they are in contact with whenever the cells are significantly active. The glial cell measures the activity present with respect to the expected activity. If the glial cell is OFF, the neurons would be generally inactive. If the cells become significantly more active than expected, the glial cell turns ON and promotes the activity of the region of cells. If the glial cell is ON and the activity falls below that expected, then the glial cell turns OFF. Interactions between glial cells that suppress one glial cell when another is active would lead to an exclusive attention mechanism.

An important part of the phenomenon of attention is that it is coupled to consciousness. We can implement this coupling by assuming that the conscious part of the brain, discussed in the previous section, controls the glial cells and thus the regional neural activity. We emphasize again that the term "glial cell" as used here might be substituted by another biological analog without changing the essence of this discussion. Moreover, it should be clear that the mechanism we have described for attention is not the only approach. It is one of the ways that are consistent with the spirit of the neural network models we have been developing.

One of the interesting outcomes of this model of attention is that it provides a mechanism for a new level of dynamics that would be related to a sequential activation of glial cells causing a sequential activation of particular regions of neurons. This can provide a missing piece in our discussion of language in the subdivided architecture. In Chapter 2 we suggested that different subdivisions of the brain are responsible for storage of distinct parts of speech. We can now suggest that a sequential firing of glial cells results in sequential activation of different parts of speech that trigger verbalized speech, are triggered by hearing speech, or represent internal dialogue in the form of organized strings of words—sentences.

3.2.14 *Summary of brain function*

We have devoted this chapter to building a relationship between our neural network models and several of the basic phenomena of brain and mind. The relationships have not only described some of the interesting phenomena, but also created a framework in which poorly understood concepts can at least be discussed. These include such diverse concepts as sleep, creativity and consciousness. To incorporate these into our model, we expanded the basic neural network to include various additional features. The comparison of many of these with the actual brain has yet to be performed. However, it is helpful to have theories that can be tested both experimentally and through simulations.

4

Protein Folding I:
Size Scaling of Time

Conceptual Outline

❚ 4.1 ❚ The simplest question about dynamics—how long does a process take?—becomes particularly relevant when the time may be so long that the process cannot happen at all. A fundamental problem associated with the dynamics of protein folding is understanding how a system of many interacting elements can reach a desired structure in a reasonable time. In this chapter, we discuss the parallel-processing idea for resolving this problem; kinetic pathways will be considered in the next chapter. Parallel processing and interdependence are at odds and must be balanced in the design of complex systems.

❚ 4.2 ❚ We use finite-size Ising type models to explore the nature of interactions that can allow a system to relax in a time that grows less than exponentially in the size of the system. These models illustrate various ways to realize the parallel-processing idea.

❚ 4.3 ❚ The simplest idealization of parallel processing is the case of completely independent spins. We discuss a two-spin model as a first example of how such a system relaxes.

❚ 4.4 ❚ Various homogeneous models illustrate some of the properties that enable systems to relax in a time that grows no more than a power law in the system size. These include ideal parallel processing, and nucleation and growth of a stable state from a metastable state. The models also illustrate cases where exponential growth in the relaxation time can prevent systems from relaxing.

❚ 4.5 ❚ Inhomogeneous models extend the range of possibilities for interaction architectures that still allow a reasonable relaxation time. Among these are space and time partitioning and preselected initial conditions. However, inhomogeneous long-range interactions generally lead to an exponential growth of relaxation time with system size.

4.1   The Protein-Folding Problem

One of the simplest questions we can ask about the dynamics of a complex system is, How long does a process take? In some cases this question presumes that we have an understanding of the initial and final state of the process. In other cases we are looking for a characteristic time scale of dynamic change. For a complex system, a particular process may not occur in any reasonable amount of time. The time that a dynamic process takes is of central importance when a system has an identifiable function or purpose. We will consider this in the context of proteins, for which this question is a fundamental issue in understanding molecular function in biological cells.

We begin by describing the structure of proteins, starting from their "primary structure." Proteins are molecules formed out of long chains of, typically, twenty different kinds of amino acids. Amino acids can exist as separate molecules in water, but are constructed so that they can be covalently bonded in a linear chain by removal of one water molecule per bond (Fig. 4.1.1). In general, molecules formed as long chains of molecular units are called polymers. Proteins, RNA and DNA, as well as other types of biological molecules (e.g., polysaccharides) are polymers. In biological cells, proteins are formed in a linear chain by transcription from RNA templates that are themselves transcribed from DNA. The sequence of amino acids forming the protein is called its primary structure (Fig. 4.1.2). The active form of proteins (more specifically

Figure 4.1.1 Illustration of the atomic composition of an amino acid. The usual notation for carbon (C), oxygen (O), nitrogen (N) and hydrogen (H) is used. R stands for a radical that is generally a hydrocarbon chain and may contain hydrocarbon rings. It is different for each of the distinct amino acids, and is the difference between them. The bottom figure is a chain of amino acids formed by removing a single water molecule and bonding one nitrogen to the carbon of the next amino acid. The sequence of amino acids is the primary structure of the protein (see Fig. 4.1.2). ∎

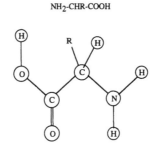

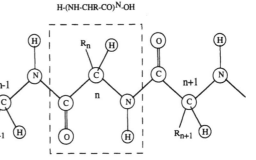

M R L N P G G Q Q Q A V E F V T G P C L V L A G A G S G K T R –
V I T N K I A H L I R G C G Y Q A R H I A A V T F T N K A A –
R E M K E R V G Q T L G R K E A R G L M I S T F H T L G L D –
I I K R E Y A A L G M K A N F S L F D D T D Q L A L L K E L –
T E F L I E D D K V L L Q Q L I S T I S N W K N D L K T P S –
Q A A A S A I G E R D R I F A H C Y G L Y D A H L K A C N V –
L D F D D L I L K P T L L L Q A N E E V R K R W Q N K I R Y –
L L V D E Y Q D T M T S Q Y E L V K L L V G S R A R F T V V –
G D D D Q S I Y S W R G A R P Q N L V L L S Q D F P A L K V –
I K L E Q N Y R S S G R I L K A A N I L I A N N P H V F E K –
R L F S E L G Y G A E L K V L S A N N E E H E A E R V T G E –
L I A H H F V N K T Q Y K D Y A I L Y R G N H Q S R V F E K –
F L M Q N R I P Y L O S G G T S F F S R P E I K D L L A Y L –
R V L T N P D D D S A F L R I V N T P K R E I G P A T L K K –
L G E W A M T R N L S M F T A S F D M G L S Q T L S G R G Y –
E A L T R F T H W L A E I Q R L A E R E P I A A V R D L I H –
G M D Y E S W L Y E T S P S P K A A E M R M K N V N Q L F S –
W M T E M L E G S E L D E P M T L T Q V V T R F T L R D M M –
E R G E S E E E L D Q V Q L M T L H A S K G L E F P Y V Y M –
V G M E E G F L P J Q S S I D E D N I D E E R R L A Y V G I –
T R A Q K E L T F T L C K E R R Q Y G E L V R P E P S R F L –
L E L P Q D D L I W E Q E R K V V S A E E R M Q K G Q S H L –
A N L K A M M A A L R G K

Common Amino Acids

Name	Notation	Name	Notation
Glycine	(gly, G)	Cysteine	(cys, C)
Alanine	(ala, A)	Methionine	(met, M)
Valine	(val, V)	Asparagine	(asn, N)
Leucine	(leu, L)	Glutamine	(gln, Q)
Isoleucine	(ile, I)	Aspartic acid	(asp, D)
Phenylalanine	(phe, F)	Glutamic acid	(glu, E)
Tyrosine	(tyr, Y)	Lysine	(lys, K)
Tryptophan	(trp, W)	Arginine	(arg, R)
Serine	(ser, S)	Histidine	(his, H)
Threonine	(thr, T)	Proline	(pro, P)

Figure 4.1.2 Amino acid sequence of the protein acetylcholinesterase — its primary structure. A list of common amino acids and their commonly used three-letter and one-letter notation is attached. ∎

Figure 4.1.3 Three-dimensional structure of the protein acetylcholinesterase. The top picture is constructed using space-filling balls that schematically portray the electron density of each atom. The bottom illustration is a simplified version showing only the backbone of the protein. Helical segments (α-helices) and regions of parallel chains (β-sheets) are visible. They are illustrated as ribbons to distinguish them from the connecting regions of the chain (turns). The α-helices, β-sheets and turns constitute the secondary structure of the protein. (Rendered on a Macintosh using RasMol [developed by Roger Sayle] and a National Institutes of Health protein databank (PDB) file) ∎

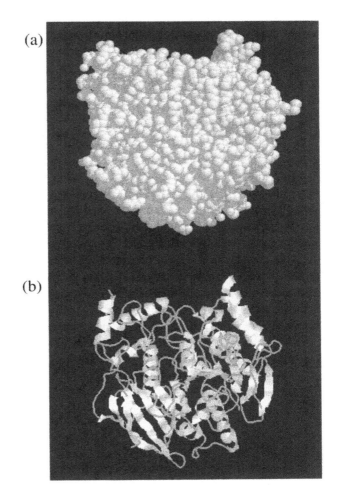

(a)

(b)

globular proteins) is, however, a tightly bound three-dimensional (3-d) structure (Fig. 4.1.3) with active sites on the surface. The active sites serve enzymatic roles, controlling chemical reactions in the cell. The transformation of the linear protein chain to the enzymatically active 3-d structure is known as protein folding. The 3-d structure arises because of additional bonding between the amino acids of the chain. These bonds are characteristically weaker than the covalent bonds along the chain. They include hydrogen bonds, van der Waals bonds and a few covalent sulfur-sulfur (disulfide) bonds. The relative weakness of the bonds responsible for the 3-d structure makes the distinction between the primary and 3-d structure meaningful.

The 3-d structure of proteins can be further analyzed in terms of secondary, tertiary and, sometimes, quaternary structure. These describe levels of spatial organization between individual amino acids and the complete 3-d structure. A plot of the protein chain backbone in space (Fig. 4.1.3 (b)) generally reveals two kinds of amino acid bonding structures known as α-helix and β-sheet. The α-helix consists of a single-chain helix, where each amino acid forms a hydrogen bond to the fourth amino acid along the chain. Each hydrogen bond attaches the N-H end of one amino acid with the C-OH end of another, resulting in 3.6 amino acids per helix turn. In this structure all such hydrogen bonds are formed, except at the ends of the helix. Thus, from the point of view of the primary chain (and without consideration of the radicals that distinguish different amino acids), this is a low-energy structure. There is a second natural way to provide hydrogen bonding. Placing two chains, or two segments of the same chain, parallel or antiparallel to each other allows a chain of hydrogen bonds. This can be extended on both sides by adding chains in a two-dimensional fashion to form a planar structure that provides complete hydrogen bonds everywhere, except at the edges. This is the β-sheet arrangement. In addition to the α-helix and β-sheet structures there are also segments of the protein, called turns, that connect different α-helix and β-sheet structures. The number of amino acids along a single α-helix typically ranges between ten and twenty-five (three to seven turns), and the number in a single strand of a β-sheet is less, only five to ten. The total number of amino acids in a region of β-sheet can be as high as fifty, divided into three to eight strands. The 3-d structure of a protein described in terms of segments of α-helix and β-sheet is known as the secondary structure of the protein. The number of different secondary-structure elements in a protein ranges from a few up to, possibly, fifty. When there are many secondary structural elements they are further grouped into intermediate structural elements. The complete 3-d structure of an individual amino acid chain is known as its tertiary structure. Several chains may be combined together to form a larger molecular aggregate that constitutes a functioning enzyme. The collective structure of the chains is the enzyme's quaternary structure. This describes the hierarchically subdivided structure of a protein. The number of components at each level of hierarchy is consistent with the generalized 7±2 rule discussed in Chapter 2. This rule is expected to apply to proteins or other complex systems that cannot be subdivided or modified locally without significant change in their global properties.

Protein folding is the transformation of a linear protein chain to the 3-d structure. The problem of understanding protein folding has achieved a separate existence from the problem of describing protein function in the cell. Many proteins can be unfolded and refolded reversibly in a test tube (*in vitro*) separate from other molecules that might otherwise be involved in the protein folding in the cell (*in vivo*). Various additives to the solution cause the protein to unfold or refold. Protein folding has attained a central significance in the effort to understand the molecular biology of the cell, because it is a key to understanding how the linear DNA code is converted into cellular function—as implemented by active enzymes. The 3-d structure of the protein is the form in which they perform enzymatic tasks.

Protein folding is an unsolved problem. What form will the solution of this problem take? One prospect is that it will be possible to predict the 3-d structure from a specified amino-acid sequence. The process of prediction may result from a complete set of rules that describe how particular sequences fold. Alternatively, the prediction may require a large-scale computer simulation of the dynamical process of folding.

Most researchers studying protein folding are concerned with determining or predicting the 3-d structure without describing the dynamics. Our concern is with the dynamics in a generalized context that applies to many complex systems.

From early on in the discussion of the protein-folding problem, it has been possible to separate from the explicit protein-folding problem an implicit problem that begs for a fundamental resolution. How, in principle, can protein folding occur? Consider a system composed of elements, where each element may be found in any of several states. A complete specification of the state of all the elements describes the conformation of the system. The number of possible conformations of the system grows exponentially with the number of elements. We require the system to reach a unique conformation—the folded structure. We may presume for now that the folded structure is the lowest energy conformation of the system. The amount of time necessary for the system to explore all possible conformations to find the lowest-energy one grows exponentially with system size. As discussed in the following paragraphs, this is impossible. Therefore we ask, How does a protein know where to go in the space of conformations to reach the folded structure?

We can adopt some very rough approximations to estimate how much time it would take for a system to explore all possible conformations, when the number of conformations grows exponentially with system size. Let us assume that there are 2^N conformations, where N is the size of the system—e.g., the number of amino acids in a protein. Assume further that the system spends only one atomic oscillation time in each conformation before moving on to the next one. This is a low estimate, so our result will be a reasonable lower bound on the exploration time. An atomic oscillation time in a material is approximately 10^{-12} sec. We should increase this by at least an order of magnitude, because we are talking about a whole amino acid moving rather than a single atom. Our conclusions, however, won't be sensitive to this distinction. The time to relax would be $2^N 10^{-12}$ sec, if we assume optimistically that each possible state is visited exactly once before the right arrangement is found.

A protein folds in, of order, 1 second. For conformation space exploration to work, we would have to restrict the number of amino acids to be smaller than that given by the equation:

$$2^N 10^{-12} \text{ sec} = 1 \text{ sec} \qquad (4.1.1)$$

or $N = 40$. Real proteins are formed from chains that typically have 100 to 1000 amino acids. Even if we were to just double our limit from 40 to 80 amino acids, we would have a conformation exploration time of 10^{12} seconds or 32,000 years. The many orders of magnitude that separate a reasonable result from this simple estimate suggests that there must be something fundamentally wrong with our way of thinking about the problem as an exploration of possible conformations. Figuring out what is a reasonable picture, and providing justification for it, is the fundamental protein-folding problem.

The fundamental protein-folding problem applies to other complex systems as well. A complex system always has a large set of possible conformations. The dynamics of a complex system takes it from one type of conformation to another type of conformation. By the argument presented above, the dynamics cannot explore all possible conformations in order to reach the final conformation. This applies to the dynamics of self-organization, adaptation or function. We can consider neural networks (Chapters 2 and 3) as a second example. Three relevant dynamic processes are the dynamics by which the neural network is formed during physiological development, the dynamics by which it adapts (is trained, learns) and the dynamics by which it responds to external information. All of these cause the neural network to attain one of a small set of conformations, selected from all of the possible conformations of the system. This implies that it does not explore all alternatives before realizing its final form. Similar constraints apply to the dynamics of other complex systems.

Because the fundamental protein-folding problem exists on a very general level, it is reasonable to look at generic models to identify where a solution might exist. Two concepts have been articulated as responsible for the success of biological protein folding—parallel processing and kinetic pathways. The concept of parallel processing suggests, quite reasonably, that more than one process of exploration may be done at once. This can occur if and only if the processes are in some sense independent. If parallel processing works then, naively speaking, each amino acid can do its own exploration and the process will take very little time. In contrast to this picture, the idea of kinetic pathways suggests that a protein starts from a class of conformations that naturally falls down in energy directly toward the folded structure. There are large barriers to other conformations and there is no complete phase space exploration. In this picture there is no need for the folded structure to be the lowest energy conformation—it just has to be the lowest among the accessible conformations. One way to envisage this is as water flowing through a riverbed, confined by river banks, rather than exploring all possible routes to the sea.

Our objective is to add to these ideas some concrete analysis of simple models that provide an understanding of how parallel processing and kinetic pathways may work. In this chapter we discuss the concept of parallel processing, or independent relaxation, by developing a series of simple models. Section 4.2 describes the approximations that will be used. Section 4.3 describes a decoupled two-variable model. The main discussion is divided into homogeneous models in Section 4.4 and inhomogeneous models in Section 4.5. In the next chapter we discuss the kinetic aspects of polymer collapse from an expanded to a compact structure as a first test of how kinetics may play a role in protein folding. It is to be expected that the evolving biology of organisms will take advantage of all possible "tricks" that enable proteins to fold in acceptable time. Therefore it is likely that both parallel processing and kinetic effects do play a role. By understanding the possible generic scenarios that enable rapid folding, we are likely to gain insight into the mechanisms that are actually used.

As we discuss various models of parallel processing we should keep in mind that we are not concerned with arbitrary physical systems, but rather with complex systems. As discussed in Section 1.3, a complex system is indivisible, its parts are interdependent. In the case of proteins this means that the complete primary structure—the sequence of amino acids—is important in determining its 3-d structure. The 3-d structure is sometimes, but not always, affected by changing a single amino acid. It is likely to be affected by changing two of them. The resulting modifications of the 3-d structure are not localized at the position of the changed amino acids. Both the lack of effect of changing one amino acid, and the various effects of changing more amino acids suggest that the 3-d structure is determined by a strong coupling between the amino acids, rather than being solely a local effect. These observations should limit the applicability of parallel processing, because such a structural interdependence implies that the dynamics of the protein cannot be separated into completely independent parts. Thus, we recognize that the complexity of the system does not naturally lead to an assumption of parallel processing. It is this conflict of the desire to enable rapid dynamics through independence, with the need to promote interdependence, which makes the question of time scale interesting. There is a natural connection between this discussion and the discussion of substructure in Chapter 2. There we showed how functional interdependence arose from a balance between strong and weak interactions in a hierarchy of subsystems. This balance can also be relevant to the problem of achieving essentially parallel yet interdependent dynamics.

Before proceeding, we restate the formal protein-folding problem in a concrete fashion: the objective is to demonstrate that protein folding is consistent with a model where the basic scaling of the relaxation time is reduced from an exponential increase as a function of system size, to no more than a power-law increase. As can be readily verified, for 1000 amino acids, the relaxation time of a system where $\tau \sim N^z$ is not a fundamental problem when $z < 4$. Our discussion of various models in this chapter suggests a framework in which a detailed understanding of the parallel minimization of different coordinates can be further developed. Each model is analyzed to obtain the scaling of the dynamic relaxation (folding) time with the size of the system (chain length).

4.2   Introduction to the Models

We will study the time scale of relaxation dynamics of various model systems as conceptual prototypes of protein folding. Our analysis of the models will make use of the formalism and concepts of Section 1.4 and Section 1.6. A review is recommended. We assume that relaxation to equilibrium is complete and that the desired folded structure is the energy minimum (ground state) over the conformation space. The conformation of the protein chain is described by a set of variables $\{s_i\}$ that are the local relative coordinates of amino acids—specifically dihedral angles (Fig. 4.2.1). These variables, which are continuous variables, have two or more discrete values at which they attain a local minimum in energy. The local minima are separated by energy barriers. Formal results do not depend in an essential way on the number of local minima for each variable. Thus, it is assumed that each variable s_i is a two-state system (Section 1.4), where the two local minima are denoted by $s_i = \pm 1$.

A model of protein folding using binary variables to describe the protein conformation is not as farfetched as it may sound. On the other hand, one should not be convinced that it is the true protein-folding problem. Protein conformational changes do arise largely from changes in the dihedral angles between bonds (Fig. 4.2.1). The

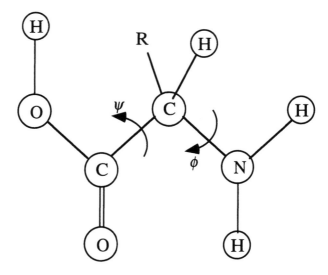

Figure 4.2.1 Illustration of the dihedral angles ψ and ϕ. These coordinates are largely responsible for the variation in protein chain conformation. Changing a single dihedral angle is achieved by rotating all of the protein from one end up to a selected backbone atom. This part of the protein is rotated around the bond that goes from the selected atom to the next along the chain. The rotation does not affect bond lengths or bond-to-bond angles. It does affect the relative orientation of the two bonds on either side of the bond that is the rotation axis. ∎

energy required to change the dihedral angle is small enough to be affected by the secondary bonding between amino acids. This energy is much smaller than the energy required to change bond lengths, which are very rigid, or bond-to-bond angles, which are less rigid than bond lengths but more rigid than dihedral angles. As shown in Fig. 4.2.1, there are two dihedral angles that specify the relative amino acid coordinates. The values taken by the dihedral angles vary along the amino acid chain. They are different for different amino acids, and different for the same amino acid in different locations.

It is revealing to plot the distribution of dihedral angles found in proteins. The scatter plot in Fig. 4.2.2 shows that the values of the dihedral angles cluster around two pairs of values. The plot suggests that it is possible, as a first approximation, to define the conformation of the protein by which cluster a particular amino acid belongs to. It might be suggested that the binary model is correct, by claiming that the variable s_i only indicates that a particular pair of dihedral angles is closer to one of the two aggregation points. However, this is not strictly correct, since it is conceivable that a protein conformation can change significantly without changing any of the binary variables defined in this way.

For our purposes, we will consider a specification of the variables $\{s_i\}$ to be a complete description of the conformation of the protein, except for the irrelevant ro-

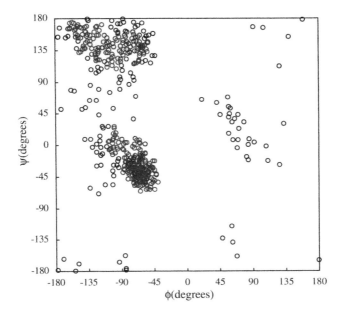

Figure 4.2.2 Scatter plot of the dihedral angle coordinates (Fig. 4.2.1) of each amino acid found along the protein acetylcholinesterase (Figs. 4.1.2–4.1.3). This is called a Ramachandran plot. The coordinates are seen to cluster in two groups. The clustering suggests that it is reasonable to represent the coordinates of the protein using binary variables that specify which of the two clusters a particular dihedral angle pair is found in. The two co-ordinates correspond to α-helix and β-sheet regions of the protein. The more widely scattered points typically correspond to the amino acid glycine which has a hydrogen atom as a radical and therefore has fewer constraints on its conformation. (Angles were obtained from a PDB file using MolView [developed by Thomas J. Smith]) ∎

tational and translational degrees of freedom of the whole protein. The potential energy, $E(\{s_i\})$, of the protein is a function of the values of all the variables. By redefining the variables $s_i \rightarrow -s_i$, when necessary, we let the minimum energy conformation be $s_i = -1$. Furthermore, for most of the discussion, we assume that the unfolded initial state consists of all $s_i = +1$. We could also assume that the unfolded conformation is one of many possible disordered states obtained by randomly picking $s_i = \pm 1$. The folding would then be a disorder-to-order transition.

The potential energy of the system $E(\{s_i\})$ models the actual physical energy arising from atomic interactions, or, more properly, from the interaction between electrons and nuclei, where the nuclear positions are assumed to be fixed and the electrons are treated quantum mechanically. The potential energy is assumed to be evaluated at the particular conformation specified by $\{s_i\}$. It is the potential energy rather than the total energy, because the kinetic energy of atomic motion is not included. Since a protein is in a water environment at non-zero temperature, the potential energy is actually the free energy of the protein after various positions of water molecules are averaged over. Nevertheless, for protein folding the energy is closely related to the physical energy. This is unlike the energy analog that was used in Chapter 2 for the attractor neural network, which was not directly related to the physical energy of the system.

In addition to the energy of the system, $E(\{s_i\})$, there is also a relaxation time τ_i for each variable, s_i. The relaxation time is governed by the energy barrier E_{Bi} of each two-state system—the barrier to switch between values of s_i. The value of E_{Bi} may vary from variable to variable, and depend on the values of the other variables $\{s_j\}_{j\neq i}$. The model we have constructed is quite similar to the Ising model discussed in Section 1.6. The primary difference is the distinct relaxation times for each coordinate. Unless otherwise specified, we will make the assumption that the time for a single variable to flip is small. Specifically, the relaxation times will be assumed to be bounded by a small time that does not change with the size of the system. In this case the model is essentially the same as an Ising model with kinetics that do not take into account the variation in relaxation time between different coordinates. In specific cases we will address the impact of variation in the relaxation times. However, when there is a systematic violation of the assumption that relaxation times are bounded, the behavior is dominated by the largest barriers or the slowest kinetic processes and a different approach is necessary. Violation of this assumption is what causes the models we are about to discuss not to apply to glasses (Section 1.4), or other quenched systems. In such systems a variable describing the local structure does not have a small relaxation time. The assumption of a short single-variable relaxation time is equivalent to assuming a temperature well above the two-state freezing transition.

Our general discussion of protein folding thus consists of assigning a model for the energy function $E(\{s_i\})$ and the dynamics $\{\tau_i\}$ for the transition from $s_i = +1$ to $s_i = -1$. In this general prescription there is no assumed arrangement of variables in space, or the dimensionality of the space in which the variables are located. We will, however, specialize to fixed spatial arrays of variables in a space of a particular dimension in many of the models. It may seem natural to assume that the variables $\{s_i\}$ occupy a space which is either one-dimensional because of the chain structure or three-dimensional because of the 3-d structure of the eventual protein. Typically, we use the dimensionality of space to distinguish between local interactions and long-range interactions. Neither one nor three dimensions is actually correct because of the many possible interactions that can occur between amino acids when the chain dynamically rearranges itself in space. In this chapter, however, our generic approach suggests that we should not be overly concerned with this problem.

We limit ourselves to considering an expansion of the energy up to interactions between pairs of variables.

$$E(\{s_i\}) = -\sum h_i s_i - \sum J_{ij} s_i s_j \qquad (4.2.1)$$

Included is a local preference field h_i determined by local properties of the system (e.g., the structure of individual amino acids), and the pairwise interactions J_{ij}.

Higher-order interactions between three or more variables may be included and can be important. However, the formal discussion of the scaling of relaxation is well served by keeping only these terms. Before proceeding we note that our assumptions imply $\sum h_i < 0$. This follows from the condition that the energy of the initial unfolded state is higher than the energy of the final folded state:

$$E(\{s_i = +1\}) - E(\{s_i = -1\}) = -2\sum h_i > 0 \qquad (4.2.2)$$

Thus, in the lower energy state s_i tends to have the same sign as h_i. We will adopt the magnetic terminology of the Ising model in our discussions (Section 1.6). The variables s_i are called spins, the parameters h_i are local values of the external field, the interactions are ferromagnetic if $J_{ij} > 0$ or antiferromagnetic if $J_{ij} < 0$. Two spins will be said to be aligned if they have the same sign. Note that this does not imply that the actual microscopic coordinates are the same, since they have been redefined so that the lowest energy state corresponds to $s_i = -1$. Instead this means that they are either both in the initial or both in the final state. When convenient for sentence structure we use UP ($\uparrow$) and DOWN ($\downarrow$) to refer to $s_i = +1$ and $s_i = -1$ respectively. The folding transition between $\{s_i = +1\}$ and $\{s_i = -1\}$, from UP to DOWN, is a generalization of the discussion of first-order transitions in Section 1.6. The primary differences are that we are interested in finite-sized systems (systems where we do not assume the thermodynamic limit of $N \rightarrow \infty$) and we discuss a richer variety of models, not just the ferromagnet.

In this chapter we restrict ourselves to considering the scaling of the relaxation time, $\tau(N)$, in these Ising type models. However, it should be understood that similar Ising models have been used to construct predictive models for the secondary structure of proteins. The approach to developing predictive models begins by relating the state of the spins s_i directly to the secondary structure. The two choices for dihedral angles generally correspond to α-helix and β-sheet. Thus we can choose $s_i = +1$ to correspond to α-helix, and $s_i = -1$ to β-sheet. To build an Ising type model that describes the formation of secondary structure, the local fields, h_i, would be chosen based upon propensities of specific amino acids to be part of α-helix and β-sheet structures. An amino acid found more frequently in α-helices would be assigned a positive value of h_i. The greater the bias in probability, the larger the value of h_i. Conversely, for amino acids found more frequently in β-sheet structures, h_i would be negative. The cooperative nature of the α and β structures would be represented by ferromagnetic interactions J_{ij} between near neighbors. Then the minimum energy conformation for a particular primary structure would serve as a prediction of the secondary structure. A chain segment that is consistently UP or DOWN would be α-helix or β-sheet respectively. A chain that alternates between UP and DOWN would be a turn. Various models of this kind have been developed. These efforts to build predictive models have met with some, but thus far limited, success. In order to expand this kind of model to include the tertiary structure there would be a need to include interactions of α and β structures in three dimensions. Once the minimum energy conformation is determined, this model can be converted to a relaxation time model similar to the ones we will discuss, by redefining all of the spins so that $s_i = -1$ in the folded state.

4.3   Parallel Processing in a Two-Spin Model

Our primary objective in this chapter is to elucidate the concept of parallel processing in relaxation kinetics. Parallel processing describes the kinetics of independent or essentially independent relaxation processes. To illustrate this concept in some detail we consider a simple case of two completely independent spins—two independent systems placed side by side. The pair of spins start in a high energy state identified as $(1,1)$, or $s_1 = s_2 = 1$. The low-energy state is $(-1,-1)$, or $s_1 = s_2 = -1$. The system has four possible states: $(1,1)$, $(1,-1)$, $(-1,1)$ and $(-1,-1)$.

We can consider the relaxation of the two-spin system (Fig. 4.3.1) as consisting of hops between the four points $(1,1)$, $(1,-1)$, $(-1,1)$, and $(-1,-1)$ in a two-dimensional plane. Or we can think about these four points as lying on a ring that is essentially one-dimensional, with periodic boundary conditions. Starting from $(1,1)$ there are two possible paths that might be taken by a particular system relaxing to $(-1,-1)$, if we neglect the back transitions. The two paths are $(1,1) \rightarrow (1,-1) \rightarrow (-1,-1)$ and $(1,1) \rightarrow (-1,1) \rightarrow (-1,-1)$. What about the possibility of both spins hopping at once $(1,1) \rightarrow (-1,-1)$? This is not what is meant by parallel processing. It is a separate process, called a coherent transition. The coherent transition is unlikely unless it is enhanced by a lower barrier (lower τ) specifically for this process along the direct path from $(1,1)$ to $(-1,-1)$. In particular, the coherent process is unlikely when the two spins are independent. When they are independent, each spin goes over its own barrier without any coupling to the motion of the other. The time spent going over the barrier is small compared to the relaxation time τ. Thus it is not likely that both will go over at exactly the same time.

There are several ways to describe mathematically the relaxation of the two-spin system. One approach is to use the independence of the two systems to write the probability of each of the four states as a product of the probabilities of each spin:

$$P(s_1, s_2; t) = P(s_1; t)P(s_2; t) \tag{4.3.1}$$

The Master equation which describes the time evolution of the probability can be solved directly by using the solution for each of the two spins separately. We have solved the Master equation for the time evolution of the probability of a two state (one spin) system in Section 1.4. The probability of the spin in state s decays or grows exponentially with the time constant τ:

$$P(s;t) = (P(s;0) - P(s;\infty))e^{-t/\tau} + P(s;\infty) \tag{4.3.2}$$

which is the same as Eq. (1.4.45). The solution of the two-spin Master equation is just the product of the solution of each spin separately:

$$P(s_1, s_2; t) = P(s_1; t)P(s_2; t) \tag{4.3.3}$$

$$= [P(s_1;0)e^{-t/\tau} + (1 - e^{-t/\tau})P(s_1;\infty)][P(s_2;0)e^{-t/\tau} + (1 - e^{-t/\tau})P(s_2;\infty)]$$

For simplicity, it is assumed that the relaxation constant τ is the same for both. This equation applies to each of the four possible states. If the energy difference between

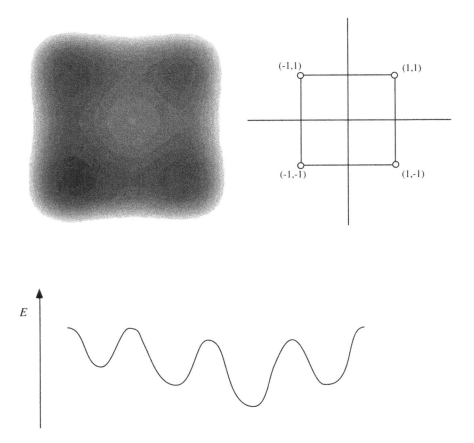

Figure 4.3.1 Illustration of a four-state (two-spin) system formed out of two independent two-state systems. The two-dimensional energy is shown on the upper left. The coordinates of the local energy minima are shown on the right. Below, a schematic energy of the system is shown on a one-dimensional plot, where the horizontal axis goes around the square in the coordinate space of the top right figure. ∎

the UP state and the DOWN state of each spin is sufficiently large, essentially all members of the ensemble will reach the $(-1,-1)$ state. We can determine how long this takes by looking at the probability of the final state:

$$P(-1,-1;t) = [(1 - e^{-t/\tau})P(-1;\infty)][(1 - e^{-t/\tau})P(-1;\infty)] \approx (1 - e^{-t/\tau})^2 \quad (4.3.4)$$

where we have used the initial and final values: $P(-1;0) = 0$, $P(-1;\infty) \approx 1$. Note that a key part of this analysis is that we don't care about the probability of the intermediate states. We only care about the time it takes the system to reach its final state. When does the system arrive at its final state? A convenient way to define the relaxation time of this system is to recognize that in a conventional exponential convergence, τ is the time at which the system has a probability of only e^{-1} of being anywhere else. Applying this condition here we can obtain the relaxation time, $\tau(2)$, of two independent spins from:

$$1 - P(-1,-1;\tau(2)) = e^{-1} \approx 1 - (1 - e^{-\tau(2)/\tau})^2 \tag{4.3.5}$$

or

$$\tau(2) = \tau[-\ln(1 - (1 - e^{-1})^{1/2})] = 1.585\tau \tag{4.3.6}$$

which is slightly larger than τ. A plot of $P(-1,-1;t)$ is compared to $P(-1;t)$ in Fig. 4.3.2.

Why is the relaxation time longer for two systems? It is longer because we have to wait until the spin that takes the longest time relaxes. Both of the spins relax with the same time constant τ. However, statistically, one will take a little less time and the other a little more time. It is the longest time that is the limiting one for the relaxation of the two-spin system.

Where do we see the effect of parallel processing? In this case it is expressed by the statement that we can take either one of the two paths and get to the minimum energy conformation. If we take the path $(1,1)\rightarrow(1,-1)\rightarrow(-1,-1)$, we don't have to make a transition to the state $(-1,1)$ in order to see if it is lower in energy. In the two-spin system we have to visit three out of four conformations to get to the minimum energy conformation. If we add more spins, however, this advantage becomes much more significant.

There may be confusion on one important point. The ability to independently relax different coordinates means that the energies of the system for different states are correlated. For example, in the two-spin system, the energies satisfy the relationship

$$E(1,1) - E(1,-1) = E(-1,1) - E(-1,-1) \tag{4.3.7}$$

If we were to assume instead that each of the four energies, $E(\pm1, \pm1)$, can be specified independently, energy minimization would immediately require a complete exploration of all conformations. Independence of the energies of different conformations for a system of N spins would require the impossible exploration of all phase

Figure 4.3.2 Comparison of the relaxation of a single spin with the relaxation of a two-spin system. The curves show the probability that the system is not in the ground state. The probability goes to zero asymptotically. The relaxation time is identified with the time when the probability is e^{-1}. ∎

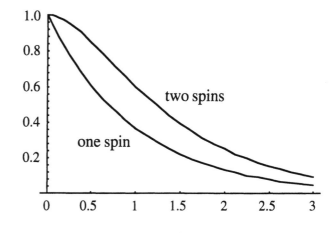

space. It is the existence of correlations in the energies of different conformations that enables parallel processing to work.

4.4   Homogeneous Systems

The models we will consider for a system of N relaxing spins $\{s_i\}$ naturally divide into homogeneous models and inhomogeneous models. For homogeneous models a transformation can be made that maps any spin s_i onto any other spin s_j, where the transformation preserves the form of the energy. Specifically, it preserves the local fields and the interactions between spins. A homogeneous model is loosely analogous to assuming that all amino acids are the same. Such a polymer is called a homopolymer. Boundary conditions may break the transformation symmetry, but their effect can still be considered in the context of homogeneous models. In contrast, an inhomogeneous model is analogous to a heteropolymer where amino acids along the chain are not all the same. Inhomogeneities are incorporated in the models by varying local fields, relaxation times or interactions between spins in a specified way, or by assuming they arise from a specified type of quenched stochastic variable.

In the homogeneous case, all sites are equivalent, and thus the local fields h_i in Eq. (4.2.1) must all be the same. However, the interactions may not all be the same. For example, there may be nearest-neighbor interactions, and different second-neighbor interactions. We indicate that the interaction depends only on the relative location of the spins by the notation J_{i-j}:

$$E(\{s_i\}) = -h\sum s_i - \sum J_{i-j}s_is_j \tag{4.4.1}$$

J_{i-j} is symmetric in $i-j$ and each pair i,j appears only once in the sum. Eq. (4.2.2) implies that h is negative. A further simplification would be to consider each spin to interact only with z neighbors with equal interaction strength J. This would be the conventional ferromagnet or antiferromagnet discussed in Section 1.6. When it is convenient we will use this simpler model to illustrate properties of the more general case. In the following sections, we systematically describe the relaxation in a number of model homogeneous systems. The results of our investigations of the scaling behavior of the relaxation time are summarized in Table 4.4.1. Each of the models illustrates a concept relevant to our understanding of relaxation in complex systems. This table can be referred to as the analysis proceeds.

4.4.1 *Decoupled*

The simplest homogeneous model is the decoupled case, where all spins are independent. Starting from Eq. (4.4.1) we have:

$$E = -h\sum s_i \tag{4.4.2}$$

This is the N spin analog of the two-spin system we considered in Section 4.3. The energetics are the same as the noninteracting Ising model. However, our interest here is to understand the dependence of kinetics on the number of spins N. The dynamics

Model	Scaling
Decoupled model	$O(\ln(N);1)$
Essentially decoupled model	$O(\ln(N);1)$
Nucleation and growth—with neutral boundaries	$O(a^{N^{(d-1)/d}};N^{-1};\ln(N);1)$
—with nucleating boundaries	$O(N^{1/d};\ln(N);1)$
Boundary-imposed ground state	$O(N^{2/d})$
Long-range interactions	$O(\ln(N);a^{N^2})$

Table 4.4.1 Summary of scaling behavior of the relaxation time of the homogeneous models discussed in Section 4.4. The notation indicates the different scaling regimes from smaller to larger systems. ∎

are defined by the individual two-state systems, where a barrier controls the relaxation rate. Relaxation is described by the exponential decay of the probability that each spin is +1.

We have to distinguish between two possible cases. When we analyzed the two-spin case we assumed that essentially all members of the ensemble reach the unique state where all $s_i = -1$. We have to check this assumption more carefully now. The probability that a particular spin is in the +1 state in equilibrium is given by the expression (Eq. (1.4.14)):

$$P(+1;\infty) = e^{-E_+/kT}/(1+e^{-E_+/kT}) \qquad (4.4.3)$$

where $E_+ = -2h$ is the (positive) energy difference between $s_i = +1$ and $s_i = -1$. If we have N spins, the average number that are +1 in equilibrium is

$$N_+ = Ne^{-E_+/kT}/(1+e^{-E_+/kT}) \qquad (4.4.4)$$

Because N can be large, we do not immediately assume that this number is negligible. However, we will assume that in equilibrium a large majority of spins are DOWN. This is true only when $E_+ \gg kT$ and $e^{-E_+/kT} \ll 1$. In this case we can approximate Eq. (4.4.4) as:

$$N_+ = Ne^{-E_+/kT} \qquad (4.4.5)$$

There are now two distinct possibilities depending on whether N_+ is less than or greater than one. If N_+ is less than one, all of the spins are DOWN in the final state. If N_+ is greater than one, almost all, but not all, of the spins are DOWN in the final state.

In the first case, $N_+ \ll 1$, we proceed as with the two-spin system to consider the growth of the probability of the final state:

$$P(\{s_i = -1\};t) = \prod_i [P(s_i = -1;0)e^{-t/\tau} + (1-e^{-t/\tau})P(s_i = -1;\infty)] = (1-e^{-t/\tau})^N$$

$$(4.4.6)$$

Defining the relaxation time as for the two-spin case we have:

$$1 - P(\{s_i = -1\};\tau(N)) = e^{-1} \approx 1 - (1-e^{-\tau(N)/\tau})^N \qquad (4.4.7)$$

or

$$\tau(N) = \tau[-\ln(1-(1-e^{-1})^{1/N})] \tag{4.4.8}$$

For large N we expand this using $a^{1/N} \sim 1 + (1/N)\ln(a)$ to obtain

$$\tau(N) \sim \tau[-\ln(1-(1+(1/N)\ln(1-e^{-1})))] = \tau[-\ln(-(1/N)\ln(1-e^{-1}))]$$
$$= \tau[\ln(N) - \ln(-\ln(1-e^{-1}))] = \tau[\ln(N) + 0.7794] \tag{4.4.9}$$

Neglecting the constant term, we have the result that the time scales logarithmically with the size of the system $\tau(N) \sim \ln(N)$. We see the tremendous advantage of parallel processing, where the relaxation time grows only logarithmically with system size rather than exponentially.

In the second case, $N \gg N_+ \gg 1$, we cannot determine the relaxation time from the probability of a particular final state of the system. There is no unique final state. Instead, we have to consider the growth of the probability of the set of systems that are most likely—the equilibrium ensemble with N_+ spins $s_i = +1$. We can guess the scaling of the relaxation time from the divisibility of the system into independent groups of spins. Since we have to wait only until a particular fraction of spins relax, and this fraction does not change with the size of the system, the relaxation time must be independent of the system size or $\tau(N) \sim 1$. We can show this explicitly by writing the fraction of the remaining UP spins as:

$$N_+(t) = \sum_i P(s_i = 1; t) = \sum_i [P(s_i = +1; 0)e^{-t/\tau} + (1-e^{-t/\tau})P(s_i = +1; \infty)]$$
$$= N\left[e^{-t/\tau} + (1-e^{-t/\tau})e^{-E_+/kT}\right] \approx N\left[e^{-t/\tau} + e^{-E_+/kT}\right] \tag{4.4.10}$$

where we use the assumption that $e^{-E_+/kT} \ll 1$. We must now set a criterion for the relaxation time $\tau(N)$. A reasonable criterion is to set $\tau(N)$ to be the time when there are not many more than the equilibrium number of excited spins, say $(1 + e^{-1})$ times as many:

$$N_+(\tau(N)) = (1+e^{-1})N_+(\infty) \tag{4.4.11}$$

This implies that:

$$N\left[e^{-\tau(N)/\tau} + e^{-E_+/kT}\right] = (1+e^{-1})Ne^{-E_+/kT} \tag{4.4.12}$$

or

$$\tau(N) = \tau(E_+/kT + 1) \equiv \tau_\infty \tag{4.4.13}$$

This relaxation time is independent of the size of the system or $\tau(N) \sim 1$; we name it τ_∞.

The $\tau(N) \sim 1$ scaling we found for this case is lower than the logarithmic scaling. We must understand more fully when it applies. In order for N_+ (Eq. (4.4.5)) to be greater than 1, we must have:

$$N > e^{+E_+/kT} \tag{4.4.14}$$

Thus N must be large in order for N_+ to be greater than 1. It may seem surprising that for larger systems the scaling is lower than for smaller systems. The behavior of the scaling is illustrated schematically in Fig. 4.4.1 (see Question 4.4.1).

There is another way to estimate the relaxation time for very large systems, τ_∞. We use the smaller system relaxation Eq. (4.4.9) at the point where we crossover into the regime of Eq. (4.4.14) by setting $N = e^{+E_+/kT}$. Because the relaxation time is a continuous function of N, at the crossover point it should give an estimate of τ_∞. This gives a similar result to that of Eq. (4.4.13):

$$\tau_\infty \sim \tau[E_+ / kT + 0.7794] \qquad (4.4.15)$$

Question 4.4.1 Combine the analysis of both cases $N_+ \ll 1$ and $N_+ \gg 1$ by setting an appropriate value of $N_+(\tau(N))$ that can hold in both cases. Use this to draw a plot like Fig. 4.4.1.

Solution 4.4.1 The time evolution of $N_+(t)$ is described by Eq. (4.4.10) for either case $N_+ \ll 1$ and $N_+ \gg 1$. The difficulty is that when $N_+ \gg 1$, the process stops when $N_+(t)$ becomes less than 1, and there is no more relaxation

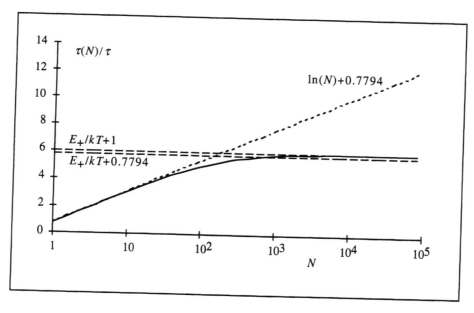

Figure 4.4.1 Relaxation time $\tau(N)$ of N independent spins as a function of N. For systems that are small enough, so that relaxation is to a unique ground state, the relaxation time grows logarithmically with the size of the system. For larger systems, there are always some excited spins, and the relaxation time does not change with system size. This is the thermodynamic limit. The different approximations are described in the text. A unified treatment in Question 4.4.1 gives the solid curve. In the illustrated example, the crossover occurs for a system with about 150 independent spins. This number is given by $e^{E_+/kT}$ so it varies exponentially with the energy difference between the two states of each spin. ∎

to be done. For this case we would like to identify the relaxation time as the time when there is less than one spin not UP. So we replace Eq. (4.4.11) by

$$N_+(\tau(N)) = (1 + e^{-1})N_+(\infty) + N_r$$

where N_r is a constant we can choose which is less than 1. When $N_+ \gg 1$, the first term will dominate and we will have the same result as Eq. (4.4.13), when $N_+ \ll 1$ the second term will dominate. Eq. (4.4.16) leads instead of Eq. (4.4.13) to:

$$\tau(N) = \tau \ln(N / N_r + e^{-1}N_+)) \qquad (4.4.17)$$

When $N_+ \ll 1$ this reduces to:

$$\tau(N) = \tau(\ln(N) - \ln(N_r)) \qquad (4.4.18)$$

which is identical to Eq. (4.4.9) if we identify

$$N_r = -\ln(1 - e^{-1}) = 0.4587 \approx .5 \qquad (4.4.19)$$

which shows that our original definition of the relaxation time is equivalent to our new definition if we use this value for the average number of residual unrelaxed spins.

The plot in Fig. 4.4.1 was constructed using a value of $E_+/kT = 5$ and Eqs. (4.4.9), (4.4.13), (4.4.15) and (4.4.17). ∎

The behavior for large systems satisfying Eq. (4.4.13) is just the thermodynamic limit where intrinsic properties, including relaxation times, become independent of the system size. In this independent spin model, the relaxation time grows logarithmically in system size until the thermodynamic limit is reached, and then its behavior crosses over to the thermodynamic behavior and becomes constant. To summarize the two regimes, we will label the scaling behavior of the independent system as $O(\ln(N);1)$ (the O is read "order").

While the scaling of the relaxation time in the thermodynamic limit is as low as possible, and therefore attractive in principle for the protein-folding problem, there is an unattractive feature—that the equilibrium state of the system is not unique. This violates the assumption we have made that the eventual folded structure of a protein is well defined and precisely given by $\{s_i = -1\}$. However, in recent years it has been found that a small set of conformations that differ slightly from each other constitute the equilibrium protein structure. In the context of this model, the existence of an equilibrium ensemble of the protein suggests that the protein is just at the edge of the thermodynamic regime. In the homogeneous model there is no distinction between different spins, and all are equally likely to be excited to their higher energy state. In the protein it is likely that the ensemble is more selective. For essentially all models we will investigate, for large enough systems, a finite fraction of spins must be thermally excited to a higher energy state. The crossover size depends exponentially on the characteristic energy required for an excited state to occur. This energy is just E_+ in the independent spin model. Because the fraction of excited states also depends exponentially on the temperature, the structure of proteins is affected by physical body temperature. This is one of the ways in which protein function is affected by the temperature.

Either the logarithmic or the constant scaling of the independent spin model, if correct, is more than adequate to account for the rapid folding of proteins. Of course we know that amino acids interact with each other. The interaction is necessary for interdependence in the system. Is it possible to generalize this model to include some interactions and still retain the same scaling? The answer is yes, but the necessary limitations on the interactions between amino acids are still not very satisfactory.

4.4.2 *Essentially decoupled*

The decoupled model can be generalized without significantly affecting the scaling, by allowing limited interactions that do not affect the relaxation of any spins. To achieve this we must guarantee that at all times the energy of the spin s_i is lower when it is DOWN than when it is UP. For a protein, this corresponds to a case where each amino acid has a certain low-energy state regardless of the protein conformation. We specialize the discussion to nearest-neighbor interactions between each spin and z neighbors—a ferromagnetic or antiferromagnetic Ising model. We also assume the same relaxation time τ applies to all spins at all times. The more general case is deferred to Question 4.4.2.

When there are interactions, the change in energy upon flipping a particular spin s_i from UP to DOWN is dependent on the condition of the other spins $\{s_j\}_{j \neq i}$. We write the change as:

$$E_{+i}(\{s_j\}_{j \neq i}) = E(s_i = +1, \{s_j\}_{j \neq i}) - E(s_i = -1, \{s_j\}_{j \neq i}) = -2h - 2\sum_{j \neq i} J_{i-j} s_j \qquad (4.4.20)$$

The latter expression is for homogeneous systems. For only nearest-neighbor interactions in both ferromagnet and antiferromagnet cases

$$E_{+i}(\{s_j\}_{j \neq i}) = -2h - 2J \sum_{nn} s_j \qquad (4.4.21)$$

where the sum is over the z nearest neighbors of s_i. Note that this expression depends on the state of the neighboring spins, not on the state of s_i. For the spins to relax essentially independently, we require that the minimum possible value of Eq. (4.4.21)

$$E_{+min} = -2h - 2z|J| \qquad (4.4.22)$$

is greater than zero. To satisfy these requirements we must have

$$|h| > z|J| \qquad (4.4.23)$$

which means that the local field $|h|$ is stronger than the interactions. When it is convenient we will also assume that $E_{+min} \gg kT$, so that the energy difference between UP and DOWN states is larger than the thermal energy.

The ferromagnetic case $J > 0$ is the same as the kinetics of a first-order transition (Section 1.6) when the local field is so large that nucleation is not needed and each spin can relax separately. Remembering that $h < 0$, the value of E_{+i} starts from its minimum value when all of the spins (neighbors of s_i) are UP, $s_j = +1$. E_{+i} then increases as the system relaxes until it reaches its maximum value everywhere when all the spins (neighbors of s_i) are DOWN, $s_j = -1$ (see Fig. 4.4.2). This means that initially the interactions fight relaxation to the ground state, because they are promoting the alignment of the spins that are UP. However, each spin still relaxes DOWN. The final state with all spins DOWN is self-reinforcing, since the interactions raise the energy of isolated UP spins. This inhibits the excitation of individual spins and reduces the probability that the system is out of the ground state. Thus, ferromagnetic interactions lead to what is called a cooperative ground state. In a cooperative ground state, interactions raise the energy cost of, and thus inhibit, individual elements from switching to a higher energy state. This property appears to be characteristic of proteins in their 3-d structure. Various interactions between amino acids act cooperatively to lower the conformation energy and reduce the likelihood of excited states.

In order to consider the relaxation time in this model, we again consider two cases depending upon the equilibrium number of UP spins, N_+. The situation is more complicated than the decoupled model because the eventual equilibrium N_+ is not necessarily the target N_+ during the relaxation. We can say that the effective $N_+(E_+)$ as

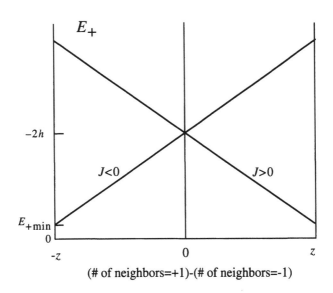

(# of neighbors=+1)-(# of neighbors=-1)

Figure 4.4.2 Illustration, for the essentially decoupled model, of the value of the single-spin energy E_+ as a function of the number of its neighbors (out of a total of z) that are UP and DOWN. At the right all of the neighbors are UP, and at the left all of the neighbors are DOWN. E_+ measures the energy preference of the spin to be DOWN. E_+ is always positive in the essentially decoupled model. The relaxation process to the ground state takes the system from right to left. For a ferromagnet, $J > 0$, the change reinforces the energy preference for the spin to be DOWN. For the antiferromagnet, $J < 0$, the change weakens the energy preference for the spin to be DOWN. Implications for the time scale of relaxation are described in the text. ∎

given by Eq. (4.4.5) changes over time. Because E_+ starts out small, it may not be enough to guarantee that all spins will be DOWN in the final state. But the increase in E_+ may be sufficient to guarantee that all spins will be DOWN at the end.

The situation is simplest if there is complete relaxation toward the ground state at all times. This means:

$$N < e^{+E_{+min}/kT} \qquad (4.4.24)$$

In this case, the relaxation time scaling is bounded by the scaling of the decoupled model. We can show this by going back to the equation for the dynamics of a single relaxing two-state system, as written in Eq. (1.4.43):

$$\dot{P}(1;t) = (P(1;\infty) - P(1;t))/\tau \qquad (4.4.25)$$

The difficulty in solving this equation is that $P(1;\infty)$ (Eq. (4.4.3)) is no longer a constant. It varies between spins and over time because it depends on the value of E_+. Nevertheless, Eq. (4.4.25) is valid at any particular moment with the instantaneous value of $P(1;\infty)$. When Eq. (4.4.24) holds, $P(1;\infty) < 1/N$ is always negligible compared to $P(1;t)$, even when all the spins are relaxed, so we can simplify Eq. (4.4.24) to be:

$$\dot{P}(1;t) = -P(1;t)/\tau \qquad (4.4.26)$$

This equation is completely independent of E_+. It is therefore the same as for the decoupled model. We can integrate to obtain:

$$P(1;t) = e^{-t/\tau} \qquad (4.4.27)$$

Thus each spin relaxes as a decoupled system, and so does the whole system with a relaxation time scaling of $O(\ln(N))$.

When Eq. (4.4.24) is not true, the difficulty is that we can no longer neglect $P(1;\infty)$ in Eq. (4.4.25). This means that while the spins are relaxing, they are not relaxing to the equilibrium probability. There are two possibilities. The first is that the equilibrium state of the system includes a small fraction of excited spins. Since the fraction of the excited spins does not change with system size, the relaxation time does not change with system size and is $O(1)$.

The other possibility is that initially the relaxation allows a small fraction of spins to be excited. Then as the relaxation proceeds, the energy differences $E_{+i}(\{s_j\}_{j \neq i})$ increase. This increase in energy differences then causes all of the spins to relax. How does the scaling behave in this case? Since each of the spins relaxes independently, in $O(1)$ time all except a small fraction χN will relax. The remaining fraction consists of spins that are in no particular relationship to each other; they are therefore independent because the range of the interaction is short. Thus, they relax in $O(\ln(\chi N))$ time to the ground state. The total relaxation time would be the sum of a constant term and a logarithmic term that we could write as $O(\ln(\chi N)+1)$, which is not greater than $O(\ln(N))$. This concludes the discussion of the ferromagnetic case.

For the antiferromagnetic case, the situation is actually simpler. Since $J < 0$, remembering that $h < 0$, the value of E_+ starts from its maximum value when all $s_j = +1$, and reaches its minimum value when all $s_j = -1$ (see Fig. 4.4.2). Thus $N_+(E_+)$ is largest in the ground state. Once again, if there is a nonzero fraction of spins at the end that are UP then the relaxation must be independent of system size, $O(1)$. If there are no residual UP spins in equilibrium, then in Eq. (4.4.25) $P(1;\infty)<1/N$ always, and the relaxation reduces directly to the independent case $O(\ln(N))$.

The ferromagnetic case is essentially different from the antiferromagnetic case because we can continue to consider stronger values of the ferromagnetic interaction without changing the ground state. However, if we consider stronger antiferromagnetic interactions, the ground state will consist of alternating UP and DOWN spins and this is inconsistent with our assumptions (we would have redefined the spin variables). Thus, nearest-neighbor antiferromagnetic interactions, as long as they do not lead to an antiferromagnetic ground state, do not affect the relaxation behavior.

When there are spin-spin interactions, we would also expect the relaxation times τ_i to be affected by the interactions. The relaxation time depends on the barrier to relaxation, E_{Bi}, as shown in the energy curve of the two-state system Fig. 1.4.1. When the energy difference E_+ is higher, we might expect that the barrier to relaxation of the two-state system will become lower. This would be the case if we raise E_+ without raising the energy at the top of the barrier. On the other hand, if the energy surface is multiplied by a uniform factor to increase E_+, then the barrier would increase. These differences in the barrier show up in the relaxation times τ_i. In the former case the relaxation is faster, and in the latter case the relaxation is slower. For the nearest-neighbor Ising model, there would be only a few different relaxation times corresponding to the different possible states of the neighboring spins. We can place a limit on the relaxation time $\tau(N)$ of the whole system by replacing all the different spin relaxation times with the maximum possible spin relaxation time. As far as the scaling of $\tau(N)$ with system size, this will have no effect. The scaling remains the same as in the noninteracting case, $O(\ln(N);1)$.

Question 4.4.2 Consider the more general case of a homogeneous model with interactions that may include more than just nearest-neighbor interactions. Restricting the interactions not to affect the minimum energy of a spin, argue that the relaxation time scaling of the system is the same as the decoupled model. Assume that the interactions have a limited range and the system size is much larger than the range of the interactions.

Solution 4.4.2 As in Eq. (4.4.20), the change in energy on flipping a particular spin is dependent on the conditions of the other spins $\{s_j\}_{j \neq i}$.

$$E_{+i}(\{s_j\}_{j \neq i}) = -2h - 2\sum_{j \neq i} J_{i-j} s_j \tag{4.4.28}$$

We assume that $E_{+i}(\{s_j\}_{j \neq i})$ is always positive. Moreover, for relaxation to occur, the energy difference must be greater than kT. Thus the energy must be bounded by a minimum energy E_{+min} satisfying:

$$E_{+i}(\{s_j\}_{j \neq i}) > E_{+min} \gg kT \tag{4.4.29}$$

This implies that the interactions do not change the lowest energy state of each spin s_i. For the energy of Eq. (4.4.1), E_{+min} can be written

$$E_{+min} = -2h - 2\sum_{j \neq i} |J_{i-j}| \qquad (4.4.30)$$

Interactions may also affect the relaxation time of each spin $\tau_i\{s_j\}_{j \neq i}$, so we also assume that relaxation times are bounded to be less than a relaxation time τ_{max}.

We assume that the parameters τ_{max} and E_{+min} do not change with system size. This will be satisfied, for example, if the interactions have a limited range and the system size is larger than the range of the interactions.

Together, the assumption of a bound on the energy differences and a bound on the relaxation times suggest that the equilibration time is bounded by that of a system of decoupled spins with $-2h = E_{+min}$ and $\tau = \tau_{max}$. There is one catch. We have to consider again the possibility of incomplete relaxation to the ground state. The scenario follows the same possibilities as the nearest-neighbor model. The situation is simplest if there is complete relaxation to the ground state at all times. This means:

$$N < e^{+E_{+min}/kT} \qquad (4.4.31)$$

which is a more stringent condition than Eq. (4.4.29). In this case the bound on $\tau(N)$ is straightforward because each spin is relaxing to the ground state faster than in the original case. Again using Eq. (1.4.43):

$$\dot{P}(1;t) = (P(1;\infty;t) - P(1;t))/\tau(t) \qquad (4.4.32)$$

This equation applies at any particular moment, with the time-dependent values of $P(1;\infty;t)$ and $\tau(t)$, where the time dependence of these quantities is explicitly written. Since $P(1;\infty;t)$ is always negligible compared to $P(1;t)$, when Eq. (4.4.31) applies, this is

$$\dot{P}(1;t) = -P(1;t)/\tau(t) \qquad (4.4.33)$$

We can integrate to obtain:

$$P(1;t) = e^{-\int_0^t \frac{dt}{\tau(t)}} < e^{-t/\tau_{max}} \qquad (4.4.34)$$

The inequality follows from the assumption that the relaxation time of each spin is always smaller than τ_{max}. Each spin relaxes faster than the decoupled system, and so does the whole system. The scaling behavior $O(\ln(N))$ of the decoupled system is a bound for the increase in the relaxation time of the coupled system.

When Eq. (4.4.31) is not true, we can no longer neglect $P(1;\infty;t)$ in Eq. (4.4.32). This means that while the spins are relaxing faster, they are not relaxing to the equilibrium probability. There are two possibilities. The first is that the equilibrium state of the system includes a small fraction of excited spins. Since the range of the interactions is smaller than the system size, the fraction of the excited spins does not change with system size and the relaxation time does not change with system size. The other possibility is that initially the values of $E_{+i}(\{s_j\}_{j \neq i})$ do not satisfy Eq. (4.4.31) and so allow a small fraction of spins to be excited. Then as the relaxation proceeds, the energy differences $E_{+i}(\{s_j\}_{j \neq i})$ increase. This increase in energy differences then causes all of the spins to relax. The relaxation time will not be larger than $O(\ln(N))$ as long as $E_{+min} \gg kT$ (Eq. (4.4.29)) holds. Because of this condition, each of the spins will almost always relax, and in $O(1)$ time all except a small fraction χN will relax. The remaining fraction consists of spins that are in no particular relationship to each other; they are therefore independent, because the range of the interaction is short, and will relax in at most $O(\ln(\chi N))$ time to the ground state. The total relaxation time would be the sum of a constant term and a logarithmic term that we could write as $O(\ln(\chi N)+1)$, which is not greater than $O(\ln(N))$. ∎

We have treated carefully the decoupled and the almost decoupled models to distinguish between $O(\ln(N))$ and $O(1)$ scaling. One reason to devote such attention to these simple models is that they are the ideal case of parallel processing. It should be understood, however, that the difference between $O(\ln(N))$ and $O(1)$ scaling is not usually significant. For 1000 amino acids in a protein, the difference is only a factor of 7, which is not significant if the individual amino acid relaxation time is microscopic.

One of the points that we learned about interactions from the almost decoupled model is that the ferromagnetic interactions $J > 0$ cause the most problem for relaxation. This is because they reinforce the initial state before the effect of the field h acts to change the conformation. In the almost decoupled model, however, the field h dominates the interactions J. In the next model this is not the case.

The almost decoupled model is not satisfactory in describing protein folding because the interactions between amino acids can affect which conformation they are in. The next model allows this possibility. The result is a new scaling of the relaxation with system size, but only under particular circumstances.

4.4.3 Nucleation and growth: relaxation by driven diffusion

The next homogenous model results from assuming that the interactions are strong enough to affect the minimum energy conformation for a particular spin:

$$E_{+min} < 0 \qquad\qquad (4.4.35)$$

From Eqs. (4.4.20) and (4.4.21) we see that this implies that the total value of the interactions exceeds the local preference as determined by the field h. Eventually, it is h that ensures that all of the spins are DOWN in the ground state. However, initially when all of the spins are UP, due to the interactions the spins have their lowest energy UP rather than DOWN. During relaxation, when some are UP and some are DOWN, a particular spin may have its lowest energy either UP or DOWN. The effect of both the external field and the interactions together leads to an effective field, $h + \sum_j J_{i-j} s_j$, that determines the preference for the spin orientation at a particular time.

The simplest model that illustrates this case is the Ising ferromagnet in a d-dimensional space (Section 1.6). The interactions are all positive, and the spins try to align with each other. Initially the local preference is for the spins to remain UP; the global minimum of energy is for all of the spins to be DOWN. The resolution of this problem occurs when enough of the spins in a particular region flip DOWN using thermal energy, to create a critical nucleus. A critical nucleus is a cluster of DOWN spins that is sufficiently large so that further growth of the cluster lowers the energy of the system. This happens when the energy lowering from flipping additional spins is larger than the increase in boundary energy between the DOWN cluster and the UP surrounding region. Once a critical nucleus forms in an infinite system, the region of down spins grows until it encounters other such regions and merges with them to form the equilibrium state. In a finite system there may be only one critical nucleus that is formed, and it grows until it consumes the whole system.

The nucleation and growth model of first-order transitions is valid for quite arbitrary interactions when there are two phases, one which is metastable and one which is stable, if there is a well-defined boundary between them when they occur side by side. This applies to a large class of models with finite length interactions. For example, there could be positive nearest-neighbor interactions and negative second-neighbor interactions. As long as the identity of the ground state is not disturbed, varying the interactions affects the value of the boundary energy, but not the overall behavior of the metastable region or the stable region. We do not consider here the case where the boundaries become poorly defined. In our models, the metastable phase consists of UP spins and the stable phase consists of DOWN spins. A system with only nearest-neighbor antiferromagnetic interactions on a bipartite lattice is not included in this section. For $J < 0$ on a bipartite lattice, when Eq. (4.4.35) is satisfied, the ground state is antiferromagnetic (alternating $s_i = \pm 1$), and we would have redefined the spins to take this into consideration.

The dynamics of relaxation for nucleation and growth are controlled by the rate of nucleation and by the rate of diffusion of the boundary between the two phases. Because of the energy difference of the two phases, a flat boundary between them will move at constant velocity toward the metastable phase, converting UP spins to DOWN spins. This process is essentially that of driven diffusion down a washboard potential as illustrated in Fig. 1.4.5. The velocity of the boundary, v, can be measured in units of interspin separation per unit time.

During relaxation, once a critical nucleus of the stable phase forms, it grows by driven diffusion and by merging with other clusters. The number of spins in a particular region of the stable phase grows with time as $(vt)^d$. This rate of growth occurs because the region of the stable phase grows uniformly in all directions with velocity v. Every part of the boundary diffuses like a flat boundary (Fig. 4.4.3). This follows our assumption that the boundary is well defined. There are two parts to this assumption. The first is that the thickness of the boundary is small compared to the size of the critical nucleus. The second is that it becomes smooth, not rough, over time. When these assumptions are satisfied, the stable region expands with velocity v in all directions.

There are several cases that must be considered in order to discuss the scaling of the relaxation time of a finite system of N spins. First we must distinguish three different ranges for the system size. The system may be smaller than the size of a critical nucleus, N_{c0}. If the system is larger than a critical nucleus, then it may be smaller than the typical distance between critical nuclei. Third, it may be larger than this distance. Finally, we must also consider the properties of the boundary of the system, specifically whether or not it promotes nucleation.

Nonnucleating boundary

We start by considering the three system sizes when the boundary of the system is either neutral or suppresses nucleation. Under these circumstances, we can neglect the effect of the boundary because relaxation depends upon nucleation and growth from the interior. The spins near the boundary join the stable phase when it reaches them. We assume throughout that the number of spins in the boundary is negligible compared to the number in the interior.

The case of the system being smaller than the size of the critical nucleus, $N < N_{c0}$, is special because the energy barrier to relaxation grows as the system size increases. The energy may be seen schematically in Fig. 4.4.4 (or Fig. 1.6.10) as a function of cluster size. The washboard-like energy rises in the region below the critical nucleus size. When the system is smaller than the size of a critical nucleus, the energy necessary to form a region of DOWN spins of roughly the size of the system controls the rate of relaxation. Because the energy barrier to forming this region increases roughly linearly with system size, the relaxation time grows exponentially with system size. We can be more precise by using an expression for how the barrier energy grows with system size. The energy of a cluster in an infinite system grows with the number of spins in the cluster as (see also Question 1.6.14):

$$E_c(N_c) = 2hN_c + bN_c^{(d-1)/d} \tag{4.4.36}$$

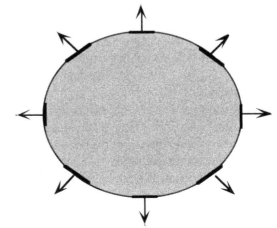

Figure 4.4.3 When a critical nucleus of a stable phase has formed in a metastable phase, the nucleus grows by driven diffusion. The motion of the boundary increases the volume of the equilibrium phase at the expense of the metastable phase. Each part of the boundary moves at a constant average velocity v. Thus, every dimension of the equilibrium phase grows at a constant rate. The number of spins in the equilibrium phase grows as $(vt)^d$ where d is the dimensionality of the space. ∎

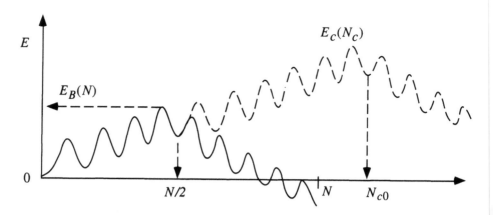

Figure 4.4.4 Schematic illustration of the energy of a cluster of DOWN spins in a metastable background of UP spins as a function of the number of spins in the cluster N_c. The corrugation of the line indicates the energy barrier as each spin flips from UP to DOWN. The dashed line illustrates the energy $E_c(N_c)$ of the cluster in an infinite system. The energy increases until it reaches the size of a critical nucleus N_{c0} and decreases thereafter as the cluster grows to become the stable phase. The solid line indicates the energy in a finite system of size $N < N_c0$. In this case the maximum energy, which is the barrier to relaxation, is located in the vicinity of $N/2$, as indicated. ∎

The first term is the bulk energy of the DOWN spins in the cluster as compared to metastable UP spins. The second term is the boundary energy, where b is a measure of the boundary energy per unit length. This expression is reasonable if the critical nucleus is large compared to the boundary width—the boundary is well defined. The critical nucleus for an infinite system is determined by the maximum value of $E_c(N_c)$. This is obtained setting its derivative with respect to N_c to zero. Aside from a factor of $(d-1)/d$, this means that both terms are equal in magnitude for the critical nucleus. If the system is smaller than the critical nucleus size, then the boundary energy must dominate the bulk energy of a cluster for all possible cluster sizes. Thus for a system with $N < N_{c0}$ we can neglect the first term in $E_c(N_c)$, leaving us with the energy $E_c(N_c) \approx bN_c^{(d-1)/d}$.

For a system with $N < N_{c0}$ that has periodic boundary conditions, the boundary of a cluster grows only as long as the cluster contains less than one-half of the spins in the system. Beyond this point, the boundary of the cluster shrinks. So the maximum cluster energy is reached when N_c is $N/2$. This is still true for a fixed boundary if the boundary is neutral. The relevant cluster may be identified by bisecting the system with UP spins on one side and DOWN spins on the other. If the boundary suppresses nucleation, then the maximum value of N_c may be greater than $N/2$, but it is not larger than N. As long as the maximum value of N_c is proportional to N, the results given below are essentially unaffected.

The cluster energy we have calculated is the energy at the bottom of a particular well in Fig. 4.4.4. It does not include the height of the corrugation E_{B0} which is the energy barrier to flipping a single spin. The energy barrier for nucleation in the system with $N < N_{c0}$ is thus given by

$$E_B(N) = E_c(N/2) + E_{B0} = b(N/2)^{(d-1)/d} + E_{B0} \qquad (4.4.37)$$

The relaxation time is given approximately by the probability that the system will reach this barrier energy, as given by a Boltzmann factor of the energy. More specifically, it is given by Eq. (1.4.44), which gives the relaxation of a two-state system with the same barrier (we neglect the back transition rate):

$$\tau(N) = v^{-1}e^{E_B(N)/kT} = v^{-1}e^{(E_{B0}+bN^{(d-1)/d}/2^{(d-1)/d})/kT} = \tau e^{bN^{(d-1)/d}/2^{(d-1)/d}/kT} \quad (4.4.38)$$

This shows the exponential dependence of the relaxation time on system size in this small system limit when $N < N_{c0}$. We note that we have neglected to consider the many possible ways there are to form a cluster of a particular size, which may also affect the scaling of the relaxation time.

The existence of a region of exponential growth of the relaxation time should be understood in a context where we compare the nucleation time with the observation time. If the nucleation time is long compared to the observation time, we would not expect to see relaxation to the ground state.

If the size of the system is much larger than the size of a critical nucleus, $N \gg N_{c0}$, we can consider each nucleus to be essentially a point object of no size, when it forms. A nucleus forms at a particular site according to a local relaxation process with a time constant we denote τ_{c0}—the nucleation time. The nuclei then grow, as discussed previously, with a constant velocity v in each direction. During the relaxation we either have one or many nuclei that form. Only one nucleus forms when the typical time for forming a nucleus in the system is longer than the time a single nucleus takes to consume the whole system. As soon as one nucleus forms, its growth is so rapid that no other nuclei form during the time it grows to the size of the whole system (Fig. 4.4.5(a)). The relaxation time is determined by the time that passes until the first nucleation event occurs in the system. For larger systems, the number of possible nucleation sites increases in direct proportion to N. Thus the time till the first nucleation event decreases, and the relaxation time actually decreases with system size. We will derive the result that $\tau(N) \sim N^{-1}$. To determine when this scaling applies we must find expressions for the nucleation time, and the time a nucleus takes to grow to the size of the system. Independent nuclei can form on every N_{c0} sites. The typical time to form a critical nucleus anywhere in the system, τ_{cN}, where $\tau_{cN} \ll \tau_{c0}$, is the time it takes any one of the possible N/N_{c0} sites to form a single critical nucleus:

$$(N/N_{c0})e^{-\tau_{cN}/\tau_{c0}} = N/N_{c0} - 1 \qquad (4.4.39)$$

expanding the exponential using $\tau_{cN}/\tau_{c0} \ll 1$ gives

$$\tau_{cN} = \tau_{c0}N_{c0}/N \qquad (4.4.40)$$

Figure 4.4.5 Several cases of the relaxation of systems by driven diffusion are illustrated. See the text for a detailed discussion. In (a) the system is larger than the size of a critical nucleus but small enough so that only one nucleation event occurs in the system. The boundary is nonnucleating. In (b) the system is large enough so that several nucleation events occur during the relaxation; the boundary is nonnucleating. In (c) the boundary nucleates the equilibrium phase. ∎

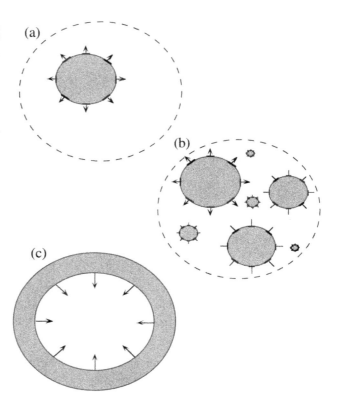

This result says that the time to form a single nucleus is inversely proportional to the size of the system. The time for a single nucleus to grow to the size of the system, τ_v, is given by

$$(v\tau_v)^d = N \qquad (4.4.41)$$

or

$$\tau_v = N^{1/d}/v \qquad (4.4.42)$$

We are neglecting numerical factors that reflect different possible locations the nucleus may form and their relationship to the boundary of the system.

The condition that a single nucleus will form $\tau_{cN} > \tau_v$ is given by combining Eq. (4.4.40) and Eq. (4.4.41) to obtain

$$(v\tau_{c0}N_{c0})^{d/(d+1)} > N \gg N_{c0} \qquad (4.4.43)$$

where we have also repeated our assumption that the size of the system is larger than the critical nucleus. Eq. (4.4.43) describes the bounds on the system size so that only one nucleus is important. Under these circumstances the relaxation time actually decreases with system size, because as the size of the system increases so do the opportunities for forming critical nuclei. The relaxation time is given by the sum of the nucleation time and the time for consumption of the whole system. The latter has been assumed to be small compared to the former:

$$\tau(N) = \tau_{cN} + \tau_v \approx \tau_{cN} = \tau_{c0} N_{c0} / N \qquad (4.4.44)$$

Thus the scaling of the relaxation time is $O(N^{-1})$.

If the system is large enough so that more than one nucleation event occurs (Fig. 4.4.5(b)), then different regions of the material may be treated as essentially decoupled. We expect from the analysis of independent systems that the scaling of the relaxation time is logarithmic. A more detailed analysis given as Question 4.4.3 shows the scaling is $O(\ln(N)^{1/(d+1)})$. While the analysis in Question 4.4.3 has interesting features, the difference between this and $O(1)$ or $O(\ln(N))$ scaling is unlikely to be significant. Finally, as with the independent spin model, the relaxation time is independent of N if N_+ is greater than 1. For convenience we assume that this occurs after the transition between the regime of Fig. 4.4.5(a) and Fig. 4.4.5(b), i.e., for systems in which there are many nucleation events.

Question 4.4.3 Calculate the scaling of the relaxation time when there are many nuclei formed in a system with N spins with boundaries that do not affect the nucleation. Assume that all spins are DOWN at the end of the relaxation. Numerical factors that do not affect the dependence of $\tau(N)$ on N may be neglected.

Solution 4.4.3 Nucleation sites occur randomly through the system and then grow and merge together. In order to find the time at which the whole system will become DOWN, we calculate the probability that a spin at a particular site will remain UP. A particular spin s_i is UP at time t only if there has been no nucleation event in its vicinity that would have grown enough to reach its site.

The probability that no critical nucleus formed at a position r_j with respect to the site s_i until the time t is given by the probability of a two-state system with a time constant τ_{c0} remaining in its high energy state or

$$e^{-t/\tau_{c0}} \qquad (4.4.45)$$

If we are looking at the spin s_i at time t, we must ask whether there was formed a nucleus at a distance r away prior to $t' = t - r_j/v$. If the nucleus formed before t' then the nucleus would arrive before time t at the site s_i. The maximum distance that can affect the spin s_i is $r_{max} = \min(vt, R)$, where $R \propto N^{1/d}$ is the size of the system. When there are many nuclei in the system, then each nucleus is much smaller than the system and $R \gg vt$, so that $r_{max} = vt$. The probability that no nucleus formed within this radius at an early enough time is given by:

$$\prod_{r_j}^{r_{max}} e^{-(t - r_j/v)/\tau_{c0}} = e^{-\sum (t - r_j/v)/\tau_{c0}} \qquad (4.4.46)$$

where the product and the sum are over all possible nucleation sites within a distance r_{max}.

The sum can be directly evaluated to give:

$$\sum_j (t-r_j/v) \propto \frac{1}{N_{c0}} \int_0^{r_{max}=vt} (t-r/v)r^{d-1}dr \propto (vt)^{d+1}/(vN_{c0}) \qquad (4.4.47)$$

where we divided by the volume of a nucleation site and neglected constants. The number of sites that remain UP is given by N times Eq. (4.4.46) with Eq. (4.4.47) substituted in:

$$N_+ = Ne^{-\chi(vt)^{d+1}/(v\tau_{c0}N_{c0})} \qquad (4.4.48)$$

The coefficient χ accounts for the numerical prefactors we have neglected. Requiring that this is a number less than 1 when $t = \tau(N)$ gives the relaxation time $\tau(N) \sim \ln(N)^{1/(d+1)}$ as indicated in the text.

If we consider this same derivation but do not substitute for r_{max} in Eq. (4.4.47) then we arrive at the expression:

$$\sum_j (t-r_j/v) \propto \frac{1}{N_{c0}} \int_0^{r_{max}} (t-r/v)r^{d-1}dr \propto (t-(d/d+1)r_{max}/v)r_{max}^d/N_{c0}$$

$$(4.4.49)$$

and

$$N_+ = Ne^{-\chi(t-(d/d+1)r_{max}/v)r_{max}^d/(\tau_{c0}N_{c0})} \qquad (4.4.50)$$

This more general expression also contains the behavior when we have only one nucleation event. We can recover this case by substituting a constant value of $r_{max} = R \propto N^{1/d}$. Then the time dependence of N_+ is given by the simple exponential dependence with the relaxation constant $\tau(N) = \tau_{c0}N_{c0}/\chi r_{max}^d \propto 1/N$. ∎

Nucleating boundary

If the boundary of the system promotes nucleation, the nucleus formed at the boundary will increase by driven diffusion. If there are no other nucleation events (Fig. 4.4.5(c)) then the relaxation-time scales as $\tau(N) \sim N^{1/d}$. Since the critical nucleus forms at the boundary, the system does not have to be larger than a critical nucleus for this to occur. If the system is large enough so that there are many nucleation events, then the behavior of the boundary is irrelevant and the same scaling found before applies.

We have found an anomaly in the intermediate regime characterized by Eq. (4.4.43). In this regime the relaxation time of a system with a nonnucleating boundary decreases, while that with a nucleating boundary increases. It should be understood that for the same microscopic parameters (except at the boundaries), the relaxation time is longer in the former case than in the latter.

Summary

In summary, a system of finite size with a driven-diffusion relaxation has a scaling of the relaxation time with system size as $O(a^{N^{(d-1)/d}}, N^{-1}; \ln(N); 1)$ for a nonnucleating boundary, and $O(N^{1/d}; \ln(N); 1)$ for a nucleating boundary. The results are illustrated in Fig. 4.4.6.

One interesting conclusion from the results in this section is that we do not have to create a very complicated model in order to find a relaxation time that grows exponentially with system size. A ferromagnetic Ising model with a large critical nucleus is sufficient. What is the significance of this result? The size of the critical nucleus N_{c0} and the nucleation time τ_{c0} are both controlled by the magnitude of h compared to the interaction strength J. When h is large the critical nucleus is small and the nucleation time is small. In this model h is the driving force for the relaxation; when this driving force is weak, the relaxation may take arbitrarily long.

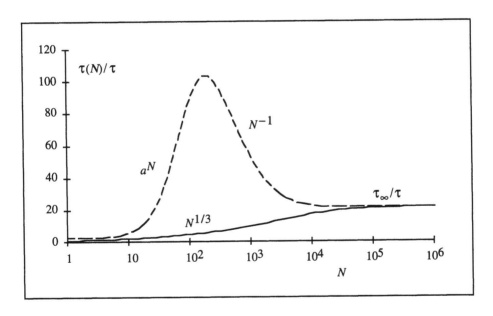

Figure 4.4.6 Schematic plot of the relaxation-time behavior for a system that equilibrates by driven diffusion (see Figs. 4.4.3–4.4.5). Two cases are shown, the solid line is for a system with a boundary that nucleates the stable phase; the dashed line is for a system with a nonnucleating boundary. When the boundary nucleates the stable phase, the stable phase grows by driven diffusion. It consumes the whole system in a time that scales with system size as $N^{1/d}$. For this plot, d is taken to be 3. When the boundary does not nucleate the stable phase, nucleation becomes harder as the system increases in size until it reaches the size of a critical nucleus. For larger systems, the relaxation time decreases because it becomes easier to form a critical nucleus somewhere. Independent of the boundary behavior, when a system becomes so large that the nucleation time, τ_{cN}, becomes equal to the time it takes for driven diffusion to travel the distance between one nucleus and another, τ_v, then the system reaches the large size (thermodynamic) limit and the relaxation time becomes constant. Logarithmic corrections that may arise in this regime have been neglected in the figure. ∎

Our assumption that the relaxation time of an individual spin is rapid should be discussed in this context. We have seen that the nucleation time can become longer than the experimental time. In overcoming the nucleation barrier, the formation of a nucleus is like the relaxation of a two-state system. What we have done, in effect, is to group together a region of spins that is the size of a critical nucleus, and treat them as if they were a single spin. This is a process of renormalization as discussed in Section 1.10. The nucleation time becomes the effective local relaxation time. Thus we see that if the field h is small enough, the effective local relaxation time increases. Even though the ultimate behavior of the system is that of growth by driven diffusion of the stable phase, the relaxation is inhibited locally. This leads to the persistence of the metastable phase. One example of a system where equilibration is inhibited by a long local relaxation time is diamond. Diamond is a metastable phase under standard conditions. The stable phase is graphite.

The second interesting conclusion is the importance of the boundary conditions for the scaling behavior. It is particularly interesting that the new scaling behavior, $N^{1/d}$, arises only for the case of nucleation by the boundary of the system. The scaling behavior of a system with nonnucleating boundaries is quite different, as discussed above.

The model of nucleation and growth of the stable phase has played an important role in conceptual discussions of protein folding. Various theoretical and experimental efforts have been directed at identifying how nucleation and growth of secondary and tertiary structure of proteins might occur. Of particular significance is that it allows interdependence through interactions, and yet can allow relaxation to proceed in a reasonable time. From our discussion it is apparent that nucleating boundaries are beneficial. Our treatment of nucleating boundaries is a mechanism for including the effect of certain system inhomogeneities. While we will consider nucleation and growth more generally in the context of inhomogeneous systems, we will not gain much further insight. The central point is that when there are predetermined nucleation sites, at a boundary or internally in the system, the relaxation of a system into the stable state can proceed rapidly through driven diffusion. This behavior occurs when the interactions in the system are cooperative, so that locally they reinforce both the metastable and stable phases. It is easy to imagine particular amino acids or amino acid combinations serving as nucleation sites around which the 3-d structure of the protein forms. In particular, the formation of an α-helix or β-sheet structure may nucleate at a particular location and grow from there.

Our discussion of nucleation and growth takes care of almost all cases of relaxation in homogeneous systems when the interactions are short-ranged and there is a well-defined ground state in the bulk—away from the boundaries. We can, however, have a well-defined ground state of the system even if the bulk ground state is not well-defined, if the boundary conditions impose the ground state. This is the subject of the next section.

4.4.4 Boundary-imposed relaxation

We have been careful to consider cases in which the energy of the state with all spins DOWN, $\{s_i = -1\}$, is lower in energy than any other state, and in particular of the initial state with all spins UP, $\{s_i = +1\}$. In a system with ferromagnetic interactions, if the energies of the initial and final states are equal, then there are two ground states. In the homogeneous model this is the case where $h = 0$. In general, the existence of two ground states is counter to our assumptions about relaxation to a unique ground state. However, we can still have a unique ground state if the boundaries impose $s_i = -1$. Such boundaries mean that the ground state is uniquely determined to be $\{s_i = -1\}$, even though $h = 0$.

In the absence of additional nucleation events, such a system would equilibrate by inward diffusion of the interface between the UP interior and the DOWN border, as in Fig. 4.4.5(c). There is no bulk driving force that locally causes the UP region to shrink. The only driving force is the interface energy (surface tension) that causes the interface to shrink. We can treat the system as performing a driven random walk in the number of UP spins. However, we must treat each part of the boundary as moving essentially independently. The rate of change of the average number of UP spins $N_+(t)$ is given by the boundary velocity times the boundary length:

$$\frac{dN_+(t)}{dt} \propto vN_+(t)^{(d-1)/d} \qquad (4.4.51)$$

The velocity of a driven random walk is (from Eq. (1.4.58) and (1.4.60))

$$v = av(e^{-\Delta E_+/kT} - e^{-\Delta E_-/kT}) = ave^{-\Delta E_+/kT}(1 - e^{(\Delta E_+ - \Delta E_-)/kT}) \qquad (4.4.52)$$

From Fig. 1.4.5 we can see that $(\Delta E_+ - \Delta E_-)$ is the energy difference between steps. A single step changes the number of UP spins by one, so

$$(\Delta E_+ - \Delta E_-) = E(N_+(t)) - E(N_+(t) - 1) \approx \frac{dE(N_+(t))}{d(N_+(t))} \propto N_+(t)^{-1/d} \qquad (4.4.53)$$

where $E(N_+(t)) \propto N_+(t)^{(d-1)/d}$ is the average surface energy for a cluster with $N_+(t)$ UP spins. Since the single-step energy difference decreases with the number of UP spins, we can assume it is small compared to kT. We can then expand the exponential inside the parenthesis in Eq. (4.4.52) and substitute the resulting expression for the velocity into Eq. (4.4.51) to obtain

$$\frac{dN_+(t)}{dt} \approx -\frac{av}{kT}e^{-\Delta E_+/kT}(\Delta E_+ - \Delta E_-)N_+(t)^{(d-1)/d} \propto -N_+(t)^{-(d-2)/d} \qquad (4.4.54)$$

The negative sign is consistent with the decreasing size of the region of UP spins. We integrate Eq. (4.4.54) to find the dependence of the relaxation time on the size of the system:

$$\tau(N) \propto \int_N^0 -N_+(t)^{(2-d)/d} dN_+(t) = N^{2/d} \qquad (4.4.55)$$

The final expression is valid even in one dimension, where the boundary executes a random walk because there is no local preference of the boundary to move in one direction or the other and $\tau(N) \propto N^2$.

In this discussion we have ignored the possible effect of the nucleation of regions of DOWN spins away from the boundary of the system. One way to understand this is to note that the size of a critical nucleus is infinite when $h = 0$. Nucleation may only change the relaxation behavior when the interface between UP and DOWN spins is not well-defined. Otherwise, nucleation does not help the relaxation, since any region of DOWN spins inside the region of UP spins will shrink. Thus, for the case where the boundary determines the ground state, the relaxation is $O(N^{2/d})$.

It is also possible to consider the case where h is positive and the bulk preference is to have all of the spins UP. However, the boundary imposes all $s_i = -1$, and the competition between the bulk and boundary energies still results in all of the spins in the ground state being DOWN. This occurs for systems where h is much smaller than $J > 0$, so that the influence of the boundary can extend throughout the system. The energy at the interface is $bN_+(t)^{d/d+1}$ where $b \propto J$ is a local measure of the boundary energy. The bulk energy is $-2hN_+(t)^d$. The latter must be smaller in magnitude than the former. As $N_+(t)$ becomes smaller, the bulk energy becomes still smaller compared to the interface energy. Thus we can neglect the bulk energy in calculating the relaxation time, which scales with N as if $h = 0$.

4.4.5 Long-range interactions

When interactions have a range comparable to the system size, the possibility of defining interior and exterior to a domain does not generally exist. If we assume a long-range ferromagnetic interaction between spins so that $J_{ij} = J$, for all i and j, the energy of the system is

$$E(\{s_i\}) = -h \sum_i s_i - J\tfrac{1}{2} \sum_{i \neq j} s_i s_j \qquad (4.4.56)$$

There is a difficulty with this expression because the energy is no longer extensive (proportional to N) since the second term grows as N^2 when all the spins are aligned. As discussed in Section 1.6, for many calculations the long-range interactions are scaled to decrease with system size, $J \sim 1/N$, so that the energy is extensive. However, it is not obvious that this scaling should be used for finite systems. If we keep h and J fixed as the system size increases, then, as shown below, one term or the other dominates in the energy expression.

We can solve Eq. (4.4.56) directly by defining the collective variable

$$M = \sum_i s_i \qquad (4.4.57)$$

Substituting this into the energy gives:

$$E(\{s_i\}) = -hM - \tfrac{1}{2}JM^2 + NJ/2 \qquad (4.4.58)$$

The term $NJ/2$, which is independent of M, accounts for the missing $i = j$ terms in Eq. (4.4.56). It does not affect any of the results and can be neglected. Adding the entropic contribution from Section 1.6 to obtain the free energy as a function of M we obtain

$$F(M) = -hM - \tfrac{1}{2}JM^2 - TNs_0(M/N) \tag{4.4.59}$$

$$s_0(x) = k\{\ln(2) - \tfrac{1}{2}(1+x)\ln(1+x) + (1-x)\ln(1-x)\} \tag{4.4.60}$$

The exact substitution of the collective variable M for the many variables s_i indicates that this system reduces to a single-variable system. The maximum value of M increases linearly with N. In the following we show self-consistently that M itself grows linearly with N, and obtain the relaxation-time scaling.

Assuming that M grows linearly with N, the first and third terms in the free energy grow linearly with N. The second term $\tfrac{1}{2}JM^2$, describing interactions, grows quadratically with N. For small enough N the interaction term will be insignificant compared to the other terms, and the system will become essentially decoupled. For a decoupled system M must grow linearly with N. The relaxation time is also the same as the relaxation time of a decoupled system.

For N larger than a certain value, the terms that are linear in N become negligible. Only the interaction term is important. All of the spins must be either UP or DOWN in order to minimize the free energy. This also implies M grows linearly with N. There is a small energy difference between UP and DOWN that is controlled by the value of h. However, to switch between them the system must pass through a conformation where half of the spins are UP and half are DOWN. The energy barrier, $F(M=0) - F(M=N) = JM^2/2$, scales as N^2. Because the barrier grows as N^2 the relaxation time grows as e^{N^2}. Thus the system is frozen into one state or the other. We can still consider raising the temperature high enough to cause the system to flip over the barrier. In this case, however, the difference in energy between UP and DOWN is not enough to force the system into the lower energy state.

Including the small system regime, where the long-range interactions are not relevant, and the large system regime, gives a relaxation-time scaling of $O(\ln(N), e^{N^2})$. We see that even simple models with long-range interactions have a relaxation time that scales exponentially with system size. Another conclusion from this section is that in the presence of long-range interactions, the relaxation-time scaling does not decrease as the system size increases. This behavior was characteristic of systems that have short-range interactions.

It is interesting to consider what would happen if we scale the interactions $J \sim 1/N$. Since all the energies are now extensive, the free-energy barrier would grow linearly in the size of the system and the relaxation time would grow exponentially with the size of the system. Starting from all of the spins UP, the system would rapidly relax to a metastable state consisting of most of the spins UP and a fraction of DOWN spins as determined by the local minimum of the free energy. This relaxation is fast and does not scale with the system size. However, to flip to the ground state with most of the spins DOWN would require an $O(e^N)$ relaxation time.

We could also consider decaying interactions of the form

$$E(\{s_i\}) = -h\sum s_i - \sum J(|r_i - r_j|)s_i s_j \tag{4.4.61}$$

$$J(x) \propto x^p \tag{4.4.62}$$

For $p < -1$ this is essentially the same as short-range interactions, where there is a well-defined boundary and driven diffusion relaxation. For $p > 1$ this is essentially the same as long-range interactions with exponential relaxation time. $p = 1$ is a crossover case that we do not address here.

4.5   Inhomogeneous Systems

In the general inhomogeneous case, each spin has its own preference for orientation UP or DOWN, determined by its local field, h_i, which may be positive or negative. This preference may also be overruled by the interactions with other spins. We begin, however, by reconsidering the decoupled or essentially decoupled model for inhomogeneous local fields and relaxation times.

4.5.1 Decoupled model—barrier and energy difference variation

There are two ways in which inhomogeneity affects the decoupled model. Both the spin relaxation time, τ_i, and the energy difference $E_{+i} = -2h_i$, may vary between spins. Analogous to Eq. (4.4.10), the average number of UP spins is given by:

$$N_+(t) = \sum_i P(s_i = -1; t) \approx \sum_i \left[e^{-t/\tau_i} + e^{-E_{+i}/kT} \right] = \sum_i e^{-t/\tau_i} + N_+(\infty) \quad (4.5.1)$$

For a distribution of relaxation times $P(\tau)$ this can be written as:

$$N_+(t) = N \int d\tau\, P(\tau) e^{-t/\tau} + N_+(\infty) \quad (4.5.2)$$

We are assuming that $P(\tau)$ does not depend on the number of spins N. The relaxation time of the system is defined so that all spins relax to their ground state. It might seem natural to define the system relaxation time $\tau(N)$ as before by Eq. (4.4.11) or Eq. (4.4.16):

$$N_+(\tau(N)) = (1 + e^{-1})N_+(\infty) + N_r \quad (4.5.3)$$

However, allowing an additional factor of e^{-1} spins that are unrelaxed can cause problems in the inhomogeneous model that were not present in the homogeneous case. When there is only one microscopic relaxation time, the existence of nonequilibrium residual UP spins can be considered as a small perturbation on the structure, if they are a smaller fraction of spins than the equilibrium UP spins. There is no special identity to the spins that have not yet relaxed. In the present case, however, the spins with longest relaxation times are the last to relax. It is best not to assume that the structure of the system is relaxed when there are specific spins that have not relaxed. This leads us to adopt a more stringent condition on relaxation by leaving out the e^{-1} in Eq. (4.5.3), $N_+(\tau(N)) = N_+(\infty) + N_r$. Combining this with Eq. (4.5.2) we have:

$$N_r = N \int d\tau\, P(\tau) e^{-\tau(N)/\tau} \quad (4.5.4)$$

where N_r is a number that should be less than one, or for definiteness we can take $N_r \approx 0.5$, as in Eq. (4.4.19).

One way to understand Eq. (4.5.4) is to let all of the spins except one have a relaxation time τ_1. The last one has a relaxation time of τ_2. We ask how does the relaxation time of the final spin affect the relaxation time of the whole system. The relaxation of the spins with τ_1 is given by the usual relaxation time of a system of N spins (Eq. (4.4.17)). If the relaxation time of the final spin is shorter than this, it does not affect the system relaxation time. If it is longer, then the system relaxation time will be determined by τ_2. Thus spins with long relaxation times, almost as long as the relaxation of the whole system, can exist and not effect the relaxation time of the system. The conclusion is more general than the decoupled model. A realization of this in protein folding is the amino acid proline. Experimental studies indicate that proline has two conformations that correspond to cis and trans isomers. The conversion of one form to the other has been found to limit the time of folding of particular proteins. We note that the temperature at which the folding is done can play a role in the relative importance of a single long relaxation time as compared to the relaxation of the rest of the system. When a single relaxation time is large in comparison to the relaxation of other spins, it becomes proportionately even larger as temperature is lowered (Question 4.5.1). The existence of the long proline relaxation time is consistent with a rule of thumb that nature takes advantage of all possibilities. Since it is possible for such a long relaxation time to exist, it does.

Question 4.5.1 Assume all of the spins in a system except one have a relaxation time of τ_1 and the last one has a relaxation time of τ_2. Show that if the last spin has the same relaxation time as the rest of the spins together, at a particular temperature, then it is slower at lower temperatures and faster at higher temperatures.

Solution 4.5.1 The key point is that the relaxation times depend exponentially on the temperature and the large relaxation time will change more rapidly with temperature than the smaller one. The ratio of relaxation times of individual spins as a function of temperature is given by:

$$\tau_2(T)/\tau_1(T) = e^{-E_{B2}/kT} / e^{-E_{B1}/kT} = e^{-(E_{B2}-E_{B1})/kT} \qquad (4.5.5)$$

where E_{B1} and E_{B2} are the barrier energies for the respective relaxation processes. In order for the relaxation time of the last spin to be relevant we must have $E_{B2} > E_{B1}$. As a function of temperature, the ratio increases exponentially with decreasing temperature:

$$\tau_2(T)/\tau_1(T) = \left(\tau_2(T_0)/\tau_1(T_0)\right)^{T_0/T} \qquad (4.5.6)$$

where T_0 is a reference temperature.

We are interested in comparing the relaxation time of $N-1 \approx N$ spins whose individual relaxation time is $\tau_1(T)$, with the relaxation of one spin whose individual relaxation time is $\tau_2(T)$. Thus we are concerned with the quantity:

$$\tau_2(T)/\tau_1(T,N) \approx \tau_2(T)/(\tau_1(T)\ln(N)) \qquad (4.5.7)$$

where we write $\tau_1(T,N)$, as the relaxation time of N spins whose individual relaxation time is τ_1. We have used the expression for this relaxation time obtained from the decoupled spin model of Section 4.4.1. This is not essential; as discussed below the result really only depends on having $\tau_2(T_0)/\tau_1(T_0) \gg 1$.

Since we are given that the last spin has the same relaxation time as the rest of the spins together at the reference temperature T_0, i.e., $\tau_2(T_0)/\tau_1(T_0,N) = 1$ evaluating Eq. (4.5.7) at $T = T_0$ we have that:

$$\left(\tau_2(T_0)/\tau_1(T_0)\right) = \ln(N) \qquad (4.5.8)$$

Considering this relaxation time ratio as an expression for $\ln(N)$, we substitute Eq. (4.5.8) and Eq. (4.5.6) into Eq. (4.5.7) to find that:

$$\tau_2(T)/\tau_1(T,N) = \left(\tau_2(T_0)/\tau_1(T_0)\right)^{(T_0/T)-1} \qquad (4.5.9)$$

which implies the desired result. For $T > T_0$ this ratio is less than one, and for $T < T_0$ this ratio is greater than one.

For the decoupled model, because the relaxation time increases only slowly with the number of spins, the ratio of the relaxation times in Eq. (4.5.8) is not very large, so that the temperature dependence of the ratio of relaxation times will also not be strong, even though it is exponential. However, Eq. (4.5.9) is more general. We can understand this by allowing the rest of the system to interact, except for the individual spin. Our conclusions hold as long as the relaxation of the interacting spins depends on a large number of hops over barriers. These interacting spins give rise to a relaxation time $\tau_1(T,N)$ that depends on the number of spins as some function of N. The consequence in the above equations would only be to replace $\ln(N)$ with this function of N. Eq. (4.5.9) would be unaffected. The ratio of individual spin relaxation times at a reference temperature, $\left(\tau_2(T_0)/\tau_1(T_0)\right)$, could even be determined empirically. Moreover, if a single barrier has a relaxation time of the same duration as the rest of the protein, the conclusion is immediate. Since microscopic relaxation times of a single degree of freedom can be as small as 10^{-10} seconds, and that of the protein is of order 1 second, the ratio between the two relaxation times is large and Eq. (4.5.9) would imply a rapid dependence of the relaxation time ratio with temperature. ∎

The more general case of an arbitrary distribution of individual spin relaxation times $P(\tau)$ in Eq. (4.5.4) can lead to arbitrary scaling of the total relaxation time with the number of spins. Intuitively, there appears to be a problem with this statement, since the spins are independent. How can the relaxation time grow arbitrarily if we only, say, double the system size? The reason that the relaxation time can grow arbitrarily is that when we increase the system size, there is a greater chance for spins with longer relaxation times to occur. It is the addition of spins in the tail of the distribution of probabilities $P(\tau)$ that controls the scaling of the relaxation time of the system. However, if we only have a few different relaxation times corresponding to a limited number of types of amino acids, then increasing the system size cannot change the relaxation time more than logarithmically with the system size. Thus, if the distribution of spin relaxation barriers is relatively narrow or is composed of a number N_τ of narrow distributions, where $N_\tau \ll N$, then we will still have the characteristic scaling $O(\ln(N); 1)$. This will be assumed for the remaining inhomogeneous models.

From Eq. (4.5.4) we see that variations in E_{+i}, while keeping τ_i fixed, do not affect the scaling of the relaxation time in the decoupled model. If we return to a consideration of the basic properties of relaxation there are two points that imply this conclusion. The first is the effect of E_{+i} on the relaxation rate of an individual spin. The relaxation rate of an individual spin can be affected only if the difference in energy between the two states becomes very small. Even in this case, the change can be at most a factor of 2 (see Eq. (1.4.44)). A factor of 2 is not particularly important when we consider relaxation-time scaling. The second point is that in general we do not allow the value of E_{+i} to become very small because of our assumption that almost all of the spins relax to their ground state. Thus the impact of variations in E_{+i} should be negligible.

Our discussion in this section of the effect of variations in τ_i and E_{+i} is valid also in the case of the essentially decoupled model, where interactions are allowed between spins as long as the interactions do not affect which of the states of a particular spin is the lowest energy. In addition to allowing variations in τ_i and E_{+i}, we can also allow inhomogeneous interactions between spins. In Section 4.4.2, in the homogeneous case, it was natural to assume that the parameters τ_{max} and E_{+min} do not change with system size. In the inhomogeneous case this assumption is less natural. However, once this assumption is made, the arguments presented in Question 4.4.2 proceed as before.

More significant for our interests is that the inhomogeneous case provides new models that retain the same relaxation-time scaling as the decoupled model. Specifically, it is possible for interactions to affect the minimum energy conformation of particular spins without changing the relaxation-time scaling. This is the topic of the next section.

4.5.2 Space and time partition (decoration of the decoupled model)

The next inhomogeneous model includes interactions that change the minimum energy state of particular spins. In the homogeneous case this led immediately to models with relaxation controlled by nucleation and growth. In the inhomogeneous case there is a richer analysis. Our first objective is to construct a generalization of the decoupled model that still relaxes with the same scaling. This can happen because, even if a few spins start out with their local equilibrium being UP, as long as the other spins have their equilibrium as DOWN the few spins will relax DOWN once the rest have. We can generalize this systematically. The idea that we will develop in this section is that an inhomogeneous system may be constructed so that it can be partitioned in space and time. The partitioning results in a finite collection of subsystems. We can then relate the relaxation of the whole system to the relaxation of each of the subsystems, and to the behavior of the subsystems as N increases. Partitioning the system in space and time is closely related to the discussion of subnetworks in Chapter 2. Partitioning in space is directly related to the discussion of subdivision in attractor networks, while partitioning in time is more loosely analogous to the discussion of feedforward networks.

It is useful to consider again the conceptual framework of renormalization discussed in Section 1.10. In essence the subsystems that we will construct are decoupled relaxing variables. They act like individual spins in a decoupled model. We can think about renormalizing the system by grouping together all of the spins in each subsystem. Each subsystem is then replaced by a single spin, with a relaxation time equal to the relaxation time of the original subsystem. The result of the renormalization is a decoupled system of spins. Another way to think about this is to invert the process of renormalization. This inverse process is called decoration. Starting from the decoupled model, we decorate it by replacing each spin with a subsystem formed out of many spins.

Space partitioning is the separation of the whole system into subsystems. We impose a much more stringent form of separation than that in Chapter 2. Within each subsystem the values of the spins may affect each other's minimum energy state but they do not affect the minimum energy state of spins in other subsystems. This does not mean that there are no interactions between spins in different subsystems, only that they are not strong enough to matter. The whole system then relaxes according to the combination of relaxation times of each subsystem combined as in the decoupled case, specifically Eq. (4.5.4). However, the distribution of relaxation times $P(\tau)$ may now depend directly upon N.

As N increases, either the number of subsystems or the size of subsystems grows. If the size of the subsystems does not grow with N, the internal behavior of each subsystem does not affect the scaling of the relaxation time of the whole system. The relaxation of the system depends only on the distribution of relaxation times of the subsystems, exactly as Eq. (4.5.4) describes the relaxation in terms of individual spins. If the number of subsystems does not change and the subsystems grow linearly with N, then the relaxation-time scaling of the whole system follows the relaxation-time scaling of the subsystem with the longest relaxation time. Unless special circumstances apply, this would correspond to the highest scaling. There are other possible ways for the growth of the system with N to be distributed between subsystem growth and growth of the number of subsystems. They can be analyzed in a similar manner.

Time partitioning implies that some spins know their equilibrium conformation from the start. When they are equilibrated, their effect on the remainder causes some of the remaining spins to relax. Then a third set of spins relax. The dynamics is like a row of dominoes. This can be illustrated first by considering only two subsystems. Let

$$W_1 = \{ s_i \mid \min(E_{+i}(\{s_j\}_{j \neq i})) > 0 \}$$

$$W_2 = \{ s_i \mid \min(E_{+i}(\{s_j\}_{j \neq i})) \leq 0 \}$$

(4.5.10)

Thus, W_2 is the set of s_i such that $E_{+i}(\{s_j\}_{j \neq i})$ can be negative. If all s_i in W_2 are in some sense independent of each other, then the relaxation of the system will still scale as $O(\ln(N);1)$. This is because the spins in W_1 relax first, then the spins in W_2 relax. The condition of independence of spins in W_2 that we need to impose has to do with which spins can affect the sign of their energy $E_{+i}(\{s_j\}_{j \neq i})$. Specifically, the spins whose state can affect the sign of $E_{+i}(\{s_j\}_{j \neq i})$ must all be in W_1, not W_2. This implies that only relatively weak interactions exist between two spins in W_2. If this is true, then consider all spins in W_1. These spins satisfy the conditions of the essentially independent model, so their relaxation takes at most $O(\ln(N);1)$ time. Once these have flipped DOWN, the remaining UP spins, all of which must be in W_2, are decoupled and therefore must satisfy $E_{+i}(\{s_j\}_{j \neq i}) > 0$. Since they satisfy the conditions of the essentially independent model, they also relax in $O(\ln(N);1)$. The total relaxation is (at most) the sum of these relaxation times and so is also $O(\ln(N);1)$. In summary, the relaxation scaling does not depend on spins that prefer to be UP for some arrangements of their neighbors, if none of their neighbors have this property at the same time as they do.

The partitioning of the system into two subsystems that relax sequentially can be generalized to a finite number of subsystems. If the spins of a system can be partitioned into a finite set of subsystems $\{W_k\}$, such that for a spin s_i of set W_k, $E_{+i}(\{s_j\}_{j \neq i}) > 0$ when all the $s_j = -1$ in sets $W_1, ..., W_{k-1}$, then the system relaxes in $O(\ln(N);1)$. This follows because the subsystems relax sequentially, each in $O(\ln(N);1)$. One may think about the subsystems as illustrated in Fig. 4.5.1. Each successive circle denotes the boundary of a subsystem. The smallest region relaxes first, followed by the next larger one. The scaling $O(\ln(N);1)$ for the whole system follows from the scaling of each subsystem in $O(\ln(N);1)$, when the number of subsystems is assumed to be independent of N. It is also possible to construct models where the number of subsystems grows with N. For specific assumptions about how the number of subsystems changes with N, the relaxation-time scaling can be determined.

A better way to describe the partitioned model uses a concept of the neighborhood of a spin. (The definition of "neighborhood" used in this section does not satisfy the conditions necessary to give a topology on the space.) For statistical fields, the physical distance is not particularly significant; it is the magnitude of the interaction between spins that determines the effective proximity. For the nearest-neighbor Ising models in Section 1.6, we determine interactions by using a spatial arrangement of spins and assign equal interactions to the nearest neighbors. For a cubic lattice, the number of nearest neighbors is directly related to the dimensionality ($z = 2d$). Other lattices give different numbers of neighbors. More generally, we can define the neighbors of a spin s_i as the spins s_j that can change the minimum energy state of the spin s_i.

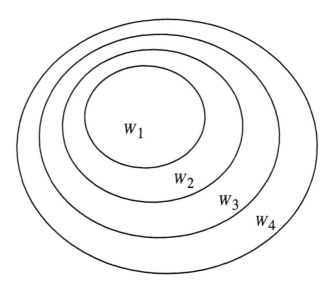

Figure 4.5.1 Illustration of the time-partitioning model. A system of N spins is partitioned so that each of the subsystems $W_1,...,W_k$ relaxes sequentially. In order for the partitions to relax sequentially, the interaction between spins must satisfy a constraint on their interactions described in the text. For example, the relaxation of a spin in W_2 can only depend on spins in W_1 and not on any others. Under these circumstances, once the spins in W_1 relax, so can the spins in W_2. There is no implied spatial arrangement of spins in this model. ∎

Let a neighbor s_j of a spin s_i be a spin that can affect the minimum energy conformation of s_i. Let the neighborhood U_i of s_i be the set of its neighbors. Then the neighborhood of an arbitrary set of spins is the union of the neighborhoods of all its members. We can summarize the results of time partitioning by recognizing that the definition of W_k implies that a spin s_i in W_k must have all of its neighbors in the subsystems $W_1,..., W_{k-1}$. Thus, time partitioning corresponds to a set of subsystems W_k such that the neighborhood of W_k is contained entirely in $W_1,..., W_{k-1}$. For such a system the relaxation time is $O(\ln(N);1)$.

We follow a chain of seemingly natural definitions. The interior W^I of a set W is the set of spins whose neighborhoods are entirely contained in W. The exterior W^E of a set W is the set of spins whose neighborhoods do not intersect W. The boundary W^B of a set W is the set of spins that are not in its interior or exterior (spins whose neighborhoods intersect but are not entirely contained within W). For the time-partitioned model, all subsystems W_k are contained in their own exterior, $W_k \subset W_k^E$. This unusual conclusion points to the difference between our neighborhoods and the usual concept of neighborhood. It is rooted in a fundamental asymmetry in our definition of "neighbor".

Time partitioning depends on an asymmetric neighbor relationship. If s_j is a neighbor of s_i, then s_i does not have to be a neighbor of s_j. This arises through inhomogeneity of the local fields h_i that make J_{ij} have a different significance for s_i than for

s_j. The spins with the largest values of h_i tend to be in W_k with lower values of k. A spin in W_1 must have a large enough h_i so that it dominates all of the interactions and there are no spins in its neighborhood.

The definition of "neighborhoods" enables us also to summarize space partitioning. The partitioning of space corresponds to a partitioning of the system into disjoint neighborhoods. The neighborhood of each subsystem does not intersect any other subsystem. Thus, in this case, we can say that each subsystem is the same set of spins as its own interior. Space partitioning can result from both inhomogeneous interactions and fields.

The model of decorated independent relaxation with both spatial and temporal subsystems is attractive as a model of the relaxation in protein folding. The existence of secondary structure, with limitations on the size of secondary-structure elements, suggests that secondary-structure elements may form first. Moreover, each of them may form essentially independently of the others. This would correspond to space partitioning. Each set of coordinates that change and cause the formation of a particular secondary-structure element would be a single subsystem. All of these together would be included in the same time partition. Then there is a second stage of relaxation that forms the tertiary structure. The coordinates that control the formation of the tertiary structure would constitute the second time partition. It is possible, however, and even likely, that during the second stage in which tertiary structure is formed, some of the secondary structure also changes.

4.5.3 Nucleation and growth in inhomogeneous fields

Diffusive equilibration can be generalized to the inhomogeneous case. General conclusions can be reached by relatively simple considerations; a complete analysis is difficult. Nucleation and growth is a model that applies when nucleation is a relatively rare event, so that only one critical nucleus forms in a large region. After the critical nucleus is formed, the region of the stable phase grows by driven diffusion of the boundary between the stable and metastable phases. In order to have a diffusive inhomogeneous system, the interactions between spins J_{ij} must be important compared to the variation in the local field, h_i, and the interactions must be essentially local and uniform. Inhomogeneities tend to enhance nucleation and inhibit diffusion of the boundaries between stable and metastable phases. Thus, increasing the inhomogeneity tends to reduce the relevance of nucleation and growth. We will discuss more specifically the effect of variations in h_i and J_{ij}, and then the effect of inhomogeneity in general, on the scaling of the relaxation time.

Inhomogeneities of the local fields h_i cause variations in the local strength of preference for the stable and metastable phases. Regions that have a larger average negative h_i will tend to nucleate before regions of a smaller average negative h_i. Since the problem is to form a nucleus somewhere, in contrast to the rare nucleation in a homogeneous system, this variation increases the rate of nucleation. The effect of variation in h_i on diffusion of a boundary between stable and metastable phases occurs through local variation in the driving force. Sites that have a larger than average negative h_i tend to increase the boundary velocity v, while sites of lower than average negative h_i tend to decrease the boundary velocity. The boundary must sweep through every site. Moreover, there is no bound on how long the boundary can be delayed, so the sites that slow it tend to trap it. Thus, on average the velocity is reduced.

Inhomogeneities of the interactions J_{ij} cause similar variations in nucleation and diffusion. Smaller values of J_{ij} make nucleation easier and the boundary diffusion slower. Conversely, larger values of J_{ij} make nucleation harder and the boundary diffusion faster. Since nucleation can occur anywhere while diffusion must sweep through everywhere, again the nucleation rate is increased while the diffusion rate is reduced.

For the case of nonnucleating boundaries, the effect on relaxation time is particularly significant. The time necessary to form a critical nucleus is apparent in the relaxation-time scaling behavior as a peak in Fig. 4.4.6. With the introduction of inhomogeneities, the peak will decrease in height. For the case of nucleating boundaries, the relaxation time is controlled by the diffusion rate and so the relaxation time will increase. For both cases, the transition to the thermodynamic limit, where the relaxation time is independent of N, will occur at smaller system sizes. This occurs because the increasing nucleation rate and decreasing diffusion rate causes the typical size to which one nucleus grows—which is the size of independently relaxing parts of the system—to decrease.

Another consideration in the discussion of diffusive relaxation in inhomogeneous fields is the structure of the boundary. In the presence of inhomogeneities, the moving boundary becomes rougher due to the inhomogeneities that slow and speed its motion. As long as the bulk energy dominates the boundary energy, it will remain smooth; however, when the variation in boundary energy becomes large enough, the boundary will become rough and the dynamic behavior of the system will change. Since we have limited ourselves to considering smooth boundaries, our discussion does not apply to this regime.

As briefly discussed in Section 4.4.3, the model of diffusion in variable fields is likely to be of relevance to understanding the local properties of protein folding in the nucleation and growth of the secondary structure. If this applies locally to each of the segments of secondary structure separately, then the scaling of this relaxation is not necessarily relevant to the folding as a whole. However, it is relevant to our understanding of the local kinetics by which secondary structural elements are formed.

4.5.4 Spin glass

There have been some efforts to describe the problem of protein folding in terms of a spin glass model and spin glass dynamics. Spin glasses are treated using models that have long-range random interactions between all spins (Section 1.6.6):

$$E[\{s_i\}] = -\frac{1}{2N}\sum_{ij} J_{ij} s_i s_j \qquad (4.5.11)$$

The difficulty with this model is that many of the properties of spin glasses do not apply to proteins. Spin glasses have many degenerate ground states, the number of which grows with the size of the system. This means that there is no unique conformation that can be identified with the folded state of the protein. Choose any conformation, the system will spend much more time in dramatically different conformations because of the essential degeneracy of ground states. Moreover, the barriers between low-lying states also grow with the size of the system. Thus, the relaxation time between any of the low-lying states grows exponentially with system size. Even the concept of equilibration must be redefined for a low temperature spin glass, since true equilibration is not possible. What is possible is a descent into one of the many low-lying energy states. If we model a particular set of interactions J_{ij} as being specified by the primary structure of a protein, there would be no correlation between low-lying states reached by different proteins with the same primary structure. This is in direct contrast to protein folding, where a unique (functional) structure of the protein must be reached.

Despite the great discrepancy between the phenomenology of spin glasses and the protein-folding problem, there are reasons for considering this model. The use of a spin glass model for protein folding is based on the understanding that many possible bonding arrangements between amino acids are possible. For a sufficiently long chain there are many compact conformations of the chain where different bonding arrangements are found. There is always an inherent frustration in the competition between different possible bonding arrangements of the amino acids. This frustration is similar to the frustration that is found in a spin glass. Because of this, in the very long chain limit, the spin glass model should become relevant. In this limit the frustration and multiple ground states are likely to be the correct description of the chain.

However, as discussed in Section 4.4.5, even when there are long-range interactions, the local fields, h_i, can be more important than the interactions, J_{ij}, for small enough systems. In an inhomogeneous system we can expand the term "local field" to include the effect of local interactions:

$$E(\{s_i\}) = -\sum h_i s_i - \frac{1}{2}\sum_{<ij>} J_{ij} s_i s_j - \frac{1}{2}\sum_{ij} J'_{ij} s_i s_j \qquad (4.5.12)$$

where the second sum describes the near-neighbor interactions and the third describes the long-range interactions. Long-range interactions that give rise to frustration may not dominate over local interactions. There are many different energies in the problem of protein folding. The analog of local interactions in Eq. (4.5.12) are the interactions between amino acids near each other along the chain, not interactions that are local in space. Hydrogen bonding between different parts of the chain, even though it is local in space, can give rise to frustration. Note that the α-helix structure is constructed entirely out of short-range interactions, while the β-sheet structure is formed out of a combination of short-range and long-range interactions.

There is a difference between bonding between different parts of the amino acid chain and long-range interactions in an Ising model. Although there are many possible hydrogen bond interactions between amino acids, these interactions are quite restricted. The number of amino acids that can interact with a particular amino acid at any one time is limited. Moreover, the chain structure restricts which combinations of amino acids can interact at one time. These limitations do not eliminate the problem of frustration for very long chains. They do, however, increase the chain length at which crossover occurs, from the regime in which local interactions dominate, to the regime in which long-range interactions dominate. It is the latter regime which is a candidate for the spin glass model.

Our discussion implies that proteins are fundamentally restricted in their length, and that a treatment of their dynamics should include this finite length restriction. From experiment we know that each element of the secondary structure has a limited number of amino acids, and the number of secondary-structure elements in the protein is also limited. These observed limitations on protein size are consistent with our discussion of the relative importance of local fields and long-range interactions. Structural frustration due to long-range interactions must limit the size of proteins to the regime in which local fields, or more generally local interaction energies, are important. It should be assumed that proteins extend up to their maximal possible size. Thus, the largest proteins are likely to be at the crossover point when both short-range and long-range interactions compete. This competition should then play an important role in the relaxation-time scaling.

The assumption of frustration in the long-range interactions appears to be the opposite of the cooperative bonding that has been found in proteins. Cooperative bonding is equivalent to long-range ferromagnetic interactions that enhance the stability of the ground state. Frustration implies that different bonds are competing with each other. It is possible to argue that the low-energy states of the spin glass represent cooperative action of many bonds and therefore constitute cooperative bonding. On the other hand, proteins are engineered, so that we would expect that bonds are designed to reinforce each other and cooperatively lower the energy of the folded state to increase its stability. This is unlike the random spin glass model. This notion of engineered cooperativity leads us to consider the engineered spin glass, which is more typically used as a model of neural network memory.

4.5.5 *Engineered spin glass—neural network*

Neural networks (Chapter 2) have been modeled as engineered spin glass systems (the attractor network) where energy minima of the system are specified. This might be considered to be analogous to the engineering of the 3-d structure of a folded protein by selection of the amino acid sequence. In the attractor network, the interactions J_{ij} determine the minimum energy states. In our discussion of protein folding in this chapter, it is largely the local fields h_i that determine the minimum energy state. One of the differences is that the attractor network cannot have a unique ground state—the inverse of a state has the same energy.

The simplest way to model the engineered spin glass is through the Mattis model (Question 1.6.12). In this model a particular state is determined to be a ground state using only interactions J_{ij} and no local fields h_i. We can redefine all of the spins in the ground state to be $s_i = -1$. Then the Mattis model is equivalent to the long-range ferromagnetic Ising model with no external field, $h = 0$, and all $J_{ij} = J$. Since it is the interactions that determine the ground state, both $s_i = -1$ and its inverse $s_i = +1$ are ground states.

Under these circumstances we cannot consider the folding transition to be from $s_i = +1$ to $s_i = -1$. We can recognize, however, that the essential point of this model is to consider the impact of the initial conditions. We therefore abandon our insistence on starting from a state where all of the spins are UP. The system will relax to the desired ground state if the initial conditions are favorable, specifically, if more of the spins in the initial state are DOWN than are UP.

One way to think about this is to look at the transition in terms of changing suddenly the interaction parameters. Indeed, this is a physically meaningful analogy, since the actual folding of proteins is achieved by changing the interaction energies of the real system. Fig. 4.5.2 illustrates several different transitions on a phase diagram of the ferromagnet that includes both the interaction J and the field h. The transition we have been considering thus far in this chapter is the transition across the first-order transition boundary shown as (A). In this section we are considering the disorder-to-order transition that is represented by (B). As long as there are a majority of DOWN

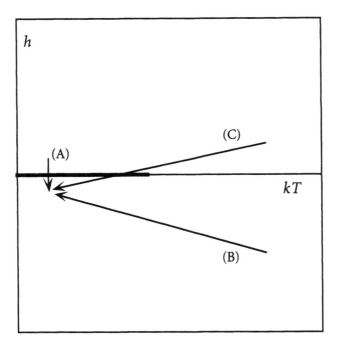

Figure 4.5.2 Illustration of transitions in a ferromagnetic Ising model that start with different initial conditions. The transitions, indicated by arrows, are superimposed on the Ising model phase diagram. The final state in each case corresponds to having all spins DOWN. (A) is a first-order transition starting from all spins UP. (B) and (C) both start from a largely random arrangement of spins but (B) starts from a majority of DOWN spins. (C) starts from a majority of UP spins. ∎

spins in the initial state, there is no need for the process of nucleation and growth to occur. The relaxation is local, and the system reduces to the decoupled model.

We can generalize the Mattis model to the attractor neural network models discussed in Chapter 2. In these models, there may be more than one energy minimum. As with the random spin glass, an arbitrary initial condition leads to any one of these low-energy states. Therefore, we cannot talk about a unique folded state in equilibrium. However, there is a difference in this case. The neural network can be designed to have only a limited number of low-energy states. Each energy state has a basin of attraction that consists of all of the states of the system that will naturally fall toward the low-energy state. The basin of attraction of a particular minimum energy state consists of initial states that have more than a certain overlap with the minimum energy state. Within this basin of attraction, the dynamics that updates the spins reaches the ground state in a finite number of steps. This can be seen to be equivalent to the time-partitioned decoupled model (Section 4.5.2). The spins that flip in a particular update correspond to a particular subsystem. The time scale for relaxation is again $O(\ln(N);1)$.

To make use of the neural network model for protein folding, we can choose an initial conformation that has a finite fraction of spins overlapping with the desired ground state. There is a lesson to be learned from this model regarding the importance of the initial conformation in protein folding. Recently there have been suggestions that the initial conformation is not arbitrary, but instead assumes one of a restricted family of conformations that are either partially folded or are related in some way to the eventual folded conformation. This would be consistent with the concept of a basin of attraction. The introduction of a limited phase space exploration, where the protein does not explore all possible conformations but is restricted from the beginning to the basin of attraction of the folded conformation, also brings us to the second mechanism for reducing the relaxation time—kinetic effects. We will discuss kinetic effects more generally in the next chapter.

The attractor neural network model may also be useful for understanding more complex protein dynamics than just protein folding. Proteins act as enzymes. However, their enzymatic efficiency may be influenced by chemical or other influences that control their function. One mechanism for this control is a change in conformation that affects the active enzymatic site. Thus a protein may respond to a variety of controlling influences by changing its conformation. This suggests that there may be two or more well-defined folded conformations that are each relevant under particular external conditions. If a change in conformation due to a particular external influence is maintained for some time after the external influence is removed, then a description of the protein in terms of multiple minimum energy conformations may become useful.

Missing from attractor neural networks is the incorporation of propagative structures, specifically, interactions that can support driven diffusion or diffusion. Thus, the equilibration of neural network spin glass systems corresponds to the decoupled model and not to any of the models that include driven diffusion or diffusion. The absence of propagative structures is not realistic either for protein folding or for the general description of neural networks. Feedforward networks are a simple approach to incorporating propagation in neural networks. More complex propagative structures are likely both in proteins and the brain.

4.6   Conclusions

In this chapter we have considered a variety of models that display a range of scaling behavior of the relaxation time with system size. There are diverse individual features of these models that can be related to properties observed in protein-folding experiments. The models also provide some insight into the nature of the relaxation time and its relationship to inter-amino-acid interactions. All of these models, however, are missing the chain structure and its relaxation in space. When a chain is spread out in space, there is an inherent scaling of the relaxation time with chain length, due to the travel time of amino acids through the space before they can bond with other amino acids. In the following chapter we show that this travel time leads to a characteristic relaxation time that scales approximately as N^2 for an expanded chain.

While the models in this chapter are general enough that they cannot be used directly as models of the kinetics of protein folding, this investigation does allow us to relate our findings to other complex systems. There are some general conclusions that can be made. First, it is not difficult to design models that cannot relax in any reasonable time. Long-range interactions, in particular, lead to exponential scaling of the relaxation time. A weak driving force for the transition may also cause problems. There are, however, systematic approaches to interactions that give rise to relaxation in a time that scales as a low power of the size of the system. One approach is partitioning in space and time; another is diffusion or driven diffusion of boundaries; a third is predisposing the system by its initial state; a fourth is dominance of local interactions. All of these are likely to occur in protein folding as well as in the dynamics of other complex systems. It should be apparent that creating a complex system where interactions cause interdependence and yet allow dynamics to proceed in a reasonable time requires a careful design. Complex systems have specific properties that are not generic to physical systems. The issues of how complex systems arise will be discussed in Chapter 6.

5

Protein Folding II:
Kinetic Pathways

Conceptual Outline

■ 5.1 ■ When kinetics limits the domain of phase space explored by a system, the scaling of the relaxation time $\tau(N)$ may be smaller than exponential. Polymers in a liquid can be in an expanded or compact form. The transition between the two—polymer collapse—is a prototype of protein folding. Using simulations, we will explore possible origins of kinetic limitations in the phase space exploration of long polymers during collapse.

■ 5.2 ■ Before we study collapse, we must understand the properties of polymers in their expanded state in good solvent. Simple arguments can tell us the scaling of polymer size, $R(N) \sim N^\nu$. The time scale of relaxation of a polymer from one conformation to another follows either Rouse $\tau(N) \sim N^{2\nu+1}$ or Zimm $\tau(N) \sim N^{3\nu}$ scaling, depending on the assumptions used.

■ 5.3 ■ Polymer simulations can be constructed in various forms. As long as they respect polymer connectivity and excluded volume, the behavior of long polymers is correctly reproduced. A two-space model where monomers alternate between spaces along the chain is a simple and convenient cellular automaton algorithm.

■ 5.4 ■ During polymer collapse monomers bond and aggregate. Simulations of collapse and scaling arguments suggest that the aggregation occurs primarily at the ends of the polymer because of the greater flexibility of polymer-end motion. Thus the aggregates at the end appear to diffuse along the polymer contour accreting monomers and smaller aggregates until they meet in the middle. This results in an aggregation process that is systematically ordered by the kinetics. The end-dominated collapse-time scales linearly with polymer length, which is faster than the usual polymer relaxation. The orderly formation of bonds in end-dominated collapse also suggests that kinetics may constrain the possible monomer-monomer bonds that are formed and thus limit the domain of phase space that is explored in protein folding.

5.1   Phase Space Channels as Kinetic Pathways

In Chapter 4 we introduced the problem of time scale: How can a system composed out of many elements reach a desired structure in a reasonable amount of time? The ability of proteins to fold into a well-defined compact structure exemplified this trait. For our first answer we assumed that the desired structure was the equilibrium state of the system. We then studied various energy functions that would enable relaxation to the equilibrium state. All of these energy functions embodied some variation on the idea of parallel processing. In this chapter we consider the possible influence of kinetics on the time scale for reaching a final structure. Considering kinetic pathways as a mechanism that enables a system to reach a desired structure is a qualitatively different idea from parallel processing. In this approach, a system follows a particular pathway through the phase space to the final desired structure. The pathway may not be unique, but it is severely limited compared to the space of possible paths in the whole space. As a result, there is no reason to expect that the system reaches the absolute minimum energy equilibrium conformation. It does, however, reach a conformation that is low in energy compared to any of the accessible conformations. Because the system only visits a limited set of conformations along the path from its initial to final state, our expectation is that the relaxation time—the time to reach the final conformation—will scale less than exponentially with the size of the system. There are a number of ways that such kinetic pathways can arise. We will discuss a few of these in this section and describe a strategy for considering the effect of kinetics in protein folding.

The simplest form of kinetic pathway can be illustrated using the model of two independent two-state systems introduced in Section 4.2. Each of the two-state systems (spins) has two states $s_i = \pm 1$, for $i = 1, 2$. Relaxation of each spin occurs independently from +1 to −1. Let the relaxation time of the first spin be extremely long compared to the second spin, $\tau_1 \gg \tau_2$. If τ_1 is long compared to relevant times (e.g., years) then only s_2 relaxes. The system starts from the state $(1, 1)$. It makes a single transition to the state $(1, -1)$ and is stuck there. The ground state $(-1, -1)$ is never reached. This, however, is fine if $(1, -1)$ is the desired state. It actually doesn't matter whether $(1, -1)$ or $(-1, -1)$ is the ground state. The long relaxation time τ_1 corresponds to a large barrier to the transition of s_1. The accessible domain of phase space includes only states that have $s_1 = 1$. More generally, this form of kinetic pathway assumes that there are energy barriers that partition phase space so that some regions are inaccessible. These regions play no role during the relaxation. For the case of protein folding, this means that certain conformations would be completely inaccessible in a transition from the initial unfolded to the final folded conformation.

An example where energy barriers limit the space of conformations during protein folding is the preservation of primary structure. The bonds between amino acids along the chain are strong bonds that have a low probability of breaking and reforming. Thus, during folding, the protein does not explore the possible arrangements of amino acids along the chain and all of their conformations. The breaking of the chain is prevented by kinetics, even though there may be other orderings of the amino acids that have lower energy structures. Breaking the protein chain would enter a different domain of phase space. It is interesting that there are specific examples where the chain is broken during protein folding (proteolytic reactions). This is done by enzymes that break the amino acid chain in specific places. The subchains that result are then formed into the final folded protein structure.

The next step is to consider how kinetic pathways might affect the space of conformations of a chain with a particular amino acid sequence. Starting from an unfolded conformation, there do not appear to be any strong bonds or energy barriers that would prevent it from reaching a large number of compact polymer conformations. The number of such compact conformations grows exponentially with the size of the polymer. Thus energy barriers do not appear to be relevant in explaining the ability of a protein to reach a definite structure. However, the kinetic barriers need not exist in the initial conformation. It is enough for them to arise during the process itself. During protein folding, new bonds form. These bonds might restrict the domain of space that is explored. A pictorial illustration of the formation of barriers is shown in Fig. 5.1.1. It shows the emergence of barriers during the kinetic process. These bar-

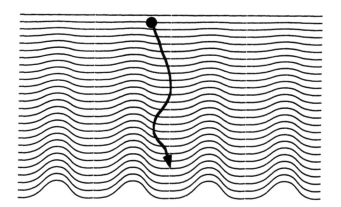

Figure 5.1.1 The simplest concept of a kinetic pathway is a path bounded by energy barriers that prevent departure from the path and thus prevent an exhaustive search of all conformations. In this figure we see that barriers may not exist initially; they may, however, develop as the relaxation proceeds. The illustration should be read as an energy landscape. Horizontal lines are plots of the energy in the horizontal direction. A vertical bias in the energy is assumed so that progressively lower lines are lower in energy. The conformation of a protein is a point in the plane of the page. A possible trajectory is illustrated. From the starting point, it appears that all three of the possible low-energy conformations at the bottom of the illustration are accessible. However, once the relaxation begins there are barriers that prevent the conformation from switching from one of the vertical channels to the other. In order for the correct final state to be chosen, the initial state must be restricted to be close to the channel that leads to the desired conformation. This conformation may or may not be the lowest-energy conformation. ∎

riers do not prevent the protein from folding into an undesirable structure. However, the conjunction of a particular initial configuration and the barriers that arise serve to limit the exploration of space and determine the ultimate conformation.

The picture of strong bonds causing large barriers that form kinetic pathways is not complete. Kinetic restrictions that limit the domain of phase space that is explored during folding arise also from entropic bottlenecks. An entropic bottleneck (Fig. 5.1.2(a)) is a narrow channel between one part of phase space and another. Because the channel is narrow, it is unlikely that the system will move from one part to the other. Thus a whole region of conformations may not arise. Another way in which entropy can be relevant is illustrated in Fig. 5.1.2(b). In this case, entropy differences in the inlets to kinetic pathways reduce the sensitivity of the final conformation to the

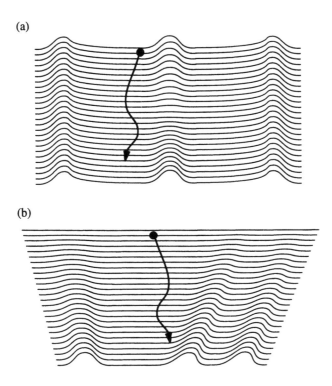

Figure 5.1.2 Entropy can play more than one role in the properties of kinetic pathways, as shown in (a) and (b). (a) illustrates an entropy bottleneck that prevents exploration of all conformations. Two regions of conformations with different energy minima are connected by a channel that is very narrow; like two valleys connected by a narrow mountain pass. It is unlikely that the system will go through the channel, because it will rarely be found by random motion, even if there is no energy barrier in the channel. A different effect of entropy is shown in (b). A wide inlet to a particular kinetic pathway causes it to be preferentially selected over a channel with a narrow inlet. This picture explains how kinetic pathways may lead to a predictable final conformation independent of the initial conformation. Such predictability would be necessary for kinetic pathways to be relevant to protein folding. ∎

initial conformation. Compare this picture with the picture illustrated in Fig. 5.1.1. For our study of protein folding, Fig. 5.1.2(b) will turn out to be relevant. Certain bonds are likely to form during the initial stages of folding. These bonds then inhibit the exploration of all conformations.

Considering energy barriers or entropic bottlenecks to the exploration of phase space are both part of a thermodynamic approach. They are relevant when diffusion dominates the kinetics. Diffusion is a random-walk process that occurs when a system is coupled to a thermal reservoir (Section 1.4.4 and Section 7.2.3). Diffusion is not important in a system that is far from equilibrium and not coupled to a thermal reservoir. The system then follows ballistic motion along a path that is determined by Newtonian equations of motion. This can give rise to other kinetic effects because a system follows a specific trajectory rather than an exploration process. Proteins are, however, embedded in a liquid that serves as a thermal reservoir. The kinetic energy is dissipated, and stochastic diffusive motion dominates. Thus, for proteins, we are amply justified in limiting ourselves to consider diffusive motion. More generally, in order for a stable final conformation to be reached, there must be dissipation of kinetic energy. This suggests that diffusion is important, but does not imply that ballistic motion plays no role.

In the previous chapter we adopted a series of models that ignored the spatial structure of polymers in favor of abstract representations in terms of Ising spin variables. This approach was helpful in developing an understanding of the issues and concepts of parallel processing for protein folding. However, in order to address kinetic limitations that may select pathways for folding, we must build a model of the polymer in space and its dynamics. Our objective is to establish the possible existence of a specific sequence of events in protein folding. Such a sequence of events would result from a particular order in which amino acids encounter each other. The encounters can then result in specific bonds being formed. To determine if a particular sequence of encounters occurs, we can consider a simplified polymer model that retains its spatial structure but does not represent the details of amino acid structure. This approach implies that we are interested in the very first part of the folding process, which might extend no more than a microsecond out of the typically one-second folding time. The potential impact of this initial time is to set the stage for later processes by forming bonds that limit the subsequent exploration of conformations, and by placing the system conformation in the vicinity of its eventual stable state.

In contrast to our studies in Chapter 4 which were essentially analytic, in this chapter we will focus on simulations (Section 1.7) as a tool for investigating the behavior of complex systems. Nevertheless, we begin in Section 5.2 by describing an analytic theory of the kinetics of long polymers. This analytic theory sets the tone for our investigations. It shows that many of the properties that are of interest do not depend on the specific microscopic structure of the polymer, but rather on the general behavior of a long chain and its many conformations in space.

To further develop an understanding of polymer kinetics, particularly the kinetics of polymer collapse, we will turn to Monte Carlo simulations. In Section 5.3 we construct a lattice Monte Carlo model of polymer dynamics. This simple model of polymer dynamics is constructed to be in the form of a cellular automaton (Section 1.5). The model dynamics serves as a simulation tool. Over many steps, the motion of the model polymer is consistent with the expected behavior of long polymers for relaxation or diffusion. This is ensured by the Monte Carlo method because only local steps in the space of polymer conformations are taken. In Section 5.4, using this model for polymer dynamics, we simulate polymer collapse and find evidence for specific kinetic pathways dictated by a preferential ordering of encounters between parts of the polymer chain. Motivated by the results of the simulations, we develop an analytic scaling theory that describes the kinetics of the transition of a polymer from an expanded to a collapsed state. This generalizes and reinforces the conclusions from the simulations. Finally, we consider a number of variations of the simulations to test the scaling theory and explore the domain of its applicability to physical polymers.

Before we proceed, we point out an example of relaxation that does not illustrate kinetic pathways, even though it appears to. The concept of a kinetic pathway implies a well-defined sequence of intermediate protein structures between the initial and the final conformation. However, the converse is not true. The existence of a well-defined sequence of intermediate protein conformations does not necessarily mean that the system is kinetically limited to this pathway. Another mechanism that may cause a well-defined pathway is the inhomogeneous decoupled model discussed in Section 4.5.1. In that model, certain degrees of freedom relax before others. If the degrees of freedom can be grouped into sets with well-separated relaxation time, then after each set of degrees of freedom relaxes, a well-defined structure arises. Kinetics does play a role because of the degrees of freedom that are frozen at any time. However, by the end of the relaxation, all degrees of freedom can and do relax. This is counter to the assumption of kinetic pathways.

5.2 Polymer Dynamics: Scaling Theory

Polymers are molecules formed out of long chains of atoms that are generally recognizable as a sequence of units (monomers) like amino acids. Biological polymers include proteins, DNA, and polysaccharides. Artificial polymers include polystyrene and polyethylene. Polymers whose monomers are all the same are known as homopolymers. Polymers that have more than one kind of monomer are called heteropolymers. Homopolymers are simpler to model, and we will concentrate on describing their dynamics, though many of the results also apply to heteropolymers. Polymers are found dissolved in liquids or embedded in composite materials. When they are dissolved in liquids, there are essentially two possible structures. Either the polymer collapses into a compact structure or the polymer is expanded. Which of these occurs depends on whether the effective interaction between monomers is attractive or repulsive. The effective interaction includes both the hard core repulsion between atoms and the longer-range attraction or repulsion. Entropy favors the expanded polymer structure over the compact structure because of the greater number of expanded conformations. However, this is generally a weaker effect than that of the energy of attraction or repulsion. In a solution with many polymers, compact polymers often aggregate to each other and precipitate. We will focus on the dynamics of a single polymer, not on polymer-polymer interactions.

When a protein is folded and unfolded in solution, it is crossing the line between compact and expanded structures. The transition may be driven by changes in temperature. However, the more usual approach is to add a chemical to the solution that affects the monomer-monomer attraction. What is relevant is the affinity of the monomers for each other compared to their affinity for the solvent. Because of the importance of the solvent for the transition, a polymer in its expanded state is said to be in a good solvent. A compact polymer is said to be in a poor solvent. The transition is called the θ-point. In this section we are concerned with the properties of a polymer in a good solvent. It is essential to understand the structure and dynamics of this state before we can study the dynamics of transition from the expanded to the compact state. We will be concerned with the scaling behavior of the properties of long polymers as a function of polymer length N. This is similar to Chapter 4 where we considered the scaling of relaxation with system size.

The scaling of the structure and dynamics of long polymers should not depend greatly on their chemical composition. The scaling theory of polymers is one of the great successes of simple concepts in understanding complex systems. A book by de Gennes, who received the 1991 Nobel Prize in physics, contains many of the elegant arguments that describe polymers simply and successfully.

A long polymer in a liquid has a local structure that is more or less flexible. The bonding of adjacent monomers controls the local polymer structure. For a specific pair of monomers, there may be several possible bonding configurations or there may be only one allowed configuration. However, a long enough polymer is always flexible, so we can start by considering it to be a random walk in space with N steps. When a polymer is modeled as a random walk, the size of a step is understood to depend on the polymer flexibility, with stiff polymers having many monomers per step and flexible polymers having few monomers per step. For convenience, we can redefine our monomers so that each step is between the new effective monomers.

Polymers are generally found in three-dimensional space. However, we generalize our discussion to d-dimensions. In d-dimensions a random walk is performed independently in each dimension. The average distance traveled along the polymer from one end to the other is called the polymer end-to-end distance R_0. We use σ to represent the root mean square distance traveled in a random walk in a single dimension. The random walk in one dimension satisfies (Section 1.2):

$$\sigma = \sqrt{<(x_N - x_1)^2>} = \sqrt{<(\sum_i (x_{i+1} - x_i))^2>} = \sqrt{\sum_i <(x_{i+1} - x_i)^2>} = \sqrt{N}a$$

(5.2.1)

where x_N and x_1 are the x coordinates of the last and first monomers respectively. More generally, x_i is the x coordinate of the ith monomer. The third equality follows from the independence of the steps. a is the distance of an elementary step in one dimension—the average distance between coordinates of adjacent monomers. Since the random walk in each dimension is independent, the end-to-end polymer length, R_0, is given by a similar expression:

$$R_0 = \sqrt{<(\mathbf{r}_N - \mathbf{r}_1)^2>} = \sqrt{<(\sum_i (\mathbf{r}_{i+1} - \mathbf{r}_i))^2>} = \sqrt{\sum_i <(\mathbf{r}_{i+1} - \mathbf{r}_i)^2>} = \sqrt{dN}a$$

(5.2.2)

where $\mathbf{r}_i$ is the d-dimensional vector position of monomer i. The result follows from the existence of dN independent terms in the sum. It could also be written as $R_0 = \sqrt{N}a'$, where $a' = \sqrt{d}a$ is the monomer-monomer distance.

A polymer is a thermodynamic system whose equilibrium size is determined by its free energy. When we consider the polymer as a random walk, we assume that all of the possible configurations have the same energy. The size is then determined by the entropic part of the free energy. As discussed in Section 1.4, the probability of a particular end-to-end polymer distance R can be found using the relationship:

$$P(R, N) = e^{-F(R, N)/kT}/Z = e^{-(E(R, N) - TS(R, N))/kT}/Z = e^{S(R, N)/k}/Z \quad (5.2.3)$$

$F(R, N)$ is the free energy, $E(R, N)$ and $S(R, N)$ are the energy and entropy respectively, and Z is a normalization constant. The final expression only applies when the energy can be neglected. We have already essentially calculated the probability $P(R, N)$ when we counted the number of random walks of a particular length in Section 1.2. From Eq. (1.2.39), the probability distribution for the distance traveled in a one-dimensional random walk is a Gaussian. In d-dimensions we take a product of the Gaussian probability in each dimension to obtain the probability for a particular end-to-end vector $\mathbf{R}$:

$$P(\mathbf{R}, N) = \frac{1}{(2\pi\sigma)^{d/2}} e^{-R^2/2\sigma^2} = \frac{1}{\left(2\pi R_0/d^{1/2}\right)^{d/2}} e^{-dR^2/2R_0^2}$$

(5.2.4)

where R is the magnitude of $\mathbf{R}$. To obtain the probability of a particular end-to-end distance, R, this must be multiplied by the $d-1$ dimensional surface area of a sphere. It turns out that none of this detail is important for what follows; however, for completeness we can write the surface area as a constant Ξ^{d-1} times R^{d-1}.

$$P(R, N) = \frac{\Xi^{d-1} R^{d-1}}{\left(2\pi R_0/d^{1/2}\right)^{d/2}} e^{-dR^2/2R_0^2}$$

(5.2.5)

From Eq. (5.2.3), the free energy of a particular end-to-end distance is given by the logarithm of the probability of a particular length times the normalization constant Z:

$$F_{random-walk}(R, N) = -kT \ln(ZP(R, N)) = F_0 + \frac{dkTR^2}{2R_0^2}$$

(5.2.6)

where we have neglected the logarithmic terms in R and N. Since only free-energy differences matter, the constant F_0 could also be neglected.

There is something unusual about this expression for the free energy. The free energy minimum does not occur at the end-to-end distance R_0 that was found before. It occurs instead at $R = 0$. Part of this problem arises because we neglected the logarithmic term $(d-1)\ln R$. However, even when we include this term, the minimum occurs at $\sqrt{((d-1)/d)}R_0$ rather than R_0. This system does not satisfy the usual property of a macroscopic system, that the probability distribution becomes sharp as the system becomes large. In the usual case we can identify the expected value of a system property as the value that maximizes the free energy. For the random walk, the free energy has the more general interpretation, discussed in Section 1.4, as the logarithm of the probability. If we were to use the free energy to evaluate the value of the average radius, we would still have to calculate the average over the probability distribution. After calculating the average we would recover the value R_0. This discussion shows the connection between the entropy, free energy and the characteristic size of a polymer. When we need to add additional terms to the free energy, such as the monomer-monomer interactions, we can recalculate the polymer size using the same expressions.

Our discussion of the random walk in Section 1.2 included a proof of the central limit theorem that allows us with some confidence to consider various random walks to have a Gaussian probability distribution. However, the polymer differs in an essential way from the random walks that were discussed there. The difference is that the steps are not uncorrelated. The stiffness of a polymer would tend to make a polymer continue in the same direction. More generally, the bonding and interactions between monomers near each other along the contour cause constraints between neighboring steps. This turns out not to be an essential problem, because the coupling between steps along a polymer decays exponentially. There is a characteristic distance along the chain after which the constraints become negligible. This means we can choose to label our polymer with random steps as long as the steps are larger than the correlation distance (persistence length). The number of these steps becomes our effective monomer number N. When we have many of them, then, and only then, can we consider the polymer to be long. Proteins in their expanded form turn out to be quite flexible and so the correlation length is short, approximately a single amino acid. On the other hand, DNA is quite stiff. For a single strand of DNA the persistence length is roughly 200 to 400 monomers. For a double strand helix, the persistence length is approximately 2,000 monomers (base pairs).

We are, however, still missing an important aspect of the interaction between monomers. This is the contact interaction between any two monomers that encounter each other. We have argued that we do not need to consider the interactions of monomers near to each other along the chain, because the correlations disappear for long enough polymers. However, the interactions that occur between any two monomers still must be considered. Since the monomers are repelling each other when the polymer is in a good solvent, we can represent the interaction between them as an excluded volume—a volume around each monomer that other monomers can-

not enter. In Section 1.10 we discussed the use of renormalization theory to understand the relevance of parameters in the macroscopic limit. It is possible to show that
for isolated polymers in good solvent, the only relevant parameters in the long length
limit are the length of the polymer and the excluded volume. This simplifies our considerations and guarantees that our results apply to any polymer, if it is long enough.
Our task is to identify the properties of a polymer that has an excluded volume. Such
abstract polymers are called self-avoiding random walks. Self avoiding walks should
be larger than random walks because of the repulsive interactions between the
monomers.

There is a scaling argument for the size of a self-avoiding random walk constructed by Flory. The argument is based on the competition between the repulsive
energy of the excluded volume that tries to expand the polymer, and the tension due
to entropy reduction when the polymer chain is stretched. Assume that we have a
polymer that occupies a volume R^d. The density of the monomers in this volume is
given by:

$$c = \frac{N}{R^d} \tag{5.2.7}$$

In this expression, and throughout, we avoid numerical coefficients, since the objective is only to understand the scaling. The energy of the monomer-monomer interactions is given by the probability that two monomers encounter each other. On average, an encounter costs an amount of energy characteristic of the thermal kinetic
energy of the monomers, kT. Once two monomers approach each other close enough
to cost this amount of energy, they cannot approach any closer. If we neglect the structure of the polymer, then we can calculate the probability of an encounter. We think
of the monomers as distributed with uniform probability in the volume R^d. We imagine placing each of the monomers at random in this volume. Each monomer has an
excluded volume V. The number of monomer-monomer overlaps (interactions) is
then proportional to the square of the concentration. More specifically, it is given by
the number of monomers times the fraction of the volume occupied by monomers.
The energy associated with the excluded volume is thus:

$$F_{excluded-volume}(R,N) = kTN\frac{NV}{R^d} = kTV\frac{N^2}{R^d} \tag{5.2.8}$$

where kT gives the units of energy. The neglect of the polymer structure is a neglect
of correlations between monomer positions. This is characteristic of a mean field approach (see Section 1.6). Thus this equation is a kind of mean field treatment of
monomer interactions.

The excluded volume energy in Eq. (5.2.8) is smaller for larger R. To this energy
we add the free energy for the random walk, Eq. (5.2.6), that neglected the excluded
volume. We obtain the free-energy expression:

$$F(R,N) = F_0 + \frac{dkTR^2}{2R_0^2} + kTV\frac{N^2}{R^d} \tag{5.2.9}$$

We minimize this expression to obtain the typical size of the polymer. We can do this as long as the result is not zero. Neglecting all coefficients we obtain:

$$\frac{R_g}{R_0^2} = \frac{N^2}{R_g^{d+1}} \tag{5.2.10}$$

From Eq. (5.2.2) we have $R_0^2 \propto N$, so:

$$R_g \sim N^\nu \tag{5.2.11}$$

where the exponent ν is given by the expression

$$\nu = \frac{3}{d+2} \tag{5.2.12}$$

The tilde, ~, is used to indicate that the result only holds in the asymptotic regime, i.e., for long enough polymers. The result we have obtained is remarkable. It is exact in one dimension where $\nu = 1$, because the excluded volume walk is a straight line. It has been shown to be exact in two dimensions where $\nu = 0.75$. In three dimensions, where $\nu = 0.6$, it is in reasonable agreement with both experiment and numerical simulation. In four dimensions, it gives the random walk result $\nu = 0.5$. In higher than four dimensions, this must continue to be the result, since it indicates that the excluded volume is irrelevant. This has also been proven to be correct. The reason that the random walk becomes correct in four or higher dimensions is that the free energy due to monomer-monomer interactions for a random walk decreases with the length of the chain (see Question 5.2.1). Thus this simple mean field scaling argument appears to give the exact result in all dimensions. Why does the mean field approach give an exact result in all dimensions? Unlike the mean field treatment of the Ising model, which was exact only in four or higher dimensions, the mean field treatment of polymers appears to benefit from a cancellation of errors.

The alert reader may note that we actually made what might seem an unreasonable step in combining the free-energy expressions in Eq. (5.2.6) and Eq. (5.2.8) to obtain Eq. (5.2.9). The definition of R used to obtain Eq. (5.2.6) was the end-to-end distance of the polymer. The definition of R used to derive the form of the excluded volume energy in Eq. (5.2.8) was the characteristic spatial size of the polymer. In effect, we assumed that all characteristic linear dimensions of the polymer behave in the same way. This is a simplification that is a reasonable first assumption to be made in constructing a scaling theory.

Question 5.2.1 Show in more than four dimensions that the monomer-monomer interactions have decreasing importance and are therefore irrelevant in the long polymer limit. However, for fewer than four dimensions they are not irrelevant.

Solution 5.2.1 In order to see whether the excluded volume is relevant, we evaluate its effect on the polymer free energy. We do so assuming the polymer has a volume given by the random walk without excluded volume. This is the maximum effect the excluded volume can have. Using the value of the polymer end-to-end distance for a random walk, the excluded volume term in the free energy scales as:

$$F_{excluded-volume} = kTV\frac{N^2}{R_0^d} \propto N^{2-d/2} \tag{5.2.13}$$

The random-walk term in the free energy is independent of polymer length. Thus for any dimension greater than four, the excluded volume interaction has decreasing relative importance with length of the polymer and will not significantly affect the asymptotic behavior of the polymer size. For fewer than four dimensions, the excluded volume term in the free energy increases with the size of the polymer and therefore is relevant. ∎

The dynamics of an isolated polymer consists of diffusion of the whole polymer and internal relaxation of its conformation. Diffusion describes the motion of the polymer center of mass. The internal relaxation of a polymer describes how the polymer changes from one conformation to another. We think of this as a relaxation process because if we know the conformation of a polymer at one time, then the ensemble of this polymer consists of many replicas of the same conformation. However, the random motions of the liquid will cause the ensemble over time to forget the initial conformation and become indistinguishable from an ensemble that started from any other conformation. This is the equilibrium ensemble. The process of relaxation to the equilibrium ensemble resembles the exponential relaxation in a two-state system in Section 1.4. The characteristic relaxation time $\tau(N)$ depends on the length of the polymer. Our objective is to determine the scaling of the relaxation time with polymer length. Dynamic scaling is generally more difficult and less universal than scaling of static quantities like the size of a polymer. Of particular significance when we consider the dynamics of polymers is our treatment of the fluid. This was not relevant when we considered the static structure of the polymer.

There are two established estimates for the scaling of the relaxation time with polymer length $\tau(N)$. The Rouse relaxation time describes the dynamics of a polymer assuming that the motion of a monomer is not correlated to the motion of monomers that are far away along the polymer chain. However, the motion of the fluid, described by hydrodynamics, couples the motion of one monomer to another when they are near each other, no matter how far apart they are along the chain. When a monomer moves, it causes a flow of fluid that in turn moves other monomers. Also, a flow of fluid that moves one monomer has a spatial extent that can move other monomers at the same time. This coupling is taken into consideration in the Zimm relaxation time. We discuss first the Rouse and then the Zimm relaxation.

The dynamics of a polymer becomes slower as the polymer length increases. The relaxation time, therefore, is controlled by the dynamics of the longest length scale—the movement of half of the polymer from one place to another. We first illustrate this using a simple elastic string model. Using this model, we derive the Rouse relaxation for a random walk that neglects the excluded volume. In the elastic string model, we assume that the distance between adjacent monomers has the same distribution as the end-to-end distance of a polymer. This means that the Gaussian distribution applies already to the intermonomer separation. We can assume this for a very long polymer because we can always relabel our monomers to be farther apart along the chain. For example, we can label every tenth or every hundredth monomer as a new "monomer." We call the chain between the new monomers "the bond" between them. Since this only changes the number of monomers by a constant factor, it will not change the scaling of properties of the polymer. However, by relabeling the monomers, the bond between two of our new monomers itself acts like a long polymer. The relabeling idea only works when we neglect excluded volume.

The free energy of the elastic string depends on the intermonomer separation as given in Eq. (5.2.6). This is the equation for a spring where the energy is proportional to the square of the distance. The force between two adjacent monomers is then proportional to the distance between them. We can write the total force on the ith monomer as:

$$K[(r_{i+1} - r_i) + (r_{i-1} - r_i)] \approx K\frac{d^2r}{di^2} \qquad (5.2.14)$$

where K is the spring constant. On the right we have taken a continuum limit with i the position along the contour of the polymer. We assume that the motion of a monomer in the fluid is overdamped, which means that the velocity of a monomer is proportional to the force upon it. Multiplying the force times the mobility of a monomer μ gives the velocity:

$$\frac{dr}{dt} \approx \mu K \frac{d^2r}{di^2} \qquad (5.2.15)$$

The solution of this equation is given by exponential relaxation of spatial waves:

$$r_i(t) \approx A\cos(ki)e^{-t/\tau_k} + B\sin(ki)e^{-t/\tau_k} \qquad (5.2.16)$$

where $k = 2\pi/\lambda$ is the wave vector of the oscillation of the elastic string. The boundary conditions at the ends restrict the wavelength λ to be no greater than the string length N. By inserting the solution into the differential equation we see that the relaxation time for a particular wavelength is given by

$$\tau_k = \lambda^2/\mu K(2\pi)^2 \qquad (5.2.17)$$

Neglecting all the numerical coefficients gives us the longest relaxation time ($\lambda \propto N$) as:

$$\tau(N) \sim N^2 \qquad (5.2.18)$$

This is the Rouse relaxation time when excluded volume is neglected. It applies much more generally than its derivation for the elastic string model would indicate. The main reason for this generality is that the longest relaxation time involves motion of essentially the whole polymer, and therefore does not depend on the local polymer properties.

We take a different approach in order to incorporate the excluded volume and the effect of hydrodynamics in the scaling of the relaxation time. This approach relates polymer relaxation to the polymer diffusion constant. Relaxation of a polymer occurs when a significant part of the polymer (say half) is able to diffuse in a random walk across the whole volume occupied by the polymer. This means that we may write the relaxation time using a random-walk expression by assuming the distance traveled is the diameter of the polymer:

$$R(N)^2 \sim D(N)\tau(N) \tag{5.2.19}$$

This is the usual relationship of distance traveled to the diffusion constant and the time (e.g., Eq. (1.4.56)). We have used the diffusion constant of the whole polymer $D(N)$ rather than $D(N/2)$ because their scaling dependence on N is the same. $R(N)$ is a characteristic length, such as the diameter or radius of the polymer.

Since we already know the size scaling of the polymer, we must derive an expression for the diffusion constant of a polymer. The diffusion of the polymer is given by the displacement of the center of mass of the polymer. In an interval of time Δt, the center of mass of the polymer r_{cm} changes according to:

$$\Delta r_{cm} = \Delta <r_i> = \frac{1}{N}\sum \Delta r_i \tag{5.2.20}$$

Assuming a mean field treatment, we neglect monomer-monomer correlations. Accordingly, the movement of each monomer is uncorrelated to other monomers. Each term in the sum Δr_i is an independent random number. The sum in Eq. (5.2.20) is like a random walk with N steps. Thus the sum over all the independent displacements of the monomers is proportional to $N^{1/2}$. The center of mass displacement is given by:

$$\Delta r_{cm} \sim N^{-1/2} \tag{5.2.21}$$

The scaling of the diffusion constant is obtained by setting this distance to be the result of a random walk of the center of mass:

$$D = \Delta r_{cm}^2 / \Delta t \sim N^{-1} \tag{5.2.22}$$

where Δt, the time for monomers to take the steps Δr_i, is independent of the polymer length.

Using this diffusion constant, we obtain the relaxation time from Eq. (5.2.19) as the time for diffusion of the polymer across its own radius:

$$\tau(N) = R(N)^2 / D(N) \sim N^{2\nu+1} \tag{5.2.23}$$

This is the generalization of Rouse dynamics to self-avoiding random walks. For two dimensions (three dimensions), it gives an exponent of 2.5 (2.2). Without excluded volume, $\nu = 1/2$, it reduces to the previous result.

In order to describe the relaxation of a polymer including hydrodynamics of the solvent, we start from the Navier-Stokes equation. We will not need to solve this equation; however, we do need to know what parameters are involved in order to construct a scaling argument. The standard complete set of hydrodynamic equations

$$\frac{\partial \mathbf{v}}{\partial t} + (\mathbf{v} \cdot \nabla)\mathbf{v} = -\nabla P/\rho + \frac{\eta}{\rho}\nabla^2 \mathbf{v}$$

$$\nabla \cdot \mathbf{v} = 0$$

(5.2.24)

describe the macroscopic behavior of the motion of an incompressible fluid. In these equations $\mathbf{v}$ is the velocity of the fluid, P is the local pressure, ρ is the density which is essentially constant in an incompressible fluid, and η is the viscosity. We do not derive these equations here, but we review the origin of each term. The top equation (Navier-Stokes equation), is Newton's law $d\mathbf{v}/dt = F/m$ applied at a particular location in the fluid. The left side of the equation is the acceleration of a fluid element. The second term accounts for the displacement of the accelerated fluid element. The right side of the equation is the force divided by the mass. This has two parts, the force due to the pressure gradient and the force due to the effects of shear. The second equation is the absence of a divergence of velocity (outflow of matter from a point) in an incompressible fluid. There are four equations, the three components of the top equation and the bottom equation, and four unknowns, the three components of the velocity and the pressure (divided by the density) $(\mathbf{v}, P/\rho)$.

The underlying assumption of our treatment of a polymer in a hydrodynamic fluid is that the polymer moves with the fluid in which it is located. Thus we think about the motion of the polymer as the motion of a spherical volume R^d of the fluid. Like the other mean field treatment, this approximation neglects the effects of monomer-monomer bonding on polymer motion. In order to obtain the diffusion constant of the polymer, we need to know which parameters it may depend on. A scaling argument follows from dimensional constraints. We imagine the diffusion of a spherical volume of fluid. At any instant, the velocity and pressure fields are solutions of the Navier-Stokes equations. There is one piece of information not contained in the Navier-Stokes equation—the size of the random thermal motion of the sphere of fluid. This is given by the thermodynamic expression for the average velocity of a particle at temperature T (Eq. (1.3.83)):

$$<v^2> \propto kT/m = kT/\rho R^d$$

(5.2.25)

The expression used for the mass of the fluid, $m = \rho R^d$, neglects the small mass of the polymer distributed within it. In addition to the characteristic velocity, there are only two other parameters that are relevant to the motion. One is the size, R, of the fluid volume that is moving. The other is the viscosity, η, which characterizes the fluid. The viscosity only appears in the Navier-Stokes equation in combination with the density as η/ρ.

The diffusion constant must be a function of only three parameters; $(<v^2>, R, \eta/\rho)$. The diffusion constant is related to the thermal velocity by the relationship:

$$D \propto <v^2> \tau'$$

(5.2.26)

This relationship is derived later in Section 7.2.3. The time τ' is not the relaxation time of the polymer. It is the characteristic time between changes in velocity of the sphere of fluid. In effect, it is the time between collisions of the sphere with the rest of the fluid. This time depends only on the remaining two parameters $(R, \eta/\rho)$. By looking at Eq. (5.2.24) the dimensions of η/ρ can be seen to be length^2/time. The only combination of η/ρ and R that has the dimensions of time is $R^2\rho/\eta$, which must therefore be proportional to τ'. Inserting this into Eq. (5.2.26) we have:

$$D \propto <v^2> R^2\rho/\eta \propto \frac{kTR^2}{\rho R^d}\frac{\rho}{\eta} = \frac{kT}{\eta R^{d-2}} \tag{5.2.27}$$

This is called Stokes' law. To concretize this result we give the diffusion constant of a sphere in three dimensions, which can be obtained by solving the Navier-Stokes equation:

$$D = \frac{kT}{6\pi\eta R} \propto \frac{1}{R} \tag{5.2.28}$$

Eq. (5.2.28) is in agreement with Eq. (5.2.27) in three dimensions, and it provides the numerical prefactor for the specific case of a sphere.

There is a problem with the result of Eq. (5.2.27) for two-dimensional systems. There are two aspects to this problem. The first issue is the result itself. In two dimensions, according to Eq. (5.2.27), there is no dependence of the diffusion constant on the size, R, of the system. It turns out that a more careful treatment yields a logarithmic correction, which is nonanalytic. The second issue is the nature of the two-dimensional system that is being modeled. Any two-dimensional system that we encounter is embedded into a three-dimensional universe. A two-dimensional Navier-Stokes equation assumes one of two scenarios. The first scenario is that we have a system formed out of very long cylinders. For a polymer this would correspond to having monomers that are very extended in one dimension. The direction in which the monomers are extended is perpendicular to the direction in which the monomers are bonded to each other. It is also perpendicular to the two dimensions in which the monomers can move. Alternatively, the two-dimensional equation would correspond to having a polymer in a solvent between two solid plates whose separation is no greater than the width of a monomer. These plates allow the polymer to diffuse without sticking. Neither of these scenarios are easy to construct. It is more relevant to consider a two-dimensional problem where a polymer is trapped at an interface. The interface might be a boundary between two liquids. In this case the polymer occupies a space like that of a flat disk in the two-dimensional interface. The Navier-Stokes equation we would solve to describe the motion of the disk is a three-dimensional equation. Even though the polymer only moves in two dimensions, the diffusion constant scales the same as in three dimensions.

In order to see this we must embellish our scaling argument slightly. R would play the role of an overall scale factor of the disk shape. The radius would be given by R, and the height of the disk would be given by ζR, where ζ is a small dimensionless number. All of the scaling statements would hold as before for three dimensions leading to Eq. (5.2.27) for $d = 3$. We are not yet done, because we assumed that the height of the disk changes with the radius, which is not true for a polymer at an interface. However, for a thin disk, the interaction between the fluid and the disk only occurs at the faces of the disk. The height is irrelevant, and we can use the same scaling for a disk whose height does not change. Thus, as in Eq. (5.2.27), the scaling of the diffusion constant is inversely proportional to the disk radius (polymer radius), $D \propto 1/R$, in the two-dimensional space. Later when we want to show a simulation of two-dimensional polymer collapse, we will choose this scaling dependence both because of its similarity to the properties of three-dimensional collapse and because it is a model of the dynamics of a polymer at an interface.

The diffusion constant of a polymer is given by either Eq. (5.2.27) or Eq. (5.2.28) with the radius given by Eq. (5.2.11). Inserting Eq. (5.2.27) into Eq. (5.2.19) gives the Zimm relaxation time:

$$\tau \sim R^d \sim N^{dv} \tag{5.2.29}$$

or for our modified scaling using Eq. (5.2.28) in both two dimensions and three dimensions:

$$\tau \sim R^3 \sim N^{3v} \tag{5.2.30}$$

The Zimm relaxation scaling in three dimensions is $3v = 1.8$, which is smaller than the Rouse relaxation result 2.2. In two dimensions it is either $2v = 1.5$ according to Eq. (5.2.29), or $3v = 2.25$ according to Eq. (5.2.30). For much of our discussion the differences between the various relaxation-time scaling exponents will not be significant.

This concludes our study of the structure and dynamics of polymers in good solvent. The next step is to introduce techniques for the simulation of polymers that enable us to investigate the properties of polymer collapse. We will return to scaling arguments for the same problem in Section 5.4.3.

5.3   Polymer Dynamics: Simulations

5.3.1 *Introduction to simulations of polymer dynamics*

In this section we describe several methods for simulating polymers that are both cellular automata (Section 1.5) and Monte Carlo algorithms (Section 1.7). They also illustrate the technique of space partitioning that can be used generally for parallel processing of spatially distributed systems. From a theoretical point of view, one of the most interesting features of these algorithms is that they allow Monte Carlo simulations of extended objects but are inherently parallel. With the advent of massively parallel computers, including cellular automaton machines, inherent parallelism can also be a practical advantage. The algorithms are also quite simple and they illustrate how a simple cellular automaton algorithm can be designed. Simplicity often makes it easier to work with and reason about models. A simple algorithm leads to small, fast programs, and small programs are easier to write, debug, optimize, execute, maintain and modify. One of the algorithms, the two-space algorithm, is particularly convenient and efficient and will form the basis of our simulations of polymer collapse in Section 5.4.

In general, polymer simulations, like other simulations, use either molecular dynamics or Monte Carlo dynamics (Section 1.7). Molecular dynamics simulations are suggestive of Newtonian dynamics and are implemented by moving all atoms with small steps (no more than 10^{-2} of a characteristic interatomic distance) according to forces calculated from modeled interatomic forces. Monte Carlo simulations represent the dynamics of an ensemble of polymers by steps that take into account transition probabilities required by thermodynamics. Both of these simulation methodologies give the same results for equilibrium ensemble properties like the polymer end-to-end distance or other average structural properties. They also give the same results for dynamical properties that involve motions on a scale that is large compared to a step of an individual monomer. Large-scale motions include polymer conformational change, relaxation and diffusion. All atoms can be moved in parallel (at the same time) in molecular dynamics, which therefore appears to be ideally suited for parallel processing. However, with a processor attached to each atom, calculation of the forces requires a large number of communications between processors. Each atom must communicate to every other atom its position. Connections between processors are the limiting feature of parallel computers. Monte Carlo simulations have a fundamental advantage in that movements of monomers can be much larger and there is no need to specify forces. It is sufficient to specify the energy for a simple model polymer system. Monte Carlo simulations also can take into consideration the thermal reservoir effect of the fluid without simulating the fluid itself. Hydrodynamics, however, is not included. We will focus on the Monte Carlo method, describe why the straightforward approach to parallelization does not work, and then construct a parallel cellular automaton algorithm that does.

In Monte Carlo simulations of polymers, a long chain of monomers is represented by the coordinates of each monomer. There are many different methods for describing the monomer-monomer interactions, the local structure of the polymer and the process of each move. Just as for real polymers, the local polymer structure should not affect the characteristic properties of long polymers, such as the scaling of the size, $R(N)$, or relaxation time, $\tau(N)$.

An example, illustrated in Fig. 5.3.1, is the ball-and-string model. The monomers are rigid balls of radius r_0 that are attached to nearest-neighbor monomers by strings of length d_0 that have no elasticity. The rigid balls are not allowed to overlap—the energy is infinite if they do. The strings are not allowed to break and simply prevent adjacent monomers from separating further than a distance d_0 apart. Smaller distances down to $d_0 - 2r_0$ are possible. To construct a Monte Carlo dynamics for this model we must specify the nature of a move (the matrix λ in Eq. (1.7.19)). The easiest specification is to allow one monomer at a time to move anywhere within a distance r_m from its current location. Then a single step of the Monte Carlo consists of two parts: (1) selecting a move, and (2) accepting or rejecting the move. Selecting a move consists of selecting at random one of the monomers, and selecting a direction and a distance to move the monomer with equal probability within the ball of radius r_m around its original location. The process of accepting or rejecting the move is explained as follows. There are two possibilities, either the final conformation is allowed or it is not allowed. It is allowed if two conditions are satisfied: there is no overlap of the monomer we chose with any other monomer, and the move did not take the

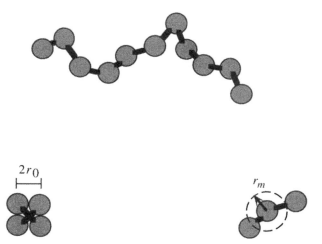

Figure 5.3.1 Illustration of an abstract polymer composed of monomers that are connected to neighbors and do not overlap "excluded" volumes. This is called the ball-and-string model. The motion of monomers is restricted so that they do not separate further than the string length, d_0, from monomers they are bonded to. Monomers have a ball radius r_0 and any two are prevented from overlapping. The strings act only as limits to the separation between monomers, and have no other physical existence. In order to ensure that the polymer cannot cross through itself, d_0 should be less than $2\sqrt{2}r_0$. As illustrated on the lower left, for a larger d_0 the polymer crosses through itself when two bonded monomers at opposite corners of the square move up out of the page and the other two bonded monomers move down into the page. In two dimensions it is enough that $d_0 < 4r_0$, preventing a monomer from passing between two other monomers.

   This model polymer can be conveniently simulated using Monte Carlo displacements of individual monomers. A monomer is moved (lower right) to a randomly selected position within a radius r_m around its original position, but only if it does not then violate the structural constraints. Unlike molecular dynamics simulations, however, there are problems in moving monomers in parallel. Moving two bonded monomers might break their bond, and moving any two monomers at the same time can lead to inadvertent overlap. ∎

monomer further than d_0 away from the neighbors to which it is attached by strings. It is not allowed if either of these conditions is violated. For any monomer move that is allowed, the energy of the polymer is unchanged. For any move that is not allowed, the energy increases to infinity. Because the energy change is either zero or infinite, the temperature of the simulation does not matter. Allowed moves are accepted, moves that are not allowed are rejected.

   We construct below several different ways of simulating long polymers. In all of them a simulation step consists of selecting a monomer, monomer i, from the polymer chain and performing a move subject to the following constraints:

1. The move does not "break" the polymer connectivity—monomer i does not dissociate itself from its nearest neighbors along the chain.

2. The move does not violate excluded volume—monomer i does not overlap the volume of any other monomer j.

These two constraints, connectivity and excluded volume, are sufficient to guarantee that the structural properties of a long polymer will be found.

In order to study the dynamics of a polymer, we must also guarantee that the steps taken are local steps in the space of polymer conformations. This is generally satisfied when monomer steps themselves are local. However, we must also be sure that the polymer cannot cross through itself. For the ball-and-string model, this limits the size of d_0 (see Fig. 5.3.1). For the types of models we will use, it is easy to verify that a polymer cannot pass through itself.

In naive parallel processing, a set of processors would be assigned one-to-one to perform the movement of a set of the monomers. A processor does not know the outcome of the movement of the other monomers; it can only know their position before the current step. With the two constraints (1) and (2) it would be impossible to perform parallel processing in this way, since moving different monomers at the same time is likely to lead to dissociation or overlap. Dissociation only restricts the parallel motion of nearest neighbors. However, the excluded volume constraint restricts the parallel motion of *any* two monomers, presenting a fundamental difficulty for parallel processing.

5.3.2 *Cellular automata for polymer simulations*

The idea of a cellular automaton is to think about simulating the space rather than the objects that are in it. This is useful for parallel simulation of polymers because, as long as there are no long-range interactions, the motion of monomers that are far apart must be independent of each other. Thus we can assign parallel processors to separated regions of space. When this concept is applied to a continuous space, we call the methodology space partitioning.

Space partitioning could be applied to the ball-and-string model. As shown in Fig. 5.3.2, the space would be partitioned into regions. For Monte Carlo simulations of the ball-and-string model, we could move at the same time monomers selected from regions separated by more than a distance of $2r_m + 2r_0$. At this distance, two monomers moving toward each other at the same time would not enter each other's excluded volume. This approach can work for other polymer models as well. However, it is simplest to implement and simulate for a cellular space of binary variables where the presence or absence of a monomer is indicated by a cell being ON or OFF.

To construct a polymer in a cellular space we could make a polymer model very similar to the ball-and-string model. Instead of a continuum of positions, the locations of monomers would be specified on a lattice. There is an algorithm, the bond-fluctuation algorithm, that implements such a ball-and-string model. However, in the design of a cellular automaton there is an additional feature to keep in mind. We would like to know which monomers are attached to each other solely by their relative

Figure 5.3.2 For the ball-and-string model of Fig. 5.3.1, or other polymer models without long range interactions, monomers sufficiently far apart may be moved independently and in parallel. The figure illustrates the use of space partitioning. If one monomer is selected from each shaded region, the selected monomers can be moved at the same time without chance of overlap. The location of the shaded regions should then be shifted so that all monomers can be moved. ∎

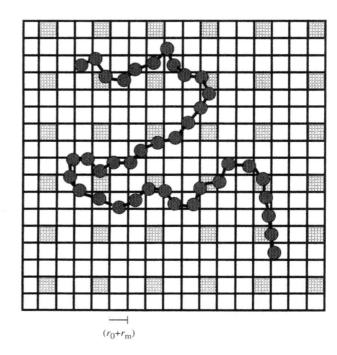

$(r_0 + r_m)$

location in space. This is unlike the ball-and-string model where neighbors attached by strings can be farther apart from each other (up to a distance of d_0) than two monomers not attached by strings. All monomers, bonded or not bonded, can approach each other to a smaller distance of $2r_0$. Because of this, we have to keep track of which monomers are bonded to which monomers by labeling the monomers. For a single polymer, the label might be the monomer sequence number along the polymer, which would tell us which monomers are neighbors along the chain and which are not. This labeling would not be convenient in a cellular automaton. The idea of a cellular automaton is that the dynamics only depends on the local spatial conformation. Bonds should be specified only by the relative position of the monomers. Thus we think about a bonded neighbor as a monomer that is closer than a certain distance, and any other monomer has to be farther away. We call the space around a monomer in which its bonded neighbors are located the bonding neighborhood. We note that, since monomers that are not bonded cannot be closer to each other than bonded monomers, in any such model, the polymer chain cannot pass through itself.

A polymer model that implements this in two dimensions is shown in Fig. 5.3.3. In this model, monomers are bonded if they are adjacent either horizontally, vertically or diagonally. The bonding neighborhood is a 3×3 region around a monomer. In three dimensions, we could use a $3 \times 3 \times 3$ cube as a bonding neighborhood. Any monomer except those that are bonded must be excluded from occupying any of these sites. We can think about this as an excluded volume that is larger than a single cell, as illustrated in Fig. 5.3.4. This excluded volume applies to all monomers, except the

Figure 5.3.3 Illustration of a cellular (lattice) polymer model. In this model, monomers are considered bonded to each other if they are touching at either faces or corners. Other nonbonded monomers are not allowed to approach this close. This enables us to distinguish bonded neighbors from other monomers just by the relative location of the monomers. This property is not satisfied by the ball-and-string model. ∎

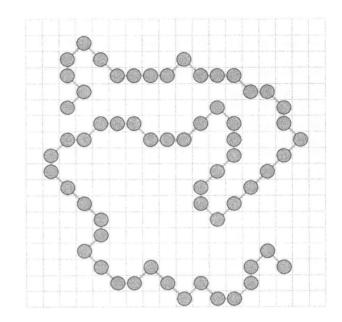

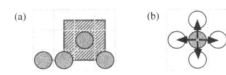

Figure 5.3.4 (a) shows the size of the effective excluded volume for the model of Fig. 5.3.3. The excluded volume is larger than a single cell. It only applies to nonbonded neighbors and prevents them from approaching the adjacent lattice sites. Note that the excluded volume illustrated does not apply to bonded neighbors which have only a one-cell excluded volume with respect to each other. (b) shows the possible moves of a monomer selected for a Monte Carlo update. There is nothing special about this choice of possible moves. We could allow diagonal moves, but the choice of possible moves must be made once and for all for a simulation. ∎

bonded neighbors. Bonded neighbors are prevented from occupying the same site, but can be adjacent. As with other variations of local polymer structure, this is not important for the structure of long polymers. If anything, this is a more realistic model for the bonding of real polymers. Bonds in real polymers are also specified by relative location of monomers—not by a labeling scheme.

The cellular space model in Fig. 5.3.3 could be simulated just like other Monte Carlo models by choosing a monomer, choosing one of the compass directions NSEW, and moving the monomer if the move does not violate either connectivity or excluded volume constraints. We can, however, turn it into a cellular automaton using the Margolus dynamics approach of updating plaquettes (see Section 1.5.6). This is necessary because we need to conserve various quantities in this dynamics: the existence of the monomers and their bonding and excluded volume constraints. Updating plaquettes enables the implementation of conservation laws and constraints in a natural way. A Margolus dynamics for this model uses a partition of the space into plaquettes that are 3×3 regions with an additional one-cell buffer between plaquettes (Fig. 5.3.5). This enables us to move monomers around in each of the plaquettes independently of other plaquettes in the space. The easiest way to do this is to move only monomers at the central sites of the plaquettes. Choosing a direction for each monomer at a central site, we move it if the constraints allow. Fig. 5.3.6 illustrates the monomer moves that are possible and the moves that are not allowed.

After updating each of the plaquettes, we shift the plaquettes around in the space so that we can move all the monomers. We must keep in mind that even as a cellular automaton this is still a Monte Carlo algorithm. In order for the Monte Carlo algorithm to satisfy detailed balance, it is important to pick the location of the plaquettes at random. Specifically, it is necessary to allow the same location of the plaquettes to be chosen in the next time step as well. This guarantees that a particular move can be reversed. More correctly, detailed balance requires that all possible moves have the same probability of occurrence in every time step, and the random selection of a location for the plaquettes guarantees this. The complete Monte Carlo algorithm consists of selecting a location for the plaquettes, selecting a direction from NSEW for each monomer at the center of a plaquette, and moving the monomer if the move is allowed.

Figure 5.3.5 In order to convert the lattice Monte Carlo algorithm to a cellular automaton we use a Margolus dynamics that consists of 3x3 plaquettes with buffer regions as illustrated. Within each of the plaquettes, we can move the monomers around without interfering with other plaquettes. The simplest way to perform moves in the 3x3 plaquettes is given in Fig. 5.3.6. The periodicity of this partition of space is 4x4. ∎

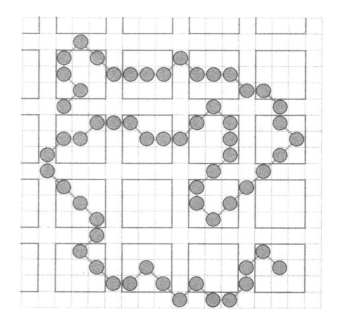

We note that the parallel version of the Monte Carlo algorithm is not, strictly speaking, a Metropolis Monte Carlo algorithm, since the parallel moves do not satisfy Eq. (1.7.19). However, it can be readily shown that a move consisting of a number of parallel independent moves, where each one of them is of the Metropolis form, satisfies detailed balance, Eq. (1.7.17). This is true because the transition matrix factors, as does the equilibrium probability distribution.

The cellular automaton model for polymer dynamics we have constructed can be readily simulated. However, it has a problem that suggests we continue to develop better algorithms. The problem is that the polymer is locally very rigid and the possible local motions of monomers are limited. One way to think about this problem is that for very long polymers, there are two types of motion that are possible: motion of monomers perpendicular to the contour of the polymer and motion along the polymer contour. The latter includes a local stretching or compression of the polymer. For our model, the local motion is always perpendicular to the polymer contour. If we take a long enough polymer, there will be various small folds of the polymer, and motion along the large-scale polymer contour would be possible. However, this means that in order to see the dynamics of very long polymers, we need a very long chain of monomers. Since our objective is to simulate long polymer behavior using as little computation time as possible, we would be better off to have a polymer model that reproduces the long polymer behavior for short polymer chains.

One way to solve this problem is to generalize the cellular automaton model by allowing the monomers that are bonded to each other to separate by one lattice space. Bonded monomers would be located in a larger bonding neighborhood—a 5×5

Figure 5.3.6 In the simplest implementation of the cellular automaton of Fig. 5.3.5, we use the usual Monte Carlo process to update a monomer located at the central site of each plaquette. As illustrated, moves are considered only in plaquettes with monomers in the middle cell of the 3x3 plaquette. Moves that are not allowed due to connectivity or excluded volume constraints are marked by an X in the target square. ∎

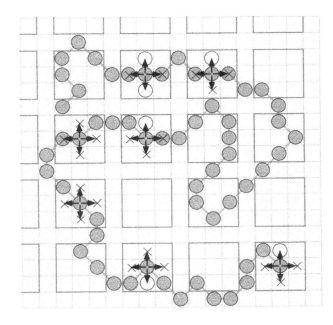

region in two dimensions, or a $5 \times 5 \times 5$ region in three dimensions. This choice of bonding neighborhood is convenient, but others could be specified as well. As before, we do not allow monomers to violate excluded volume by entering a bonding neighborhood, and we do not allow monomers to break a bond by leaving. A monomer move is accepted if monomers are not removed from nor added to the bonding neighborhood by the move. The larger bonding neighborhood allows more flexibility to the motion because adjacent monomers can move toward and away from each other, enabling local contraction and expansion of the polymer. We call this algorithm the one-space algorithm in order to contrast it with the two-space algorithm discussed next.

The problem of polymer flexibility also has a second solution—the two-space algorithm—that has some additional advantages. The simplest way to describe the two-space algorithm in two dimensions is to consider a polymer on two parallel planes (Fig. 5.3.7). The monomers alternate between the planes so that odd-numbered monomers are on one plane and even-numbered monomers are on the other. The neighbors of every monomer reside in the opposite space. We define a 3×3 region of cells around each monomer in the opposite space as its bonding neighborhood. This is the region of cells in which its neighbors reside and no other monomers are allowed to enter. To construct a polymer we place successive monomers so that each monomer has its nearest neighbors along the contour in its bonding neighborhood. The dynamics is defined, as before, by requiring that the motion of a monomer be allowed only if its movement to a new position (selected at random from NSEW directions) does not add or remove monomers from its bonding neighborhood. This preserves both connectivity, preventing loss of a neighbor, and excluded volume, preventing the addition of a neighbor (Fig. 5.3.8).

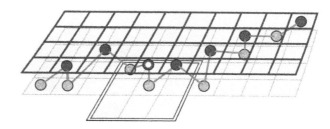

Figure 5.3.7 Schematic illustration of a two-space polymer. In two dimensions, the two spaces are parallel planes. Monomers on the upper plane are shown as circles with dark shading; monomers on the lower plane are shown as circles with light shading. Along the polymer, the monomers alternate spaces so that odd monomers are in one space (the light space) and even monomers in the other space (the dark space). Bonds are indicated by line segments between monomers. Monomers are bonded only to monomers in the other space. The "bonding neighborhood" of each monomer is a 3x3 region of cells located in the opposite plane. The bonding neighborhood of the dark monomer marked with a white dot is shown by the region with a double border. The two neighbors of this monomer, both light monomers, are located in the bonding neighborhood. ∎

Using this model an additional flexibility is achieved, because neighboring monomers can be "on top of each other," so that even the 3×3 bonding neighborhood allows local expansion and contraction. Even more interesting, it is possible in this dynamics to move all of the monomers in one space at the same time without concern for interference, because both connectivity and excluded volume are implemented through interactions with the other space. This allows 1/2 of the monomers to be updated in parallel. The simple parallelism of the two-space algorithm lends it to implementation on a variety of computer architectures. Because we can update 1/2 of the polymer at a time, there are two different ways to implement parallelism: space partitioning and polymer partitioning. Space partitioning is the usual cellular automaton assignment of processors to different regions of space. Polymer partitioning is the assignment of processors to different parts of the polymer. Spatial assignment is particularly convenient when a simulation is performed with a high density of monomers. For example, there is considerable interest in simulations of entangled polymers at high densities (polymer melts). Polymer assignment is convenient when the polymer occupies only a small fraction of the space. This is the case for expanded isolated polymers, or problems that might include a polymer moving in a static matrix.

To show that all the monomers in one space can be moved independently, we must show that their motion cannot result in either breaking the polymer or violating excluded volume. Since each monomer move preserves its bonded neighbors, the polymer cannot be broken. Excluded volume is different for two monomers within a space and for two monomers in opposite spaces. For two monomers in opposite spaces, the excluded volume is implemented by preventing monomers from entering

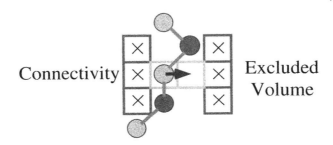

Connectivity   Excluded Volume

Figure 5.3.8 Illustration of the movement of a monomer in the two-space polymer algorithm. The movement of a light monomer requires checking connectivity and excluded volume in the dark space. The picture illustrates a move where the light monomer is to be moved to the right. To ensure that connectivity is not broken, we check that no monomers are left behind. This is equivalent to checking that there are no dark monomers in the three cells marked with Xs on the left. To ensure that excluded volume is not violated is equivalent to checking that there are no dark monomers in the three cells marked with Xs on the right. If there are no monomers in these cells, then no monomers are removed from or added to the bonding neighborhood of the light monomer as a result of the move. In the picture the move is allowed. ∎

each other's bonding neighborhood. The excluded volume between nonbonded monomers is the same as that shown for the first cellular automaton in Fig. 5.3.2. For bonded neighbors there is no excluded volume. For two monomers in the same space, excluded volume is just the requirement that two monomers do not move onto the same site. They can be adjacent, since they are not within each other's bonding neighborhood. In a proof by contradiction that two monomers cannot move onto the same site, assume that two monomers were to move to the same site. In this state they will have the same bonded neighbors. Since they start with different bonded neighbors and our algorithm explicitly prevents any two monomers from changing their bonded neighbors, this can not happen. There is only one exception, which we may simply avoid (or treat specially). For a polymer of length three, the two end monomers both have the same neighbor and they are not prevented by the algorithm from landing on the same site.

How do we choose the next monomer to move in the two-space dynamics? In order to preserve detailed balance, we must choose which of the two spaces to update at random. This ensures that all possible moves have equal probability in each step. Alternating spaces would not satisfy detailed balance. The order of updates of monomers within one of the spaces does not matter and may be done sequentially rather than randomly.

The two-space algorithm may be implemented in three dimensions by considering the polymer to be in a double space with a $3 \times 3 \times 3$ bonding neighborhood, and a similar generalization of possible monomer moves. If it was desired, we could also remap all of the monomers into a single space with an unusual implementation of excluded volume. As before, the local properties do not affect the long polymer scaling.

To test the algorithm, we can implement and simulate it and measure various structural properties as a function of time. The simulations we perform for these tests are in two dimensions. Rather than measuring the polymer end-to-end distance, we choose to measure the characteristic size of the polymer as given by the radius of gyration $R_g(N;t)$:

$$R_g(N;t)^2 = \langle(r(t)-\langle r(t)\rangle)^2\rangle = \frac{1}{N}\sum_i(r_i(t)-\langle r(t)\rangle)^2$$

$$\langle r(t)\rangle = \frac{1}{N}\sum_i r_i(t)$$

(5.3.1)

This is just the standard deviation of monomer positions in space. As indicated, the averages are over monomers rather than over time. To initialize the simulation, we start from a straight stretched polymer that alternates from space to space. This is the easiest way to lay out a polymer initially. Simulating the polymer dynamics then results in Fig. 5.3.9. We see that after some number of steps, the polymer relaxes and fluctuates around an average polymer size that we can calculate as a time average. The value of the time average, $R_g(N)$, is indicated on Fig. 5.3.9. It is better to leave out the first part of the simulation in calculating this average in order to eliminate the effect of the improbable first configuration. For a long enough simulation, this correction is unimportant.

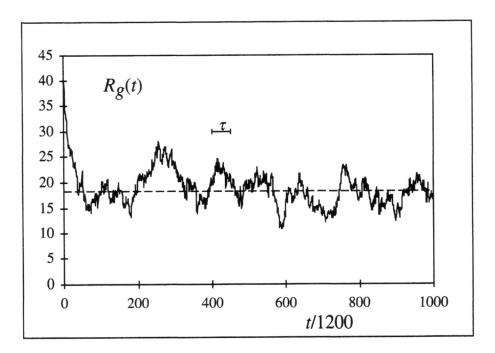

Figure 5.3.9 Plot of the characteristic polymer size, the radius of gyration, $R_g(t)$, as a function of time in a Monte Carlo simulation of the two-space algorithm. The two-dimensional polymer simulated has $N = 140$ monomers. The simulation starts from a completely straight conformation which has an unusually large size. After relaxation, the radius of gyration fluctuates around the average value, $R_g = 18.39$, indicated by the horizontal dashed line. The characteristic time over which the polymer conformation relaxes τ is the correlation time of the radius of gyration indicated by the horizontal bar. The values plotted of the radius of gyration are sampled every 1200 plane updates. There are about 50 samples in a relaxation time. ∎

In order to see the scaling properties of $R_g(N)$ we calculate the average radius of gyration for polymers of different lengths. On a log-log plot (Fig. 5.3.10) the radius of gyration is a straight line for large N. This means that it follows a power-law behavior where the slope of the line is the value of the exponent. The value of the exponent is in agreement with the expected scaling result $R_g(N) \sim N^{0.75}$ from Eq. (5.2.12) in two dimensions. We note that rather than plot $R_g(N)$ as a function of the number of monomers N, the horizontal axis of the plot is $N-1$, which is the contour length of the polymer. For long polymers, the difference is not significant. For short polymers, this causes the results to follow more closely the long-polymer scaling behavior. The long-polymer behavior is reached for remarkably small polymer chains of only a few monomers. Since our objective is to simulate long polymers, this is a desirable result.

The second test is to evaluate the dynamics of relaxation of the polymer. We can see from Fig. 5.3.9 that there is a characteristic time over which the polymer forgets the value of its radius of gyration. Values of $R_g(N;t)$ fluctuate with a characteristic time

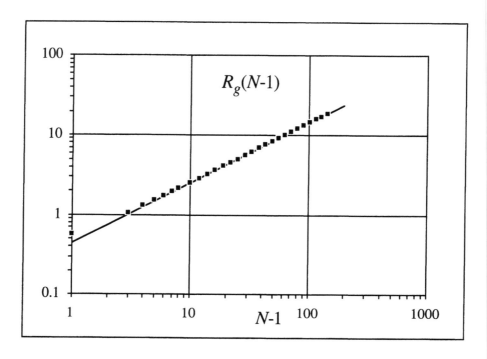

Figure 5.3.10 Plot of the average (root mean square) radius of gyration as a function of polymer length for the two-space algorithm in two dimensions. The average values are obtained by simulations like that shown in Fig. 5.3.9 using 100,000 samples and without including the first 100 samples. The horizontal axis is the number of links, $N - 1$, in the chain. The line in the figure is fitted to the data above $N = 10$ and has a slope of 0.756. This is close to the exact asymptotic scaling exponent for long polymers, $v = 0.75$. ∎

shown by the horizontal bar on the plot. Over shorter times than this, values of $R_g(N;t)$ are correlated. Over longer times, they are essentially independent. This characteristic time is the relaxation time, $\tau(N)$. To find a value for the relaxation time we study the correlation over time of $R_g(N;t)$ (for simplicity the dependence on N is not indicated):

$$A[R_g(t)](\Delta t) = \frac{<(R_g(t+\Delta t)-R_g)(R_g(t)-R_g)>}{<(R_g(t)-R_g)^2>}$$

(5.3.2)

$$R_g = <R_g(t)>$$

This correlation function measures the relationship between the value of $R_g(t)$ and $R_g(t+\Delta t)$, and is a function of Δt. The averages are over time. The overall behavior of the correlation function can be readily understood. For $\Delta t = 0$ it is normalized to 1. For large Δt, where the value of $R_g(t+\Delta t)$ is independent of $R_g(t)$, the average of the product in the numerator would be the product of the averages of the two factors performed independently. Since the average of either factor is zero, the value of the cor-

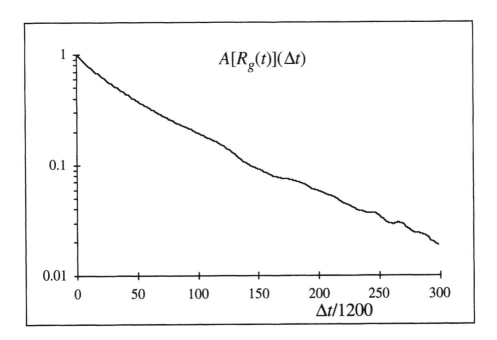

Figure 5.3.11 Plot of the autocorrelation function of the radius of gyration for a polymer of length $N = 140$. The correlation decays approximately exponentially, so the logarithm of the autocorrelation function is roughly linear in time. As in Fig. 5.3.9, the horizontal axis is marked in units of samples taken. The correlation time, τ, is the time at which the autocorrelation function drops to the value $1/e$. Using the integral method described in the text, the correlation time is $\tau = 49.09$ samples of 1200 updates each, or $\tau = 58900$ updates. Only the first 50 values shown of the autocorrelation function determine the relaxation time. ∎

relation function is zero. A plot of the correlation as a function of time is shown in Fig. 5.3.11. If the relaxation of the polymer were simple, the value of the correlation function would be an exponential in time. Since Fig. 5.3.11 is a semilog plot, it would appear as a straight line. The plot is somewhat curved, indicating that it is not a simple exponential decay. Our objective is limited to finding a characteristic relaxation time, $\tau(N)$. We can do this by finding the time at which the correlation falls to $1/e$ of its initial value. However, a better way to measure $\tau(N)$, which reduces the effect of statistical errors, is to integrate the correlation function. If we integrate out to a value $A_0 \approx 1/e$ then we can estimate the relaxation time using

$$\tau \approx \frac{\int_0^{\Delta t(A_0)} d(\Delta t)\, A[R_g](\Delta t)}{(1 - A_0)} \tag{5.3.3}$$

where $\Delta t(A_0)$ is the time interval at which $A[R_g](\Delta t) = A_0$. This formula would be exact if the correlation were exponential and without statistical error.

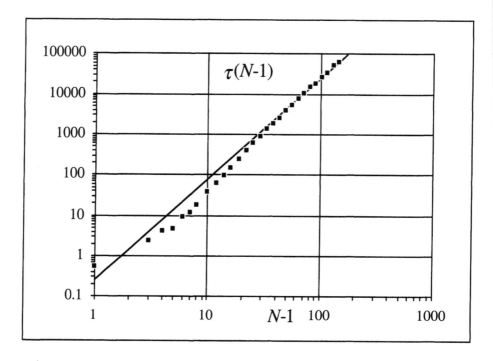

Figure 5.3.12 Plot of the relaxation time τ of polymers as a function of the number of links $N - 1$ for the two-space algorithm in two dimensions. The line in the figure is fitted to the last four points that are relaxation times for polymers longer than $N = 100$. The slope of this line is 2.51, which is close to the asymptotic scaling exponent expected for Rouse relaxation, $2v + 1 = 2.5$. ∎

A plot of the characteristic relaxation time $\tau(N)$ as a function of the polymer length is shown in Fig. 5.3.12. We see that $\tau(N)$ increases with length, and for long enough polymers it agrees with the Rouse power law prediction, $\tau(N) \sim N^{2.5}$. Since we have not included hydrodynamics, the Rouse scaling is to be expected, not the Zimm scaling. Short polymers do not have this behavior, but the asymptotic scaling of long polymers is satisfied for polymers longer than approximately 50 monomers. This is still quite short for a polymer, and it suggests that we can effectively simulate long polymer behavior using this algorithm. This concludes our tests of the scaling behavior of the two-space algorithm. In order to apply it to the simulation of polymer collapse, we have to modify it so that monomer-monomer aggregation can occur.

A different Monte Carlo algorithm that enables the study of the radius of gyration or other static properties, but not the dynamics of polymers, is described in Question 5.3.1.

Question 5.3.1 In Section 1.7 we discussed the possibility of considering Monte Carlo algorithms that were nonlocal. These algorithms would provide the correct equilibrium ensemble, but the dynamics would be different from that of local Monte Carlo algorithms. There is an interesting non-local algorithm for polymer simulations. It can be used for various polymer models, including both the one- and two-space polymers. A step of the non-local Monte Carlo moves a monomer from one end of the polymer to the other. To achieve such a change in conformation of the polymer by local monomer steps would require many steps. However, the final conformation is allowed, and so in a Monte Carlo process we can enable this transition. A complete specification of the algorithm is: Select one of the two polymer ends. Delete the end monomer. Select one of the possible neighboring locations of the monomer at the opposite end at random. If the addition does not violate excluded volume constraints, accept the move. Otherwise reject it. Convince yourself that this algorithm satisfies the requirements of a Monte Carlo process. Program and simulate it and see that the radius of gyration does satisfy the standard scaling behavior, but the relaxation does not. ∎

5.4 Polymer Collapse

5.4.1 *Introduction to polymer collapse*

The objective of this section is to develop an understanding of the kinetics of collapse. We do this first through simulations, then a scaling argument, then some more simulations. One of the goals we achieve is to obtain the scaling of the collapse time of a polymer as a function of polymer length. Our primary objective, however, is achieved when we find that the kinetic process of collapse can systematically and reproducibly restrict the possible conformations that are explored. This implies that kinetics of collapse may reproducibly lead to a desired folded conformation without exploring all of the possible conformations of the polymer.

The collapse of a polymer is controlled by the difference of the temperature $\Delta T = \theta - T$ from the θ-point temperature. The lower the temperature the more rapid the collapse and the more important the kinetic effects. This, it may be noted, could be used in experiments to determine the significance of kinetic effects during collapse. If kinetics play an important constructive role, then collapse that occurs too close to the θ-point might not result in properly folded conformations. Of course, the collapse under these circumstances is also slower.

The process of collapse involves many encounters between monomers that form weak bonds to each other, like hydrogen bonds. Some of these bonds might break and others form instead. The bonds that are formed build larger and larger aggregates. If we are concerned about the kinetic effects, then we don't have to be overly concerned about the bonds that are broken; we can focus only on the formation of bonds. We then imagine a process of irreversible sticking of monomers. One way to think about this is that the key to kinetic effects for polymers is the process of first encounter—those monomers that find each other first. A second way to think about this is that we are considering only large values of ΔT, where the energy of a single bond becomes large compared to the temperature T and the chance of breaking it is small. This picture becomes increasingly valid for longer polymers. For long polymers, we can think about a coarse graining process that groups monomers and bonds together. The local formation and breaking of a single bond is less relevant than the formation of clusters. The possibility of breaking up a cluster becomes less and less likely for larger and larger clusters because the bonding energy is larger and larger. It is possible to prove that for long polymers we can always think about the process of collapse as if it occurs for large values of ΔT. This is demonstrated formally in the following paragraph.

We can discuss the thermodynamics of polymer collapse using arguments similar to those in Section 5.2, by including an additional term in the free energy that describes the interaction of the polymer with the solvent. The energy of polymer-solvent interaction is given by the energy of a monomer-solvent interaction times the number of such interactions. In a mean field picture where correlations are ignored, this would be written as:

$$F_{polymer-solvent} = \chi kTN(1 - \frac{NV}{R^d})$$ (5.4.1)

The prefactor χkT gives the energy of interaction of monomer with adjacent solvent. The rest is the number of monomers times (in parenthesis) the probability that solvent is found adjacent to a monomer. This probability is 1 minus the volume fraction of the solution occupied by monomers. Adding Eq. (5.4.1) to Eq. (5.2.9) we have

$$F(R) = F_0 + \frac{dkTR^2}{2R_0^2} + (1 - \chi)kTV\frac{N^2}{R^d}$$ (5.4.2)

We see that the interaction with the solvent acts to change the effective monomer-monomer interaction. If the coefficient is negative, then the polymer self-attracts and collapses. When this happens, the free energy we have written down is not sufficient, because it has no terms that stop the radius from decreasing to a point. We need to add a term that increases with increasing monomer concentration and can stop the collapse. To do this we treat the free energy as an expansion in the concentration and add a positive term with one higher power of the concentration $c = N/R^d$:

$$F(R) = F_0 + \frac{dkTR^2}{2R_0^2} + (1 - \chi)kTV\frac{N^2}{R^d} + fkT\frac{N^3}{R^{2d}}$$ (5.4.3)

where f is known as the third virial coefficient. It is convenient to rewrite this in terms of the ratio of the radius to the random-walk radius $y = R/R_0$ giving

$$F(R) = F_0 + \frac{dkT}{2}y^2 + (1 - \chi)kTV\frac{N^{2-d/2}}{y^d} + fkT\frac{N^{3-d}}{y^{2d}}$$ (5.4.4)

To find the expected value of y, we take the derivative and set the result equal to zero to obtain an equation:

$$0 = y^{d+2} - (1 - \chi)VN^{2-d/2} - 2f\frac{N^{3-d}}{y^d}$$ (5.4.5)

We know that $(1 - \chi)$ has to be positive for T greater than θ and negative for T less than θ. We can substitute a linear dependence for $(1 - \chi) = c\Delta T$. Limiting ourselves to three dimensions, we can write Eq. (5.4.5) as:

$$0 = y^{d+2} - cV(\Delta TN^{1/2}) - 2f\frac{1}{y^d} \qquad (5.4.6)$$

Looking at this equation, we see that a large value of ΔT has the same effect as a large value of N, since the only relevant parameter that controls the collapse is $\Delta TN^{1/2}$. This shows that for long polymers, large values of N, we can always think about the collapse as occurring at low effective temperatures, or large $\Delta T_{eff} = \Delta TN^{1/2}$. This argument provides a formal justification of our treatment of collapse using monomers that encounter each other and stick (bond) irreversibly.

In Section 5.4.2 we describe simulations that indicate a possible relevance of kinetics to polymer collapse. They motivate a scaling argument, described in Section 5.4.3, which generalizes the results. Section 5.4.4 contains a discussion of the implications of the results for protein folding and other systems. Additional simulations in Section 5.4.5 explore the sensitivity of the results to the details of the polymer structure.

5.4.2 *Two-space simulations of collapse*

Using the cellular automaton Monte Carlo algorithms developed in Section 5.3 we can study the problem of polymer collapse. The simulation of polymer collapse starts from a set of equilibrium (expanded) polymer configurations in good solvent generated by Monte Carlo simulations. To generate these conformations, we use either the two-space algorithm with the local monomer moves or the nonlocal Monte Carlo described in Question 5.3.1. The nonlocal Monte Carlo is faster than the local Monte Carlo dynamics and yields the same equilibrium polymer conformations. However, because it is nonlocal, we cannot use it for simulating dynamics such as the collapse itself.

In order to simulate collapse of a polymer, we must allow monomers to stick to each other and form aggregates. Once aggregates form, we have to track their shape, move them as a unit, and allow continued aggregation at their boundaries as they move. This will complicate our simulations substantially. Before we try this, is there an easier way? Aggregation would be much simpler if we allowed monomers to move onto the same site of the lattice. Then the aggregate would look the same as a monomer for the simulation. We can make the simulations a little more realistic by keeping track of the aggregate mass—the number of monomers that have accumulated. Since this is easy to do, we might try to simulate collapse and see what results we find. Later we can make tests to verify or correct the results. This approach is quite similar to the way scaling relations were derived in Section 5.2—use the simplest method possible, then check if it makes sense and verify it with a more complete analysis. Whether the simple method works or not, we will have learned something valuable about what is important and what isn't. If the simple approach works we learn that the results are general and robust. If the simple approach does not work, then by investigating what details change the results we learn what aspects of the problem or parameters play a role. An example of this approach is the treatment of the scaling of polymer size in Section 5.2. The random-walk model was not quite enough to give the correct exponent. Incorporating excluded volume was necessary and also sufficient. For simulations of polymer collapse, a simple approach that doesn't quite work is described in Question 5.4.1.

Question 5.4.1 In terms of our polymer simulation algorithms, aggregation is easiest to think about as the elimination of the excluded volume constraint. This allows one monomer to approach another and bond by entering its bonding neighborhood. Once they bond we would not allow them to separate. This means we continue to impose the connectivity constraint by not allowing a monomer to leave behind another monomer. We might simulate collapse by keeping track of each individual monomer, even if the monomers occupy the same site. However, there is a problem with this approach. Consider what happens to the diffusion constant of an aggregate. Show that the diffusion constant of an aggregate is not realistic and that this must distort the outcome of the simulations.

Solution 5.4.1 In the proposed method, monomers aggregate by moving into each other's bonding neighborhoods. As time goes on, aggregates form with many monomers bonded to each other in their mutual bonding neighborhoods. The problem with diffusion is that for an aggregate with N monomers, the diffusion constant of the aggregate becomes exponentially small with N. In order to see this, we can focus on the motion of the bonding neighborhood associated with a monomer. Assume we begin with a bonding neighborhood located at a particular place in the lattice. All of the bonded monomers are located in this bonding neighborhood. In order for the bonding neighborhood to displace by one square to the right, none of the monomers must be in the leftmost set of cells of the bonding neighborhood. Any monomer that stays on the left would veto the motion to the right. Thus, in effect, in order to move to the right all of the monomers must move to the right at the same time. Since each monomer is moving independently, the probability of this occurring decreases exponentially with the number of monomers in the aggregate. This is unrealistic. In a simulation, once aggregation starts to occur, the continued motion of an aggregate is too slow. This problem can be solved as discussed in the text, by moving aggregates as a unit. ∎

The example described in Question 5.4.1 illustrates an important feature of computer simulation and research in general. One of the most important and yet most difficult skills to learn is to distinguish a correct from an incorrect result by simple criteria. This capability is crucial when performing computer simulations because, to one degree or another, the computer acts as a black box. We are unable to verify the performance of the simulation ourselves directly. This is a problem both for the presence of an error—bug—in the computer program, as well as an error in the methodology or approach to simulation. The latter is illustrated by Question 5.4.1. For some students, the problem of telling whether a simulation is correct may seem an impossible one. However, this is a skill that we develop. An example is the ability to determine if two numbers have been multiplied correctly. After the multiplication we can check whether the order of magnitude is correct and whether the number is even or odd. We can perform these and other tests independent of the multiplication itself. By verifying that these aspects of the multiplication are correct, we increase the likelihood that the entire multiplication is correct. When computer simulations are performed, one of our best tools to determine whether it is valid is to view the simulation as a movie. By viewing it, our qualitative concepts about the simulation can be evaluated and compared with what is observed. For this reason, it is important to build a mental model of how the simulations should appear. When the simulations and mental model do not agree, we can either correct the mental model or the simulations based upon further inquiry. This approach is not guaranteed to work, since we might have an error that falsely makes the simulation agree with our mental concept. However, performing such tests does increase the reliability of the results. In effect, it is one way we can compare two independent models of the same process. Whenever we can compare more than one model of the same process, we gain an understanding of the reliability and robustness of results.

We begin to investigate the behavior of polymer collapse using the two-space lattice Monte Carlo algorithm. Starting from an initial equilibrium conformation, polymer collapse is simulated by eliminating the excluded volume constraint. We discuss below why excluded volume may not be necessary during collapse even though it is necessary for the original polymer conformation. Once the excluded volume constraint is eliminated, the usual monomer Monte Carlo steps are taken. Monomers are no longer prevented from entering the neighborhood of another monomer; however, they continue to be required not to leave any neighbors behind. This enables monomers of the same type (odd or even) to move on top of each other. Once one monomer moves onto another monomer, they lose separate identity and become an aggregate. Aggregates are moved as a unit. This avoids the problem we found in Question 5.4.1. We keep track of the mass M of an aggregate, which is the total number of monomers that reside on the same site. We can assign a diffusion constant to the aggregate which depends on the mass of the aggregate. The most natural choice is to set the diffusion constant according to Stokes' law: $D(M) \sim 1/M^x$. As discussed at the end of Section 5.2, we use a diffusion constant in two and three dimensions that scales as $D \sim 1/R$, so in d-dimensions $x = 1/d$. By incorporating Stokes' law into the collapse, we have introduced hydrodynamics into the simulation. Hydrodynamics is not generally part of a lattice simulation. However, by explicitly setting the diffusion constant, we have incorporated its primary effect when there are aggregates present.

What role does the diffusion constant play in the simulations? The diffusion constant is proportional to the rate at which an aggregate hops from a lattice site to a lattice site (Eq. (1.4.55)). When we perform a Monte Carlo simulation, steps occur in discrete time. The rate of hopping is represented as a probability of making a hop in a single time step. Thus we can implement the diffusion constant by controlling the probability of making a hop when an aggregate is selected to move. We choose our time scale and normalize the probabilities by setting to one the probability that a single monomer will move when chosen. All moves, of course, may be rejected if they violate other constraints.

The polymer dynamics are then simulated by selecting at random an aggregate (monomers are included as aggregates of mass 1), and moving the aggregate in one of four compass directions with a probability given by the diffusion constant, and only if connectivity constraints allow—the aggregate does not leave any neighbors behind. In order to move the aggregate with a probability given by its diffusion constant, a random number ranging between zero and one is compared with the diffusion constant. The monomer is moved only if the random number is smaller than the diffusion constant.

In order to describe the time dependence of the collapse, we must keep track of the passage of time in the simulation. Time is normally measured in a Monte Carlo simulation of polymers by choosing randomly N monomers to move in a single time interval. On average, each monomer is moved once in a time interval. During the collapse we must do a similar counting, where one time interval of the simulation consists of performing a number of aggregate moves equal to the number of remaining aggregates. Since the number of aggregates can change during the time interval, it is arbitrarily taken to be the number at the end of the time interval. As monomers are moved, a counter is incremented and compared with the number of aggregates remaining. When the number of moves exceeds the number of aggregates, a new time interval is started.

When we simulate collapse, we introduce effective interactions between monomers in the same space. This interaction is the aggregation itself. We therefore do not consider parallel processing in the simulation of collapse. While we don't take advantage of parallelism, other aspects of the two-space algorithm make it convenient for these simulations.

A sequence of frames from a simulation in two dimensions is shown in Fig. 5.4.1. Each aggregate is shown by a dot. The area of the dot is the mass of the aggregate. Most striking in these pictures is that the ends of the polymer have a special role in the collapse. The ends diffuse along the contour of the polymer, eating up monomers and smaller aggregates until the two end aggregates meet in the middle. Along the contour, away from the ends, the polymer becomes progressively smoother. The polymer becomes more and more like a dumbbell. One way to think about this process is that the polymer collapse becomes essentially a one-dimensional process along the polymer contour. We call this process end-dominated collapse.

5.4.3 Scaling theory of collapse

In this section we develop a scaling theory that describes the results found in the simulations. Before proceeding, we summarize a mean field model for the kinetics of polymer collapse that is relevant to conditions where the collapse occurs slowly be-

Figure 5.4.1 Frames, "snapshots," of the collapse of a single homopolymer of length $N = 500$ monomers in two dimensions. The plot is constructed by placing dots of area $M^{1/2}$ for an aggregate of mass M. This does not reflect the excluded volume of the aggregates, which is zero during this collapse simulation. Simulations that include excluded volume during collapse demonstrate similar results (Section 5.4.5). Successive snapshots are taken at intervals of approximately one-quarter of the collapse time. The initial configuration is shown at the top. The results demonstrate the end-dominated collapse process, where the ends diffuse along the contour of the polymer accreting small aggregates. ∎

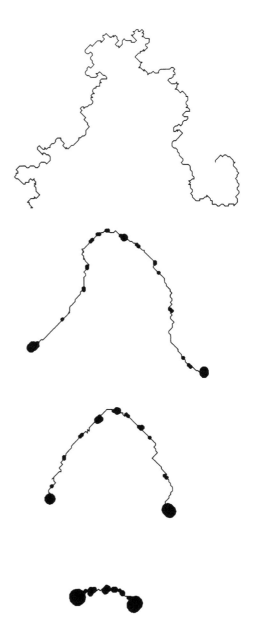

cause T is very close to the θ-point. This means that monomers bind and unbind many times during the collapse. Under these conditions, the behavior of the polymer in solution is like the behavior of two liquids trying to separate because they are immiscible. This is called phase separation. The slowest process is the longest length scale separation. The final structure of the system is a sphere of polymer, so the longest length scale relaxation is the contraction of a prolate spheroid, a sausage shape, to a sphere. The longest time scale is set by the diffusion that causes the sausage to thicken as the ends contract. Despite the difference in the nature of this process from our usual polymer relaxation, the scaling of the collapse time is essentially consistent with the Zimm relaxation of a polymer above the θ-point, $\tau(N) \sim N^{3\nu}$.

In contrast to the mean field scenario, in order to investigate the kinetic effects in polymer collapse we must consider larger departures from the θ-point. As discussed in Section 5.4.1, a model in which monomers stick irreversibly becomes valid progressively closer to the θ-point as the polymer length increases. Below the θ-point, monomers attract each other. However, unless the monomers are charged, for a long polymer the interactions are short range. Thus we will consider monomer-monomer interactions only when they come into contact. When two monomers come into contact, they stick to each other and do not separate. Therefore, the first step in understanding the kinetics of polymer collapse is identifying the order in which monomers encounter each other. In completely irreversible collapse (sticking), every encounter causes a kinetic barrier to arise, as described in Fig. 5.1.1, that does not allow the two monomers to separate. In a more realistic model there would be reversibility; however, the final conformation will evolve from a conformation established by the initial encounters. This picture is convenient for our analysis because the initial encounters between monomers occur when the polymer is expanded, and therefore we can understand it beginning from the theory we have developed for the polymers in good solvent.

When we consider qualitatively the process of polymer collapse, we realize that encounters of monomers that are distant from each other along the contour of the chain are unlikely because they are also, on average, distant from each other in space. We can therefore begin by limiting ourselves to consider aggregation as primarily a local process where a monomer forms an aggregate with neighboring monomers along the contour. This kind of aggregation is, however, inhibited by the existing bonds. Aggregation occurs when two monomers that are near each other in space move close enough to form a new bond. The easiest aggregation would occur if a monomer could move to aggregate with one of its neighbors, however, the neighbor on the other side prevents this because stepping away from the other neighbor would break an existing bond. Without curvature in the chain, the monomer is unable to move to aggregate with either neighbor, because it is bonded to the neighbor on the other side. If there is some curvature, then monomers can aggregate. The aggregation would cause the curvature to decrease, and further aggregation becomes more difficult. The same argument applies if we consider a monomer moving to bond to its second or third neighbors along the contour. These problems do not occur at the ends of the polymer. The ends, because they have only one neighbor, can move to aggregate with the monomers near them along the contour. Thus during collapse, the aggregates at the ends grow more rapidly than aggregates along the contour and eventually the polymer looks like a dumbbell. This is what was found in the polymer collapse simulations (Fig. 5.4.1).

To develop a scaling argument for irreversible collapse that distinguishes only between the ends and the average collapse along the contour, we assume that we can summarize the collapse using two variables $M(t)$ and $M_0(t)$. $M(t)$ is the average mass of an aggregate along the polymer at time t, but it does not include the end aggregates. $M_0(t)$ is the mass of the aggregate at either end of the polymer. At time t the end aggregate has grown to size $M_0(t)$ and the average mass along the contour is $M(t)$. In a small time interval, each end aggregate has a probability proportional to its diffusion constant of collecting more mass by moving toward and accreting its immediate neighbor aggregate. This neighbor has an average mass $M(t)$ and is a distance a away that does not depend on time. Thus, on average $M_0(t)$ grows according to

$$\frac{dM_0(t)}{dt} \propto M(t)D_0(t)/a^2 \tag{5.4.7}$$

By Stokes' law (see the discussion at the end of Section 5.2) the diffusion constant $D_0(t)$ of the end aggregate decreases in time as its mass increases according to the relationship

$$D_0(t) \propto 1/M_0(t)^{1/d} \tag{5.4.8}$$

We assume that the two quantities $M(t)$ and $M_0(t)$ follow a power law scaling with time:

$$M_0(t) \propto t^{s_0}$$
$$M(t) \propto t^{s} \tag{5.4.9}$$

Inserting Eq. (5.4.9) and Eq. (5.4.8) into Eq. (5.4.7), we can ignore prefactors and set the exponents of the time on both sides equal.

$$t^{s_0-1} \propto t^{s}/t^{s_0/d} \tag{5.4.10}$$

Solving for s_0 in terms of s we obtain:

$$s_0 = (s+1)d/(d+1) \tag{5.4.11}$$

The two exponents are equal when $s = s_0 = d$. This would correspond to the case of uniform collapse, where there is no difference between collapse at the ends and collapse along the chain. We will find instead that s is much smaller than d, and therefore s_0 is much larger than s.

The value of s may be obtained from a second scaling argument by considering collapse of a polymer with fixed ends at their average equilibrium separation. This removes the dynamics of the end motion from the problem. The collapse of this fixed-end polymer would result in a straight rod of aggregates with an average mass $M = N/R$, where N is the number of original monomers and $R \sim N^{\nu}$ is the average end-to-end distance of the original polymer, which is also the length of the resulting rod. Since we have eliminated the special effects of the ends, the time over which the collapse occurs can be approximated roughly by the usual dynamics of a polymer with $\tau \sim N^z$, where $z = 3\nu$ or $z = 2\nu + 1$ for Zimm or Rouse relaxation respectively. We think about the fixed-end polymer collapse as a simple model for the collapse of the original polymer along the contour away from the ends. We are assuming that the polymer can be approximated locally by polymer segments whose ends are pinned. Over time, progressively longer segments are able to relax to rods. The average mass of the aggregates at a particular time t is then determined by the maximal segment length $N(t) \sim t^{1/z}$ that relaxes by time t. This means that

$$M(t) \sim N(t)/R(t) \sim N(t)^{1-v} \sim t^{(1-v)/z} \tag{5.4.12}$$

or

$$s = (1-v)/z \tag{5.4.13}$$

The value of s obtained from this argument (see Table 5.4.1) is small. Using Eq. (5.4.11) we also find that s_0 is much larger than s. This means that the ends will play a special role in the collapse of long polymers. The mass of the ends increases more rapidly than the average mass along the chain and eventually dominates the aggregation. We have seen this result in the simulations of collapse.

The idea that the polymer becomes straighter with time, because regions of higher curvature aggregate more rapidly, can be made more precise. To characterize this behavior it is convenient to compare the distance, r, between two designated monomers (not necessarily the polymer ends) with the contour length, l, of the polymer connecting them. The contour length is the number of bonds between them along the chain. When aggregation occurs, the small aggregates that form, appearing like beads on the chain, decrease the effective contour length of the polymer. We can define the effective contour length by counting the minimum number of monomer-monomer bonds that one must cross in order to travel the polymer from one designated monomer to the other. Bonds formed by aggregation allow us to bypass the usual polymer contour. In this way the effective number of links in the chain decreases over time.

We consider the scaling of $r(l, t)$ as a function of l. Before collapse begins ($t = 0$), the scaling is given by $r \sim l^v$. This is just the usual scaling of the end-to-end distance of a self-avoiding random walk, Eq. (5.2.11), because the number of links is essentially the number of monomers. If the polymer becomes straighter, the scaling exponent will increase over time. At long enough times, the scaling will approach that of a straight line ($r \sim l$). However, the smoothing occurs first at the shortest length scales. The characteristic time over which a particular length of polymer becomes straight is the relaxation time of the polymer segment. We can approximate this as in the previous paragraph using the usual relaxation, $\tau \sim l^z$.

We can summarize the behavior of the polymer over time using a universal scaling function. This function describes how the end-to-end distance depends on the contour length as a function of time. The essential idea of the scaling function is that the behavior on different length scales can be described by the same function. Specifically we have a relationship of the form:

$$r(l,t) = l f\left(t/l^z\right) \tag{5.4.14}$$

This relationship summarizes our previous discussion through properties of the universal scaling function $f(w)$. $f(w)$ is a constant for large values of its argument (long times) because at long times $r \sim l$. We also know that it scales as $w^{(1-v)/z}$ for small values of its argument in order to ensure the correct scaling of the self-avoiding random walk, $r \sim l^v$, at $t \to 0$. The crossover between one behavior and the other occurs at a particular value of the argument, $w = w_0$, so that the relaxation time satisfies $t = \tau = w_0 l^z \sim l^z$.

This concludes our analytical study of irreversible collapse. This analysis has given us the tools to discuss the simulations in a way that will show us important features of the collapse. In particular we can study the behavior of the quantities $M(t)$, $M_0(t)$, r and l. Before we do so we address one of the questions we asked before about the simulations.

In our simulations, during polymer collapse we eliminated the excluded volume. Isn't the excluded volume important for collapse? We know that excluded volume is relevant to the initial polymer conformation in good solvent. Moreover, excluded volume is relevant to the final collapsed state of the polymer—without excluded volume the polymer collapses to a point. However, we see that excluded volume does not enter in the scaling argument leading to the relationship between s and s_0, Eq. (5.4.11). This argument describes the kinetics of collapse itself. Thus, we do not expect excluded volume to affect the behavior of the collapse. In particular, we do not expect it to affect the relationship between s and s_0. According to Eq. (5.4.11) this relationship depends only on the dimension d of the space. On the other hand, the value of s derived in Eq. (5.4.13) is dependent on the values of the exponents v and z. This means that we can expect the precise values of s and s_0 to be somewhat more sensitive to the presence of an excluded volume. However, the range of possible values (Table 5.4.1) indicates that the overall behavior of the collapse should not be affected. In Section 5.4.5 we will describe simulations that have excluded volume, and find that the results are indeed similar. We can understand why the excluded volume is not important because during collapse the kinetics is primarily affected by the net attractive interaction between monomers, rather than the hard core repulsive part.

We can analyze the simulations to compare with the scaling argument by studying the average mass of the polymer except the ends, $M(t)$, and the mass of the ends, $M_0(t)$. The result of averaging many collapse simulations are shown in Fig. 5.4.2 for two and three dimensions. As indicated, two lengths of polymers were simulated in each case. Longer polymers follow the collapse of the shorter polymers but extend the curves to longer times. This is consistent with the picture of end-dominated collapse, where the only effect of a longer polymer is increasing the length of time till the end aggregates meet in the middle.

$s = (1 - v)/z$	$z = 3v$	$z = 2v + 1$
$2-d$: $v = 0.75$	0.11	0.10
$3-d$: $v = 0.6$	0.22	0.18
$v = 0.5$	0.33	0.25

Table 5.4.1 Values of the exponent s from the scaling relation Eq. (5.4.13) using different assumptions for v and z. $v = 0.5$ would occur for a random walk without excluded volume in any dimension. The other values of v are for self-avoiding random walks. $z = 3v$ is for Zimm relaxation that includes hydrodynamics. $z = 2v + 1$ is Rouse relaxation that does not include hydrodynamics. All of these values are small and indicate that s_0 is much larger. ∎

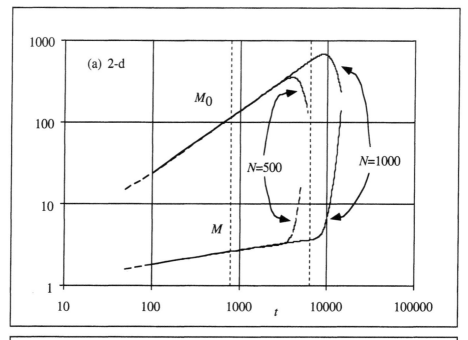

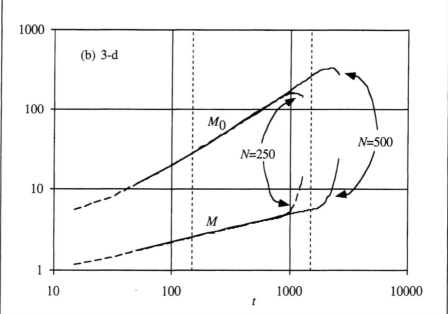

Figure 5.4.2 Plot of the time evolution during polymer collapse of the average total mass of the polymer ends $M_0(t)$, and of the average mass $M(t)$ of aggregates not including the ends. (a) shows collapse in two dimensions of polymers of length $N = 1000$ (averaged over 500 samples), and $N = 500$ (1000 samples). (b) shows collapse in three dimensions of polymers of length $N = 500$ (500 samples), and $N = 250$ (1000 samples). Scaling exponents fitted to the longer polymer collapse for times between the vertical dashed lines are given in Table 5.4.2 and discussed in the text. ∎

Both $M(t)$ and $M_0(t)$ follow a power-law scaling behavior. Exponents from the lines in Fig. 5.4.2 are given in Table 5.4.2. There are two results from our scaling analysis that we can compare with. The more reliable derivation is that of the relationship between s and s_0. A comparison is made in Table 5.4.2 by calculating the expected value of $s_0(s)$ using the scaling relation, Eq. (5.4.11), and the measured value of s. This is compared with the measured value of s_0. The agreement with the scaling relationship is striking since there are corrections which may be expected due to the neglect of the effects of small rings, or changes over time in curvature and compression of the polymer. The statistical errors are smaller than the quite small difference between the expected and measured value of s. This difference, in principle, might be real but could also be due to systematic error from the use of polymers that are not long enough to determine this level of precision.

The comparison of the value of s obtained from the simulation with the values predicted by the scaling argument (Eq. (5.4.13)) show that there is qualitative but not quantitative agreement. The value in two dimensions does not agree with that expected—it is halfway between the values expected in two and three dimensions. The value in three dimensions is very close to that expected without excluded volume ($v = 0.5$) and with Zimm relaxation. There are two ways to discuss this: one is to downplay the success and the other to downplay the failure. All the values of the exponent s are small, and this suggests that the general discussion in the scaling argument is essentially correct. We could also excuse the disagreement in two dimensions because we are using Stokes' law to move aggregates, and hydrodynamics is not well behaved in two dimensions (see the end of Section 5.4.2). The estimate given for three dimensions using Zimm relaxation and without excluded volume $s = 1/3$ is in coincidence with the simulations. It is hard to believe that this result can be justified except as a coincidence. While it is true that we implement hydrodynamics by performing moves of aggregates according to Stokes' law, and that we have performed the simulation without excluded volume, nevertheless, the initial conformation of the polymer should set the distances that control the collapse time. This initial polymer conformation is for a polymer with excluded volume. This would suggest that a

	s	s_0	$s_0(s)$ [Eq. (5.4.11)]	$s_0 - s_0(s)$
1-d	0	0.5	0.5	0
2-d	0.154±0.001	0.7734±0.0006	0.7695±0.0006	0.004±0.001 (0.5%)
3-d	0.337±0.002	0.982±0.005	1.003±0.002	−0.021±0.005 (2%)

Table 5.4.2 Power-law exponents for the scaling of end mass (s_0) and mass along the contour (s) during polymer collapse. The first column gives the dimension of space. The second and third columns are fitted to the simulation results between the dashed lines in Fig. 5.4.2. Fits were chosen to minimize standard errors. Errors given are only statistical—they reflect the standard deviation of the simulation data around the fitted line. The simulation results are compared to the scaling relation, Eq. (5.4.11), in columns four and five. Results in one dimension are exact, since there is no possibility of collapse along the contour, $s = 0$. ∎

smaller value of s should be expected. For our purposes, the precise value of s is not essential, but the central result of the end-dominated collapse is.

Finally, to investigate the validity of the universal scaling of polymer smoothing, Eq. (5.4.14), we plot in Fig. 5.4.3 values of r/l against the rescaled time, $w = t/l^z$, with $z = 3v$. This is a way of showing directly the function $f(w)$. What is important in this figure is that all the different values lie along the same curve. This is the significance of the universal scaling function. The generally good coincidence of the different curves confirms that the simulation obeys Eq. (5.4.14). Moreover, we see that the function $f(w)$ approaches the expected value, 1, at large values of w, consistent with the polymer becoming straighter with time. If we try to change the assumptions we find a poorer fit. For example, assuming $r \sim l^{0.95}$ at long times or changing slightly the value of z would lead to visibly poorer coincidence of the curves. The fit in two dimensions is somewhat less precise. This may be due to the difficulties with modeling hydrodynamics in two dimensions, or it may be due to other effects we neglected such as ring formation during collapse.

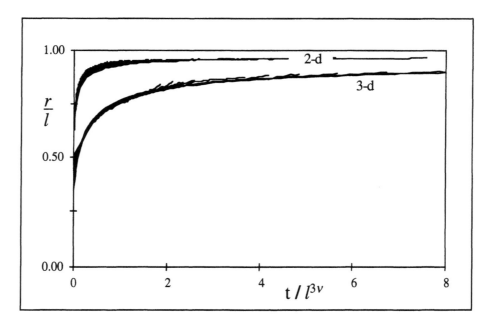

Figure 5.4.3 Plot of the rescaled end-to-end polymer segment distance, r/l, as a function of the rescaled time, t/l^{3v}. Results for both two and three dimensions are shown. In both cases the polymer contained 500 monomers and results were averaged over 200 collapses. The many curves in each case arise from different values of l. The coincidence of the curves is consistent with validity of the universal scaling relationship, Eq. (5.4.14). The plot shows the universal scaling function, $f(w)$. This function describes the structure of the polymer contour, not including the ends. For large w it approaches one, consistent with the expectation that the polymer contour becomes straight at long times. ∎

5.4.4 *Implications of end-dominated collapse*

We have found that polymer collapse in both two and three dimensions appears to reduce essentially to the behavior of a one-dimensional collapse. In one dimension we
would have a starting conformation of a set of monomers along a line. Then, without
excluded volume, the ends in each time step can move to occupy the same site of, and
aggregate with, their neighbor. In each time step the end has equal probability to try
to move away (it cannot) or to try to move on top of its neighbor. Thus, 50% of the
steps, it aggregates with its neighbor. Collapse thus proceeds by the driven diffusion
of the polymer ends. The driving force is the aggregation of the monomers. The only
modification of simple diffusion is the change in the diffusion constant due to the
accretion of mass onto the ends. In two and three dimensions, the results are similar.
Ends diffuse along the contour accreting monomers and smaller aggregates.

We can analyze our results to give the scaling of the collapse time $\tau(N)$. In end-
dominated collapse this is the time that passes until the end aggregates meet in the
center. By this time the end aggregate has reached one-half of the mass of the whole
polymer:

$$M_0(\tau(N)) \propto N \qquad (5.4.15)$$

where we need to write only that the end mass is proportional to the number of
monomers. Substituting Eq. (5.4.9) we obtain

$$\tau(N) \sim N^{1/s_0} \qquad (5.4.16)$$

From the simulations we find that $1/s_0 = 1.293 \pm 0.001$ in two dimensions and
1.018 ± 0.005 in three dimensions (errors are statistical). Thus the collapse time is predicted to scale linearly with polymer length in three dimensions. This indicates that
kinetic effects through end-aggregation accelerate the collapse from the usual relaxation time scaling of $\tau(N) \sim N^z$.

Even without a value of s_0 from the simulations, we can see that end-aggregation
must accelerate the collapse from the equilibrium relaxation. The scaling relation
Eq. (5.4.11) gives a minimum possible value for the exponent s_0. s_0 is a monotonic increasing function of s. The minimum value of s_0 results from setting $s = 0$ in the scaling relation. This corresponds to collapse without any aggregation along the contour.
The only process that occurs is the increasing mass of the ends. The minimum value
$s_0 = 2/3(3/4)$ in two (three) dimensions gives the slowest possible collapse, or the
largest collapse-time scaling. Thus the maximum possible collapse-time scaling exponent $(1/s_0)$ is $3/2$ in two dimensions and $4/3$ in three dimensions. These exponents
are still significantly smaller than the usual polymer relaxation-time scaling exponent,
$z \approx 2$. They are also not dramatically different from the values we found using s_0 from
the simulations.

When we simulate a system like a polymer, we are always concerned that our simulation results are characteristic only of the size simulated and not of the regime we
are interested in. Our objective is to understand the properties of long polymers. Thus
we must ask the question: Are the polymers simulated long enough to show the correct collapse mechanism for very long polymers? We can address this question

because we know how the collapse-time scales with the size of the system for end-dominated collapse. If another process were to become the dominant process for longer polymers, it would have to scale as an even lower power of N than the end-dominated collapse process. Only then could it become important for longer polymers. However, the scaling of the collapse time we found is significantly lower than other known collapse mechanisms. The mean field collapse scaling near the θ-point is similar to the usual polymer relaxation. Thus it is unlikely that for longer polymers a faster process will dominate.

How can we relate our discussion of polymer collapse to the problem of protein folding? Our results on the end-dominated collapse of polymers indicated that collapse can be faster than would be expected from the usual polymer dynamics. Instead of a scaling of $O(N^2)$ we have a scaling of $O(N)$. This appears to be a significant reduction in the time; however, when we consider this in light of our discussions in the previous chapter, we see that this is not as significant as we might expect. Both scaling exponents are reasonable for a polymer collapse time since there is enough time for either scaling to reach its conclusion. We allowed, in principle, for exponents up to $O(N^4)$. Even if $O(N^4)$ is generous, $O(N^2)$ appears quite possible.

Where does the difficulty then arise? The difficulty is that we have considered all of the collapsed polymer conformations as equally acceptable. All we have shown is that we can make a transition from expanded to collapsed polymers in a reasonable amount of time. We have not shown that we can select the right compact form of the polymer. This is where the simulations can provide their most important clue to the benefits of kinetics. End-dominated collapse does not give an arbitrary compact polymer. The process of sequential accretion of monomers and small aggregates by the ends should reproducibly yield a particular compact structure.

Unlike a general uniform collapse of the polymer, the end-dominated collapse proceeds by an orderly process of sequential monomer encounters. These encounters build up the aggregate compact structure (globule) in a manner that is not random. A consequence of this orderly kinetic process is that the resulting globules may be expected to be selected from a limited subset of all possible globules. This is precisely what we have been looking for—a kinetic process that might enhance the process of arriving at a specific folded protein structure. One of the most striking implications of the end-dominated collapse process is that the order of monomer encounters is essentially independent of the initial conformation of the polymer. It is not easy at this point to see what the precise nature of the globules that are formed are. However, we can note that they are likely to be formed out of two parts corresponding to the aggregate formed from one end and the aggregate formed from the other. A more subtle feature of this process is that the globule is likely to contain fewer knots than would be generally found in a globule. This is because the diffusive end motion tends to unknot the polymer, since the ends are passed through any knots rather than closing or tightening them. To discuss this formally would require defining knots in a polymer with free ends, which is a feasible but tricky task.

End-dominated collapse is also consistent with a model, called the molten globule model, that has been proposed for the kinetics of protein folding. It suggests that there is a fast initial process of forming a compact globule followed by a rearrangement of the globule to form the final folded protein. Our simulations and scaling results describe the fast process by which a polymer makes a transition from an expanded form to a compact globule. The rearrangement process should take more time and is likely to be the limiting step in the formation of the protein. Unlike the collapse, this process requires segments of polymer to move around each other, which is a much more difficult dynamic process. The significance of the end-dominated collapse is that by preselecting the initial compact globule, the rearrangement process is shortened and does not necessarily explore all possible compact conformations of the protein before settling in the desired state.

There is another interesting feature of the end-dominated collapse relevant to proteins. Proteins, when they are formed, are not formed all at once. Instead the chain emerges sequentially from a ribosome. This process is called extrusion. It is thus quite likely that the protein in the cell (*in vivo*) performs much of the folding sequentially as it is formed. In many ways this is similar to the sequential process of end-dominated collapse. Thus the polymer during extrusion has a natural sequential process for forming a definite final folded structure. The appearance of end-dominated collapse in polymer simulations may simply reveal why folding also works when the protein is unfolded and refolded *in vitro*. One possible difference between the two environments is that the collapse process *in vitro* occurs from both ends while during extrusion it occurs from one end only.

We note that it has yet to be demonstrated experimentally that kinetics plays a significant role in protein folding, or in other processes like DNA aggregation. There is an interesting consequence of end-dominated collapse that has relevance to experimental tests. End-dominated collapse represents a significant departure from the usual rule of thumb that linear and ring polymer dynamics are similar. The primary other exception to this rule is reptation in polymer melts. When polymers are placed together at high density, the resulting fluid is called a melt. The motion of polymers in the melt is inhibited by entanglements. Essentially the only way a polymer can change position is to move along its own contour by local stretching and contraction. This is a process, called reptation, that is possible for polymer chains. For rings it is not. Other processes must become relevant for rings, and motion should be much slower. The simulations and scaling argument we have described in this chapter indicates that ring collapse should be significantly slower than linear polymer collapse. This is one of the possible ways that the predictions of these simulations could be tested by experiment.

5.4.5 *Variations in the polymer microstructure*

In this section we discuss additional simulations of homopolymer and heteropolymer collapse. The objective is to investigate how robust are the results we have found in the simple simulations discussed in Section 5.4.2 and the scaling argument in Section 5.4.3. If the results are sensitive to the choice of model, then we should doubt their applicability to real polymers. On the other hand, if the results are robust then we can feel confident that they will also be relevant to real polymers. For a variety of systems the results suggest that collapse of long polymers is dominated by diffusion of the polymer ends, which accrete monomers and small aggregates. Collapse does not proceed uniformly along the polymer. However, in a model where *only* pairwise bonding is allowed, the collapse is uniform, since more flexible end motion does not result in continued end accretion.

All of the simulations in this section are in three dimensions and are based on the one-space cellular automaton algorithm described in Section 5.3. The one-space algorithm uses a $5 \times 5 \times 5$ bonding neighborhood for monomers. It allows adjacent monomers to separate by one lattice site providing flexibility in the polymer dynamics. Motion of monomers is performed by Monte Carlo steps that satisfy the polymer constraints.

We summarize briefly the general process of simulation of collapse, then discuss each of several models that test various aspects of the dependence of the results on the local properties of the polymer. As with the two-space simulations, the simulation of collapse starts from a set of equilibrium polymer configurations. Each of the models consists of a particular scenario for monomer-monomer sticking whereby aggregates are formed from individual monomers. In all models, once formed, aggregates are moved as a unit. The collapse simulations include only aggregation, and not disaggregation. As discussed in Section 5.4.3, the primary effect of hydrodynamics is included by scaling the diffusion constant of aggregates by Stokes' law. For the three-dimensional simulations described here, $D \sim 1/M^{1/3}$. Polymer dynamics are simulated by selecting an aggregate (monomers are included as aggregates of mass 1) and moving the aggregate in one of the four compass directions with a probability given by the diffusion constant, and only if connectivity constraints allow—the aggregate does not leave any neighbors behind. One time interval consists of performing a number of aggregate moves equal to the number of remaining aggregates, taken to be the number at the end of the time interval.

Six different models of polymer microstructure are described in the following numbered paragraphs. These simulations are also compared with the two-space collapse simulations that did not include any form of excluded volume during collapse. All of the six models were simulated using polymers of length $N = 250$. The first two models explore variations in the collapse of homopolymers. The third and fourth models explore heteropolymer collapse. The results of these four models are shown in Fig. 5.4.4. The last two models only allow pairwise bonding. Results of these two models are shown in Fig. 5.4.5.

1. The first version of one-space collapse explores the significance of excluded volume. Since excluded volume is essential for the final structure of the polymer as well as the initial structure, it might be expected to be relevant to collapse. This was completely neglected in the two-space simulations. In these one-space simulations, during collapse, monomers are allowed to enter the bonding neighborhood. However, excluded volume is maintained during collapse by preventing monomers from occupying the same lattice site. Monomers aggregate by moving adjacent (NSEW) to other monomers. As before, aggregates are moved as a unit

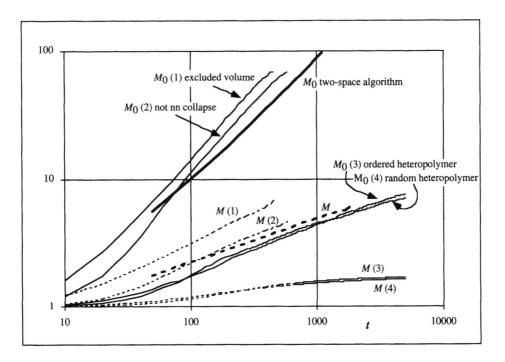

Fig. 5.4.4 Plot of the time evolution during polymer collapse in three dimensions of the average total mass of the polymer ends, $M_0(t)$, and of the average mass, $M(t)$, of aggregates not including the ends. The different lines correspond to the different models described in the text. The end mass evolution is shown by a solid line and the mass along the contour is shown as a dashed line. The bold curves are for the two-space algorithm (Fig. 5.4.2). All of the others are for variations on the one-space algorithm for polymers of length $N = 250$ and are averaged over 300 simulations: (1) for the one-space polymer including excluded volume; (2) same as (1) but preventing aggregation of nearest neighbors along the contour; (3) a model for heteropolymer collapse where only the odd monomers aggregate; (4) is similar to (3) but monomers that aggregate are selected at random. ∎

using a step probability governed by Stokes' law. This means that we must keep track of each aggregate's structure in space, and move it as a unit. After a step we must check all sites at the boundary of the aggregate in order to perform additional aggregation. From the figure we see that the inclusion of excluded volume appears to increase the rate of collapse. Conceptually, we might think about this as resulting from a decrease in the distance monomers need to travel in order to aggregate. Both exponents s and s_0 increase slightly. However, we see that excluded volume does not affect overall behavior and does not even dramatically change the values of the exponents (Fig. 5.4.4 and Table 5.4.3).

2. One of the strange features of model (1) is that monomers aggregate by directly attaching to their neighbors along the chain. Thus, much of the aggregation

	s	s_0	$s_0(s)$	$s_0 - s_0(s)$
1-d	0	0.5	0.5	0
2-d	0.154±0.001	0.773±0.001	0.769±0.001	0.004±0.001 (0.5%)
3-d	0.337±0.002	0.982±0.005	1.003±0.002	-0.021±0.005 (2%)
3-d (1)	0.484±0.002	1.102±0.004	1.113±0.002	-0.011±0.004 (1%)
3-d (2)	0.453±0.002	1.079±0.004	1.090±0.002	-0.011±0.004 (1%)
3-d (3)	0.061±0.001	0.363±0.001	0.796±0.001	
3-d (4)	0.050±0.001	0.293±0.001	0.787±0.001	

Table 5.4.3 Power-law behavior exponents fitted to the simulation results of Fig. 5.4.5. The results from the two-space algorithm (Table 5.4.2) are included for comparison in the first three lines. Fits were chosen to minimize standard errors. Errors given are only statistical. Results are compared to the scaling relation given by Eq. (5.4.11). As discussed in the text, the heteropolymer collapse results are not in agreement with this scaling relation. ∎

occurs between monomers that previously were already bonded as neighbors. This would be a particularly convenient process for end aggregation and may exaggerate its importance. The aggregated structure of model (1) is formed out of rodlike structural components. These rods arise because the attachment to nearest neighbors follows the contour of the polymer. A more realistic model would exclude such nearest-neighbor aggregation. The second version of one-space collapse is similar to model (1) except that nearest neighbors along the contour are prevented from aggregating to each other. Monomers or aggregates are forced to move around their nearest neighbor to bond to a monomer further along the chain. This prevents the simplest end monomer accretion of nearest neighbors. Simulations show, however, that not only does the end-domination persist but (Fig. 5.4.4 and Table 5.4.3) that this change does not change significantly the exponent values. Presumably this is because the need to move around neighbors to aggregate affects collapse along the contour similarly to its affect at the ends, making both more difficult. The only apparent effect is an overall slowing of the collapse as seen by the shift of the $M(t)$ and $M_0(t)$ curves to longer times. The following simulations (3)–(6) are based on model (2).

3. We now consider heteropolymers. Thus far our discussions, both in simulations and in scaling arguments, have not distinguished between different monomers. However, there are significant differences in the bonding of different monomers in heteropolymers. In order to investigate the effect of such variation, we take an extreme case where there are some monomers that bind and some that do not bind at all. This is a simple model of proteins that takes into account the difference between hydrophobic and hydrophilic monomers. In our language, this is the same as monomers that want to collapse by aggregation and monomers that do not. The third one-space model of collapse includes both kinds of monomers. Using the one-space algorithm as in (2), only odd monomers are allowed to aggregate. This is an ordered 50% hydrophilic version of this model of hy-

drophilic/hydrophobic collapse. The way the collapse works is that monomers that do not want to bind at all must be on the outside of an aggregate. The aggregate has an interior filled with bonding monomers and a surface of nonbonding monomers. These nonbonding monomers prevent further aggregation and thus slow the continued formation of larger aggregates. The process of increasing shielding is not included in our scaling arguments, so Eq. (5.4.11) should not be expected to apply, and it doesn't. Despite the shielding of continued collapse by nonbonding monomers, the collapse is still dominated by end motion for the length and time scale simulated. The overall collapse is significantly slowed—the exponent s_0 is reduced to a third of its value for the homopolymer case. There is also some indication that the collapse would completely saturate in this case and would not go through to completion. The overall fraction of hydrophobic (nonbonding) monomers must be reduced to reach complete aggregation. A reduction in the proportion of hydrophobic monomers would make the simulation results more like the previous homopolymer models.

4. The fourth version of one-space collapse is similar to the third version but tests the relevance of the order in model (3). Instead of alternating the hydrophobic and hydrophilic monomers along the chain, they are placed at random along the chain. The collapse behavior is almost the same as in (3). Opportunistic collapse, which results from convenient local arrangement of bonding monomers, speeds the collapse at first. However, the scaling of the masses is slightly lower, and eventually collapse is slightly slower at later times.

5. In order to make end-dominated collapse as unfavorable as possible, we must eliminate the advantage that is gained by the high mobility of the ends. We can do this by eliminating entirely the continued aggregation. The fifth version of one-space collapse starts from the same conditions as (2); however, collapse only includes pairwise bonding. Once two monomers are bonded, other monomers that become adjacent are not aggregated. This is not like protein folding, since each amino acid can form two hydrogen bonds with other amino acids. Also, there is additional bonding due to van der Waals forces. Nevertheless, we can consider the pairwise bonding model as a version of collapse. End dominance in this case would correspond to a progressive pairing from the ends inward. A plot of the pair-density along the chain (Fig. 5.4.5) shows that, despite a tendency toward more rapid pairing at the ends, the collapse is essentially uniform. Thus in this model we do not have end-dominated collapse.

6. The sixth and final version of collapse (Fig. 5.4.5) is like the fifth version where pairwise bonding is allowed; however, only even-odd monomer combinations are allowed for pairwise bonding. The results do not differ significantly from (5).

We have learned from these simulations that in heteropolymer collapse the values of the exponents s and s_0 may change; however, only when we go to an extreme and unrealistic model of polymer aggregation do we lose the end-dominated collapse entirely.

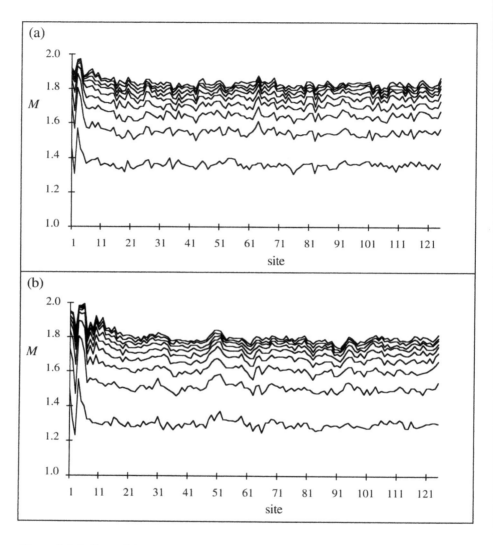

Figure 5.4.5 Plots of the average aggregate mass at every site along the contour when collapse only allows pairwise bonding. Each curve shows the averages at a particular time, with later times having higher masses. Collapse is uniform except for a tendency for the monomers near the end to pair up first. The minimum mass is 1 and the maximum mass is 2. The collapsing polymer has a length $N = 250$. By symmetry only the first 125 sites are shown. (a) shows the case (5) of arbitrary monomer bonding with the exception of no nearest-neighbor bonding; (b) shows the case (6) of only odd to even monomer bonding. Progressively later times are shown separated by (a) 50 updates and (b) 100 updates. For clarity only the first ten times are shown. ∎

The scaling time $\tau(N) \sim N^{1/s_0}$ can also be calculated for the first four one-space models of collapse. For homopolymer collapse $1/s_0 = 0.907 \pm 0.003$ in model (1) and, 0.927 ± 0.003 in model (2). Thus the collapse time still scales approximately linearly with polymer length in three dimensions. In contrast, the time for heteropolymer collapse with nonbonding monomers scales as $1/s_0 = 2.75 \pm 0.007$ in model (3) and 3.41 ± 0.008 in model (4). This is much slower than the heteropolymer collapse. It also approaches the limit of our allowed exponents for protein folding discussed at the beginning of Chapter 4.

Question 5.4.2 One way to think about the effect of various properties of the microstructure of a polymer is as a change in how the diffusion constant of an aggregate grows with its mass. For example, if there are monomers that do not aggregate at all, they become like a surface coating on an aggregate that prevents further aggregation. Rather than model this as a limitation in aggregation, we could simplify the effects by modeling them as a progressively larger diffusion constant that would also limit continued aggregation. Find the relationship between s and s_0 for different values of x in $D \sim 1/M^x$. Then simulate the collapse of a polymer for different values of x and see if the scaling relationship between s and s_0 continues as before. How should this affect the value of s?

Solution 5.4.2 The generalized form of Eq. (5.4.11) is:

$$s_0 = (s+1)/(x+1) \qquad (5.4.17)$$

Simulations of collapse when x is varied are shown in Fig. 5.4.6. The end-dominated collapse occurs for all values of x that are simulated. The scaling relationship continues to be satisfied.

The size of s appears to be nearly constant. A quite reasonable value for s_0 may be obtained from a single value for s in each dimension and use of the scaling relation Eq. (5.4.17). In three dimensions, s appears to decrease slowly with increasing x. ∎

5.4.6 Conclusions

In this chapter our objective was to understand how kinetic processes could accelerate the formation of a selected final polymer structure. We found that there is a distinctive kinetic process that occurs in the initial stages of polymer collapse that results in a characteristic order of monomer-monomer encounters. This suggests that it might be possible to design a sequence of amino acids that fold into a particular structure using this order of events. The polymer might not reach the desired structure if the polymer were to explore all possible conformations. We have not demonstrated that this process applies specifically to proteins, but the robustness of the end-dominated collapse to variations in models suggests that it should play some role. Our analysis has been specific to polymers. How can we generalize this discussion to apply more generally to complex systems? The most important feature of these simulations

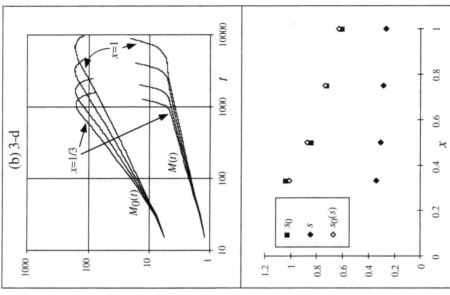

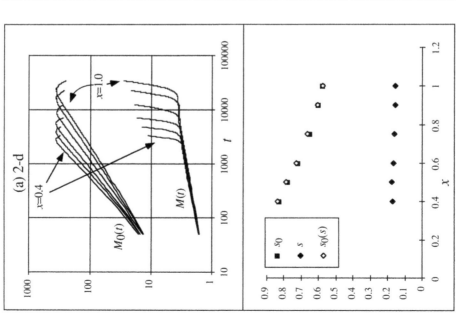

Figure 5.4.6 Exploration of the variation of polymer collapse with x, where the diffusion constant scales as $D \sim 1/M^x$. In the top panels of (a) and (b) are plots of the time evolution during polymer collapse of the average total mass of the polymer ends $M_0(t)$ and of the average mass $M(t)$ of aggregates not including the ends. (a) shows collapse in two dimensions of polymers of length $N = 250$ (500 samples) for $x = \{0.4, 0.5, 0.6, 0.75, 0.9, 1.0\}$. (b) shows collapse in three dimensions of polymers of length $N = 250$ (500 samples) for $x = \{0.5, 0.75, 1\}$. The value at $x = 1/3$ is from polymers of length $N = 500$. In the bottom panels the scaling exponents s_0 and s obtained from fits to the plots of $M_0(t)$ and $M(t)$ are plotted as a function of x. The value of $s_0(s)$ obtained from the scaling relation Eq. (5.4.16) is also plotted. It is in good agreement with the values of s_0 found in the simulations. ∎

is the recognition that there may be a natural sequentiality to events. Such sequentiality does not necessarily mean that a desired structure will be attained; however, it provides an opportunity for control over the final structure. This leads us to the topic of self-organization and organization by design, which we address in the following two chapters.

6

Life I: Evolution—
Origin of Complex Organisms

Conceptual Outline

■ 6.1 ■ We are interested in developing an understanding of organisms in their environments. The primary focus is on the evolutionary dynamics of populations of complex organisms, rather than on how they act collectively.

■ 6.2 ■ Evolution is a general approach to the formation of complex organisms through incremental change. The phenomenology of life is rich and motivates the discussion of evolution. Conceptual models of incremental evolutionary processes include monotonic evolution on a fitness incline, divergence of traits and extinction.

■ 6.3 ■ In theories of evolution, fitness is the only property of the organism which determines the evolutionary dynamics. The fitness can be described as a function of the genome, but it is more directly related to the phenome. Variations in the fitness as a function of changes in either phenome or genome may sometimes be large and may sometimes be insignificant. Conventional evolutionary theory is, however, based upon gradual changes in fitness.

■ 6.4 ■ A model of organisms evolving by diffusing on a fitness landscape is equivalent to particles moving on an energy surface. In the context of this model, many aspects of evolution can be discussed. However, it is far from trivial to account in a robust way even for basic phenomena such as the existence of groups of organisms with well-separated traits.

■ 6.5 ■ The use of dynamical equations that model reproduction, competition for resources, and predation can model a variety of dynamic phenomena in populations. They illustrate how various organism properties contribute to fitness. Moreover, the dynamics of such models is fundamentally different from that of the models discussed in Section 6.4. However, like the models in Section 6.4, these models cannot account for the existence of groups of organisms with well-separated traits. To overcome this problem requires introducing a variety of resources, with their own dynamics.

■ 6.6 ■ Returning to the consideration of collective behavior of components, we find that sexual reproduction takes advantage of composite patterns to form high-fitness organisms. A mean field approach that neglects correlations in the genome does not apply when organism types diverge due to correlations imposed by selection and reproduction. Moreover, a discussion of altruism and aggression in evolution is relevant to understanding the existence of organisms formed out of components, or social groups of organisms, that exhibit collective behaviors.

■ 6.7 ■   Various systems, including the immune system and artificial computer software, can be used as laboratories for developing an understanding of evolution.

6.1   Living Organisms and Environments

The study of living organisms, their behavior and evolution, using mathematical tools, is one of the rapidly developing areas in the study of complex systems. In this chapter we discuss the collective evolutionary dynamics of living organisms. There is an essential difference between this endeavor and the study of neural networks (Chapters 2 and 3) or protein folding (Chapters 4 and 5). In discussing spatial substructure and temporal relaxation in these systems, we were able to construct models from the interactions of simple elements. Even though these elements were drastic simplifications of neurons or amino acids, meaningful questions were addressed. They were meaningful because our focus was on the complex collective behavior. When we discuss life in general, we are not interested in the collective behavior of the organisms, but in the behavior of complex individual organisms in interaction with their environment. It is still possible to consider the emergent collective behavior of many individuals; however, it is not clear that this behavior is complex. In contrast, the individual behavior is often complex.

The contrast can be illustrated by two examples. The first example is related to the concept of subdivision discussed in Chapter 2. Collections of animals or plants do not generally satisfy the conditions that were considered necessary for a complex organism. Flocks of animals or collections of plants can be diminished in size without essentially affecting their collective behavior. Indeed "diminished in size" would not be the natural phrase in the previous sentence. We would say instead "diminished in number." While there are collective effects, they are not sufficient to satisfy our criteria for a complex organism. The second example is related to the development of life over time, analogous to our discussion in Chapter 4 of protein folding. The development of life is generally described as evolution. Evolution is the development of capabilities of the individual organism—specifically, an increase in its complexity. While there is also a development of species and ecosystems, evolution is not considered primarily the development of a collective behavior of many organisms. We will discuss this more fully later in this chapter. Our primary focus, however, is to explore the consequences of the shift in emphasis from the collective system to the individual.

As a result of the shift in emphasis, in discussing models of life we cannot use a two-state variable to represent the elements of our system. And yet, in the construction of models, the use of simple elements cannot be avoided. In order to design models of complex organisms, they must themselves be composed out of suitably designed simple elements that capture as much of the complexity of their behavior as we are interested in.

As we consider the construction of models of life, it must be understood that any model of life includes a model of the environment. Behaviors of individual organisms are generally measured in response to external stimuli. The relationship between the capabilities of the individual and the demands of the environment plays an important role in the description of the organism. The environment also plays a central role in the dynamics of evolutionary change. As we discuss both here and in Chapter 8, the complexity of an organism and the complexity of its environment are often closely related. Moreover, while the behavior of individual organisms is central to the discussion of life, much of the interest in describing life is in the interaction of an organism with other organisms. This interaction may take the form of competition, cooperation, reproduction, communication, exploitation, consumption, etc.

Since our objective is to model aspects of the evolutionary dynamics of populations, we can adopt quite abstract models of organism behavior that do not relate directly to their biology. Nevertheless, these models provide insight into population dynamics and interactions. The models describe an organism and its behavior as a coordinate in an abstract configuration space. In general, we are not concerned with the mapping of this coordinate to specific behavioral attributes. Any concrete computational model of behavior must be represented by a set of parameters that we consider to be our abstract configuration space. In this way we ensure that our discussion is relevant to behavioral models of organisms. The generality of the representation we use for organisms may be argued on the basis of universality of computation and information theories described in Sections 1.8 and 1.9. However, any choice of representation emphasizes particular aspects of a complex system. There is no claim that these models address all of the questions of interest in discussions of life.

There are important connections between this chapter and Chapter 7. In both chapters we are considering processes associated with heritable organism traits. The heritable physiological and behavioral traits are called the "phenome." These traits also exist in an encoded form called the "genome," which is commonly associated with DNA, though some other heritable molecular and cellular structures should be included. In this chapter, we are concerned with the joint evolution of the phenome and genome, which are linked together. In Chapter 7 we are concerned with the process of expression of the genome—the process of development which connects the genome with the phenome. This connection is essential to our understanding of evolution. Moreover, the same central question is present in both: How are complex systems formed? In this chapter we discuss concepts relevant to self-organization. In Chapter 7 we discuss concepts relevant to organization by design. Also significant is the connection between the processes that we consider. In both cases it will turn out that we are considering pattern formation. In this chapter we consider patterns of organisms in the space of possible organisms. In Chapter 7 we consider pattern formation in cell populations and physiology.

This chapter is divided into five major sections. In Section 6.2 we review briefly the phenomenology of life that motivates evolutionary theory. Section 6.3 sets a general tone for discussions of evolution by considering the representation space in which organisms evolve, and the concept of fitness which is central to the theory of evolution. Section 6.4 presents, mostly through conceptual discussion, a Monte Carlo random-walk model of evolution. Fundamental limitations of this model motivate the introduction in Section 6.5 of a collection of models of evolution by reproduction and selection. These models also turn out to have fundamental limitations that become apparent through simulation and analysis. While the models of Sections 6.4 and 6.5 have difficulties, they provide various insights into evolution. Moreover, through recognizing the difficulties, we are forced to develop a better concept of the aspects of organism-environment and organism-organism interactions that must be incorporated in more complete models of evolution. Finally, in Section 6.6 we return to discuss the relationship between components and collective behavior. We discuss sexual reproduction and social behaviors (altruism and aggression) to make connections between the behavior of genes (genome components), molecules, cells, organisms and populations of organisms.

6.2   Evolution Theory and Phenomenology

6.2.1 *The theory of evolution*

For modern biologists, evolution evokes the rich phenomenology of life on earth. Evolution is considered to be a universal process (dynamics) that gives rise to the nonuniversal (diverse) phenomenology of life. The nature of evolution from a biologist's perspective has also been modified over the years. In particular, the relationship of evolution to organism complexity has been bypassed almost entirely in recent years. This arises in part from the recognition that the process of evolutionary change need not give rise to more complex organisms. However, for our purposes it is essential that evolution is a process that *can* give rise to more complex organisms, whether or not it does so under particular circumstances. Thus we focus on the concepts that have been developed in biology to understand the change in organisms as a part of a theory of evolution that not only pertains to the phenomena of life but also indicates quite generally how complex systems can arise.

In this context, the objective of the theory of evolution is to explain the existence of complex life on earth. The need for an explanation arises because it is assumed that the earth began in a state devoid of life. Since living organisms today are complex, they are highly improbable combinations of the building blocks of nature—atoms. An explanation of their existence is necessary. Traditionally, the scope of evolution is divided into two parts. This separation is in recognition of the essential role played by organism self-replication (reproduction). The first part is the formation of relatively simple self-replicating organisms from molecules. The second part is the formation of complex organisms from simple organisms. While our discussion will focus on the latter, the dividing line is not fundamental. Conditions exist in which various molecules can replicate, and thus a theory of evolution can apply to molecules and the formation of cells as complex molecular structures, as well as to organisms. Our discussion of this point will be delayed to the end of the chapter in Question 6.6.6.

Evolutionary theory was introduced as an alternative to two older theories. The first of these is the theory of creation by a prior being capable of creating life. This theory is manifest in some form in most mythologies and religions. Typically this model assumes that life was created at a particular time in a form similar to that we see today. The creation model is difficult to accept because it assumes an external agent that has not been observed. It also gives no explanation for the phenomenology that exists in life discussed below. The second theory is that of spontaneous generation. This theory assumes that life can form spontaneously under certain conditions that arise naturally. In an experimental context it was discussed as a reason for the formation of maggots in rotting meat, until it was shown that without parent organisms that could lay eggs this would not happen. The difficulty with the model of spontaneous generation is precisely our original problem, that the spontaneous formation of a complex system is highly improbable.

Evolutionary theory provides an alternative to these models by proposing that incremental changes over many generations of organisms led to increasing complexity. Spontaneous changes by themselves are assumed to be random, but organism selection through interaction with the environment can lead to a process that systematically increases the complexity of the organisms. The selection process is the driving force in evolution that replaces the physical force in systems governed by classical mechanics. A relevant image is that of biased random walk or biased diffusion similar to that discussed in Section 1.4. In evolution, the biased diffusion occurs in the space of possible organisms. Selection is a consequence of differences in fitness, which plays the role of the energy. Fitter organisms survive at the expense of less fit organisms. We say that the organisms compete for survival, though the intentionality in the term "compete" may be an unnecessary anthropomorphism. The concept of incremental changes leaves many details of the theory unspecified. The importance of the theory is that it provides a framework in which we can understand the appearance of complex systems through a dynamic pathway. The incremental changes are understood to be encoded largely in the genome, which transfers information from generation to generation.

Evolutionary theory is powerful because it describes a large variety of phenomena in life. Darwin's articulation of the theory of evolution preceded the discovery of DNA and its role in preserving traits from generation to generation, and many other relevant discoveries that have given a firm basis for the concept of incremental changes, which is necessary for evolutionary theory to hold. Nevertheless, as a theory of the origin and phenomena of complex life on earth, there are missing pieces, because it is not easy to verify whether processes articulated conceptually, but not reproduced experimentally are sufficient to explain the phenomena of life on earth. For some there is a belief that evolutionary theory requires only verification; others suggest that major new concepts are likely to be discovered that will modify qualitatively our global understanding of evolution. There are also key unanswered questions related to the incremental concept of change.

There is a connection to be made between the study of evolution and the problem of protein folding considered in Chapters 4 and 5. Both deal with dynamics of organization. Chapters 4 and 5 assumed that a unique folded structure had to be reached by a process that selected one out of many possible structures. We can articulate the problem of the formation of life in a similar manner. The problem is the formation of a biological organism from atoms. The developmental process from, e.g., a fertilized egg, can be included as part of this process. Like the protein-folding problem, an attempt to search all possible arrangements of atoms is impossible on any reasonable time scale (e.g., the lifetime of the earth). Even the formation of a single protein out of its atomic components is a much more difficult problem than folding the same protein. Much more difficult still is the formation of long DNA chains found in living organisms. The protein is an engineered system with a specified amino acid sequence. We assumed that it was designed to lead to a special conformation, and discussed the properties of the energy that were necessary in order to enable this to occur. For the formation of life on earth, there is no readily apparent analog to the initial amino acid sequence that served as a template for the formation of life. Thus we have a much more difficult problem with fewer tools. The opportunity present is that, unlike protein folding, we are not required to succeed every time. A process can be designed where many attempts are made. A successful attempt may be reproduced and can be the starting point for successive developments.

While the process of incremental change is conceptually powerful, we will encounter fundamental difficulties in our attempt to understand the overall process of the development of life. Perhaps one of the key issues that underlies these difficulties is that the theory of evolution assumes that the emergence of complex organisms is reducible to understanding incremental changes. We have found in previous chapters that a system composed out of many components cannot be understood in a reductionist manner as trivially related to the behavior of components. Instead, it is necessary to understand their interactions and how these interactions result in collective behavior. Similarly, the process from atoms to organisms cannot be understood as a direct result of a few elementary incremental evolutionary processes. This becomes apparent in this chapter as we attempt to construct a global representation of fitness that can account for evolution by an incremental model.

6.2.2 Fitness—what is being optimized?

We should pause and consider the fundamental justification for use of a fitness property in the dynamics of organisms. Our study of thermodynamics specified that the state of a system is determined by maximizing the entropy of an isolated system or minimizing the free energy at a particular temperature. What gives us the freedom to postulate an alternate dynamical process—a process in which organisms increase their fitness rather than decrease their free energy? The key point is that the earth is not in equilibrium. It receives energy from the sun and emits lower energy photons to black space. This energy flow implies that the second law of thermodynamics does not apply, and it enables the existence of nonequilibrium structures that themselves consume energy and emit waste heat. Without the energy flow, this would be impossible. Having said that such structures are possible does not necessarily mean that they must occur in the form of living organisms. However, as an underlying concept, the idea that this nonequilibrium circumstance can lead to nonequilibrium entities then enables us to ask constructively what entities or organisms will be in existence. The answer provided implicitly by evolutionary theory is that in some sense the organisms that exist are those that optimize some function of the energy flow. Ultimately this is the nature of fitness.

To reach the fitness of organisms, the discussion also requires an additional step that would relate the overall process of energy flow not to the whole system, but rather to individual organisms. The overall nonequilibrium conditions create local nonequilibrium conditions in which organisms exist. The availability of energy in various derivative forms different from that provided by the sun, as well as the availability of heat sinks of other derivative forms, enables the local process of a living organism to proceed. Within this local circumstance, the organisms that exist are the result of some dynamic process that need not optimize the free energy, but may optimize some function of the energy flow. When we can interpret each organism as optimizing the cost function separately, the resulting cost function is the fitness. The process of optimization causes incremental changes to appear in the system. The assumption of independent optimization by particular organisms bears resemblance to parallel processing in the protein-folding problem, where components of the system act, in part, independently.

6.2.3 *Phenomenology of life*

The phenomenology of life is rich and diverse. There is a lot of specific information that is known, and general observations that can be made. The general observations should be addressed by a complete theory of evolutionary dynamics. In each of the following paragraphs, we summarize some of the general observations to motivate aspects of our discussion of evolution.

Existence of life—Aside from the observation that without life we wouldn't be here to talk about it, the existence of life tells us that in principle it is possible to have life. What it doesn't tell us is whether it is a highly improbable or a probable occurrence, and whether there are other forms we have not encountered.

Existence of variety of life—Not only do we find that life exists, we also find that various forms of life exist. The variety is remarkable: animals and plants, living beings that can exist in various environments, animals that can swim, walk and fly. Conventional life-forms range in size from single-celled organisms to whales. We can also say that life exists in many different degrees of complexity. This indicates that not only can life exist, but that there are many varied forms it can take.

Existence of distinct traits—Organisms living today are grouped together in various ways. Certain animals are similar and others are quite different. There is no continuum of organism traits at the present time. Instead there are groupings of organisms that are more similar and less similar to other organisms. For example, there is no continuum of organisms between a giraffe and a spider or a giraffe and a grass plant. There are many different forms of variation that appear to exist, and others that do not. One might wonder, for example, why domestic dogs come in a wide range of sizes, while domestic cats do not.

Existence of shared traits—Various quite different organisms share similar traits. For example, bony fishes, sharks and dolphins share a superficial body form. Butterflies, birds and bats share certain attributes of wing structure. Organisms share similar numbers and roles of appendages. Mice and men share common organs, tissues and chemical processes.

Existence of fossils—Fossils illustrate various changes in traits of organisms over geologic times. Specifically:

a. *Changing of traits*—Many existing organism traits did not exist in previous times. The rate of change is not uniform over history. Slow changes occur at some times, and rapid changes at others.

b. *Extinction*—One example of dramatic change is extinction. Large dinosaurs are the most prominent example of disappearance of a set of complex organisms. Other organisms also have disappeared at particular times.

c. *Persistence of species*—While some organisms have disappeared, others have persisted over long times. Single-celled organisms similar to those existing today are found in fossils at the beginning of recorded life 3×10^9 years ago. Among the longest continuously existing animals are cockroaches, horseshoe crabs and certain sharks.

d. *Systematic changes (evolutionary progress)*—Among the changes that are shown by fossils are examples of incremental monotonic changes from an initial form to another form. A classic example is the horse for which a sequence of progressively larger fossils was found.

Trait persistence from generation to generation by reproduction—Organisms that reproduce (asexually or sexually) pass traits from generation to generation. Dogs don't give birth to plants. More specific traits ranging from size to color are correlated from generation to generation, though a detailed description of inheritance must include mixing in sexual reproduction and various statistical correlations rather than deterministic relationships.

Death—All organisms appear to die. Death is due to various circumstances including accidents, disease, hunger and predation. Barring other causes, for some organisms there appears to be a "natural life span" following which death occurs by senescence, i.e., "old age."

Migrations and domains—Organisms may migrate from place to place on the earth. Generally a species exists within a certain domain. Other places it is not found.

Trait change by human (artificial) selection—When human beings select members of a domesticated species to reproduce, this can cause progressive changes in organism traits. Eventually, different varieties can be formed. Various properties can be modified, such as size, disease resistance, or quantity of a product (milk, eggs, meat, grain, etc.). This is apparent in both plants and animals that have been domesticated over many years.

Apparent competition for resources—Studies of organisms suggest that they compete for resources such as food, territory and mating rights.

Apparent interdependence and reliance—Organisms are also interdependent. There are parasitic relationships, symbiotic relationships between species and social behaviors within a single species.

Food web—There is a food web that corresponds to organisms consuming other organisms as food. There is specialization in this food web. There are herbivores and carnivores as well as omnivores. Other aspects of specialization in the consumption of resources are also apparent. Different herbivores consume different plants, or different parts of the same plant. Different carnivores consume different animals.

Reproduction—All organisms reproduce. There exists asexual and sexual reproduction among all major groups of living organisms. Sexual reproduction appears to become more prevalent among more complex organisms.

Relevance of organism size—Among organisms with larger body sizes there are characteristically fewer individuals, fewer progeny and also fewer species.

Role of DNA—Many traits have been traced to DNA sequence. Various genetic associations of traits and generational transfer of traits as well as the direct manipulation of DNA have established DNA as a source of information that determines hereditary physiological traits of plants and animals.

6.2.4 *Life and reproduction*

The existence of life relies upon a diversity of molecules and molecular types. Polymers, discussed in Chapters 4 and 5, formed of several types of units are essential. Proteins appear to serve primarily as enzymes. DNA and RNA, formed from chains of nucleotides (bases), appear to serve primarily as repositories of information. This information is represented by the particular sequence which is composed of four distinct molecular units. For DNA the units are adenine, cytosine, thymine and guanine. For RNA the thymine is replaced by uracil and all have a systematic modification that changes deoxyribose to ribose forms. DNA and RNA could also be used as catalysts, but this does not appear to be their primary function in cells. Polysaccharides and lipids are polymers that serve both structural functions, and for storage of energy.

A number of polymers are involved in the formation of two-dimensional membranes that are self-organizing molecular assemblies. A membrane is formed when certain polymers having both hydrophobic and hydrophilic ends are present. Under a certain range of conditions, the polymers form a planar double layer consisting of internal hydrophobic ends that avoid water, and external hydrophilic ends that seek water. Once a membrane is formed, other molecules can be added to modify its behavior. The formation of a membrane is, more than the existence of complex molecules, the boundary of living and nonliving. It bridges from molecular systems to organisms because it establishes a distinction between the interior and exterior of a system. Ultimately a membrane enables the interior environment to be controlled so that, in turn, a variety of molecular processes can be controlled.

The hierarchy of living organisms is now understood to be classified into largely single-cell prokaryotes and largely multicell eukaryotes. By number, most of the organisms on earth are in the category of prokaryotes. Prokaryotes, which include bacteria, are simpler and, according to fossil records, arose earlier (3×10^9 years ago) than eukaryotes (1.5×10^9 years ago). Prokaryotic cells consist of a cellular membrane with molecules or molecular aggregates inside. In contrast, eukaryotic cells have additional internal levels of structure in the form of membrane-bound organelles. These can include a nucleus, mitochondria, chloroplasts, endoplasmic reticulum, Golgi apparatus and food vacuoles. Eukaryotic cells are typically at least 10 times larger than prokaryote cells. Prokaryotes can form colonies but never achieve the highest levels of organization seen in eukaryotes. Eukaryotic cells can be either single-celled organisms or part of multicellular organisms including plants and animals.

Prokaryotic cells reproduce by replication. The DNA in a prokaryotic cell replicates as materials are available. It consists of a single DNA double helix. In the bacterium E. coli it is 2×10^6 bases long. Eukaryotic cells undergo a more complex process of reproduction. Individual cells reproduce by mitosis, which involves DNA duplication and then separation in an organized fashion to ensure proper grouping of multiple DNA strands, each of which is called a chromosome.

Multicellular organisms frequently reproduce by sexual reproduction, which involves two processes. The first is the formation of gametes consisting of cells that contain half of the full set of chromosomes. This occurs by a process of cell division called meiosis, during which a mixing (recombination) of the parent chromosomes occurs. The second is a developmental process that occurs once two gametes from different organisms are combined. This developmental process creates a new multicellular organism by cell division, growth, differentiation, locomotion and changes in shape and function of cells. The developmental process is the topic of Chapter 7.

Cell replication involves both molecular replication and cell growth. DNA can be replicated because the nucleotides preferentially bind in specific pairs: adenine with thymine, cytosine with guanine. This enables a complementary chain to be readily formed with the help of additional molecular machinery—a polymerase. The reaction of replication can be performed in a test tube. The test-tube version is called the polymerase chain reaction, and it is the basis of modern methods for determining the sequence of DNA nucleotides and other uses of DNA. What does it mean to replicate the molecule? Since this is the most basic biological replication process, it is helpful to ask exactly what is being replicated. The atoms are not replicated, thus it is better to think about DNA replication as a replication of the information in the sequence of the polymer. More generally, polymers can grow by selective addition of monomers facilitated by catalysts. Cell growth occurs by selective addition of molecules to the cell. This generally requires consumption of energy in order to execute the selection against the influence of entropy.

Replication is central to the process of evolution, which involves changes in the organism type over many generations. Incremental changes occur through processes that we generically call mutation. The simplest of these is the change of a single base— a transcription error. There are other processes that change the genome from generation to generation. The main process is that involved in sexual reproduction. In this process, the DNA of male and female parent are mixed, typically by taking half of the chromosomes from each. In order for this to make sense, there must be some way to ensure that all essential functions of the cell and multicellular organism will be represented in the final combination. This is accomplished by the presence of two (or more) homologous chromosomes in each organism that would perform similar function but are different. When the chromosomes are separated into zygotes, which have only half of the chromosomes, the process of meiosis is designed to ensure that one of each of the homologues ends up in each of the zygotes. The mating of zygotes then gives a new DNA sequence formed out of the DNA of the parents. In human beings there are 23 homologous pairs of chromosomes. With such a process, the number of distinct individuals would be the product of the number of distinct homologue chromosomes. During meiosis, however, there is also a process that encourages crossover between the homologue chromosomes. Segments of DNA are transferred between them. The locations of the DNA segments can also be rearranged. This results in a much larger set of possible variations in the DNA of offspring. The basic functional parts of chromosomes are called genes. In the simplest picture, a single gene contains the code for a single protein. Finally, other processes, such as extra chromosome duplication, can change the number of chromosomes in the cell. Note that genetic mixing by sexual reproduction is quite analogous to the formation of composite states discussed in Section 2.4 and similarly assumes partial independence of component function.

In our discussion of evolution, we will assume the existence of a basic cell with DNA and replication machinery. Processes that led to such a system would involve molecular evolution that can be discussed in a similar framework. The difficulty in discussing molecular evolution is that organisms that involve other types of polymers might be possible and we have no grasp of the space of possibilities. Even restricting the organisms to those primarily encoded by DNA and ancillary molecules, we know little about the enormous space of possibilities. Our discussion of proteins at the beginning of Chapter 4 counted the large number of conformations for a protein. The number of possible DNA sequences which are no longer than human DNA is $4^{10^{10}}$. Our interest is in this space of possibilities. We note that the number of possibilities in *Star Trek*'s domain—"Space, the final frontier"—pale in comparison to the space of possibilities that is being explored by nature through the process of evolution, in which exploration human beings participate.

6.2.5 *Qualitative incremental dynamics of evolution*

The theory of evolution is based upon two processes, mutation and selection, that are assumed to give rise to incremental changes in organisms. We discuss here qualitatively the processes and the types of incremental changes, and then we address their properties more systematically through mathematical models that turn out to be more subtle than the qualitative picture would suggest. The approach we take here illustrates the dangers of qualitative models, how more quantitative models can be constructed, and some of the problems, as well as the benefits, of doing so.

Mutation, as previously mentioned, is used as a generic term for heritable variation in an organism largely through changes in the genome. Specific processes that result in changes in organisms from generation to generation include point mutations, rearrangement, mixing by sexual reproduction and gene duplication. Mutation increases the variety of organisms, thus enabling selection to effect a change in the overall population of organisms.

Selection in a qualitative sense is differential reproduction. Organisms reproduce. The type of organism that is more likely to be around in the future is one that has more offspring. The ability to have offspring requires survival and reproduction. This can be prevented by various problems—nonviability, nonfertility, lack of food, death by predators, death by disease, lack of a mate—leading to death without offspring. Fitness is, by definition, the quality that is selected for.

Equilibrium—To illustrate conceptually the processes of selection and mutation it is helpful to consider first a condition in which there is no net change in the population of organisms from generation to generation. We will often call this an equilibrium, but it is actually a steady-state condition, because resources are consumed. We imagine a population of organisms (Fig. 6.2.1) that has a distribution of some property and whose population is described by a Gaussian distribution. Without considering many implicit assumptions in this picture, we suggest that mutation is a process

Figure 6.2.1 Conceptual illustration of an equilibrium that results from mutation and selection. The top figure shows the increase in diversity of a population of organisms as a result of mutation. The middle figure shows the decrease in diversity due to the action of selection. When these processes are balanced, there is no net change. The relative normalization of the curves is chosen only for convenience. The horizontal axis is some heritable property of the organisms. The bottom figure illustrates the variation of the fitness with this heritable property. In order to retain consistency with dissipative physical systems that move to lower energy, we plot the negative of the fitness. By this convention, equilibrium occurs in a valley. ∎

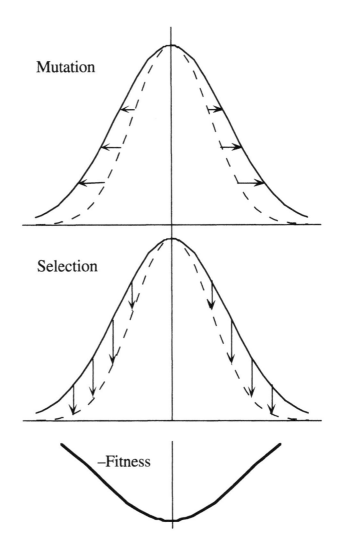

that would, by itself, increase the width of the distribution, while selection decreases the width, thus leading to equilibrium.

Evolution—The central process of evolutionary change is the displacement of a population along some coordinate. Qualitatively this is illustrated in Fig. 6.2.2, where motion on a slanted fitness surface as a function of some property is assumed to give rise to population evolution. This motion results from a combination of mutation and selection, where the mutation increases the width of the distribution, and selec-

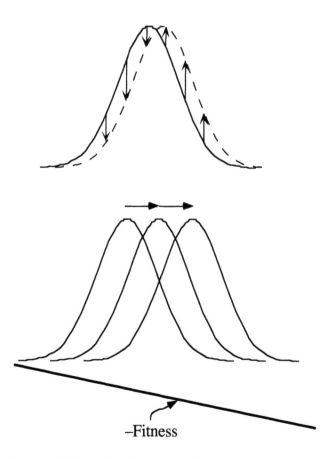

Figure 6.2.2 Conceptual illustration of a process of incremental evolution by biased selection. In the top figure, nonuniform selection is shown acting on a population of organisms. Selection occurs by preferential death of organisms and/or by preferential reproduction of organisms. In the figure, only the effect of both is shown. This results in a net movement of the population. The bottom figure illustrates progress down a fitness slope. Only one of the subtleties that arise in this illustration is that we did not appear to need mutation. Mutation is necessary because the population is discrete rather than continuous and therefore does not have an arbitrarily long tail. At every step, mutation must create the forward tail of the distribution, which is then increased by reproduction. ∎

tion causes the preferential selection of the more fit organisms. We say that selection is a force driving the evolutionary process.

Trait divergence and speciation—In order to account for the existence of a variety of organism types, it is necessary to have a possibility of splitting a single population into two populations with distinct traits. For this to occur it is assumed that under some circumstances there is a process of divergence of the organism traits. The beginning of this process is called disruptive selection. This is illustrated in Fig. 6.2.3. The opposite process of true convergence of two populations is not often considered, for reasons discussed below. The formation of groups of organisms with distinct traits may also lead to the formation of distinct species—organisms that cannot interbreed.

Extinction—Finally, organism types can disappear through extinction (Fig. 6.2.4). It is also possible for an organism type to increase dramatically in population over a short period of time. Of the four processes illustrated in Figs. 6.2.1–6.2.4 this process is the one that most clearly suggests that what is shown is only part of the picture, since extinction is a strictly time-dependent phenomenon. This implies that something external to the organism—its environment—is changing. We note that the nonreversibility of evolution is manifest in extinction, since the reverse process is spontaneous generation.

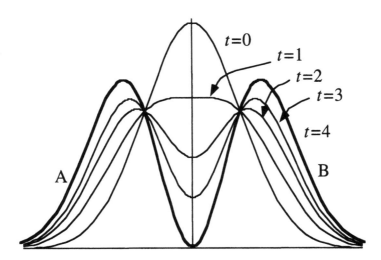

Figure 6.2.3 Conceptual illustration of a process that may result in trait divergence — the formation of two populations with distinct traits starting from one population. Beginning from a population that is located at a particular value of a heritable trait, the population separates into two parts by disruptive selection that broadens the distribution and then forms two peaks that separate over time. In order for this to occur, the fitness in the center must be smaller than at the sides of the distribution. A question that immediately arises is, Why did the initial single population peak form? Resolution requires some additional features that must be included in a model but are not contained in this picture. ∎

Figure 6.2.4 Conceptual illustration of extinction where a population of organisms disappears for reasons that are not apparent from this picture. Influences of the environment are responsible but their nature is not specified. ∎

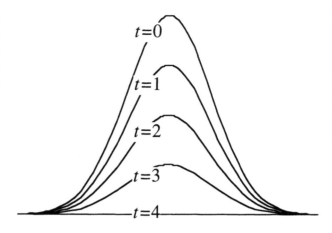

6.3   Genome, Phenome and Fitness

6.3.1 *Complex to simple: a single behavioral parameter*

The theory of biological evolution is predicated on the assumption of a measure of fitness and its relationship to survival, reproduction and competition for resources with other organisms. Fitness is considered to be a single valued function of the parameters, s, describing the organism. Depending on one's viewpoint, the parameters describing the organism may refer to the genetic code or to physiology and behavior. It may be easiest to imagine the organism described by the genome as an explicit list containing the sequence of DNA base pairs. For example $s = (ATCGAAGCT...A)$. The genome should also be understood to include a description of other parameters such as molecules used in transcription and inherited cellular materials. Alternatively, we can consider s as a representation of the phenome: physical and behavioral characteristics such as height, weight, speed, instinctive behavior patterns, disease resistance, etc. These attributes appear more closely related to capabilities of the organism. The genomic space is related to the phenomic space through development. The converse relationship, where behavior and physiological traits affect the genome, is a consequence of evolution. Our primary objective here is not to relate the genome to the phenome, but rather to discuss the generational dynamics of s as a representation of the processes of evolution. When we want to emphasize the distinction between phenome and genome, we will use w for the former and s for the latter.

The fundamental assumption of evolutionary theory is that the fitness can in principle be expressed as a unique function of the organism. More specifically, we have a real number $K(s)$, which is the fitness. It will turn out that the models that we will develop to describe evolution will have quite distinct roles for the fitness. These models are a random walk model and a collection of differential reproduction models. Before we discuss these models, we will describe in Sections 6.3.2 and 6.3.3 the basic properties of the genomic and phenomic space and the sources of fitness.

6.3.2 *Genome and phenome*

Before we consider various models for fitness and evolution, we must first clarify the nature of the variable *s*. The space of possibilities of *s* can always be enumerated discretely as a set, but this is not necessarily helpful. The space of *s* generally indicates the connectivity of the space—what values are able to make transitions to what values. A transition is a change in *s* from generation to generation. The spatial structure of values of *s* thus represents our expectations of changes that are likely (states that are close together in the space) and changes that are less likely in a single step but might occur through a number of steps (states that are far apart in the space). We illustrate using a few relevant examples.

The first case is a binary variable $s = \pm 1$, similar but not necessarily with the same dynamics as the two-state system of Section 1.4. This model could be a simplified single base—considered as two possibilities rather than four. In sexual reproduction, a binary variable can represent a gene with two possibilities (alleles) in a population. The alleles might correspond to particular traits—for example, brown or blue eyes among humans. We are concerned with the time dependence of the relative proportion of the two traits. A standard picture of evolution would indicate the growth and eventual dominance of one trait over the other. We can expand the two-state system to a larger set of discrete possibilities. As long as there are only a few, a discussion of the dynamics is similar to that for the two-state system.

The second case is a one-dimensional continuum. For our purposes, the continuum and a set of discrete possibilities associated with the integers is the same. How does such a model relate to the genomic or the phenomic space? The phenome appears to have continuum parameters such as the height or weight of an organism. For animal breeding, we might consider chickens with a larger egg, cows with more milk, or faster horses. A conventional picture of evolution would include the classic example of incremental growth in size of the horse over time. How does this relate to the genome? The natural continuum parameter of the genome is its length. We could consider the possibility that the height of the organism is related to genome length—the addition of bases in a particular part of the sequence increases the height. It would be more reasonable, however, to assume that a number of discrete modifications of the genome lead to a larger animal. How can discrete modifications lead to a continuum? If we assume that each modification is independent and their effects are additive, then the height, *w*, is determined by an expression of the form:

$$w = w_0 + \sum_i s_i \delta w_i \qquad (6.3.1)$$

where w_0 is a constant added for convenience. We assume a representation $s = \{s_i\}$, where s_i is either ± 1. δw_i is a number that determines the effect of s_i on *w*. A very rough but useful first approximation is to consider each s_i to correspond to a single gene or even base. It is important to recognize that this expression has implications for the natural distribution of heights, which is different from the distribution expected in a continuum. If we assume that all possible genomes are equally represented in a population (selection does not apply) then there is a natural distribution of heights. This would correspond to setting the values of s_i at random.

The simplest case to consider is when all δw_i have the same value, $\delta w_i = \delta w$. Then a random distribution of s_i implies that the heights have a Gaussian distribution with a width given by $\delta w \sqrt{N}$, where N is the number of independent terms. This is unlike a usual continuum, where a random distribution would correspond to equal probability of all heights. This also implies that there are bounds to the values of possible heights, $\pm N\delta w$, though the bounds are much wider than the distribution. There is another way that this model is relevant to any discussion of evolution. When we apply a selective force to such a continuum parameter, it gives rise to a bias in the probabilities of values of the s_i. However, the phase space of possible genetic representations decreases with increasing deviation of the height from the unbiased average.

The distribution of values of δw_i for a more realistic model should be considered carefully. A single value might be replaced by a power-law distribution, or a distribution which implies that particular mutations have a large effect and others have a small effect. For example, eye color, while dominated by the blue/brown dichotomy, also has more subtle distinctions. Moreover, the additive representation might be replaced by a contingent representation where one mutation can occur only after another has happened. Complicated distributions of traits among organisms that are not subject to selection reflect features of the underlying genomic representation. As usual, in modeling such distributions it is reasonable to build a preliminary discussion upon simple models which illustrate features of more complete models.

The existence of an underlying genomic representation also has direct impact on the dynamics of the continuum model. There is no reason to believe that the probability of a mutation to the right is the same as to the left. These probabilities will vary at different location in the continuum. The simplest example is an organism that starts with a genome of the form $(-1,-1,-1,-1,...,-1)$ and we study evolution of an ensemble of these organisms where the only trait we measure is the total number of 1s, which is the relevant phenomic property. Then the genome mutates at random in every generation. Initially, every mutation changes one of the digits to a 1. This would look like a constant drift in the value of the phenome. Then as time goes on there are fewer -1s, and the mutations may change either -1s to 1s or 1s to -1s. Without any selection bias it will eventually set the digits equally on average to 1s and -1s with a distribution that extends from $N/2 - \sqrt{N}$ to $N/2 + \sqrt{N}$. This distribution is not changed by mutation. We will discuss the effect of a bias due to fitness selection in the following sections.

There are many other possible spaces to consider in addition to a binary and a continuum space. We can consider various d-dimensional continuum spaces and combinations of continuum and discrete spaces. We could also consider a set of N binary variables, an Ising-like model, representing the various bases or alleles of genes.

The final case we consider is a more direct representation of the genome as a space of strings, $s = (s_1 s_2 s_3 ... s_l ...)$ where all characters s_i for $i > l$ are zero, and l is the genome length. The s_i might be taken to be bases or genes with a prespecified alphabet of possibilities. One kind of transition in this space alters the characters but does not change the string length. For example, point mutations change the value of a single character. Another kind of transition changes the length of the genome—for example, adding or deleting a single base at the end, or inserting or deleting a base somewhere in the middle. If we limit ourselves to considering only these transitions, we can consider the process of genome extension as distinct from the process of changing the genome for a specific length. The rate of point mutation need not be the same as the rate of mutations that change the genome length. Later, we will discuss the implications of this model for the problem of generating organisms of higher complexity.

In considering the space of possible organisms, it is essential to consider the independence or dependence of parameters. Only when parameters are independent is it possible to consider a phenomic trait or a particular gene as the subject of evolutionary study. Similar to the discussions in Chapters 2 and 4, there are likely to be parameters that are partially independent. For example, the structure of the digestive system is largely independent of the mating behavior or the absolute size of the organism. Because of the independence of certain physiologic or behavioral parameters, we can consider evolutionary change in the different parameters separately. Even when they are coupled, partial independence implies that there are organisms that share one trait but vary in another. This variation can allow for evolutionary selection in one and not in another. When chickens are bred for increased size, this can be done largely independent of the color of the chicken. The independence of phenomic traits should, however, be carefully considered in the context of the underlying genomic representation. The essential point is illustrated by considering the representation described in Eq. (6.3.1) and allowing for two different traits w,v to rely upon the same representation:

$$w = w_0 + \sum_i s_i \delta w_i$$

$$v = v_0 + \sum_i s_i \delta v_i$$

(6.3.2)

As long as we are considering values of w and v that can be represented by a large number of possible genomes, then w and v may act independently. However, when one of the phenome parameters is pushed by selection to an extreme limit, then the space of available genomes becomes reduced and the possible values for the other phenome parameter also becomes reduced. For example, if all (or nearly all) s_i in Eq. (6.3.2) are selected to have the same sign as δw_i to achieve the maximal value of w then we are restricted to a particular value (or limited set of values) of v which is (are) unlikely to be optimal. Systematically, we can say that a high selection pressure increases the coupling between various phenomic parameters. Moreover, this shows that it is progressively difficult to optimize multiple phenomic parameters at the same time. This is important when we consider several different traits that superficially are independent, such as chicken size, the number of eggs laid per day, and the resistance to a particular disease. Starting from an unbiased distribution, by selection we may be able to change them independently. Once they are strongly selected they often become coupled to each other.

The idea of trait independence, and direct coupling of a single trait to a single gene, has attained a popular following that is reflected in the searches for individual genes responsible for a variety of human physiological and behavioral traits. It is not unreasonable to suggest that this view arises largely out of our ignorance of the complex interplay of genome and phenome.

In a more global context, the coupling of attributes is a motivation for diversity of life even if the possible organisms are continuous. We can consider different organisms that optimize particular capabilities and not others: sensory acuity or large size or quickness. Each of these can provide an opportunity for fitness improvement, but eventually to the exclusion of improvements in other properties. This suggests that different organisms could survive by optimizing different traits. However, in order for this to be the case, there must be a nonlinear relationship between fitness and the phenomic properties. A linear optimization would still mean that a particular combination of characteristics wins over the others, and diversity would not result.

6.3.3 *Fitness sources*

The variation of fitness as a result of mutation may range from large to insignificant. A large variation in fitness may result in offspring that aren't viable—that are unable to survive or reproduce. An insignificant variation in fitness means that the difference doesn't affect selection; such mutations are called neutral. In a historical controversy it has been debated whether neutral mutations dominate the space of possible mutations. The controversy is relevant, because if neutral mutations dominate, then random changes (diffusion) of the genome, rather than selection, would cause evolutionary changes. Conventional evolution by selection occurs when changes in fitness are significant but gradual, so that populations change over many generations. On the other hand, the proportion and spatial distribution of nonviable organisms may result in boundaries to the course of evolution that are likewise important to understand. In this section we discuss some of the possible reasons for large variations in fitness, or a lack of variation in fitness, that can give insight into these issues. Remarks in later sections will clarify the neutralist/selectionist controversy.

Fitness may be considered as a function of the genome or phenome space. The function $K(s)$ has different properties depending on what s represents. Contributions to fitness variation are considered for each case in the following paragraphs.

We begin with the contributions to the fitness for the genomic space. We focus on complex multicellular organisms and their viability. A specific genome may not be viable because it does not provide for its own reproduction, or for effective expression of its information. The genome contains markers that indicate where to begin and where to end transcription so that proteins are formed. Eliminating or adding such markers may be readily understood to cause nonviability of the genome. Moreover, the genome acts as a set of instructions that lead to development of a multicellular organism. It may fail due to inconsistent instructions. It may also describe a nonviable organism where necessary physiological functions do not exist, or where organs or systems are improperly connected or sited. We might also distinguish between organisms that are viable under some circumstances, but not the circumstances that prevail. For example, a fish born on land. We could introduce the concept of a domain of viability as the set of conditions under which a particular organism is viable. Evidence for a large number of nonviable genomes exists. It is believed that approximately one-third of successful impregnations in human beings result in early (first trimester) miscarriages that are often unnoticed. A significant proportion of these are believed to be due to nonviability of the genome. It should be understood that this occurs even though the possible genomes are very selectively chosen due to their origination from functional genomes of the parents with limited types of variation. This also suggests that developmental viability is a major constraint on fitness. Finally, some organisms are not fertile even when properly developed, with the classic example being the mule.

In contrast to the reasons for nonviability, there are also reasons that mutations are neutral. A significant amount of DNA in cells does not appear to code for proteins and, at least to a first approximation, does not directly affect the system function. Discussions of the role of this "junk DNA" have yet to resolve whether it has a functional role, such as in the structure of the DNA molecule or as latent coding DNA, or if it has no functional role at all. At the present time it is reasonable to assume that a significant fraction of changes in this part of the DNA are neutral with respect to selection. We can also discuss changes in the coding parts of the DNA. Changing a single base that codes for a particular protein may not change the amino acid that it codes for. This is because the mapping of DNA to amino acids is not one-to-one. Even if a base change does change the amino acid, changing one amino acid generally does not affect the structure of a protein or its enzymatic activity. Even when a protein is changed so that its activity is compromised, the change may be compensated by other cellular or physiological mechanisms. This suggests that many changes in the genome do not affect the phenome and thus do not affect the fitness in a conventional way.

When we consider the fitness as a function of the phenome, we would consider as coordinates various properties of an organism such as height, weight, bone structure or speed of locomotion. However, a central problem with this description is that it is not clear whether it is possible to create an organism with a particular set of physiological properties from a genomic description. We can describe various organisms in terms of their traits, but they may be impossible to create. We could protect ourselves from this problem by considering only organisms that are known to exist and comparing their fitness. However, this simplification does not allow us to address basic questions that we want to understand regarding the reasons for the existence of organisms in the form and with the evolutionary history that is found. Moreover, from a practical point of view it is important to understand what are the factors that prevent horses from running faster, chickens from laying more eggs or cows from giving more milk. Thus nonviability has meaning in this context as nonfeasibility. Feasibility can be divided into several categories according to which constraint prevents the formation of the organism—physical or representational. We also can discuss generally the effect of population interactions.

Physical constraints—In the context of particular external circumstances, physical law places various constraints on possible organisms. There are requirements on strength of bone in order to support an organism in a certain gravity. There are also constraints on senses—ears and eyes are limited in their sensitivity by quantum mechanics. External conditions such as temperature, air pressure and composition impose additional constraints. For example, the visibility in air is limited to a window of frequencies that are relied upon by the eye. The composition of the atmosphere places constraints on the organisms that can exist in it. The terrain places constraints on the nature of locomotion and the limbs that may be useful for it. The ocean and its composition imposes quite different constraints. The cycle of day and night results in other constraints. The qualitative differences between organisms in the water and on the land, or even between fresh and salt water and between different land climates, and specifically the lack of viability of one organism in another environment, suggests the importance of physical constraints on fitness.

Representation constraints—Even if certain traits are possible within physical law they might not be possible when we consider their implementation using DNA encoding. The process of developmental biology does not allow all systems to be formed. For example, automobiles are possible, but it appears likely that developmental biology cannot create a car directly using DNA encoding (indirectly, of course, it has). There may also be limitations in the structures that can be formed because of the use of particular chemical processes. This is not due to physical law but rather to the mechanisms of coding. A milder form of the encoding constraints exists in the form of coupled traits. Thus certain physiological/behavioral traits may be coupled to others due to their representation in the genome. Such constraints are difficult to consider without understanding the processes involved in developmental biology. They will become somewhat clearer as we discuss these processes in Chapter 7.

Population interactions—Population interactions might be thought to be significant only after issues of viability cease. This is not entirely true because, for example, parenting can enable organisms to be viable when they would not otherwise be. There are various interactions between different organisms that are important for fitness. There are interactions between organisms that are distant from each other in the genomic space. Examples include: interactions between plants and animals, interactions between bacteria and multicellular organisms and interactions between parasites and their hosts. There are also interactions between organisms that are proximate to each other in the genomic or phenomic space. They can include competition for the same resource, cooperation in group protection, reproductive interactions and parental attention.

Is the variation in fitness dominated by interactions between organisms or by inherent (physical or genomic) limitations? This question is superficial in that it is quite clear that both contribute in essential ways to the determination of fitness. Moreover, physical considerations, such as the composition of the atmosphere, due in part to the balance of plants and animals, may also reflect indirect interactions between organisms. Interactions and their effect on fitness are also related to physical considerations. Nevertheless, the issue of the relative importance of interactions and physical causes of fitness is important because it is relevant to questions that are at the heart of evolutionary theory. For example, it is relevant to the importance of randomness and historical accident in determining the course of evolution.

It is tempting to consider all variation in phenomic traits as significantly affecting fitness so that there are no neutral variations. The more practical aspect of this approach is to understand the existing variation of phenomic properties of a particular population of organisms. The central issue becomes whether the variation in phenome represents a diversity that is being acted upon by selection and therefore certain traits will eventually be forced to disappear in favor of others, or whether the variation is neutral with respect to selection and will persist. This limits the scope of the question from the space of all possibilities to the space of extant organisms. Even in this context, the controversy between neutralists and selectionists is not easy to resolve. The issue is still more complicated since populations of organisms may not act solely to select individual properties but also properties of the whole population. In this case variation may reflect the effects of selection. This will be discussed in Section 6.6.2.

From the discussion in this section, we see that there are a wide variety of contributions to the fitness of an organism. These factors change in time due to various events that range from change of weather to fluctuations in populations of other organisms. Since we are describing the evolution of organisms due to a fitness that itself depends on the existence of organisms, we are describing a self-consistent process. Such self-consistency was discussed in Section 1.6 in the simpler context of the Ising model for magnets. In essence, the concept of fitness itself represents a mean field approach. The assumption is that at any time, an average over influences that affect fitness is a meaningful concept, and that evolution takes place in the context of this average fitness. This is one of the central assumptions in evolutionary theory, not just in the models we will be discussing. Whenever the fitness is discussed as a fixed external parameter independent of the changes in the population, this simplification is being made. Corrections to the mean field approximation can be included in various ways; however, it is not clear how well it serves as a first approximation.

Question 6.3.1 If large regions of the space represent nonviable organisms, this might prevent evolution from one part of the space to another. The relevant question is the degree of isolation of regions of the space, like valleys in a mountain range. What phenomenological evidence suggests that the phenomic space is connected?

Solution 6.3.1 One observation that suggests that the space is connected is the existence of various widely different classes of animals such as land animals, winged animals, and water animals. The existence of flying insects, flying birds and flying mammals also suggests that there are multiple pathways between widely separated parts of phenomic space, as does the existence of different kinds of swimming animals. It is difficult, however, to rule out the possibility that other classes of organisms do not occur because evolution is unable to reach them. ∎

6.4   Exploration, Optimization and Population Interactions

6.4.1 Exploration and optimization on a fitness landscape

At the root of evolutionary theory is the concept of optimization. It is not to be assumed that the optimum has been, or ever will be, reached. However, incremental evolutionary processes increase the fitness. Thus, it makes sense that a first mathematical model of evolution relies upon our understanding of the dynamics of optimization. Optimization problems can generally be written as a moving point on a landscape representing the cost function. The prototype optimization problem is the motion of a particle on an energy landscape where dissipation of kinetic energy causes it to move to lower potential energy. A nonzero temperature causes the particle to bounce around, enabling movement up in potential energy, but the tendency is to settle in lower regions. This system was introduced in Section 1.4 for simple energy landscapes and discussed in Section 1.7 in the context of Monte Carlo computer simulations. It was also the basis of our discussion of the relaxation of proteins to their folded conformation. For evolution, we modify this picture by allowing the existence of more than one organism performing the optimization at the same time. Interactions between the organisms change the nature of the optimization. We first introduce and motivate a conventional optimization picture, and later discuss how the interactions affect it.

A central difficulty in constructing mathematical theories of evolution is the necessity of describing reproductive proliferation of a single organism, a variable population size and population interactions. On the other hand, survival pressure is based on the concept of a population of limited size. In a simplified form, one organism replaces another due to limited resources. It is not unreasonable to model this first by using the dynamics $s(t)$ of an organism that reproduces and dies immediately after giving birth to a single mutated offspring. Thus, as a basis for our discussion we can consider a single mutating organism in a fixed size population—an ensemble. This is a Monte Carlo random walk model (Section 1.7.2).

In the Monte Carlo random walk model we begin with a population of N noninteracting organisms identified by their locations $\{s^i\}$ on the fitness landscape. The organisms are called walkers. In each time interval, every walker attempts to take a step. Steps correspond to changes in the value of s^i. The step of an organism is selected at random from all changes in s that are allowed by organism mutation. The probability of a mutation is represented by a matrix $\lambda(s'|s'')$ which gives the probability of a mutation from s'' to s' in a particular step. The move is accepted or rejected according to the fitness $K(s)$. A convenient, but by no means unique, way to do this uses the Metropolis form, which says that if the new fitness is higher, the step is taken. If the new fitness is lower, then the step may still be taken but with a reduced probability given by the ratio of fitnesses: $K(s')/K(s'')$. The lower the fitness is at s', the smaller is the chance the step will be taken. When the step is not taken, the walker stays in its original location. We can think about this process in terms of the competition for survival. Starting with an organism at s'', we perform a mutation to s'. We think about this as the momentary existence of two organisms at s'' and at s'. Then we perform selection. Either the new or the old organism survives and the other one disappears. In the Metropolis form there is an asymmetry between the selection of the old and new organisms. If the mutation leads to an organism that is more fit, the new organism always survives. If the mutation leads to an organism that is less fit, then there is still a probability that the new organism will survive. This probability is given by the fitness ratio. We could also choose a selection rule that treats the new and old organisms the same. This would not change the overall evolutionary behavior in this model.

Quite generally, the stochastic dynamics of an ensemble with walkers that do not interact can be written as a Markov chain (Section 1.2). The ensemble is represented by the probability $P(s; t)$ of finding a particular organism s at time t. This probability changes with time due to mutation, reproduction and death. The probabilities of organisms at one time determine the probability of organisms after an interval of time by a linear matrix equation (Eq. (1.2.5)) which we rewrite here:

$$P_s(s'; t) = \sum_{s''} P_s(s' \mid s'') P_s(s''; t-1) \tag{6.4.1}$$

The matrix $P_s(s' \mid s'')$ is the probability an organism at s'' will go to s' in the next step. It is specified by the matrix $\lambda(s' \mid s'')$ and the fitness $K(s)$. The precise expression for $P_s(s' \mid s'')$ is not essential for our discussion, but for the Metropolis form it is given by Eq. (1.7.19) as:

$$
\begin{aligned}
P_s(s' \mid s'') &= \lambda(s' \mid s'') & K(s')/K(s'') \geq 1 \quad s'' \neq s' \\
P_s(s' \mid s'') &= \lambda(s' \mid s'') K(s')/K(s'') & K(s')/K(s'') < 1 \quad s'' \neq s' \\
P_s(s'' \mid s'') &= 1 - \sum_{s' \neq s''} P_s(s' \mid s'')
\end{aligned}
\tag{6.4.2}
$$

It is important to note that Eq. (6.4.1) is linear in the organism population. It applies when there are no explicit interactions between organisms. Because the equations are linear, we can consider evolution by starting from a population located at a single point, and apply superposition to obtain the evolutionary behavior of any initial set of organisms.

There is also one more point that we must consider—the granularity of the ensemble. Eq. (6.4.1) uses continuous values of the probability $P(s; t)$. We should not use a continuum model to describe populations, because a subunit population makes no sense biologically. The Monte Carlo random walk has granularity built in. However, as long as the model is linear this granularity is not essential. When we consider interactions between organisms that make the model nonlinear, we can do so in the context of the random walk model.

Since we are discussing the properties of a system for which we do not actually know the landscape, we should review what we know about the general properties of Markov chains that are true for any landscape. We know that after enough time has passed, a Markov chain in a connected finite space will reach equilibrium. This is true about the model populations independent of whether the parameters of the model are derived by assuming nonequilibrium organisms and nonequilibrium processes of birth, consumption and death. The extensibility of the genomic space suggests it may not be finite. However, any limit on the ultimate length of the genome implies that over long enough time the system must reach equilibrium. The time to reach equilibrium may be much longer than any reasonable amount of time (e.g., the lifetime of the universe) but our discussion does not depend upon this, since in this chapter (contrast Chapter 4) we are concerned about the dynamics of the ensemble, not the time scale to reach equilibrium.

The overall behavior of the Markov chain is that of a relaxation process of $P(s; t)$ to the target (equilibrium) probability distribution $P(s)$ which we recognize as the fitness $P(s) = K(s)$. This follows from our use of the fitness to determine the probabilities of taking a step in the random walk. We may choose to represent the fitness as:

$$K(s) = P(s) = e^{-E(s)/kT} \tag{6.4.3}$$

where $E(s)$ is determined as a function of the fitness using

$$E(s) = -kT \log(K(s)) \tag{6.4.4}$$

We call $E(s)$ the energy since it plays a similar role to the energy in particle motion on an energy landscape. However, the energy as it is defined here is not the actual energy or energy consumption of the system—it is only a way of writing the fitness. High energy implies low fitness, and low energy implies high fitness. The energy landscape is the landscape for motion of particles representing the genome of organisms. In principle, the parameter kT plays no essential role and could be set to 1. If the temperature kT is kept as a tunable parameter, it is an overall scale factor that changes how flat the fitness landscape is. This reflects the influence of chance in the selection process. For low kT the chance of a higher-energy organism surviving is insignificant. For higher kT higher-energy organisms are more likely to survive.

The identification of the fitness with the target probability distribution of the organisms enables us to think about the evolutionary process directly. The concept of selection appears in the target population distribution, since the higher the fitness, the greater the target population of the organism. Even though the target distribution $K(s)$ is not the same as the distribution at a particular time $P(s; t)$, under some circumstances the relative populations between organisms given by $K(s)$ may be the same as in $P(s; t)$. We will discuss the conditions under which this is true below.

There is one aspect of this model that may already be troubling. If the fitness is directly related to the current population of organisms, this would strongly favor microbes over insects and insects over human beings. The difficulty here is not superficial and is important for the understanding of evolutionary theory.

Our objective will be to discuss generally the consequences of the random walk model. In particular, we will focus on building an understanding of the relationship between general biological phenomena and the motion of populations on the fitness landscape. We will also consider the effect of the environment through the shape of the landscape and the implications of interaction between organisms. Eventually we will find this model to be quite limited and will discuss ways that it must be improved to account for the phenomena we hope to describe.

Quite generally, the dynamics of a Monte Carlo random walk on a landscape is an exploration of the space with longer times spent in regions of lower energy. As we discussed in Section 6.3.2, the nature of the space of possibilities s can be used to describe the possible mutations. The coordinates of organisms that can mutate to each other are close in space, and those that require several steps are further apart. Once we have determined the nature of the space, we must provide values for the fitness. Then we can tentatively apply our intuition to the dynamics of a population that appears to diffuse in the space.

6.4.2 *Shape of the landscape*

In a theory of evolution based upon fitness, there is only one mathematical entity—the fitness. Thus we must satisfy ourselves that using only the fitness landscape we can account for all of the phenomena of life. If we knew the landscape, we could analyze it to arrive at these conclusions. Alternatively, we may analyze the requirements that the landscape must have in order to satisfy these properties. At this time, the latter phenomenological approach is appealing, since we have yet to develop a systematic approach to obtaining the actual landscape. A systematic determination of the landscape would require us not only to know the fitness of specific organisms, but also their genome or phenome in order to map the fitness space. It would be necessary to know this both for organisms that are found on earth and organisms that might be created by genomes that do not exist. Using the phenomenological approach, we can relate general properties of the landscape to the phenomena of life. Various experiments have more specific bearing on the nature of the landscape.

When we consider mathematical models for the landscape of the fitness there are several generic possibilities:

Flat—The landscape may be essentially flat, corresponding to the neutralist perspective.

Smooth—When there are variations it might be smooth, so that fitness varies continuously with changes in the organism. It suggests that there are only a few widely separated minima.

Rugged/random—A rugged landscape implies that the fitness of one organism is uncorrelated with the fitness nearby. The fitness might be selected at random from a distribution.

Locally correlated—If there is local correlation then the fitness is correlated within a limited distance, and becomes random for larger distances.

Locally rugged with long-range correlations—If the landscape is random over short distances, it may still be smooth if we average the values of fitness over neighboring sites, or look at the minimum of the values of fitness over neighboring sites and consider the longer-range variation.

Complex—A truly complex landscape implies that there exists structure on every scale. There might be power-law correlations between different locations as a function of distance. Different regions may be smooth or rough. This allows for many different kinds of evolutionary behavior. In this case we should not infer from one or two phenomenological examples what the general behavior is like. However, the existence of various possibilities does not necessarily mean they are relevant to the global behavior of evolution.

6.4.3 Evolution—local landscape dynamics

We start by considering incremental pictures of the dynamics of a population of organisms analogous to Section 6.2.5. These should be related to smooth fitness surfaces for continuum s. The first picture of equilibrium as a balance between mutation and selection (Fig. 6.2.1) can be readily understood as the behavior of a population in a valley. The equilibrium distribution $P(s) = K(s) = e^{-E(s)/kT}$ is realized with the energy a parabola $E(s) \sim \alpha(s - s_0)^2$ to first order in arbitrarily many (continuous) dimensions. This picture works. In the dynamics of Monte Carlo walkers, mutation increases the diversity of the population, while selection reduces it.

We run into some trouble with the second picture (Fig. 6.2.2), of population motion on a linear slope. In the model that we are considering the population does diffuse down the slope, but the distribution broadens (Fig. 6.4.1). What happens when we add more dimensions? When the landscape is smooth, there is only one direction in which the fitness is increasing (the steepest descent direction of the energy $\nabla K \propto -\nabla E$), and all orthogonal directions have no change in fitness. This is a property of a

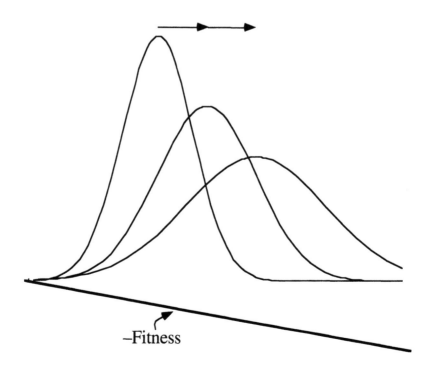

Figure 6.4.1 Schematic illustration of evolution on a linear fitness slope in the Monte Carlo random-walk model. Equivalent to the problem of diffusion, the model always results in a spreading of the population unless it is confined in a well. Thus the population spreads as it translates in average location. This is unlike the conceptual illustration of Fig. 6.2.2, and is unlike the models that will be described in Section 6.5. ∎

smooth function, and does not require any special conditions. This should be recalled in the context of the selectionist/neutralist controversy. To first order (i.e., valid for incremental evolution) only one out of many dimensions of variation of the mean of a population of organisms can be affected by selective pressure. In the other dimensions the population will spread out until it reaches second-order changes in the fitness.

In general, the landscape model readily accounts for the spreading of a population throughout space. We might argue that this is a favorable outcome for the explanation of the diversity of life. However, there is greater difficulty in accounting for confinement of the population. Confinement is evident when a population of organisms has a limited range of traits. It can be confined in a valley; however, a population evolving as a whole cannot be in a valley. In order to confine an evolving population, it is necessary to assume that the evolution is in one dimension only and that other dimensions are confined as in a channel. Even in this case, from Fig. 6.4.1 we see that spreading occurs in the direction of evolution.

Trait divergence requires the confinement of population traits, since two populations of organisms must be separated from each other. To understand the formation of two groups of organisms with distinct traits, as illustrated in Fig. 6.4.2 we would consider a spreading population encountering a ridge that will separate the population at later times. As long as the landscape is smooth, the population will be continuous. Only when there arise barriers will the population separate into different parts. While there is need for a cause for the separation, there is no need for a cause for the broadening of the distribution.

Another way to understand the existence of groups of organisms with distinct traits is through local minima in the landscape. A rough or correlated landscape has multiple minima and barriers over which walkers must cross to reach them. Starting at a point within a particular valley, the population spreads and becomes a Gaussian distribution at the minimum. Over time, the population will escape from the valley to find other valleys. For two valleys this is just the two-state model of Section 1.4. The population evolves by changing the relative probability of the two states until an equilibrium is established between them. There is a characteristic time for this equilibration. As shown in Fig. 6.4.3, this can serve as a model for trait divergence or extinction.

When there are many valleys, we can characterize the population at any time by a quasi-equilibrium that applies to the region of space which has been reached by the population. In this region of space the relative population of different organisms is given by their relative fitness. Thus, in this region the population approximates the expression

$$P(s;t) = \frac{K(s)}{\sum K(s)} = \frac{e^{-E(s)/kT}}{\sum e^{-E(s)/kT}} \tag{6.4.5}$$

where the values of s in the sums are also limited to the region of space that the population has reached. It is important to note that a feature of this model is that the population of organisms in a well does not evolve together. Instead, individual organisms explore space and accumulate at valleys that are then identified as groups of organisms with similar traits. Since there is no interaction between the organisms, there is no reason for them to evolve together. This is related to the problem of con-

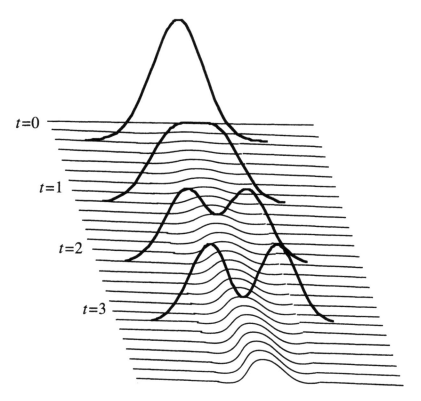

Figure 6.4.2 Schematic illustration of trait divergence in the Monte Carlo random-walk model. The process requires at least two dimensions in which fitness is varying. In the first direction, a linear fitness slope causes the population to translate over time. This dimension is indicated in the figure by successive closely spaced curves displaced towards the bottom right of the page. The second dimension is shown by the curves themselves. As the population evolves down the slope, it encounters a fitness ridge in the second dimension which causes the population to split into two parts. ∎

finement. The problem related to confinement is most apparent when we consider whether the population maintains a structure of separated groups of organisms or continues to disperse until each organism is isolated.

On a rough landscape in a one-dimensional space, the population of walkers is confined to a limited region of space, because barriers prevent it from expanding to fill the space. However, in higher dimensions the population can generally escape around barriers to explore ever larger regions of space and therefore also find progressively lower minima if they exist. We note that in order to account for the phenomena of life, the landscape must be constructed in such a way that groups of organisms continue to exist: there isn't complete accumulation in one valley, and at the same time there isn't complete dispersal.

We can see that describing a landscape that enables the creation of distinct organism types without causing complete dispersal is difficult in this model. The need for this kind of balance is not healthy in a generic model, because it compels us to pro-

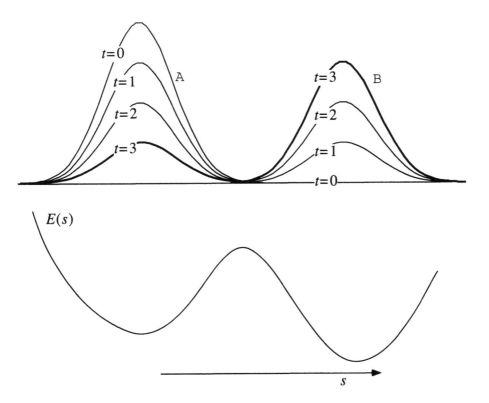

Figure 6.4.3 Evolution of population on a landscape with two wells, similar to the time evolution of the two-state system in Section 1.4. Starting from a population in one of the two wells, the population in the other well grows until equilibrium is reached. This can be a model for trait divergence if organisms of both types continue to exist in equilibrium, or if multiple wells are being filled and emptied, as in a washboard energy with progressively lower wells. It is also a model for extinction, when a well becomes completely depopulated. Note that for trait divergence we could also start from a population in the lower well and create a smaller population of new organisms by occupying the upper well till it reaches equilibrium. ∎

vide some reason that the landscape is so constructed. No reason is readily apparent. We will try to solve this problem with interactions between organisms in Section 6.4.5, but we will be only partially successful.

In the random-walk model there is a natural way to discuss the effect of the mapping of genome to phenome. The genomic representation can be accounted for by writing an effective phenome fitness in terms of the genome fitness as:

$$K(w) = \sum_s \delta_{w(s),s} K(s) \qquad (6.4.6)$$

which says that the fitness of the phenome coordinate is the sum over the fitness of the respective genomes that give rise to this phenome. This makes sense, because the phenome target population is the sum over the respective genome target populations for all genomes that give rise to this phenome. Eq. (6.4.6) is the same as treating the fitness using a free energy for the phenome coordinate.

$$K(w) = \sum_s \delta_{w(s),s}\, e^{-E(s)/kT} = e^{-F(w)/kT} \tag{6.4.7}$$

The free energy, defined as in Eq. (1.4.27),

$$F(w) = -kT\ln(\sum_s \delta_{w(s),s}\, e^{-E(s)/kT}) \tag{6.4.8}$$

plays the same role for the phenome as the energy did for the genome. It contains the effects of the different number of possibilities of the genome for each value of the phenome.

The free energy can also be written in terms of an entropy with the usual relationship between energy, entropy and free energy. Assuming the fitness is only a function of the phenome w means that the energy $E(s)$ can be written as $E(w)$ and can be removed from the sum in Eq. (6.4.7) to obtain:

$$F(w) = E(w) - kT\ln(\sum_s \delta_{w(s),s}) = E(w) - TS(w) \tag{6.4.9}$$

The sum in the logarithm is the distribution of possible values of $w(s)$.

For the phenome representation of Eq. (6.3.1) with $\delta w_i = \delta w$, the use of a phenome fitness takes into account the larger number of possibilities of the distribution being near w_0. For random s_i (no selection), $w(s)$ is a random walk in the variables s_i. Thus the distribution is a Gaussian (Eq. (1.2.39)) and the free energy is a quadratic in w (constant terms due to the normalization of the Gaussian can be neglected):

$$F(w) = E(w) + (kT/2N\delta w^2)(w - w_0)^2 \tag{6.4.10}$$

Thus the maximum of the phenome fitness is at w_0. A similar calculation was done at the beginning of Chapter 5.

The use of a phenome fitness enables us to perform the Monte Carlo walk in the phenomic space without considering the genomic space. In general we have to be concerned about the possible transitions in w as a result of mutations in s. For this simple case where $\delta w_i = \delta w$, mutations can change w by only ± 1. As discussed in Section 6.3.2, when there is no fitness bias in the underlying genome representation, $E(w) = 0$, there is still a fitness bias in the phenome representation $F(w) \neq 0$. Changes in w are linear in time if we start sufficiently far away from w_0. Every step toward w_0 is accepted and every step away is rejected. Near w_0 steps are random. Eventually the population reaches equilibrium in the Gaussian distribution.

To consider a fitness bias and selection that would lead to a phenome that has, for example, taller horses, we would write the free energy as:

$$F(w) = -\alpha w + (kT/2N\delta w^2)(w - w_0)^2 \tag{6.4.11}$$

where the linear energy $E(w) = -\alpha w$ is the phenome fitness bias. The new equilibrium value of the phenome is obtained by minimizing the free energy and is given by

$$w_0' = w_0 + \alpha N \delta w^2 / kT \qquad (6.4.12)$$

The equilibrium distribution $e^{-F(w)/kT}$ is a Gaussian of the same width as before. Because it is displaced from the center of the genomic space, there are fewer distinct genomes that comprise the population. This reduction can be estimated by the number of genomes at the peak location w_0' which is reduced from the number at w_0 by a factor:

$$\frac{\sum_s \delta_{w_0',s}}{\sum_s \delta_{w_0,s}} = e^{(S(w_0')-S(w_0))/k} = e^{-(kT/2N\delta w^2)(w_0'-w_0)^2} = e^{-\alpha^2 N \delta w^2 / kT} \qquad (6.4.13)$$

This illustrates the effect of selection which, by definition, decreases the number of possible organisms in the population.

> **Question 6.4.1** Consider a genome that consists of the values of all $\delta w_i = \delta w$ except for one mutation s_0 which has the value of $\delta w_0 = N \delta w$. Start from an equilibrium distribution without selection. Discuss strategies for artificially selecting organisms for obtaining large w.

Solution 6.4.1 The initial distribution of w consists of two Gaussian peaks located at $w_0 \pm \delta w_0$. It is clear that the organisms that are optimal all have $s_0 = 1$. The key point in performing selection, however, is realizing that in addition to the gross effect of the single mutation, the best organisms also have many small effects due to $s_i = 1$ for $i \neq 0$ that accumulate to reach the optimal w. To achieve a population of such organisms, the best approach is not to select the upper peak of the equilibrium distribution but rather the upper tails of both peaks. The upper tail of the lower peak is only one mutation away from the upper tail of the upper peak. In contrast, organisms in the lower tail of the upper peak are many mutations away from the upper tail of the upper peak.

Consider the problem of developing a selection strategy for a more complex distribution of δw_i. Also, does the answer change for sexual reproduction? ∎

6.4.4 *Complexity increase*

The increase of complexity of organisms is tied to the increasing length of the genomic representation. For now we can consider this as an intuitive relationship which will be clarified somewhat during the discussion. A more careful formulation of the relationship of genome length and complexity is deferred to Chapter 8.

A simple model of the process of genome extension can be constructed out of the genome-space model consisting of strings of characters where point mutations, insertions and deletions are allowed. There is a difficulty with this model that will become apparent in a moment. Consider first a model where the fitness is the same for all possible genomes. We are interested in the time dependence of genome length $l(s)$ of the population when we start from a population of organisms with a short genome, which we might for simplicity take to be length zero. As a result of mutation, the population will spread out in genomic space. It will radiate outward from short genomes to longer ones. Without any fitness bias, there is equal probability of reaching any of the genomes of a particular length. We can treat the genome probability as a function of the length $P(l(s))$. Over time the characteristic length of the genome increases. However, it does not behave like a usual random walk in one dimension. One reason for this is that the genome lengths must be nonnegative, so steps to negative lengths are rejected. More significantly, since we have an expanding number of possibilities for longer-length genomes (Fig. 6.4.4), each step of an organism in length has a larger probability of increasing than decreasing its length. The number of ways to increase the genome is $q(l+1)$ (if we assume there are $l+1$ places to insert q possible bases) and only l possible deletions to decrease it. This leads to a bias toward longer sequences.

The bias corresponds to an entropy (and free energy) difference between strings of length l and $l+1$. The number of possible organisms of length l is q^l. The effective entropy of strings of length l is $S(l)/k = l\ln(q)$. The free energy difference between strings of length l and $l+1$ is $-kT\ln(q)$. Thus, without any underlying fitness bias, the increasing number of possibilities (phase space) for longer genomes creates a bias in the diffusion. The bias would result in an average genome length that grows linearly with time. Does this mean that it is easy to create more complex organisms? The increasing number of organisms that are more complex appears to cause a bias in favor of their creation. There is a basic problem with this argument, however, because we have ignored the entropy loss associated with adding a base to the genome from the fluid that surrounds it. Under normal circumstances we would assume that bases in free solution have a higher entropy than bases in a long chain. Adding a nucleotide to the end of a chain decreases the overall physical entropy even though it increases the entropy in the genomic space. This would cause a counterbias against the creation of longer genomes. A more complete analysis would include the energy and entropy in the free energy difference for adding the base to the chain. An even more complete analysis would also include the nonequilibrium conditions of chemical energy sources in the cell that drive such processes as DNA replication. The main lesson to be learned is that a simulation of the genomic space cannot ignore the physical free energy, because this neglect can give rise to an unphysical bias to the formation of longer chains.

We still have to address the question of the bias from a different point of view. Is it sufficient to argue that for a particular set of conditions the genome may be driven to longer lengths to explain evolution? Should we argue that the conditions in the cell may be such as to form longer genomes, and that this is responsible for the increasing complexity of organisms? To answer this question we must consider again, and more carefully, the complexity of organisms. If there is a bias to the addition of more bases, does this really create more complex organisms? No. It is only when the longer DNA is used for some purpose that the organism is more complex.

The problem is that if all possible genomes are created, then the description of the population is simple. It is the selection of organisms by some criteria that makes them complex (Question 6.4.2). This arrives at the crux of the evolution of complexity. It is the selection of an organism from a large number of possibilities that makes

Figure 6.4.4 Illustration of the expanding space of genome possibilities that starts from a single base on the left. Lengthening the genome by a single base moves one step to the right and multiplies the number of possibilities by four. The space is only schematically indicated after three bases. Many different steps are possible between genome lengths if we allow deletion or insertion of bases. If we only consider the space available, starting from an organism of a particular length genome, and without any selection, the genome will lengthen by diffusion because of the much larger number of longer genomes. This, however, does not take into account the actual free energy for adding a base. ∎

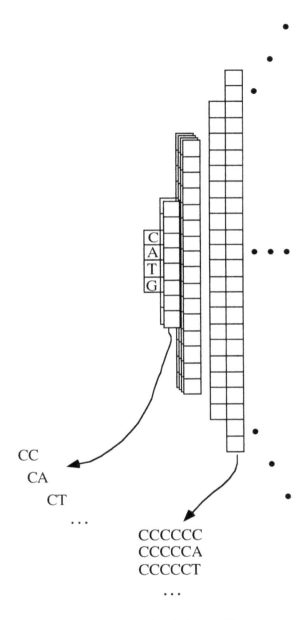

CC
CA
CT

...

CCCCCC
CCCCCA
CCCCCT

...

it complex. In the theory of evolution, the selection criterion is fitness. The assumption is that longer genomes are systematically able to represent fitter organisms. Since there are many more possible organisms of longer genome length, this enables selection of more specific traits that correspond to higher fitness. It is presumed that the highest fitness of a particular length genome

$$K(l) = \max_{l(s)=l} K(s) \qquad (6.4.14)$$

is a monotonically increasing function of the genome length.

We can see the advantage of lengthening the genome if we look at our phenome/genome relationship in Eq. (6.3.1). This relationship describes a phenomic trait in terms of the available genomic parameters. As long as the optimal trait exists within this representation, there is no problem. However, if it does not, the addition of extra parameters in the form of possible genome coordinates increases the possible options for a particular trait or for combinations of traits. This increase in the phase space of possibilities is exactly the motivation for increase in genome and organism complexity.

Our discussion of selection and organism complexity is also relevant to the neutralist/selectionist controversy. If neutral mutations dominate the space of possible organisms, then we are left with the circumstance of a large number of possible organisms with selection not playing a significant role in evolution. This is unsatisfactory as an explanation of the evolution of complex organisms that, without selection, have no mechanism by which to arise. Thus, even if neutral mutations account for many of those that are possible, it is the mutations that do affect fitness that account for the part of evolutionary changes that we are most interested in.

Question 6.4.2 We noted that selection is what causes an organism to be complex. What is wrong with the following statement: "If you create all organisms of length 10^{10} base pairs, you will also create human beings, and therefore you will have created complex organisms"?

Solution 6.4.2 A part of the problem with this statement is the number of organisms that would have to be created, which is $4^{10^{10}}$. However, this is not yet a complete answer. Another problem with the argument is that in order to see that you have also created human beings, you must have some way of pointing them out among the large (huge) number of other organisms. It is pointing them out which is equivalent to selection. Otherwise we can only see a typical organism out of this set, which would not be a complex organism. ∎

Question 6.4.3 (for further thought) If there are a larger variety of complex organisms, then why are there fewer distinct types of complex organisms than simple organisms currently on earth?

6.4.5 Interactions

In this section we consider interactions between organisms—reproduction, consumption, predation, symbiosis, parasitism—which affect fitness. To understand the effects of interactions in the random-walk model we treat fitness as a property of the entire population of organisms rather than of a particular organism. We write the fitness as $K(\{N(s)\})$, where $N(s)$ is the number of organisms of genome s. Using this fitness of the whole population, we can still treat evolution as an optimization of fitness. We can also define the fitness of a particular organism as the difference in the fitness of the total collection of organisms minus the fitness when the organism is not present:

$$K(s) = K(\{N(s') + \delta_{s',s}\}) - K(\{N(s')\}) \qquad (6.4.15)$$

We can see that our original fitness landscape already included interactions. However, they were included only in a time-independent average (mean field) way. To get back to our original picture, we would write the mean field landscape of a single organism in terms of the fitness of the population as

$$K(s) = K(\{N_0(s') + \delta_{s',s}\}) - K(\{N_0(s')\}) \qquad (6.4.16)$$

for a reference population $\{N_0(s)\}$. This assumes that variations that occur in the populations of organisms do not significantly affect the fitness of a particular organism. This tends to be valid when the population of organisms is large and unchanging. For smaller populations that change on the time scale relevant to the evolutionary dynamics (this would seem to be a tautology), we must include the interactions explicitly. This means that the existence of a particular organism affects the fitness of other organisms. From Eq. (6.4.15), the fitness landscape changes with time along with the changes in populations.

It is important to recognize, however, that as soon as we assume a fitness which is only a function of the population, $K(\{N(s)\})$, we also have a symmetry of interaction. When an organism at s' lowers (raises) the fitness of an organism at s'', then an organism at s'' lowers (raises) the fitness of an organism at s'. This symmetry is shown in Question 6.4.4. If we want to model asymmetric interactions, then we must use entirely different models discussed in Section 6.5. Within the Monte Carlo random-walk model there are thus only two types of interactions, interactions between organisms that raise their fitness and interactions that decrease their fitness.

Question 6.4.4 Prove that "When an organism at s' lowers (raises) the fitness of an organism at s'', then an organism at s'' lowers (raises) the fitness of an organism at s'." This assertion is true whenever we have a model that assigns a unique fitness to the collection of organisms $K(\{N(s)\})$, where $N(s)$ is the number of organisms with genome s.

Solution 6.4.4 The only difficulty is translating the English into an equation. The statement is an answer to the question, How does adding an organism to s' affect the fitness of an organism at s''? Start from a set of organisms $\{N(s)\}$. The change in fitness due to adding an organism at s'' before adding an organism at s' is:

$$K(\{N(s) + \delta_{s,s''}\}) - K(\{N(s)\}) \qquad (6.4.17)$$

After adding an organism at s' it is:

$$K(\{N(s) + \delta_{s,s'} + \delta_{s,s''}\}) - K(\{N(s) + \delta_{s,s'}\}) \qquad (6.4.18)$$

The difference between these two is:

$$\left(K(\{N(s) + \delta_{s,s'} + \delta_{s,s''}\}) - K(\{N(s) + \delta_{s,s''}\}) \right) - \left(K(\{N(s) + \delta_{s,s'}\}) - K(\{N(s)\}) \right)$$
$$= K(\{N(s) + \delta_{s,s'} + \delta_{s,s''}\}) + K(\{N(s)\}) - K(\{N(s) + \delta_{s,s'}\}) - K(\{N(s) + \delta_{s,s''}\})$$

$$(6.4.19)$$

which is symmetric in s' and s'', so the assertion is proven. ∎

A convenient way to think about the interactions is that adding an organism at one place in the phase space (genome or phenome) changes the landscape for other organisms by raising or lowering their fitness. A uniform raising or lowering of the landscape does not affect anything; only the differential effect on the fitness matters. The simplest interactions are those that raise the fitness of all nearby organisms, or those that lower the fitness of all nearby organisms. The effect is assumed to decrease with distance. When the fitness of organisms is raised (the energy is lowered), a depression (energy well) is created around the organism that causes other organisms to be drawn toward it—an attraction between organisms. If the fitness is lowered, other organisms tend to move away—a repulsion between organisms.

When there is an attraction, the energy well may cause a self-consistent trapping, effectively binding the organisms in a group. This trapping causes the organisms to move together on the landscape rather than as individual organisms. This is the effect we need in order to account for the confinement of populations discussed in Section 6.4.3.

For evolution down an incline, Fig. 6.4.2, the spreading of the organisms would be limited. For more than one dimension, the mutual attraction automatically creates a channel. Then the co-moving organisms would appear to be analogous to our understanding of evolutionary change in Fig. 6.2.2. There still is a difficulty with this picture because the local interactions become less relevant as the dimension of space increases. In particular, in four or more dimensions, short-range interactions are irrelevant. Intuitively, this is because in a large dimensional space there are too few encounters between organisms for their interactions to matter. Alternatively, the reason is that the mean field theory becomes exact in four or more dimensions. Thus in the apparent high number of dimensions of the phenomic or genomic space, the attractions should be irrelevant. There are two possible flaws in this argument. The first is that the number of relevant dimensions in distinguishing between organisms may not be as large as the number of apparent dimensions. The second is that the way we are modeling interactions is inadequate. The latter would again force us into a different class of models.

Even if attractions help with confinement, they do not as readily help with trait divergence (Fig. 6.2.3). The picture of a ridge, Fig. 6.4.3, would be difficult to justify except as a low-probability occurrence. On a smooth landscape, the likelihood that a self-attracting population is precisely at a location where a ridge occurs (as opposed to on one side or the other) is small. Speciation would be more readily understood as a process where a self-attracting population splits by chance into two populations by random processes. The most likely scenario is when a small population separates itself from the whole. This is just the escape of one (or a few) organisms from the energy well created by the large population. If several individual organisms escape, they may encounter each other and aggregate to form a co-moving group. Once again, however, this scenario requires a delicate balance between the tendencies of organisms to disaggregate and aggregate, which cannot be expected to apply generically.

Before leaving the topic of attraction, we consider, for future reference, the rate of change of a self-attracting population in two cases. When the collection of organisms moves together, random motion on a flat landscape is slower than the motion of an individual organism. In Chapter 5 the same problem was discussed for a polymer. The diffusion constant was shown to decrease with the number of monomers as $D \sim 1/N$ and the distance traveled as $\Delta r_{cm} \propto N^{-1/2}$. On an incline, the speed of travel of the self-attracting population would be the same as for a single organism, because on average each organism of the group feels the effect of the bias.

Thus far we have discussed attraction. An organism that repels other organisms would move on the landscape in isolation. If, because of a valley, the organisms accumulate, they would tend to escape more readily and rapidly from it than without the repulsion.

Thus far we have discussed the effects of either attraction or repulsion separately. In order to understand how attraction and repulsion affect evolutionary behavior, we must recognize that the attraction and repulsion are properties of each place in the space, not of the walkers themselves. The primary effect of the interactions is to cause organisms to bunch in regions of space where they are attracting each other. Regions where they repel would tend to be empty—all other things being equal. Since the properties of attraction and repulsion vary from place to place in the space, an evolving population may encounter both aggregation and dispersal. We might consider creating a model using this variation to account for trait divergence and other properties of evolution. It is important, however, to recognize that this kind of model is a significant departure from the model that tries to explain evolution from a single fitness function of individual organisms.

Finally, we discuss long-range interactions. Long-range interactions between organisms can cause circumstances where the fitness valley in which one type of organism exists depends on the existence of another organism at some other location in the space. This is most simply illustrated by the dependence of animals on plants in general, or by specific relationships between predator and prey. Note that these relationships are largely untreated in this model of an energy landscape (for example, they are not symmetric). However, in general we can recognize that relationships of mutual dependence exist. Changes that occur in one organism thus result in changes in the fitness landscape of other organisms and thus changes in the other organisms as well. This leads us to a recent innovation in evolutionary theory—the concept of avalanches in evolutionary change. Over time, a system of dependencies is developed which can be disrupted when one organism undergoes a change in population that is sufficiently severe, whether due to evolutionary change or external influence, such as environmental change. The change causes a cascade of changes in other organisms. Depending on the nature of the mutual dependencies, the cascade of changes can be large or small. The modeling of such phenomena is outside of the fitness landscape model because it is dependent on interactions that are only being added as secondary effects to the fitness landscape. A model that has been used to think about such processes is a sandpile model. In this model, grains of sand fall at random onto a surface. They cause piles to grow which are formed out of grains supporting themselves on other grains. The addition of a single grain can destabilize a pile and cause an avalanche that can move many grains of sand. The sandpile model has a power-law distribution of avalanche sizes. This model has been used as an analog of what may happen in mutually interacting networks of organisms.

There is now substantial evidence that evolutionary change has undergone periods of rapid change in many organisms (e.g., the Precambrian explosion) after long periods of slow change. Known as the model of punctuated equilibria, it may be possible to describe this by a model of avalanches. The idea of a mutually consistent network of dependencies is also a model for the sudden extinction of large dinosaurs after their extended existence. The fitness of individual organisms was high due to mutual interactions. However, when a sufficient disturbance occurred (possibly due to impact of a comet) then the self-consistent network of dependencies was disrupted. Once this occurred, other organisms that were less fit under the original circumstance (not just for climactic reasons) were able to increase and dominate the population of organisms.

We will return to consider interactions between organisms in Section 6.5 in the context of a different class of models. We will also discuss the impact of interactions such as altruism and aggression and the formation of collective behaviors in Section 6.6.

6.4.6 Evolution—global landscape dynamics

There are conceptual problems in understanding a global fitness landscape that includes microorganisms and man. After several billion years of evolution, we might expect that the relative populations of microorganisms, insects and man would reflect evolutionary progress and fitness. On one hand, the fossil record suggests that evolution proceeded from microorganisms through insects to mammals. On the other hand, the numbers appear to have remained in favor of the smaller and simpler organisms that arose earlier in evolution.

If the fitness is directly related to the number of organisms according to Eq. (6.4.3) where $P(s) = K(s)$, then fitness would strongly favor microbes over insects and insects over human beings. There is, however, an alternative definition that is equally valid. We could set the fitness to be the mass times the organism population, $P(s)M(s) = K(s)$. $K(s)$ would still be the limiting distribution of the ensemble which represents mass rather than organisms. A Monte Carlo walker would represent a unit of organism mass. This definition gives a much higher relative fitness for large organisms. We do not want to argue which definition is correct, but to understand why there is an ambiguity. The key point is that the use of an ensemble is predicated on the existence of a conserved total number of elements in the ensemble. The total number of organisms is not conserved. Neither is the total mass of the organisms, though this might seem to be a better approximation. Since such quantities are not conserved, we do not have a well-defined ensemble. This is only one of the problems that give rise to a difficulty in defining the relative fitness of widely different organisms. With this in mind, we discuss three scenarios for the global behavior of evolution on the fitness landscape. The first is evolution downhill, the second is evolution uphill, and the third is an alternative that relies upon a changing landscape.

One traditional view of evolution is most easily considered as starting from the pinnacle (or somewhere on the side) of a long hill (Fig. 6.4.5). The motion of the population consists of a descent downward. Grooves in the hill and self-attraction are essential to account for trait divergence causing a separation into droplets. Unlike flowing water on a hill that typically converges upon a single channel, the flow is an outward branching that results in a treelike structure of different subpopulations with distinct traits. The branching is an assumption about the dynamics; it is more reasonable for a large-dimensional landscape than the usual two-dimensional one. Of course, we must also explain why there isn't a complete dispersal into very few organisms per subpopulation. Starting from this picture, however, another key question would be, Why do there persist primitive organisms such as single-celled organisms or insects that were formed earlier and thus higher on the hill? One possible answer is that these organisms continued to evolve and increased their fitness without dramatically changing form. The problem with this picture is that improving fitness would seem to require manifest changes in phenome that are not evident. A second possible answer is that nonconservation of organisms enables the microorganisms to continue to regenerate even though they are high on the hill. However, the ability to regenerate populations alters radically the assumptions of selection according to the fitness landscape. Any such new model requires its own analysis. Thus, while the picture of evolution downhill is consistent with the view that fitness propels forward the process of evolution, it is difficult to reconcile this with the population ratios that

Figure 6.4.5 A first model of the global fitness landscape considers all of evolution to occur on a single long fitness slope upon which the evolution of organisms consists of progress to increasing fitness, like the flow of water downhill in energy, but with the added assumption of outward branching. ∎

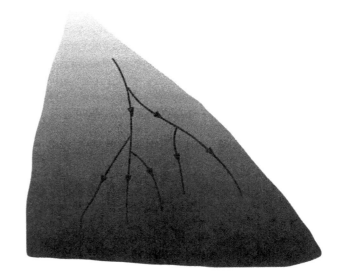

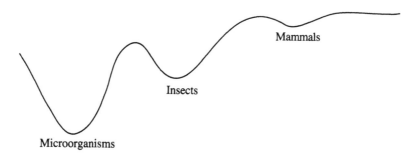

Figure 6.4.6 A second model of the global fitness landscape assumes that organisms are found today in rough proportion to their fitness, and therefore that the fitness landscape consists of energy valleys that are lowest for microorganisms, higher for insects and higher still for mammals. ∎

continue to retain most of life as microorganisms, virtually independent of the existence of the higher forms of life.

An alternative view would adopt the existing population ratios as a model for the probability function $P(s)$ and use this probability function to define fitness (Fig. 6.4.6). In this model evolution started with microorganisms that are low-energy, high-fitness organisms. What is the driving force for the existence of higher forms of life? The answer is that random mutations provide a possibility of moving upward. The landscape of mountains and valleys is then responsible for the observed pattern of organism traits and species. The smaller numbers of complex organisms that occurred later in evolution reflect their lower fitness. A low fitness does not preclude their existence, because there are only a few of them compared to the high-fitness microorganisms. The fossil record is explained by the motion of populations upward to overcome an obstacle, and downward into the subsequent valley. Thus far the model seems to account for observations, but this quickly breaks down with further thought. Extinctions are a problem. They might be explained by temporary occupation of local minima that are higher in energy than minima currently occupied. However, extinctions would not be permanent, since such valleys are likely to be repopulated later. A more serious problem is that organisms would be expected to regularly evolve from valley to valley, both forward and backward in evolutionary order. This conclusion follows, because once a time scale of moving from one valley to another is reached, migration between them continues to occur. It is possible that existing experiments missed such processes, but it would require a dramatic revision of prevailing thought. Another controversial conclusion that follows from this model is that organism populations are close to equilibrium. Even though randomness through mutation plays the essential role of causing evolutionary change, because populations are close to equilibrium, they must be independent of history. However, the main problem with this model is that it does not agree with the overall size of the phase space of organisms. If we are close to equilibrium, then essentially all possible organisms would be

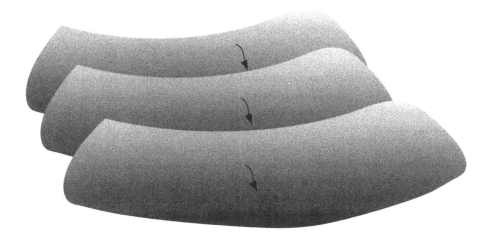

Figure 6.4.7 A third model of the global fitness landscape is a dynamic landscape that consists of expanding waves upon which organisms always evolve toward decreasing energy, like surfers on ocean waves. ∎

represented in existing organisms. Since the genomic space is so large, we would have to assume that almost none of the genomic space is viable, but that organisms evolved anyway. This is not reasonable.

There is a third alternative for the fitness landscape that is consistent with the existence of a diversity of organism types at different stages of evolutionary progress but maintains the evolutionary pressure of fitness. This approach emphasizes that the long-range structure of the landscape must include the long-range interactions between organism types discussed in the previous section. For simplicity we can set the inherent fitness to be the same everywhere. The existence of a long-range interaction implies that a particular organism type promotes the existence of another organism type that may be far away on the landscape. For example, the fitness of a sheep would not be high without plants. A picture that takes advantage of this property of the landscape considers evolution as a process of surfing on self-consistent expanding waves where the large populations of simpler organisms are responsible for the fitness waves on which higher organisms evolve (Fig. 6.4.7). This picture is consistent with a local evolutionary pressure of fitness, and the persistence of primitive organisms. A specific mathematical model that is related to this picture will be introduced at the end of Chapter 7, because it is in the class of models of pattern formation in developmental biology. In Section 6.5.4 we also address related issues.

6.4.7 *Randomness and determinism*

Many of the questions often articulated about evolution, such as

- How does the formation of life depend on the conditions?

- What is the likelihood of the formation of life and how likely is life to be found elsewhere?
- How sensitive is the present form of life to chance, and how determined is it?
- Is the form of life on earth unique or are there viable alternatives?

have to do with randomness and determinism. One of the difficulties, however, is understanding what we mean by these terms. The statement that existing forms of life were determined "by chance events" and therefore were not predetermined is not a sufficiently clear statement since there are multiple issues that must be resolved. First, it must be recalled (Section 1.1) that chaotic behavior is deterministic. The reason for chaotic behavior is that the system is sensitive to initial conditions—small differences become amplified over time. Second, a stochastic system (Section 1.2) is a system where external influences affect the system behavior and are presumed random. Finally, in the study of thermodynamic systems (Section 1.3) randomness played a crucial role in the dynamics, but the equilibrium state of a system is completely and uniquely determined and is stable and insensitive to initial conditions.

Our objective here is to clarify rather than answer the fundamental questions. We separate the discussion into two issues. The first issue is whether life that exists on earth is representative of what would arise from any evolutionary process under a wide range of initial and intermediate conditions. If it is representative, then life is essentially determined in the same sense that equilibrium states are determined. On the other hand, if life is not a typical outcome of evolutionary processes, then the second issue is to determine which influences were important in determining the existing form of life. Are these the effect of microscopic thermal vibrations or macroscopic influences? Do the macroscopic or microscopic effects trace themselves to the initial conditions or to persistent external influences such as solar radiation?

We have noted several times that stochastic iterative dynamics of an ensemble should lead to equilibrium. The equilibrium state is not affected by the specific and possibly random path it took to get there. Thus, the existence of randomness in the dynamics does not necessarily mean that the outcome is not determined. Thermodynamics uses an ensemble in the same way that we are thinking about the collection of organisms on earth. There is, however, a basic assumption in thermodynamics which need not be true about the collection of organisms. An ensemble is an arbitrarily large collection of systems. The sense in which it must be arbitrarily large is that the number of systems is larger than the number of system states. Even if this is not satisfied, at least a significant fraction of distinct high-likelihood possibilities must be represented. The reason that an ensemble is not affected by randomness is that whenever one of the systems takes a step, another takes a step in the opposite direction.

Let us consider the organisms on earth as a limited set of examples of possible organisms. We know that the genomic space is very large and therefore we could easily argue that even with all of the organisms on earth, there is unlikely to be a representative sampling of genomic space. However, we may not care about sampling the genome space, but rather the phenome space. It is harder to tell if we have representatives of all viable phenomes. To take this discussion further, we should include the interactions between organisms. A self-attracting population that forms a species which moves together on the fitness landscape becomes correlated, and therefore a single system rather than a collection of systems. This both reduces the number of independent samples present on earth and increases the time scale over which random changes occur. More generally we know that any interaction that causes interdependence of organisms reduces the effective number of independent systems present on earth. However, we may still focus on the set of organisms and try to determine if they are representative.

Can we arrive at any conclusions from the general phenomenology of life? On one hand the existence of a wide variety of organisms suggests that many are possible; on the other hand this wide variety might be sampling all possibilities. The persistence of certain organisms since early in evolution suggests that the correlation time for organisms is very long, and therefore that independent samples of all possibilities may not have been realized. On the other hand, their persistence suggests that there may not be many other alternatives. A better source of evidence is the artificial breeding of organisms. By demonstrating the existence of many varieties of organisms that differ from those found in nature, we can conclude that the naturally occurring organisms are not representative of the possibilities. The large dinosaurs also provide an important piece of evidence through their persistence and complete disappearance. The more different they are from current living organisms and the longer their persistence on earth, the better is the argument that there are many possible living organisms and the present samples of life on earth are only a few nonrepresentative examples.

Our discussion indicates that the ratio between the number of organisms to the space of possible organisms is important in determining whether the existing population is representative. This suggests that microorganisms might be effectively in equilibrium even if multicellular organisms are not. Thus we might not want to ask whether equilibrium applies, but rather at what level of organism complexity it applies. If it turns out that the simplest prokaryotes are not fully represented, then we may conclude that this is also true about more complex organisms (Question 6.4.5).

Let us now assume, reasonably, that the existing organisms on earth are not representative of all possible living organisms. Then it becomes relevant to discuss the nature of the pathways of evolution, and the role of initial conditions or external perturbations. Either becomes important when there are multiple options at a particular moment for incremental evolutionary changes, of which only one can be chosen. The question becomes how the path is chosen, and our interpretation of randomness or determinism in this context. Of course it is not enough that different pathways exist; they must also not converge at later times. The evolutionary path taken by an organism, or a collection of self-attracting organisms, must be distinct at all later times from other possible paths in order for the choice to be important. The expanding phase space for ever more complex organisms is the best argument in favor of a lack of converging pathways. A model (or phenomenology) that shows that over time organisms are always exploring new regions of space would be relevant.

The main point to understand in this discussion is the relevance of selection. The whole idea of selection is that there are multiple possibilities of which only a few are selected. This is also the nature of what we mean by a complex organism—that it is differentiated from other possible organisms by many details that must be selected. Thus we must further ask whether the pathway taken in evolution is selected by fitness or whether other effects are significant. If fitness is the primary selective force, then we are attributing the selection to macroscopic environmental effects external to an organism and not to microscopic or macroscopic randomness. Another possibility is that, of the possible mutations that might occur, only a small subset do occur. In this case selection due to fitness can only apply to the possibilities that occurred, and microscopic randomness is relevant. Moreover, if the survival of a particular organism is not determined by fitness but only statistically related to fitness through random occurrences, then macroscopic randomness plays a role. Here again we must be careful to recognize that if fitness selection eventually forces the organisms to reach a particular place in genome space, then all prior divergences in paths due to randomness are irrelevant.

Q **uestion 6.4.5** Discuss evidence that microorganisms involved in diseases are not in equilibrium.

Solution 6.4.5 Equilibrium implies that all possible microorganisms exist. If this were true, the eradication of disease (by natural or artificial methods) and the appearance of new diseases would both be impossible. Since both appear to be possible, it seems reasonable to assume that microorganisms are far from equilibrium. ∎

6.4.8 *Space and time*

Thus far our fitness landscape has been discussed as a function of genome or phenome. We must also include the dimensions of space in our considerations. Assuming a well-defined fitness landscape as a function of the phenome or genome also assumes that spatial variation in the landscape is smooth or that its effect can be averaged over. Rapid spatial variations are likely to have a significant impact on the model properties. However, even relatively smooth variations in space and time have important effects. The main effect of a spatial dimension is the existence of populations of organisms that exist simultaneously in time and can evolve in part independently. This would be a valid statement even if the fitness landscape is the same in different locations. However, the situation is more interesting because the fitness landscape is different in different locations. Including ocean and land environments, different climates as well as the existence of distinct combinations of organisms causes the fitness landscape to vary greatly on earth. All organisms survive only in a limited range of conditions, and have restricted spatial regions in which they are found on earth. The dynamics of the entire system become more interesting when we consider coupling the different environments through migrations. Migrations enable organisms evolving in one location to encounter alternate environments. An important realization is that this gives rise to an additional type of selection—selection by the organism of its environment. Thus, organisms are not necessarily subject to a unique fitness criteria. By migrating they may be able to select an environment to which they are well suited.

Time-dependent variations in the fitness can also cause a variety of effects. An example is the mass extinction attributed to a comet that changed climactic conditions on earth and led to the demise of large dinosaurs. A more current example might be a forest fire. In either case it is easy to understand how such events might be disastrous for evolution if they happen too often or too severely. However, the comet may have been responsible for a large step forward in evolution by enabling other animals (mammals) to emerge. As discussed earlier, the disruption of an existing network of organisms may enable other organisms to arise and cause rapid evolutionary changes. The smaller example of a local forest fire is now understood to provide opportunities for the survival of organisms that would not have a chance in well-developed forests. One way to think about this is through the fitness landscape, where valleys are formed by interactions that cause self-trapping of the population. When the existing organisms are reduced in number, these valleys may also not be as deep. This enables more rapid movement of organisms on the landscape.

When large variations in the fitness landscape occur frequently, there are other effects that may occur, including the development of organisms that are better suited to these variations either through self-imposed genetic diversity (requiring collective behaviors discussed in Section 6.5) or adaptability.

Question **6.4.6** Why can't we just think about the dimensions of space as additional dimensions for the fitness landscape?

Solution 6.4.6 The nature of steps in the spatial dimension is radically different from the nature of steps in the genome dimension. Nevertheless, within the context of the model of Monte Carlo steps on a fitness landscape, we can consider the different types of steps in the same way. The only place we run into trouble is when the steps in space are directed rather than random. Specifically, when an organism can identify which direction it should move in, then there is a violation of the assumptions of the model. ∎

Question **6.4.7** Discuss the relevance of spatial dimension to the problem of walkers exploring the space of possible organisms.

Solution 6.4.7 The fitness landscape may involve obstacles that consist of regions that are not viable under particular environmental circumstances. Thus we can expect that the connectivity of the phenomic or genomic space is very poor if we consider only one particular environment. The existence of a spatial dimension with different environments enables organisms to move around obstacles in the genomic space because there are more possible ways to have organisms in a variety of environments. For example, it is not clear that whales could develop from fish directly. However, according to the current view, by a process of moving from water to land and back to water it became possible for whales to appear. ∎

Question 6.4.8 Discuss, from the point of view of a fitness landscape, the process of organisms crossing a mountain range to an isolated valley.

Solution 6.4.8 This scenario contains a number of important elements. First it is assumed that a population of organisms evolved in one region of the land but was not found across a particular mountain range. The mountain range is a barrier in both physical and fitness space because it is assumed that the organisms do not live easily on the mountain. By crossing the mountain, a group of organisms becomes independent of the original set of organisms of which it was a part. They participate in the evolutionary process in the isolated valley. The distinct evolutionary pressures or random influences that affect this small population also change its position in genomic space. Some time later the organisms may recross the barrier, but the two populations that evolved separately are now at different positions in genome space. This is significant, because we expect that an attractive interaction between organisms of similar type prevents the separate evolution of subpopulations. Thus, physical separation is an additional mechanism for the formation of distinct organism types. ∎

6.4.9 Adaptive organisms

An adaptive organism responds to its environment in a manner that adjusts behavior to improve fitness. We could more generally state that an adaptive organism has a phenome that depends on its environment. However, this does not affect the fundamental relationships of genome or phenome and fitness. Specifically, given a genome of an adaptive organism, the fitness is still as well defined a quantity as it is without the adaptation. However, by making additional assumptions, we can try to understand the effect of adaptation.

One perspective is that there is no special ability that adaptation provides over nonadaptive organisms. However, an adaptive organism can approximate the behavior of more than one nonadaptive organism. It cannot do so exactly, because adaptation carries its own cost. However, it can do so well enough to reach close to their fitness. This is an advantage when the fitness landscape is spatially or temporally varying, because the adaptive organism can survive in varied conditions. We might write that:

$$K(s,e) = \min_{s'} K(s',e) + \varepsilon \qquad (6.4.20)$$

where s represents an adaptive organisms and s' varies over a set of nonadaptive organisms. We have written the fitness as a function of the environment e explicitly. ε represents the inherent cost to fitness of adaptation, because the optimal behavior is not automatically realized for a particular environment.

A consequence of this view is that the fitness landscape due to changes in genome (that do not affect the adaptive ability) for adaptive organisms tends to be flatter than for nonadaptive organisms. As the genome of an adaptive organism changes, if the domain of s' in Eq. (6.4.20) does not change, then neither will its fitness. Even if the domain of s' in Eq. (6.4.20) changes, the variation in the fitness will be more gradual than for a nonadaptive organism. The organism, in effect, reduces barriers to evolution by using adaptation to move around them. This picture, however, also implies that the ultimate benefit of adaptation becomes small for a relatively static fitness landscape, since it is advantageous for adaptation to disappear in favor of the genetically determined optimal behavior pattern.

A different perspective suggests that it may be possible that adaptation can enable certain phenomes to exist that cannot be described directly by the genome-to-phenome developmental relationship. Thus, for example, certain behavioral patterns may not be possible to specify genetically and can only arise through adaptation. In this case, adaptation becomes an extension of the physiological developmental process in creating the resulting phenome.

6.4.10 *Limitations of the fitness landscape*

We have discussed many limitations of the fitness landscape in previous sections and have introduced some ways to work around them. Here, however, we recall the most basic ways in which the fitness-landscape model breaks down, to motivate a different approach taken in the following section. Ultimately, the main problem in the fitness-landscape model is the use of a conserved population that forces a particular treatment of reproduction and death. Let us think what this means in terms of the model behavior. When an organism reproduces or dies, it causes a change in the local population of organisms at a particular genome or phenome. The random-walk model treats this by assuming that reproduction of a single offspring and death, either of the offspring or the parent, are directly linked. If we do not do this, but still require the conservation of population, then the birth of one organism is tied to the death of another organism. However, the death and birth may be at very different locations in the genomic space. Thus we are forced to consider various nonlocal moves. Including nonlocal moves is not, however, sufficient, because a Monte Carlo move is possible or impossible independent of the population itself. The birth of one organism and the death of another forces a particular nonlocal move that would not be possible without the existence of the parent organism. Specifically, we can imagine an organism that gives birth to many offspring as a process of ingathering of organisms from various other regions of space. This type of nonlocal move is not readily treated in Monte Carlo and another approach is necessary.

One illustration of how reproduction can affect the behavior of evolution on a fitness landscape is a hybrid picture in which we think about organisms evolving on a landscape but with reproduction and a nonconserved population. By processes of mutation, an organism might overcome a fitness barrier and end up in a fitness well. There is no need for other organisms to follow over the barrier, since the organism can reproduce, increasing the population in the well that is reached. Even if we include sexual reproduction, there is only need for a reproducing population to cross the barrier. The decoupling of the population in one well from the population in another well is a problem for the model we have been discussing. To enable us to think about this picture, we must develop different tools. We might note, however, that this image suggests additional problems for a single global fitness function tied to the independence of growth and death of different populations.

Consistent with this discussion, we recognize that the Markov chain, even in its most general form, does not allow the transition rate from one location in space to another to depend on the population. This is because the description is limited to that of an ensemble of independently evolving systems treated statistically. To progress, we must write a time dependence of the organism population $N(s;t)$, as given by a non-linear dependence on its population and other populations $\{N(s;t)\}$. We therefore abandon the Markov chain formalism in favor of a more explicit discussion of reproduction, death and selection and the interaction of organisms.

Finally, by abandoning the Markov chain formalism we can also eliminate the use of a target limiting distribution for the dynamics. This inherently prevented us from considering many possible dynamical behaviors of population evolution. For example, fluctuations in populations driven by predator-prey relationships. The lack of such dynamics is related to the impossibility of including asymmetric interactions between organisms that increase the fitness of one and decrease the fitness of the other. It should be noted that this is a limitation that is often assumed in evolutionary theory even without the assumption of a Markov chain or limiting distribution, because fluctuating populations would be represented by fluctuations in the fitness function with time.

Our efforts to understand the random-walk model were not in vain. It is a difficult and valuable accomplishment to demonstrate that an entire class of models is not adequate, and to understand in what way it is not adequate. Moreover, we have discussed many important issues and gained insights that will also show us limitations in the seemingly better models that we will proceed to investigate.

6.5   Reproduction and Selection by Resources and Predators

The objective of this section is to present and discuss several mathematical models for the process of incremental evolutionary change in a population of reproducing organisms. We will see that there are subtleties that arise in such models that may initially be counterintuitive, and this will lead to a better understanding of evolution. In these models we often assume two or more types of reproducing organisms and follow their relative populations as a function of time. Our attention will be focused on understanding what parameters control selection—the survival of one type of organism at the expense of the other. One common model for evolution relates fitness directly to reproductive rates. Organisms with more offspring are more likely to survive and therefore more fit than organisms with fewer offspring. We will see by analyzing a few more detailed models that this is too simplified and incomplete a picture.

The models we will use directly describe the behavior of a population of organisms $N(s;t)$ in terms of an iterative map:

$$N(s;t+1) = f_s(\{N(s;t)\};t) \tag{6.5.1}$$

or in terms of a differential equation:

$$\frac{dN(s;t)}{dt} = f_s(\{N(s;t)\};t) \tag{6.5.2}$$

In either iterative map or differential equation forms, the models of the last section would account for any case where the function f_s is linear and population conserving. We will rapidly depart from this in our efforts to describe reproduction, death, resources and predators.

6.5.1 *Reproduction, resources and selection*

We start with a simple model for population growth. An organism that reproduces at a rate of $\lambda > 1$ offspring per individual per generation has a population growth that is exponential. Using an iterative equation (Section 1.1) this is written as:

$$N(t) = \lambda N(t-1) \tag{6.5.3}$$

In the simplest interpretation, this represents synchronous generations with death following reproduction, but the behavior is more general. We can also write a differential equation that represents similar growth:

$$\frac{dN(t)}{dt} = \lambda' N(t) \tag{6.5.4}$$

where $\lambda' > 0$. If we have two organisms whose populations grow exponentially, the faster growing population will eventually dominate the slower one. However, both organisms continue to exist.

We obtain a standard model for fitness and selection by taking two equations of the form Eq. (6.5.3) for two populations $N_1(t)$ and $N_2(t)$ with λ_1 and λ_2 respectively, and normalize the population at every step so that the total number of organisms remains fixed at N_0. We have that

$$N_1(t) = \frac{\lambda_1 N_1(t-1)}{\lambda_1 N_1(t-1) + \lambda_2 N_2(t-1)} N_0$$

$$N_2(t) = \frac{\lambda_2 N_2(t-1)}{\lambda_1 N_1(t-1) + \lambda_2 N_2(t-1)} N_0 \tag{6.5.5}$$

Because we did not change the relative dynamics of the two populations, and only the total population is affected by the normalization, we know that the faster-growing population will dominate the slower-growing one. If we call λ_i the fitness of the ith organism we see that according to this model the organism populations grow at a rate that is determined by the ratio of their fitness to the average fitness of the population. This model is similar in form, but not behavior, to the two-state system of Section 1.4, which is a prototype for the model of evolution discussed in Section 6.4. Question 6.5.1 addresses the similarities and differences of this population model and the two-state system.

Question 6.5.1 We can choose to write Eq. (6.5.5) in terms of the probability of having each organism type by writing $P_1(t) = N_1(t)/N_0$ and similarly for $P_2(t)$. Compare the qualitative behavior of Eq. (6.5.5) with the behavior of the two-state system that also describes the dynamics of two probabilities.

Solution 6.5.1 The most dramatic difference between the behaviors of the two models is that the two-state system, at any particular energy difference and temperature, equilibrates at a particular ratio of the two different populations. In Eq. (6.5.5), unless the fitnesses are exactly equal, the lower fitness population will eventually disappear no matter what the relative fitnesses are. The relative fitness only controls the rate of disappearance. ∎

The model for selection in Eq. (6.5.5) is useful in that it provides an alternative dynamics to the two-state model. However, we would like to develop an understanding of the process by which population size is limited. The model of Eq. (6.5.5) does not represent population limits directly. Instead it simply normalizes the population size. In order to have a better model for the interaction between organisms that gives rise to selection, we should directly limit the number of organisms and then see how one organism grows at the expense of the other. A standard way to limit the population growth is to use a differential equation of the form:

$$\frac{dN(t)}{dt} = \lambda N(t)(1 - N(t)/N_0)$$
(6.5.6)

This equation appears similar to the quadratic iterative map discussed in Section 1.1, but this differential equation is not the same (Question 6.5.3) and it has a relatively simple behavior. Eq. (6.5.6) can be solved analytically or integrated numerically to obtain the behavior shown in Fig. 6.5.1 (Question 6.5.2). Starting from a small population, the population grows exponentially, then saturates at the value N_0. The qualitative behavior can be understood directly from Eq. (6.5.6) because the factor $(1 - N(t)/N_0)$ reduces the growth rate to zero as $N(t)$ approaches N_0.

Question 6.5.2 In this section we use both iterative maps and differential equation models when convenient. It is simplest to integrate the differential equations by converting them to an iterative map, as long as it is well behaved, by the straightforward method of converting an equation of the form

$$\frac{dN(t)}{dt} = f(N(t))$$
(6.5.7)

to

$$N(t) = N(t-dt) + f(N(t-dt))dt$$
(6.5.8)

and reducing dt until the results are insensitive to it.
Try this for Eq. (6.5.6) and plot the results.

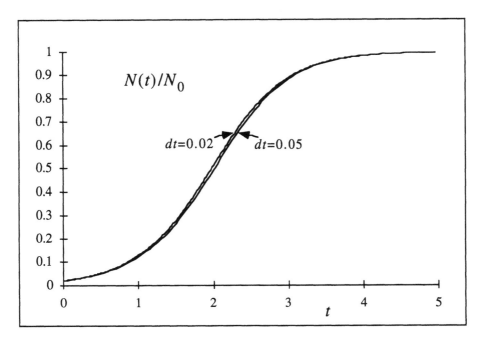

Figure 6.5.1 Solution of the logistic equation (Eq. 6.5.6) with $\lambda = 2$ using an iterative map to perform the integration. When starting from low values, the population increases and saturates at the value N_0. The two curves are for different time increments in the integration (see Question 6.5.2). ∎

Solution 6.5.2 See Fig. 6.5.1. ∎

Q **uestion 6.5.3** Show analytically that Eq. (6.5.6), unlike the quadratic iterative map, should not have chaotic behavior.

Solution 6.5.3 The iterative map corresponds to the equation:

$$N(t+1) = N(t) + \lambda N(t)(1 - N(t)/N_0)dt$$
$$= (1 + \lambda dt)N(t) - (\lambda dt/N_0)N(t)^2 \qquad (6.5.9)$$
$$= (1 + \lambda dt)N(t)(1 - cN(t))$$

where

$$c = \frac{\lambda dt}{N_0(1 + \lambda dt)} \qquad (6.5.10)$$

Defining $s(t) = cN(t)$ we have the same quadratic map as in Section 1.1:

$$s(t+1) = (1 + \lambda dt)s(t)(1 - s(t)) \qquad (6.5.11)$$

where the coefficient can be made incrementally greater than one, which is in the stable regime. ∎

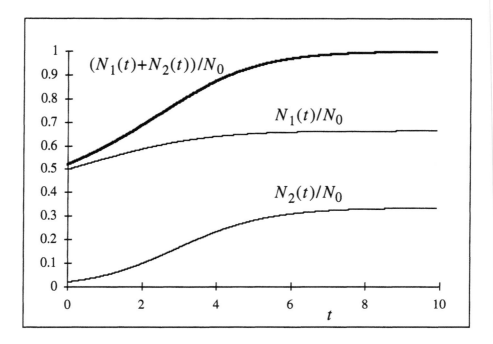

Figure 6.5.2 A model of competition based on Eq. (6.5.6) where two types of organisms are limited to have a total population less than N_0. The first organism has a reproduction rate λ_1 = 0.2 and the second $\lambda_2 = 10\lambda_1$. The initial conditions are set so that the first organism with $N_1 = 0.5N_0$ dominates the second $N_2 = 0.02N_0$. The concept of evolution by selection suggests that the second organism should grow in number and eventually dominate the first organism. However, the figure shows that both populations grow so that the first organism continues to dominate the second. ∎

In order to consider selection between two organisms, we use two equations that describe the growth of each of the populations with the same form as Eq. (6.5.6) but with different growth-rate parameters λ_1 and λ_2:

$$\frac{dN_1(t)}{dt} = \lambda_1 N_1(t)\left(1 - \frac{N_1(t) + N_2(t)}{N_0}\right)$$

$$\frac{dN_2(t)}{dt} = \lambda_2 N_2(t)\left(1 - \frac{N_1(t) + N_2(t)}{N_0}\right)$$

(6.5.12)

To couple the equations, we have assumed that the limitation on the number of organisms applies to both of them together. In solving these equations, our intuitive assumption is that one type of organism will dominate over the other and grow to have most of the population regardless of the initial starting point. However, when we look more closely we see that this cannot be true.

We notice first that if at any time the total population $N_1(t) + N_2(t)$ is N_0, then regardless of the mix of organisms, the number of organisms of each type does not change, because the expression in parenthesis is zero. So we consider instead starting the organisms with a total population below N_0. In this case both populations are monotonically increasing as long as the total population is smaller than N_0. This means that whatever our initial conditions are, the lower growth-rate type of organism will never have fewer than its starting number. This is illustrated in Fig. 6.5.2, where the population of the lower growth-rate type of organism starts at $0.5\,N_0$ and the population of the higher growth-rate type of organism starts at $0.02\,N_0$. We see that it is not possible for the organisms with the higher growth rate to overcome the organisms with the lower growth rate. This does not correspond to our intuition about selection. According to this equation, an organism type that exists cannot be superseded by a newcomer even if the newcomer is reproducing more rapidly.

To try and overcome this problem we might consider the possibility of adding noise that would cause the total population sometimes to be greater than N_0 and sometimes to be less than it. This would cause the populations of the organisms alternately to grow and shrink. Then we might expect to see the higher growth-rate type of organism dominate. In a numerical integration this would look like:

$$N_1(t) = \lambda_1 N_1(t-dt)\left(1 - \frac{N_1(t-dt)+N_2(t-dt)}{N_0}\right)dt + N_1(t-dt) + \delta \bullet (\Xi(t) - 0.5)$$

$$N_2(t) = \lambda_2 N_2(t-dt)\left(1 - \frac{N_1(t-dt)+N_2(t-dt)}{N_0}\right)dt + N_2(t-dt) + \delta \bullet (\Xi(t) - 0.5)$$

$$(6.5.13)$$

where $\Xi(t)$ is a random number in the range 0 to 1 and δ controls the impact of the noise. If we simulate this problem many times, we will find that the faster growing population does not usually dominate. If δ is large enough, there are large fluctuations, and one or the other population might become extinct, but it is the population that starts out with the greater number that survives on average. The reason for this is that Eq. (6.5.12) assumes that the factors λ_1 and λ_2 control the population increase when the total population is less than N_0, and they also control the population decline when the total population is greater than N_0. Thus the faster-growing population is also the faster-declining population when there are too many organisms, and this prevents it from dominating the slower-growing one.

We are now faced with an interesting situation where we have several options. The model as we have constructed it has a built-in assumption about the relationship between the population growth and the population decline of an organism. We could argue that this relationship might not be correct, and introduce a model where there are two parameters; one describing the population growth and one describing the population decline. While this can work, we should learn something more significant: that the rate of population growth in a circumstance of plenty is not the factor that controls the fitness of the organism from an evolutionary perspective. The necessity of introducing an additional parameter demonstrates this. If we introduce another parameter, then an interplay between the two different parameters controls the fitness. Thus, according to our analysis, the reproduction rate by itself does not determine the fitness.

Rather than pursuing a model with a new parameter for population decline, we can consider instead whether there is a different model that better captures what we have in mind when we consider selection. The real difficulty with the model in Eq. (6.5.12), and Eq. (6.5.6) upon which it is based, is the way the limitation on population is implemented.

A more natural model for selection represents organisms in competition for a resource. Instead of limiting the population directly, the population is limited by the resource necessary for reproduction. This resource could be food—e.g., grass that regrows to a limited height after being grazed—or space—e.g., nesting sites that are limited in number but are available again after offspring are grown. We will call this model the renewable-resource model. The amount of resource is measured in elementary units, each of which is sufficient to enable an organism to reproduce. We let $r(t)$ be the amount of resource available at time t. This amount is determined by resource renewal as well as by the amount that is consumed by organisms. If there are no organisms, the amount of resource reaches a maximum value r_0. The resource that is available at time t is assumed to be given on average by:

$$r(t) = r_0 - N(t-1)P(t-1) \qquad (6.5.14)$$

where the available resource has been reduced by the product of the number of organisms at the previous time $N(t-1)$, times the probability that any one of them will consume the resource $P(t-1)$.

Each type of organism is assigned an effectiveness κ, which is the probability that the organism can consume the resource if there is only one available. The probability that it consumes the resource when there are $r(t)$ available is:

$$P(t) = (1 - (1 - \kappa)^{r(t)}) \approx 1 - e^{-\kappa r(t)} \qquad (6.5.15)$$

The latter expression is valid when $r(t)$ is large and κ is small. It is not a very limiting assumption, though we will not need to use it. Finally, the number of organisms at time t is given by:

$$N(t) = \lambda N(t-1)P(t-1) \qquad (6.5.16)$$

which means that each organism that consumes a resource produces λ progeny for the next generation, and then dies. The model described by the three Eqs. (6.5.14)–(6.5.16) is an iterative map that can be used to represent competition for a resource. For a single type of organism, the population grows like the solution of Eq. (6.5.6). This is shown in Fig. 6.5.3. The organism grows until it reaches an equilibrium. However, when we have two organisms, the behavior is quite different from what we found before. Question 6.5.4 describes the construction of equations that generalize this model for two organisms. The results of a simulation show that if we have one organism at equilibrium and add a single organism of a type that has a

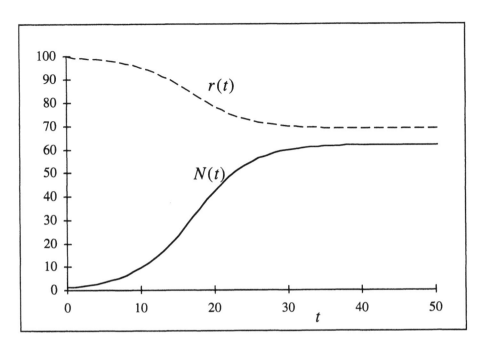

Figure 6.5.3 Renewable-resource model of population growth described by Eqs. (6.5.14)–(6.5.16). The organism population, $N(t)$, grows and saturates in a similar manner to Fig. 6.5.1. The limitation in population growth arises, however, from a reduction in the amount of resources, $r(t)$, consumed by the organism. The parameters used for this simulation are $r_0 = 100$, $\lambda = 2$, and $\kappa = 0.01$, and the initial population is $N(0) = 1$. An incremental version of the model discussed in Question 6.5.5 gives similar results. For other values of the parameters, e.g. higher values of λ, the incremental model is necessary due to chaotic behavior in the original equations. ∎

slightly higher effectiveness κ, or a slightly higher reproduction rate λ, then the new organism will grow and the original organism will become extinct (Fig. 6.5.4).

Question 6.5.4 Write the equations for two types of organisms and simulate their behavior for various initial conditions and parameter values.

Solution 6.5.4 Instead of Eq. (6.5.14) the resource left is:

$$r(t) = r_0 - N_1(t-1)P_1(t-1) - N_2(t-1)P_2(t-1) \qquad (6.5.17)$$

The other two equations are the same as before for each of the organisms:

$$P_1(t) = (1 - (1 - \kappa_1)^{r(t)})$$
$$P_2(t) = (1 - (1 - \kappa_2)^{r(t)})$$
$$N_1(t) = \lambda_1 N_1(t-1)P_1(t-1)$$
$$N_2(t) = \lambda_2 N_2(t-1)P_2(t-1)$$

$(6.5.18)$

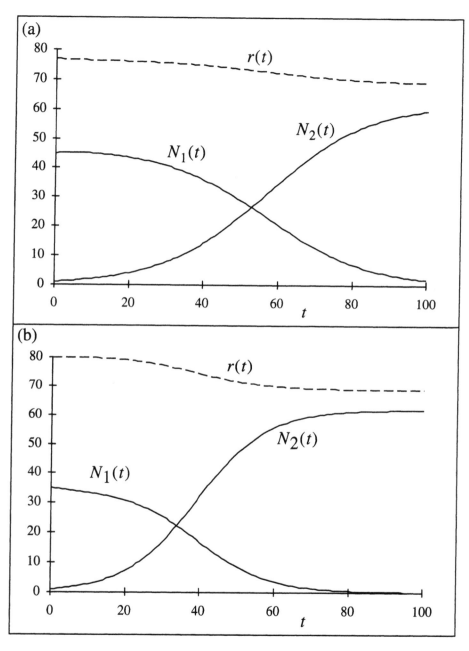

Figure 6.5.4 Renewable-resource model of competition between two organisms showing how the second organism population grows and dominates the first organism. The two figures illustrate different reasons for selection of the second organism over the first. In both cases the second organism has the same parameter values as in Fig. 6.5.3 ($\lambda = 2$, $\kappa = 0.01$). For (a) the first organism has a lower consumption effectiveness, $\kappa = 0.009$. For (b) the first organism has a lower number of offspring per resource consumption $\lambda = 1.8$. The initial conditions are close to, but not equal to, the steady-state value for the first organism. The initial population of the second organism is $N_2(0) = 1$. The baseline resource is set to $r_0 = 100$. ∎

See Fig. 6.5.4 for two simulations for organisms with different values of the parameters. ∎

Question 6.5.5 Eqs. (6.5.14)–(6.5.16) together constitute an iterative map with a tendency to chaotic behavior. The reason for this is that the whole population is being updated at once. We can, however, use a model where both population growth and consumption of the resource occur incrementally. Set up an incremental analog of the iterative map. Hint: The difficulty is in determining how the resource should behave.

Solution 6.5.5 One way to do this is to assume a continuously growing resource that grows in proportion to the amount that is missing:

$$r(t) = r(t-dt) + (r_0 - r(t-dt))dt - N(t-dt)P(t-dt)dt \quad (6.5.19)$$

Eq. (6.5.15) requires no modification and Eq. (6.5.16) becomes:

$$N(t) = N(t-dt) + (\lambda P(t-dt) - 1)N(t-dt)dt \quad (6.5.20) \ \blacksquare$$

We see from Fig. 6.5.4 that this model displays an intuitive behavior of selection of one organism over another. The reason for this behavior can be found by considering the nature of the population control exercised by a resource. For a single organism, the equilibrium population is reached when there is no change in the value of $N(t)$. We can solve the equations in this case directly. Using Eq. (6.5.16) we find that $N(t) = N(t-1)$ implies:

$$1 = \lambda P(\infty) \quad (6.5.21)$$

and from Eq. (6.5.15) that:

$$1/\lambda = (1 - (1-\kappa)^{r(\infty)}) \quad (6.5.22)$$

We can solve this for the amount of resource that is available in equilibrium as:

$$r(\infty) = \frac{\log(1-1/\lambda)}{\log(1-\kappa)} \approx \frac{1}{\lambda\kappa} \quad (6.5.23)$$

The latter expression applies when λ is large and κ is small. The meaning of $r(\infty)$ is that when this amount of resource is available, the population is self-sustaining. This implies that the probability of consumption is enough to generate the same number of organisms in the next generation. We can also conclude that if the amount of resource is less than $r(\infty)$ the population of the organism will fall; if the amount of resource is greater than $r(\infty)$ the population of the organism will grow. The product of the effectiveness of the organism and the reproduction rate sets this equilibrium value of the resource, and the resource controls the population.

Consider what happens when we have two organisms that are competing for the resource. The relevant parameter of each one is their respective $r(\infty)$. This reflects the efficiency of utilization of the resource. The more efficient the organism is, the smaller is $r(\infty)$. The population of the organism that has a higher efficiency will grow at the equilibrium concentration of resource of the organism that is less efficient, while the

population of the organism that is less efficient will shrink at the equilibrium concentration of resource of the organism that is more efficient. Thus the less-efficient
organism must disappear while the more efficient one must increase in number and
dominate the population. Thus, in this model fitness is given by the efficiency of resource utilization:

$$K = 1/r(\infty) \approx \lambda \kappa \qquad (6.5.24)$$

To see how the fitness is distinct from the population of the organism in equilibrium, we can write down the equilibrium population of each type of organism by itself. This is given by:

$$N(\infty) = \lambda(r_0 - r(\infty)) \qquad (6.5.25)$$

This means, reasonably, that the population is the reproduction rate times the amount
of resource that is consumed. We can think about the case where the efficiency of organisms is high so that the residual resource $r(\infty)$ is much smaller than r_0. Then the
population of a type of organism is directly proportional to its reproduction rate λ.
However, this is not the same as the fitness in Eq. (6.5.24). Thus we have found that
starting from a first organism type with an equilibrium population $N_1(\infty)$ we can introduce a second organism type that grows and dominates the first organism type because it has a higher fitness $K_2 > K_1$. But even after the first organism is entirely eliminated, and the second organism has reached its equilibrium population $N_2(\infty)$, we
find that $N_2(\infty) < N_1(\infty)$. Specifically, when $\kappa_2 > \kappa_1$ then the fitness can increase, even
though the total number of organisms decreases because $\lambda_2 < \lambda_1$.

6.5.2 Predators and selection

The discussion of the previous section leads us to consider what happens when one
evolving organism serves as a resource for another organism. A first model that considers a reproducing organism as a resource is the Lotka-Volterra predator-prey
model. This model is a pair of coupled differential equations that describes the exponential growth of a population of prey whose population is limited only by its consumption by a predator. The predator population is limited by the availability of prey,
without which it declines. For convenience we write the prey population as $a(t) =
N_a(t)$ and the predator population as $b(t) = N_b(t)$. The equations are:

$$\frac{da}{dt} = \lambda_a a - \gamma ab$$

$$\qquad (6.5.26)$$

$$\frac{db}{dt} = -\mu b + \lambda_b \gamma ab$$

The parameters are the reproduction rate of the prey λ_a, the probability that predators meeting prey consume them γ, the rate of death of predators in absence of prey
μ, and the number of offspring produced by predators after consumption of prey λ_b.
Solutions of these equations display oscillations as shown in Fig. 6.5.5. These oscillations result from the interplay between the effects of growth of the two organisms.

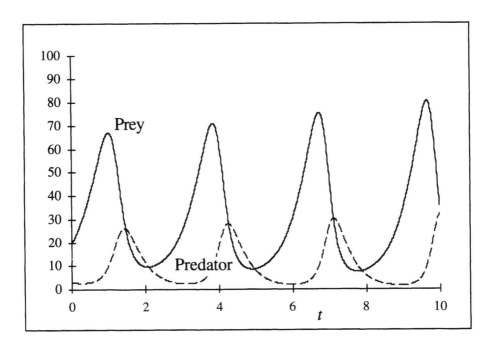

Figure 6.5.5 Simulation of the predator-prey model described by Eq. (6.5.26). The predator and prey populations undergo periodic oscillations as discussed in the text. The parameters are $\lambda_a = 2$, $\gamma = 0.2$, $\mu = 3$, $\lambda_b = 0.5$ and the initial conditions are $a(0) = 20$ prey, and $b(0) = 3$ predators. It is important to recognize that the progressive increase in the height of the peaks is an artifact due to the numerical integration of these equations using Eq. (6.5.8) and a time increment of $dt = 0.01$. A solution using smaller values of time increment would be more closely periodic. An analytic solution of the equations is exactly periodic. This is an illustration of the inherent sensitivity of the predator-prey model to perturbations. ∎

When the prey increases in population, the predator population increases so much that it decreases the prey population, which then results in a decrease in predator population. We can add a second type of prey to this model and see how the fitness selection of the two types of prey would work:

$$\frac{da_1}{dt} = \lambda_1 a_1 - \gamma_1 a_1 b$$

$$\frac{da_2}{dt} = \lambda_2 a_2 - \gamma_2 a_2 b \qquad (6.5.27)$$

$$\frac{db}{dt} = -\mu b + \lambda_b (\gamma_1 a_1 + \gamma_2 a_2) b$$

The result is simulated in Fig. 6.5.6 for several variations in parameters. We see that the prey which has either a higher reproduction rate (larger λ) or a better avoidance

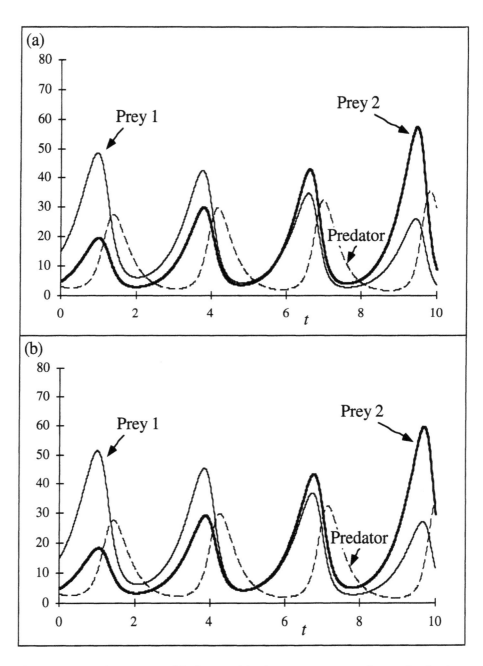

Figure 6.5.6 Predator-prey model of competition between two types of prey, showing how the second type of prey population grows and dominates the first type of prey. The two figures illustrate different reasons for selection of the second type of prey over the first. In both cases the first type of prey has the same parameter values as in Fig. 6.5.5 ($\lambda_a = 2$, $\gamma = 0.2$). For (a) the second type of prey has a higher reproduction rate, $\lambda_a = 2.2$. For (b) the second type of prey has a lower probability of being eaten $\gamma = 0.18$. The initial population of the first and second type of prey are 15 and 5 respectively. ∎

of being eaten (lower γ) will survive and therefore is the fitter organism. It is a combination of these two traits that is the important criteria for fitness. Question 6.5.6 describes a method for obtaining the longer time dynamics of the evolutionary process from these equations. It is significant that in this model, as in the renewable-resource model, it is not just the population growth by itself that is important.

Question 6.5.6 When there are two or more different types of prey whose parameters (λ, γ) differ by a small amount, they together undergo oscillations in population. As this occurs, one of them increases in population at the expense of the others. This longer-time evolutionary dynamics can be separated from the short-time oscillations. Write a differential equation for the longer-time dynamics of the ratio of the populations of two types of prey with incrementally different parameters. Determine the unique parameter that controls the fitness.

Solution 6.5.6 We write the density of the second prey in Eq. (6.5.27) as

$$a_2(t) = \psi(t)a_1(t) \tag{6.5.28}$$

so that $\psi(t)$ is the population ratio. Inserting in Eq. (6.5.27) we obtain:

$$\psi\frac{da_1}{dt} + a_1\frac{d\psi}{dt} = \lambda_2 a_1\psi - \gamma_2 a_1\psi b \tag{6.5.29}$$

Substituting the first prey equation from Eq. (6.5.27) we have:

$$\psi\left(\lambda_1 a_1 - \gamma_1 a_1 b\right) + a_1\frac{d\psi}{dt} = \lambda_2 a_1\psi - \gamma_2 a_1\psi b \tag{6.5.30}$$

or:

$$\frac{d\psi}{dt} = \Delta\lambda\psi - \Delta\gamma b\psi \tag{6.5.31}$$

where $\Delta\lambda = \lambda_2 - \lambda_1$ and $\Delta\gamma = \gamma_2 - \gamma_1$. This equation has the same form as the differential equations describing the prey population. However, since the parameters $\Delta\lambda$ and $\Delta\gamma$ are small, we know that the change in ψ is small, and so we can average the coefficient of ψ on the right over the time that b is fluctuating. This shows that the population ratio changes at a rate controlled by:

$$\Delta\lambda - \Delta\gamma < b > \tag{6.5.32}$$

which means, quite intuitively, that the fitness is controlled by the difference in the reproduction rate minus the average probability that an organism will be eaten over time. ∎

We can consider a similar question to that asked about the renewable-resource model. If a particular prey is replaced by a fitter organism, would the eventual total population of the prey be larger after the change? The result of Question 6.5.6 contained in Eq. (6.5.32) might be wrongly interpreted to mean that with a higher reproduction rate and/or a lower consumption rate, the population of the prey would necessarily increase. However, this is not the case. The average population of the prey is not determined solely by the parameters; it is very sensitive to the initial conditions—how many predator and prey are present at a particular time. Since there is no unique stable equilibrium toward which the equations lead, we cannot define the average prey population directly. We can, however, make some relevant remarks.

For the one steady-state solution of Eq. (6.5.26) obtained by setting the time derivative to zero,

$$a = \mu / \gamma \lambda_b$$
$$b = \lambda_a / \gamma \quad\quad\quad (6.5.33)$$

there is an increase in the value of a with lower consumption rate γ, but there appears to be no effect of its own reproduction rate λ_a. The reason is that the predator population is affected by the rate of increase of the prey population which then affects the prey population. Moreover, for a particular set of initial conditions, it is possible to show (by simulations or by solving the differential equations) that the average prey population does not increase with its reproduction rate.

The predator-prey model with evolving prey can be readily expanded to consider what happens when both the predator and the prey can evolve. This process of evolution of coupled organisms is called coevolution. Its study is a step toward developing an understanding of the network of interdependence discussed in Section 6.4.5. An essential parameter in the fitness of both the prey and the predator is the ability of the predator to eat the prey. Changes in one organism are echoed by changes in the fitness criteria for the other organism, which in turn drive its selection.

The results we have found from the models in this and the previous section contribute to our understanding of fitness and evolution on a more global scale. An important conclusion was the decoupling of fitness from the equilibrium or average number of organisms. As discussed in Section 6.4.6, a relationship between fitness and population, e.g., $P(s) = K(s)$, is in conflict with the idea that selection resulted in evolution to larger, more complex organisms. We know that the number of small, relatively simple organisms greatly exceeds the number of complex organisms. This might suggest that the fitness of the smaller organisms is greater. However, the results that we have found indicate that fitness is not directly related to the number of organisms. In these models, parameters such as the efficiency of resource utilization as well as reproduction rate control the fitness rather than the equilibrium number of organisms. We are still left with the problem of understanding why the presumably less fit small organisms continue to exist in the presence of the more fit complex organisms. This will be addressed in Section 6.5.4.

6.5.3 *Mutation*

In the discussion of selection in the previous sections, we assumed the existence of two types of organisms and investigated the consequences. In this section we consider the process by which changes in organisms occur through mutation. Our objective is to consider implications of the existence of many possible mutations that can occur in an organism. In the context of a simple evolutionary model, we would categorize the effect of these mutations in terms of their effect on fitness. Some mutations improve the fitness, others decrease it. In general, it is also important to allow mutations that do not change the fitness. Moreover, once a mutation has occurred, the organism has changed and the effect of subsequent mutations is contingent on the mutation that has already occurred. We will start, however, by considering only mutations that increase or decrease the fitness by a fixed amount. Of particular significance is the fundamental assumption that mutations occur at random. Mutations occur with a probability that is not affected by the contribution of the mutation to the fitness. This does not mean, however, that mutations that improve fitness are equally likely to those that decrease it.

We simplify the problem by considering what happens if there is a fixed proportion $1/\Lambda$ of mutations that increase fitness for any organism. Moreover, all mutations change the fitness by the same amount up or down. With these assumptions there is no significant difference between two organisms that have distinct genomes or phenomes but the same fitness. Organisms that have the same fitness will coexist and their population will grow or decline together. We can consider together the class of organisms of the same fitness—a fitness class. Our concern is to understand how the population in a fitness class changes with time through the effect of mutation and selection. We have chosen to write $1/\Lambda$ for the proportion of fitness-improving mutations because, due to prior fitness selection, it is less likely to have a mutation that increases fitness to one that decreases it. Thus we expect and assume that Λ is significantly greater than one.

For definiteness we consider an organism that reproduces while consuming a renewable resource as given by Eqs. (6.5.14)–(6.5.16) or, better, their incremental analog (Question 6.5.5). We introduce a certain rate μ at which mutations can occur that change the fitness class of offspring. Each fitness class is identified by its limiting resource $r_i(\infty)$. For simplicity we will consider only variations in the resource utilization effectiveness κ which will be taken to have the value:

$$\kappa_i = g^{i-1}\kappa_1 \tag{6.5.34}$$

where i is the fitness class and g is the ratio of the value of κ_i from one class to the next, assumed to follow a geometric sequence. This is a convenient choice because we will find that the ratio of κ_i determines the relative growth of the population of a fitness class.

The simulation must be performed in such a way that a fractional organism is not allowed to reproduce or mutate. The use of a differential equation can cause problems when care is not taken with this granularity. A set of incremental equations that do account for the granularity are developed in Question 6.5.7.

Question 6.5.7 Write a set of incremental equations based on those in Question 6.5.4 that account for granularity and allow for mutation between a set of fitness classes.

Solution 6.5.7 Two of the relevant equations are:

$$r(t) = r(t - \delta t) + (r_0 - r(t - \delta t))\delta t - \sum_i N_i(t - \delta t)P_i(t - \delta t)\delta t \quad (6.5.35)$$

$$P_i(t) = (1 - (1 - \kappa_i)^{r(t)}) \quad (6.5.36)$$

A subtlety in setting up the equation for the number of organisms in a class is realizing that mutation into a class should be treated probabilistically. Specifically, at any step there is a certain probability of mutation. When a mutation occurs, one organism moves from one class to another. If we naively try to make the continuum equations deterministic, we would introduce a fractional transfer of organisms. This can be treated by accumulating fractional organisms inside a class but not using them for reproduction or mutation. When the fraction of an organism reaches a whole organism, then we do use it. This corresponds, on average, to the moment at which one organism in a stochastic process would have reached there.

The number of offspring that would arise in a single generation of the organism in class i is given by:

$$O_i(t) = \lambda_i P_i(t) \lfloor N_i(t) \rfloor \quad (6.5.37)$$

where $\lfloor x \rfloor$ indicates the integer part of x. Some of these offspring will mutate to another class—specifically, μO_i will. To write an incremental model we assume that only a fraction dt of the organisms reproduce at once and we have:

$$
\begin{aligned}
N_i(t) = N_i(t - dt) + (O_i(t - dt) - \lfloor N_i(t - dt) \rfloor)dt \\
+ \mu\left(\frac{1}{\Lambda + 1}O_{i-1}(t - dt) + \frac{\Lambda}{\Lambda + 1}O_{i+1}(t - dt) - O_i(t - dt)\right)dt
\end{aligned}
\quad (6.5.38)
$$

The subtraction of $\lfloor N_i(t - dt) \rfloor$ in the first line corresponds to the assumption that the parent dies when the offspring are born. The second line describes the effect of mutation, where a fraction μ of the offspring of class i mutate and leave the fitness class. Of these $\mu/(\Lambda + 1)$ go to the next higher fitness class and $\mu\Lambda/(\Lambda + 1)$ go to the next lower one. The equation is written in terms of the changes in the ith fitness class due to mutations from the $i + 1$ and $i - 1$ classes. We can see that this part of the equation corresponds to a biased diffusion of population in fitness classes. ∎

A simulation of the model of mutation is shown in Fig. 6.5.7. As mutations occur, the fitness class of the organisms increase. We might imagine this process as accounting for some of the historical fossil record where over many years an organism changes monotonically from one form to another. However, since there is no specific trait or traits assumed to be associated with the mutations, this is also a general description of evolutionary progress.

There are several interesting features of this model that we can understand by considering the effect of various parameters. First we should recognize that there is a finite range of possible fitness classes. This range is set by the total amount of the re-

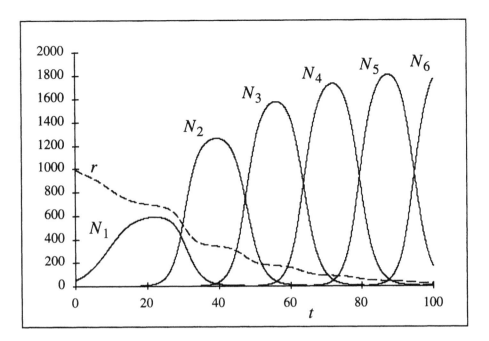

Figure 6.5.7 Model of evolutionary progress by mutation and selection based upon the renewable-resource model. Mutation enables organisms to move from one fitness class to another. The improved resource utilization by the higher fitness classes causes their population to increase and dominate the lower fitness classes, as the amount of resource available, r, declines due to its utilization. In this simulation the base resource is $r_0 = 10^3$, all fitness classes have $\lambda = 2$ offspring per unit of consumed resource, the first fitness class has a resource utilization effectiveness $\kappa_1 = 10^{-3}$, the resource utilization effectiveness of each successive fitness class is multiplied by $g = 2$, the ratio of fitness improving mutations to fitness reducing mutations is $1/\Lambda = 1/4$, and the mutation rate is $\mu = 10^{-3}$. The first fitness class starts with 50 organisms and all others start with none. The time increment for integration is $dt = 0.05$. ■

source r_0. The smallest value of κ which makes sense is $\kappa_{min} = 1/(r_0\lambda)$. A smaller value results in an organism that is not viable with this amount of resource. Recognizing that there is a lower bound to the viability of an organism is important. It is related to the problem of creating the first viable organism. On the other end of the scale there is also a maximum fitness that arises when there is only one unit of resource left. This gives $\kappa_{max} = 1/\lambda$. It is impossible for an organism to improve further because there is no resource to be consumed. Thus, a finite amount of resource leads to a bound on how much improvement in fitness is possible.

We can gain additional insight into the behavior of this model by determining the rate of evolutionary progress—the time for fitness class $i-1$ to be replaced by fitness class i. There are two parts to this process the first is the time τ_1 till a first organism appears in class i and the second is the time τ_2 till its population becomes dominant. We can make a complete analysis when τ_1 is longer than τ_2. In this case equilibrium is reached in class $i-1$ and it dominates the population of organisms before mutation creates organisms in class i.

To study the condition of equilibrium, we use Eq. (6.5.38) to describe the time dependence of class $i-1$ by shifting i to $i-1$ everywhere. We can then impose the equilibrium condition, $N_{i-1}(t) = N_{i-1}(t-1)$. The resulting equation simplifies because when class $i-1$ is dominant the population of other classes is negligible; also, we don't need to take the integer part of $N_{i-1}(t)$. We find:

$$N_{i-1}(t) = (1-\mu)O_{i-1}(t) \qquad (6.5.39)$$

which says that the offspring that do not mutate replace their parents. From Eq. (6.5.37) we have:

$$1 = (1-\mu)\lambda_{i-1}P_{i-1}(t) \qquad (6.5.40)$$

From the resource equation Eq. (6.5.35), with $r(t) = r(t-1)$, we can obtain a value for $N_{i-1}(t)$:

$$N_{i-1}(t) = (r_0 - r(t))/P_{i-1}(t) = (1-\mu)\lambda_{i-1}(r_0 - r(t)) \approx (1-\mu)\lambda_{i-1}r_0 \qquad (6.5.41)$$

The latter approximation holds unless the organism is just marginally viable.

The time τ_1 to create a first organism in class i is determined by Eq. (6.5.38) with all of the terms equal to zero $(\lfloor N_i(t)\rfloor = 0)$ except for the contribution by mutation from class $i-1$:

$$N_i(t) - N_i(t-1) = \frac{\mu}{\Lambda+1}O_{i-1}(t-dt)dt \approx \frac{\mu\lambda_{i-1}r_0}{\Lambda+1}dt \qquad (6.5.42)$$

This equation is linear, so the time to reach a single organism τ_1 is:

$$\tau_1 = (\Lambda+1)/\mu\lambda_{i-1}r_0 \qquad (6.5.43)$$

This expression says that the time to obtain a single organism in class i is proportional to the difficulty in finding a fitness-improving mutation, and inversely related to the number of mutated offspring per generation produced by fitness class $i-1$.

Once class i has an organism, we can neglect mutation from class $i-1$, because $N_i(t)$ grows by reproduction. Moreover, now that class i has more than one organism, it is not essential to take the integer part of $N_i(t)$. $N_i(t)$ grows according to (Eq. 6.5.38):

$$N_i(t) - N_i(t-dt) = ((1-\mu)\lambda_iP_i(t) - 1)N_i(t-dt)dt \qquad (6.5.44)$$

To solve this we recognize that the amount of resource available during the growth of $N_i(t)$ is determined by the equilibrium resource of the fitness class $i-1$. It is essentially independent of time, and therefore so is $P_i(t)$. From Eq. (6.5.36) the equilibrium resource of fitness class $i-1$ is:

$$r(t) = \frac{\ln(1-1/(1-\mu)\lambda_{i-1})}{\ln(1-\kappa_{i-1})} \approx \frac{1}{(1-\mu)\lambda_{i-1}\kappa_{i-1}} \qquad (6.5.45)$$

Then we have:

$$(1-\mu)\lambda_iP_i(t) \approx (1-\mu)\lambda_i(1-(1-\kappa_i)^{r(t)}) \approx (1-\mu)\lambda_i\kappa_ir(t) \approx \frac{\lambda_i\kappa_i}{\lambda_{i-1}\kappa_{i-1}} = g \qquad (6.5.46)$$

where we have used approximations to simplify the form of the result. Using this in Eq. (6.5.44) we have exponential population growth in class i:

$$N_i(t) \propto e^{t/(g-1)} \qquad (6.5.47)$$

τ_2 is the time for the population of class i to grow from a single organism to the equilibrium population of class $i-1$. This is given by:

$$N_{i-1}(t) = e^{\tau_2/(g-1)} \qquad (6.5.48)$$

or:

$$\tau_2 = \ln(N_{i-1})/\ln(g-1) = \ln((1-\mu)\lambda_{i-1}r_0)/\ln(g-1) \qquad (6.5.49)$$

We conclude that the total time for a change of fitness class is given by (setting $\lambda_i = \lambda$):

$$\tau = \tau_1 + \tau_2 = (\Lambda+1)/r_0\lambda\mu + \ln((1-\mu)\lambda r_0)/\ln(g-1) \qquad (6.5.50)$$

This is the evolution time between fitness classes. It becomes invalid when the second term becomes large enough compared to the first that significant growth of class i occurs before the growth of the class $i-1$ is completed.

We can develop an understanding of Eq. (6.5.50) by realizing that the first term is large compared to the second term when the mutation rate μ is small or the probability of finding a fitness-improving mutation is small (Λ is large). In this case, the organisms evolve in distinct stages where a fitness class replaces the one immediately preceding it. If Λ is not too large and the mutation rate becomes high enough (it cannot be greater than one), τ_1 may become shorter than τ_2. In this case there are several overlapping classes that exist at the same time, and Eq. (6.5.50) is no longer valid. Fig. 6.5.8 illustrates the latter case, where at any time there is a heterogeneous population of organisms undergoing selection.

The model of mutation and selection appears in its overall behavior to be similar to the Monte Carlo random-walk model of downhill diffusion that was discussed in Section 6.4. However, there are a number of differences between these two models. The most important difference is the role of the rarity of fitness-improving mutations (phase space). In the Monte Carlo model we can analyze its role through the properties of equilibrium. In equilibrium the number of organisms that mutate from class $i-1$ to class i is the same as from class i to class $i-1$. The relative number of organisms in equilibrium in the different classes is set by this condition. We can calculate the number of mutating organisms in the random-walk model using the parameters of the mutation and selection model. In each time step, a walker chooses one of the possible mutations. The proportion of these that improve the fitness is $1/(\Lambda+1)$, while the proportion that decrease it is $\Lambda/(\Lambda+1)$. All of the mutations that improve the fitness are accepted, but only $K(i-1)/K(i) = 1/g$ of those that decrease the fitness are accepted. This means that in equilibrium the proportion of the population in class $i-1$ and class i is given by:

$$N_i/N_{i-1} = g/\Lambda \qquad (6.5.51)$$

This means that the population of the lower fitness class will be larger if the number of fitness-improving mutations is sufficiently small. If we think about dynamics,

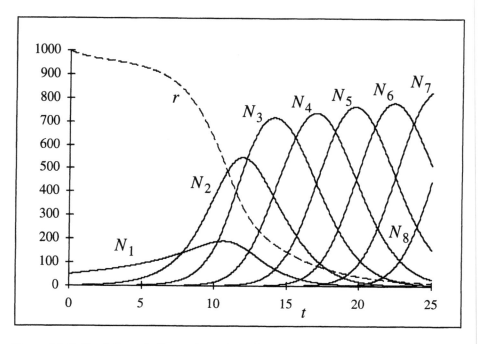

Figure 6.5.8 Simulations similar to those shown in Fig. 6.5.7. The only difference is that a higher mutation rate $\mu = 10^{-1}$ was used. The fitness classes overlap because each one does not reach a steady state population before the next one arises. ∎

under these circumstances the evolution will progress uphill rather than downhill. The reason for this is that the balance between entropy and energy is being won by the entropy of the much greater number of lower-fitness organisms.

This conclusion is not true for the reproduction and selection model. The time that it takes to improve the fitness Eq. (6.5.50) increases with increasing Λ. However, for any value of Λ the fitness increases. This is an important result for our understanding of evolution. It means that selection with reproduction is more powerful than entropy. Our understanding becomes more complete if we recognize that the advance in fitness does stop when the resource is scarce—when the fitness reaches κ_{max} so that the amount of resource is a single unit. Thus it is the nonequilibrium driving force of resource consumption that plays a different and more powerful role than a difference in energy or entropy.

A related difference between the two models arises when we consider the possibility that an individual organism will move counter to evolutionary progress—downward in fitness or upward in energy. In a reproduction and selection model, the possibility of movement to a significantly lower fitness class is vanishingly small. This is because steps downward in fitness become progressively more and more difficult. A step downward consists of two parts, a mutation downward and a successful reproduction in the lower fitness class. The first part does not depend on which fitness class we start from. However, the second decreases for lower-fitness classes because the reproduction rate is controlled by the available resource which is controlled by the dominant fitness class. In contrast, Monte Carlo steps upward in energy have the same probability no matter what the starting energy is. If we think about the fitness landscape as formed of valleys and ridges, this difference in model behavior is directly relevant to the possibility of climbing over ridges to find other valleys. In the Monte Carlo model, it may be possible. In the reproduction and selection model, it is very unlikely. This is also related to the observation that in the Monte Carlo model the population tends to spread out on an incline. In the reproduction and selection model this is not the case.

Finally, there is also a difference in the effect of the absolute population size in the two models. In a simple Monte Carlo model where each walker moves independent of the others, the population size does not enter in any way. When there is self-attraction of the organisms (Section 6.4.5), it plays a role in random movement on a flat landscape—the motion is faster for smaller populations. However, population size does not play a role in the rate of evolution on an incline. On the other hand, in the reproduction and selection model, the probability of finding a rare mutation per generation increases with population size. Thus the rate of evolution increases almost linearly with population size when the probability of finding the right mutation is small. We can also think about the reproduction and selection model as a kind of fitness optimization algorithm. The linear increase in rate of evolution with population size implies that it works as an efficient parallel algorithm where each processor (organism) contributes to the optimization.

Qualitatively our conclusions from this section are that the process of reproduction and selection is effective at finding rare fitness-improving mutations and therefore is effective at forcing evolutionary progress against the influence of entropy. This is precisely what is needed to generate complex organisms. However, we also find that reproduction and selection tend to drastically confine the exploration of possible organisms to a steepest descent in the fitness landscape. Thus, evolutionary progress should become stuck in the first fitness valley that is encountered, and organism change will no longer be possible. This problem leads to even more dramatic consequences when we consider it in the context of trait divergence in the next section.

6.5.4 *Trait divergence, extinction and the tree of life*

The incremental process of evolution by mutation and selection described in the previous section must be accompanied by a discussion of trait divergence in order to account for the phenomenology of life. As discussed before and illustrated in Fig. 6.5.9, it is assumed that all organisms, ranging from single-celled organisms to plants and animals, originated from the same microorganisms early in evolutionary history. This requires a process of divergent evolution. At various moments in time, originally similar organisms evolved in different ways to create (at least) two types of organisms from the original one. The evidence in support of this picture includes the similarity of various organisms in their various levels of structure (chemical, cellular, physiological) and the experience of breeding where distinct varieties can be generated. The

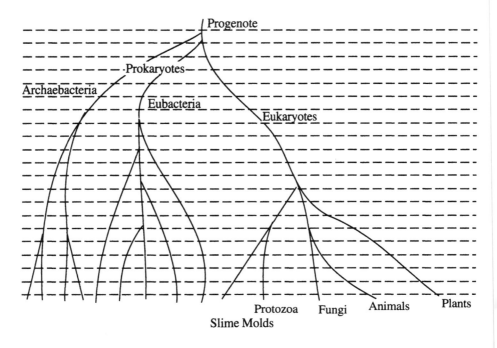

Figure 6.5.9 Schematic illustration of the tree of life formed by evolution and trait divergence, resulting in all of the diversity of life, and originating in one type of single-celled organism. As discussed in the text, a model of reproduction and selection that requires competition based on fitness between all organisms would not be able to account for trait divergence, because at every time all of the species must have essentially the same fitness. This would require all to evolve in precise lockstep. In order to account for trait divergence, selection must be understood to apply separately to organisms that consume distinct resources—identified both by resource type and geographical location. ∎

relative proximity of organisms on the tree of life has been studied by assuming that the number of differences in the genome reflects their distance on the evolutionary tree. In Section 6.4.6 we discussed several global models, flow downhill, uphill motion and expanding waves, which might be considered in the context of Monte Carlo models for speciation. What can we say about the reproduction-and-selection models discussed in this section?

Remarkably, we can point out that any of the reproduction-and-selection models discussed in this section are inconsistent with trait divergence. For definiteness, consider the reproduction-and-selection model based upon resource utilization in Section 6.5.3. We can introduce branching channels for organisms in a pattern consistent with that shown in Fig. 6.5.9. Progress along any channel is directly mapped onto mutations that increase fitness. Thus we identify a particular fitness class with the horizontal lines in Fig. 6.5.9. Starting from an organism at the beginning of the tree, the organism evolves downward to the first branching. Then some organisms move to one channel and some to the other. The problem with this model arises in the interplay between organisms evolving in the two channels. Assume that organisms in one channel, even by chance, progress to a higher class slightly before they do so in the second channel. Then the organisms in the second channel will be rapidly suppressed by selection. This suppression is an accelerating process, because once there are fewer organisms, they have less chance to reach the subsequent fitness class. Thus, unless we could demonstrate a reason for all organisms to evolve in lockstep, there is no possibility that the reproduction-and-selection model will allow the coexistence of distinct types of organisms. This is the literal conclusion of "survival of the fittest"—only one type of organism can exist. This conclusion applies not only to the renewable-resource model but also to the simplest reproduction-and-selection model in Eq. (6.5.5), and to the predator-prey model for the evolution of prey (or predators) in Section 6.5.2.

Question 6.5.8 Contrast the formation of diverse organism traits in the Monte Carlo model with its absence in the reproduction-and-selection models. What is the key difference?

Solution 6.5.8 The key difference is the scope of selection. In the Monte Carlo model, selection only occurs between an organism and its one mutated offspring. In the reproduction-and-selection model, selection occurs between all organisms at the same time. In the first case, we can have different traits for every walker. In the second, essentially only one organism type is possible. ∎

Question 6.5.9 Consider neutral mutations in the model of the last section. These mutations consist of additional genome or phenome dimensions in which the organisms can change without affecting their fitness. How does the population evolve in these dimensions? For simplicity, consider the case where Λ is large.

Solution 6.5.9 The trick is to recognize that evolution in the direction of increasing fitness affects the population movement in the neutral dimensions as well. Thus we cannot describe the evolution of the population as a diffusion in the neutral dimensions.

The case where Λ is large means that fitness-improving mutations are rare. As a result there is essentially only one organism that mutates to a higher fitness class. This organism reproduces to form the entire population at the next fitness class. Thus the population of every fitness class begins from a unique genome. During the proliferation of the organisms in this class, they diffuse in the neutral dimensions to form a more diverse population. Then a mutation takes one organism to the next fitness class, and the process of spreading starts over. Thus, the width of the distribution in the neutral dimensions is limited to be the width of a random walk that occurs in the time to reach the next fitness class; i.e., it is proportional to $\sqrt{\tau}$. In our model where Λ does not change, the width of the distribution also does not change as the organisms evolve. This is consistent with the observation that this model precludes the coexistence of distinct organism types.

However, there is a change over time in the average location of the population in the neutral dimensions. At the time a fitness-improving mutation occurs, the population consists of a distribution of width $\sqrt{\tau}$. From this distribution, one organism mutates to the higher fitness class. Thus in the next class the average location of the population in the neutral dimensions will be different from that in the previous class by a distance proportional to $\sqrt{\tau}$. We see that the population as a whole undergoes a random walk in the neutral dimensions. The typical distance traveled in the random walk is proportional to $\sqrt{\tau n} = \sqrt{t}$, where n is the total change in fitness class. This means that the typical distance traveled in the neutral dimensions is independent of τ. ∎

Where have we neglected an essential element in our models that would enable evolutionary coexistence of organism types? By considering natural phenomena, we recognize that the main problem in this picture is that all organisms do not compete directly for the same resource. Instead, there are many different resources that organisms are consuming. The primary resource is the energy that is arriving from the sun and radiated as heat into space. This resource is converted by interaction with the physical world, as well as with the biological organisms that exist, to other forms. The utilization of this resource by one organism type (e.g., plants) leads to another resource type (oxygen, sugar) that can be used by other organisms. The interplay of this process with physical climate and geologic conditions also leads to variations in the form the resource takes and conditions under which it can be utilized. In addition to their distinct forms, resources may also be distinct through spatial physical isolation—the separation of two different areas by physical obstacles that prevent easy migration from one to the other. This variation and isolation of resources leads to distinct channels of evolution related to their utilization.

We conclude that it is the existence of different resources that enables distinct organism types. Thus, to simulate the formation of different types in a model similar to that in Section 6.5.3, we must modify it to allow for the existence of distinct resources, say r_1 and r_2. These may represent different sides of the same mountain range, different types of grass, or grass and leaves or even sunlight and plants. To allow the formation of different types of organisms, mutation (or migration) must then allow creation of organisms that pursue these distinct resources. Construction of a model of this kind is not difficult. We assume a number of organism types indexed by i that consume the two resources with efficiencies κ_{i1} and κ_{i2}. Trait divergence would occur when organisms consume the resources in a manner that would systematically cause one type of evolving organism to consume one resource, while the other consumes the other resource. In the simpler model of Eq. (6.5.5) the same effect would be modeled by introducing more than one fitness parameter per organism. We will not expend effort to build and simulate such models here. In Chapter 7, models will arise naturally that contain the essential features discussed here.

Question 6.5.10 Let us try one more time to create a global fitness model using an equation similar to Eq. (6.5.4) to define what we mean by fitness in the most direct and natural way. The fitness K is the rate of increase in the population of the organism defined as:

$$\frac{dN}{dt} = KN \qquad (6.5.52)$$

In what way might this be a useful definition? In what way is it not?

Solution 6.5.10 The concept of a fitness assumes that it can be expressed in terms of the organism properties—in particular, as a function of the genome s. To the extent that we can obtain $K(s)$ independent of the evolutionary dynamics we wish to describe, it is useful. If we use the definition in Eq. (6.5.52) in the context of the model of Section 6.5.3, we see that for any fitness class, $K(s)$ starts out greater than one, passes through unity when it is populated and becomes smaller than unity as it disappears. This shows that the fitness defined in this natural way is a strong function of time through changes in the environment, which also consists of other organisms.

Using Eq. (6.5.52) as a model for global evolution, we imagine a fitness landscape $-K(s)$ (the negative sign is for consistency of up and down with the energy model). Organisms do not diffuse downward on the landscape; instead the landscape itself moves upward. This is not an essential difference. What is different is that all the organisms at any time are points located on a band near unit fitness. Organisms increasing in number have slightly higher fitnesses (are lower on the landscape) and those decreasing have slightly lower fitnesses. This picture would be appealing and simple if the landscape were rigid. It would then correspond to all organisms evolving uniformly in lockstep. However, if we consider an organism that persists for long times and organisms that undergo dramatic evolutionary changes during the same time, we see that the landscape itself is changing shape (morphing). Regions of the landscape where organisms are evolving move quickly, while other regions remain fixed in place. This illustrates how the existence of multiple resources manifests itself in this model. ∎

The necessity of considering multiple resources in the study of multiple organism types is an indication that an essential problem in studying evolution is understanding the dynamics of the resources and their categorizations and distinctions. Since most of the time, aside from sunlight and space, the resources that are consumed are themselves organisms or related to the existence of organisms, this creates an interdependence between the evolution of one organism and the evolution of others. For our limited purposes here, we need only recognize that without a model for the dynamics of the resources, a dynamics of organism types cannot be understood. Within such a model, formation of distinct organism types can be readily understood.

Another process that may be understood within this picture is extinction. There are various times in evolutionary history where organism types that existed become extinct. We can understand this by assuming that after multiple organism types are created, there is parallel evolution of organisms that are consuming different resources. Some time later, one organism type may have a mutation that enables it to better consume a resource that the second organism type depends upon. This leads to extinction of the second organism type.

A process that is rarely considered is true convergent evolution: two organisms evolve by mutation and become the same organism. This is the opposite of divergent evolution. There are several reasons why it is not likely. The first is the large space of possible organisms. Moving around in this space, two evolutionary tracks are unlikely to encounter each other. Another reason that this is not likely can be seen from the multiple-resource model. As two organisms become similar they become competitors for the same resource. However, as this occurs they are still different enough so that one should have a lower fitness than the other and will become extinct.

In order to understand the role of different resources in causing multiple organism types, it is helpful to consider the notion of organism complexity, which will be developed further in Chapter 8. We can think about a particular environment and resource as establishing a particular demand on organisms. As discussed above, associated with an amount of resource is a minimum fitness that corresponds to the first viable organism that can survive by utilizing it. This also means that there is a minimal complexity for viable organisms. This is the minimum complexity of organisms that have sufficient fitness to survive in the environment by consuming this resource. For example, it appears that there are photosynthetic organisms that can exist in the ocean which are simpler than photosynthetic life on land. We might imagine a map of the minimal complexity of viable organisms at every location on the two-dimensional surface of the earth, for each of multiple resources present there. We can view the process of evolutionary progress as the creation of new organisms that are complex enough to exist in a certain environment and consume a particular resource. Progress enables organisms to spread from one environment to another. One example is the often discussed emergence of life from ocean to land. Once life exists that consumes a particular resource in a particular place, evolution continues until it reaches the maximal fitness for the resource. This maximal fitness is the lowest fitness that enables organisms to consume all of the resource. It also corresponds to a maximum complexity that would be reached by evolution in this environment. This picture may be described as the evolution of organisms to fill ecological niches that exist due to the presence of resources. We can intuitively understand that in order for an organism to be able to consume more than one type of resource it must be more complex than an organism that consumes only one type of resource. Qualitatively, this explains why evolution created progressively more complex organisms but did not systematically eliminate the previous organisms. Organisms that existed and filled ecological niches remain. This also includes the prokaryote single-celled organisms that may have initiated the process of organism evolution.

This description of evolution enables us to compare and contrast human beings with the most closely related species—the apes. Apes occupy small ecological niches and consume limited resources. In contrast, human beings utilize a significant fraction of many different resources in many distinct environments. This suggests that apes evolved in order to be able to utilize resources that can only be consumed by such highly complex organisms. Human beings break this pattern by being sufficiently complex to meet the challenges of consuming a large variety of resources. Within this context, we can understand the potential and actual extinction of various organisms as a result of the actions of mankind in consuming resources. Human consumption of plants and animals is in part a predator-prey situation where overconsumption can lead to decline of the predator (human beings) as well as the prey. On the other hand, the direct competition for resources (prey or space) causes extinction or danger of extinction to many animals and plants. This is consistent with considering human beings as part of the evolutionary process where the increased fitness of human beings through their ability to consume resources is an advantage over animals and plants and may cause extinction of the latter. Of course, an explanation for extinction does not mean that it is in the best interest of human beings. However, we see that the widespread ability to cause extinction for many different organisms signals a qualitative change in what is more typical of evolution. Organisms typically evolve and are complex enough to occupy only specific ecological niches.

The discussion of multiple resources allowed us to introduce spatial variation in resources and thus in fitness. It is also relevant to discuss temporal variation in resources. We consider the possibility of a fluctuating base line resource $r_0(t)$ for the renewable-resource model. This can be seen to cause a variation in the selection pressure. Without any variation, the selection pressure is great due to consumption of the available resource by the fittest organisms present. If, however, the resource suddenly becomes more plentiful, then the amount of resource $r(t)$ is more than the equilibrium value $r_i(\infty)$ for organisms in lower fitness classes. The organisms in these classes then increase in numbers. It is less significant that the fittest organisms multiply faster. The increase in population of lower-fitness organisms allows, in principle, for the possibility of escape from valleys in the fitness landscape. This solves a major problem of describing evolutionary progress in the reproduction-and-selection model. An illustrative example is the effect of forest fires. Originally thought to be solely harmful, the occasional loss of old trees is now understood to have many benefits. We discussed the effects of forest fires in the context of interdependent networks of organisms in Section 6.4.5. Here, the effect of such catastrophes is direct. We can recognize that a fire enables a larger variety of plant life and other organisms to grow in the plentiful resource (sunlight) whose consumption is not dominated by a particular type of optimal-fitness tree. This increase in diversity of organisms participates in the process of evolution through creation of variety that can then be the subject of selection as the forest matures.

One final temporal consideration is important when we consider the competition that causes extinctions. If we consider competition between two species, we see that the organism type that evolves faster has an advantage when competition for the same resource occurs. Thus, fitness of a species can depend not only on the fitness of a particular organism but also of the rate of organism evolution. This can be directly affected by the lifetime of a generation that sets the time scale for the models of evolutionary progress. We see that selection may cause organisms to have short life spans—a possible reason for senescence. Even more significantly, considering these effects causes us to look beyond the individual to the effects of selection on species, since the rate of evolutionary progress is not part of the fitness of a particular organism. This will be the topic of the following section.

6.6   Collective Evolution: Genes, Organisms and Populations

Our objective in this section is to reconsider the properties of components in complex systems and their relevance to evolution. Sexual reproduction involves an entire species in an evolutionary process rather than each individual organism. Interestingly, it is possible to take another approach in which sexual reproduction decouples the evolution to a process that pertains to individual genes that are parts of the organism. We will discuss this approach in Section 6.6.1, developing an understanding of when (and in what sense) it is valid and when it is invalid, since we are interested in the nature of interdependence of components of complex systems. In Section 6.6.2 we will approach more generally the problem of understanding why systems built out of components are formed in evolution. Why don't the components just fend for themselves? This question is related to philosophical questions about selfish and altruistic individuals in a society, and conceptual problems of understanding the appearance of altruism in evolution. By addressing these questions, we will gain an intuitive understanding of the process of formation of collectives in evolution.

6.6.1 *Genetic mixing in sexual reproduction*

One of the interesting phenomena of biology is the existence of sexual reproduction in all but relatively simple organisms. Sexual reproduction mixes hereditary traits. This mixing poses serious philosophical problems for the understanding of evolution. Simply stated, the problem is that in sexual reproduction the organism that is selected for is not the same as its offspring. How is this consistent with the concept of selection as a mechanism of evolution?

One of the approaches that has been taken to deal with this issue is the gene-centered view of evolution. In this view there are assumed to be indivisible elementary units of the genome (often thought of as individual genes) that are preserved from generation to generation. Different versions of the gene (alleles) compete and mutate rather than the organism as a whole. Thus the genes are the subject of evolution. We will show below that this view is precisely equivalent to a mean field approach (Section 1.6) where correlations between the different genes are ignored. Each gene evolves in an effective environment formed within the organism and its environment. This effective environment is an average environment (mean field) within a sexually reproducing population (species). By showing that the gene-centered view of evolution is a mean field approach, we can recognize why it is useful and we can also recognize when it is invalid—when correlations between genes are significant.

Correlations between genes arise when the presence of one allele in one place in the genome affects the probability of another allele appearing in another place in the genome. One of the confusing points about the gene-centered theory is that there are two stages in which correlations must be considered: selection and sexual reproduction (gene mixing). Correlations occur in selection when the probability of survival favors certain combinations of alleles, rather than being determined by a product of terms given by each allele separately. Correlations occur in reproduction when parents are more likely to mate if they have certain combinations of alleles. If correlations only occur in selection and not in reproduction, the mean field approach continues to be at least partially valid. However, if there are correlations in both selection and sexual reproduction, then the mean field approach and the gene-centered view becomes completely invalid. It is sufficient for there to be very weak correlations in sexual reproduction for the breakdown to occur. This turns out to be particularly relevant to trait divergence of populations.

In order to understand the gene-centered view we will study a simple model of the process of sexual reproduction that explicitly eliminates correlations in reproduction. Two specific examples will be worked out in some detail. Then we will discuss a more complete theory showing that the simple theory is a mean field approach. Later we will present reasons for the existence of sexual reproduction. It is helpful to recall that during sexual reproduction an offspring obtains half of the chromosomes of nuclear DNA from each parent. The chromosomes are paired in function—homologous pairs. Each homologue chromosome of the offspring is formed in a parent by a process (crossover during meiosis) that combines segments of DNA from both of the parents' homologues.

A first model of sexual reproduction begins by assuming that recombination of the genome components during sexual reproduction results in a complete mixing of the possible alleles, not just in the organism itself but rather throughout the species— the group of organisms that is mating and reproducing. Thus the offspring represent all possible combinations of the genomes from reproducing organisms. From the point of view of a particular allele at a particular gene, the complete mixing means that at all other genes, alleles will be present in the same proportion that they appear in the population—there are no allele correlations after reproduction.

In the simple model, selection operates on the entire genome of the organism. Thus, after selection there may be correlations in the allele populations. It is assumed that the reproducing organisms are the ones that have successfully survived the process of selection. If we would further simplify this model by assuming that each gene controls a particular phenomic trait for which selection occurs independent of other genes, then each gene would evolve independently; a selected gene reproduces itself and its presence within an organism is irrelevant. The existence of a gene as part of an organism means, however, that selection occurs on the genome, not on individual genes, and allele correlations after selection will occur. This means that fitness depends not on individual genes but rather on gene combinations. As the proportion of one allele in the population changes due to evolution, the fitness of another allele at a different gene will be affected. However, due to the assumption of complete mixing in sexual reproduction, only the average effect (mean field) of one gene on another is relevant. We could consider the organism to be part of the changing environment in which the gene evolves. The following two examples will help us examine this more carefully.

The first example we discuss is the special case of interdependence of two homologue genes. This is special because the same alleles are found in both genes. We allow there to be only two different alleles . The evolutionary dynamics describes the proportion of genes with each allele in the population. The proportion of the alleles is given by $P_1(t)$ and $P_{-1}(t) = 1 - P_1(t)$. An individual organism has two homologue genes and may be either homozygous with both of the same kind or heterozygous with one of each. Using our assumption of random mixing during reproduction, offspring represent the ensemble of possible combinations of the alleles; the specific composition of the parent generation cannot matter to the composition of the offspring. Thus the offspring organisms are in proportions:

$$P_{1,1}(t) = P_1(t-1)^2$$
$$P_{1,-1}(t) = 2P_1(t-1)(1 - P_1(t-1)) \qquad (6.6.1)$$
$$P_{-1,-1}(t) = (1 - P_1(t-1))^2$$

where $P_{i,j}$ is the proportion of an organism with i and j alleles. If there is no selection bias, these organisms will reproduce to form the subsequent generation. We confirm that the proportion of alleles is unchanged from generation to generation (Hardy-Weinberg theorem). The proportion of one allele is given by

$$P_1(t) = \frac{1}{2}\left(2P_{1,1}(t) + P_{1,-1}(t)\right) \qquad (6.6.2)$$

where the prefactor of 1/2 comes from normalization of the probability because there are two alleles per organism. From Eq. (6.6.1) this is:

$$P_1(t) = \frac{1}{2}\left(2P_1(t-1)^2 + 2P_1(t-1)(1 - P_1(t-1))\right)$$
$$= P_1(t-1) \qquad (6.6.3)$$

To introduce selection we assume that it acts on the organisms, and assign a fitness to each of the organisms, not each of the alleles. We use the simplest selection model of Eq. (6.5.5) where number of offspring λ determines fitness. The parameters are indexed by the two alleles λ_{11}, $\lambda_{1,-1}$, and $\lambda_{-1,-1}$. We have a two-step dynamics consisting of reproduction and selection. In generation t the population proportions after selection (indicated by primes) are:

$$P'_{i,j}(t) = \frac{\lambda_{i,j}}{<\lambda>}P_{i,j}(t)$$
$$<\lambda(t)> = \sum_{ij}\lambda_{i,j}P_{i,j}(t) = \lambda_{1,1}P_1(t)^2 + 2\lambda_{1,-1}P_1(t)(1 - P_1(t)) + \lambda_{-1,-1}(1 - P_1(t))^2 \qquad (6.6.4)$$

The reproduction step that determines $P_1(t)$ is given by Eq. (6.6.2) with primed probabilities:

$$P_1(t) = \frac{1}{2}\left(2P'_{1,1}(t-1) + P'_{1,-1}(t-1)\right)$$

$$= \frac{1}{<\lambda(t-1)>}\left(\lambda_{1,1}P_1(t-1)^2 + \lambda_{1,-1}P_1(t-1)(1-P_1(t-1))\right) \tag{6.6.5}$$

We can find the steady state where $P_1(t) = P_1(t-1) = P_1$. Multiplying Eq. (6.6.5) by $<\lambda>$ gives the equation:

$$\lambda_{1,1}P_1^3 + 2\lambda_{1,-1}P_1^2(1-P_1) + \lambda_{-1,-1}P_1(1-P_1)^2 = \lambda_{1,1}P_1^2 + \lambda_{1,-1}P_1(1-P_1) \tag{6.6.6}$$

We have two trivial solutions $P_1 = 0, 1$. Dividing this equation by $P_1(1-P_1)$—the easiest way is to combine together the first term on both sides—enables us to obtain the third root, which is given by:

$$2\lambda_{1,-1}P_1 + \lambda_{-1,-1}(1-P_1) = \lambda_{1,1}P_1 + \lambda_{1,-1} \tag{6.6.7}$$

or:

$$P_1 = \frac{\lambda_{1,-1} - \lambda_{-1,-1}}{2\lambda_{1,-1} - \lambda_{-1,-1} - \lambda_{1,1}} \tag{6.6.8}$$

The two trivial solutions occur when either $\lambda_{1,1}$ or $\lambda_{-1,-1}$ is the highest fitness. In this case we can say that one of the alleles is more fit than the other. If $\lambda_{1,-1}$ is the highest fitness, then the third solution that corresponds to a mixed population results. This is the circumstance where an organism with one allele of one type and one allele of the other type is most fit. A well-known example is the sickle-cell allele which, when combined with a normal allele, has higher fitness in the presence of malaria. To see how this mixed solution functions, we can assume that the fitness for homozygous organisms is zero, so that none of them reproduce. Then we have $\lambda_{1,1} = \lambda_{-1,-1} = 0$. From Eq. (6.6.8) $P_1 = 1/2$. Even though the organisms that are homozygous do not reproduce, they still exist in every generation and comprise half of the population at birth. Thus, selection in favor of heterozygotes creates a mixed population.

In a sense, this is a straightforward example of the creation of correlations between alleles that might be expected to violate a mean field theory. Selection imposes a correlation by requiring the existence of different alleles at the two genes. However, if we only consider the composition of offspring, then the alleles become uncorrelated due to sexual mixing. The dependence of one allele on the other allele for survival is obscured by the averaging due to reproductive mixing. In Question 6.6.1 the replacement of the fitness of an organism with an effective fitness of an allele is discussed. Note also, that the way the model is formulated so that the population is always normalized, obscures the need to overcome selection by having greater numbers of offspring.

This example can be generalized. We could consider two genes with two alleles each, where the only reproducing organism has a combination of all different alleles. Each of the alleles on each gene would be present half of the time and the reproducing organism occurs only one-quarter of the time. Three-quarters of the organisms do not reproduce. We can see that survival would become poor for organisms if there is such severe selection for particular combinations of genes. Since there are estimated to be of order 10^5 genes in mammals, this is an unlikely scenario. Specifically, if many individual genes strongly affect selection, then the number of organisms surviving to reproduce becomes very small. This problem will become more significant when we consider correlations between nonhomologue genes.

As a second example, we consider a case of selection in favor of a particular combination of alleles on nonhomologue genes. Specifically, when allele A_1 appears in one gene, allele B_1 must appear on a second gene, and when allele A_{-1} appears on the first gene, allele B_{-1} must appear on the second gene. We can write these high fitness organisms with the notation $(1,1)$ and $(-1,-1)$; the organisms with lower fitness (for simplicity, $\lambda = 0$) are $(1,-1)$ and $(-1,1)$. It is clear that there are two stable states of the population with $(1,1)$ or with $(-1,-1)$. If we start with exactly 50% of each allele, then there is an unstable steady state. In every generation, 50% of the organisms reproduce and 50% do not. Any small bias in the proportion of one or the other will cause there to be more and more of one type over the other, and the population will eventually have only one set of alleles.

We can solve the example directly. It simplifies matters to realize that the reproducing parents must contain the same proportion of the correlated alleles (A_1 and B_1) so that:

$$P_{1,1}(t) + P_{1,-1}(t) = P_{1,1}(t) + P_{-1,1}(t) = P_1(t)$$

$$P_{-1,1}(t) + P_{-1,-1}(t) = P_{1,-1}(t) + P_{-1,-1}(t) = P_{-1}(t) = (1 - P_1(t))$$

(6.6.9)

The reproduction equations are:

$$P_{1,1}(t) = P_1(t-1)^2$$

$$P_{1,-1}(t) = P_{-1,1}(t) = P_1(t-1)(1 - P_1(t-1))$$

$$P_{-1,-1}(t) = (1 - P_1(t-1))^2$$

(6.6.10)

The proportion of the alleles in the generation t is given by the selected organisms:

$$P_1(t) = \left(P_{1,1}'(t) + P_{1,-1}'(t) \right)$$

(6.6.11)

Since the less fit organisms $(1,-1)$ and $(-1,1)$ do not reproduce this is described by:

$$P_1(t) = P_{1,1}'(t) = \frac{1}{P_{1,1}(t) + P_{-1,-1}(t)} P_{1,1}(t)$$

(6.6.12)

This gives the update equation:

$$P_1(t) = \frac{P_1(t-1)^2}{P_1(t-1)^2 + (1 - P_1(t-1))^2}$$

(6.6.13)

which has the behavior described above and shown in Fig. 6.6.1. This problem is reminiscent of the ferromagnet at low temperature as studied in Section 1.6. Starting from a nearly random state with a slight bias in the number of UP and DOWN spins, the spins align, becoming either all UP or all DOWN.

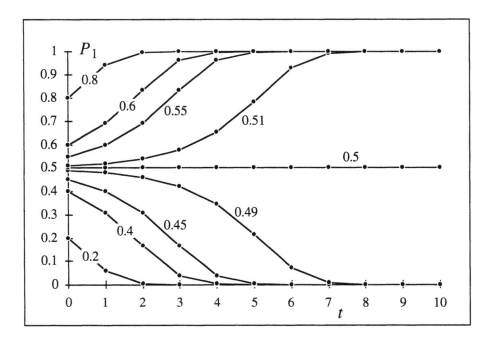

Figure 6.6.1 Time evolution of the allele population in sexual reproduction when selection enforces a correlation between alleles on two genes, Eq. (6.6.13). The proportion of the first coupled pair of alleles $P_1(t)$ in successive generations is shown by the dots, and connecting lines are included for clarity. The starting proportions $P_1(0)$ are indicated by the labels. The figure shows that the pair of alleles which starts with a larger proportion eventually dominates. For $P_1(t) = 0.5$ there is an unstable equilibrium. The text discusses a species that has nonuniform composition in physical space. In one area one allele pair dominates and in another area the second allele pair dominates. Under the influence of correlated selection linked to correlated reproduction, two distinct populations arise. ∎

In order to relate the examples and assumptions we have used to a more general formulation of sexual reproduction, we write a two-step general model for sexual reproduction:

$$\{N(s;t)\} = R[\{N'(s;t-1)\}] \tag{6.6.14}$$

$$\{N'(s;t)\} = D[\{N(s;t)\}] \tag{6.6.15}$$

The first equation describes reproduction. The number of offspring $N(s;t)$ having a particular genome s is written as a function of the reproducing organisms $N'(s;t-1)$ from the previous generation. The second equation describes selection. The reproducing population is written as a function of the offspring. The brackets on the left indicate that each of these equations represents a set of equations for each value of the genome. The brackets within the functions indicate, for example, that each of the offspring populations depends on the entire set of parent populations.

A mean field approximation is performed by assuming that the reproduction step depends only on the proportion of alleles and not on their specific combinations in the reproducing population. This proportion can be written as:

$$P'(s_i;t) = \frac{1}{N_0'(t)} \sum_{\{s_j\}_{j \neq i}} N'(s;t) \tag{6.6.16}$$

where we use $s = (s_1, ..., s_N)$ to represent the genome in terms of alleles, and the sum is over all gene alleles keeping s_i fixed. $N_0'(t)$ is the total reproducing population at time t. According to our assumption about reproduction, the same offspring would be achieved by a population with a number of reproducing organisms given by

$$\tilde{N}'(s,t) = N_0'(t) \prod_i P'(s_i;t) \tag{6.6.17}$$

since this has the same proportions as Eq. (6.6.16). The form of this equation indicates that the probability of a particular genome is a product of the probabilities of the individual genes—they are independent. Thus complete reproductive mixing assumes that:

$$R[\{\tilde{N}'(s;t)\}] \approx R[\{N'(s;t)\}] \tag{6.6.18}$$

Once this is assumed, then a complete step including both reproduction and selection can also be written in terms of the allele probabilities in the whole population. The update of an allele probability is:

$$P'(s_i;t) \approx \frac{1}{N_0'(t)} \sum_{\{s_j\}_{j \neq i}} D[R[\{\tilde{N}'(s;t-1)\}]] \tag{6.6.19}$$

Given the form of Eq. (6.6.17) we could write this as an effective one-step update

$$P'(s_i;t) = \tilde{D}[\{P'(s_i;t-1)\}] \tag{6.6.20}$$

which describes the allele population change. Thus the assumption of complete mixing allows us to write the evolution of a single allele in this way. However, because Eq. (6.6.20) is a function of all the allele populations, the fitness of an allele is coupled to the evolution of other alleles.

Eq. (6.6.17) describes the neglect of allele correlations in reproduction consistent with a mean field approximation. It should be apparent that this is only a rough first approximation. It is valid only when the gene appears with sufficiently many different combinations of other genes so that correlations are unimportant. In more realistic models, correlations between genes affect both reproduction and selection.

We can provide a specific example of breakdown of the mean field approximation using the previous example of selection of gene combinations leading to Eq. (6.6.13). In this example, if there is a spatial variation in the organism population that results in a starting population that has more of the alleles represented by 1 in one region and more of the alleles represented by −1 in another region, and reproduction is correlated by region, then we will form patches of organisms that have (1,1) and (−1,−1) after several generations. This symmetry breaking, like in the ferromagnet, is the usual breakdown of the mean field approximation. Here, we see that it creates correlations in the genetic makeup of the population. When the correlations become significant, then the species has a number of types. The formation of organism types depends on the existence of correlations in reproduction that are, in effect, a partial form of speciation—what is important is whether interbreeding occurs, not whether it is possible.

Thus we see that the most dramatic breakdown of the mean field approximation occurs when multiple organism types form. This is consistent with our understanding of ergodicity breaking, phase transitions and the mean field approximation from Section 1.6. Interdependence at the genetic level is echoed in the population through the development of subpopulations. We should emphasize, that this symmetry breaking required both selection and reproduction to be coupled to gene correlations. Moreover, if there is a small bias in the fitness of (1,1) over (−1,−1) then the formation of the two varieties will not persist due to competition between them over many generations. Thus we still need the existence of multiple resources, as discussed in the previous section, to enable the distinct types to persist.

Even if we assume that there exist multiple resources, interbreeding of organisms may continue to mix and force them to remain a single type with diverse individuals. However, the more correlations are important during selection, the more ineffective this becomes. When viable organisms are a small subset of the organisms formed by reproduction, a large number of offspring are required in order to maintain the population. Specifically, the number of offspring grows exponentially with the number of genes whose alleles are coupled to each other in selection. Strong genomic correlations in selection eventually make reproductive mixing impossible. The actual situation is not quite so extreme, because meiosis does not result in complete mixing of parent DNA. Only limited crossover occurs, so that chromosomes do not loose much of the preexisting allele correlations.

The existence of sexual reproduction implies that from a fitness perspective it is beneficial. Our next objective is to understand how it might be beneficial and what this says about fitness. In sexual reproduction, organisms are paired and their offspring are not copies of the original organisms but rather composites of them. We use the term "composites" in the same sense as used in Chapter 2, and its significance there is related to its significance in the present context. The implication is that organisms are composites of partially independent components, designed to correspond to partially independent aspects of the fitness. Specifically, distinct physiological or behavioral attributes have varying degrees of interdependence in selection. The approach of trying composite states of previously successful combinations applies in this case, as it did in the learning of patterns in neural networks. This, however, does not entirely explain the existence of sexual reproduction.

We have discussed various aspects of the problem of sexual reproduction however, these have not indicated why sexual reproduction itself improves organism fitness. By conventional argument, sexual reproduction should be a physiological or behavioral attribute that increases the ability of an organism to produce surviving off-spring. On the face of it this seems ludicrous. Asexual reproduction does not depend on the existence of a mate, and therefore an organism that can reproduce asexually is more likely to reproduce. Moreover, asexual reproduction seems to require a much smaller overhead in terms of physiological machinery. This physiological machinery uses resources that could be utilized for other purposes. In other words, an organism that had all of the physiological traits of a sexually reproducing organism but could reproduce itself asexually (e.g., by cloning) would seem to have a fitness advantage.

There are difficulties with this picture that illustrate problems with simple for-mulations of the theory of evolution. The first is that the ability to produce surviving offspring depends on a time-varying fitness landscape rather than a static one. Of particular significance in this variability is the evolution of competing organisms. An organism that evolves more slowly has a lower probability of producing surviving off-spring than an organism that evolves more rapidly. This sentence, however, does not make sense, since an organism does not evolve, only a population of organisms evolve. In order to understand this we must improve the language that we use to de-scribe evolution.

We often say that fitness-based selection implies that an organism exists because of its ability to survive and have offspring. The problem with this statement is that an organism does not exist because of what happens in the future but rather what hap-pened in the past. The two are only the same when every generation is the same. Thus the organisms that exist at any one time are offspring from organisms that survived, who in turn were offspring of a set of organisms that survived. Thus we must write an iterative equation of the form

$$N(s, t) = R[D[R[D[R[D[R[D[...\{N(s; 0)\}...]$$ (6.6.21)

representing the selection of organisms at every generation. Because selection applies not to a particular organism but to a chain of ancestors, the rate of fitness improve-ment is essential to the selection. We see that this causes conceptual problems, because selection is no longer based only upon the fitness of an organism but rather on the rate of fitness change, which is a property of generations of organisms.

The process of sexual reproduction accelerates evolution because of the same rea-son that composite states are useful in pattern recognition. As long as the genes on an organism cause partially independent effects, it is advantageous to attempt possible composites and establish more definite relationships only as the correlations are es-tablished by selection. This process, however, describes not selection of an organism but selection of the collectively evolving species. Thus we return to the discussion at the end of the last section, where selection acts not upon an organism but upon species that evolve in parallel and compete for resources. This should not be overly troubling because, after all, sexual reproduction does inherently involve the coupling of past and future generations of any organism with the other organisms that are re-productively coupled. We will return to this issue in the next section.

Our remarks about the relevance of sexual reproduction to fitness suggest that evolution is a process that is far from equilibrium, which is quite reasonable. In such a far-from-equilibrium process, we are not only selecting for the static properties but also for the dynamic properties. Earlier our treatments assumed that selection applied to a persistent property rather than a dynamical property. Thus, to include the dynamic property in Section 6.4 we would assume that different parts of the landscape have walkers that take steps at different rates. In section 6.5.3 we would assume that organisms have higher or lower rates of mutation as determined by part of the genome itself. An analogy to a car race may be helpful. Our previous discussion assumed that the race is won by the car that is farther along on the road, as opposed to the car that is faster. Of course it makes sense that the faster car is farther along the road, since we assumed in evolution, like in a usual car race, that all started from the same place. The point is that the nature of selection causes the properties of the organism not only directly through fitness but also self-consistently through the process of selection itself.

Question 6.6.1 Criticize the following statement: Since we can define the proportion of a gene in generation t and in generation $t+1$, we can always write an expression for the allele evolution in the form:

$$P(s_i;t) = \lambda_{s_i} P(s_i;t-1)$$
$$\sum_{s_i} \lambda_{s_i} = 1 \qquad (6.6.22)$$

so that we always have evolution that can be described in terms of genes. Does it matter if Eq. (6.6.17) applies?

Solution 6.6.1 The difficulty lies in the dependence of the coefficients λ on time t through its dependence on the changing population. In steady state, λ values would not change. Of course, in steady state there is no need to describe the dynamics. The equation is only useful as a description of the dynamics if the values of λ are slowly varying in time compared to the changes in P. When Eq. (6.6.17) is not valid, neglecting the subpopulation correlations is formally equivalent to considering the average gene dynamics over all of the organisms on earth despite their differences and the lack of mixing of species. This is the ultimate form of the gene-centered view of evolution. Eq. (6.6.17) applies to the two examples given in the text. The coefficients λ can be written explicitly. They can be seen to vary significantly when the gene population is changing. For the first example we obtain the value $\lambda_1(t)$ = $\lambda_{1,1}P_1(t) + \lambda_{1,-1}(1 - P_1(t))$ from Eq. (6.6.5). For the second example we read the value $\lambda_1(t) = P_1(t)$ from Eq. (6.6.13). ∎

Question 6.6.2 Discuss the existence of nonreproducing organisms such as mules in light of Eq. (6.6.21).

Solution 6.6.2 The problem with nonreproducing organisms is understanding why they should exist, since they do not have offspring and therefore by usual concepts are completely unfit—nonviable. Eq. (6.6.21) solves the formal problem of their existence by indicating logically that the existence of a mule depends only on the existence of its parents. However, it does not explain why evolutionary changes have not caused horses and donkeys to avoid coupling. By coupling, their ability to produce reproducing offspring and therefore their fitness would seem to decrease. In this light we can only remark that the existence of mules suggests the importance of the dynamics of the evolutionary process over the equilibrium view. ∎

There are several concepts that relate to sexual reproduction that we briefly mention here. An implicit issue in sexual reproduction is identifying the region of genome or phenome space that can interbreed, which defines the boundaries of possible spread of a single species. We generally assume that for each species this is a well-defined domain separate from other species. In principle, the domains might interleave; however, the large size of phase space suggests otherwise.

A related issue is the relevance of sexual reproduction to our earlier discussion of interactions between organisms in the space. We first recognize that sexual reproduction represented in terms of walkers in the genome space on a fitness landscape corresponds to a step that starts by selecting two walkers at different locations and creates new walkers that are related to the original walkers in that some coordinates are from one and some from the other. The notion of composite states suggests that the landscape is quite rough; however, the selection of composites tends to place organisms into valleys that may be separated from each other by high ridges. Such steps are efficient nonlocal Monte Carlo moves. In effect these steps enable the population to move and spread in what is likely to be a fragmented space. Indeed, if the space were smooth the formation of composites would not be an improvement over standard Monte Carlo steps.

We can think about the impact of sexual reproduction on the effective interactions such as attraction and repulsion. Because of the boundaries of the domain of sexual reproduction, the species inherently evolves together. This corresponds to an attractive interaction within this domain and a repulsive interaction outside of it. Organisms that are similar but located outside the domain of reproduction would be impeded from reproducing if they are present in small numbers because of the prevalence of organisms with which they cannot reproduce. The need to identify organisms with which one can reproduce may be a motivation for the creation of patterns on animal skins discussed in Chapter 7, as well as bird calls and mating behavior.

6.6.2 *Genes, organisms and groups—*
the evolution of interdependence

The partial independence of genic evolution in the evolution of sexually reproducing species has led to a gene centered view of evolution where the gene, rather than the organism of which it is a part, is the "entity of interest." One way to react to this is to consider this approach as the result of a too seriously taken mean field approximation. However, there are important philosophical issues that are raised in discussions of this point that have direct relevance to our understanding of complex systems, so we will spend some time discussing them.

The fundamental philosophical difficulty that appears to give impetus to the focus on gene evolution is the concept of interdependence. Ironically, interdependence that is so essential to the concept of a complex system appears to be at odds with the evolutionary concepts of competition and selection that are supposed to create them. If we consider phenomenologically the evolution of molecular fragments (e.g., genes), or molecules, or cells, or organisms, we find the formation of collections of interdependent individuals. The real problem is understanding the reason for the emergence of interdependence in the context of evolution. Why, after all, would the selection of fit individuals give rise to the appearance of a collective interdependence? We see this problem in discussions of altruism and selfish or even aggressive social behavior. Why would organisms develop altruistic behaviors in a competition for survival? It appears straightforward to assume that selfish or aggressive social behavior provides selective advantage and altruism selective disadvantage when competition for a resource determines survival. The same problem arises at the molecular fragment level when we consider why genes should assemble into interdependent chromosomes. We might easily imagine that the gene that is responsible for coding a replicase could replicate itself many more times with the available resources without the other genes of a complex organism. Until this point we have been concerned with the processes of incremental modification of organisms and speciation. In order to develop an understanding of evolution, we must also understand the formation of interdependent communities of organisms.

In order to make progress, we use the language of organisms evolving on a fitness landscape and the interactions between them as discussed in Section 6.4.5. We considered interactions between organisms that caused either a mutual lowering or a mutual raising of fitness. We argued that organisms would aggregate in regions of space where there was a mutual raising of fitness. We can understand this simply in the context of genic interactions or human social interactions: when there are similar organisms, that help each other, this mutual assistance increases their fitness or success. This model of a mutually supportive community of organisms is not, however, complete, as we now discuss.

Let us imagine introducing into this community a single selfish or aggressive individual (gene, cell, organism or human). The selfish individual benefits from the help that others give, while not providing help to them. Instead it utilizes the additional effort for self-benefit. It is important to recognize that in the language of interactions this is an asymmetric interaction. The existence of the selfish individual does not improve the success of the other individuals, while their existence improves the success of the selfish individual. Similarly, an aggressive biological organism is assumed to decrease the fitness of the other organisms, while they do not decrease its fitness. We might also add that an altruistic individual is assumed to increase the fitness of other organisms while decreasing its own.

This interaction asymmetry means that analysis of this problem does not fit within the framework of energies and equilibria. Instead it must be analyzed as a dynamical system such as a predator-prey problem. This conclusion has important consequences for the idea that selfish or aggressive behavior cause benefits for the individual. It is not generally reasonable to analyze a dynamical system with the concepts of a model of energies and equilibrium. In the latter we assign definite steady state properties to the entities involved. In the former this is not generally possible. For example, in what way can we compare the fitness of a predator and its prey? However, just because we cannot analyze the model in the same way does not mean that we can avoid dealing with this problem.

With this in mind, let us continue our introduction of selfish individuals. Assuming there is benefit to the individual in being selfish, then such individuals will proliferate. The fitness and success of mutually supportive individuals will decrease. Moreover, as the proportion of the selfish individuals in the population increases they encounter each other, and their fitness and success will also decrease even as their advantage over the mutually supportive individuals continues to exist. We should not analyze this as a static situation but rather introduce a dynamic model that describes possible fluctuations in the numbers of the different types of individuals. However, we can take a step further without considering the dynamic aspect of the model.

We know from direct consideration of the interactions that a completely mutually-supportive collection of individuals consists of individuals that are more fit than the mixed community that has been created by introduction of selfish individuals. Nevertheless, we also know that the former is unstable to the introduction of such selfish individuals. Thus, if we can modify the community of mutually supportive individuals in such a way that can stabilize it to the introduction of selfish individuals, then the mutually supportive group will be more successful win—in a competition with the mixed group.

What is completely essential to realize in this discussion is that we have made the step to the competition between communities of organisms rather than competition between individuals. It is only at the level of community competition that the success of mutual support is sufficient to eliminate the selfish individuals. Thus, it is only when we have several communities of organisms, some of which have selfish individuals in them and some of which do not, that we will have selection in favor of the community of mutually-supportive individuals over the communities with selfish individuals.

How can we stabilize the community of mutually supportive individuals to the introduction of selfish or aggressive individuals? Social behaviors that prevent the introduction of selfish individuals are not necessarily individually based, because they only arise in the context of the selection of collections of individuals. However, they may be individually-based behaviors such as recognition and rejection (ostracism) of selfish individuals. As has often been noted, this rejection causes extrinsic consequences to the selfish behavior. Rejection is not inherent in the selfish behavior. Its presence results from feedback through the process of community selection that promotes such extrinsic consequences. In the context of the individual, these consequences may not appear causally related to the behavior itself but are manifest as indirect social consequences. One way to think about this is that the action of selection at the higher level of organization modifies the influence of selection at the lower level of organization. This is because inherently the environment of the individual includes the social environment, and selection cannot be considered independently of this social environment.

We can see such social mechanisms at all levels of collective organization. Cellular systems provide for regulation of gene expression or replication. Physiological systems provide for prevention of cancerous growths. More generally we can consider the immune system as combating selfish or aggressive individual cells that appear as infections. Social systems ranging from spousal selection in sexual reproduction to ostracism to human legal systems provide for regulation or reward of individual behaviors.

Once a collective community identity is formed and it is competing for resources and survival at the next level of organization, then various other properties of the collective entity may be introduced through incremental evolutionary change. These include specialization and more elaborate forms of interdependence that are found in complex systems. These arise as improvements in the capabilities of the community.

In this discussion we have given support to the emergence of mutually supportive behaviors and the elimination of selfish or aggressive behaviors. However, we must emphasize that in our argument these behaviors arise only when there is competition and selection at the higher level of organization. Specifically, only when there is competition between communities of organisms. There is no direct mechanism to suppress selfish or aggressive behaviors at the highest level of organization present at a particular time. Ultimately, this is also tied to the statement that evolution applies to members of a population—a single organism cannot evolve by itself.

Questions 6.6.4 and 6.6.5 focus on these concepts in the context of human societies. This follows from the recognition that social evolution has similar properties to biological evolution through the existence of heritable behavioral traits that are transmitted by education. This should not be too much of a surprise, since biological selection is also based on information. It is not the atoms that are selected, but rather the sequence of base pairs—information—that is selected.

Question **6.6.3** What is wrong with this seemingly logical statement: "Survival of the fittest is based upon a competition for survival. Therefore an organism that competes by eliminating its competitors is more likely to survive"?

Solution 6.6.3 The solution was articulated in this section as a difficulty with analyzing a dynamical system with equilibrium concepts. The analysis shows that a system of mutual support is unstable to the introduction of an organism or individual that acts aggressively. The problem is that this situation is not stable either. From an equilibrium model we can only say that mutually supporting individuals are better off than mutually destructive individuals. Once there are asymmetric relationships, the system has a dynamic population that may, for example, display population fluctuations like the predator-prey model unless the system is stabilized by additional interactions that create social consequences for the behavior of "eliminating its competitors." ∎

Question 6.6.4 Consider bravery and altruism in the context of human societies. Discuss these behaviors in the context of the discussion of the appearance of collective social behaviors. How do they persist from generation to generation? For simplicity, define "altruism" as a behavior that increases the reproductive success of others at the expense of reproductive success of the altruistic individual.

Solution 6.6.4 Bravery is actually an example of altruistic behavior, since it implies facing danger to oneself for a cause that is generally of collective benefit. The soldier facing danger in war is a prime example. Such behavior causes a lower probability of individual survival. It should thus be selected against and disappear from civilization. As discussed in Chapter 3, human behavior has both a genetic and a learned component. In order to promote the existence of bravery, there appear to be social behaviors that promote its presence. If we assume a persistence of bravery, the social behaviors that promote bravery must be sufficiently strong to compensate for both the genetic losses (death before reproduction) and the loss due to experience (cessation of bravery due to learning). In general, spousal selection may promote altruistic behaviors by increasing the probability of reproduction or average number of offspring of subgroups that are less likely to survive. Socialization in pedagogy may serve to compensate for the effects of experience, and social rewards for altruistic behavior may mitigate their effects. As discussed previously, these must manifest their benefit in the action of competition and selection of groups. ∎

Question 6.6.5 Consider war in the context of human societies. Is war necessary for the creation of altruistic behavior?

Solution 6.6.5 It is not clear that competition between groups requires war. However, war is a mechanism of group competition that, by the arguments presented here, should promote interdependence within societies. Thus it can also promote altruistic behavior. Some other mechanisms such as mass starvations due to competition for resources are not more appealing. Our arguments suggest that whatever the mechanism, the emergence of complex social behaviors must arise as a result of competition of populations that can only be manifest in the (selective) demolishing of whole populations. In this regard we note that wars are manifest in ants, which have complex social structures, sometimes called superorganisms. When social behaviors are learned rather than genetically based, then other possibilities for competition arise. This is because one form of reproduction is the transfer of learned social behaviors. Defeat in economic competition can cause one group to adopt social behaviors of a second group, which is equivalent to its reproduction. ∎

Question 6.6.6 Evolution is often separated into two parts; the formation of self-reproducing organisms from molecules and the evolution of these organisms into their present form. Argue that a distinction is not necessary. Why would it seem a natural distinction if it is not really one?

Solution 6.6.6 Under certain environments, molecules can also be self-reproducing organisms. However, these conditions are much more limited than the conditions under which individual cells can self-reproduce. This is because the cell membrane creates a distinction between inside and outside environments that allows an artificial environment for reproduction of molecules to exist. If the internal environment of the cell was typical of the fluid in which the cells were found, then the cell membrane would not be necessary. We can imagine that at some time and place there was such an environment that was conducive to molecular reproduction. Molecules evolved within this environment, forming various molecular types. At some point the molecules formed collective entities that were able to move out of this limited environment in the same way that organisms left the ocean for the land. Several stages of such processes gave rise to cells. The formation of cells expanded greatly the possible environments/resources that could be consumed by self-reproducing organisms. ∎

We have focused on the formation of collectives in order to describe interdependence. In a sense, the interdependence was already explained previously in this chapter in two steps. First we noted the necessity of increasing complexity through increasing the genomic space. This provided a selective advantage for long genomes but did not explain the existence of subcomponents. The second step was arguing that there are partially independent traits of organisms and therefore that a composite genome would be effective in accelerating evolution. This completes our argument, because we now have both an argument for the creation of collective communities and an argument for the retention of substructure.

The development of higher levels of organization might be thought to create additional problems for our discussion of the global evolutionary process in the context of a single fitness parameter. Do we need to introduce a separate parameter to consider the fitness of the collective? Actually this is not the case. Any parameter that describes replicative proliferation of a collective organism which has a well-defined number of components is the same parameter as that of any one of its components. In a sense this manifests exactly what we mean by a complex system that is collectively and individually interdependent.

6.7   Conclusions

Our primary effort in this chapter has been to develop an understanding of the context in which incremental evolutionary processes relate to global phenomena of evolution. A problem we encountered is that it is hard if not impossible to define fitness so that it can compare *E. coli* and human beings. We began with a model of Monte Carlo walkers on a fitness landscape. While there were problems with this model, we were able to discuss incremental changes including trait divergence and augment the model to include population interactions. Difficulties with global phenomena were compounded, however, when we introduced models of reproduction and selection through competition. Specifically, we were left with the problem of understanding trait divergence when competition would eliminate distinct organisms. We came to the conclusion that in order to think about fitness we must limit the scope of selection by considering multiple resources. When selection occurs with multiple resources, it acts more broadly than in the Monte Carlo model and in a more limited fashion than in a single-resource model. The introduction of multiple resources suggests that in order to understand global evolution we must have a clear understanding of the properties of resources as well as organisms, and a consideration of the latter separately is not adequate. At every stage of our discussion we found that interactions between organisms were an essential part of understanding evolution. Mean field approaches that may be used in incremental evolutionary theory break down and are deceptive if we want to understand global evolution. In particular, in the context of discussing sexual reproduction and collective behavior we developed an understanding of the creation of various levels of structure and interdependence in organisms.

How can we develop a better understanding of the processes associated with the evolution, and thus creation, of complex systems? In recent years several additional examples of evolution have been studied. In our immune system, an evolutionary process enhances the recognition and removal of antigens—foreign and/or harmful cellular or molecular entities. The immune system creates molecular receptors and antibodies that bind antigens and enable them to be eliminated. To achieve this, receptors undergo reproduction and selection for high-affinity binding. This process includes rapid genetic mutation of immune cell DNA that codes for the receptors. Our understanding of this process and the development of mechanisms for rapid replication of DNA in a test tube have led to recent implementation of artificial molecular evolution. In this process DNA or RNA is itself used as a reacting molecule or enzyme. The desired molecular action is obtained by repeated test tube selection and replication. There is hope that molecular evolution will enable the formation of targeted medical drugs. Finally, there are increasing efforts to implement evolutionary processes for the creation of software algorithms. This requires representation of algorithms in a manner that allows them to undergo mutation and selection. The immune system maturation, test tube molecular evolution and software evolution all provide opportunities for further study and for application of knowledge gained about evolution.

7

Life II: Developmental Biology—
Complex by Design

Conceptual Outline

■ 7.1 ■ Developmental biology strives to understand the sequence of events by which a single cell becomes a system of many differentiated interacting cells. This process involves placing different structures in particular locations and interconnecting them.

■ 7.2 ■ To model differentiation we focus on the formation of color patterns on animal skins that have a variety of forms. Cellular automaton models show the relevance of local activation and long-range inhibition of pigment production to the formation of patterns. Chemical reaction-diffusion systems illustrate similar patterns using slow- and fast-diffusing species.

■ 7.3 ■ Other elements of the tool kit for developmental processes include mechanisms for changes in cell structure, cell motion, timing and counting. Of particular interest are sequential steps (programs) that can form branching structures.

■ 7.4 ■ Theoretical modeling can better complement phenomenological studies of biological systems if the different objectives of theory and experiment are recognized.

■ 7.5 ■ The approach of developmental biology to the design of complex systems may be a useful framework for considering the design of complex artificial systems.

■ 7.6 ■ Models of pattern formation may be better suited to discussions of global properties of the evolution of organisms than the models discussed in Chapter 6.

7.1 Developmental Biology: Programming a Brick

Reproduction in multicellular organisms, animals and plants, occurs through a process of development from a single cell. The fundamental objective of developmental biology is to understand how an individual cell through cell division, differentiation and growth results in a complex physiology. The controls for this process of

621

development are present within the initial cell and also in the environment in which the cell develops.

Our concern in this chapter is largely with the cellular behavior in development rather than with the internal functioning of the cell. However, in the following paragraphs we discuss briefly models for the mechanisms that exercise control over the developmental process as part of the internal functioning of the cell.

It is generally believed that the design of plant or animal physiology is contained within the nuclear DNA of the cell. DNA is often called the blueprint for the biological organism. However, it is clear that DNA does not function like an architect's blueprint because the information does not represent the structure of the physiology in a direct way—there is no homunculus there. For our purposes it is convenient to think about the DNA blueprint as a program that specifies the interaction between a cell and its environment, including cells in its vicinity, as well as the internal functioning of the cell. However, in describing DNA as a program we are implicitly subsuming the functions and description of the entire cellular machinery in the DNA. For our abstract purposes, there is no difference in various sources of information, as there is no essential difference between information that is found on the tape of a Turing machine and information in the table of the read-write head (see Section 1.9.4). There are, however, other conceptual issues to address.

First, we must clarify the nature of DNA function within the cell. DNA serves at least in part as a collection of templates (genes) that may be thought of as blueprints for protein chains. These templates are sometimes being transcribed (active) and sometimes not being transcribed (inactive). Thus, the role of DNA at a particular time is described by a set of transcription activities. The activity of a particular gene depends on the activity of other genes. Thus, a useful analogy may be a neural network model where the transcription activities are analogous to the neuronal activities in the network. Like the synapses of the network, the molecular machinery of the cell mediates the activities (and performs the transcriptions) of the DNA. The patterns of activity of the transcription of DNA are a part of the patterns of activity of the cell as a whole which constitute possible behaviors of the cell. Thus it may be reasonable to consider the relevance of attractors of patterns of activity, as in the neural network models discussed in Chapter 2, to the study of cellular function. The development of an organism consists of a temporal sequence of such patterns of cellular function.

Second, we must clarify the relationship of information and behavior. It is likely that the DNA in a cell contains a large proportion of the information needed to describe the function of the cell, the developmental dynamics and the physiological function of the organism. However, this does not mean that the DNA should be thought of as controlling the processes in the conventional sense of the term "control." A useful analogy is the role of a library in society. It is quite likely that most of the information about the function of society in one way or another may be found in the Library of Congress. However, this does not mean that the library controls this function. It may, indeed, be better to think about the molecules in the cell as akin to a society of entities that act upon each other and respond to external stimuli. DNA then serves this society as a source of information—in part as a repository of blueprints for the manufacture of cellular machinery.

In this regard it may be helpful, though somewhat subtle, to recognize that DNA is not by itself a complex organism. It does not satisfy our criteria of nondivisibility, since its structure and behavior (including transcription) is essentially local. It is only when the information in DNA takes form in the context of cellular or organismal behavior that the behavior is itself complex, and the system as a whole satisfies the conditions of a complex organism. Incidentally, this is also a reason that the structure of DNA does not satisfy the 7±2 rule—there are 23 pairs of homologue chromosomes in most human beings, and a wide variation in the number of chromosomes in other organisms.

Returning to our central focus in this chapter, for our purposes development is a largely deterministic sequence of cellular states that results in a multicellular organism. In this sense the organism can be described as the result of a program, since all deterministic processes can be so described. The program is largely contained in the original cell. It is essential to recognize that all cells of an organism begin from one cell in a unique state, and therefore inherit all or parts of the same set of information, and thus the same program.

The central problem of developmental biology is to describe how the cells differentiate in such a way as to place particular functions in particular locations in the body—not to describe the specific eventual function of each cell. Part of this problem is to describe how cells become interconnected by necessary structures formed out of individual cells such as long branching neurons, or many-celled structures such as blood vessels. This must be achieved by the program that specifies the sequence of cell states and cell interactions. The overall process of development is shown in Fig. 7.1.1.

Biological development is a systematic approach to the very difficult problem of designing complex systems. It enables the creation of a large variety of systems. In studying this approach it may be helpful to think about designing a building in a similar manner. Allowing some imagination, we might consider writing a program for a brick. The program describes how a brick should move and interact with other bricks in its vicinity. Providing the same program for each brick in a pile, we walk away and return to find the whole building, with windows, ducts, and utilities in place. Cells, unlike bricks, are themselves like organisms in consuming resources and producing waste; they are self-reproducing and mobile. They also have the ability to change shape. Through shape change and changes in chemical processes they can adopt a large variety of functions in a multicellular organism. Even if we endow bricks with similar abilities, it still requires careful thought to understand how the design of a complex structure can arise from a program describing their interactions.

It is significant that this approach balances design with self-organization. In Chapter 6 on evolution, we assumed a self-organizing process that occurred by chance and external selection. In contrast, organism development should reliably achieve a desired outcome from a preexisting (internal) design. Nevertheless, the built-in design directs a dynamic process where mutually interacting entities self-organize into the desired complex structure.

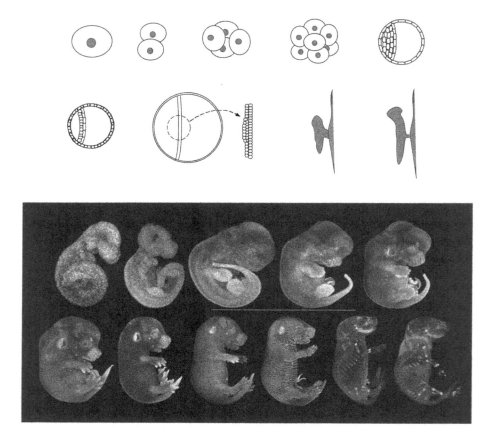

Figure 7.1.1 Illustration of some of the stages in the development of an animal. The top two rows are schematic illustrations of the initial stages where a single fertilized cell undergoes multiple divisions to form a spherical shell with a membrane separating its internal cavity into two parts that become the primary yolk sack and the amniotic cavity. Cells from part of the internal membrane then form the growing fetus. The bottom two rows are magnetic resonance microscopy images of mouse fetal development from 9 days to birth. All images are shown at the same size despite a 10-fold increase in fetal dimensions from the first to last image of this sequence. The multicellular structure of the organism arises through a set of programmed steps originating in a single cell. The identification of processes and mechanisms for this development is the subject of developmental biology (magnetic resonance microscopy images are courtesy of Brad Smith, Elwood Linney and the Center for In Vivo Microscopy at Duke University [A National Center for Research Resources, NIH]). ∎

For many who have had occasion to contemplate a newborn, development is miraculous. From a scientific point of view there are at least two reasons that this reaction arises. First, the relationship between process and outcome is emergent—the relationship between individual parts of the dynamics and the whole is difficult to understand. This is the nature of a complex self-organizing process. Second, designing a dynamic process that can reliably arrive at a specific complex outcome is difficult. When a process involves many steps and an error in any step may give rise to failure, the likelihood that the process will be successful is vanishingly small. Our analogy with a computer program is telling, since a single bit error in computer hardware or software would generally cause failure. It is useful to compare this with our discussion of protein folding in Chapters 4 and 5, where we were also concerned about arriving at a definite final structure. In Chapter 4 we considered exploration of conformation space to find an energy minimum. As long as the dynamics could reach the energy minimum, its identity was not in question. In Chapter 5 we argued that directed sequential steps could arrive at a desired final structure. Here we recognize that in a strictly directed (deterministic) process, there must be no error in the dynamics so that there will be no error in the eventual structure. What is particularly remarkable is that the dynamic process must at the same time be stable to many perturbations, and yet modifiable through mutations that enable evolutionary changes. To understand how this is possible we must eventually recognize that the dynamics as a whole must be formed out of a sequence of attractors that are sufficiently stable to be the outcome of a variety of intermediaries. In this way the nonequilibrium dynamics and its outcome may be relatively stable to perturbations.

From the most basic complex-systems point of view, the problem of developmental biology is composed out of two parts: first, to identify general and specific processes that cause a homogeneous set of cells to differentiate in a controlled fashion so that specific structures are located in specific locations with respect to each other; and second, to identify mechanisms for creating structures that interconnect or support various functional regions of the system. Much of the quantitative modeling of such processes is relatively recent. In this chapter we focus on the problem of differentiation. In Section 7.2 we describe models of the formation of patterns on animal skins. This problem captures an essential aspect of differentiation and structure. The advantage of such patterns is that their structure is not very specific and therefore lends itself to a simpler analysis. However, the interplay of such patterns with specific boundary conditions can give rise to well-defined structures when they are necessary in development. In Section 7.3 we describe some more tools necessary for developmental formation of physiological systems. Of particular emphasis is the formation of branching structures found in plants and animals in the lungs, nervous and vascular systems. In Section 7.4 we discuss some of the general objectives and methodologies of theory and mathematical modeling of biological systems. In Section 7.5 we discuss the general properties of organization by design in biological complex systems and contrast it with the conventional approaches used in human design and engineering. Finally, in Section 7.6 we return to consider the implications of the models of pattern formation in this chapter for the problem of evolution discussed in Chapter 6.

7.2 Differentiation: Patterns in Animal Colors

7.2.1 *Introduction to pigment patterns*

Many animals have patterns of coloration on their external surfaces. Color is the result of pigment produced in cells. Often the patterns are composed of only two different colors, but in some cases there are more. For our purposes, the examples that are convenient to think about are the patterns on the fur of both predator and prey mammals. Zebras, giraffes, tigers, leopards and many others have distinctive patterns as a species. These patterns also vary in more subtle ways from individual to individual. Other kinds of patterns are present in some insects—particularly butterflies—fish—particularly tropical fish—and birds—particularly tropical birds.

The functional relevance of patterns or brilliant coloration for animals is an interesting topic of study. We can try to understand the reasons for coloration through the concepts of evolution discussed in the previous chapter. Evolutionary theory suggests that such physiological attributes arise from a survival advantage. It is a common practice to offer explanations for the existence of physiologic or behavioral features based on this premise. The ultimate difficulty is that these explanations, no matter how well reasoned, are rarely subject to direct experimental test. However, there appears little doubt that a uniform color for some animals is used for camouflage within a well-defined environment. This is characteristic of various green, brown or black insects and lizards that are found on leaves, various tree trunks, or the ground. Patterns of coloration, whether of black and white or of brilliant colors, appear to be directly counter to this purpose. Alternative explanations rely upon some form of social or collective behavior. The coloration of prey such as zebras and giraffes might serve to confuse predators because, in the context of a herd of animals, it inhibits the distinction of one individual from another. The boundaries between animals become less distinct than the internal coloration boundaries. Since the herd as a whole is not readily attacked, the individual disguised as part of a larger system is protected. This is consistent with the general discussion in Chapter 6 about the nature of collective behavior. However, this does not explain the coloration in the predators—tigers, leopards, cheetahs, etc. The careful distinctness of the patterns of different species, however, suggests that they serve as identification. The ability to identify animals of the same species either for herding (animals finding their way back to the herd or smaller group) or for mating may be more important for survival than camouflage. It may also be that individuals—mates, young or others—are identified through the specific distinctions between individual coloring patterns. Regardless, the functional purpose of colors is not directly relevant to the problem of determining a process that can give rise to them—the topic of this chapter.

Why are color patterns interesting as a problem in developmental biology? It would seem that they are quite incidental to more important problems such as the formation of limbs, the development of organs and the formation of networks, neural or vascular. While coloration appears to be superficial, it captures a basic feature necessary for many of the other processes—differentiation. A central problem in development is to assign distinct tasks. In order for limbs to develop, at some point in time there must be an identification of which cells are to proliferate in such a way as to give rise to the limbs, and which cells are not to proliferate. This requires the formation of a pattern in the initially undifferentiated cells. Only after a pattern has been established can the processes associated with differential function of the cells proceed. In a more general context, understanding pattern formation as a form of spatial and temporal structure is a central issue in the formation and function of complex systems in general.

Our objective is to construct mathematical models that can result in the formation of patterns such as those present on the skins of mammals (Fig. 7.2.1). The essentially two-dimensional animal surfaces enable us to illustrate more readily the models than if they were in three dimensions. The models might use a cellular space with a variable representing the color of each cell in an array. Since many of these animals have essentially two colors, we can use a binary variable s_i. This type of model is suggestive of a simple cellular automaton (CA, Section 1.5) where the individual cell determines its state (the color at that location) as a consequence of interactions with neighboring cells. Indeed, the process of intercellular influence in biology is generally suggestive of a CA—as long as communication between cells is local, and we do not consider migration of cells or changes in their shape. The most direct model represents each biological cell by a lattice cell; however, we can also consider a homogenous region of biological cells to be represented by a single lattice cell. Such CA are often natural models for processes that take us from the behavior of an individual cell (or homogenous region) to the inhomogenous behavior of a collection of cells. On a finer scale we can model the diffusion and reaction of chemical messengers between cells and their effect on pigmentation. This provides an additional level of detail to models of such patterns. In Section 7.2.2 we will consider CA models for pattern formation. In Sections 7.2.3 and 7.2.4 we introduce mathematical treatments of chemical diffusion and reaction. Section 7.2.5 describes pattern formation in reaction-diffusion systems. Section 7.2.6 discusses the coupling of a patterned chemical to additional chemical processes. Finally, Section 7.2.7 describes patterns that might form in vertebrates during development by diffusion of pigment cells from their origin along the spinal cord. A discussion of the relative benefits of CA and reaction-diffusion approaches is included later, in Section 7.4.

As will become apparent in the following sections, creating an interacting system that evolves to a pattern requires us to specify interactions that satisfy various constraints. Since systems evolve toward equilibrium, the principle issues are not dynamic, but rather revolve around constructing a model with a complex pattern as its equilibrium or steady-state structure. In simple systems the equilibrium is homogeneous and has no distinguishable or controllable features. The ability to make patterns requires the specification of a system that behaves in an unconventional manner in equilibrium or steady state.

7.2.2 *Activation and inhibition in pattern formation: CA models*

We begin by thinking about the equilibrium behavior of some simple models. For a CA, the equilibrium is generally described by a stochastic field such as the Ising model

Figure 7.2.1 Photographs showing examples of pigment patterns on animal skins. From top left by row: Grant's zebra, South African cheetah, Grevy's zebra, Uganda giraffe, reticulated giraffe and Masai giraffe. These patterns arise from a process that requires differentiation between regions that contain pigment-producing cells and those that do not. The study of such patterns captures one of the essential processes involved in various stages of development that require differentiation in order to form structures and organs that form a functioning physiology. ∎

(Section 1.6). Since the developmental process leads to a long-lived pattern that remains as the color of the animal, this seems a reasonable starting point. Are there indications that such models can give rise to patterns? The seeds of pattern formation are present in the behavior of an antiferromagnet on a square lattice (Fig. 1.6.7) with alternating values of the variables s_i in its equilibrium state. This pattern arises from simple interactions between neighbors that compel adjacent cells to have opposite values of the spin variable. Considered as a color pattern, it is a checkerboard—the simplest of two color patterns (there are only two such patterns). Is there a way to generalize this to form more elaborate patterns characteristic of animal colors? The most basic feature of the color patterns of animals that is not captured by the checkerboard is the existence of a new length scale. This length scale, the size of dots or bands of color, is characteristic of the pattern. It is not given by the size of the cells or by the size of the animal but rather is a characteristic length scale of its own. It is important to consider how such a length scale can arise. An alternating black and white pattern on the scale of individual cells would appear gray on the scale of the organism.

A straightforward method for creating a new length scale in CA is to extend the range of the interactions between cells. We will take this approach and investigate the consequences. Before we do this let us consider what this means from the point of view of biological cells. It might seem that biological cells interact only with adjacent cells. This interaction occurs by emitting chemicals into the intercellular fluid. The chemicals are then detected by the adjacent cells. Such interactions, however, are not necessarily local, because the distance over which the chemicals travel is controlled by their diffusion constant and lifetime in the intercellular fluid or, more correctly, in the matrix of cells and intercellular fluid. Thus an individual cell can interact with a region of cells in its vicinity, where the size of this region is controlled by the diffusion constant of the chemical as well as reactions that might affect it. More direct modeling of diffusion is discussed in the following section. Here we consider only the effective interaction that results between cells.

In order to generate patterns that consist of a large number of cells that are either all black or all white in regions of a characteristic size, we use interactions that extend a distance typical of the linear dimension of the regions. There are two possible types of pairwise interactions between cells. When a cell producing pigment causes other cells to produce pigment we say that the interaction is activating. When a cell causes others not to produce pigment we say that it is inhibiting. As with the discussion of nerve cell interactions in Chapter 2, the terminology and mathematics of activation and inhibition is similar to the use of ferromagnetic and antiferromagnetic interactions that cause the spins in an Ising model to align or antialign. Spins that are UP are producing pigment, while spins that are DOWN are not. Loosely speaking, a ferromagnetic interaction corresponds to mutual activation. An antiferromagnetic interaction corresponds to inhibition.

How can we design an Ising type model that will give rise to domains of locally aligned spins (either ON or OFF) but will have large scale variation so that adjacent to a region of ON cells there will be a region of OFF cells? The interactions must achieve two effects. First, they must cause the cells that are nearby to have a bias toward having the same color so that the regions of color are formed. Second, they must have the effect of causing regions that are farther away to have the opposite color. This suggests a short-range interaction that is mutually activating and a long-range inhibiting interaction or, in magnetic language, a short-range ferromagnetic interaction and a long-range antiferromagnetic interaction. This is the model we will be using to obtain various pigment patterns.

It turns out that the magnetic analogy is not without practical application. Real magnetic materials form magnetic domains. The reason for these magnetic domains is that the short-range ferromagnetic interaction between spins is a local effect of quantum mechanics. However, the long-range interaction between spins is through the magnetic field that tries to antialign the spins—an antiferromagnetic interaction. This gives rise to domains of magnetization that form a pattern of regions of UP and DOWN spins that has a large scale compared to the atomic distances. When a piece of iron is magnetized, it is forced into a metastable state by aligning these magnetic domains. After long enough time, it demagnetizes by returning to its equilibrium state. Modern use of patterns of magnetization appears in magnetic bubble memories that vary external fields to manipulate the patterns of magnetic domains very much in the manner described below.

We will adopt the Ising model terminology of spin variables to describe pattern formation. In Fig. 7.2.2 the spin-spin interaction for a model of pattern formation is plotted as a function of distance. The energy of the system would be written as:

$$E[\{s_i\}] = -h\sum_i s_i - \frac{1}{2}\sum_{i,j} J(r_{ij})s_i s_j \qquad (7.2.1)$$

where $s_i = \pm 1$ is ON and OFF respectively. $J(r_{ij})$ is the interaction as a function of distance r_{ij} between spins. This is similar to Eq. (1.6.52) but includes only a uniform bias field h that controls how likely a cell is to have pigment (ON) as opposed to no pigment (OFF). Writing explicitly the interaction in terms of two parameters $J_1 > 0$ and $J_2 < 0$ we have:

$$E[\{s_i\}] = -h\sum_i s_i - \frac{1}{2}J_1 \sum_{r_{ij}<R_1} s_i s_j - \frac{1}{2}J_2 \sum_{R_1<r_{ij}<R_2} s_i s_j \qquad (7.2.2)$$

What should we expect from the equilibrium structure of this model? The main point is that the existence of long-range antiferromagnetic interactions should cause patches of color. However, we have already found in some cases that the presence of antiferromagnetic interactions causes many low-energy states rather than only a single unique one. While this was true in Section 1.6 only for nonbipartite lattices, we anticipate that it will be true for this more complicated model. Thus we will avoid trying to describe directly the equilibrium states of this model and focus instead on what we are more interested in anyway—the outcome of its dynamics. For convenience, we take a square lattice and start from a random set of values with half of the cells ON. We construct a dynamics for the system, then run it until there are no changes and record the resulting pattern.

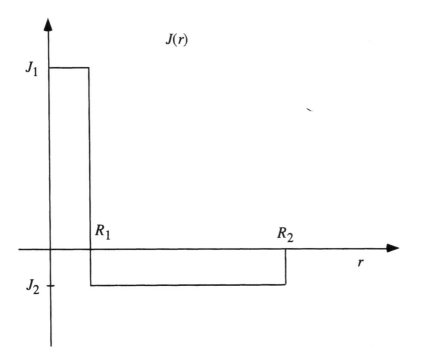

Figure 7.2.2 A CA model of pattern formation uses interactions that cause short-range activation and long-range inhibition of pigment. The interaction as a function of distance $J(r)$ in this model is illustrated. The same model describes interactions that are locally ferromagnetic and long-range antiferromagnetic in a magnetic system. ∎

The dynamics is the same as that used in Chapter 2 for the neural network—zero-temperature Glauber or Monte Carlo dynamics. We choose a particular cell to update and set it ON or OFF according to which gives the lower energy. Stated differently, the cell is set ON if the net interaction of the cells causes it to be ON. The total influence of the other cells is given by the effective field:

$$h_i = h + J_1 \sum_{r_{ij} < R_1} s_j + J_2 \sum_{R_1 < r_{ij} < R_2} s_j \tag{7.2.3}$$

We thus set the value of $s_i(t)$ to be:

$$s_i(t) = \mathrm{sign}\left(h + J_1 \sum_{r_{ij} < R_1} s_j(t-1) + J_2 \sum_{R_1 < r_{ij} < R_2} s_j(t-1) \right) \tag{7.2.4}$$

This equation is quite similar to the equation describing the update of neural cells Eq. (2.2.4). The difference between Eq. (7.2.4) and Eq. (2.2.4) is how we set the values of the interactions between the spin variables, and the presence of a bias h. The pigment cells are locally interacting, while in Chapter 2, neural cells were interconnected throughout by interactions J_{ij}. In Chapter 2 we considered the update of cells to be synchronous because in the presence of random interactions this does not generally cause different results. Here, it is better to update the system asynchronously by selecting cells to update sequentially at random. This avoids oscillations that can occur when all the cells are updated simultaneously.

There are five parameters in this model: the two interaction ranges R_1 and R_2, the two interaction strengths J_1 and J_2 and the bias field h. Since we have not yet chosen the scale of the interaction strength, we can choose it so that one of them takes a convenient value. We set $J_1 = 1$. It is positive, as required for a ferromagnetic interaction. J_2 takes a negative value. It makes sense to choose a value of J_2 smaller than J_1 because J_2 acts over a larger area. We choose $J_2 = -0.1$. We set the value of R_1, the range of the short-range interaction, to a nearest-neighbor distance or $R_1 = 1$. Distance is measured in the cellular space by cell size. The range of R_2 should have something to do with the size of the pattern elements that result. We set this to a value of $R_2 = 6$ to have a large enough value that will be distinct from the nearest-neighbor distance and small enough to be comfortably within the space we simulate, which will be 60 × 60 cells. We start by setting the value of the bias field $h = 0$ and vary it to create patterns with more or less ON or OFF cells. Fig. 7.2.3 illustrates the generation of a pattern from a random starting configuration of the cells. The computer program used to generate these patterns is similar to those used in Sections 1.5 and 1.6 to investigate the dynamics of cellular automata and the Ising model respectively. We can see that long-range inhibition gives rise to alternating regions of colors at a characteristic separation distance.

Fig. 7.2.4 illustrates the variation in patterns that can be generated as a result of changing the value of the bias field h. All of the patterns on the left result from the same initial random array of dots. Similarly, all of the patterns on the right result from a single but different random initial array of dots. Considering the left patterns and the right patterns separately, we can see how the change in the bias field affects the eventual pattern that is formed. At one extreme there are black dots in a sea of white. The dots are not regularly spaced or shaped. They are variable in size and some are elongated. As the value of h is increased, more dots elongate and connect forming bands that interconnect and eventually become the black background in which white dots exist. These patterns are reminiscent of some animal color patterns.

In Fig. 7.2.5 we investigate the effect of increasing the value of R_1 (right panels) and R_2 (left panels). The most obvious changes occur with R_2. The characteristic size of the pattern increases and is directly controlled by R_2. Increasing R_1 does not affect the size of the pattern but rather the shape of the boundaries between regions of ON and OFF cells. Increasing R_1 ensures that the boundaries of dots and stripes are smoother, with more gradual changes in curvature. This is particularly apparent in our simulations because the size of R_1 is comparable to the size of cells. For more realistic animal color patterns, the size of R_1 should be larger than the size of cells, to avoid sharp corners.

The initial conditions of the simulation can be important. We started these simulations with 50% of the cells set ON at random. The effect of the initial random

Figure 7.2.3 A simulation of a CA model of pattern formation. ON cells (black) produce pigment and OFF cells (white) do not. The initial conditions assign cells to be ON or OFF at random with probability 1/2. Five updates are shown and then the unchanging stable limit that is reached after about 20 updates. The parameters are $R_1 = 1$, $R_2 = 6$, $J_1 = 1$, $J_2 = -0.1$, and $h = 0$. ∎

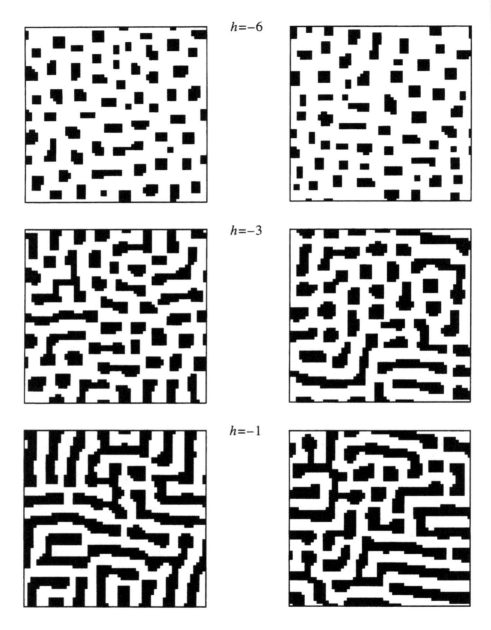

Figure 7.2.4 Additional simulations of the CA model that illustrate the effect of varying the bias field h; other parameters are the same as Fig. 7.2.3. All patterns shown are the unchanging stable limit of a simulation. h biases the system to have more or less ON cells. Varying h results in patterns with black spots on a white background, white and black stripes, or white spots on a black background. The left and right panels differ only in the initial conditions of the simulation. All of the left panels start with the initial condition shown

$h=1$

$h=3$

$h=6$

in Fig. 7.2.3. The right panels begin from a different random initial condition. We see that left and right panels share general characteristics but are different in detail. While both initial conditions have a probability of 1/2 that cells are ON, qualitative aspects of the final patterns are not sensitive to the initial probability, since they are determined by the ensemble of stable states of the system. ∎

$R_2=6.0$ $R_1=1.0$ $h=0$

$R_2=6.0$ $R_1=1.5$ $h=0$

$R_2=7.0$ $R_1=1.0$ $h=0$

$R_2=6.0$ $R_1=1.5$ $h=-3$

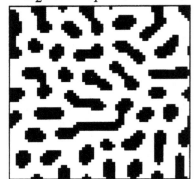

$R_2=8.0$ $R_1=1.0$ $h=0$

$R_2=6.0$ $R_1=1.5$ $h=-6$

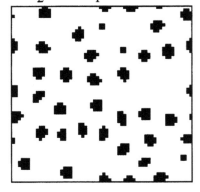

Figure 7.2.5 Changing other parameters in the CA model. Each pattern is the steady state of a simulation with the parameters indicated, and for all cases $J_1 = 1$ and $J_2 = -0.1$. The left panels show the variation in the spatial scale of the pattern that results from changing the range R_2 of the antiferromagnetic interaction. Simulations for three different values of R_2 are shown. There is a direct relationship between R_2 and the size of the stripes. The right panels show patterns that arise when the range R_1 of the ferromagnetic interaction is increased. Simulations with three values of h are shown all with the same increased value of R_1. The top right panel should be compared with the top left panel. The other two right panels should be compared to the panels of Fig. 7.2.4 with the same value of h. The effect of the increase in R_1 is to round the corners of the spots and stripes. ∎

configuration is apparent from the nonuniform nature of the pattern, and the two different results shown in left and right panels of Fig. 7.2.4. Changing the initial proportion of ON cells has very little effect on the qualitative behavior of the model because the resulting pattern is essentially an equilibrium pattern—one of many with similar number and shapes of color regions. However, the specific pattern of dots and their shapes is sensitive to the precise starting pattern of ON and OFF cells. If we consider this as a theory of the origin of animal color patterns, it suggests that individual differences may be due to randomness rather than genetic control, while the overall characteristics are controlled by the underlying mechanism, which is genetic and species specific. In this case the particular pattern is not heritable, and even identical twins would have different patterns. This should not be the case with many other characteristics.

Question 7.2.1 In the model we have just simulated, patterns appear to arise in equilibrium. We have argued in Section 1.3 that equilibrium systems have simple behavior. Why doesn't this apply in our case?

Solution 7.2.1 The thermodynamic limit discussed in Section 1.3 applies when we take the limit of large enough system size. The results in this section do not apply when we take this limit, since then the system would appear uniform and homogenous, because the size of the spots that we are discussing would be so small as to be irrelevant. When this limit is not used, then the conclusions also do not apply.

A more thorough discussion would note that there are actually two conditions that are not met by these systems, consistent with the discussion in Section 1.3.6.

First, the ergodic theorem does not apply. This means that the ensemble of possible states of the system is not being explored. This is apparent when we consider that the system iterates to a steady state and that this steady state is a unique state that is unchanging even though there are many such possible states. Moreover, when the ergodic theorem applies, the initial state is irrelevant to the final equilibrium state. The reason that this model breaks the ergodic theorem in such a direct way is that we are modeling it at zero temperature—the temperature is so low that no random changes occur. The only changes are those dictated by energy reduction. In a system where temperature causes random changes, there would be a time dependence to the pattern. If our observations of such a system were averaged over a long time, we would not see any specific pattern, but only a homogeneous average. If our observations were averaged over a shorter time, we would see an individual pattern.

Second, in these simulations a correlation length exists that is not small on the scale of the whole system. The length scale that is relevant is the characteristic length scale of the pattern. In some patterns it may actually be larger than the size of the stripes or dots, since the positions of stripes can be correlated with each other. We can see the relevance of the pattern length scale by considering what would happen if we observed a system that was much larger than this length scale. Then the pattern would become irrelevant and the color would be gray on the scale of our observations.

The relevance of temporal and spatial scale to the complexity of a system will be discussed further in Chapter 8. ■

Question 7.2.2 Consider a model using variables $\bar{s}_i = 0,1$ to represent unpigmented and pigmented cells. We will use overbars to indicate all quantities in this model. Set the update rule to be similar to that in Eq. 7.2.4 but with the bias field $\bar{h} = 0$. When the effective local field $\bar{h}_i$ is negative, the cell is set to 0; when it is positive the cell is set to 1. This is a more intuitive representation of activation and inhibition since both are effected by cells that are ON. Cells that are OFF have no influence on the pigment production in other cells. It is also assumed that there is no tendency for cells to spontaneously become pigment producing. Simulate this model and vary the strength of the inhibition $\bar{J}_2$ to obtain various patterns. How can this model be transformed back to that given in the text?

Solution 7.2.2 Using the same parameters as the text except for the value of J_2 and h, the results of simulations are shown in Fig. 7.2.6.

To transform this model to that in the text, we can perform the substitution $\bar{s}_i \rightarrow (s_i + 1)/2$ so that $0,1 \rightarrow -1,+1$. Once this substitution is performed in Eq. (7.2.3) we can recognize the parameters that would give the same patterns in the original model:

$$J_1 = \bar{J}_1/2$$
$$J_2 = \bar{J}_2/2$$
$$h = \bar{J}_1/2 \sum_{r_{ij}<R_1} 1 + \bar{J}_2/2 \sum_{R_1<r_{ij}<R_2} 1$$

$$(7.2.5) ■$$

Question 7.2.3 Consider what will happen if $R_1 = 0$, i.e., there is no activation in the model of the text and $\bar{R}_1 = 0$ for the model of Question 7.2.2.

Figure 7.2.6 Using a different parametrization of the CA model for pattern formation with $\bar{s}_i$ = 0, 1 and $\bar{h}$ = 0 in Eq. 7.2.3, we generate patterns that are similar to Fig. 7.2.3 by varying the strength of the inhibition $\bar{J}_2$. The other parameters were taken to be $\bar{R}_1 = 1$, $\bar{R}_2 = 6$, $\bar{J}_1 = 1$. Left and right panels use different initial random conditions similar to Fig. 7.2.4 (see Question 7.2.2). ∎

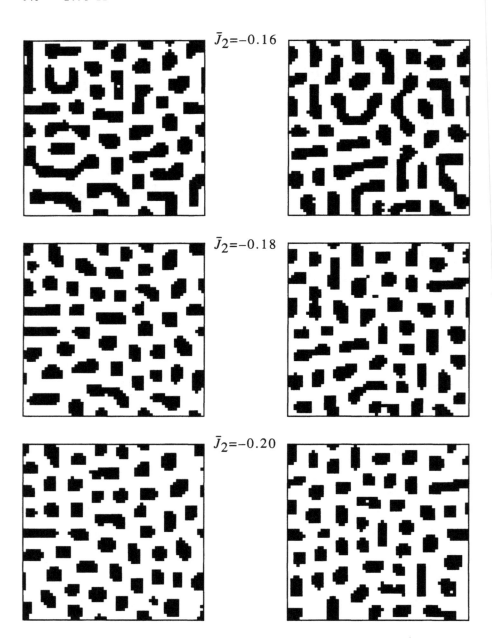

$\bar{J}_2=-0.16$

$\bar{J}_2=-0.18$

$\bar{J}_2=-0.20$

Figure 7.2.6 (*continued*)

Solution 7.2.3 Since the pattern formation seems to depend on local activation and long-range inhibition, we might think that setting $R_1 = 0$ would eliminate pattern formation. However, in the case of the model discussed in the text, patterns still form. In order for the long-range antiferromagnetic interaction to lower the system energy, it is necessary for regions to be locally ferromagnetic. Another way to understand this is that the long-range interaction controls the long-range properties of the pattern while, as seen in Fig. 7.2.5, only the boundary shapes are controlled by the short-range interaction. We could even have a short-range antiferromagnetic interaction and still have patterns, as long as the short-range interaction is not too strong. However, when we consider the model of Question 7.2.2 we realize that in order for cells to turn ON there must be a local activation, otherwise cells can only turn OFF in the dynamics. From Eq. (7.2.5) because $\bar{J}_2 < 0$, $\bar{J}_1 = 0$ would correspond to a large negative h in the model discussed in the text. The large negative h would likewise prevent any cell from turning ON. ∎

The patterns that can be generated using the activation-inhibition CA model suggest that the variability between different species can be readily achieved by variations in parameters of such models. However, this particular set of patterns does not capture the appearance of many of the common animals. One example is the giraffe (specifically the Uganda giraffe), which has patterns of coloration characterized by regions of pigment separated by relatively narrow and straight lines without pigment. We will discuss an approach to generating such patterns which illustrates that there may be other mechanisms for generating patterns of a certain scale. The method begins by noting (Fig. 7.2.7) that giraffe patterns appear to be similar to patterns generated in two steps. First we choose a sparse set of initial dots. Then we divide the plane into regions associated with each dot. The region associated with a dot consists of all points that are closer to it than any other dot. Then the boundaries of these regions are not colored, while the interiors are.

To generate this pattern, we use a CA that grows regions of color from isolated points, which are cells initialized ON. The growing regions then stop when they reach the proximity of another region that is growing. In this rule, the characteristic size of the pattern is given by the density of the initial ON cells. This would be similar to a process of nucleation and growth (Section 1.6.8), where nucleation creates the isolated points that expand rapidly compared to the nucleation time. The CA rule we use is similar to those described in Sections 1.5.2 and 1.5.3, for the condensation model and Conway's Game of Life. We will construct the rule step by step.

To allow regions to grow, a cell is set ON at time t if at time $t - 1$ there were more than zero cells ON in its neighborhood. This results in growth from a point expanding into the space in a uniform fashion. Because of our square lattice, there is a problem in the shape of growth—it is not circular as would be expected in a physiological system. By expanding the range of influence of a single cell, which corresponds to increasing the size of the neighborhood, we can make it more circular, as illustrated in Fig. 7.2.8.

Figure 7.2.7 The patterns of coloration on a giraffe can be understood geometrically. They appear to be generated by dividing the two-dimensional surface according to their distance from a sparse selected set of points. In this figure the selected set of points are indicated by circles. Line segments connecting them are shown as thin lines. By coloring areas that are close to each of the selected points, but not points that are approximately the same distance from two or more points, we can generate patterns similar to those found on some species of giraffe. ∎

In order to leave uncolored the regions between growing dots, cells must recognize when the two growing regions meet. When the pigment grows from a point, the shape of the ON region is convex. We can identify a cell in the encounter region because it has more ON cells around it than a cell at the boundary of the growing region. A cell that has more than a certain number of neighbors with pigment must be in the encounter region and should not turn ON. Thus, the CA sets a cell ON if it has some but not too many ON neighbors. Note that in this model we are not allowing cells that are ON to turn OFF. This is important, because otherwise cells in the interior of a spot would turn OFF once we impose the condition that stops the growth.

We start the growth by setting cells ON at random with a probability of 1 in 100. The result of this simulation, illustrated in Fig. 7.2.9, is not very satisfactory. Some of the cells at the boundaries of growing regions do not turn ON. However, these do not form continuous lines. To overcome this problem we need to have wider regions of

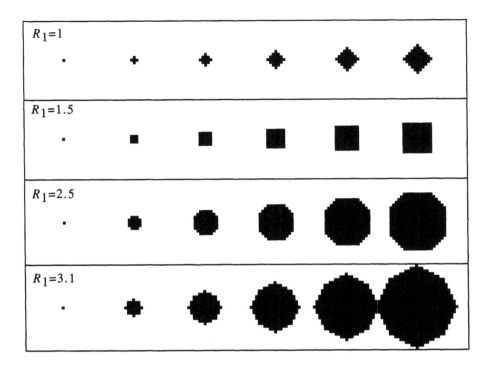

Figure 7.2.8 Starting from a seed pigment cell we can grow outward using a rule that sets a cell ON if there are any ON cells in its neighborhood. However, the shape of the growing region on a square lattice depends on the way we grow it. Here, growth of a region is shown for various sizes of the neighborhood given by its radius R_1. A larger R_1 leads to more circular pigmented areas. ∎

OFF cells. To achieve the desired result, we can take a clue from the previous model of pattern formation and set up two distances, a distance R_1 over which the growth is determined, and a distance R_2 over which the stopping is determined. Thus, using binary variables $s_i = 0,1$ we turn a cell ON when the value of Eq. (7.2.3) is positive, with parameter values of $R_1 = 2.5$, $R_2 = 4.3$, $J_1 = 1$ and $J_2 = -.5$. The values of these parameters can be adjusted by trial and error.

The patterns generated in Fig. 7.2.10 using this approach are reminiscent of the patterns of giraffes; however, they are not entirely satisfactory. While some of the regions follow the convex shape that we expect, other regions are more convoluted. By looking carefully at the patterns, we see that this occurs because the separations between the initial ON cells vary in distance. This would not occur if the starting points were more regularly spaced. There are many ways to consider placing the points at more regular intervals. A reasonable approach for this case is to use the previous method of creating patterns using activation-inhibition to generate a pattern of spots such as those shown in Fig. 7.2.4 and then to apply the growth starting from these spots. This is illustrated in Fig. 7.2.11, where the initial pattern is generated

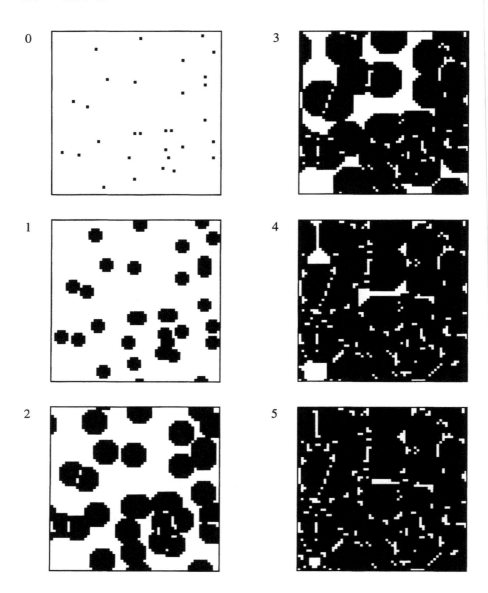

Figure 7.2.9 A first attempt at forming a color pattern similar to that of a giraffe. The initial conditions are obtained by setting cells to be ON at random with a small probability, here taken to be 1 in 100. The algorithm updates the cells synchronously and sets them ON if the number of cells in a neighborhood of radius $R_1 = 2.5$ is nonzero, but also less than 10. The color grows out from the initial ON cells. When growing regions meet, there are some cells that do not turn ON because of the limiting condition on the number of cells in the neighborhood. However, these regions of residual OFF cells are not continuous. ∎

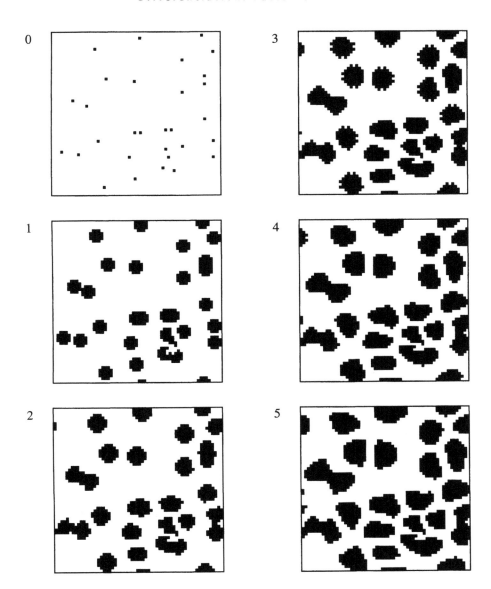

Figure 7.2.10 Better simulations of the formation of giraffe patterns than those in Fig. 7.2.9 result if we use a larger region to set the condition for stopping growth. The parameters, adjusted by hand, are inspired by the activation-inhibition model. The growth results from activation of cells adjacent to cells that are ON, while too many ON cells in a larger region (long-range inhibition) cause cells not to turn ON. Here a particular simulation is shown from its initial condition for seven updates, and then the final stable result. Three outcomes starting from other random initial conditions are shown in the rightmost column. All initial conditions were set with a probability 1 in 100 of cells being ON. These simulations are not entirely satisfactory because many of the spots have unusual shapes. ∎

6

7

∞

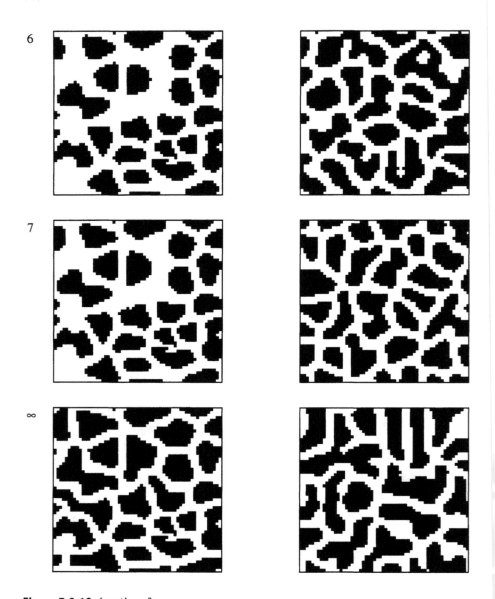

Figure 7.2.10 (*continued*)

from a CA activation-inhibition model, resulting in more regular but still randomly placed spots. By growing out into the OFF regions we form a pattern that is closer to the patterns on the giraffe coats. More specifically, this coloration is similar to that of the Uganda giraffe (Fig. 7.2.1). Two other kinds of giraffe—the reticulated giraffe and the Masai giraffe—would require additional tuning of parameters. The reticu-

$R_2=12.0$ $R_1=1.0$ $h=-25$

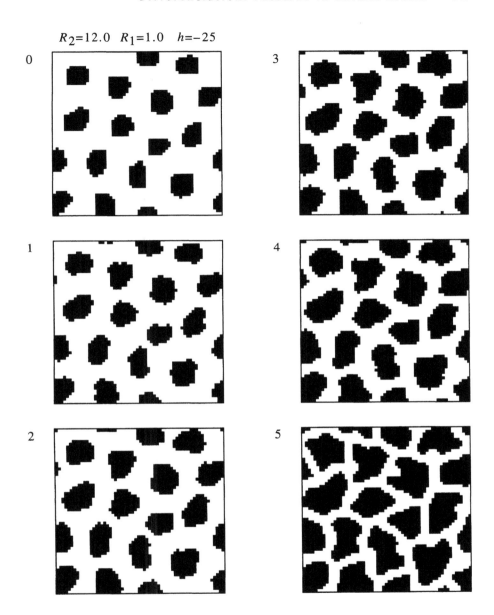

Figure 7.2.11 The giraffe color patterns generated in Fig. 7.2.10 can be improved by start-ing from points that are more regularly spaced in the plane. They might be placed more reg-ularly by several processes, one of which is illustrated here. The initial conditions result from an activation-inhibition CA model simulation with parameters as indicated on the upper left. This is the starting conformation for the growth outward of pigmented regions. The subse-quent frames show updates using the same algorithm as Fig. 7.2.10. This results in a more regular pattern reminiscent of the Uganda giraffe. Other patterns can be generated by vary-ing the parameters. ∎

lated giraffe would be generated by a smaller ratio of the line width to the size of the spots. This requires a finer mesh of points but could be simulated by the same algorithm. The third kind of giraffe, the Masai giraffe, has spots that are blotches with fingering. Such fingering can also be achieved by varying the parameters in this algorithm.

7.2.3 *Chemical diffusion*

We can add an additional layer of detail in our models by considering more directly the properties of molecules produced in cells and their motion through the matrix of cells and intercellular fluid. Molecules generally move by a random walk that is not directed but results from the random thermal motion of the liquid, mostly water, in which they are located. A single molecule undergoing a random walk travels a characteristic distance proportional to the square root of the time, or $\sqrt{Dt}$, where D is the diffusion constant. The probability distribution of the behavior of a single molecule also describes what happens to a density of weakly interacting molecules. If there is a localized density of molecules at one place, it will spread over time and the distribution will approximate a Gaussian that broadens and flattens over time (Section 1.2). This molecular motion, diffusion, in the continuum limit is described by a differential equation (Section 1.4.4) that represents the changes in density $n(x;t)$ with time when it is sufficiently smooth:

$$\frac{dn(\mathbf{x};t)}{dt} = D\nabla^2 n(\mathbf{x};t) \qquad (7.2.6)$$

This discussion suggests that we consider pattern formation arising from a differential equation representing the evolution of molecular density. This approach was taken by Turing, more generally known for the invention of Turing machines discussed in Section 1.9.4. The resulting color patterns are known as Turing patterns.

The CA approach in the previous section treated diffusion as an incidental process which was summarized by an effective interaction between the cells. This simplified the study of the process of pattern formation so that the activation and inhibition were readily apparent. In this and the following section we construct two essential parts of the differential equation approach—the diffusion and reaction of molecules. Then we discuss and simulate specific sets of equations that give rise to patterns.

We derived the diffusion equation (Eq. (7.2.6)) in Section 1.4.4 from the motion of a particle in a periodic set of wells. It is more usually derived from the motion of a low-density "gas" of molecules that have a varying density profile as a function of position as illustrated in Fig. 7.2.12(a). We consider the current $J(x)$ of molecules at a particular position x and relate this current to the variation of the density with position $n(x)$. In order to obtain the current, we make use of simplifying assumptions. The result is more general than the assumptions suggest. We assume that molecules undergo instantaneous collisions with a fluid or matrix in which they are embedded. The characteristic time between collisions is τ. In between collisions, particles have a characteristic velocity v and travel a distance $l = v\tau$. v is determined by thermal motion—

(a)

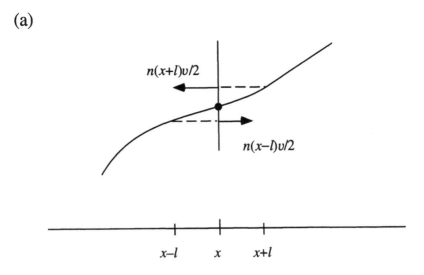

(b)

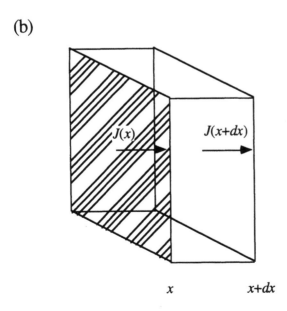

Figure 7.2.12 We derive the diffusion equation using a model consisting of a weakly inter-acting nonuniform density of particles embedded in a medium. The derivation relates the change in density with time to the spatial variation in density. It takes two steps: (a) the particle current at a point x is related to the spatial variation in density, and (b) the change in density with time is related to the spatial variation in the current. Consult the text for details. ∎

it is controlled by the temperature—and τ is related to the interactions with the fluid or matrix, so neither depend on the density $n(x)$.

When we look at a position x we see molecules traveling to the right and to the left. These molecules originated a distance l to the left and a distance l to the right respectively. At these locations their density was $n(x - l)$ and $n(x + l)$ respectively. Since we expect half of the molecules from $n(x - l)$ to be traveling to the right and half at $n(x + l)$ to be traveling to the left, we infer that the current at x is given by:

$$J(x) = \frac{v}{2}(n(x-l) - n(x+l)) \approx -lv\frac{dn(x)}{dx} = -v^2\tau\frac{dn(x)}{dx} \qquad (7.2.7)$$

where we have expanded in a Taylor series keeping the first term, and thus assuming that l is small compared to distances over which the density varies significantly.

We want to describe the changes in $n(x)$ as a function of time. To do this we also need the continuity equation that relates the current to the change in density. From Fig. 7.2.12(b) describing the change of density in a small box in terms of the currents at two faces, this is given by

$$\Gamma\Delta x \frac{dn(x;t)}{dt} = (-J(x+\Delta x/2;t) + J(x - \Delta x/2;t))\Gamma = -\Gamma\Delta x \frac{dJ(x;t)}{dx} \qquad (7.2.8)$$

where Γ is the area of a face and Δx is the length of the side. Combining Eq. (7.2.7) and Eq. (7.2.8) we have the diffusion equation:

$$\frac{dn(x;t)}{dt} = v^2\tau\frac{d^2n(x;t)}{dx^2} \qquad (7.2.9)$$

This is generalized to Eq. (7.2.6) when the density varies in three dimensions.

The many assumptions in this derivation can be avoided if we consider Eq. (7.2.6) as an expansion in the density and its derivatives (Question 7.2.4). The right side is the lowest-order term that is not excluded by symmetries of the problem. It controls the longest spatial and temporal behavior. This is the reason for the applicability of the diffusion equation under a large variety of circumstances.

Question 7.2.4 We want to write a differential equation describing the time dependence of the density

$$\frac{dn(x;t)}{dt} = \cdots \qquad (7.2.10)$$

in terms of various spatial derivatives—local properties—of the density. Consider including terms that involve up to three derivatives (in three dimensions) of $n(x;t)$:

$$\frac{d}{dx}n(x;t), \quad \frac{d^2}{dx^2}n(x;t), \quad \frac{d^2}{dxdy}n(x;t), \quad \frac{d^3}{dx^3}n(x;t), \quad \frac{d^3}{dx^2dy}n(x;t), \quad \frac{d^3}{dxdydz}n(x;t)$$

$$(7.2.11)$$

Argue:

a. That of these terms only the second term can be used.

b. There are additional terms involving four derivatives that can be used.

c. That terms of the form

$$n(x;t)\frac{d^2}{dx^2}n(x;t), \quad \left(\frac{d}{dx}n(x;t)\right)^2 \qquad (7.2.12)$$

become smaller than the second term in Eq. (7.2.11) when the density is small enough.

d. That terms that do not involve derivatives—a function of $n(x;t)$ itself or a constant—cannot be included if the number of molecules is conserved.

Solution 7.2.4

a. $dn(x;t)/dt$ does not change when we invert any of the spatial coordinates, for example by setting $x \to -x$. Thus any term on the right-hand side of Eq. (7.2.10) must also not change. Since the inversion of x changes the sign of dx no odd derivatives are admissible and only the second term is possible.

b. Additional fourth-order terms are of the form:

$$\frac{d^4}{dx^4}n(x;t), \quad \frac{d^4}{dx^2dy^2}n(x;t) \qquad (7.2.13)$$

These terms are corrections to the diffusion equation and must be used if the spatial variations in the density are large enough, or we are concerned about behavior on a small enough length scale.

c. Consider multiplying the density by a factor λ. The terms listed in Eq. (7.2.13) vary as λ^2 while those in Eq. (7.2.11) vary as λ. Thus at low enough density these terms are insignificant.

d. Consider the case of a uniform density $n(x;t) = n_0$. Any function of $n(x;t)$ that does not involve derivatives will give a changing density that must be the same everywhere. A uniform changing density does not conserve the number of molecules. Thus we cannot include such terms. We are implicitly assuming that x itself does not appear in the equation—points in space are indistinguishable before molecules are placed there. Otherwise this argument would not be valid. ∎

Diffusion causes molecules on average to move from higher density regions to lower density regions. This can be readily understood from the random-walk behavior of the molecules and the discussion in Section 1.2. This motion leads to a more uniform density profile. Thus if there is a nonuniform pattern of molecular density initially imposed on a system, diffusion leads to a loss of the pattern through the smoothing of the density. The key problem in discussing color patterns is identifying how we can cause nonuniform densities to arise out of diffusing molecules. As we remarked before, this is related to the fundamental problem that equilibration generally causes uniformity and lack of structure.

The solution to this problem is through the interaction of more than one type of molecule. Recognizing this was central to the contribution of Turing. The interactions are chemical reactions that change the local densities of molecules. In addition to the reacting molecules, the reactions may involve catalysts that accelerate them. Of particular importance are autocatalyzing reactions where molecules that are reacting are also catalysts. Autocatalysis causes a nonlinear dependence of the reaction rate on the densities. Systems of reacting and diffusing molecules are called reaction-diffusion systems.

7.2.4 *Chemical reactions*

Chemical reactions cause molecular densities to change with time even when there is no diffusion. A reaction may combine different molecules, decompose a molecule into parts or just change the structure of a molecule. We write the general dynamic behavior of the molecular densities using a set of coupled equations of the form:

$$\frac{dn_i(x;t)}{dt} = D\nabla^2 n_i(x;t) + R_i(\{n_j(x;t)\})$$

(7.2.14)

where $R_i(\{n_j(x;t)\})$ is the rate of change in the concentration of a molecular species i due to generation or annihilation in reactions that involve other molecular species. In order to solve such equations, it is necessary to have an expression for $R_i(\{n_j(x;t)\})$ in terms of the densities of the molecules present.

As with diffusion, a discussion of reaction rates requires some simplifying assumptions. In writing Eq. (7.2.14) we have already assumed that the density is not too rapidly varying in space, so that the local reaction rate depends only on the local densities and not their gradients. We will also assume that the diffusion time of a molecule between reactions is large compared to the time of a reaction. This assumption implies that the limiting step in the rate of reaction is the rate at which molecules encounter each other. In order to satisfy this assumption, we need three conditions: that interactions between molecules are short range, that the molecular densities are low and that once the molecules encounter each other the reaction is fast. For simplicity we can think of this as a low density limit. As in the discussion of diffusion, violations of the assumptions can be incorporated in the equations when necessary.

Under these assumptions the rate of a reaction involving molecules A, B and C (with molecular densities n_A, n_B and n_C of the form

$$A + B \rightarrow C$$

(7.2.15)

is proportional to the probability of encounter of the reagents—it is proportional to:

$$n_A n_B$$

(7.2.16)

This follows from our assumptions because each molecule diffuses and reacts independently of other molecules of the same type. Thus, the probability of a reaction is proportional to the reactant concentrations. Since one molecule of A and of B disappears for every reaction, and a molecule of C appears, the rate of change of the densities, due to this reaction, are given by:

$$\frac{dn_A}{dt} = -k_1 n_B n_A$$

$$\frac{dn_B}{dt} = -k_1 n_B n_A \quad\quad\quad (7.2.17)$$

$$\frac{dn_C}{dt} = k_1 n_B n_A$$

where k_1 is positive, and called a reaction constant.

The reverse reaction

$$A + B \leftarrow C \quad\quad\quad (7.2.18)$$

has a rate which is proportional to n_C. Including this in Eq. (7.3.17) results in the equations:

$$\frac{dn_A}{dt} = -k_1 n_B n_A + k_2 n_C$$

$$\frac{dn_B}{dt} = -k_1 n_B n_A + k_2 n_C \quad\quad\quad (7.2.19)$$

$$\frac{dn_C}{dt} = k_1 n_B n_A - k_2 n_C$$

Thus, reactions give rise to differential equations coupling the densities of different molecules.

It is important to emphasize that reactions we write in the form of Eq. (7.2.15) and Eq. (7.2.18) are to be considered elementary reactions that reflect actual molecular encounters. In chemistry, the same notation is often used to describe the net consequence of many reactions. The reaction then reflects only the proportions of molecules involved (stoichiometry). The rate of the reaction is not proportional to reactant density, and therefore must be determined separately.

There are three approximations that can be used to simplify the equations resulting from chemical reactions. These are the condition of quasi-equilibrium, the extreme kinetic regime and the quasi-static regime.

If the two reactions Eq. (7.2.15) and Eq. (7.2.18) are in equilibrium, then the density of A no longer changes with time and we can set Eq. (7.2.19) equal to zero. This gives a relationship between the densities:

$$n_B n_A = k_2' n_C \quad\quad\quad (7.2.20)$$

where $k_2' = k_2/k_1$. When a reaction is close to equilibrium and we disturb the conditions by adding one of them, then the reaction will act to change the densities of the other chemicals to restore equilibrium. If this were the only reaction we were interested in, then the equilibrium would describe all of the dynamics. However, the molecules might be involved in additional reactions that are slower. Then the fast reaction that restores equilibrium always maintains the chemicals involved in a quasi-equilibrium. Under these conditions we may use the relationship of Eq. (7.2.20) to simplify the system of equations.

The second simplifying circumstance is when the densities of molecules are far from equilibrium. Then one of the two terms in Eq. (7.2.19) will be much larger than the other. In this case we may consider a reaction as proceeding only in one direction. This is the kinetic regime of the reaction, where equilibrium is essentially irrelevant.

The third simplifying circumstance is the quasi-static regime. It is applicable when a quantity is slowly varying on the time scale of observations. The simplest way this can occur is for one of the molecules in a reaction to have a much larger density than the others. Then the change in its density, as compared to the density itself, can be negligible. For example, if the density of C in Eq. (7.2.19) is very large compared to the other molecules, and the value of k_2 is not too large, we may be able to approximate the second term and write, for example:

$$\frac{dn_A}{dt} \approx -k_1 n_B n_A + k_2'' \tag{7.2.21}$$

where $k_2'' = k_2 n_C$ assumes that n_C is approximately constant. This describes a constant source of the molecule A implicitly originating from molecule C. n_C need not be explicitly written when it is essentially constant.

We will be interested in two sets of chemical reactions. The first is the activator-inhibitor system. It represents activation and inhibition more directly, and can be described by

$$A \to 0$$
$$B \to 0$$
$$2A + D \to 2A + B \tag{7.2.22}$$
$$2A + C \to 3A + C$$
$$C + B \leftrightarrow E$$

The second is the activator-substrate system. It is simpler and implements the properties of activation and inhibition in a more indirect way to be explained later, and can be described by:

$$A \to 0$$
$$0 \to B \tag{7.2.23}$$
$$2A + B \to 3A$$

We discuss simplifications of our treatment of the reactions using the methods discussed above. The discussion will justify the functional form of the differential equations used in the next section.

In both sets of reactions, we have used the convention that 0 represents a chemical species whose density is not of relevance to our discussion. When 0 produces a relevant molecule (e.g., $0 \to B$) it has a large enough density so that any change is insignificant over the time of observation. This is the quasi-static approximation. We also denote by 0 a molecule produced by a reaction that is inert (A → 0). This is one way the extreme kinetic limit manifests itself in the reactions. Another way it does so is in all the reactions that have only one direction indicated. The reverse direction is assumed to be irrelevant. There may be other molecules involved in reactions that are not indicated at all. For example we could also write the third reaction in Eq. (7.2.23) as $2A + B + 0 \rightarrow 3A + 0$, where the two 0s indicate molecules whose density is unchanging (the first) or that are inert (the second). Indeed, one of the reactions would definitely not make sense without additional reactants. (Which one?) We will also use the quasi-equilibrium approximation to describe the last reaction in Eq. (7.2.22). Before we describe this, we will discuss the nonlinear reactions that appear in these systems.

Both activator-inhibitor and activator-substrate systems have reaction rates that depend in a nonlinear fashion on molecular densities. The simplest example of a nonlinear dependence is a molecule that reacts with itself:

$$2A \rightarrow B \tag{7.2.24}$$

which would give rise to two coupled equations of the form:

$$\frac{dn_A}{dt} = -k_3 n_A^2$$
$$\frac{dn_B}{dt} = k_4 n_A^2 \tag{7.2.25}$$

The value of k_3 would be twice as large as k_4 because of the loss of two molecules of A per reaction.

More complex examples of nonlinear dependence result from autocatalysis. First we describe simple catalysis. A catalyst accelerates reactions by creating an additional pathway for the reaction. An example would be:

$$A + B + D \rightarrow C + D \tag{7.2.26}$$

where D is a catalyst since, regardless of intermediate stages, it reappears at the end of the reaction. The density of the catalyst affects the rate of the reaction, but the reaction does not affect the density of the catalyst. An example that appears in Eq. (7.2.22) with A as a catalyst is:

$$2A + D \rightarrow 2A + B \tag{7.2.27}$$

This would give rise to two coupled equations of the form:

$$\frac{dn_D}{dt} = -k_5 n_A^2 n_D$$
$$\frac{dn_B}{dt} = k_5 n_A^2 n_D \tag{7.2.28}$$

Since there is no change in number of A molecules, there is no effect on dn_A/dt.

In autocatalyzed reactions one of the reactants also acts as a catalyst. An example from Eq. (7.2.23) is:

$$2A + B \rightarrow 3A \tag{7.2.29}$$

Each reaction results in the gain of a molecule of A and the loss of a molecule of B. A also acts as a catalyst. The related differential equations take the form:

$$\frac{dn_A}{dt} = k_3 n_A^2 n_B$$

$$\frac{dn_B}{dt} = -k_3 n_A^2 n_B$$

(7.2.30)

If two new molecules of A appeared in the reaction, $2A + B \to 4A$, we would still have the same functional dependence $n_A^2 n_B$. However, the coefficients in the two equations would differ by a factor of 2.

We now consider the last reaction in Eq. (7.2.22) that we will treat using a quasi-equilibrium condition. Some care must be exercised in simplifying equations based upon the interplay between fast processes and the dynamics we are observing. This example is relatively simple because the density of C is only affected by the last reaction. It acts as a catalyst in the second to last reaction. We assume that the last reaction is rapid and therefore maintains a relationship between n_C, n_B and n_E similar to that in Eq. (7.2.20):

$$n_B n_C = k_2' n_E$$

(7.2.31)

To simplify matters further, we assume that n_E is always very large and effectively constant. Then n_C would be inversely proportional to n_B.

$$n_C = \frac{k_2'''}{n_B}$$

(7.2.32)

We can use this relationship in other equations. To illustrate this we write an equation for the time dependence of n_A from the first reaction and the second to last reaction in Eq. (7.2.22):

$$\frac{dn_A}{dt} = -k_1 n_A + k_3 n_A^2 n_C \approx -k_1 n_A + k_3' n_A^2 / n_B$$

(7.2.33)

We see that increasing the density of B reduces the rate of the reaction of A through mediation by C. We can say that B inhibits the reaction that affects the density of A. The next problem is to describe the rate of change of n_B. Since B is affected by several reactions in addition to the fast, quasi-equilibrium one, this is more complicated. We can think about the problem as writing a set of equations that no longer contains the variable n_C. While it is not overly difficult to do this, we can simplify matters further by assuming conditions that decouple the behavior of n_B from the quasi-equilibrium equation. This requires that n_B is significantly greater than n_C ($n_C \ll n_B$). To see how this works we write the complete equations (from Eq. (7.2.22)) that affect n_B and n_C:

$$\frac{dn_B}{dt} = -k_4 n_B n_C + k_5 n_E + (k_3 n_A^2 n_D - k_2 n_B)$$

(7.2.34)

$$\frac{dn_C}{dt} = -k_4 n_B n_C + k_5 n_E \qquad (7.2.35)$$

The density n_C changes only through the reaction that is in quasi-equilibrium. n_B has the same terms, but also the terms in parenthesis reflecting the additional reactions that include B. If $n_C \ll n_B$ then any change in n_C is also much smaller than n_B. Thus we can neglect the first two terms in the rate of change of n_B which are the same as the rate of change of n_C. Then we are left with only the terms in parenthesis:

$$\frac{dn_B}{dt} = k_3 n_A^2 n_D - k_2 n_B \qquad (7.2.36)$$

A more complete treatment is discussed in Questions 7.2.5 and 7.2.6.

Question 7.2.5 Write an expression instead of Eq. (7.2.32) for the dependence of n_C on n_B when we cannot assume that n_E is unchanging.

Solution 7.2.5 In order to obtain the more general form of Eq. (7.2.32) we must recognize that the sum $n_E + n_C$ is conserved in the reactions of Eq. (7.2.22). We can define this sum to be n_0 and write the quasi-equilibrium condition Eq. (7.2.31) as:

$$n_B n_C = k_2'(n_0 - n_C) \qquad (7.2.37)$$

or

$$n_C = k_2' n_0 / (n_B + k_2') \qquad (7.2.38)$$

We see that as long as n_B is larger than k_2' this correction can be ignored. The correction will be important when we do simulations later because it is unphysical that the rate of change of n_A given in Eq. (7.2.33) diverges when the density of n_B is small. ∎

Question 7.2.6 Derive an equation instead of Eq. (7.2.36) that incorporates an approximate quasi-equilibrium relationship but doesn't assume $n_C \ll n_B$. Assume n_E is essentially constant.

Solution 7.2.6 We start from the quasi-equilibrium relationship Eq. (7.2.31). To use this relationship we recognize that the equality is not exact, but holds approximately. The difference between the two sides, which appears in Eq. (7.2.35), ensures that n_C changes when n_B does. Any change in n_B must be matched by a change in n_C to maintain the quasi-equilibrium relationship itself. Thus an incremental change of Eq. (7.2.31) can be written:

$$n_B dn_C + n_C dn_B \approx 0 \qquad (7.2.39)$$

where we have assumed n_E is essentially constant. Dividing by a time increment dt and using the approximate quasi-equilibrium relationship we relate the rate of change of n_C to that of n_B:

$$\frac{dn_C}{dt} \approx -\frac{k_2''' \, dn_B}{n_B^2 \, dt} \tag{7.2.40}$$

We use this expression instead of the first two terms on the right side of Eq. (7.2.34):

$$\frac{dn_B}{dt} \approx -\frac{k_2''' \, dn_B}{n_B^2 \, dt} + (k_3 n_A^2 n_D - k_2 n_B) \tag{7.2.41}$$

or

$$\left(1 + \frac{k_2'''}{n_B^2}\right) \frac{dn_B}{dt} \approx (k_3 n_A^2 n_D - k_2 n_B) \tag{7.2.42}$$

or finally:

$$\frac{dn_B}{dt} \approx \frac{1}{1 + k_2''' n_B^{-2}} (k_3 n_A^2 n_D - k_2 n_B) \tag{7.2.43}$$

The precise conditions under which this equation is valid can be understood by recognizing that Eq. (7.2.40) can be obtained from the time derivative of Eq. (7.2.35) by neglecting the second time derivative of n_C. ∎

7.2.5 Pattern formation in reaction-diffusion systems

The combination of reaction and diffusion terms in a differential equation can give rise to pattern formation under particular circumstances. Ultimately the source of the pattern formation may be the same as that of the CA rules—short-range activation and long-range inhibition. However, this is not as transparent in the differential equation form. There are two ways to think about the formation of differential equation patterns. The first is a conceptual one that connects with activation and inhibition. The second is through the analytic properties of the differential equations that can give rise to a pattern. To understand the conceptual approach, we must relate the notion of action at a distance of the CA model to the reaction-diffusion model. The influence of one molecular species over a distance is achieved by diffusion. Typically, when diffusion is faster the influence is longer range. Since there are two processes—activation and inhibition—that occur over different ranges, it makes sense to consider the effects of two types of molecules, one with a short-range influence corresponding to a small diffusion constant, and one with a long-range influence corresponding to a large diffusion constant. Activation is a process by which one cell produces a signal molecule that causes other cells around it to produce the same molecule. From the point of view of the molecules, this is a self-catalyzing reaction that causes a nonlinear increase in the molecule density. Thus we expect that the molecule with a small diffusion constant autocatalyzes a reaction that increases its own density. Cell pigment is then coupled to its density. The second molecule, with a longer-range influence, must perform an inhibition of the reaction that forms the first molecule. We will see that the equations developed to demonstrate pattern formation have these properties.

Efforts have been made to construct models of actual physiological reaction-diffusion processes. It is to be expected that such systems involve more than two types of molecules, though quasi-static, kinetic and quasi-equilibrium approximations may allow their description to be simplified. We will discuss two sets of equations that are not specifically obtained from the physiology of pattern formation but are used to illustrate how the patterns can form. The equations have only two types of molecules A and B whose density $n_A(x,y;t)$ and $n_B(x,y;t)$ in space and time we write for simplicity as $a(\mathbf{x};t)$ and $b(\mathbf{x};t)$. The molecules diffuse with different diffusion constants D_a and D_b. The differential equations describing their behavior can be written generally as

$$\frac{da(\mathbf{x};t)}{dt} = D_a\nabla^2 a(\mathbf{x};t) + f(a(\mathbf{x};t),b(\mathbf{x};t))$$

$$\frac{db(\mathbf{x};t)}{dt} = D_b\nabla^2 b(\mathbf{x};t) + g(a(\mathbf{x};t),b(\mathbf{x};t))$$

(7.2.44)

The functions $f(a,b)$ and $g(a,b)$ reflect the effects of chemical reactions. They describe the time dependence of the densities when the density is uniform.

We now write down and simulate two sets of equations that form patterns. The first set of equations may be obtained from the activator-inhibitor reactions in Eq. (7.2.22) as discussed in the previous section (see Question 7.2.7). They are described by:

$$f(a,b) = k_1 a^2/b - k_2 a$$

$$g(a,b) = k_3 a^2 - k_4 b$$

(7.2.45)

The first term $k_1 a^2/b$ describes the autocatalytic formation of the activator A which is inhibited by the presence of B. The inhibitor B is produced by A in the term $k_3 a^2$. If the molecules of B are rapidly diffusing, the creation of B in regions where a is large causes long-range inhibition of the formation of A. The densities of both A and B are limited by decay processes responsible for the second term in each equation. Patterns formed from this set of equations are shown in Figs. 7.2.13 and 7.2.14. We will discuss the methodology of these simulations in greater detail below.

The second set of equations that we use to generate patterns may be obtained from the activator-substrate reactions in Eq. (7.2.23) (see Question 7.2.8):

$$f(a,b) = k_1 a^2 b - k_2 a$$

$$g(a,b) = k_3 - k_4 a^2 b$$

(7.2.46)

In this set of reactions, the presence of B is necessary for the autocatalytic reaction that creates A, as is evident in the expression $k_1 a^2 b$. The same reaction causes the disappearance of B and the formation of A. B is spontaneously created by a process, given by k_3, that is independent of the density of A or B. Finally, the density of A is limited by decay, as evident in the term $-k_2 a$. We can consider the autocatalytic increase of A as local self-activation. Long-range inhibition arises when the diffusion constant of B

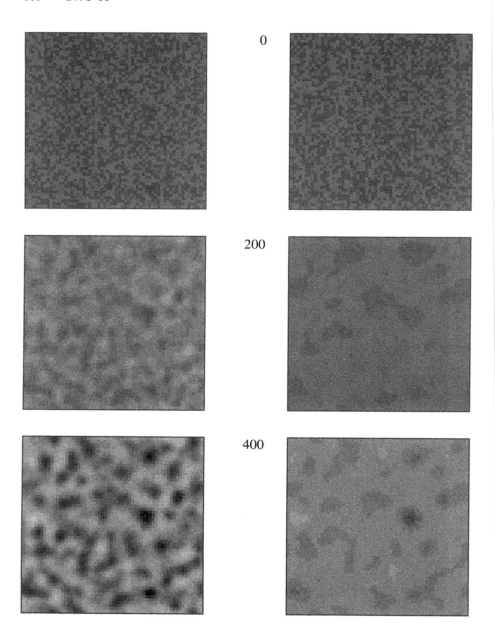

Figure 7.2.13 Simulations of the first set of reaction-diffusion equations, the activator-inhibitor system. At each time two panels are shown. The left panel shows the density a of the activator A. The right panel shows the density b of the inhibitor B. The parameters were chosen as described in the text with $k_1 = k_2 = k_3 = k_4 = 1$. The initial conditions, shown as the first panel, consist of density values either of 1 or of 1.3 placed randomly with equal probability. The same initial conditions are used for Figs. 7.2.14 - 7.2.16. Note that since B is created by A they both have maxima and minima at the same locations. The more rapid diffu-

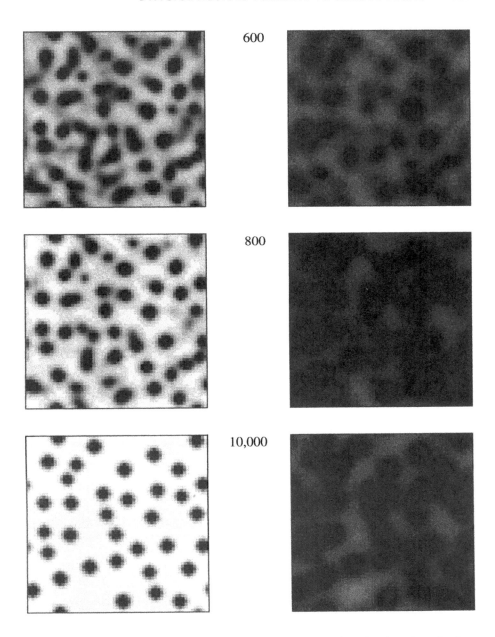

600

800

10,000

sion of B causes the regions around the maxima to be depleted of A. The plots show the density using a representation (gray scale) that uses gray values ranging from white for 0 to black for 2. The figures are labeled by the time in units of updates. Since our convention is that the time per update is $\Delta t = 0.01$ the frame marked 200 would correspond to $t = 2$. A steady state is essentially reached by 10,000 updates. This was verified and used throughout for the other reaction-diffusion simulations in Figs. 7.2.14–7.2.16. Note the difference between this and the number of updates (20) necessary for the CA models of Section 7.2.2. ∎

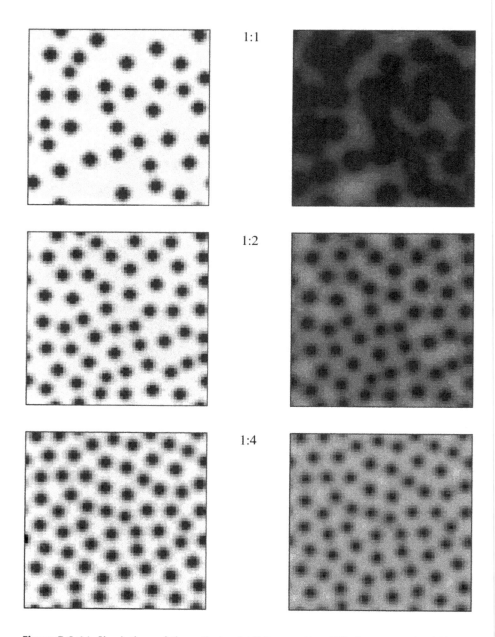

Figure 7.2.14 Simulations of the activator-inhibitor reaction-diffusion system for different values of the reaction constants. The left panels show the density a of the activator A. The right panels show the density b of the inhibitor B. All frames show the steady-state result after 10,000 updates. The parameters for the frames are $k_1 = k_2 = 1$ and $k_3 = k_4 = 1, 2, 4$ respectively. The parameters for the top frames are the same as Fig. 7.2.13 and reproduce the last time step of that figure. ∎

is much larger than the diffusion constant of A. Because B moves rapidly and is consumed by reaction with A, the density of B is depleted not only where A is high in density, but also in the surrounding region. Since B is necessary for the creation of A, this inhibits the formation of A in this larger region. Patterns formed from this set of equations are shown in Fig. 7.2.15. In the activator-inhibitor set of reactions, the maxima of b occur at the same locations as the maxima of a. In the activator-substrate system, the minima of b are at the maxima of a.

Q **uestion 7.2.7** Identify the relevant equations and approximations from Section 7.2.4 used to obtain Eq. (7.2.45) from the reactions in Eq. (7.2.22).

Solution 7.2.7 The two most directly relevant equations are Eq. (7.2.33) and Eq. (7.2.36). The approximations leading to them are relevant, including the quasi-equilibrium approximation for the last reaction in Eq. (7.2.22). The only additional modification is that n_D, which plays no real role in the discussion of the last section, is assumed to be essentially unchanging. ∎

Q **uestion 7.2.8** Identify the approximations used to obtain Eq. (7.2.46) from the reactions in Eq. (7.2.23). There is an inconsistency between the reactions and the equations. In the activator-substrate system the reaction

$$2A + B \rightarrow 3A \qquad (7.2.47)$$

is the only reaction that is responsible for two terms in the differential equations. These two terms in Eq. (7.2.46) have the coefficients k_1 and k_4. This would mean that $k_1 = k_4$, since one A is gained and one B is lost. Describe a modified reaction in which $k_1 = 2k_4$ (easy) and a reaction in which $k_4 = 2k_1$ (hard). One of our assumed cases in the simulations corresponds to the latter.

Solution 7.2.8 For the case $k_1 = 2k_4$ we produce twice as many A in each reaction as B is lost, this can be done using:

$$2A + B \rightarrow 4A + C \qquad (7.2.48)$$

The difficulty in the case $k_4 = 2k_1$ is that the left side of the equation must have only one B, but we want to make twice as many B disappear as A appears. To do this we need to have A be a composite molecule formed by a fast reaction from two equivalent parts (ligands) that bind together to form a complete A. We call each part D, then we have the reactions:

$$2A + B \rightarrow 2A + D + C$$
$$\qquad\qquad\qquad\qquad (7.2.49)$$
$$2D \rightarrow A$$

where the second is a fast reaction. This combination gives the desired result.

Another possible solution is to use two catalyzed reactions. One causes A to appear and is catalyzed by B. The second causes B to disappear and is catalyzed by $2A$. Then the coefficients can be set independently. This suggests some of the subtlety necessary to create actual pattern-forming systems. ∎

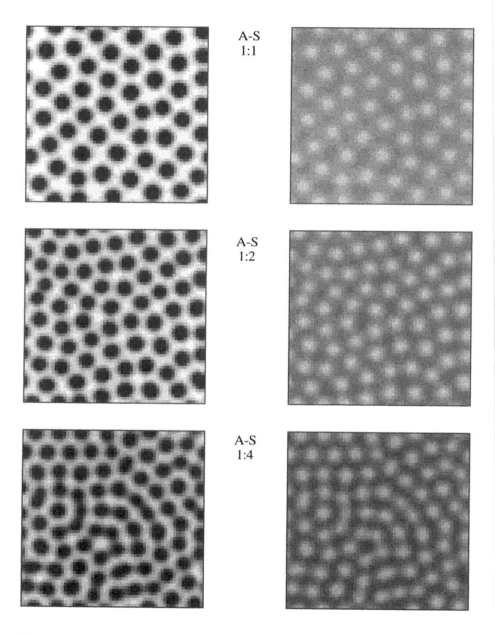

A-S
1:1

A-S
1:2

A-S
1:4

Figure 7.2.15 Similar to Fig. 7.2.14, but for the activator-substrate system. The right panels show the substrate B, which is consumed by the activator A. B is depleted and has its minima where A has its maxima. Due to the more rapid diffusion of B, it is depleted in a region around maxima of A. Thus, the growth of A is inhibited in regions surrounded by maxima of A. ∎

Eqs. (7.2.45) and (7.2.46) involve many parameters, six different ones including the two diffusion constants and four reaction constants. Exploring the six-dimensional parameter space would involve much effort. Exploring large dimensional spaces to discover particular pattern-forming regions of the space can give insight into the difficulty of evolutionary processes that form these systems. However, we can significantly simplify our problem mathematically by recognizing that each of the densities a and b and the variables x and t can be measured in convenient units. By normalizing these variables we do not change the form of the pattern, only its scale. Full use of this would reduce the number of independent parameters to only two. For our simulations, the time and length scale must be related to the time step and lattice size. However, we can conveniently scale the densities a and b.

It is easier to scale the densities if we make use of the observation that these equations have a solution that is uniform and does not change in time. This solution, obtained by setting $f(a,b) = g(a,b) = 0$, is unstable in the parameter domain in which patterns form. Adding any small perturbation leads to the formation of a pattern. We will discuss this in more detail at the end of this section. In the meantime we use the uniform solution to choose coefficients—patterns typically consist of positive and negative excursions from the unstable uniform solution. By normalizing the coefficients, we can set the uniform solution so that it is $a = b = 1$. For both sets of equations, this imposes the same relationships between the coefficients:

$$k_1 = k_2$$
$$k_3 = k_4$$

(7.2.50)

Using these relationships also makes it easier to display simulations, since we can use a consistent scale for all plots of the densities. All of the figures showing density plots of the patterns are formed using a scale that begins with white at 0 and ends with black at 2.

In simulating the behavior of these differential equations, we can use a finite difference representation of the diffusion operator:

$$\frac{d^2 a(x)}{dx^2} \to \frac{1}{\Delta x^2}(a(i+1) + a(i-1) - 2a(i))$$

(7.2.51)

or in two dimensions:

$$\frac{d^2 a(\mathbf{x})}{dx^2} + \frac{d^2 a(\mathbf{x})}{dy^2} \to \frac{1}{\Delta x^2}(a(i+1,j) + a(i-1,j) + a(i,j+1) + a(i,j-1) - 4a(i,j))$$

(7.2.52)

The time derivative is represented as a time difference:

$$\frac{da(t)}{dt} \to \frac{1}{\Delta t}(a(t) - a(t-1))$$

(7.2.53)

where we use t also as the discrete time index. These substitutions return us to a CA consistent with a random-walk model of molecular motion. It has the form:

$$a(i,j;t+1) = a(i,j;t) + \Delta t\, f(a(i,j;t),b(i,j;t))$$

$$+\frac{\Delta t}{\Delta x^2} D_a \big(a(i+1,j;t) + a(i-1,j;t) + a(i,j+1;t) + a(i,j-1;t) - 4a(i,j;t)\big)$$

$$b(i,j;t+1) = b(i,j;t) + \Delta t\, g(a(i,j;t),b(i,j;t)) \tag{7.2.54}$$

$$+\frac{\Delta t}{\Delta x^2} D_b \big(b(i+1,j;t) + b(i-1,j;t) + b(i,j+1;t) + b(i,j-1;t) - 4b(i,j;t)\big)$$

The choice of Δt and Δx is coupled to the choice of the remaining coefficients—the reaction constants k_1 and k_3 and the value of the two diffusion constants D_a and D_b. Their value determines the characteristic time to equilibration and the length scale of the pattern that is found. The time scale must be set so that significant changes do not happen in a single increment, because otherwise the differential equation is not being correctly approximated, and oscillatory or chaotic dynamics of iterative maps may occur. The inherent time scale of the system is set by the amount of time it takes for a typical molecular density to change significantly. If we assume that our reaction constants k_1 and k_3 are set approximately equal to one, and we have already chosen the characteristic density of both reactants to be one, then the time for the characteristic density to change is also one. We must ensure that this is much larger than the time interval Δt so we choose $\Delta t = 0.01$. For the simulations we choose the length scale to be approximately one lattice constant, so we set $\Delta x = 1$.

How should we choose the values of the diffusion constants D_a and D_b? We can set their relative values by noting that the range of diffusion $\sqrt{Dt}$ is proportional to the square root of the diffusion constant. In the CA models, we used a ratio of activation range to inhibition range of 6:1. Thus we would like the diffusion constants $D_b : D_a$ to be approximately 36:1 with D_a approximately 1. For the simulations, a ratio of 40:1 was used with $D_a = 0.5$ and $D_b = 20$. $D_b = 20$ was used instead of $D_b = 40$, because for this value the coefficient of $b(i,j;t)$ in Eq. (7.2.54) is greater than 1 (it is -1.6) which causes numerical instabilities (see also Question 7.2.9).

With most of the parameters determined, the only remaining choice is the relative values of the reaction constants k_1 and k_3 with both approximately one. We fix $k_1 = 1$, and vary k_3. Not all values of k_3 produce patterns. In Fig. 7.2.13 patterns formed from the activator-inhibitor system are shown for $k_3 = 1,2,4$. For smaller values of k_3, the spots become sparser as is evident already from the behavior at $k_3 = 1$. For higher values of k_3 ($k_3 = 8$), the pattern disappears and a uniform solution of the differential equation becomes the steady-state result. Simulations for the same values of $k_3 = 1,2,4$ are shown in Fig. 7.2.14 for the activator-substrate system. However, in this case we see that at low values of k_3 the spots become slightly bigger but not significantly sparser. For still lower values of k_3, the simulations, as described above, become unstable and do not arrive at a steady-state result. For higher values of k_3 ($k_3 = 8$), a uniform solution becomes stable. An analytic approach to understanding the pattern-forming range of k_3, and incidentally why both sets of equations have a similar pattern-forming behavior, is described in Questions 7.2.7 and 7.2.8.

The finite difference form of the differential equations in Eq. (7.2.54) is a CA. This CA is both simpler and more elaborate than the CA in Section 7.2.2. Here the interactions between cells are nearest neighbor and the variables at every site are two real numbers—a major part of the pattern-forming behavior arises from the on-site part of the rule. In Section 7.2.2 the interactions were longer range and each cell had only a single binary variable—the pattern formation arose from the interactions. We note that CA rules that are derived from differential equations are designed to be studied in the limit where the cell size is small enough that granularity does not affect the result. This is not necessarily the case with all CA rules; however, in the case of pattern formation, a similar limit should be taken where the cell size is small compared to the typical size of the pattern.

Question 7.2.9 The parameters of the differential equations that give rise to patterns must, in biological systems, arise out of the properties of the molecules involved. If we assume that simple diffusion applies, the diffusion constant arises largely from the volume of the molecule, so the slow diffuser must be much larger than the fast diffuser. Discuss the practicality of the activator-inhibitor or activator-substrate systems simulated here.

Solution 7.2.9 Using Stokes' law (see also Section 5.2) for spherical molecules, the diffusion constant is inversely proportional to the cube root of the volume. For simplicity we can assume the volume is approximately proportional to the mass. Since the diffusion constants were set to have a ratio of 40:1, the masses must have a ratio of 64,000 or approximately 10^5. Recall that the characteristic distance traveled is proportional to the square root of the diffusion constant, which is inversely proportional to the cube root of the mass. Thus the characteristic distance traveled is inversely proportional to the sixth root of the mass—a very weak function of the mass.

Since the fast diffuser must be complicated enough to participate in well-defined ways in specific reactions, we can not expect it to be easy to design a small molecule to do this. If the small molecule is itself large, the large molecule must be huge. Thus, either the slow diffuser must be a behemoth or some other approach must be taken. One solution is that the slow diffuser is actually a cell rather than a molecule (see Section 7.2.7). Another possibility is that other effects, such as reactions that temporarily bind molecules, reduce its diffusion rate. ∎

Using the reaction-diffusion equations and the chosen parameters, the patterns formed are those of spots. We have seen from the discussion of the CA activation-inhibition models in Section 7.2.2 that there are several ways to cause such patterns to form stripes. One way (Question 7.2.2) is to change the relative strength of the inhibition compared to the activation. In the reaction-diffusion systems, the same terms in the differential equations are responsible for both activation and inhibition (k_1 in Eq. (7.2.44) and k_1, k_4 together in Eq. (7.2.45)). Thus it does not appear possible to

control separately the activation and inhibition. However, these terms describe activation when a is large and inhibition when a is small. Thus we can vary their relative
strength by introducing an additional density dependence to these terms that reduces
the activation at high values of a and maintains the inhibition at low values of a. For
the first set of equations (activator-inhibitor):

$$f(a,b) = k_1 a^2 / b(1 + k_5 a^2) - k_2 a$$
$$g(a,b) = k_3 a^2 - k_4 b \qquad\qquad (7.2.55)$$

For the second set (activator-substrate):

$$f(a,b) = k_1 a^2 b / (1 + k_5 a^2) - k_2 a$$
$$g(a,b) = k_3 - k_4 a^2 b / (1 + k_5 a^2) \qquad\qquad (7.2.56)$$

While we do not discuss the possible chemical origins of this modification in detail,
we can understand it as a saturation (effectively an inhibition) of the autocatalytic reaction in the presence of high densities of the activator. It could be caused by an additional inhibitor whose density is tied to a. Patterns formed from these equations in
Fig. 7.2.16 show the formation of stripes.

In summary, we see that the conditions under which patterns can be generated
include cases where there are two types of molecules, one diffusing rapidly and the
other slowly. The slow diffuser A autocatalyzes a reaction that increases its own density. The fast diffuser B reacts with the slow diffuser and decreases the density of A in
the vicinity of a high-density region of A. This results in patterns like that of the
activation-inhibition CA model in the previous section. The primary difference between the two sets of differential equations is that the fast diffuser B acts to inhibit in
two distinct ways, in the activator-inhibitor system through its presence, and in the
activator-substrate system through its absence (depletion).

The discussion of these equations in terms of activation and inhibition can be
augmented by a discussion of their analytic properties. Diffusion in the absence of reactions causes the density to become uniform and patterns are not possible. What are
the mathematical conditions under which patterns will form when there are reactions? Central to our understanding of the formation of patterns is the recognition
that a uniform solution of the equations continues to exist even when patterns are
formed. However, this uniform solution is unstable. This means that adding a small
nonuniform density (perturbation) to the uniform solution will cause the system to
evolve to a pattern such as those shown in the figures. An analytic study of the stability of the uniform solution is known as linear stability analysis. Using a linear expansion of the equations around the uniform solution, we can determine if it is stable.
When it is not stable then the quadratic terms become important in determining the
solution, which may be a nonuniform pattern.

We can take the analysis one step further by recalling that a key aspect of the pattern is the existence of a length scale characteristic of the distance between spots. This
length scale arises even though a differential equation (unlike the CA) has no cellular
length scale; it is also independent of the size of the system. The characteristic length

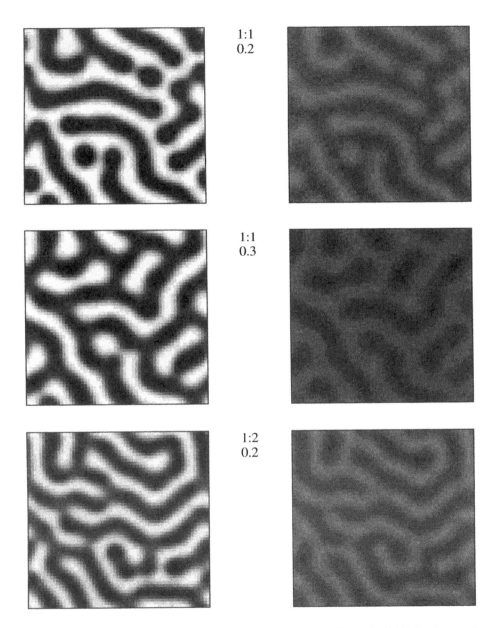

1:1
0.2

1:1
0.3

1:2
0.2

Figure 7.2.16 The addition of a parameter that causes the rate of growth of A to be decreased at high density of A and increased at low density causes the formation of stripes in both the activator-inhibitor (this page) and activator-substrate (p. 670) systems. The parameter values shown are: for all cases $k_1 = k_2 = 1$, for top $k_3 = k_4 = 1$, $k_5 = 0.2$, for middle $k_3 = k_4 = 1$, $k_5 = 0.3$ and for bottom $k_3 = k_4 = 2$, $k_5 = 0.2$. ∎

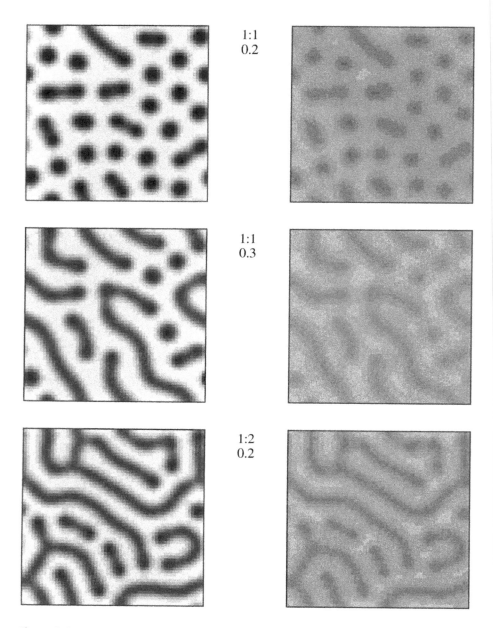

1:1
0.2

1:1
0.3

1:2
0.2

Figure 7.2.16 (*continued*)

scale arises because of the nature of the instability of the uniform solution. Instead of being unstable to all perturbations, the system is only unstable to perturbations of a range of length scales that characterize the patterns. Using the linear expansion around the stable solution, we can identify the range of length scales over which it is unstable to perturbations, and thus identify whether a pattern will form (or for what range of parameters a pattern may be expected to form), and what its characteristic length scale should be. This analysis is discussed in Questions 7.2.10 and 7.2.11.

Question 7.2.10 Patterns are generated when a differential equation has a uniform steady-state solution that is unstable to perturbations at the length scale of the pattern. The instability means that a small addition to the uniform solution grows in time until it is stopped by a process that limits the continued growth. To perform an analytical investigation of the reaction-diffusion equations, expand the reaction part of the reaction-diffusion equations $f(a,b)$, $g(a,b)$ around the uniform solutions in the form

$$a = 1 + u$$
$$b = 1 \mp v$$

(7.2.57)

using the $-(+)$ sign for the activator-inhibitor (activator-substrate) system of equations. Then write differential equations for the time evolution of u and v. It is only necessary to keep the linear terms.

Solution 7.2.10 We expand $f(a,b)$, $g(a,b)$ to second order. We use only the first-order terms, but the second-order terms will illustrate a point. Inserting Eq. (7.2.57) and expanding the activator-inhibitor set of equations gives:

$$f(1+u,1-v) = k_1(1+u)^2/(1-v) - k_1(1+u) +$$
$$= -k_1(1+u) + k_1(1+2u+u^2)(1+v+v^2+...)$$
$$= -k_1(1+u) + k_1(1+2u+u^2+v+2uv+v^2+...)$$
$$= k_1(u+v) + k_1(u+v)^2 + ...$$
$$g(1+u,1-v) = k_3(1+u)^2 - k_3(1-v)$$
$$= k_3(2u+v) + k_3u^2$$

(7.2.58)

For the activator-substrate set of equations, we have:

$$f(1+u,1+v) = k_1(1+u)^2(1+v) - k_1(1+u)$$
$$= -k_1(1+u) + k_1(1+2u+u^2+v+2uv+u^2v)$$
$$= k_1(u+v) + k_1(u^2+2uv) + ...$$
$$g(1+u,1+v) = k_3 - k_3(1+u)^2(1+v)$$
$$= -k_3(2u+v) - k_3(u^2+2uv) + ...$$

(7.2.59)

The differential equations for u and v are obtained by inserting Eq. (7.2.57) and Eq. (7.2.58) or Eq. (7.2.59) into Eq. (7.2.44). After substitution we switch signs in the second equation, when necessary, to obtain:

$$\frac{du(x;t)}{dt} = D_a \nabla^2 u(x;t) + f(1+u(x;t),1 \mp v(x;t)) \approx D_a \nabla^2 u(x;t) + k_1(u(x;t)+v(x;t))$$

$$\frac{dv(x;t)}{dt} = D_b \nabla^2 v(x;t) \mp g(1+u(x;t),1 \mp v(x;t)) \approx D_b \nabla^2 v(x;t) - k_3(2u(x;t)+v(x;t))$$

$$(7.2.60)$$

We have written only the first-order terms, which are the same in both sets of equations. The second-order terms are not the same. The equivalence of the first-order terms in part explains the similarity in the results obtained by simulating the two sets of equations. The inequivalence of the second-order terms is responsible in large part for the differences. ∎

Question 7.2.11 Eq. (7.2.60) consists of two coupled linear differential equations. For a uniform solution where u and v are independent of x, the solution must either be a growing exponential or a decaying exponential. The two possibilities correspond to an unstable or stable uniform solution of the original equations. We can also consider nonuniform solutions by using the trial solutions:

$$u(x;t) = u_0 e^{\lambda t} \sin(\kappa x + \phi)$$
$$v(x;t) = v_0 e^{\lambda t} \sin(\kappa x + \phi)$$

$$(7.2.61)$$

A one-dimensional spatial variation has been assumed, since there is no y dependence. Substitute and find possible values of λ. Plot the real part of λ as a function of κ for the parameter values used in the simulations above. When the real part of λ is positive, the uniform solution is unstable; when the real part is negative, the uniform solution is stable.

Solution 7.2.11 Substituting Eq. (7.2.61) into Eq. (7.2.60) we have:

$$\lambda u_0 = -\kappa^2 D_a u_0 + k_1(u_0 + v_0) = (-D_a \kappa^2 + k_1)u_0 + k_1 v_0$$
$$\lambda v_0 = -\kappa^2 D_b v_0 - k_3(2u_0 + v_0) = -2k_3 u_0 + (-D_b \kappa^2 - k_3)v_0$$

$$(7.2.62)$$

To determine the solutions, we must find eigenvectors and eigenvalues of the matrix:

$$\begin{pmatrix} -D_a \kappa^2 + k_1 & k_1 \\ -2k_3 & -D_b \kappa^2 - k_3 \end{pmatrix}$$

$$(7.2.63)$$

The eigenvalues, which are the possible values of λ, can be obtained with some algebra:

$$\lambda_{\pm} = \frac{1}{2} \left((-\kappa^2(D_a + D_b) + k_1 - k_3) \pm \sqrt{(-\kappa^2(D_a - D_b) + k_1 + k_3)^2 - 8k_1 k_3} \right)$$

$$(7.2.64)$$

We could find the solutions (values of u_0 and v_0). Our objective, however, is only to consider the eigenvalues λ_+. Their real part determines whether the solutions grow or decay. If they decay, then the uniform solution of the original equations $a = b = 1$ is stable and no pattern will form. If one of the solutions grows, then the system will form a pattern. Without analyzing these eigenvalues in great detail, we can plot their values for the parameters used in the simulations to form patterns as a function of $1/\kappa$, which is proportional to the length scale of the perturbation. This is done in Fig. 7.2.17. We see that the real part of λ_+ is positive for a range of values around unity but is negative both at $1/\kappa = 0$ and $1/\kappa = \infty$. This means there is a limited range of length scales at which the equations are unstable, and this range determines the size of the pattern that is formed. ∎

7.2.6 *Cellular switches*

The patterns of molecular density discussed in the previous two sections may describe the behavior of patterns of pigment. More generally, in developmental biology it is

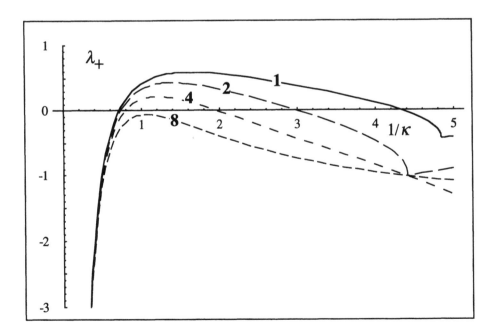

Figure 7.2.17 Plots of the real part of the eigenvalue λ_+ as a function of $1/\kappa$ as obtained in Question 7.2.8. The real part of λ_- is always negative for the parameters chosen. The plots are for parameter values: $D_a = 0.5$, $D_b = 20$, and $k_1 = 1$. The value of $k_3 = 1, 2, 4, 8$ is indicated on each curve. We see that the range of length scales over which the uniform solution is unstable decreases with increasing k_3 and eventually vanishes, causing the uniform solution to become stable at $k_3 = 8$. This is consistent with the simulations for $k_3 = 1, 2, 4$ shown in Figs. 7.2.14 and 7.2.15. The uniform solution (not shown in Figs. 7.2.14 and 7.2.15) was indeed found to be the result of simulations at $k_3 = 8$. ∎

necessary to use such patterns to activate certain cells to perform specific functions, change shape or initiate another stage of pattern formation. For any of these to occur, a chemical process inside a cell must be initiated. The chemical process should persist independent of the original pattern of molecular density. Then the pattern itself need not persist as the system further develops. This requires a one-way chemical switch that can then activate additional cellular functions.

In order to realize the behavior of a one-way switch, what is needed is a chemical system that has two stable states and can be switched from one to the other by a pre-specified concentration of the patterned molecule. The prespecified concentration is genetically encoded to achieve the desired control. We require a new reaction equation that depends on the concentration a of the patterned substance A and controls the concentration c of a substance C:

$$\frac{dc(t)}{dt} = h(a,c) \tag{7.2.65}$$

$h(a,c)$ must have the property that as a function of c it can have at least two solutions c_{-1} and c_1 (in a moment we will see that it must have three) of

$$h(a,c) = 0 \tag{7.2.66}$$

which are the steady-state conditions in which c does not change. These solutions are functions of a, and can be assumed to vary smoothly with a. However, above a speci-fiable density of a, one of the two solutions, say c_{-1}, disappears. This causes the den-sity of C to switch to c_1.

We can analyze the properties of $h(a,c)$ that are necessary and suggest specific forms it might take. In order for c_{-1} and c_1 to be stable solutions of Eq. (7.2.66), the derivative of $h(a,c)$ must be negative at these values:

$$\left.\frac{dh(a,c)}{dc}\right|_{c_{\pm1}} < 0 \tag{7.2.67}$$

This means that a small positive increment results in a negative dc/dt (see Eq. (7.2.65)) while a small negative increment leads to a positive dc/dt. In either case $c = c_{\pm1}$ is restored.

The burden of creating a switch is on the density c, so we represent simply the ef-fect of A on C as direct production ($A{\rightarrow}C$), or catalysis of production ($A{\rightarrow}A{+}C$) leading to the form:

$$h(a,c) = k_1 a + \tilde{h}(c) \tag{7.2.68}$$

We can now design $\tilde{h}(c)$ with the desired properties and consider how it can be gen-erated using chemical reactions. For simplicity, we set $\tilde{h}(c)$ to have its first steady state at $c_{-1} = 0$ so that there is no constant term in $\tilde{h}(c)$. Since it must have a negative de-rivative at $c_{-1} = 0$ we have $\tilde{h}(c) = -k_2 c + \ldots$ where the ellipsis represents higher-order terms.

In order to have two solutions of Eq. (7.2.66) with negative derivatives, $\tilde{h}(c)$ must have a form like that illustrated in Fig. 7.2.18(a). In particular, there must also be a

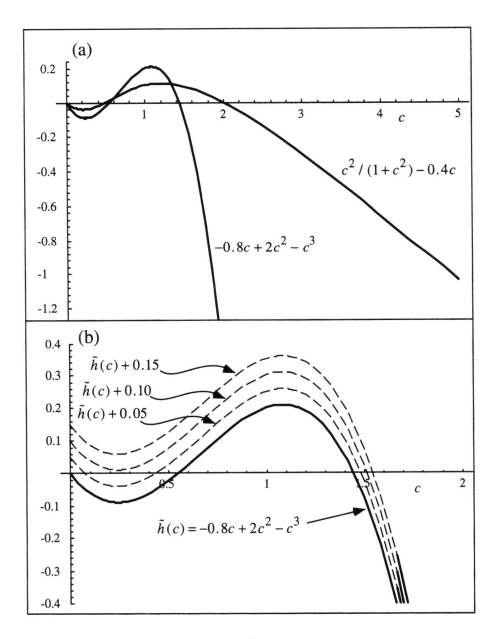

Figure 7.2.18 (a) Plots of two possible forms $\tilde{h}(c)$. This function describes the rate of change of the density c used to create a chemical switch. There are two stable solutions (low and high density) and one unstable solution of $\tilde{h}(c) = 0$. (b) When a is added to the system the curve is displaced upward, as shown by the dashed lines. For a high enough value of a only the high density solution is left. If we start with the low-density solution and raise the density of a the density of c will rise gradually and then switch to the high-density value. When a is lowered back down, c stays at the high-density solution. This sequence describes turning the switch ON. ■

third solution of Eq. (7.2.66) with a positive derivative. This can be achieved using a polynomial of the form:

$$h(a,c) = k_1 a - k_2 c + k_3 c^2 - k_4 c^3 \qquad (7.2.69)$$

For the last term, we can use any power of c that is greater than 2. Writing down reactions that in principle would lead to this form is not difficult. However, they may appear overly contrived.

Another way to satisfy the conditions is to make use of a system that has the structure:

$$h(a,c) = k_1 a - k_2 c + \frac{k_3 c^2}{1 + k_4 c^2} \qquad (7.2.70)$$

The third term has the interpretation that it consists of a molecule produced in a sigmoidal fashion—it increases quadratically by autocatalysis and then saturates at a maximum value. In both Eq. (7.2.69) and Eq. (7.2.70) the second term represents a process of molecular decay. Fig. 7.2.18(b) shows the switching action when there is a change in the concentration of a.

7.2.7 Pigment cell diffusion

The study of the formation of patterns in Sections 7.2.2 and 7.2.5 considered systems where the initial conditions provided pigment placed at random throughout the space. The dynamics then caused these pigment molecules to bunch together to form the pattern. Experiment suggests, however, that vertebrates create pigment patterns by the migration of pigment-producing cells (melanophores). Early in fetal development, the melanophores are formed on the line that eventually becomes the spinal cord and from there migrate across the surface and aggregate into a pattern that becomes the pigment pattern. The number of these pigment-producing cells need not be conserved during this process, however, they must arise in most regions by migration, rather than by initial seeding or by spontaneously being produced by other cells.

Thus, we must consider a model where the initial conditions place pigment only in a limited part of the space, and from there the pigment diffuses through the space to form the pattern. We consider this process in the context of the reaction-diffusion systems described in Section 7.2.5. The slow diffusing species is the melanophore, while the fast species is assumed to be a molecule (Question 7.2.9). In both the activator-inhibitor and activator-substrate systems, the slow-diffusers (A) are not spontaneously generated—some of A is required in order to make more of A—consistent with the properties of melanophore reproduction. However, both of the models must be modified to allow the initial conditions to consist of only a single initial band of A and B (Fig. 7.2.19 (top)).

For the activator-inhibitor set of equations, the problem with the initial conditions arises in regions where b is zero. The first term in Eq. (7.2.45) diverges. This occurs not just because of the initial conditions but also because B is generated by the presence of A, which is limited in space by our assumptions. Thus, as discussed in Question 7.2.5 (Eq. 7.2.38), we introduce an additional constant k_6:

$$f(a,b) = k_1 a^2 / (b+k_6)(1+k_5 a^2) - k_2 a$$
$$g(a,b) = k_3 a^2 - k_4 b \qquad\qquad (7.2.71)$$

The results of simulations are not very sensitive to the value of k_6, which was chosen to be 0.1.

For the activator-substrate equations (Eq. (7.2.45)), the problem arises from the uncontrolled growth of B in the regions where A has not yet reached. This eventually causes the simulation to break down as the gradients in B become too large to be integrated using the parameters chosen. It makes sense to limit the spontaneous generation of B using an additional parameter k_6 in the following way:

$$f(a,b) = k_1 a^2 b / (1+k_5 a^2) - k_2 a$$
$$g(a,b) = k_3 (1-k_6 b^2) - k_4 a^2 b / (1+k_5 a^2) \qquad\qquad (7.2.72)$$

A quadratic term rather than a linear term was used so that the first-order expansion of the function would not be affected. The first-order terms, as discussed in Questions 7.2.7 and 7.2.8, play an essential role in the existence of patterns, while the higher-order terms are less crucial. A value of $k_6 = 0.1$ was found to be reasonable and was used for the simulations. It limits the growth of b to no more than $\sqrt{10}$.

The simulations of these two systems are quite distinct. Simulations of the activator-inhibitor system are shown in Figs. 7.2.19 and 7.2.20. For certain values of the parameters, the pigment does not expand out of the region in which it started. This can be understood when we think about how this system functions. The pigment cells A produce the fast diffuser B which inhibits the formation of A. Since the highest concentration of B is in the immediate vicinity of high concentrations of A, it becomes difficult if not impossible for A to move into additional areas. For other values of the parameters, the initial line of pigment is unstable to bending, and the pigment expands to fill the space in spots or stripes, or combinations of spots and stripes. An example is shown in Fig. 7.2.20.

In contrast, pigment in the activator-substrate system (Figs. 7.2.21 and 7.2.22) generally expands to fill the space. This occurs because the fast diffuser B, which enables A to increase in concentration, is readily available in regions away from regions of high concentration of A. The melanophores A diffuse outward and increase in numbers due to the availability of B. It is helpful to recall that inhibition in the growth of A arises only when regions of high density of A surround a region of low density. In the central region, A cannot grow because the density of B is maintained at a low level due to reaction with the surrounding A.

One of the patterns that appears in these simulations are stripes that run parallel to each other. In the activator-inhibitor model (Fig. 7.2.20), they form by extension of each stripe and they are essentially perpendicular to the originating line (spine). In contrast, the stripes in the activator-substrate system (Fig. 7.2.22) are formed sequentially and are parallel to the originating line. Depending on the parameters, the whole space may become stripes, or stripes may give way to spots.

Many animals have stripes that are better described by these results than by the patterns formed from random initial conditions of Sections 7.2.2 and 7.2.5. Some have stripes that run head to tail. These are more easily accounted for by the activator-substrate model. In particular, patterns with two stripes along the spine and dots below are found (e.g., genets) similar to Fig. 7.2.22. In other animals, such as the zebra, stripes run perpendicular to the spine. This could be generated by a version of the activator-inhibitor model where the stripes originate along the spine. Alternatively, if the pigment cells only originate at the skull, the activator-substrate model might form the stripes sequentially. We can identify which model is reasonable from the pattern by noting whether the stripes are continuous across the spine. In the activator-substrate model the stripes would be continuous across the spine, while in the activator-inhibitor model the stripes would be broken at the spine.

7.3   Developmental Tool Kit

In this chapter we have focused on the modeling of pattern formation as a fundamental aspect of developmental biology. In this section we briefly describe other processes that are important or essential for the process of development. The ability to cause these processes to occur provides a tool kit for the formation of organisms with functioning interdependent organs and physical structures designed for particular tasks.

The formation of physical structures including organs, limbs and tissues involves various processes that occur both inside and between cells that change the number, shape and location of cells. Growth in absolute size of the system occurs by cellular replication (growth and division). Once cells have differentiated in function due to patterning, diffusion or directed motion of cells in chemical gradients plays an important role in the relative location of cell types. Programmed cell death also plays a role in the formation of structures. Changes in external and internal structure of the organism also arise from changes in the structure of individual cells, particularly the cell membrane. Oriented adhesion of cells also results from cell membrane behavior. These processes involve changes at the molecular level in the cellular membrane and cytoplasm. They are developmental processes within the cell that contribute to the development of the whole organism. Among the physiological structures that are formed are spheroids, balls, membranes, tubes and branching systems. In some areas intercellular spaces also become filled with various excretions of cells to form support structures for the cells and the whole organism.

For the study of patterns in growth, the formation of treelike branching structures (Section 1.10.2) is particularly interesting. In plants, these include external structures—branches and roots. Internal branching structures occur in plants (veins in leaves) and animals (veins, nerves, air passages in lungs, and duct systems in certain organs). Most of these are multicell systems that may be formed by elongation of tubelike structures through cellular division and growth, then a periodic or occasional

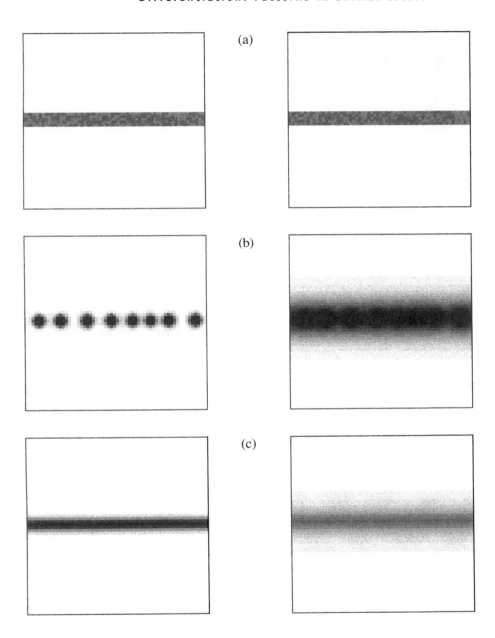

Figure 7.2.19 The reaction diffusion activator-inhibitor system simulated starting from initial conditions of a single linear strip of A (left frame) and B (right frame). (a) illustrates the initial conditions which are similar to that used in Fig. 7.2.13–7.2.16 but are restricted to a linear strip as shown. (b) shows the steady-state result of a simulation with parameter values $k_1 = k_2 = 1$, $k_3 = k_4 = 1$, $k_5 = 0$, and $k_6 = 0.1$. (c) shows a simulation with the same parameters except $k_5 = 0.1$. For the parameter values of both (b) and (c) the pigment remains confined to its initial line. ∎

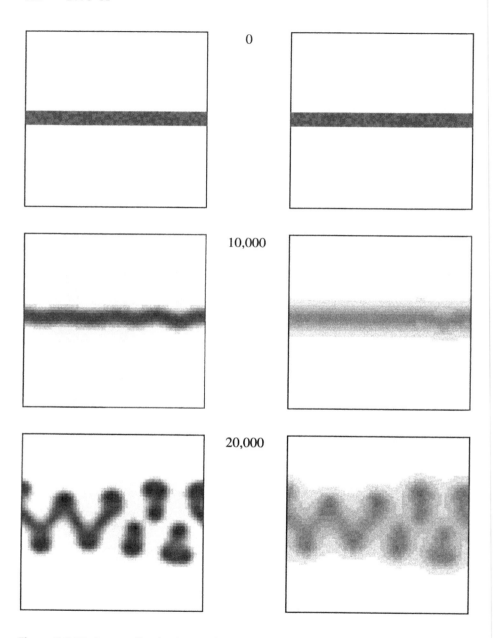

Figure 7.2.20 Frames of a simulation of the activator-inhibitor system with parameter values $k_1 = k_2 = 1$, $k_3 = k_4 = 2$, $k_5 = 0.3$, and $k_6 = 0.1$. The results are unlike the simulations shown in Fig. 7.2.19, for the same system but using different parameter values. In this case the initial line becomes unstable and the pattern of pigment expands to fill the space. Note that the lines of pigment are extended at their ends into the empty space. They are largely perpendicular to the line found in the initial conditions. Note also the long simulation time. The activator-substrate model has different behavior, as shown in Figs. 7.2.21 and 7.2.22. ∎

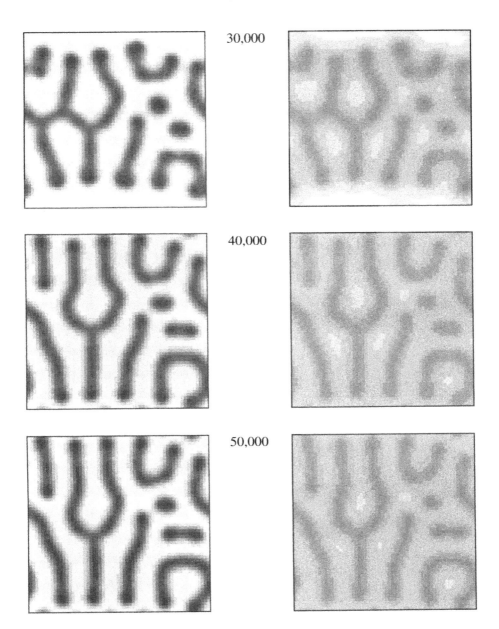

30,000

40,000

50,000

Figure 7.2.20 (*continued*)

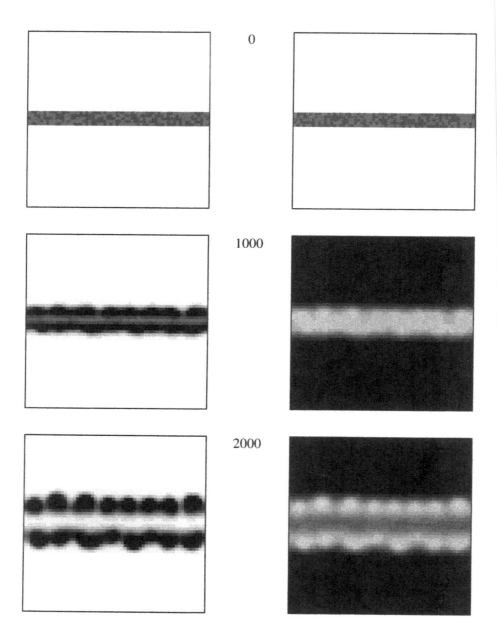

Figure 7.2.21 Frames of a simulation of the activator-substrate system with parameter values $k_1 = k_2 = 1$, $k_3 = k_4 = 2$, $k_5 = 0$, and $k_6 = 0.1$. The pigment expands to fill the space with spots using a process of spot splitting and diffusion. Compare Fig. 7.2.19 for the activator-inhibitor system. This model may also be relevant to evolution and trait divergence as discussed in Section 7.6. ∎

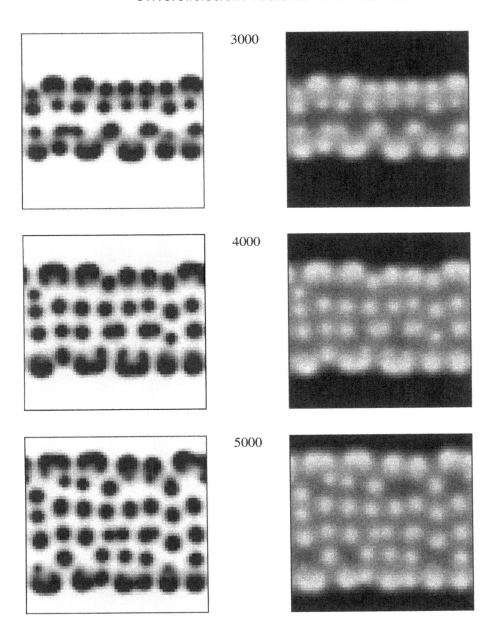

Figure 7.2.21 (*continued*)

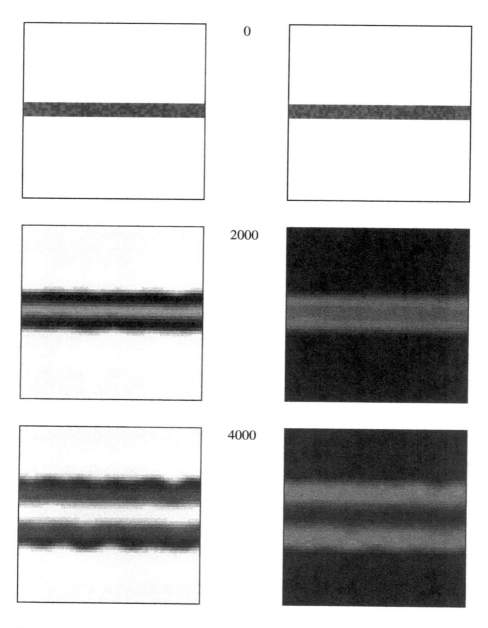

Figure 7.2.22 Similar to Fig. 7.2.21 except for the parameter $k_5 = 0.2$. We see that spots are formed when the original stripe diffuses outward. Then new stripes form parallel to the initial stripe of pigment by merging of the spots. For these parameter values, the spots continue to form into lines until the lines fill the whole space (not shown). All lines formed run parallel to the initial line of pigment. Compare Fig. 7.2.20 for the activator-inhibitor system. ∎

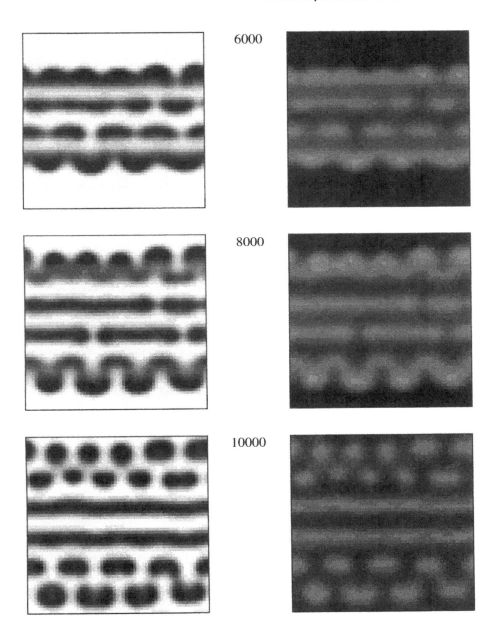

Figure 7.2.22 (*continued*)

initiation of branching. Some networks, such as the plant leaf veins, may also be formed by direct chemical patterning similar to the patterns shown in section 7.2. Branching nerve cells discussed in Chapter 2 are an example of individual cells that branch using molecular changes. Cell elongation and branching must be controlled through the addition of molecules to the cell membrane. Both multicellular and single-cell branching structures also, in general, must include a form of target tracking that imposes some overall systematic behavior on the branching system. This target tracking may cause the network to fill space more or less uniformly so that all regions are served (veins or ducts). Alternatively, a directional bias to growth may be important, such as provided by sunlight on tree branches, or chemical gradients causing a bias in growth direction that results in interconnection of organs inside an animal.

The mathematical modeling of branching structures would seem to be natural, since it is only necessary to specify an algorithm by which the branching occurs. However, there is a conceptual difficulty in representing such systems, because we generally think about the storage of information about a system in terms of storage locations that are themselves given by a linear string—a nonbranching data structure. When a branching structure grows, cells replicate at many sites, forming new cells whose existence and state must be specified. A better approach for describing tree structures, known as L-systems, has been developed by Lindenmayer based upon concepts originating in treelike hierarchies in linguistics. This approach uses a character string representation, but there are delimiters that indicate branching. Moreover, the dynamics allows the insertion of multiple characters at any site. These dynamics are specified by operators that act upon all the characters in a string. Each character can be considered as representing the state of a particular cell. We will illustrate this using a very simple example of a tree-generating algorithm.

We assume there are three states of a cell indicated by A, B and C. The update algorithm is specified by state transitions of cells that include the possibility of replication to two cells. Branchings are indicated by delimiters (brackets). A simple state transition table is:

$$A \rightarrow A$$
$$B \rightarrow AC \tag{7.3.1}$$
$$C \rightarrow [B][B]$$

The first few updates of a string are as follows:

$$[B]$$
$$[AC]$$
$$[A[B][B]] \tag{7.3.2}$$
$$[A[AC][AC]]$$
$$[A[A[B][B]][A[B][B]]]$$

This illustrates the representation of a tree with binary branches. The nongrowing part of the tree are cells in the state A. Cells in the state B replicate to extend the length of their branch, and cells in state C replicate to form two new branches. By further elaborating such an algorithm, it is possible to specify geometric information that can fully describe a treelike structure. Various natural structures have been modeled in this way.

The formation of limbs through budding (including arms, legs, tail, head and fingers) may appear to be similar in many ways to the formation of branching structures. It may also be related to the formation of pigment patterns that specify the location of limbs to be formed. However, there is an essential difference between limb formation in animals and the other types of patterns. Both color patterns and branching structures can be treated primarily as a statistical process that allows significant variation between specific realizations. However, limb budding must be a reproducible process with definite outcome so that the number and type of limbs is consistent and directly controlled by the developmental process. Small-scale patterns involving only a few limbs would be much more reproducible and controllable than the large number of spots in patterns discussed in Section 7.2. The precise form of small-scale patterns is controlled by the boundary conditions that are imposed through the size (or number of cells) of the organism or the internal system in which the pattern is being formed. Our discussion in Chapter 2 of the 7±2 rule may be relevant here as well, suggesting a limit to the number of limbs that can be created reliably through such patterns.

Strictly repeating patterns such as faceted eyes of certain insects are another class of patterns that are different from those discussed in Section 7.2. Such patterns can be formed by sequential addition of elements. It is less reasonable to use a chemical patterning as a template to achieve strictly periodic order extending over a large number of elements. The main difference between periodic patterns and limb budding is not that there are few or many, but that in limb budding there are differences between the limbs that are important and the total number is well defined, while in a periodic structure all of the components are essentially the same and a few less or more doesn't really matter.

There are several other processes in addition to pattern formation and physical changes in cells that are important. These processes control the timing or order of developmental stages. We have already discussed in Section 7.2.6 the operation of one-way chemical switches that can serve to couple different processes. The presence of one chemical density above a threshold causes a second chemical to be produced. The second chemical continues to be produced even when the first is removed. Irreversible processes like chemical switches are an important component of timing mechanisms that count regularly spaced events. Timing mechanisms may be used for processes within a cell, including setting the time between cell divisions. Timing mechanisms are also necessary across several cell divisions. For this purpose, one way to monitor time is to count the number of cell divisions. This would require a sequential process (such as chemical switches) that can serve as a counter. It is believed that a certain number of bases at one end of DNA chains do not replicate in normal cell division (telomere shortening). The bases may be added later; alternatively, the progressive shortening of the chain may serve as a counter of cell divisions to control development and aging.

Our discussion of cellular processes in developmental biology is far from comprehensive, though a few of the important processes have been mentioned. A further level of detail could be added to the internal cellular processes. This level would include: the transmission of signals through membranes via cellular receptors, the transfer of such signals from the cellular membrane to the cell nucleus, the coupling of chemical processes to the activity of gene expression, the production of various enzymes, and the transmission of signals from their production sites out of the cell and into the intercellular fluid. Discussions of these processes are relevant to considering the cell as a complex system in its own right.

7.4   Theory, Mathematical Modeling and Biology

We have used various techniques to model pattern formation in biological systems. The primary tools were simulations of CA and differential equations. There is a need to develop some perspective on the utility of mathematical models for the study of biological systems. Biology is largely a phenomenological science. It is dominated by the experimental study of systems, their description and classification. This is to be contrasted with the physical sciences, where theory and mathematical modeling play a more integral role. At least for some biologists, the use of mathematical models misses the essence of the study of biological systems. Aside from the usual political/sociological issues that can affect such perspectives, there is validity to the concern that mathematical modeling may not capture the processes that are important in biological systems. It is important, therefore, to understand more systematically the objectives of theory in general and mathematical modeling in particular.

The role of theory in science rests on three legs—description, explanation and prediction. Description implies that a theory has the ability to describe the existing observations and phenomena. Explanation implies that the theory has a comparatively simple set of concepts and relationships that capture the system behavior more concisely than the phenomena that are described. This is tied to the notion of simplicity of scientific theory and Occam's razor, which requires a theory to be as simple as possible. Prediction is linked to the ability to describe existing phenomena but demands that clear and testable predictions be possible. In particular, a theory is considered poor if it cannot be falsified by direct experimental test. In essence the theory must distinguish between possible outcomes of an experiment whose implications would not otherwise be known. In a certain sense, the more unanticipated (surprising) are its predictions, the more useful is the theory.

From the point of view of experimentalists interested in further elucidating the phenomena of biological systems, the most important role of theory is the suggestion and prediction of the results of experiments. Indeed, every experiment that is performed is based upon some concept of what phenomena are important to measure, and therefore reflects a conceptual theoretical framework in which the experiment is performed. In biology, much of this theoretical framework is not based upon quantitative theory. As a consequence, there has been little expectation that significant quantitative predictions are possible. Recent efforts have demonstrated that constructive and predictive theories are possible, and the role of theory in biology is expanding.

In order to clarify the role of theory further, it is important to distinguish it from that of experiment. Experiment has a responsibility to uncover truths in the measurement of actual systems. Theorists are often assumed to have the role of proving truths through inference. However, their actual role is to propose assumptions—the theory itself—and correctly derive from these assumptions various predictions. Only when experiment tests the predictions can the assumptions themselves be tested and truth be determined. Because of the different objectives of theory and experiment, it is not appropriate to evaluate the contribution of theory by the goals of experiment. This is just as true about the evaluation of experiment by its contribution to the goals of theory. For example, in most cases experiment does not provide a general understanding, only an understanding of specific phenomena.

Increasingly, two additional roles of theory have arisen that cause more confusion about its ultimate responsibility. The first of these is the appearance of *ab-initio* calculations of system properties. This approach is most often applied to the study of solid or molecular systems. These studies extend the traditional objective of providing theoretical predictions for experimental results. However, the assumption in these studies is that the underlying theory has been so fully tested that the result of a proper calculation is as correct as an experiment that is performed on the same system. The challenge for the theorist is to ensure that the calculation is correct, if this is satisfied then the results are assumed valid. In this way it is like an experiment. The concept of *ab-initio* calculations has limitations in that there are no calculations that have perfect accuracy, and their implementation always requires assumptions about the relationship of the computer model with actual systems. Such limitations also apply to laboratory experiments and the relationship of the experimental condition to other circumstances. The objective of developing *ab-initio* methodologies is a positive one. However, it should not be confused with the more traditional objective of proposing fundamental simplifications and their experimental consequences.

The second additional role of theory is the mathematical modeling of experimental phenomena. This is known as phenomenological theory and represents a significant part of theoretical work in biology as well as in the physical sciences. Much of this chapter is rooted in phenomenological theory. While such theory is generally strong on description, it is weak on explanation and prediction. The reason for these difficulties is that a particular observation may be described by many distinct phenomenological theories. Thus we have seen that color patterns can be obtained from several different sets of differential equations and from CA. The general term for this feature of modeling is universality. The concept of universality implies that in many systems only a few aspects of the properties of a system are important in determining its characteristic (simple) behavior. The reason for this should be apparent: a simple behavior arising from a complex system cannot depend on all of the properties of the complex system. If it did, it would in turn be complex. Thus only a few features of the underlying system must be relevant, and many models should give rise to the same behavior.

A phenomenological model can be expanded to a more complete theoretical effort in order to provide additional information. One possible approach is to study directly the universality of the models. This means that we develop an understanding of the essential properties of models that can give rise to a particular phenomenology.

This approach takes us beyond the particular model and toward a general framework (theory) that provides a more systematic understanding of the origins of a phenomenon. One step in this direction is the articulation in this chapter of the principles of (a) activation and inhibition, and (b) fast and slow diffusers. Another is the analytic expansion of the differential equations in Questions 7.2.10 and 7.2.11. More formal approaches to universality are also possible (see Section 1.10.5).

A second possible approach is to discuss distinctions between different phenomenological models in order to provide contrasting predictions that can then be tested by experiment. This enables the phenomenological model to become more predictive and suggest experiments that can increase our understanding of the underlying causes of a phenomenon. The underlying causes themselves may not be readily accessible to experiment. For example, the discussion of diffusing melanophores and the difference between the activator-inhibitor and activator-substrate models in Section 7.2.7 provides a mechanism for distinguishing between the two models of pattern formation without direct knowledge of the actual processes involved. Without such a discussion there would be no way to tell which of the models applied for a particular animal except to study the molecular processes, and little would be gained from the theory. In general, the more independent tests are performed on a phenomenological model (or theory), the more it can be relied upon to describe new circumstances.

Most important for the consideration of the success or failure of theoretical modeling in biology is the recognition that complex phenomena require, by their nature, a complex model to generate them. This means that we cannot expect simple models to generate truly complex behavior. Thus, a basic skepticism about the ability of theory to describe biological phenomena can be justified. What is missing, however, is an ability to know, a priori, what are truly complex phenomena and what properties of complex organisms can be attributed to simple universal behaviors. Through a number of examples in this text, various approaches to the description of aspects and attributes of complex systems have been illustrated using relatively simple concepts and models. This is ultimately an important objective of the field of complex systems.

We conclude this section with a discussion of the relative utility of the CA, reaction-diffusion and other models of pattern formation as an illustration of the use of simulations in the study of biological systems. There are various biases regarding the use of particular forms of equations and this discussion is designed to illustrate that the form to be used should be dictated by the nature of the question that is to be addressed. We have seen that the CA models introduced in Section 7.2.2 were convenient for developing a basic understanding of activation and inhibition as a simplest model of pattern formation. The differential equation models in Section 7.2.5 provided a more microscopic view of these same processes in the sense that they modeled the chemical processes that might underlie the activation and inhibition. It would be important to recognize, however, that the particular differential equations used are not necessarily indicative of the actual processes in a biological system. They are thus only particular realizations of systems that embody the activation and inhibition phenomenon. These equations show us that reaction-diffusion systems can form patterns. To achieve a yet more microscopic view of the processes, we might turn to another type of CA—the lattice gas—that would describe the diffusive motion of molecules directly rather than using the average density of the molecules as its essential variable. To be even more microscopic, we could use a particle model that includes Newtonian mechanics. This would require modeling the medium in which the particles are located. These examples are designed to suggest that there should be no inherent bias toward one approach. The bias is generated by the nature of the questions and answers that are desired.

7.5   Principles of Self-Organization as Organization by Design

In previous sections we discussed the dynamics of systems that achieve a complex structure through the interaction of their components. The context is our effort to understand developmental processes that are involved in the reproduction of multicellular biological organisms. There are, however, other processes that can result in reproduction. When a cell reproduces, it recreates itself by direct duplication. Each of the components is duplicated or simply divided in two parts and they are grouped to form two daughter cells. Does direct duplication play a role in the reproduction of multicellular organisms? Many plants can reproduce by growing new plants directly from a mature plant. Despite the connection to the parent plant, and its provision of nourishment, this is not duplication. Instead, differentiation and pattern formation occur in the creation of roots, stem and leaves. The other mode of reproduction, through a seed, is essentially independent of the parent plant. Thus the entire process is developmental. In animals, the interaction between parents and offspring is also secondary to the inherent developmental process. Fertilized eggs may be warmed by birds and the young may be fed and trained. Mammals have a more direct relationship, initially through a controlled uterine environment, then through nurture. Nevertheless, the specification of the process of physiological development is understood to be largely self-contained in the initial cell. It would be remarkable if it were found that some structures are transmitted directly from mother to fetus in-utero by migration of differentiated cells. However, the basic developmental phenomenology appears independent. The question that arises is, Why do biological multicellular organisms reproduce using a process of development? What benefit is there in this process?

In order to understand why a developmental process is desirable we should consider the general task of creating a complex system, and specifically the problems with duplication. The first aspect of duplication that we might consider is rooted in the difference between individual cells and multicellular organisms—the existence of more levels of structure. In order to duplicate a multicellular system, we would duplicate each cell and then we would have to disentangle the two resulting organisms. This problem is linked to the spatial structure of the organism. An essentially two-dimensional organism in three dimensions would not have this problem. Individual cells are able to overcome this problem when organelles in the cell are replicated and separated, though this is a complex process. By lining up DNA strands along a single two-dimensional plane, this part of the system is reduced to two dimensions. We might consider whether there are ways to do this for multicellular organisms. Instead, we will look for other reasons for a developmental process that are also fundamental.

Another mechanism for duplicating a system would involve a process that is more akin to our manufacturing processes, where many copies of a system are produced. Note that for many multicellular organisms there actually is mass production of offspring, so that this is not an unreasonable model. Starting from a prototype or a description (representation) of the whole system, we create a process that produces and then places each of the components in its proper location. There are various problems with this for an interdependent complex system. One of these is that the system must be maintained in partial form. Sustaining the various components separately creates an additional burden on the manufacturing process. This problem exists in actual manufacturing, since structures that become self-supporting must be maintained during construction. Extrinsic supportive structures (scaffolding) may be necessary during construction that are later removed. For a very complex interdependent system such scaffolding would be much more difficult to design. Even for a developmental process, the problem exists. It is manifest in the support systems in a reptile or bird egg, and in a mammalian uterus. However, the internal organs are still largely maintained by self-consistent systems that develop into the mature systems of the multicellular organism.

While the two problems discussed in the previous paragraphs are important, there is another way to understand the reason for a developmental process, which will be particularly relevant for our understanding of the design of complex systems. It relates not to the structure itself but to the problem of specifying the structure. Any design process implicitly assumes that a description of the system exists before the system itself does. This description, generically called a blueprint, is in many ways like a model of the system. We can better appreciate how science and engineering are related when we recognize that the relationship between system and description plays an important though different role in both. Ultimately, it is the interplay between system and description that science is investigating. A key difference between science and engineering is that science can advance by using partial descriptions, while for engineering a useful description must be sufficiently complete. Our concern here is to understand the relevance of representation to developmental biology. In particular, What is the advantage of a representation that describes the developmental process of formation rather than the final system itself? This reformulation of our question suggests an answer: the developmental process can be more concisely described.

We can understand this answer when we think about the existence of various relationships between different parts of a complex system as well as the different activities of the parts. If we take advantage of these relationships, we can reduce the amount of information necessary to describe the whole system. More correctly, we use the relationship when we create the program of development that constructs the system. The program of development allows us to have less duplication of information. If the same basic structure is relevant to several components, we have them undergo the same developmental processes and then modify them later to accommodate differences. Even after two components are different, the same developmental process may be used to achieve incremental modifications that are common to the two parts. The process that we are describing is the creation of an algorithm from which the system is to arise. The reason it is useful is because the explicit blueprint is inherently compressible. The algorithm appears to contain fewer pieces of information than the final form, even though they both ultimately contain the same information. We will enter further into a discussion of system representation and information theory in Chapter 8 (See also Section 1.8 and 1.9). The ideas articulated here are parallel to the idea of algorithmic complexity, where the notion of reducing the length of a description to its smallest possible representation (compressing a character string) is investigated. Here we are considering the applications of these ideas to design.

When we consider developmental biology in this context, we must expand our understanding of compression from the usual notion that allows only deterministic compression algorithms. Randomness or noise is available from molecular motion in biological systems. Information that is essentially arbitrary can be provided from this randomness. An example may be found in the pattern formation discussed in this chapter. To describe the patterns formed in all of their detail would require many pieces of information. However, for the animal skins, the specific details of the pattern are not essential—we can vary them and still have patterns with the same properties. As long as we are interested in the generic properties, such as size and overall shape of dots, then the details can be provided by randomness. In the simulations, this is provided either by the initial conditions or in the update process when there is a random selection of cells to update. To think about this more clearly we must recognize that the eventual state of the system is selected from an ensemble which results from the influence of randomness. As long as we are interested in properties that are generic in the ensemble, this is satisfactory. However, if we want to select a particular feature that is rare in the ensemble then we must specify it a priori as part of the design.

There is another source of information that may be used in the process of forming the system. This is the existence of specific well-defined influences of the environment. The environmental influences are in addition to the support structures and nutrition provided in the seed, egg or uterine development environment. We can illustrate this by another example. As mentioned in Chapter 3, the development of basic neural connections in the visual system of mammals is influenced by stimulation by light. This is not really a form of adaptation to the environment, it is instead the use of specific external stimuli as part of the developmental process. The algorithm is taking advantage of persistent information about the external environment—the existence of light.

We thus find that the process of development is convenient because it allows the system design to be more concisely represented. This answer is not complete. We must still explain why it is advantageous to have a more concise description of the complex system. The advantages of a concise design become particularly meaningful when the design is to be modified. Modifications should preserve many of the essential relationships encoded in the description. This reduces the possibility of design errors. In a complex system, a major source of errors is inconsistencies in the design. By definition, an inconsistency reflects the violation of a constraint or relationship that is necessary in the final system. If the design automatically incorporates the required relationship in its compressed form, then the inconsistency cannot arise. More directly, if the description is shorter, there are fewer places errors can be made—the space of possible systems is reduced. This advantage is readily apparent when we consider the range of sizes and structures of mammals that contain similar internal organs with mutually consistent function and interconnections. It is also apparent when we consider the variety of cars that are produced and realize that systematic (algorithmic) relationships exist in the structure and placement of different components. A drawn blueprint cannot describe the interrelationships of engine size to car mass. Instead, the burden of applying such relationship is typically placed on human beings who know them as design rules. A developmental approach would incorporate these rules in an algorithm that could be modified to produce cars with various features and sizes. If the algorithmic description is sufficiently concise, then even random changes will still lead to viable designs. The importance of modification of design in biology is apparent in our discussions in Chapter 6 about sexual reproduction and the importance of random variation in evolution.

There are also disadvantages of a concise design. One arises when the design must be precisely duplicated. Without any redundancy in the information, copying errors may be introduced. We can see that the advantages of a developmental approach are most important when a design is to be modified frequently, and less so when it is to be duplicated many times without modification. Still, the problem of duplication can be largely overcome. This is done through compression with a limited number of redundancies that enable error detection.

Without further elaboration of these matters we can recognize the central issues that have been raised. The connections that we want to make in this discussion are between the biological developmental process, the design of complex systems, and the field of information and computation theory which is more commonly discussed in the context of computer algorithms.

There are various ways in which interrelationships or algorithms are used in the design of man-made products, whether these are physical entities such as cars and airplanes or computer programs. One common methodology is the use of modular design. For example, in the construction of apartment buildings or housing developments, identical units need not be individually described. If modules differ, however, they must generally be separately described. In order to execute the design, it is not sufficient to describe only the modifications. In software design, compilers or interpreters translate from a more concise, higher-level language into a form suitable for execution. We could also consider human elaboration of a design in a similar manner. The various stages in design development elaborate a concise specification. The first design might be an overall concept which is very concise. This concise description is elaborated in a process that can involve many human beings. Such a process is a kind of development when we include the human beings as part of the system. Thus we see that there are various ways in which algorithmic relationships are incorporated into the design of man-made systems. However, it is apparent that the systematic use of algorithms is not yet well developed.

How can we further incorporate the concept of self-organization or algorithmic description in the design of man-made products? It is hard to imagine a developmental process that could create houses. However, it is not hard to imagine a computer-aided design system that can apply various modifications to a design and automatically incorporate design rules. Computer aided design in general can be understood as a process of elaboration of concise descriptions. In its present form it is not developmental in approach. Because the design description in a developmental process is more concise, the application of these concepts to design is a strategy for reducing the complexity of the design and engineering task. Our discussion of developmental biology suggests that it will become progressively important as the systems that are being designed become more complex and are modified more frequently.

7.6   Pattern Formation and Evolution

In Chapter 6 we considered various models of evolution as an undirected process that can give rise to complex systems. The essential concepts are the formation of diversity and selection from this diversity. In this chapter we considered models for pattern formation in developmental biology. The types of models we used are quite different. Here we point out that the mathematical models of pattern formation may also be relevant to the problem of evolution. Contact between these two problems arises from the pattern of organism populations in genomic or phenomic space and in physical space. The existence of a particular organism corresponds to a density $n(s)$ in this space. Species or trait groups correspond to patches of high density that are surrounded by regions that do not contain organisms. We assume that the pattern of populations is formed by evolution.

Evolution considered as diffusion in genomic space, including interactions, has resemblance to a reaction-diffusion system with some important modifications. We consider first the origin of short-range activation and long-range inhibition that may have given rise to the pattern of spots separated by unoccupied regions. This was already discussed in Chapter 6; we summarize here only a few points. There are various mechanisms for activation. The most direct is reproduction. Various social behaviors such as flocking are also mechanisms of short-range activation. Long-range inhibition must have a range that is reflected by the gaps between species. An important cause of inhibition is the consumption of resources. Similar organisms typically consume similar resources. Thus the existence of an organism causes inhibition of organisms over a range of genomes or phenomes. We can reasonably assume that the range of organisms that consume similar resources is larger than the range of organisms that are enhanced by reproduction. There may also be even longer-range interactions, but these we might include in a mean field treatment for the pattern-forming model.

With this motivation, we consider evolution modeled by a reaction-diffusion system with two components formed from the organism and its resource. The second set of reaction-diffusion equations (Eq. (7.2.45)) is a natural model where the substrate b is the resource and the activator a is the organism. Organism diffusion is a consequence of mutation if s represents genomic or phenomic space. It is physical migration if s represents physical space. Resource diffusion need not be taken literally—the same effect (long-range inhibition) may be achieved due to organism behavior in consuming resources of various related types, or at various physical locations in the vicinity of its domicile.

For plant evolution we consider resources to be sun, water, nutrients and space. For herbivore evolution we consider plants and space to be the primary resources. For carnivore evolution we consider herbivores to be the primary resource. According to Eq. (7.2.45), the resource grows spontaneously but the organism reproduces when consuming the resource. The quadratic dependence of reproduction on organism population a in the terms $k_1 a^2 b$ and $k_4 a^2 b$ is a nonlinear or cooperative effect in consumption and reproduction. Sexual reproduction by itself only gives rise to a nonlinear dependence if the probability of mate encounter is small. If the probability of encounter is not small then reproduction is linear, since organisms are often limited to a certain number of offspring. The nonlinear dependence is suggestive of the cooperativity of effective consumption (e.g., a wolf pack or a lion pride) and resulting reproduction. The other term $k_2 a$ is the rate of organism death. From our studies of the behavior of reaction-diffusion systems, there are various modifications of this system that would still give rise to pattern formation, however, not all systems will result in patterns, and the pattern character varies.

There are two additional differences between an evolutionary model and the pattern-forming model: first, the existence of a fitness, or fitness-related parameters, that control the growth of population at a particular genome, and second, the existence of a higher-dimensional space than the two-dimensional space that we considered for patterns. To implement these modifications the equations would take the form:

$$f(a(s),b(s)) = k_1(s)a(s)^2 b(s) - k_2(s)a(s)$$
$$g(a(s),b(s)) = k_3(s) - k_4(s)a(s)^2 b(s)$$

(7.6.1)

where we have just included the state dependence of all of the constants. They are all genome or phenome and location dependent, because the resources appropriate for a particular organism have their own dynamics, as do the organisms.

The essential behavior of this model without the species dependence of the parameters has already been simulated in the context of the pattern formation through diffusion of pigment cells. The modeling of diffusion of the pigment from a line in Section 7.2.7 is particularly relevant. We saw how patterns of spots can be formed that, in a model of evolution, would be interpreted as species or trait groups. The species closer to the starting line would correspond to simpler and more primitive organisms, while those far away would correspond to more complex organisms formed at a later stage of evolutionary history. We could readily imagine that such patterns will form in higher-dimensional spaces and with various species-dependent parameters. The degree to which variability of parameters would affect the relevance of such a model is still to be studied.

There are several advantages of a reaction-diffusion model for evolution that are appealing when contrasted with the models used in the previous chapter. The reaction-diffusion model gives insight into the reason that organisms continue to exist at different scales and at different stages of the evolutionary tree, including the coexistence of simple and complex organisms. We find this in the pattern-forming model, because the pattern continues to have populations in all regions of the space. The underlying reason for this is that the model inherently assumes that there is a variety of resources that are consumed by different organisms. A more complex organism that occurs later in evolution does not consume the same resources that a simpler organism does. To return to a model of the competition for a single resource, we would simply replace the many variables $b(s)$ with a single variable b. Or, more properly, we would expand the range of inhibition (by increasing D_b) to include the whole space. This would be similar to the renewable-resource model with only one resource. In this case, we have argued in Chapter 6 that only one type of organism would survive.

The model of a pattern-forming evolutionary process is also interesting in that competition is no longer the primary reason for the creation of complex organisms. Instead, the creation of complex organisms is due to the existence of resources that cannot be consumed by simple organisms. We might call these complex resources. Through mutation, organisms are formed that can consume the complex resources. Competition for resources causes the pattern of species or trait groups, but is not responsible for the existence of complex organisms.

We can modify this model to incorporate competition more fully by considering the space of resources and the space of organisms to be related in a more elaborate manner. Specifically, that organisms that are far apart in genome or phenome might consume the same resource. In order to know which organisms would be in competition, we consider the phenome space as projecting onto the resource space in a many-to-one map. As evolution proceeded, there would come instances in which organisms at different phenome locations but the same resource location would coexist, and the fitter organism would survive while the less fit would become extinct. However, there would still be a variety of resources giving rise to a variety of organisms at any stage of evolution.

In summary, the phenomenological existence of diverse species suggests that a reaction-diffusion model of pattern formation, with distinct resources for different organisms, is more realistic than a model that assumes a single resource for all organisms. The persistence of organisms over varied periods of evolutionary history and particularly the continued existence of organisms that originally appeared at much earlier stages of evolutionary history is suggestive of such a model. It is also consistent with the wide variety of resources found in nature.

The notion that the pattern of species is analogous to a developmental process of pattern formation also brings into focus the recognition that all life on earth is interrelated and in some sense is a single complex system. Loosely, by analogy, we might consider the collection of organisms on earth to be a collective organism similar to the collection of cells in a particular organism. This is relevant to the study of ecosystems and their behavior. We will address this from a more specific point of view in

Chapter 9 when we discuss the possibility that the collection of human beings on earth should be considered as a single complex system. This discussion will also have consequences for our understanding of the relationship between evolution and developmental biology. Before we do so we introduce and discuss in greater detail the concept of complexity in order to better evaluate the complexity of the global system of organisms on earth. Our focus, for various reasons, will be the global human civilization, but extending this discussion to include other organisms on earth is natural.

8

Human Civilization I:
Defining Complexity

Conceptual Outline

■ **8.1** ■ Our ultimate objective is to consider the relationship of a human being to human civilization, where human civilization is considered as a complex system. We use this problem to motivate our study of the definition of complexity.

■ **8.2** ■ The mathematical definition of the complexity of character strings follows from information theory. This theory is generalized by algorithmic complexity to allow all possible algorithms that can compress the strings. The complexity of a string is defined as the length of the shortest binary input to a universal Turing machine, such that the output is the string.

■ **8.3** ■ The use of mappings from strings onto system states allows us to apply the concepts of algorithmic complexity to physical systems. However, the complexity of describing a microstate of the system is not really what we mean by system complexity. We define and study the complexity profile, which is the complexity of a system observed with a certain precision in space and time.

■ **8.4** ■ We estimate the complexity of various systems, focusing on the complexity of a human being. Our final estimate is based upon a combination of the length of descriptions in human language, genetic information in DNA, and component counting.

8.1 Motivation

8.1.1 *Human civilization as a complex system*

The subject of this and the next chapter is human civilization—the collection of all human beings on earth. Our long-term objective is to understand whether and how we can treat human civilization as a complex system and, more particularly, as a complex organism. In biology, collections of interacting biological organisms acting together are called superorganisms. At times, we will adopt this convention and refer to civilization as the human superorganism. Much of what we discuss is in early stages of development and is designed to promote further research.

This subject is distinct from the others we have considered. The primary distinction is that we have only one example of human civilization. This is not true about the systems we have discussed in earlier chapters, with the exception of evolution considered globally. The uniqueness of the human superorganism presents us with questions of fundamental interest in science, related to how much we can know about an individual system. When there are many instances, we can use information provided by various examples and the statistics of their properties. When there is only one system, to understand its properties or predict its behavior we must apply fundamental principles that are valid for all complex systems. Since the field of complex systems is dedicated to uncovering such principles, the subject of the human superorganism should be considered a premiere area for application of complex systems research. Central questions are: How can we characterize this complex system? How can we determine its properties? What can we tell about its dynamics—its past and future? We note that as individuals we are elements of the human superorganism, thus our spatial and temporal experience may very well be more limited than that appropriate for analyzing the human superorganism.

The study of human civilization is guided by historical records and contemporary news. In contrast to protein folding, neural networks, evolution and developmental biology there are few reproducible laboratory experiments. Because of the irreproducibility of historical or contemporary events, these sources of information are properly not considered part of conventional science. While this can be a limitation, it is also apparent that there is a large amount of information available. Our task is to develop systematic methods for considering this kind of information that will enable us to approach questions about the nature of human civilization as a complex system. Various aspects of these problems have been studied by historians, anthropologists and sociologists.

Why consider human civilization as a single complex system? The recently discussed concept of a global economy, and earlier the concept of a global village, suggest that we should consider the collective economic behavior of human beings and possibly the global social behavior as a single system. Considering civilization as a single entity we are motivated to ask various questions about it. These questions relate to all of the topics we have covered in the earlier chapters: spatial and temporal structure, evolution and development. We would also like to understand the interaction of human civilization with its environment.

In developing an understanding of human civilization, we recognize that a widespread view of human civilization as a single entity is relatively new and driven by contemporary developments. At least superficially, the historical epoch described by the dominance of nation-states appears to be quite different from the present global economy. While recent events appear to be of particular significance to the global view, our questions must be addressed in a historical context. Thus we should include a discussion of the transition to a global economy. We postpone this historical discussion to the next chapter because of the groundwork that we would like to build in order to target a particular objective for our analysis—that of complexity classification.

We are motivated to understand complexity in the context of our effort to understand the nature of the human superorganism, or the nature of the global economy. We would like to identify the type of complex system it is—to classify it. The first distinction that we might make is between a complex material or a complex organism (see Section 1.3.6). Could part of the global system be modified without affecting the whole? From historical evidence discussed in the next chapter, the answer appears to be no. This indicates that human civilization is a complex organism. The next question we would like to ask is: What kind of complex organism is it? By analogy we could ask: Is it like a protein, a cell, a plant, an insect, a frog, a human being? What do we mean by using such analogies? At least in part the problem is to describe the complexity of an entity's behavior. Intuitively an insect is a simpler organism than a human being, and this is of qualitative importance for our understanding of their differences. The degree of complexity should provide a scale that can distinguish between the many different complex systems we are familiar with.

Our objective in this chapter is to develop a quantitative definition of complexity and behavioral complexity. We then apply the definition to various complex systems. The focus will be on the complexity of an individual human being. Once we have established our complexity scale we will be in a position to apply it to human civilization. We will understand formally why a collection of complex systems (human beings) may be, but need not be, complex. Beyond recognizing human civilization as a complex system, it is far more significant to identify the degree of its complexity. In the following brief sections we establish some additional context for the importance of measuring complexity using both unconventional and conventional examples of organisms whose complexity should be evaluated.

8.1.2 *Scenario: alien encounter*

The possibility of encountering alien life has been debated within the scientific community. In popular literature, such encounters have been portrayed in various forms ranging from benevolent to catastrophic. The scientific debate has focused thus far on topics such as the statistics of planet formation and the likelihood that planets contain life. The presence of organic molecules in meteorites and interstellar gasses has been interpreted as suggesting that alien life is likely to exist. Efforts have been made to listen for signs of alien life in radio communications and to transmit information to aliens using the Voyager spacecraft, which is leaving the solar system marked with information about human beings. Thus far there has been no scientifically confirmed evidence for the existence of alien life. Even a single encounter would change the human perspective on humanity's place in the universe.

Let us consider one possible scenario for an encounter. An object that flashes light intermittently is found in orbit around one of the planets of the solar system. The humans encountering this object are faced with the question of determining whether the object is: (a) a signal device—specifically a recording, (b) a communication device, or (c) a living organism. The central problem can be seen to revolve around determining whether, and in what way, the device is responsive to external phenomena. Do the flashes of light occur without regard to the external environment in a predetermined sequence? Are they random? If the flashes are sensitive to the environment, then what are they sensitive to? We will see that these questions are equivalent to the question of determining the complexity of the object's behavior.

The concept of life in biology is often defined, or better yet, characterized, in terms of consumption, excretion and reproduction. As a definition, these characteristics are well known to be incomplete, since there are life-forms that do not reproduce, such as the mule. Furthermore, a particular individual is still considered alive even if it/he/she does not reproduce. Moreover, there are various physical systems such as crystals and fire that have all these characteristics in one form or another. Moreover, there does not appear to be a direct connection between these biological characteristics and other characteristics of life such as sentience and self-awareness. When considering behavior, the biological perspective emphasizes the survival instinct as characteristic of life. There are exceptions to this, since there exist life-forms that are at times suicidal, either individually or collectively. The question of whether an organism actively seeks life or death does not appear to be a characterization of life but rather of life-forms that are likely to survive. In our discussions, we may be developing an additional characterization of life in terms of behavioral complexity. Definitions of life are often considered in speculating about the rights of and treatment of real or imagined organisms—injured or unconscious humans, robots, or aliens. The degree of behavioral complexity is a characterization of life-forms that may ultimately play a role in informing our ethical decisions with respect to various biological life-forms, whether terrestrial or (if found) alien, and artificial life-forms that we create.

8.1.3 *Scenario: blood cells*

One of the areas briefly touched upon in Chapter 6, which is at the forefront of complex systems research, is the study of the immune system. Blood cells, unlike other cells in the body, are mobile on a length scale that is large compared to their size. In this characteristic they are more similar to independent organisms than to the other cells of the body. By their migration they might be said to "choose" to associate with other cells of the body, or with foreign chemicals and cells. It is fair to say that our understanding of the behavior of immune cells remains primitive. In particular, the variety of possible chemical interactions between cells has only begun to be mapped out. These interactions involve a variety of chemical messengers. More direct cell-to-cell interactions where parts of the membrane or cellular fluid are transferred are also possible.

One of the interesting questions that can be asked is whether, or at what level of complexity, the interactions become identifiable as a form of language. It is not difficult to imagine, for example, that a chemical communication originating from one cell might be transferred through a chain of cell interactions to a number of other cells. In the context of the discussion in Section 2.4.5, the question of existence of a language might be formulated as a question about the possibility of messages with a grammar—a combinatorial composition of parts that are categorized like parts of speech. Such combinatorial mechanisms are known to exist even at the molecular level in the DNA coding of antibody receptors that are a composite of different parts of the genome. It remains to be seen whether intercellular communication is also generated in this fashion.

In the context of this chapter we can reduce the questions about the immune cells to a single one—What is the degree of complexity of the behavior of the immune cells? By its very nature this question can only be answered once a complete understanding of immune cell behavior is reached. A limited understanding establishes a lower bound for the complexity of the behavior. It should also be understood that different types of cells will most likely have quite different levels of behavioral complexity, just as different animals and man have differing levels of complexity. Our objective in this chapter is to show that it is possible to quantify the concept of complexity in a way that is both natural and useful. The practical application of these definitions is a central challenge for the field of complex systems.

8.1.4 *Complexity*

Mathematical definitions of the complexity of systems are based upon the theories of information and computation discussed in Sections 1.8 and 1.9. In Section 8.2 they will be used to treat complexity in the context of mathematical objects such as character strings. To develop our understanding of the complexity of physical systems requires that we relate these concepts to those of thermodynamics (Section 1.3) and various extensions (e.g., Section 1.4) that enable the treatment of nonequilibrium systems. In Section 8.3 we discuss relevant concepts and tools that may be used for this purpose. In Section 8.4 we use several semiquantitative approaches to estimate the value of the complexity of specific systems.

Our use of the word "complexity" is specified as an answer to the question, How complex is it? We say, Its complexity is <number><units>. Intuitively, we can make a connection between complexity and understanding. When we encounter something new, whether personally or in a scientific context, our objective is to understand it. The understanding enables us to use, modify, control or appreciate it. We achieve understanding in a number of ways, through classification, description and ultimately through the ability to predict behavior. Complexity is a measure of the inherent difficulty to achieve the desired understanding. Simply stated, *the complexity of a system is the amount of information necessary to describe it.*

This is descriptive complexity. For dynamic systems the description includes the changes in the system over time. We will also discuss the response of a dynamic system to its environment. The amount of information necessary to describe this response is a system's behavioral complexity. To use these definitions of complexity we will introduce mathematical expressions based upon the theory of information.

The quantitative definition of information (Section 1.8) is relatively abstract. However, it can be measured in familiar terms such as by the number of characters in a text. As a preliminary exercise in the discussion of complexity, the reader is invited to exercise intuition to estimate the complexity of a number of systems. Question 8.1.1 includes a list of systems that are designed to stimulate some thought about complexity as a quantitative measure of the behavior of a system. The reader should devote some thought to this question before proceeding with the rest of the text.

Question 8.1.1 Estimate the complexity of some of the systems in the following list. For this question use an intuitive definition of complexity—the amount of information that would be required to describe the system or its behavior. We use units of bits to measure information. However, to make it easier to visualize, you may use other convenient units such as words or pages of text. So, we can paraphrase the question as, How much would you have to write to describe the system behavior? A rough conversion factor of 1 bit per character can be used to convert these estimates to bits. It is not necessary to estimate the complexity of all the systems on the list. Considering even a few of them is sufficient to develop an understanding of some of the issues that arise. Indeed, for some of these systems a rough estimate is far from trivial. Answers to this question will be given in the text in the remainder of this chapter.

Hint You may find that you would use different amounts of information depending on what aspects of the system you are describing. In such cases try to give more than one estimate or a range of values.

Physical Systems:

Ideal gas (1 mole at $T = 0°K$, $P = 1$atm)

Water in a glass

Chemical reaction

Brownian particle

Turbulent flow

Protein

Virus

Bacterium

Immune system cell

Fish

Frog

Ant

Rabbit

Cow

Human being

Radio

Car

IBM 360

Personal Computer (PC/Macintosh)

The papers on your desk

A book

A library

Weather

The biosphere

Nature

Mathematical and Model Systems:

A number

Iterative maps (growth, bifurcation to chaos)

1-D random walk

>   short time

>   long time

Ising model (ferromagnet)

Turing machine

Fractals

>   Sierpinski gasket

>   3-D random walk

Attractor neural network

Feedforward neural network

Subdivided attractor neural network ∎

8.2   Complexity of Mathematical Models

Complexity is a property of the relationship between a system and various representations of the system. Our objective is to understand the complexity of systems composed of physical entities such as atoms, molecules or cells. Abstract representations of such systems are described in terms of characters or numbers. It is helpful to preface our discussion of physical systems with a discussion of the complexity of the characters or numbers that we use to represent them.

8.2.1 *Information, computation and algorithmic complexity*

The discussion of Shannon information theory in Section 1.8 was based on strings of characters that were generated by a source. The source generates each string, s, by selecting it from an ensemble. The information from a particular string was defined as

$$I = -\log(P(s)) \tag{8.2.1}$$

where $P(s)$ is the probability of the string in the ensemble. If all strings have equal probability then this is the logarithm of the number of distinct strings. The source itself (or the ensemble) was characterized by the average information of a large number of strings

$$<I> = -\sum_{s} P(s)\log(P(s)) \tag{8.2.2}$$

It was also possible to consider a more general source that selected characters to form a Markov chain. The probabilistic coupling between sequential characters reduced the information content of the string. It was possible to compress the string using a reversible coding algorithm (computation) that would enable the same information to be represented in a more compact form. The length of the shortest binary compact form is equal to the average information in a string.

Information theory suggests that we can define the complexity of a string of characters by the information content of the string. The information content is the same as the length of the shortest binary encoding of the string. This is intuitive—since the original string can be obtained from its shortest representation, the same information must be present in both. Within standard information theory, the encodings would be limited to compression using a Markov chain model. However, more generally, we could use any possible algorithm for encoding (compressing) the string. Questions about all possible algorithms are precisely the domain of computation theory. The definition of Kolmogorov (algorithmic) complexity of a string makes use of computation theory to describe what we mean by "any possible algorithm." Allowing all algorithms is the same as allowing more general models for the string than a Markov chain. Our objective in this section is to develop an understanding of algorithmic complexity beginning from the theory of computation.

Computation theory (Section 1.9) describes the operations of logic and computation on symbols. All the operations are deterministic and are expressible in terms of a few elementary operations. The concept of universality of computation is based on the understanding that a particular type of conceptual machine/computer—the universal Turing machine (UTM)—can perform all possible computations if the instructions are properly encoded as a finite string of characters serving as the UTM input. Since we have no absolute definition of computation, there is no complete proof. The existing proof shows that the UTM can perform all computations that can be done by a much larger class of machines—the Turing machines (TM). Other models for computation have been shown to be essentially equivalent to these TM. A TM is defined by a table of elementary operations that act on the input string. The word "program" can be used either to refer to the TM table or to its input and so its use is best avoided in this context.

We would like to define the algorithmic complexity of a string, s, as the length of the shortest possible binary TM input, such that the output is s. The relationship of this to the encoding and decoding of Shannon should be apparent. In order to use this as a definition, there are several matters that must be cleared up. To summarize: There are actually two sources of information when we use a TM, the input string and the table. We need to take both of them into account to define the complexity. There are many ways to define complexity; however, we can prove that any two definitions of complexity differ by no more than a constant. We will also show that no matter what definition we use, most strings cannot be compressed.

In order to motivate the logic of the following discussion, it is helpful to think about how we might approach compressing various strings of characters. The shortest compression should then be the complexity of the string. One string might be formed out of a long substring of zeros followed by a long substring of ones. This is convenient to write by indicating how many zeros followed by how many ones: N_0N_1. We would make a binary string notation for N_0N_1 and write a program that would read this input and then output the original string. Another string might be a representation of the Fibonacci numbers $(1,1,2,3,5,8,\ldots)$, starting from the N_0st number and ending at the N_1st number. We could write this using a similar notation as the previous one, but the program that we would write to generate the string is quite different. Both programs would be quite simple. Now imagine that we want to communicate one of the original strings to someone else. If we want to communicate it in compressed form, we would have to send the program as well as the input. If there were many strings, we might be clever and send the programs only once. The problem is that with only the input string, the recipient would not know which program to apply to obtain the original string. We need to send an additional piece of information that indicates which program to apply. The simplest way to do this is to assign numbers to each of the programs and preface the program input with the program number. Once we do this, the string that we send uniquely determines the string we wish to communicate. This is necessary, because if the interpretation of the transmitted string is not unique, then it would be impossible to guarantee a correct interpretation. We now develop these thoughts using a more formal notation.

In what follows, the operation of a TM or a UTM will be indicated by functional notation. The string that results from its application to a tape is indicated by $U(s)$ where s is the nonblank portion of the tape (input string), U is the identifier of the TM, and the initial position of the TM head is assumed to be at the leftmost nonblank character.

In order to define the complexity of a string, we identify a particular UTM U. Then the complexity $C_U(s)$ of the string s is defined as the length of the shortest string r such that $U(r) = s$. We call an input string r to U that generates s a representation of s. Thus the length of the shortest representation is $C_U(s)$. The central theorem of algorithmic complexity relates the complexity according to one UTM U and another UTM U'. Before we state and prove the theorem, we discuss several incidental matters.

We first ask whether we need to use a UTM and not just any TM in the definition. The answer is that the use of a UTM is convenient, and we cannot significantly improve the ability to compress strings by allowing the larger class of TM to be used in the definition. Let us say that we have a UTM U and a TM V, we define a new UTM W by:

$$W(0s) = V(s)$$
$$W(1s) = U(s)$$

(8.2.3)

—the first character indicates whether to use the TM V or the UTM U on the rest of the input. Since the complexity according to the UTM W is at most one more than the complexity according to the TM V, $C_W(s) \le C_V(s) + 1$, we see that using the larger class of TM to define complexities can not improve our results for any particular string by more than one bit, which is not significant for long complex strings.

We may be disturbed that the definition of complexity does not indicate that the complexity of an incompressible string is the same as the string length itself. Indeed the definition does not require it. However, if we wanted to impose this as an auxiliary condition, we could define the complexity of a string using a slightly different construction. Given a UTM U, we define a new UTM V such that

$$V(0s') = s'$$
$$V(1s') = U(s')$$

(8.2.4)

—the first character indicates whether the string is compressed. We then define the complexity $C_U(s)$ of any string s as one less than the length of the shortest string r such that $V(r) = s$. This is not quite a fair definition, because if we wanted to communicate the string s we would have to indicate all of r, including its first bit. This means that we should define the complexity as the length of r, which would be a sacrifice of at most one bit for incompressible strings. Limiting the complexity of a string to be no longer than the string itself might seem a natural idea. However, we note that the Shannon information, Eq. (8.2.1), is related only to the probability of a string, and may be larger than the original string length for a particular string.

Returning to our basic definition of complexity, we have described the existence of a shortest possible representation of any string s, and a single machine U that can reconstruct each s from this representation. The key theorem that we need to prove relates the complexity defined using one UTM U to the complexity defined using another UTM U'. The theorem is: the complexity C_U based on U and the complexity $C_{U'}$ based on U' satisfy:

$$C_U(s) \le C_{U'}(s) + C_U(U')$$

(8.2.5)

where $C_U(U')$ is independent of the string s. The proof of this expression results from the ability of the UTM U to simulate U'. To prove this we must improve slightly our definition of complexity, or equivalently, we have to limit the UTM that are allowed. This is discussed in Questions 8.2.1–8.2.3. It is shown there that we can preface binary strings input to the UTM U' with a prefix that will make them generate the same output when input to U. We might call this prefix $r_{U,U'}$ a translation program, it satisfies the property that for any string r, $U(r_{U,U'}r) = U'(r)$. Let $r_{U'}$ be a minimal representation for U' of the string s. Then $r_{U,U'}r_{U'}$ is a representation for U of the string s. The length of this string must be greater than or equal to the length of the minimum string r_U necessary to produce the same output:

$$C_U(s) = |r_U| \le |r_{U,U'}r_{U'}| = |r_{U'}| + |r_{U,U'}| = C_{U'}(s) + C_U(U')$$

(8.2.6)

$C_U(U') = |r_{U,U'}|$ is the length of the translation program. We have proven the inequality in Eq. (8.2.5).

Question 8.2.1 Show that there exists a UTM U_0 such that for any TM U that accepts binary input, there is a string r_U so that for all s and r satisfying $s = U(r)$, we have that $s = U_0(r_U r)$.

Hint One way to do this is to use a modified form of the construction given in Section 1.9. The new construction requires modifying the nature of the UTM—i.e., a trick.

Solution 8.2.1 We call the UTM described in Section 1.9, $\tilde{U}_0$. We can simulate the UTM U using $\tilde{U}_0$; however, the form of the input string would not quite satisfy the conditions of this theorem. $\tilde{U}_0$ has an input that looks like $r_U r_t(r)$, where the right part is only a function of the input string r and the left part is only a function of the UTM U. However, the tape part of the representation $r_t(r)$ uses a doubled binary form for characters and markers between them so that it is not the same as the original tape. We must replace the tape part of the representation with the original string in order to have an input string of the form $r_U r$.

Both $\tilde{U}_0$ and U have binary input strings. This means that we might try to use the tape of U without modification in the tape part of the representation given in Section 1.9. Then there would be no delimiters between characters and no doubled binary representation. There is, however, one difficulty. The UTM U_0 must keep track of where the current position of the UTM U would be during the same calculation. This was accomplished in Section 1.9 by converting one of the M_1 markers to M_6 at the current location of the UTM U. There are a number of ways to overcome this problem, but all require us to introduce something new. We will do this by allowing the UTM U_0 to have a counter that can keep track of the current position of the UTM U. There are two ways to argue this. One is to allow, by proclamation, a counter that can reach arbitrarily high numbers. The other is to recognize that the longest string we might conceivably encounter is smaller than the number of particles in the known universe, or very roughly $10^{90} = 2^{300}$. This means that we can use an internal memory of 300 bits to represent such a counter. This counter is initialized to 0 and set to the current location of the UTM U at every step of the calculation. This construction gives us the desired UTM U_0. ∎

Question 8.2.2 Using the result of Question 8.2.1, prove Eq. (8.2.5). See the text for a hint.

Solution 8.2.2 The problem is that Eq. (8.2.5) is not actually correct for all UTM (see Question 8.2.3) so we need to modify our conditions. In a sense, the modification is minor because we only improve the definition slightly. We do this by defining the complexity $C_U(s)$ for an arbitrary UTM as the minimum length of r such that $W(r) = s$ where W is defined by:

$$W(0s) = U_0(s)$$
$$W(1s) = U(s)$$

$$(8.2.7)$$

—the first bit specifies whether to use U or the special UTM U_0 constructed in Question 8.2.1. $C_U(s)$ defined this way is at most one bit more than our previous definition, for any particular string. It might be significantly smaller. This should not be a problem, because our objective is to find short representations of strings. By using our special UTM U_0 in this definition, we guarantee that for any two UTM U and U', whose complexity is defined in terms of W and W' by Eq. (8.2.7), we can write $W(r_{WW'}r_{W'}) = W'(r_W)$. This is possible because W inherits the properties of U_0 when the first character of its input string is 0. ∎

Question 8.2.3 Show that some form of qualification of Eq. (8.2.5) is necessary by demonstrating that there exists a UTM that does not satisfy this inequality. Therefore, Eq. (8.2.5) cannot be extended to all UTM.

Solution 8.2.3 One possibility is to have a UTM that uses only certain characters in its input string. Specifically, define a UTM U that acts the same as a UTM U' but uses only every other character in its input string: $U(r) = U'(r')$ if r is any string whose odd characters are the characters of r'. The complexity of a string according to U is twice the complexity according to U' and therefore Eq. (8.2.5) is invalid in this case. With the modified definition of complexity given in Question 8.2.2 this is no longer a problem. ∎

Switching U and U' in Eq. (8.2.5) gives a similar inequality with a constant $C_{U'}(U)$. Defining the larger of the two translation program lengths to be

$$C_{U,U'} = \max(C_U(U'), C_{U'}(U)) \tag{8.2.8}$$

we have proven that complexities defined by the UTM differ by no more than $C_{U,U'}$:

$$|C_U(s) - C_{U'}(s)| \le C_{U,U'} \tag{8.2.9}$$

Since this constant is independent of the complexity of the string s, it becomes insignificant for large enough complexities. Thus, for strings that are complex enough, it doesn't matter which UTM we use to define its complexity. The complexity defined by one UTM is the same as the complexity defined by another UTM. This consistency—universality—in the complexity of a string is essential in order for it to be well defined. We will use a few examples to illustrate the nature of universality provided by this definition.

The first example illustrates the relationship of algorithmic complexity to string compression. Given a string s we can ask what methods of compression are useful for the string. A useful compression algorithm corresponds to a pattern in the characters of the string. A string might have many repetitive digits, or cyclically repeating digits. Alternatively, it might be a sequence that can be generated using simple mathematical operations such as the Fibonacci series, or the digits of π. There are many such patterns that are relevant to the compression of strings. We can choose a finite set of N algorithms $\{V_i\}$, where each one is represented by a TM that reconstructs a string s from a shorter string r by taking advantage of properties of the pattern. We now construct a new TM U which is defined by:

$$U(r_i r') = V_i(r') \tag{8.2.10}$$

where r_i is a binary representation of the number i, having $\log(N)$ bits. This is a UTM if any of the V_i is a UTM or it can be made into a UTM by Eq. (8.2.3). We use U to define the complexity $C_U(s)$ of any string as described above. This complexity includes both the length of r' and the number of bits $(\log(N))$ in r_i that together constitute the length of the input r to U. Once it is defined, this complexity is a measure of the complexity of all strings. We do not use different TM to define the complexity of each string; one UTM is used to define the complexity of all strings.

Despite the message of the last example, let us assume that we are evaluating the complexity of a particular string s. We define a new UTM U_s by:

$$U_s(0s') = s$$
$$U_s(1s') = U(s)$$
(8.2.11)

—the first character tells U_s if the string is s. We can use this new UTM to define the complexity of all strings and for this definition the complexity of s is one. How does this relate to our theorem about the universality of complexity? The point is that in this case the translation program between U and U_s contains the complete information about s and therefore must be at least as long as $C_U(s)$. What we have done is to take the particular string s and insert it into the table of U_s. We see in this example how universality is tied to an assumption that the complexities that are discussed are longer than the TM translation programs or, equivalently, the information in their tables. Conceptually, we would say that universality of complexity is tied to an assumption of lack of specific knowledge on the part of the recipient (represented by the UTM) of the information itself. The choice of a particular UTM might be dictated by an implicit understanding of the set of strings that we would like to represent, even though the complexity of a string is defined without reference to an ensemble of strings. However, this apparent relativism of the complexity is limited by our basic theorem that relates the complexity of distinct UTM, and by additional results about the impossibility of compressing most strings discussed in the following paragraphs.

We have gained an additional result from the construction of a single UTM that generates all strings from their compressed forms. This is that a representation r only represents one string s. We can now prove that the probability that a string of length N can be compressed is very small. The proof proceeds from the observation that the number of possible strings decreases very rapidly with decreasing string length. A string s of length $|s| = N$ compressed by k bits is represented by a particular string r of length $|r| = C(s) = N - k$. Since there are only 2^{N-k} strings of length $N - k$, at most 2^{N-k} strings of length 2^N can be compressed by k bits. The fractional compression is k/N. For example, among all strings of length 10^6 bits, at most 1 string in $2^{100} = 10^{30}$ can be compressed by 100 bits or .01% of the string length. This is not a very significant compression. Even so, this estimate of the average number of strings that can be compressed is much too large, because strings that are not of length N, e.g., strings of length $N - 1$ $N - 2$, ..., $N - k$, would also be represented by strings of length $N - k$. Thus most strings are incompressible. Moreover, selecting a string at random will yield an incompressible string.

Question 8.2.4 Calculate a strict lower bound for the average complexity of strings of length N.

Solution 8.2.4 We assume that strings of length N are compressed so that they are represented by all of the shortest strings. One string is represented by the null string (length 0), two strings are represented by a single bit (length 1), and so on. The relationship:

$$2^N = \sum_{l=0}^{N-1} 2^l + 1 \tag{8.2.12}$$

means that we will fill all of the possible strings up to length $N-1$ and then have one string left of length N. The average representation length for any complexity measure must then satisfy:

$$<C(s)> \geq \frac{1}{2^N}\left(\sum_{l=0}^{N-1} l2^l + N\right) \tag{8.2.13}$$

The sum can be evaluated using a table of sums or:

$$\sum_{l=0}^{N-1} l2^l = \frac{1}{\ln(2)}\frac{d}{d\alpha}\sum_{l=0}^{N-1} 2^{\alpha l}\bigg|_{\alpha=1} = \frac{1}{\ln(2)}\frac{d}{d\alpha}\frac{2^{\alpha N}-1}{2^\alpha -1}\bigg|_{\alpha=1} = N2^N - 2(2^N-1) \tag{8.2.14}$$

giving:

$$<C(s)> \geq (N-2)+\frac{1}{2^N}(N+2) > N-2 \tag{8.2.15}$$

Thus the average complexity of strings of length N cannot be reduced by more than two bits. This strict lower bound applies to all measures of complexity. ∎

We can also interpret this discussion to mean that the best UTMs to use to define complexity are those that are invertible—they have a one-to-one mapping of strings to representations. In this case we have a mapping $r(s)$ which gives the unique representation of a string. The reason that such UTM are better is that there are only a limited number of representations shorter than N; if we use up more than one of them for a particular string, then we will have fewer representations to use for others. Such UTM are closely analogous to our understanding of encoding and decoding as described in information theory. The UTM is the decoder and the mapping of the string onto its representation is the encoding.

Because most strings are incompressible, we can also prove that if we have an ensemble of strings defined by the probability $P(s)$, then the average algorithmic complexity of these strings is essentially the same as the Shannon information. In particular, the ensemble of all of the strings of length N have a Shannon information of N bits and an average algorithmic complexity which is the same. The catch is recognizing that to specify $P(s)$ itself requires an algorithm whose complexity must enter into the discussion. The proof follows from the discussion in Section 1.8. An ensemble defined by a probability $P(s)$ can be encoded in such a way that the average string length is given by the Shannon information. We now realize that to define the string complexity we must include the description of the decoding operation:

$$\sum_{s} P(s)C(s) = \sum_{s} P(s)I_s + C(P) \qquad (8.2.16)$$

where the expression $C(P)$ represents the complexity of the decoding operation for the universal computer U for the ensemble given by $P(s)$. $C(P)$ depends in part on the algorithm used to specify the ensemble probability $P(s)$. For the average ensemble complexity to be essentially equal to the average Shannon information, the specification of the ensemble must itself be simple.

For Markov chains a similar result applies—the Shannon information of a string representing a Markov chain is the same as the algorithmic complexity of the same string, as long as the algorithm specifying the Markov chain is simple.

A general consequence of the definition of algorithmic complexity is a limitation on what TM can do. No TM can generate a string more complex than the input string that it is provided with, plus the information in its table—otherwise we would have redefined the complexity of the output string to take this into consideration. This is a key limitation of TM: TM (and computers that are realizations of this model) cannot generate new information. They can only process information they are given. As discussed briefly in Section 1.9.7, this limitation can be overcome by a TM that is given a string of random bits as input. The infinitely complex input means the limitation does not apply. It remains to be demonstrated what tasks such a TM can perform that are not possible for a conventional TM. If such tasks are identified, there will be important implications for computer design. In this context, it may also be suggested that some forms of creativity might be linked to the availability of randomness (see Section 1.9.7). We will return to this issue at the end of the chapter.

While the definition of complexity using UTM is appealing, there is a profound difficulty with this proof. It is nonconstructive. No method is given to determine the complexity of a particular string. Indeed, it can be proven that this is a fundamentally difficult task—the time necessary for a TM to determine $C(s)$ grows exponentially with the length of s. At least this is true when there is a bound on the complexity, e.g., by Eq. (8.2.4). Otherwise the complexity is noncomputable. We find the complexity of a string by trying all input strings in the UTM to see which one gives the necessary output. If the complexity is not bounded, then the halting problem implies that we cannot tell if the UTM will halt on a particular input, thus it is noncomputable. If the complexity of the string is bounded, then we only try strings up to this bound, and it is possible to determine if the UTM will halt for members of this bounded set of strings. Nevertheless, trying each string requires a time that grows exponentially with the bound, and therefore is not practical except for a few very simple strings. The process of finding the complexity of a string is akin to a process of trying models for the string. A model is a TM that might, when given the proper input, generate the string. It is possible to try many models. However, to determine the actual compressed string may not be practical in any reasonable time. With any particular set of models, we can, however, find an upper bound on the complexity of a string. One of the possible models is that of a Markov chain as used by Shannon information theory. Algorithmic complexity allows more general TM models. However, by our discussion it is improbable that a randomly chosen string will be compressible by any algorithm.

In summary, the universality of complexity is a statement that the use of different UTMs in the definition of complexity affects the result by no more than a constant. This constant is the length of the program that translates the input of one UTM to the other. Significantly, the more complex the string is, the more universal is the value of its complexity. This follows because the length of translation programs becomes less and less relevant for longer and longer descriptions/representations. Since we are interested in properties of complex systems whose descriptions are long, we can, with caution, rely on the universality of their complexity. This is not the case with simple systems whose descriptions and therefore complexities are "subjective"—they depend on the conventions for description. These conventions, in our mathematical definition, are represented by the choice of UTM used to define complexity. We also showed that most strings are not compressible and that the Shannon information measure is the same as the average algorithmic complexity for all concisely describable ensembles. In what follows, unless otherwise mentioned, we assume a particular definition of complexity $C(s)$ using the UTM U.

8.2.2 *Mathematical systems: numbers and functions*

One of the difficulties in discussing complexity is that many elementary mathematical constructs have unusual properties when considered from the point of view of complexity. Philosophers have been troubled by these points, and they have been extensively debated over the centuries. Most of the problems revolve around various forms of infinity. Unlimited numbers and infinite precision often simplify symbolic mathematical discussions; however, they are not well behaved from the point of view of complexity measures. There appears to be a paradox here that will be clarified when we distinguish between the complexity of a set of numbers and the complexity of an element of the set.

Let us consider the complexity of specifying a single integer. The difficulty with integers is that there are infinitely many of them. Using an information theory point of view, assigning equal probability to all integers would imply that any particular integer would have no probability of occurring. If I ask you to give me a positive integer, from 1 to infinity with equal probability, there is no chance that you will give me an integer below any particular cutoff value, say N. This means that you will need arbitrarily many digits to specify the integer, and there is no limit to the information required. Thus the complexity of specifying a single integer is infinite. However, if we allow only integers between 1 and a large positive number—say $N = 10^{90}$, roughly the number of elementary particles in the known universe—the complexity of specifying one of the integers is only $\log(N)$, about 300 bits. The drastic difference between the complexity of specifying an arbitrary integer (infinite) and the complexity of an enormously large number of integers (300 bits) suggests that systems that are easy to define may be highly complex. The whole field of number theory has shown that integers are not as simple as they first appear. The measure of complexity of specifying a single integer may appear to be far from more abstract discussions like those of the halting problem or Gödel's theorem (Section 1.9.5), however, they are related. This is apparent since these theorems do not apply to finite sets.

In what sense are integers simple? We can consider the length of a UTM input string that can generate all the positive integers. As discussed in the last section, this is similar to the definition of their Kolmogorov or algorithmic complexity. The program would, starting from zero and keeping a list, progressively add one to the preceding integer. The problem is that such a program never halts, and the task is not complete. We can generalize our definition of a Turing machine to allow for this case by saying that, by definition, this simple program is generating all integers. Then the algorithmic complexity of the integers is quite small. Another way to do this is to consider the complexity of recognizing an integer—the recognition complexity. Recognizing an integer is trivial if we are considering only binary strings, because all of them represent integers. The point, however, is that we can expand the space of possible characters to include various symbols: letters, punctuation, mathematical operations, etc. The mathematical operations might act upon integers. We then ask how long is a TM program that can recognize any integer that appears as a combination of such characters. The length of such a program is also small.

We see that we must distinguish between the complexity of elements of a set and the set itself. A program that recognizes integers is concerned with the attributes of the integers required to define them as a set, rather than the specification of a particular integer. The algorithmic complexity of the set of all integers is small even though the information contained in a single integer can be arbitrarily large. This distinction between the information contained in an element of a set and the information necessary to define the set will also be important when we consider the complexity of physical systems.

The complexity of a single real number is also infinite. Specifying an arbitrary real number requires infinitely many digits. However, if we confine ourselves to any reasonable precision, the complexity becomes very manageable. For example, the most accurately known fundamental constant in science is the electron magnetic moment in Bohr magnetons

$$\mu_e/\mu_B = 1.001159652193(10) \tag{8.2.17}$$

where the parenthesis indicates the error estimate, corresponding to 11 accurate decimal digits or 37 binary digits. If we consider $1 - \mu_e/\mu_B$ we immediately lose 3 decimal digits. Thus, similar to integers, the practical complexity of a real number is not very large.

The discussion of integers and reals suggests that under practical circumstances a single number is not a highly complex object. Generally, the complexity of a system arises because of the presence of a large number of parameters that must be specified.

However, there is only reason to consider them collectively as a system if they are coupled to each other.

The next category of mathematical objects that we consider are functions. To specify a function $f(s)$ we must either describe its operation by a formula or specify its action on each possible argument. We consider Boolean functions (functions with binary output, see Section 1.9.2), $f(s) = \pm 1$, of a binary string, $s = (s_1 s_2 \dots s_{N_e})$. The number of arguments of the function—input bits—is N_e. There are 2^{N_e} possible values of the input string. For each of these there are two possible outcomes (output values). All Boolean functions may be specified by listing the binary output for each possible input state. Each possible output is independent. The number of different Boolean functions is the number of possible sets of outputs which is $2^{2^{N_e}}$. Assuming that all of the possible Boolean functions are equally likely, the complexity of a Boolean function (the amount of information necessary to specify it) is the logarithm of this number or $C(f) = 2^{N_e}$. The representation of a Boolean function in terms of $C(f)$ binary variables can also be made explicit as a string representing the presence or absence of terms in the disjunctive normal form described in Section 1.9.2.

A binary function with N_a outputs is the same as N_a independent Boolean functions. If we assume that all possible combinations of Boolean functions are equally likely, then the total complexity is the sum of the complexity of each, or

$$C(f) = N_a 2^{N_e} \tag{8.2.18}$$

The asymmetry between input and output is a fundamental one. It arises because we need to specify for each possible input which of the possible outputs is output. Specifying "which" is a logarithmic operation in the number of possibilities, and therefore the influence of the output space on the complexity is logarithmic compared to the influence of the input. This discussion will be generalized later to consider a physical system that acts in response to its environment. The environment will be specified by a number of binary variables (environmental complexity) N_e, and its actions will be specified by a number of binary variables (action complexity) N_a.

8.3   Complexity of Physical Systems

In order to apply our understanding of the complexity of mathematical constructs to physical systems, we must develop a fundamental understanding of representations. The complexity of a physical system is to be defined as the length of the shortest string s that can represent its properties—the results of possible measurements/observations. In Section 8.3.1 we discuss the relationship between thermodynamics and information theory. This will enable us to define the complexity of ergodic and nonergodic systems. The resulting information measure is essentially that of Shannon information theory. When we consider algorithmic complexity, we can ask whether this is the smallest amount of information that might be used. This is discussed in Section 8.3.2. Section 8.3.3 introduces the complexity profile, which measures the complexity as a function of the scale of observation. Implications of the time scale of observation, for chaotic dynamics, are discussed in Section 8.3.4. Section 8.3.5

discusses examples and properties of the complexity profile. Sections 8.3.1 through 8.3.5 are based upon descriptive complexity. To better account for the behavior of a system in response to its environment we consider behavioral complexity in Section 8.3.6. This turns out to be closely related to descriptive complexity. Other issues related to the role of the observer are discussed in Section 8.3.7.

8.3.1 *Entropy and the complexity of physical systems*

The definition of complexity of a system requires us to develop an understanding of the relationship of information to the physical properties of a system. The most direct relationship is the relationship of entropy and information. At the outset, it should be understood that these are very different concepts. Entropy is a specific physical property of systems that are in equilibrium, or are in well-defined ensembles. Information is not a unique physical property. Instead it is related to representations of digits. Information can be a property of a time sequence or any other set of degrees of freedom. For example, the information content of a set of characters written on a piece of paper can be given. The entropy, however, would be largely a property of the paper or the ink. The entropy of paper is difficult to determine precisely, but simpler substances have entropies that have been determined and are tabulated at specific temperatures and pressures. We also know that entropy is conserved in reversible adiabatic processes and increases in irreversible ones.

Despite the significant conceptual difference between information and entropy, the formal definition of information discussed in Section 1.8 appears very similar to the definition of entropy discussed in Section 1.3. Thus, it makes sense that the two are related when we develop an understanding of complexity. It is helpful to review the definitions. The entropy was defined first for the microcanonical ensemble, which specifies the macroscopic energy U, number of particles N, and volume V, of the system. We assume that all states (microstates) of the system with this energy, number of particles and volume are equally likely in the ensemble. The entropy was written as

$$S = k \ln \Omega(U, N, V) \qquad (8.3.1)$$

where $\Omega(U, N, V)$ is the number of such states. The coefficient k is defined so that the units of entropy are consistent with units of energy and temperature for the thermodynamic relationship $T = dU/dS$.

Information was defined for a string of characters. Given the probability of the string of characters, the information is defined by Eq. (8.2.1). The logarithm is taken to be base 2 so that the information is measured in units of bits. We see that the information content is related to selecting a single state out of an ensemble of possibilities.

We can relate the two definitions in a mathematically direct but conceptually significant way. If we want to specify a particular microstate of a thermodynamic system, we must select this microstate from the whole ensemble. The probability of this particular state is given in the microcanonical ensemble by $P = 1/\Omega$. If we think about the state of the system as a message containing information, we can use Eq. (8.2.1) to give the amount of information as:

$$I(\{x,p\} | (U, N, V)) = S(U, N, V)/(k \ln 2) \qquad (8.3.2)$$

This expression should be understood as the amount of information contained in a microstate $\{x,p\}$, when the system is in the macrostate specified by U, N, V—it is also the information necessary to describe precisely the microstate. This is the fundamental relationship we are looking for. We review its meaning in terms of the description of a particular idealized physical system.

If we want to describe the microstate of a system, like a gas of particles in a box, classically we must specify all of the positions and momenta of the particles $\{x_i, p_i\}$. If N is the number of particles, then there are $6N$ coordinates, 3 position and 3 momentum coordinates for each particle. To specify exactly the position of each particle appears to require arbitrary precision in these coordinates. If we had to specify even a single position exactly, it would take an infinite number of binary digits. However, quantum mechanics is inherently granular, thus there is a smallest distance Δx within which we do not need to specify one position coordinate of a particle. The particle location is uniquely given once it is within a region Δx. More correctly, the particle must be located within a region of position and momentum of $\Delta x \Delta p = h$, where h is Planck's constant. The granularity defines the precision necessary to specify the positions and momenta, and thus also the amount of information (number of bits) needed in order to describe completely the microstate. The definition of the entropy takes this into account, otherwise the counting of possible microstates of the system would be infinite. The complete calculation of the entropy (which also takes into account the indistinguishability of the particles) is given in Question 1.3.2. We now recognize that the calculation of the entropy is precisely a calculation of the information necessary to describe the microstate.

There is another way to think about the relationship of entropy and information. It follows from the recognition that the number of states of a string of $I(\{x,p\}|(U,N,V))$ bits is the same as the number of states of the system. If we consider a mapping of system states onto strings, the strings enumerate or label the system states. If there are $I(\{x,p\}|(U,N,V))$ bits in each string, then there is a one-to-one mapping of system states onto the strings, and a string uniquely identifies a system state. We say that a string represents a system microstate.

We thus identify the entropy of a physical system as the amount of information necessary to identify a single microstate from a specified macroscopic ensemble. For an ergodic macroscopic system, this definition is a robust one. It does not matter if we consider a typical or an average amount of information. What happens if the system is nonergodic? There are two kinds of nonergodic systems we will discuss: a magnet with a well-defined magnetization below its ordering phase transition (see Section 1.6), and a glass where there are many frozen coordinates describing the local arrangements of atoms (see Section 1.4). Many of these coordinates do not change during the time of a typical experiment. Should we include the information necessary to specify the frozen variables as part of the entropy? We would like to separate the discussion of the frozen variables from the fast ones that are in equilibrium. We use the entropy S to refer to the fast ensemble—the enumeration of the kinetically accessible states of the system. The same function of the frozen variables we will call C.

For the magnet, the amount of information contained in frozen variables is small. For the Ising model of a magnet (Section 1.6), below the magnetization transition only a single binary variable is necessary to specify if the system magnetization is UP or DOWN. We treat the magnet by giving the information about the magnetization explicitly as part of the ensemble description. The amount of information is insignificant compared to the information in the microstate of a system, and therefore is generally ignored.

In contrast, for a glass, the amount of information that is included in the frozen variables is large. How does this information relate to the thermodynamic treatment of the system? The conventional thermodynamic theory of phase transitions does not consider the existence of frozen information. It is designed for systems like the magnet, where this information is insignificant, and thus it does not apply to the glass transition. A different theory is necessary which includes the change from an ergodic to a nonergodic system, or a change from information in fast variables to information in frozen variables. Is there any relationship between the frozen information and the entropy? If they are related at all, there are two intuitive possibilities. One is that we must specify the frozen variables as part of the ensemble, and the amount of information necessary to describe the fast variables is just as large as if there were no frozen variables. The other is that the frozen variables balance against the fast variables so that when there is more frozen information there is less information in the fast variables. In order to determine which is correct, we will need to consider an experiment that measures both. As long as an experiment is being performed in which the frozen variables never change, then the amount of information in the frozen variables is fixed. Thermodynamic experiments only depend on entropy differences. We will need to consider an experiment that changes the frozen variables—for example, heating up a glass until it becomes a liquid or cooling it from a liquid to a glass. In such an experiment the frozen information must be accounted for. The difficulty with a glass is that we do not have an independent way to determine the amount of frozen information. Fortunately, there is another system where we do.

There is an intermediate example between a magnet and a glass that is of considerable interest. The structure of ice has a glasslike frozen disorder of its hydrogen atoms below approximately $100°K$. The simplest way to think about this disorder is that it arises from a choice of orientations of the water molecule around the position of the oxygen atom. This means that there is a macroscopic amount of information necessary to specify the static structure of ice. The amount of information associated with this disorder can be calculated directly using a model for the structure of ice that takes into account the correlations between molecular orientations that are needed to form a self-consistent hydrogen structure within the oxygen lattice. A first estimate is based on an average of $3/2$ orientations per molecule or $C = Nk \ln(3/2) = 0.806$ cal/moleK. A review of better calculations is given in a book by Fletcher. The best is $C = 0.8145 \pm 0.0002$ cal/mole°K. The other calculation we need is the amount of entropy in steam. This can be obtained using a slight modification of the ideal gas calculation, that takes into account the rotational and internal vibrational motion of the water molecule.

The key experiment is to measure the change in the entropy of the system as a function of temperature as it is heated from ice all the way to steam. We find the entropy using the standard thermodynamic relationship (Section 1.3)

$$q = TdS \tag{8.3.3}$$

where q is the heat added to the system. At close to a temperature of zero degrees Kelvin ($T = 0K$) the entropy is zero because all motion stops, and there is only one possible state of the system. Thus we would expect

$$S(T) = \int_0^T q/T \tag{8.3.4}$$

—the total amount of entropy added to the system as it is heated up should be the same as the entropy of the gas. However, experimentally there is a difference of 0.82 ± 0.05 cal/moleK between the two. This is the amount of entropy in the gas that was not added to the system as it was heated. The coincidence of two numbers—the amount of entropy missing and the calculation of the information in the frozen structure of the hydrogen atoms, suggests that the missing entropy was present in the original state of the ice.

$$S(T) = C(T = 0) + \int_0^T \frac{q}{T} \tag{8.3.5}$$

This in turn implies that the information in the frozen degrees of freedom was transferred (but conserved) to the fast degrees of freedom. Eq. (8.3.5) is not consistent with the standard thermodynamic relationship in Eq. (8.3.3). Instead it should be modified to read:

$$q = Td(S + C) \tag{8.3.6}$$

This should be understood as implying that adding heat to a system increases the information either of the fast or frozen variables. Adding heat (e.g., to ice) increases the temperature of the system, so that fewer variables are frozen. In this case C decreases and S increases more than would be given by the conventional relationship of Eq. (8.3.3). When heat is not added to a system, we see that there can be processes that change the number of fast degrees of freedom and the number of static degrees of freedom while leaving their sum the same. We will consider this further in later sections.

Eq. (8.3.6) is important enough to present it again from a different perspective. The discussion will help demonstrate its validity by using a theoretical argument (Fig. 8.3.1). Rather than considering it from the point of view of heating ice till it becomes steam, we consider what happens either to ice or to a glass when we cool it down through the transition where degrees of freedom become frozen. In a theoretical description we start, above the freezing-in transition, with an ensemble of systems. As we cool the system we remove heat, and this is reflected in a decrease in the number of possible states of the system. We think of this as a shrinking of the number of elements of the ensemble. However, as we go through the freezing-in transition, the

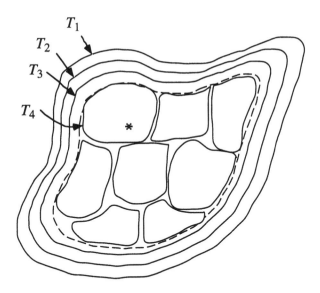

Figure 8.3.1 Schematic illustration of the effect on motion in phase space of cooling through a glass transition. Above the glass transition (T_1, T_2 and T_3) the system is ergodic — it explores the entire phase space. Cooling the system causes the phase space to shrink smoothly. The entropy, the logarithm of the volume of phase space, decreases. Below the glass transition, T_4, the system is no longer ergodic and the phase space breaks up into pieces. A particular system explores only one of the pieces. The total amount of information necessary to specify a particular microstate (e.g. indicated by the *) is the sum of $C/k \ln(2)$, the information necessary to specify which piece, and $S/k \ln(2)$, the information necessary to specify the particular state within the piece. ∎

ensemble breaks up into disjoint pieces that can not make transitions to each other. Any particular material must be in one of the disjoint pieces. Thus for a particular material we must track only part of the original ensemble. For an incremental decrease in temperature due to an incremental removal of heat, the information needed to identify (describe) a particular microstate is the sum of the information necessary to describe which of the disjoint parts of the ensemble the system is in, plus the information needed to specify which of the microstates the system is in once its ensemble fragment has been specified. This is the meaning of Eq. (8.3.6). The information to specify the ensemble fragment was transferred from the entropy S to the ensemble information C. The reduction of the entropy, S, is not reflected in the amount of heat that is removed.

We are now in a position to give a first definition of complexity. In order to describe a system and its behavior over time, we must describe the ensemble it is in. This information is given by $C/k \ln(2)$. If we insist on describing the microstate of the system, we must add the information contained in the fast degrees of freedom $S/k \ln(2)$. The question is whether we should insist on describing the microstate. Typically, the whole point of describing an ensemble is that we don't need to specify the particular microstate. We will return to address this question in greater detail later. However, for now it is reasonable to consider describing the system to be specifying just the ensemble. This implies that the information in the frozen variables $C/k\ln(2)$ is the complexity. For a thermodynamic system in the microcanonical ensemble, the complexity would be given by the (small) number of bits in the specification of the three variables (U, N, V) and the number of bits necessary to specify the type of element (atom, molecule) that is present. The actual amount of information seems not to be precisely defined. For example, we have not identified the number of bits to be used in specifying (U, N, V). As we have seen in the discussion of algorithmic complexity, this is to be expected, since the conventions of how the information is specified are crucial when there is only a small amount.

We have learned from this discussion that for a nonergodic system, the complexity (the frozen ensemble information) is bounded by the sum over the number of fast and static degrees of freedom $(C + S > C)$. For material systems, we know in principle how to measure this. As in the case of ice, we heat up the system to the vapor phase where the entropy can be calculated, then subtract the entropy added during the heating process. This gives us the value of $C + S$ at the temperature from which the heating began. If we know that $C \gg S$, then the result is the complexity itself. In order for this technique to work at all, the complexity must be large enough so that experimental accuracy can enable its measurement. Estimates we will give later imply that complexities of biological organisms are too small to be measured in this way.

The concept of frozen degrees of freedom immediately raises the question of the time scale in which the experiment is performed. Degrees of freedom that are frozen on one time scale are not on sufficiently longer ones. If our time scale of observation would be arbitrarily long, we would always describe systems in equilibrium. The entropy would then be large and the complexity would be negligible. On the other hand, if our time scale of observation was extremely short so that microscopic motions were detected, then our complexity would be large and the entropy would be negligible. This motivates the introduction of the complexity profile in Section 8.3.3.

Question 8.3.1 Calculate the information necessary to specify the microstate of a mole of an ideal gas at $T = 0°C$ and $P = 1$atm. Use the mass of a helium or neon atom for the mass of the ideal gas particle. This requires a careful investigation of units. A table of fundamental physical constants is given on the following page.

Solution 8.3.1 The entropy of an ideal gas is found in Section 1.3 to be:

$$S = kN[\ln(V/N\lambda(T)^3) + 5/2] \qquad (8.3.7)$$

$$\lambda(T) = (h^2/2\pi mkT)^{1/2} \qquad (8.3.8)$$

The information content of a microstate is given by Eq. (8.3.2).

Each of the quantities must be evaluated numerically from appropriate tables. A mole of particles is

$$N_0 = 6.0221367 \times 10^{23} \text{ /mole} \tag{8.3.9}$$

At the temperature

$$T_0 = 0 \text{ °C} = 273.13 \text{ °K} \tag{8.3.10}$$

$$kT_0 = 0.0235384 \text{ eV} \tag{8.3.11}$$

and pressure

$$P_0 = 1 \text{atm} = 1.01325 \times 10^5 \text{ Pascal} = 1.01325 \times 10^5 \text{ Newton/m}^2 \tag{8.3.12}$$

the volume (of a mole of particles) of an ideal gas is:

$$V = N_0 kT/P_0 = 22.41410 \times 10^{-3} \text{ m}^3/\text{mole} \tag{8.3.13}$$

the volume per particle is:

$$V/N = 37219.5 \text{ Å}^3 \tag{8.3.14}$$

At the same temperature we have:

$$\lambda(T) = (2\pi mkT/h^2)^{-1/2} = m[AMU]^{-1/2} \times 1.05633 \text{Å} \tag{8.3.15}$$

This gives the total information for a mole of helium gas at these conditions of

$$I = N_0 (18.5533 + 3/2 \ln(m[AMU])) = 1.24 \times 10^{25} \tag{8.3.16}$$

Note that the amount of information per particle is only of order 10 bits. ∎

$hc = 12398.4 \text{ eV Å}$
$k = 1.380658 \times 10^{-23} \text{ Joule/°K}$
$R = kN_0 = 8.3144 \text{ Joule/°K/mole}$
$c = 2.99792458 \; 10^8 \text{ Meter/second}$
$h = 6.6260755 \; 10^{-34} \text{ Joule second}$
$e = 1.60217733 \; 10^{-19} \text{ Coulomb}$
ProtonMass $= 1.6726231 \times 10^{-27} \text{ kilogram}$
1 AMU $= 1.6605402 \times 10^{-27} \text{ kilogram} = 9.31494 \times 10^9 \text{ eV}$
$M[\text{Helium}] = 4.0026 \text{ AMU}$
$M[\text{Neon}] = 20.179 \text{ AMU}$
$M[\text{Helium}] \; c^2 = 3.7284 \times 10^9$
$M[\text{Neon}] \; c^2 = 1.87966 \times 10^{10}$

Table 8.3.1 Fundamental constants ∎

8.3.2 *Algorithmic complexity of physical systems*

The complexity of a system is designed to measure the amount of information necessary to describe it, or its behavior. In this section we address the key word "necessary." This word suggests that we are after the minimum amount of information. The minimum amount of information depends on our capabilities of inference from a smaller amount of information. As discussed in Section 8.2.2, logical inference and computation lead to the definition of algorithmic complexity. However, for an ensemble that can be described simply, the algorithmic complexity is no different than the Shannon information.

Since we have established a connection between the complexity of physical systems and representations in terms of character strings, we can apply these results directly to physical systems. A physical system in equilibrium is represented by an ensemble. At any particular time, it is in a single microstate. The specification of this microstate can be compressed by encoding in certain rare cases. However, on average the compression cannot lead to an amount of information significantly different from the entropy (divided by $k \ln(2)$) of the system. This conclusion follows because the microcanonical (or canonical) ensemble can be concisely described. For a nonergodic system like a glass, the microstate description has been separated into two parts. It is no longer true that the ensemble of dynamically accessible states of a particular system is concisely describable. The information in the frozen degrees of freedom is precisely the information necessary to specify the ensemble of dynamically accessible states. The total information, $(C + S)/k \ln(2)$, represents the selection of a microstate from a simple ensemble (microcanonical or canonical). Since the total information cannot be compressed, neither can either of the two parts of the information—the frozen degrees of freedom that we have identified with the complexity, or the additional information necessary to specify a particular microstate. Thus the algorithmic complexity is the same as the information for either part.

We can now, finally, explain the experimental observation that an adiabatic process does not change the entropy of a system (Section 1.3). The algorithmic description of an adiabatic process requires only a few pieces of information, e.g., the size of a force applied over a specified distance. If a new microstate of the system can be described by the original microstate plus the process of adiabatic change, then the amount of information in the microstate has not been changed, and the adiabatic process does not change the microstate algorithmic complexity—the entropy of the system. Like other aspects of statistical mechanics (Section 1.3), this should not be understood as a proof but rather as an explanation of the relationship of the thermodynamic observation to the microscopic properties. Using this explanation, we can identify the nature of an adiabatic process as one that is described microscopically by a small amount of information.

This becomes clearer when we compare adiabatic and irreversible processes. Our argument that an adiabatic process does not change the entropy is based on considering the information necessary to describe an adiabatic process—slowly moving a piston to expand the space available to a gas. An irreversible process could achieve a similar expansion, but would not be thermodynamically the same. Take, for example, the removal of a partition that separates the gas from a second, initially empty, chamber. The irreversible process of expansion of the gas results in a final state which has a higher entropy (see Question 1.3.4). The removal of a partition in itself does not appear to require a lot of information to describe. One moment after the partition is removed, the entropy of the system is the same as before. To understand how the entropy increases, we must consider the nature of irreversible dynamics.

A key ingredient in our understanding of physical systems is that the time evolution of an isolated system can be obtained from the simple laws of mechanics (classical or quantum). This means that the dynamics of an isolated system conserves the amount of information as well as the energy. Such dynamics are called conservative. If we consider an ensemble of systems starting in a particular region of phase space, the phase space position evolves in time, but the volume of the phase space that is occupied—the entropy—does not change. This conservation of phase space can be understood from our discussion of algorithmic complexity: since the deterministic dynamics of a system can be computed, the algorithmic complexity of the system is conserved. Where does the additional entropy come from for the final equilibrium state after the expansion?

There are two parts to the process of proceeding to a true equilibrium state. In the first part the distinction between the nonequilibrium and equilibrium state is obscured. At first there is macroscopically observable information—the particles are in one half of the chamber. This information is converted to microscopic correlations between atomic positions and momenta. The conversion occurs when the gas expands to fill the chamber, and various currents that follow this expansion become smaller and smaller in extent. The microscopic correlations cannot be observed on a macroscopic scale, and for standard observations the system is indistinguishable from an equilibrium state. The transfer of information from macroscopic to microscopic scale is related to issues of chaos in the dynamics of physical systems, which will be discussed later.

The second part to the process is an actual increase in the entropy of the system. The additional entropy must come from outside the system. In macroscopic physical processes, we are not generally concerned with isolating the system from information transfer, only with isolating the system from energy transfer. Thus we can surmise that the expansion of the gas is followed by an information transfer that enables the entropy to increase to its equilibrium value without changing the energy of the system. Many of the issues related to describing this nonequilibrium process will not be addressed here. We will, however, begin to address the topic of the scale of observation at which correlations appear using the complexity profile in the following section.

8.3.3 *Complexity profile*

General approach In this section we discuss the relationship of microscopic and macroscopic complexity. Our objective is to develop a consistent language for discussing complexity as a function of length scale. In the following section we will discuss the complexity as a function of time scale, which generalizes the discussion of frozen and fast degrees of freedom in Section 8.3.1.

When we describe a system, we are not generally interested in a microscopic description of the positions and velocities of all of the particles. For a thermodynamic system there are only a few macroscopic parameters that we use to describe the system. This is indeed the reason we use entropy as a summary of the many hidden parameters of the system that we are not interested in. The microscopic parameters change too fast and over too small distances to matter for our macroscopic measurements/experience. The same is true more generally about systems that are not in equilibrium: a macroscopic description does not require specifying the position of each atom. This implies that we must develop an understanding of complexity that is not tied to the microscopic description, but is relevant to observations at a particular length and time scale.

This point lies at the root of a conceptual problem in thinking about the complexity of systems. A gas in equilibrium has a large entropy which is its microscopic complexity. This is counter to our understanding of complex systems. Systems in equilibrium are intuitively simpler than nonequilibrium systems such as a human being. In Section 8.3.1 we started to address this problem by identifying the complexity of a nonergodic system as the information necessary to specify the frozen degrees of freedom. We now discuss a more systematic approach to dealing with macroscopic observations.

In order to consider the macroscopic complexity, we have to define what we mean by macroscopic in a formal sense. The concept of macroscopic must be understood in relation to a particular observer. While we often consider experimental results to be independent of the observer, there are various ways in which the observer is essential to the observation. In this context, in which we are concerned with the meaning of macroscopic, considering the observer is essential.

How do we characterize the difference between a microscopic and a macroscopic observer? The most crucial difference is that a microscopic observer is able to distinguish between all inherently distinguishable states of the system, while a macroscopic observer cannot. For a macroscopic observer, many microscopically distinct states appear the same. This is related to our understanding of complexity, because the macroscopic observer need only specify which of the macroscopically distinct states the system is in. The microscopic observer must specify which of the microscopically distinct states the system is in. Thus the macroscopic complexity must always be smaller than the microscopic complexity of a system. Instead of considering a unique macroscopic observer, we will consider a sequence of observers with a progressively poorer ability to distinguish microstates. Using these observers, we will define the complexity profile.

Ideal gas   These ideas can be directly applied to the ideal gas. We generally think about a macroscopic observer as having an inability to distinguish fine-scale distances. Thus we expect that the usual uncertainty in particle position Δx will increase for a macroscopic observer. However, we learn from quantum mechanics that a unique microstate of the system is defined using an uncertainty in both position and momentum, $\Delta x \Delta p = h$. Thus for the macroscopic observer to confuse distinct microstates, the product $\Delta x \Delta p$ must be larger than its minimum value—an observation of the system provides measurements of the position and momentum of each particle, whose uncertainty has a product greater than h. We can label our observers by this uncertainty, which we call $\bar{h}$.

If we retrace our steps to the calculation of the entropy of an ideal gas (Question 1.3.2), we can recognize that essentially the same calculation applies to the complexity with the uncertainty $\tilde{h}$. An observer with the uncertainty $\tilde{h}$ will determine the complexity of the ideal gas according to Eq. (8.3.7) and Eq. (8.3.8), with h replaced by $\tilde{h}$. Thus we define the complexity profile for the ideal gas in equilibrium as:

$$C(\tilde{h}) = S - 3kN \ln(\tilde{h}/h) \qquad\qquad \tilde{h} > h \qquad\qquad (8.3.17)$$

This equation describes a complexity that decreases as the ability of the observer to distinguish states decreases. This is as we expected. Despite the weak logarithmic dependence on $\tilde{h}$, $C(\tilde{h})$ decreases rapidly because the coefficient of the logarithm is so large. By the time $\tilde{h}$ is about 100 times h the complexity profile has become negative for the ideal gases described in Question 8.3.1.

What does a negative complexity mean? It actually means that we have not done the calculation quite right. The counting of states we did for the ideal gas assumed that the particles were well separated from each other. If they begin to overlap then we must count the possible states differently. This overlap is significant precisely when Eq. (8.3.17) becomes negative. If the particles really overlapped then quantum statistics becomes important; the gas is said to be degenerate and satisfies either Fermi-Dirac or Bose-Einstein statistics. In our case the overlap arises only because the observer cannot distinguish different particle positions. In this case, the counting of states is appropriate to a classical ideal gas, as we now explain.

To calculate the complexity as a function of $\tilde{h}$ for an equilibrium state whose entropy is S, we start by calculating the number of microstates that the observer cannot distinguish. The logarithm of this number of microstates, which we call $S(\tilde{h})/k \ln(2)$, is the amount of information necessary to specify a microstate, if the macrostate is known. Thus we have that:

$$C(\tilde{h}) = S - S(\tilde{h}) \qquad\qquad (8.3.18)$$

To count the number of microstates that the observer cannot distinguish, we note that the possible microstates of a particular particle are grouped together by the observer into bins (regions or cells of position and momentum) of size $(\Delta x \Delta p)^d = \tilde{h}^d$, where $d = 3$ is the dimensionality of space. The observer determines only that a particle is within a certain region. In the classical ideal gas each particle moves independently, so more than one particle may occupy the same microstate. However, this is unlikely. As $\tilde{h}$ increases it becomes increasingly likely that there is more than one particle in a region. If the number of particles in a certain region is n_i, then the number of distinct microstates of the bin that the observer does not distinguish is:

$$\frac{g^{n_i}}{n_i!} \qquad\qquad (8.3.19)$$

where $g = (\tilde{h}/h)^d$ is the number of microstates within a region. This is the product of the number of states each particle may be in, corrected for particle indistinguishability. The number of microstates of the whole system that appear to the observer to be the same is the product of such terms for each region:

$$\prod_i \frac{g^{n_i}}{n_i!} \tag{8.3.20}$$

From this we can determine the complexity of the state determined by the observer as:

$$C(\tilde{h}) = S - S(\tilde{h}) = S - k\ln(\prod_i \frac{g^{n_i}}{n_i!}) \tag{8.3.21}$$

If we consider this expression when $g = 1$—a microscopic observer—then n_i is almost always either zero or one and each term in the product is one (a more exact treatment requires treating the statistics of a degenerate gas). Then $C(\tilde{h})$ is S, which means that the microstate complexity is just the entropy. For $g > 1$ but not too large, n_i will still be either zero or one, and we recover Eq. (8.3.17). On the other hand, using this expression it is possible to show that for a large value of g, when the values of n_i are significantly larger than one, the complexity goes to zero.

We can understand this by recognizing that as g increases, the number of particles in each bin increases and becomes closer to the average number of particles in a bin according to the macroscopic probability distribution. This is the equilibrium macrostate. By our conventions we are measuring the amount of information necessary for the observer to specify its observation in relation to the equilibrium state. Therefore, when the average number of particles in a bin becomes close enough to this distribution, there is no information that must be given. To write this explicitly, when n_i is much larger than one we apply Sterling's approximation to the factorial in Eq. (8.3.21) to obtain:

$$C(\tilde{h}) = S - k\sum_i n_i\left(\ln(g/n_i) + 1\right) = S + k\sum_i gP_i \ln(P_i) - kN \tag{8.3.22}$$

where $P_i = n_i/g$ is the probability a particle is in a particular state according to the observer. It is shown in Question 8.3.2 that $C(\tilde{h})$ is zero when P_i is the equilibrium probability for finding a particle in region i (note that i stands for both position and momentum (x, p)).

There are additional smaller terms in Sterling's approximation to the factorial that we have neglected. These terms are generally ignored in calculations of the entropy because they are not proportional to the number of particles. They are, however, relevant to calculations of the complexity:

$$C(\tilde{h}) = S - k\sum_i n_i\left(\ln(g/n_i) + 1\right) + k\sum_i \ln(\sqrt{2\pi n_i}) \tag{8.3.23}$$

The additional terms are related to fluctuations in the density. This will become apparent when we analyze nonuniform systems below.

We will discuss additional examples of the complexity profile below. First we simplify the complexity profile for observers that measure only the positions and not the momenta of particles.

Question 8.3.2 Show that Eq. (8.3.22) is zero when P_i is the equilibrium probability of locating a particle in a particular state identified by momentum p and position x. For simplicity assume that all g states in the cell have essentially the same position and momentum.

Solution 8.3.2 We calculate an expression for $P_i \rightarrow P(x,p)$ using Boltzmann probability for a single particle (since all are independent):

$$P(x,p) = NZ^{-1}e^{-p^2/2mkT} \tag{8.3.24}$$

where Z is the one particle partition function given by:

$$Z = \sum_{x,p} e^{-p^2/2mkT} = \int \frac{d^3x\, d^3p}{h^3} e^{-p^2/2mkT} = \frac{V}{\lambda^3} \tag{8.3.25}$$

We evaluate the expression:

$$-k\sum_i gP(x,p)\ln(P(x,p)) + kN \tag{8.3.26}$$

which, by Eq. (8.3.22), we want to show is the same as the entropy. Since all g states in cell i have essentially the same position and momentum, this is equal to:

$$-k\sum_{x,p} P(x,p)\ln(P(x,p)) + kN = k\sum_{x,p}\left(\ln(V/N\lambda^3) + p^2/2mkT\right)\left(N\lambda^3/V\right)e^{-p^2/2mkT} \tag{8.3.27}$$

which is most readily evaluated by recognizing it as:

$$kN + kNZ^{-1}\left(\ln(V/N\lambda^3) - \frac{1}{\beta}\frac{d}{d\beta}\right)Z = kN\left(\ln(V/N\lambda^3) + 5/2\right) \tag{8.3.28}$$

which is S as given in Eq. (8.3.7). ∎

Position without momentum The use of the scale parameter $\Delta x \Delta p$ in the above discussion should trouble us, because we do not generally consider the momentum uncertainty on the macroscopic scale. The resolution of this problem arises because we have assumed that the system has a known energy or temperature. If we know the temperature then we know the thermal velocity or momentum:

$$\Delta p \approx \sqrt{mkT} \tag{8.3.29}$$

It does not make sense to have a momentum uncertainty of a particle that is much greater than this. Using $\Delta x \Delta p = h$ this means there is also a natural uncertainty in position which is the thermal wavelength λ given by Eq. (8.3.8). This is the maximal quantum position uncertainty, unless the observer can distinguish the thermal motion of individual particles. We can now think about a sequence of observers who do not distinguish the momentum of particles (they have a larger uncertainty than the thermal momentum) but have increasing uncertainty in position given by $L = \Delta x$, or $g = (L/\lambda)^d$. For such observers the equilibrium momentum probability distribution is to be assumed. In this case the number of particles in a cell n_i contributes a term to the entropy that is equal to the entropy of a gas with this many particles in the volume L^d. This gives a total entropy of:

$$S(L) = k \sum_i n_i \left(\ln(L^d / n_i \lambda^3) + 5/2 \right) \qquad (8.3.30)$$

and the complexity is:

$$C(L) = S - k \sum_i n_i \left(\ln(g/n_i) + 5/2 \right) \qquad (8.3.31)$$

which differs in form from Eq. (8.3.22) only in the constant.

While we generally do not think about measuring momentum, we do measure velocity. This follows from the content of the previous paragraph. We consider observers that measure particle positions at different times and from this they may infer the velocity and indirectly the momentum. Since the observer measures n_i, the determination of velocity depends on the observer's ability to distinguish moving spatial density variations. Thus we consider the measurement of $n(x,t)$, where x has macroscopic meaning as a granular coordinate that has discrete values separated by L. We emphasize, however, that this description of a space- and time-dependent density assumes that the local momentum distribution of the system is consistent with an equilibrium ensemble. The more fundamental description is given by the distribution of particle positions and momenta, $n_i = n(x,p)$. Thus, for example, we can also describe a rotating disk that has no macroscopic changes in density over time, but the rotation is still macroscopic. We can also describe fluid flow in an incompressible fluid. In this section we continue to restrict ourselves to the description of observations at a particular time. The time dependence of observations will be considered in Section 8.3.5.

Thus far we have considered systems that are in generic states selected from the equilibrium ensemble. Equilibrium systems are uniform on all but very microscopic scales, unless we are exactly at a phase transition. Thus, most of the complexity disappears on a scale that is far smaller than typical macroscopic observations. This is not necessarily true about nonequilibrium systems. Systems that are in states that are far from equilibrium can have nonuniform densities of particles. A macroscopic observer will see these macroscopic variations. We will consider a couple of different examples of nonequilibrium states to illustrate some properties of the complexity profile. Before we do this we need to consider the effect of algorithmic compression on the complexity profile.

Algorithmic complexity and error   To discuss macroscopic complexity more completely, we turn to algorithmic complexity as a function of scale. The complexity of a system, particularly a nonequilibrium system, should be defined in terms of the algorithmic complexity of its description. This means that patterns that are present in the positions (or momenta) of its particles can be used to simplify the description.

Using this discussion we can reformulate our understanding of the complexity profile. We defined the profile using observers with progressively poorer ability to distinguish microstates. The fraction of the ensemble occupied by these states defined the complexity. Using an algorithmic perspective we say, equivalently, that the observer cannot distinguish the true state from a state that has a smaller algorithmic complexity. An observer with a value of $g = 2$ cannot distinguish which of two states each particle occupies in the real microstate. Let us label the single particle states using an index that enumerates them. We can then imagine a checkerboard (in six dimensions of position and momentum) where odd indexed states are black and even ones are white. The observer cannot tell if a particle is in a black or a white state. Thus, no matter what the real state is, there is a simpler state where only odd (or only even) indexed states of the particles are occupied, which cannot be distinguished from the real system by the observer. The algorithmic complexity of this state with particles in odd indexed states is essentially the complexity that we determined above, $C(g = 2)$—it is the information necessary to specify this state out of all the states that have particles only in odd indexed states. Thus, in every case, we can specify the complexity of the system for the observer as the complexity of the simplest state that is consistent with the observations—by Occam's razor, this is the state that the observer will use to describe the system.

We note that this is also equivalent to defining the complexity profile as the length of the description as the error allowed in the description increases. The total error as a function of g for the ideal gas is

$$\frac{1}{2}\log\left(\prod \Delta x_i \Delta p_i / h\right) = \frac{1}{2}N\log(g) \tag{8.3.32}$$

where N is the number of particles in the system. The factor of $1/2$ arises because the average error is half of the maximum error that could occur. This approach is helpful since it suggests how to generalize the complexity profile for systems that have different types of particles. We can define the complexity profile as a function of the number of errors that are made. This is better than using a particular length scale, which implies a different error for particles of different mass as indicated by Eq. (8.3.8). For conceptual simplicity, we will continue to write the complexity profile as a function of g or of length scale.

Nonequilibrium states   Our next objective is to consider nonequilibrium states. When we have a nonequilibrium state, the microstate of the system is simpler than an equilibrium state to begin with. As we mentioned at the end of Section 8.3.2, there are nonequilibrium states that cannot be distinguished from equilibrium states on a macroscopic scale. These nonequilibrium states have microscopic correlations. Thus, the microscopic complexity is lower than the equilibrium entropy, while the macroscopic complexity is the same as in equilibrium:

$$\begin{aligned} C(g) < C_0(g) = S_0 \qquad & g = 1 \\ C(g) = C_0(g) \qquad & g \gg 1 \end{aligned} \tag{8.3.33}$$

where we use the subscript 0 to indicate quantities of the equilibrium state. We illustrate this by an example. Using the indexing of single particle states we just introduced, we take a microstate where all particles are in odd indexed states. The microstate complexity is the same as that of an equilibrium state at $g = 2$, which is less than the entropy of the equilibrium system:

$$C(g=1) = C_0(g=2) < C_0(g=1)$$

However, the complexity of this system for scales of observation $g \geq 2$ is the same as that of an equilibrium system—macroscopic observers do not distinguish them.

This scenario, where the complexity of a nonequilibrium state starts smaller but then quickly becomes equal to the equilibrium state complexity, does not always hold. It is true that the microscopic complexity must be less than or equal to the entropy of an equilibrium system, and that all systems have the same complexity when L is the size of the system. However, what we will show is that the complexity of a nonequilibrium system can be higher than that of the equilibrium system at large scales that are smaller than the size of the system. This is apparent in the case, for example, of a nonuniform density at large scales.

To illustrate what happens for such a nonequilibrium state, we consider a system that has nonuniformity that is characteristic of a particular length scale L_0, which is significantly larger than the microscopic scale λ but smaller than the size of the system. This means that n_i is smooth on finer scales, and there is no particular relationship between what is going on in one region of length scale L_0 and another. The values of n_i will be taken from a Gaussian distribution around the equilibrium value n_0 with a standard deviation of σ. We assume that σ is larger than the natural density fluctuations, which have a standard deviation of $\sigma_0 = \sqrt{n_0}$. For convenience we also assume that σ is much smaller than n_0.

We can calculate both the complexity $C(L)$, and the apparent entropy $S(L)$ for this system. We start by calculating them at the scale L_0. $C(L_0)$ is the amount of information necessary to specify the density values. This is the product of the number of cells V/L^d times the information in a number selected from a Gaussian distribution of width σ. From Question 8.3.3 this is:

$$C(L_0) = k\frac{V}{L_0^d}(\frac{1}{2}(1+\ln(2\pi))+\ln\sigma) \tag{8.3.34}$$

The number of microstates consistent with this macrostate at L_0 is given by the sum of ideal gas entropies in each region:

$$S(L_0) = -k\sum_i n_i \ln(n_i/g) + (5/2)kN \tag{8.3.35}$$

Since σ is less than n_0, this can be evaluated by expanding to second order in $\delta n_i = n_i - n_0$:

$$S(L_0) = S_0 - k\sum_i \frac{(\delta n_i)^2}{2n_0} = S_0 - \frac{kV\sigma^2}{2L_0^d n_0} \tag{8.3.36}$$

where S_0 is the entropy of the equilibrium system, and we used $<\delta n_i^2> = \sigma^2$. We note that when $\sigma = \sigma_0$ the logarithmic terms in the complexity reduce to the extra terms found in Eq. (8.3.23). Thus, these terms are the information needed to describe the equilibrium fluctuations in the density.

We can understand the behavior of the complexity profile of this system. By construction, the minimum amount of information needed to specify the microstate is $C(\lambda) = S(L_0) + C(L_0)$. This is the sum over the entropy of equilibrium gases with densities n_i in volumes L_0^d, plus $C(L_0)$. Since $S(L_0)$ is linear in the number of particles, while $C(L_0)$ is logarithmic in σ and therefore logarithmic in the number of particles, we conclude that $C(L_0)$ is much smaller than $S(L_0)$. For $L > \lambda$ the complexity profile $C(L)$ decreases like that of an equilibrium ideal gas. The term $S(L_0)$ is eliminated at a microscopic length scale larger than λ but much smaller than L_0. However, $C(L_0)$ remains. Due to this term the complexity crosses that of an equilibrium gas to become larger. For length scales up to L_0 the complexity is essentially constant and equal to Eq. (8.3.34). Above L_0 it decreases to zero as L continues to increase by virtue of the effect of combining the different n_i into fewer regions. Combining the regions results in a Gaussian distribution with a standard deviation that decreases as the square root of the number of terms $\sigma \to \sigma(L_0/L)^{d/2}$. Thus, the complexity and entropy profiles for $L > L_0$ are:

$$C(L) = k\frac{V}{L^d}\frac{1}{2}(1+\ln(2\pi)) + \ln\sigma\left(L_0/L\right)^{d/2})$$

$$S(L) = S_0 - \frac{kV\sigma^2}{2(LL_0)^{d/2}n_0}$$

(8.3.37)

This expression continues to be valid until there is only one region left, and the complexity goes to zero. The precise way the complexity goes to zero is not described by Eq. (8.3.37), since the Gaussian distribution does not apply in this limit.

There are several comments that we can make that are relevant to understanding complexity profiles in general. First we see that in order for the macroscopic complexity to be higher than that in equilibrium, the entropy at the same scale must be reduced $S(L) < S_0$. This is necessary because the sum $S(L) + C(L)$—the total information necessary to specify a microstate—cannot be greater than S_0. However, we also note that the reduction in $S(L)$ is much larger than the increase in $C(L)$. The ratio between the two is given by:

$$\frac{\delta S(L)}{\delta C(L)} = -\frac{\sigma^2}{2n_0}\frac{L^{d/2}}{L_0^{d/2}}\frac{1}{\ln(\sigma/\sigma_0)}$$

(8.3.38)

For $\sigma > \sigma_0 = \sqrt{n_0}$ this is greater than one. We can understand this result in two ways. First, a complex macroscopic system must be far from equilibrium, and therefore must have a much smaller entropy than an equilibrium system. Second, a macroscopic observer makes many errors in determining the microstate, and therefore if the microstate is similar to an equilibrium state, the observer cannot distinguish the two and the macroscopic properties must also be similar to an equilibrium state. For every bit of information that distinguishes the macrostate, there must be many bits of difference in the microstate.

In calculating the complexity of the system at a particular scale, we assumed that the observer was in error in obtaining the position and momentum of each particle. However, we assumed that the number of particles within each bin was determined exactly. Thus the complexity we calculated is the information necessary to specify the number of particles precise to the single particle. This is why even the equilibrium density fluctuations were described. An alternative, more reasonable, approach assumes that particle counting is also subject to error. For simplicity we can assume that the error is a fraction of the number of particles counted. For macroscopic systems this fraction is much larger than the equilibrium fluctuations, which therefore need not be described. This approach also modifies the form of the complexity profile of the nonuniform gas in Eq. (8.3.37). The error in measurement increases as $n_0(L) \propto L^d$ with the scale of observation. Letting $\gamma m_0(L)$ be the error in a measurement of particle number, we write:

$$C(L) = k\frac{V}{L^d}(\frac{1}{2}(1+\ln(2\pi)) + \ln\frac{\sigma L_0^{d/2}}{m_0(L)L^{d/2}}) \approx k\frac{V}{L^d}\ln\frac{\sigma L_0^{3d/2}}{m_0(L_0)L^{3d/2}} \qquad (8.3.39)$$

The consequence of this modification is that the complexity decreases somewhat more rapidly as the scale of observation increases. The expression for the entropy in Eq. (8.3.37) is unchanged.

Question 8.3.3 What is the information in a number (character) selected from a Gaussian distribution of standard deviation σ?

Solution 8.3.3 Starting from a Gaussian distribution (Eq. 1.2.39),

$$P(x) = \frac{1}{\sqrt{2\pi\sigma}}e^{-x^2/2\sigma^2} \qquad (8.3.40)$$

we calculate the information (Eq. 8.2.2):

$$I = -\int dx P(x)\log(P(x)) = \int dx P(x)\left(\log(\sqrt{2\pi}\sigma) + \ln(2)x^2/2\sigma^2\right) \qquad (8.3.41)$$
$$= \log(\sqrt{2\pi}\sigma) + \ln(2)/2$$

where the second term in the integral can be evaluated using $<x^2> = \sigma^2$.

We note that this result is to be interpreted as the information in a discrete distribution of integral values of x, like a random walk, that in the limit of large σ gives a Gaussian distribution. The units that are used to measure σ define the precision to which the values of x are to be described. It thus makes sense that the information to specify an integer of typical magnitude σ is essentially $\log(\sigma)$. ∎

8.3.4 *Time dependence—chaos and the complexity profile*

General approach   In describing a system, we are interested in macroscopic observations over time, $n(x, t)$. As with the uncertainty in position, a macroscopic observer is not able to distinguish the time of observation within less than a certain time interval $T = \Delta t$. To define what this means, we say that the system is represented by an ensemble with probability $P_{L,T}(n(x;t))$, or more generally $P_{L,T}(n(x,p;t))$. The different microstates that occur during the time interval T are all part of this ensemble. This may appear different than the definition we used for the spatial uncertainty. However, the definitions can be restated in a way that makes them appear equivalent. In this restatement we recognize that the observer performs measurements that are, in effect, averages over various possible microscopic measurements. The average measurements over space and time represent the system (or system ensemble) that is to be described by the observer. This representation will be discussed further in Section 8.3.6. The use of an ensemble is convenient because the observer may only measure one quantity, but we can consider various quantities that can be measured using the same degree of precision. The ensemble represents all possible measurements with this degree of precision. For example, the observer can measure correlations between particle positions that are fixed over time. If we averaged the density $n(x,t)$ over time, these correlations could disappear because of the movement of the whole system. However, if we average over the ensemble, they do not. We define the complexity profile $C(L,T)$ as the amount of information necessary to specify the ensemble $P_{L,T}(n(x,t))$. A description at a finer scale contains all of the information necessary to describe the coarser scale. Thus, $C(L,T)$ is a monotonic decreasing function of its arguments. A direct analysis is discussed in Question 8.3.4. We start, however, by considering the effect on $C(L,T)$ of prediction and the lack of predictability in chaotic dynamics.

Predictability and chaos   As discussed earlier, a key ingredient in our understanding of physical systems is that the time evolution of an isolated system (or a system whose interactions with its environment are specified) can be obtained from the simple laws of mechanics starting from a complete microscopic description of the position and momenta of the particles. Thus, if we use a small enough L and T, so that each particle can be distinguished, we only need to specify $P_{L,T}(n(x,t))$ over a short period of time (or the simultaneous values of position and momentum) in order to predict the behavior over all subsequent times. The laws of mechanics are also reversible. We describe the past as well as the future from the description of a system at a particular time. This must mean that information is not lost over time. Systems that do not lose information over time are called conservative systems.

However, when we increase the spatial scale of observation, L, then the information loss—the complexity reduction—also limits the predictability of a system. We are not guaranteed that by knowing $P_{L,T}(n(x,t))$ at a scale L we can predict the system behavior. This is true even if we are only concerned about predicting the behavior at the scale L. We may need additional smaller-scale information to describe the time evolution of the system. This is precisely the origin of the study of chaotic systems discussed in Section 1.1. Chaotic systems take information from smaller scales and bring it to larger scales. Chaotic systems may be contrasted with dissipative systems that take information from larger scales to smaller scales. If we perturb (disturb) a dissipative system, the effect disappears over time. Looking at such a system at a particular time, we cannot tell if it was perturbed at some time far enough in the past. Since the information on a microscopic scale must be conserved, we know that the information that is lost on the macroscopic scale must be preserved on the microscopic scale. In this sense we can say that information has been transferred from the macroscopic to the microscopic scale. For such systems, we cannot describe the past from present information on a particular length scale.

The degree of predictability is manifest when we consider that the complexity of a system $C(L, T)$ at a particular L and T depends also on the duration of the description—the limits of $t \in [t_1, t_2]$. Like the spatial extent of the system, this temporal extent is part of the system definition. We typically keep these limits constant as we vary T to obtain the complexity profile. However, we can also characterize the dependence of the complexity on the time limits t_1, t_2 by determining the rate at which information is either gained or lost for a chaotic or stable system. For complex systems, the flow of information between length scales is bidirectional—even if the total amount of information at a particular scale is preserved, the information may change over time by transfer to or from shorter length scales. Unlike most theories of currents, information currents remain relevant even though they may be equal and opposite. All of the information that affects behavior at a particular length scale, at any time over the duration of the description, should be included in the complexity.

It is helpful to develop a conceptual image of the flow of information in a system. We begin by considering a conservative, nonchaotic and nondissipative system seen by an observer who is able to distinguish $2^{C(L)/k\ln(2)} = e^{C(L)/k}$ states. $C(L)/k\ln(2)$ is the amount of information necessary to describe the system during a single time interval of length T. For a conservative system the amount of information necessary to describe the state at a particular time does not change over time. The dynamics of the system causes the state of the system to change over time among these states. The sequence of states could be described one by one. This would require

$$N_T C(L)/k\ln(2) \qquad (8.3.42)$$

bits, where $N_T = (t_2 - t_1)/T$ is the number of time intervals. However, we can also describe the state at a particular time (e.g., the initial conditions) and the dynamics. The amount of information to do this is:

$$(C(L) + C_t(L,T))/k\ln(2) \qquad (8.3.43)$$

$C_t(L, T)/k\ln(2)$ is the information needed to describe the dynamics. For a nonchaotic and nondissipative system we can show that this information is quite small. We know from the previous section that the macrostate of the system of complexity $C(L)$ is consistent with a microstate which has the same complexity. The microstate has a dynamics that is simple, since it follows the dynamics of standard physical law. The dynamics of the simple microstate also describes the dynamics of the macrostate, which must therefore also be simple. Therefore Eq. (8.3.43) is smaller than Eq. (8.3.42) and the complexity is $C(L, T) = C(L) + C_t(L, T) \approx C(L)$. This holds for a system following conservative, nonchaotic and nondissipative dynamics.

For a system that is chaotic or dissipative, the picture must be modified to accommodate the flow of information between scales. From the previous paragraph we conclude that all of the interesting (complex) dynamics of a system is provided by information that comes from finer scales. The observer does not see this information before it appears in the state of the system—i.e., in the dynamics. If we allow ourselves to see the finer-scale information we can track the flow of information that the observer does not see. In a conventional chaotic system, the flow of information can be characterized by its Lyaponov exponents. For a system that is described by a single real valued parameter, $x(t)$, the Lyaponov exponent is defined as an average over:

$$h = \ln((x'(t) - x(t))/(x'(t-1) - x(t-1))) \qquad (8.3.44)$$

where unprimed and primed coordinates indicate two different trajectories. We can readily see how this affects the information needed by an observer to describe the dynamics. Consider an observer at a particular scale, L. The observer sees the system in state $x(t-1)$ at time $t-1$, but he determines $x(t-1)$ only within a bin of width L. Using the dynamics of the system that is assumed to be known, the observer can determine the state of the system at the next time. This extrapolation is not precise, so the observer needs additional information to specify the next location. The amount of information needed is the logarithm of the number of bins that one bin expands into during one time step. This is precisely $h/\ln(2)$ bits of information. Thus, the complexity of the dynamics for the observer is given by:

$$C(L,T) = C(L) + C_t(L,T) + N_T kh \qquad (8.3.45)$$

where we have used the same notation as in Eq. (8.3.43).

A physical system that has many dimensions, like the microscopic ideal gas, will have one Lyaponov exponent for each of $6N$ dimensions of position and momentum. If the dynamics is conservative then the sum over all the Lyaponov exponents is zero,

$$\sum_i h_i = \log(\prod_i \Delta x_i(t)\Delta p_i(t)/\prod_i \Delta x_i(t-T)\Delta p_i(t-T)) = 0 \qquad (8.3.46)$$

where $\Delta x_i(t) = x'_i(t) - x_i(t)$ and $\Delta p_i(t) = p'_i(t) - p_i(t)$. This follows directly from conservation of volumes of phase space in conservative dynamics. However, while the sum over all exponents is zero, some of the exponents may be positive and some negative. These correspond to chaotic and dissipative modes of the dynamics. We can imagine the flow of information as consisting of two streams, one going to higher scales and one to lower scales. The complexity of the system is given by:

$$C(L,T) = C(L) + C_t(L,T) + N_T k \sum_{i:h_i>0} h_i \qquad (8.3.47)$$

As indicated, the sum is only over positive values.

Two cautionary remarks about the application of Lyaponov exponents to complex physical systems are necessary. Unlike many standard models of chaos, a complex system does not have the same number of degrees of freedom at every scale. The number of independent bits of information describing the system above a particular scale is given by the complexity profile, $C(L)$. Thus, the flow of information between scales should be thought of as due to a number of closed loops that extend from a particular lowest scale up to a particular highest scale. As the scale increases, the complexity decreases. Thus, so does the maximum number of Lyaponov exponents. This means that the sum over Lyaponov exponents is itself a function of scale. More generally, we must also be concerned that $C(L)$ can be time dependent, as it is in many irreversible processes.

The second remark is that over time the cycling of information between scales may bring the same information back more than once. Eq. (8.3.47) does not distinguish this, and therefore may include multiple counting of the same information. We should understand this expression as an upper bound on the complexity.

Time scale dependence   Once we have chaotic behavior, we can consider various descriptions of the time dependence of the behavior seen by a particular observer. All of the models we considered in Chapter 1 are applicable. The state of the system may be selected at random from a particular distribution (ensemble) of states at successive time intervals. This is a special case of the more general Markov chain model that is described by a set of transition probabilities. Long-range correlations that are not easily described by a Markov chain may also be important in the dynamics.

In order to discuss the complexity profile as a function of T, we consider a Markov chain model. From the analysis in Question 8.3.4 we learn that the loss of complexity with time scale occurs as a result of cycles in the dynamics. These cycles need not be deterministic; they may be stochastic—cycles that do not repeat indefinitely but rather can occur one or more times through the probabilistic selection of successive states. Thus, a high complexity for large T arises when there is a large space of states with low chance of repetition in the dynamics. The highest complexity would arise from a deterministic dynamics with cycles that are longer than T. This might seem to contradict our previous conclusion, where the deterministic dynamics was found to be simple. However, a complex deterministic dynamics can arise if the successive states are specified by information from a smaller scale.

Question 8.3.4 Consider the information in a Markov chain of N_T states at intervals T_0 given by the transition matrix $P(s'|s)$. Assume the complexity of specifying the transition matrix—the complexity of the dynamics—$C_t = C(P(s'|s))$, is itself small. (See Question 8.3.5 for the case of a complex deterministic dynamics.)

a. Show that the more deterministic the chain is, the less information it contains.

b. Show that for an observer at a longer time scale consisting of two time steps $(T = 2T_0)$ the information is reduced. Hint: Use convexity of information as described in Question 1.8.8, $f(\langle x \rangle) > \langle f(x) \rangle$, for the function $f(x) = -x\log(x)$.

c. Show that the complexity does not decrease for a system that does not allow 2-cycles.

Solution 8.3.4 When the complexity of the dynamics is small, then the complexity of the Markov chain is given by:

$$C = C(s) + C_t + N_T k \ln(2) I(s'|s) \qquad (8.3.48)$$

where the terms correspond to the information in the initial state of the system, the information in the dynamics and the incremental information per update needed to specify the next state. The relationship between this and Eq. (8.3.47) should be apparent. This expression does not hold if C_t is large, because if it is larger than $N_T C(s)$, then the chain is more concisely described by specifying each of the states of the system (see Question 8.3.5).

The proof of (a) follows from realizing that the more deterministic the system is, the smaller is $I(s'|s)$. This may be used to define how deterministic the dynamics is.

To analyze the complexity of the Markov chain for an observer at time scale $2T_0$, we need to combine successive system states into an unordered pair—the ensemble of states seen by the observer. We use the notation $\{s', s\}$ for a pair of states. Thus, we are considering a new Markov chain of transitions between unordered pairs. To analyze this we need two probabilities: the probability of a pair and the transition probability from one pair to the next. The latter is the new transition matrix. The probability of a particular pair is:

$$P(\{s_1, s_2\}) = \begin{cases} P(s_1|s_2)P(s_2) + P(s_2|s_1)P(s_1) & s_2 \neq s_1 \\ P(s_1|s_1)P(s_1) & s_2 = s_1 \end{cases} \quad (8.3.49)$$

where $P(s)$ is the probability of a particular state of the system and the two terms in the upper line correspond to the probability of starting from s_1 to make the pair, and starting from s_2 to make the pair. The transition matrix for pairs is given by

$$P(\{s_1', s_2'\}|\{s_1, s_2\}) = [(P(s_1'|s_2')P(s_2'|s_1) + P(s_2'|s_1')P(s_1'|s_1))P(s_1|s_2)P(s_2)$$
$$+ (P(s_1'|s_2')P(s_2'|s_2) + P(s_2'|s_1')P(s_1'|s_2))P(s_2|s_1)P(s_1)]/P(\{s_1, s_2\})$$
$$(8.3.50)$$

which is valid only for $s_1 \neq s_2$ and for $s_1' \neq s_2'$. Other cases are treated like Eq. (8.3.49). Eq. (8.3.50) includes all four possible ways of generating the sequence of the two pairs. The normalization is needed because the transition matrix is the probability of $\{s_1' \neq s_2'\}$ occurring, assuming the pair $\{s_1, s_2\}$ has already occurred.

To show (b) we must prove that the process of combining the states into pairs reduces the information necessary to describe the chain. This is apparent since the observer loses the information about the state order within each pair. To show it from the equations, we note from Eq. (8.3.49) that the probability of a particular pair is larger than or equal to the probability of each of the two possible unordered pairs. Since the probabilities are larger, the information is smaller. Thus the information contained in the first pair is smaller for $T = 2$ than for $T = 1$. We must show the same result for each successive pair. The transition probability can be seen to be an average over two terms in the round parenthesis. By convexity, the information in the average is less than the average information of each term. Each of the terms is a sum over the probabilities of two possible orderings, and is therefore larger than or equal to the probability of either ordering. Thus, the information needed to specify any pair in the chain is smaller than the corresponding information in the chain of states.

Finally, to prove (*c*) we note that the less the order of states is lost when we combine states into pairs, the more complexity is retained. If transitions in the dynamics can only occur in one direction, then we can infer the order and information is not lost. Thus, for $T = 2$ the complexity is retained if the dynamics is not reversible—there are no 2-cycles. From the equations we see that if only one of $P(s_1|s_2)$ and $P(s_2|s_1)$ can be nonzero, and similarly for $P(s_1'|s_2')$ and $P(s_2'|s_1')$, then only one term survives in Eq. (8.3.49) and Eq. (8.3.50) and no averaging is performed. For arbitrary T the complexity is the same as at $T = 1$ if the dynamics does not allow loops of size less than or equal to T. ∎

Question 8.3.5 Calculate the maximum information that might in principle be necessary to specify completely a deterministic dynamics of a system whose complexity at any time is $C(L)$. Contrast this with the maximum complexity of describing N_T steps of this system.

Solution 8.3.5 The number of possible states of the system is $2^{C(L)/k\ln(2)}$. Each of these must be assigned a successor by the dynamics. The maximum possible information to specify the dynamics arises if there is no algorithm that can specify the successor, so that each successor must be identified out of all possible states. This would require $2^{C(L)/k\ln(2)}C(L)/k\ln(2)$ bits.

The maximum complexity of N_T steps is just $N_T C(L)$, as long as this is smaller than the previous result. Which is generally a reasonable assumption. ∎

A simple example of chaotic behavior that is relevant to complex systems is that of a mobile system—an animal or human being—where the motion is internally directed. A description of the system behavior, even at a length scale larger than the system itself, must describe this motion. However, the motion is determined by information contained on a smaller length scale just prior to its occurrence. This satisfies the formal requirements for chaotic behavior regardless of the specifics of the motion involved. Stated differently, the large-scale motion would be changed by modifications of the internal state of the system. This is consistent with the sensitivity of chaotic motion to smaller scale changes.

Another example of information transfer between different scales is related to adaptability, which requires that information about the external environment be represented in the organism. This generally involves the transfer of information between a larger scale and a smaller scale. Specifically, between observed phenomena and their representation in the synapses of the nervous system.

When we describe a system at a particular moment of time, the complexity of the system at its own scale or larger is zero—or a constant if we include the description of the equilibrium system. However, when we consider the description of a system over time, then the complexity is larger due to the system motion. Increasing the scale of observation continues to result in a progressive decrease in complexity. At a scale that is larger than the system itself, it is the motion of the system as measured by its location at successive time intervals that is to be described. As the scale becomes larger, smaller scale motions are not observed, and a simpler description of motion is possible. The observer only notes changes in position that are larger than the scale of observation.

A natural question that can be asked in this context is whether the motion of the system is due to external influences or due to the system itself. For example, a particle moving in a fluid may be displaced by the motion of the fluid. This should be considered different from a mobile bacteria. Similarly, a basketball in a game moves through its trajectory not because of its own volition, but rather because of the volition of the players. How do we distinguish this from a system that moves due to its own actions? More generally, we must ask how we must deal with the environmental influences for a system that is not isolated. This question will be dealt with in Section 8.3.6 on behavioral complexity. Before we address this question, in the next section we discuss several aspects of the complexity profile, including the relationship of the complexity of the whole to the complexity of its parts.

8.3.5 *Properties of complexity profiles of systems and components*

General properties   We can readily understand some of the properties that we would expect to find in complexity profiles of systems that are difficult to calculate directly. Fig. 8.3.2 illustrates the complexity profile for a few systems. The paragraphs that follow describe some of their features.

For any system, the complexity at the smallest values of L, T is the microscopic complexity—the amount of information necessary to describe a particular microstate. For an equilibrium state this is the same as the thermodynamic entropy, which is the entropy of a system observed on an arbitrarily long time scale. This is not true in general because short-range correlations decrease the microstate complexity, but do not affect the apparent macroscopic entropy. We have thus also defined the entropy profile $S(L,T)$ as the amount of information necessary to determine an arbitrary microstate consistent with the observed macrostate. From our discussion of nonergodic systems in Section 8.3.1 we might also conclude that at any scale L, T the sum of the complexity $C(L,T)$ and the entropy $S(L,T)$ of the system (the fast degrees of freedom) should add up to the microscopic complexity or macroscopic entropy

$$C(0,0) \approx S(\infty,\infty) \approx C(L,T) + S(L,T) \tag{8.3.51}$$

However, this is valid only under special circumstances—when the macroscopic state is selected at random from the ensemble of macrostates, and the microstate is selected at random from the possible microstates. A glass may satisfy this requirement; however, other complex systems need not.

For a typical system in equilibrium, as L,T is increased the system rapidly becomes homogeneous in space and time. Specifically, the density of the system is

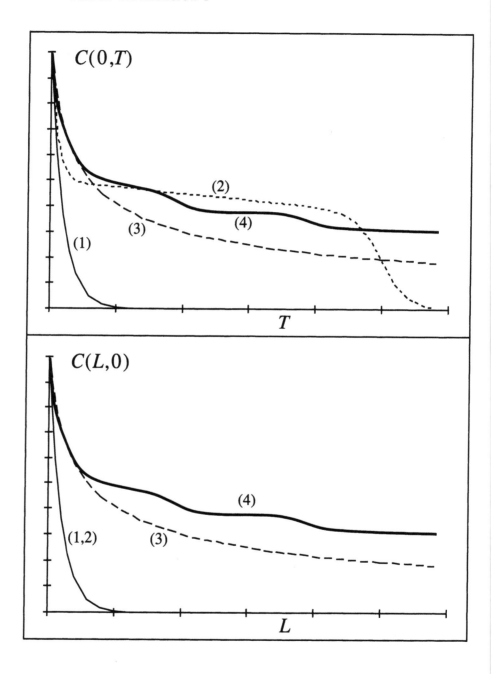

Figure 8.3.2 Schematic plots of the complexity profile $C(L,T)$ of four different systems. $C(L,T)$ is the amount of information necessary to describe the system ensemble as a function of the length scale, L, and time scale, T, of observation. Top panel shows the time scale dependence, bottom panel shows the length scale dependence. (1) An equilibrium system has a complexity profile that is sharply peaked at $T = 0$ and $L = 0$. Once the length or time scale is beyond the correlation length or correlation time respectively, the complexity is just the macroscopic complexity associated with thermodynamic quantities (U, N, V), which vanishes on any reasonable scale. (2) For a glass the complexity profile as a function of time scale $C(0,T)$ decays rapidly at first due to averaging over atomic vibrations; it then reaches a plateau that represents the frozen degrees of freedom. At much longer time scales the complexity profile decays to its thermodynamic limit. Unlike $C(0,T)$, $C(L,0)$ of a glass decays like a thermodynamic system because it is homogeneous in space. (3) A magnet at a second-order phase transition has a complexity profile that follows power-law behavior in both length and time scale. Stochastic fractals capture this kind of behavior. (4) A complex biological organism has a complexity profile that should follow similar behavior to that of a fractal. However it has plateau-like regions that correspond to crossing the scale of internal components, such as molecules and cells. ∎

uniform in space and time, aside from unobservable small fluctuations, once the scale of observation is larger than either the correlation length or the correlation time of the system. Indeed, this might be taken to be the definition of the correlation length and time—the scale at which the microscopic information becomes irrelevant to the properties of the system. Beyond the correlation length, the average behavior characteristic of the macroscopic scale is all that remains, and the complexity profile is constant at all length and time scales less than the size of the system.

We can contrast the complexity profile of a thermodynamic system with what we expect from various complex systems. For a glass, the complexity profile is quite different in time and in space. A typical glass is uniform if L is larger than a microscopic correlation length. Thus, the complexity profile of the glass is similar to an equilibrium system as a function of L. However, it is different as a function of T. The frozen degrees of freedom that make it a nonergodic system at typical time scales of observation guarantee this. At typical values of T the temporal ensemble of the system includes the states that are reached by vibrational modes of the system, but not the atomic rearrangements characteristic of fluid motion. Thus, the atomic vibrations cannot be observed except at microscopic values of T. However, a significant part of the microscopic description remains necessary at longer time scales. Correspondingly, a plateau in the complexity profile extends up to characteristic time scales of human observation. At a temperature-dependent and much longer time scale, the complexity profile declines to its thermodynamic limit. This time scale, the relaxation time, is accessible near the glass transition temperature. For lower temperatures it is not. Because the glass is uniform in space, the plateau should be relatively flat and end abruptly. This is because spatial uniformity indicates that the relaxation time is essentially a local property with a narrow distribution. A more extended spatial coupling would give rise to a grading of the plateau and a broadening of the time scale at which the plateau disappears.

More generally, for a complex system we expect that many parameters will be required to describe its properties at all length and time scales, at least up to some fraction of the spatial and temporal scale of the system itself. Starting from the microscopic complexity, the complexity profile should not be expected to fall smoothly. In biological organisms, we can expect that as we increase the scale of observation, there will be particular length scales at which details will be lost. Plateaus in the profile are related to the existence of well-defined levels of description. For example, an identifiable level of cellular behavior would correspond to a plateau, because over a range of length scales larger than the cell, a full accounting of cellular properties, but not of the internal behavior of the cell, must be given. There are many cells that have a characteristic size and are immobile. However, because different cell populations have different sizes and some cells are mobile, the sharpness of the transition should be smoothed. We can at least qualitatively identify several different plateaus. At the shortest time scale the atomic vibrations will be averaged out to end the first plateau. Larger atomic motions or molecular behavior will be averaged out on a second, larger scale. The internal cellular behavior will then be averaged out. Finally, the internal behavior of tissues and organs will be averaged out on a still longer length and time scale. It is the degrees of freedom that remain relevant on the longest length scale that are key to the complexity of the system. These degrees of freedom manifest the concept of emergent collective behavior. Ultimately, they must be traceable back to the microscopic degrees of freedom. Describing the connection between the microscopic parameters and macroscopically relevant parameters has occupied our attention in much of this book.

Mathematical models that best capture the complexity profile of a complex system are fractals (see Section 1.10). Mathematical fractals with no granularity (no smallest length scale) have infinite complexity. However, if we define a smallest length scale, corresponding to the atomic length scale of a physical system, and we define a longest length scale that is the size of the system, then we can plot the spatial complexity profile of a fractal-like system. There are two quite distinct kinds of mathematical fractals, deterministic and stochastic fractals. The deterministic fractals are specified by an algorithm with only a few parameters, and thus their algorithmic complexity is small. Examples are the Kantor set or the Sierpinski gasket. The algorithm describes how to create finer and finer scale detail. The only difficulty in specifying the fractal is specifying the number of levels to which the algorithm should be iterated. This information (the number of iterations) requires a parameter whose length grows logarithmically with the ratio of the size of the system to the smallest length scale. Thus, a deterministic fractal has a complexity profile that decreases logarithmically with observation length scale L, but is very small on all length scales.

Stochastic fractals are qualitatively different. In such fractals, there are random choices made at every scale of the structure. Stochastic fractals can be based upon the Kantor set or Sierpinski gasket, by including random choices in the algorithm. They may also be systems representing the spatial structure of various stochastic processes. Such a system requires information to describe its structure on every length scale. A stochastic fractal is a member of an ensemble, and its algorithmic as well as ensemble complexity will scale as a power law of the scale of observation L. As L increases, the amount of information is reduced, but there is no length scale smaller than the size of the system at which it is completely lost. Time series that have fractal behavior—that have power-law correlations—would also display a power-law dependence of their complexity profile as a function of T. The simplest physical model that demonstrates such fractal properties in space and time is an Ising model at its second-order transition point. At this transition there are fluctuations on all spatial and temporal scales that have power-law behavior in both. Observers with larger values of L can see the behavior of the correlations only on the longer length scales. A renormalization treatment, discussed in Section 1.10, can give the value of the complexity profile. These examples illustrate how microscopic information may become irrelevant on larger length scales, while leaving collective information that remains relevant at the longer scales.

The complexity profile enables us to consider again the definition of a complex system. As we stated, it seems intuitive that a complex system is complex on many scales. This strengthens the identification of the fractal model of space and time as a central model for the understanding of complex systems. We have also gained an understanding of the difference between deterministic and stochastic fractal systems. We see that the glass is complex in its temporal behavior, but not in its spatial behavior, and therefore is only a partial example of a complex system. If we want to identify a unique complexity of a system, there is a natural space and time scale at which to define it. For the spatial scale, L_s, we consider a significant fraction of the system—one-tenth of its size. For the temporal scale, T_s, we consider the relaxation (autocorrelation) time of the behavior on this same length scale. This is essentially the maximal complexity for this length scale, which would be the same as setting $T = 0$. However, we could also take a natural time scale of $T_s = L_s/v_s$ where v_s is a characteristic velocity of the system. This form makes the increase in time scale for larger length scales (systems) apparent. Leaving out the time scale, since it is dependent on the space scale, we can write the complexity of a system s as

$$C_s = C_s(L_s) = C_s(L_s, L_s/v_s) \approx C_s(L_s, 0) \tag{8.3.52}$$

In Section 1.10 we discussed generally the scaling of quantities as a function of the precision to which we describe the system. One of the central questions in the field of complex systems is understanding how complexity scales. This scaling is concretized by the complexity profile. One of the objectives is to understand the ultimate limits to complexity. Given a particular length or time scale, we ask what is the maximum possible complexity at that scale. One could say that this complexity is limited by the thermodynamic entropy; however, there are further limitations. These limitations are established by the nature of physical law that establishes the dynamics and interactions of the components. Thus it is unlikely that atoms can be attached to each other in such a way that the behavior of each atom is relevant to the spatiotemporal behavior of an organism at the length and time scale relevant to a human being. The details of behavior must be lost as we observe on longer length and time scales; this results in a loss of complexity. The complexity scaling of complex organisms should follow a line like that given in Fig. 8.3.2. The highest complexity of an organism results from the retention of the greatest significance of details. This is in contrast to thermodynamic systems, where all of the degrees of freedom average out on a very short length and time scale. At this time we do not know what limits can be placed on the rate of decrease of complexity with scale.

Components and systems   As we discussed in Chapter 2, a complex system is formed out of a hierarchy of interdependent subsystems. Thus, relevant to various questions about the complexity profile is an understanding of the complexity that may arise when we bring together complex systems to form a larger complex system. In general it is not clear that bringing together many complex systems must give rise to a collective complex system. This was discussed in Chapter 6, where one example was a flock of animals. Here we can provide additional meaning to this statement using the complexity profile. We will discuss the relationship of the complexity of components to the complexity of the system they are part of. To be definite, we can consider a flock of sheep. The example is chosen to expand our view toward more general application of these ideas. The general statements we make apply to any system formed out of subsystems.

Let us assume that we know the complexity of a sheep, $C_{sheep}(L_{sheep})$, the amount of information necessary to describe the relevant behaviors of eating, walking, reproducing, flocking, etc., at a length scale of about one-tenth the size of the sheep. For our current purposes this might be a lot of information contained in a large number of books, or a little information contained in a single paragraph of text. Later, in Section 8.4, we will obtain an estimate of the complexity as, of order, one book or 10^7 bits.

We now consider a flock of N sheep and construct a description of this flock. We begin by taking information that describes each of the sheep. Combining these descriptions, we have a description of the flock. This information is, however, highly redundant. Much of the information that describes one sheep can also be used to describe other sheep. Of course there are differences in size and in behavior. However, having described one sheep in detail we can describe the differences, or we can describe general characteristics of sheep and then specialize them for each of the individual sheep. Using this strategy, a description of the flock will be shorter than the sum of the lengths of the descriptions of each of the sheep. Still, this is not what we really want. The description of the flock behavior has to be on its own length scale L_{flock}, which is much larger than L_{sheep}. So we shift our observation of behavior to this longer length scale and find that most of the details of the individual sheep behavior have become irrelevant to the description of the flock. We describe the flock behavior in terms of sheep density, grazing activity, migration, reproductive rates, etc. Thus we write that:

$$C_{flock} = C_{flock}(L_{flock}) \ll C_{flock}(L_{sheep}) \ll N C_{sheep}(L_{sheep}) = N C_{sheep} \quad (8.3.53)$$

where N is the number of sheep in the flock. Among other conclusions, we see that the complexity of a flock may actually be smaller than the complexity of one sheep.

More generally, the relationship between the complexity of the collective complex system and the complexity of component systems is crucially dependent on the existence of coherence and correlations in the behavior of the components that can arise either from common origins for the behavior or from interactions between the components. We first describe this qualitatively by considering the two inequalities in Eq. (8.3.53). The second inequality arises because different sheep have the same behavior. In this case their behavior is coherent. The first inequality arises because we change the scale of observation and so lose the behavior of an individual sheep. There is a trade-off between these two inequalities. If the behaviors of the sheep are independent, then their behavior cannot be observed on the longer scale. Specifically, the movement of one sheep to the right is canceled by another sheep that starts at its right and moves to the left. Thus, only correlated motions of many sheep can be observed on a longer scale. On the other hand, if their behaviors are correlated, then the complexity of describing all of them is much smaller than the sum of the separate complexities. Thus, having a large collective complexity requires a balance between dependence and independence of the behavior of the components.

We can discuss this more quantitatively by considering the example of the nonuniform ideal gas. The loss of information for uncorrelated quantities due to combining them together is described by Eq. (8.3.37). To construct a model where the quantities are correlated, we consider placing the same densities in a region of scale $L_1 > L_0$. This is the same model as the previous one, but now on a length scale of L_1. The new value of σ is $\sigma_1 = \sigma(L_1/L_0)^d$. This increase of the standard deviation causes an increase in the value of the complexity for all scales greater than L_1. However, for $L < L_1$ the complexity is just the complexity at L_1, since there is no structure below this scale. A comparative plot is given in Fig. 8.3.3.

We can come closer to considering the behavior of a collection of animals by considering a model for their motion. We start with a scale L_0 just larger than the animal, so that we do not describe its internal structure—we describe only its location at successive intervals of time. The characteristic time over which a sheep moves a distance L_0 is T_0. We will use a model for sheep motion that can illustrate the effect of coherence of many sheep, as well as the effect of coherent motion of an individual sheep over time. To do this we assume that an individual sheep moves in a straight line for a distance qL_0 in a time qT_0 before choosing a new direction to move in at random. For simplicity we can assume that the direction chosen is one of the four compass directions, though this is not necessary for the analysis. We will use this model to calculate the complexity profile of an individual sheep. Our treatment only describes the leading behavior of the complexity profile and not various corrections.

For $L = L_0$ and $T = T_0$, the complexity of describing the motion is exactly 2 bits for every q steps to determine which of the four possible directions the sheep will move next. Because the movement is in a straight line, and the changes in direction are at well-defined intervals, we can reconstruct the motion from the measurements of any observer with $L < qL_0$ and $T < qT_0$. Thus the complexity is:

$$C(L,T) = 2N_T/q \qquad\qquad L < qL_0,\ T < qT_0 \qquad (8.3.54)$$

Once the scale of observation is greater than qL_0, the observer does not see every change in direction. The sheep is moving in a random walk where each step has a length qL_0 and takes a time qT_0, but the observer does not see each step. The distance traveled is proportional to the square root of the time, and so the sheep moves a dis-

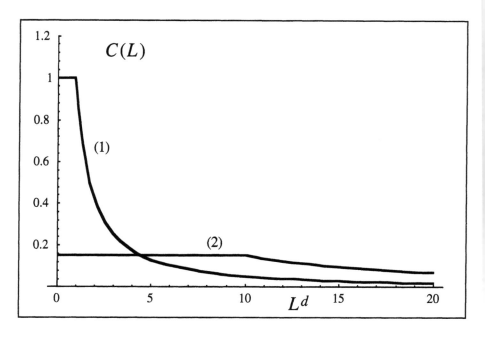

Figure 8.3.3 Plot of the complexity of a nonuniform gas (Eq. (8.3.37)), for two cases. The first (1) has a correlation in its nonuniformity at a scale L_0 and the second (2) at a scale $L_1 > L_0$. The magnitude of the local deviations in the density are the same in the two cases. The second case has a lower complexity at smaller scales but a higher complexity at the larger scales. Because the complexity decreases rapidly with scale, to show the effects on a linear scale L_1 was taken to be only $\sqrt[3]{10}L_0$, and the horizontal axis is in units of L^3 measured in units of L_0^3. Eq (8.3.39) would give similar results but the complexity would decay still more rapidly. ∎

tance L once in every $(\sigma_0/L)^2$ steps, where $\sigma_0 = qL_0$ is the standard deviation of the random walk in each dimension. Every time the sheep travels a distance L we need 2 bits to describe its motion, and thus we have a complexity:

$$C(L,T) = 2\frac{N_T}{q}\frac{\sigma_0^2}{L^2} = 2N_T\frac{qL_0^2}{L^2} \qquad L > qL_0, T < qT_0 \qquad (8.3.55)$$

We note that at $L = qL_0$ Eq. (8.3.54) and Eq. (8.3.55) are equal.

To obtain the complexity profile for long times scales $T > qT_0$, but short length scales $L < qL_0$, we use a simplified "blob" picture to combine the successive positions of the sheep into an ensemble of positions. For T only a few times qT_0 we can expect that the ensemble would enable us to reconstruct the motion—the complexity is the same as Eq. (8.3.54). However, eventually the ensemble of positions will overlap and form a blob. At this point the movement of the sheep will be described by the movement of the blob, which itself undergoes a random walk. The standard deviation of this random walk is proportional to the square root of the number of steps:

$\sigma = \sigma_0 \sqrt{T/qT_0}$. Since this is larger than L, the amount of information is essentially that of selecting a value from a Gaussian distribution of this standard deviation:

$$C(L,T) = 2\frac{N_T}{q}\min(1,\frac{qT_0}{T}(1+\log(\frac{\sigma}{L}))$$

$$= 2\frac{N_T}{q}\min(1,\frac{qT_0}{T}(1+\log(\frac{L_0}{L}\sqrt{\frac{qT}{T_0}})) \qquad L < \sigma, T > qT_0 \qquad (8.3.56)$$

There are a few points to be made about this expression. First, we use the minimum of two values to select the crossover point between the behavior in Eq. (8.3.54) and the blob behavior. As we mentioned above, the blob behavior only occurs for T significantly greater than qT_0. The simplest way to identify the crossover point is when the new estimate of the complexity becomes lower than our previous value. The second point is that we have chosen to adjust the constant term added to the logarithm so that when $L = \sigma$ the complexity matches that given by Eq. (8.3.55), which describes the behavior when L becomes large. Thus the limit on Eq. (8.3.55) should be generalized to $L > \sigma$. This minor adjustment enables the complexity to be continuous despite our rough approximations, and does not change any of the conclusions.

We can see from our results (Fig. 8.3.4) how varying q affects the complexity. Increasing q decreases the complexity at the scale of a sheep, $C(L,T) \propto 1/q$ in Eq. (8.3.54). However, it increases the complexity at longer scales $C(L,T) \propto q$ in Eq. (8.3.55). This is a straightforward consequence of increasing the coherence of the motion over time. We also see that the complexity at long times decays inversely proportional to the time but is relatively insensitive to q. The value of q primarily affects the crossover point to the long time behavior.

We now use two different assumptions to calculate the complexity of the flock. If the movement of all of the sheep is coherent, then the complexity of the flock for length scales greater than the size of the flock is the same as the complexity of a sheep for the same length scales. This is apparent because describing the movement of a single sheep is the same as describing the entire flock. We now see the significance of increasing q. Increasing q increases the flock complexity until qL_0 reaches L_1, where L_1 is the size of the flock. Thus we can increase the complexity of the whole at the cost of reducing the complexity of the components.

If the movement of sheep are independent of each other, then the flock displacements—the displacements of its center of mass—are of characteristic size $\sigma/\sqrt{N}$ (see Eq. 5.2.21). We might be concerned that the flock will disperse. However, as in our discussions of polymers in Section 5.2, interactions that would keep the sheep together need not affect the motion of their center of mass. We could also introduce into our model a circular reflecting boundary (a moving pen) around the flock, with its center at the center of mass. Since the motion of the sheep with this boundary does not require additional information over that without it, the complexity is the same. In either case, the complexity of flock motion ($L > L_1$) is obtained as:

$$C(L,T) = 2N_T\frac{qL_0^2}{NL^2} \qquad L > \sigma \qquad (8.3.57)$$

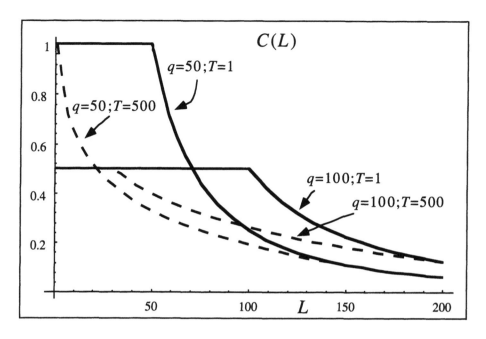

Figure 8.3.4 The complexity profile is plotted for a model of the movement of sheep as part of a flock. Increasing the distance a sheep moves in a straight line (coherence of motion in time), q, decreases the complexity at small length scales and increases the complexity at large length scales. Solid lines and dashed lines show the complexity profile as a function of length scale for a time scale $T = 1$ and $T = 500$ respectively. ∎

This is valid for all L if σ is less than L_1. If we choose T to be very large, Eq. (8.3.56) applies, with σ replaced by $\sigma/\sqrt{N}$. We see that when the motion of sheep are independent, the flock complexity is much lower than before—it decreases inversely with the number of sheep when $L > \sigma$. Even in this case, however, increasing q increases the flock complexity. Thus coherence in the behavior of a single sheep in time, or coherence between different sheep, increases the complexity of the flock. However, the maximum complexity of the flock is just that of an individual sheep, and this arises only for coherent behavior when all movements are visible on the scale of the flock. Any movements of an individual sheep that are smaller than the scale of the flock disappear on the scale of the flock. Thus even for coherent motion, in general the flock complexity is smaller than the complexity of a sheep.

This example illustrates the effect of coherent behavior. However, we see that even with coherent motion the complexity of a flock at its scale cannot be larger than the complexity of the sheep at its own scale. This is a problem for us, because our study of complex systems is focused upon systems whose complexity is larger than their components. Without this possibility, there would be no complex systems. To obtain a higher complexity of the whole we must modify this model. We must assume more generally that the motion of a sheep is describable using a set of patterns of behavior. Coherent motion of sheep still lead to a similar (or lower) complexity. To increase the complexity, the motion of the flock must have more complex patterns of motion. In order to achieve such patterns, the motions of the individual sheep must be neither independent nor coherent—they must be correlated motions that combine patterns of sheep motion into the more complex patterns of flock motion. This is possible only if there are interactions between them, which have not been included here. It should now be clear that the objective of learning how the complexity of a system is related to the complexity of its components is central to our study of complex systems.

Question 8.3.6 Throughout much of this book our working definition of complex systems or complex organisms as articulated in Section 1.3 and developed further in Chapter 2 was that a complex system has a behavior that is dependent on all of its parts. In particular, that it is impossible to take part of a complex organism away without affecting the behavior of the whole and behavior of the part. How is this definition related to the definition of complexity articulated in this section?

Solution 8.3.6 Our quantitative concept of complexity is a measure of the information necessary to describe the system behavior on its own length scale. If the system behavior is complex, then it must require many parameters to describe. These parameters are related to the description of the system on a smaller length scale, where the parts of the system are manifest because we can distinguish the description of one part from another. To do this we limit $P_{L,T}(n(x,t))$ to the domain of the part. The behavior of a system is thus related to the behavior of the parts. The more these are relevant to the system behavior, the greater is the system complexity. The information that describes the system behavior must be relevant on every smaller length scale. Thus, we have a direct relationship between the definition of a complex system in terms of parts and the definition in terms of information. Ultimately, the information necessary to describe the system behavior is determined by the microscopic description of atomic positions and motions. The more complex a system is, the more its behavior depends on smaller scale components. ∎

Question 8.3.7 When we defined interdependence we did not consider the dependence of an animal on air as a relevant example. Explain.

Solution 8.3.7 We can now recognize that the use of information as a characterization of behavior enables us to distinguish various forms of dependency. In particular, we see that the dependence of an animal on air is simple, since the necessary properties of air are simple to describe. Thus, the degree of interdependence of two systems should be measured as the amount of information necessary to replace one in the description of the other. ∎

8.3.6 Behavioral complexity

Our ability to describe a system arises from measurements or observations of its behavior. The use of system descriptions to define system complexity does not directly take this into account. The complexity profile brought us closer by acknowledging the observer in the space and time scale of the description. By acknowledging the scale of observation, we obtained a mechanism for distinguishing complex systems from equilibrium systems, and a systematic method for characterizing the complexity of a system. There is another approach to reaching the complexity profile that incorporates the observer and system relationship in a more satisfactory manner. It also enables us to consider directly the interaction of the system with its environment, which was not included previously. To introduce the new approach, we return to the underpinning of descriptive complexity and present the concept of behavioral complexity.

In Shannon's approach to the study of information in communication systems, there were two quantities of fundamental interest. The first was the information content of an individual message, and the second was the average information provided by a particular source. The discussion of algorithmic complexity was based on a consideration of the information provided by a particular message—specifically, how much it could be compressed. This carried over into our discussion of physical systems when we introduced the microscopic complexity of a system as the information contained in a particular microscopic realization of the system. When all messages, or all system states, have the same probability, then the information in the particular message is the same as the average information, and we can write:

$$I(\{x,p\}|(U,N,V)) = -\log P(\{x,p\}) = -\sum_{\{x,p\}} P(\{x,p\})\log(P(\{x,p\})) \quad (8.3.58)$$

The expression on the right, however, has a different purpose. It is a quantity that characterizes the ensemble rather than the individual microstate. It is a characterization of the source rather than of any particular message.

We can pursue this line of reasoning by considering more carefully how we might characterize the source of the information, rather than the messages. One way to characterize the source is to determine the average amount of information in a message. However, if we want to describe the source to someone, the most essential information is to give a description of the kinds of messages that will be received—the ensemble of possible messages. Thus to characterize the source we need a description of the probability of each kind of message. How much information do we need to describe these probabilities? We call this the behavioral complexity of the source.

A few examples in the context of a source of messages will serve to illustrate this concept. Any description of a source must assume a language that is to be used. We assume that the language consists of a list of characters or messages that can be received from the source, along with their probabilities. A delimiter (:) is used to separate the messages from their probability. For convenience, we will write probabilities in decimal notation. A second delimiter (,) is used to separate different members of the list. A source that gives zeros and ones at random with equal probability would be described by {1:0.5, 0:0.5}. It is convenient to include the length of a message in our description of the source. Thus we might describe a source with length $N = 1000$ character messages, each character zero and one with equal probability, as: $\{1000(1:0.5, 0:0.5)\}$. The message complexity of this source would be given by N, the length of a message. However, the behavioral complexity is given by (in this language): two decimal digits, two characters $(1, 0)$, the number representing N (requiring $\log(N)$ characters) and several delimiters. We could also specify an ASCII language source by a table of this kind that would consist of 256 elements and the probabilities of their occurrence in some database. We see that the behavioral complexity is quite distinct from the complexity of the messages provided by a source. In particular in the above example it can be larger, if $N = 1$, or it can be much smaller, if N is large.

This definition of the behavioral complexity of a source runs into a minor problem, because the probabilities are real numbers and would generally require arbitrary numbers of digits to describe. To overcome this problem, there must be a convention assumed about the limit of precision that is desired in describing the source. In principle, this precision is related to the number of messages that might be received. This convention could be part of the language, or could be defined by the specification itself. The description of the source can also be compressed using the principles of algorithmic complexity.

As we found above, the behavioral complexity can be much smaller than the information complexity of a particular message—if the source provides many random digits, the complexity of the message is high but the complexity of the source is low because we can characterize it simply as a source of random numbers. However, if the probability of each message must be independently specified, the behavioral complexity of a source is much larger than the information content of a particular message. If a particular message requires N bits of information, then the number of possible messages is 2^N. Listing all of the possible messages requires $N2^N$ bits, and specifying each probability with Q bits would give us a total of $(N + Q)2^N$ bits to describe the source. This could be reduced if the messages are placed in an agreed-upon order; then the number of bits is $Q2^N$. This is still exponentially larger than the information in a particular message. Thus, the complexity of an arbitrary source of messages of a particular length is much larger than the complexity of the messages it sends.

We are interested in the behavioral complexity when our objective is to use the messages that we receive to understand the source, rather than to make use of the information itself. Behavioral complexity becomes particularly useful when it is smaller than the complexity of a message, because it enables us to anticipate or predict the behavior of the source.

We now apply these thoughts about the source as the system of interest, rather than the message as the system of interest, to a discussion of the properties of physical systems. To make the connection between source and system, we consider an observer of a physical system who performs a number of measurements. We might imagine the measurements to consist of subjecting the system to light at various frequencies and measuring their scattering and reflection (looking at the system), observations of animals in the wild or in captivity, or physical probes of the system. We consider each measurement to be a message from the system to the observer. We must, however, take note that any measurement consists of two parts, the conditions or environment in which the observation was performed and the behavior of the system under these conditions. We write any observation as a pair (e,a), where e represents the environment and a represents a measurement of system properties (action) under the circumstances of the environment e. The observer, after performing a number of measurements, writes a description of the observations. This description characterizes the system. It captures the properties of the list of measurements, rather than of one particular measurement. It may or may not explicitly contain the information of each measurement. Alternatively, it may assign probabilities to a particular measurement. We would like to define the behavioral complexity as the amount of information contained in the observer's description. However, we must be careful how we do this because of the presence of the environmental description e.

In order to clarify this point, and to make contact between behavioral complexity and our previous discussion of descriptive complexity, we first consider the physical system of interest to be essentially isolated. Then the environmental description is irrelevant, and an observation consists only of the system measurement a. The list of measurements is the set $\{a\}$. In this case it is relatively easy to see that the behavioral complexity of a physical system is its descriptive complexity—the set of all measurements characterizes completely the state of the system.

If the entire set of measurements is performed at a single instant, and has arbitrary precision, then the behavioral complexity is the microstate complexity of the system. The result of any measurement can be obtained from a description of the microstate, and the set of possible measurements determines the microstate.

For a set of measurements performed over time on an equilibrium system, the behavioral complexity is the ensemble complexity—the number of parameters necessary to specify its ensemble. A particular message is a measurement of the system properties, which in principle might be detailed enough to determine the instantaneous positions and momenta of all of the particles. However, the list of measurements is determined by the ensemble of states the system might have. As in Section 8.3.1, we conclude that the complexity of an equilibrium system is the complexity of describing its ensemble—specifying (U, N, V) and other parameters like magnetization that result from the breaking of ergodicity. For a glass, the ensemble information is the information in the frozen coordinates previously defined as the complexity. More generally, for a set of measurements performed over an interval of time T—or at one instant but with time determination error T—and with spatial position determination errors given by L, we recover the complexity profile.

We now return to consider a system that is not isolated but subject to an environmental influence so that an observation consists of the pair (e,a) (Fig. 8.3.5). The complexity of describing such messages also contains the complexity of the environment e. Does this mean that our system description must include its environment and that the complexity of the system is dependent on the complexity of the environment? Complex systems or simple systems interact and respond to the environment in which they are found. Since the system response a is dependent on the environment e, there is no doubt that the complexity of a is dependent on the complexity of e. Three

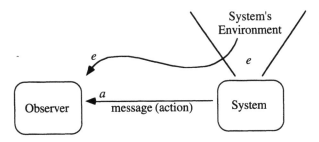

Figure 8.3.5 The observation of system behavior involves measurements both of the system's environment, e, and the system's actions, a, in response to this environment Thus we should characterize a system as a function, $a = f(e)$, where the function f describes its actions in response to its environment. It is generally simpler to describe a model for the system structure, which is also a model of f, rather than a list of all of its environment-action (e,a) pairs. ∎

examples illustrate how the environmental influence is important. The tail of a dog has a particular motion that can be described, and the complexity can be characterized. However, we may want to attribute much of this complexity to the rest of the dog rather than to the tail. Similarly, the motion of a particle suspended in a liquid follows Brownian motion, the description of which might be better attributed to the liquid than to the particle. Clearer yet is the example of the behavior of a basketball during a basketball game. These examples generalize to the consideration of any system, because measuring the properties of a system in an environment may cause us to be measuring the influence of the environment, rather than the system. The observer must describe the system behavior as a response to a particular environment, rather than just the behavior itself. Thus, we do not characterize the system by a list of actions $\{a\}$ but rather by the list of pairs $\{(e,a)\}$ where our concern is to describe f the functional mapping $a = f(e)$ from the environment e to the response a. Once we realize this, we can again affirm that a full microscopic description of the physical system is enough to give all system responses. The point is that the complexity of a system should not include the complexity of the influence upon it, but just the complexity of its response. This response is a property of the system and is determined by a complete microscopic description. Conversely, a full description of behavior subject to all possible environments would require complete microscopic information.

However, within a range of environments and with a desired degree of precision (spatial and temporal scale) it is possible to provide less information and still describe the behavior. We consider the ensemble of messages (measurements) to have possible times of observation over a range of times given by T and errors in position determination L. Describing the ensemble of responses gives us the behavioral complexity profile $C_b(L,T)$.

When the influence of the environment is not important, $C(L,T)$ and $C_b(L,T)$ are the same. When the environment matters, it is also important to characterize the information that is relevant about the environment. This is related to the problem of prediction, because predicting the system behavior in the future requires information about the environment. As we have defined it, the descriptive complexity is the information necessary to predict the behavior of the system over the time interval $t_2 - t_1$. We can characterize the environmental influence by generalizing Eq. (8.3.47) to include a term that describes the rate of information transfer from the environment to the system:

$$C(L,T) = C_b(L,T) + N_T C_e(L,T)$$

$$C_b(L,T) = C_b(L) + C_t(L,T) + N_T k \sum_{i:h_i>0} h_i \qquad (8.3.59)$$

where $C_e(L)/k\ln(2)$ is the information about the environment necessary to predict the state of the system at the next time step, and $C_b(L)$ is the behavioral complexity at one time interval. Because the system itself is finite, the amount of information about the universe that is relevant to the system behavior in any interval of time must also be finite. We note that because the system affects the environment, which then affects the system, Eq. (8.3.59) as written may count information more than once. Thus, this expression as written is an upper bound on the complexity. We noted this point also with respect to the Lyaponov exponents after Eq. (8.3.47).

This use of behavior/response rather than a description to characterize a system is related to the use of response functions in physics, or input/output relationships to describe artificial systems. The response function can (in principle) be completely derived from the microscopic description of a system. It is more directly relevant to the system behavior in response to environmental influences, and thus is essential for direct comparison with experimental results.

Behavioral complexity suggests that we should consider the system behavior as represented by a function $a = f(e)$. The input to the function is a description of the environment; the output is the response or action. There is a difficulty with this approach in that the complexity of functions is generically much larger than that of the system itself. From the discussion in Section 8.2.3 we know that the description of a function would require an amount of information given by $C_f = C_a 2^{C_e}$, where C_e is the environmental complexity, and C_a is the complexity of the action. Because the environmental influence leads to an exponentially large complexity, it is clear that often the most compact description of the system behavior will give its structure rather than its response to all inputs. Then, in principle, the response can be derived from the structure. This also implies that the behavior of physical systems under different environments cannot be independent. We note that these conclusions must also apply to human beings as complex systems that respond to their environment (see Question 8.3.8).

Q **uestion 8.3.8** Discuss the following statements with respect to human beings as complex systems: "The most compact description of the system behavior will give its structure rather than its response to all inputs," and "This implies that the behavior of physical systems under different environments cannot be independent."

Solution 8.3.8 The first statement is relevant to the discussion of behaviorism as an approach to psychology (see Section 3.2.8). It says that the idea of describing human behavior by cataloging reactions to environmental stimuli is ultimately an inefficient approach. It is more effective to use such measurements to construct a model for the internal functioning of the individual and use this model to describe the measured responses. The model description is much more concise than the description of all possible responses.

Moreover, from the second statement we know that the model can describe the responses to circumstances that have not been measured. This also means that the use of such models may be effective in predicting the behavior of an individual. Specifically, that reactions of a human being are not independent of past reactions to other circumstances. A model that incorporates the previous behaviors may have some ability to predict the behavior to new circumstances. This is part of what we do when we interact with other individuals—we construct models that represent their behavior and then anticipate how they will react to new circumstances.

The coupling between the reaction of a human being under one circumstance to the reaction under a different circumstance is also relevant to our understanding of human limitations. Optimizing the response through adaptation to a set of environments according to some goal is a process that is limited in its effectiveness due to the coupling between responses to different circumstances. An individual who is effective in some circumstances may have qualities that lead to ineffective behavior under other circumstances. We will discuss this in Chapter 9 in the context of considering the specialization of human beings in society. This point is also applicable more generally to living organisms and their ability to consume resources and avoid predators as discussed in Chapter 6. Increasing complexity enables an organism to be more effective, but the effectiveness under a variety of circumstances is limited by the interdependence of responses. This is relevant to the observation that living organisms generally consume limited types of resources and live in particular ecological niches. ∎

8.3.7 *The observer and recognition*

The explicit existence of an observer in the definition of behavioral complexity enables us to further consider the role of the observer in the definition of complexity. What assumptions have we made about the properties of the observer? One of the assumptions that we have made is that the observer is more complex than the system. What happens if the complexity of the system is greater than the complexity of the observer? The complexity of an observer is the number of bits that may be used to describe the observer. If the observer is described by fewer bits than are needed to describe the system, then the observer will be unable to contain the description of the system that is being observed. In this case, the observer will construct a description of the system that is simpler than the system actually is. There are several possible ways that the observer may simplify the description of the system. One is to reject the observation of all but a few kinds of messages. The other is to artificially limit the length of messages described. A third is to treat complex variability of the source as random—described by simple probabilities. These simplifications are often done in our modeling of physical systems.

An inherent problem in discussing behavioral complexity using environmental influence is that it is never possible to guarantee that the behavior of a system has been fully characterized. For example, a rock can be described as "just sitting there," if we want to describe the complexity of its motion under different environments. Of course the nature of the environment could be changed so that other behaviors will be realized. We may, for example, discover that the rock is actually a camouflaged animal. This is an inherent problem in behavioral complexity: it is never possible to characterize with certainty the complexity of a system under circumstances that have not been measured. All such conclusions are extrapolations. Performing such extrapolations is an essential part of the use of the description of a system. This is a general problem that applies to quantitative scientific modeling as well as the use of experience in general.

Finally, we describe the relevance of recognition to complexity. The first comment is related to the recognition of sets of numbers introduced briefly in Section 8.2.3. We introduced there the concept of recognition complexity of a set that relies upon a recognizer (a special kind of TM called a predicate that gives a single bit output) that can identify the system under discussion. Specifically, when presented with the system it says, "This is it," and when presented with any other system it says, "This is not it." We define the complexity of a system (or set of systems) as the complexity of the simplest recognizer of the system (or set of systems). There are some interesting features of this definition. First we realize that this definition is well suited to describing classes of systems. A description or model of a class of systems must identify common attributes rather than specific behaviors. A second interesting feature is that the complexity of the recognizer depends on the possible universe of systems that it can be presented with. For example, the complexity of recognizing cows depends on whether we allow ourselves to present the recognizer with all domestic animals, all known biological organisms on earth, all potentially viable biological organisms, or all possible systems. Naturally, this is an important issue in the field of pattern recognition, where the complexity of designing a system to recognize a particular pattern is strongly dependent on the universe of possibilities within which the pattern must be recognized. We will return to this point later when we consider the properties of human language in Section 8.4.1.

A different form of complexity related to recognition may be abstracted from the Turing test of artificial intelligence. This test suggests that we will achieve an artificial representation of intelligence when it becomes impossible to determine whether we are interacting with an artificial or actual human being. We can assume that Turing had in mind only a limited type of interaction between the observer "we" and the systems being observed—either the real or artificial representation of a human being. This test, which relies upon an observer to recognize the system, can serve as the basis for an additional definition of complexity. We determine the minimal possible complexity of a model (simulated representation) of the system which would be recognized by a particular observer under particular circumstances as the system. The complexity of this model we call the substitution complexity. The sensitivity of this definition to the nature of the observer and the conditions of the observation is manifest. In some ways this definition, however, is implicit in all of our earlier definitions. In all cases, the complexity measures the length of a representation of the system. Ultimately we must determine whether a particular representation of the system is faithful. The "we" in the previous sentence is some observer that must recognize the system behavior in the constructed representation.

We conclude this section by reviewing some of the main concepts that were introduced. We noted the sensitivity of complexity to the spatial and temporal scale relevant to the description or response. The complexity profile formally takes this into account. If necessary, we can define the unique complexity of a system to be its complexity profile evaluated at its own scale. A more complete characterization of the system uses the entire complexity profile. We found that the mathematical models most closely associated with complexity—chaos and fractals—were both relevant. The former described the influence of microscopic information over time. The latter described the gradual rather than rapid loss of information with spatial and temporal scale. We also reconciled the notion of information as a measure of system complexity with the notion of complex systems as composed out of interdependent parts. Our next objective is to concretize this discussion further by estimating the complexity of particular systems.

8.4   Complexity Estimation

There are various difficulties associated with obtaining specific values for the complexity of a particular system. There are both fundamental and practical problems. Fundamental problems such as the difficulty in determining whether a representation is maximally compressed are important. However, before this is an issue we must first obtain a representation.

One approach to obtaining the complexity of a system is to construct a representation. The explicit representation should then be used to make a simulation to show that the system behavior is reproduced. If it is, then we know that the length of the representation is an upper bound on the complexity of the system. We can hope, however, that it will not be necessary to obtain explicit representations in order to estimate complexities. The objective of this section is to discuss various methods for estimating the complexity of systems with which we are familiar. These approaches make use of representations that we cannot simulate, however, they do have recognizable relationships to the system.

Measuring complexity is an experimental problem. The only reason that we are able to discuss the complexity of various systems is that we have already made many measurements of the properties of various systems. We can make use of the existing information to construct estimates of their complexity. A specific estimation method is not necessarily useful for all systems.

Our objective in this section is limited to obtaining "ballpark" estimates of the complexity of systems. This means that our errors will be in the exponent rather than in the number itself. We would be very happy to have an estimate of complexity such as $10^{3\pm1}$ or $10^{7\pm2}$. When appropriate, we keep track of half-decades using factors of three, such as in 3×10^4. These rough estimates will give us a first impression of the degree of complexity of many of the systems we would like to understand. It would tell us how difficult (very roughly) they are to describe. We will discuss three methods—(1) use of intuition and human language descriptions, (2) use of a natural representation tied to the system existence, where the principle example is the genome of living organisms, and (3) use of component counting. Each of these methods has flaws that will limit our confidence in the resulting estimates. However, since we are trying to find rough estimates, we can still take advantage of them. Consistency of different methods will give us some confidence in our estimates of complexity.

While we will discuss the complexity of various systems, our focus will be on determining the complexity of a human being. Our final estimate, $10^{10\pm2}$ bits will be obtained by combining the results of different estimation techniques in the following sections. The implications of obtaining an estimate of human complexity will be discussed in Section 8.4.4. We start, however, by noting that the complexity of a human being can be bounded by the physical entropy of the collection of atoms from which he or she is formed. This is roughly the entropy of a similar weight of water, about 10^{31} bits. This is the value of $S/k\ln2$. As usual, we have assumed that there is nothing associated with a human being except the material of which he or she is formed, and that this material is described by known physical law. This entropy is an upper bound to the information necessary to specify the complete human being. The meaning of this number is that if we take away the person and we replace all of the atoms according to a specification of 10^{31} bits of information, we have replaced microscopically each atom where it was. According to our understanding of physical law, there can be no discernible difference. We will discuss the implications for artificial intelligence in Section 8.4.4, where we consider whether a computer could simulate the dynamics of atoms in order to simulate the behavior of the human being.

The entropy of a human being is much larger than the complexity estimate we are after, because we are interested in the complexity at a relevant spatial and temporal scale. In general we consider the complexity of a system at the natural scale defined in Section 8.3.5, one-tenth the size of the system itself, and the relaxation time of the behavior on this same length scale. We could also define the complexity by the observer. For example, the maximum visual sensitivity of a human being is about 1/100 of a second and 0.1 mm. For either case, observing only at this spatial and temporal scale decreases dramatically the relevance of the microscopic description. The reduction in information is hard to estimate directly. To estimate the relevant complexity, we must consider other techniques. However, since most of the information in the entropy is needed to describe the position of molecules of water undergoing vibrations, we can guess that the complexity is significantly smaller than the entropy.

8.4.1 *Human intuition—language and complexity*

The first method for estimation of complexity—the use of human intuition and language—is the least controlled/scientific method of obtaining an estimate of the complexity of a system. This approach, in its most basic form, is precisely what was asked in Question 8.2.1. We ask someone what they believe the complexity of the system is. It is assumed that the person we ask is somewhat knowledgeable about the system and also about the problem of describing systems. Even though it appears highly arbitrary, we should not dismiss this approach too readily because human beings are designed to understand complex systems. It could be argued that much of our development is directed toward enabling us to construct predictive models of various parts of the environment in which we live. The complexity of a system is directly related to the amount of study we need in order to master or predict the behavior of a system. It is not accidental that this is the fundamental objective of science—behavior prediction. We are quite used to using the word "complexity" in a qualitative manner and even in a comparative fashion—this is more complex or less complex than something else. What is missing is the quantitative definition. In order for someone to give a quantitative estimate of the complexity of a system, it is necessary to provide a definition of complexity that can be readily understood.

One useful and intuitive definition of complexity is the amount of information necessary to describe the behavior of a system. The information can be quantified in terms of representations people are familiar with—the amount of text/the number of pages/the number of books. This can be sufficient to cause a person to build a rough mental model of the system description, which is much more sophisticated than many explicit representations that might be constructed. There is an inherent limitation in this approach mentioned more generally above—a human being cannot directly estimate the complexity of an organism of similar or greater complexity than a human being. In particular, we cannot use this approach directly to estimate the complexity of human beings. Thus we will focus on simpler animals first. For example, we could ask the question in the following way: How much text is necessary to describe the behavior of a frog? We might emphasize for clarification that we are not interested in comparative frogology, or molecular frogology. We are just interested in a description of the behavior of a frog.

To gain additional confidence in this approach, we may go to the library and find descriptions that are provided in books. Superficially, we find that there are entire books devoted to a particular type of insect (mosquito, ant, butterfly), as there are books devoted to the tiger or the ape. However, there is a qualitative difference between these books. The books on insects are devoted to comparative descriptions, where various types of, e.g., mosquitoes, from around the world, their physiology, and/or their evolutionary history are described. Tens to hundreds of types are compared in a single book. Exceptional behaviors or examples are highlighted. The amount of text devoted to the behavior of a particular type of mosquito could be readily contained in less than a single chapter. On the other hand, a book devoted to tigers may describe only behavior (e.g., not physiology), and one devoted to apes would describe only a particular individual in a manner that is limited to only part of its behaviors.

Does the difference in texts describing insects and tigers reflect the social priorities of human beings? This appears to be difficult to support. The mosquito is much more relevant to the well-being of human beings than the tiger. Mosquitoes are easier to study in captivity and are more readily available in the wild. There are films that enable us to observe the mosquito behavior at its own scale rather than at our usual larger scale. Despite such films, there is no book-length description of the behavior of a mosquito. This is true despite the importance of knowledge of its behavior to prevention of various diseases. Even if there is some degree of subjectivity to the complexity estimates obtained from the lengths of descriptions found in books, the use of existing books is a reasonable first attempt to obtain complexity estimates from the information that has been compiled by human beings. We can also argue that when there is greater experience with complexity and complexity estimation, our ability to use intuition or existing texts will improve and become important tools in complexity estimation.

Before applying this methodology, however, we should understand more carefully the basic relationship of language to complexity. We have already discussed in Section 1.8 the information in a string of English characters. A first estimate of 4.8 bits per character could be based upon the existence of 26 letters and 1 space. In Question 1.8.12, the best estimate obtained was 3.3 bits per character using a Markov chain model that included correlations between adjacent characters. To obtain an even better estimate, we need to have a model that includes longer-range correlations between characters. The most reliable estimates have been obtained by asking people to guess the next character in an English text. It is assumed that people have a highly sophisticated model for the structure of English and that the individual has no specific knowledge of the text. The guesses were used to establish bounds on the information content. We can summarize these bounds as 0.9 ± 0.3 bits/character. For our present discussion, the difference between high and low bounds (a factor of 2) is not significant. For convenience we will use 1 bit/character for our conversion factor. For larger quantities of text, this corresponds to values given in Table 8.4.1.

Our estimate of information in text has assumed a strictly narrative English text. We should also be concerned about figures that accompany descriptive materials. Does the conventional wisdom of "a picture is worth a thousand words" make sense? We can consider this both from the point of view of direct compression of the picture, and the

Amount of text	Information in text	Text with figures
1 char	1 bit	-
1 page = 3000 char	3×10^3 bit	10^4
1 chapter = 30 pages	10^5 bit	3×10^5
1 book = 10 chapters	10^6 bit	3×10^6

Table 8.4.1 Information estimates for straight English text and illustrated text. ∎

possibility of replacing the figure by descriptive text. A thousand words corresponds to 5×10^3 characters or bits, about two pages of text. Descriptive figures such as graphs or diagrams often consist of a few lines that can be concisely described using a formula and would have a smaller complexity. Photographs are formed of highly correlated graphical information that can be compressed. In a black and white photograph 5×10^3 bits would correspond to a 70×70 grid of completely independent pixels. If we recall that we are not interested in small details, this seems reasonable as an upper bound. Moreover, the text that accompanies a figure generally describes its essential content. Thus when we ask the key question—whether two pages of text would be sufficient to describe a typical figure and replace its function in the text—this seems a somewhat generous but not entirely unreasonable value. A figure typically occupies half of a page that would be otherwise occupied by text. Thus, for a highly illustrated book, on average containing one figure and one-half page of text on each page, our estimate of the information content of the book would increase from 10^6 bits by a factor of 2.5 to roughly 3×10^6 bits. If there is one picture on every two pages, the information content of the book would be doubled rather than tripled. While it is not really essential for our level of precision, it seems reasonable to adopt the convention that estimates using descriptions of behavioral complexity include figures. We will do so by increasing the previous values by a factor of 3 (Table 8.4.1). This will not change any of the conclusions.

There is another aspect of the relationship of language to complexity. A language uses individual words (like "frog") to represent complex phenomena or systems (like the physical system we call a frog). The complexity of the word "frog" is not the same as the complexity of the frog. Why is this possible? According to our discussion of algorithmic complexity, the smallest possible representation of a complex system has a length in bits which is equal to the system complexity. Here we have an example of a system—frog—whose representation "frog" is manifestly smaller than its complexity.

The resolution of this puzzle is through the concept of recognition complexity discussed in Section 8.3.7. A word is a member of an ensemble of words, and the systems that are described by these words are an ensemble of systems. It is only necessary that the ensemble of words be matched to the ensemble of systems described by the words, not the whole ensemble of possible systems. Thus, the complexity of a word is not related to the complexity of the system, but rather to the complexity of specifying the system—the logarithm of the number of systems that are part of the shared experience of the individuals who are communicating. This is the central point of recognition complexity. For a human being with experience and memory of only a limited number of the set of all complex systems, to describe a system one must identify it only in comparison with the systems in memory, not with those possible in principle.

Another way to think about this is to consider a human being as analogous to a special UTM with a set of short representations that the UTM can expand to a specific limited subset of possible long descriptions. For example, having memorized a play by Shakespeare, it is only necessary to invoke the name to retrieve the whole play. This is, indeed, the essence of naming—a name is a short reference to a complex system. All words are names of more complex entities.

In this way, language provides a systematic mechanism for compression of information. This implies that we should not use the length of a word to estimate the complexity of a system that it refers to. Does this also invalidate the use of human language to obtain complexity estimates? On one hand, when we are asked to describe the behavior of a frog, we assume that we must describe it without reference to the name itself. "It behaves like a frog" is not a sufficient description. There is a presumption that a description of behavior is made to someone without specific knowledge. An estimate of the complexity of a frog would be much higher than the complexity of the word "frog." On the other hand, the words that would be used to describe a frog also refer to complex entities or actions. Consistency in different estimates of the amount of text necessary to describe a frog might arise from the use of a common language and experience. We could expand the description further by requiring that a person explain not only the behavior of the frog, but also the meaning of each of the words used to describe the behavior of the frog. At this point, however, it is more constructive to keep in mind the subtle relationship between language and complexity as part of our uncertainty, and take the given estimates at face value. Ultimately, the complexity of a system is defined by the condition that all possible (in principle) behaviors of the same complexity could be described using the same length of text. We accept the possibility that language-based estimates of complexity of biological organisms may be systematically too small because they are common and familiar. We may nevertheless have relative complexities estimated correctly.

Finally, we can argue that when we estimate the complexity of systems that approach the complexity of a human being, the estimation problems becomes less severe. This follows because of our discussion of universality of complexity given in Section 8.2.2. Specifically, that the more complex a system is, the less relevant specific knowledge is, and the more universal are estimates of complexity. Nevertheless, ultimately we will conclude that the inherent compression in use of language for describing familiar complex systems is the greatest contributor to uncertainty in complexity estimates.

There is another approach to the use of human intuition and language in estimating complexity. This is by reference to computer languages. For someone familiar with computer simulation, we can ask for the length of the computer program that can simulate the behavior of the system—more specifically, the length of the program that can simulate a frog. Computer languages are generally not very high in information content, because there are a few commands and variables that are used throughout the program. Thus we might estimate the complexity of a program not by characters, but by program lines at several bits per program line. Consistent with the definition of algorithmic complexity, the estimate of system complexity should also include the complexity of the compiler and of the computer operating system and hardware. Compilers and operating systems are much more complex than many programs by themselves. We can bypass this problem by considering instead the size of the execution module—after application of the compiler.

There are other problems with the use of natural or artificial language descriptions, including:

1. Overestimation due to a lack of knowledge of possible representations. This problem is related to the difficulty of determining the compressibility of information. The assumption of a particular length of text presumes a kind of representation. This choice of representation may not be the most compact. This may be due to the form of the representation—specifically English text. Alternatively, the assumption may be in the conceptual (semantic) framework. An example is the complexity of the motion of the planets in the Ptolemaic (earth-centered) representation compared to the Copernican (sun-centered) representation. Ptolemy would give a larger complexity estimate than Copernicus because the Ptolemaic system requires a much longer description—which is the reason the Copernican system is accepted as "true" today.

2. Underestimation due to lack of knowledge of the full behavior of the system. If an individual is familiar with the behavior of a system only under limited circumstances, the presumption that this limited knowledge is complete will lead to a complexity estimate that is too low. Alternatively, lack of knowledge may also result in too high estimates if the individual extrapolates the missing knowledge from more complex systems.

3. Difficulty with counting. Large numbers are generally difficult for people to imagine or estimate. This is the advantage of identifying numbers with length of text, which is generally a more familiar quantity.

With all of these limitations in mind, what are some of the estimates that we have obtained? Table 8.4.2 was constructed using various books. The lengths of linguistic descriptions of the behavior of biological organisms range from several pages to several books. Insects and fish are at pages, frogs at a chapter, most mammals at approximately a book, and monkeys and apes at several books. These numbers span the range of complexity estimates.

We have concluded that it is not possible to use this approach to obtain an estimate of human complexity. However, this is not quite true. We can apply this method by taking the highest complexity estimate of other systems and using this as a close lower bound to the complexity of the human being. By close lower bound we mean that the actual complexity should not be tremendously greater. According to our

Animal	Text length	Complexity (bits)
Fish	a few pages	3×10^4
Grasshopper, Mosquito	a few pages to a chapter	10^5
Ant (one, not colony)	a few pages to a chapter	10^5
Frog	a chapter or two	3×10^5
Rabbit	a short book	10^6
Tiger	a book	3×10^6
Ape	a few books	10^7

Table 8.4.2 Estimates of the approximate length of text descriptions of animal behavior ∎

experience, the complexity estimates of animals tend to extend up to roughly a single book. Primates may be estimated somewhat higher, with a range of one to tens of books. This suggests that human complexity is somewhat larger than this latter number—approximately 10^8 bits, or about 30 books. We will see how this compares to other estimates in the following sections.

There are several other approaches to estimating human complexity based upon language. The existence of book-length biographies implies a poor estimate of human complexity of 10^6 bits. We can also estimate the complexity of a human being by the typical amount of information that a person can learn. Specifically, it seems to make sense to base an estimate on the length of a college education, which uses approximately 30 textbooks. This is in direct agreement with the previous estimate of 10^8 bits. It might be argued that this estimate is too low because we have not included other parts of the education (elementary and high school and postgraduate education) or other kinds of education/information that are not academic. It might also be argued that this is too high because students do not actually know the entire content of 30 textbooks. One reason this number appears reasonable is that if the complexity of a human being were much greater than this, there would be individuals who would endure tens or hundreds of college educations in different subjects. The estimate of roughly 30 textbooks is also consistent with the general upper limit on the number of books an individual can write in a lifetime. The most prolific author in modern times is Isaac Asimov, with about 500 books. Thus from such text-based self-consistent evidence we might assume that the estimate of 10^8 bits is not wrong by more than one to two orders of magnitude. We now turn to estimation methods that are not based on text.

8.4.2 *Genetic code*

Biological organisms present us with a convenient and explicit representation for their formation by development—the genome. It is generally assumed that most of the information needed to describe the physiology of the organism is contained in genetic information. For simplicity we might think of DNA as a kind of program that is interpreted by decoding machinery during development and operation. In this regard the genome is much like a Turing machine tape (see Section 1.9), even though the mechanism for transcription is quite different from the conventional Turing machine. Some other perspectives are given in Section 7.1. Regardless of how we ultimately view the developmental process and cellular function, it appears natural to associate with the genome the information that is necessary to specify physiological design and function. It is not difficult to determine an upper bound to the amount of information that is contained in a DNA sequence. Taken at face value, this provides us with an estimate of the complexity of an organism. We must then inquire as to the approximations that are being made. We first discuss the approach in somewhat greater detail.

Considering the DNA as an alphabet of four characters provided by the four nucleotides or bases represented by A (adenine) T (tyrosine) C (cytosine) G (guanine), a first estimate of the information contained in a DNA sequence would be $N \log(4) = 2N$. N is the length of the DNA chain. Since DNA is formed of two com-

Organism	Genome length (base pairs)	Complexity (bits)
Bacteria (E. coli)	10^6–10^7	10^7
Fungi	10^7–10^8	10^8
Plants	10^8–10^{11}	3×10^8–3×10^{11}
Insects	10^8–7×10^9	10^9
Fish (bony)	5×10^8–5×10^9	3×10^9
Frog and Toad	10^9–10^{10}	10^{10}
Mammals	2×10^9–3×10^9	10^{10}
Man	3×10^9	10^{10}

Table 8.4.3 Estimates of complexity based upon genome length. Except for plants, where there is a particularly wide range of genome lengths, a single number is given for the information contained in the genome, because the accuracy does not justify more specific numbers. Genome lengths and ranges are representative. ∎

plementary nucleotide chains in a double helix, its length is measured in base pairs. While this estimate neglects many corrections, there are a number of assumptions that we are making about the organism that give a larger uncertainty than some of the corrections that we can apply. Therefore as a rough estimate, this is essentially as good an estimate as we can obtain from this methodology at present. Specific numbers are given in Table 8.4.3. We see that for a human being, the estimate is nearly 10^{10} bits, which is somewhat larger than that obtained from language-based estimates in the previous section. What is more remarkable is that there is no systematic trend of increasing genome length that parallels our expectations of increasing organism complexity based on estimates of the last section. Aside from the increasing trend from bacteria to fungi to animals/plants, there is no apparent trend that would suggest that genome length is correlated with our expectations about complexity.

We now proceed to discuss limitations in this approach. The list of approximations given below is not meant to be exhaustive, but it does suggest some of the difficulties in determining the information content even when there is a clear first numerical value to start from.

a. A significant percentage of DNA is "non-coding." This DNA is not transcribed for protein structures. It may be relevant to the structural properties of DNA. It may also contain other useful information not directly relevant to protein sequence. Nevertheless, it is likely that information in most of the base pairs that are non-coding is not essential for organism behavior. Specifically, they can be replaced by many other possible base pair sequences without effect. Since 30%–50% of human DNA is estimated to be coding, this correction would reduce the estimated complexity by a factor of two to three.

b. Direct forms of compression: as presently understood, DNA is primarily utilized through transcription to a sequence of amino acids. The coding for each amino acid is given by a triple of bases. Since there are many more triples ($4^3 = 64$) than amino acids (twenty) some of the sequences have no amino acid counterpart, and there are more than one sequence that map onto the same amino acid. This redundancy means that there is less information in the DNA sequence. Taking this into account by assigning a triple of bases to one of twenty characters that represent amino acids would give a new estimate of $(N/3)\log(20) = 1.4N$. To improve the estimate further, we would include the relative probability of the different amino acids, and correlations between them.

c. General compression: more generally, we can ask how compressed the DNA encoding of information is. We can rely upon a basic optimization of function in biology. This might suggest that some degree of compression is performed in order to reduce the complexity of transmission of the information from generation to generation. However, this is not a proof, and one could also argue in favor of redundancy in order to avoid susceptibility to small changes. Moreover there are likely to be inherent limitations on the compressibility of the information due to the possible transcription mechanisms that serve instead of decompression algorithms. For example, if a molecule that is to be represented has a long chain of the same amino acid, e.g., asp-asp-asp-asp-asp-asp-asp-asp-asp-asp-asp-asp-asp-asp-asp-asp-asp, it would be interesting if this could be represented using a chemical equivalent of (18)asp. This requires a transcription mechanism that repeats segments—a DNA loop. There are organisms that are known to have highly repetitive sequences (e.g., 10^7 repetitions) forming a significant fraction of their genome. Much of this may be non-coding DNA.

Other forms of compression may also be relevant. For example, we can ask if there are protein components/subchains that can be used in more than one protein. This is relevant to the general redundancy of protein design. There is evidence that the genome does uses this property for compression by overlapping the regions that code for several different proteins. A particular region of DNA may have several coding regions that can be combined in different ways to obtain a number of different proteins. Transcription may start from distinct initial points. Presumably, the information that describes the pattern of transcriptions is represented in the noncoding segments that are between the coding segments. Related to the issue of DNA code compression are questions about the complexity of protein primary structure in relation to its own function—specifically, how much information is necessary to describe the function of a protein. This may be much less than the information necessary to specify its primary structure (amino acid sequence). This discussion is approaching issues of the scale at which complexity is measured—at the atomic scale where the specific amino acid is relevant, or at the molecular scale at which the enzymatic function is relevant. We will mention this limitation again in point (d).

d. Scale of representation: the genome codes for macromolecular and cellular function of the biological organism. This is much less than the microscopic entropy, since it does not code the atomic vibrations or molecular diffusion. However, since our concern is for the organism's macroscopic complexity, the DNA is likely to be coding a far greater complexity than we are interested in for multicellular organisms. The assumption is that much of the cellular chemical activity is not relevant to a description of the behavior on the scale of the organism. If the DNA were representing the sum of the molecular or cellular scale complexity of each of the cells independently, then the error in estimating the complexity would be quite large. However, the molecular and cellular behavior is generally repeated throughout the organism in different cells. Thus, the DNA is essentially representing the complexity of a single cellular function with the additional complication of representing the variation in this function. To the extent that the complexity of cellular behavior is smaller than that of the complete organism, it may be assumed that the greatest part of the DNA code represents the macroscale behavior. On the other hand, if the organism behavior is comparatively simple, the greater part of the DNA representation would be devoted to describing the cellular behavior.

e. Completeness of representation: we have assumed that DNA is the only source of cellular information. However, during cell division not only the DNA is transferred but also other cellular structures, and it is not clear how much information is necessary to specify their function. It is clear, however, that DNA does not contain all the information. Otherwise it would be possible to transfer DNA from one cell into any other cell and the organism would function through control by the DNA. This is not the case. However, it may very well be that the description of all other parts of the cell, including the transcription mechanisms, only involves a small fraction of the information content compared to the DNA (for example, 10^4–10^6 bits compared to 10^7–10^{11} bits in DNA). Similar to our point (d), the information in cellular structures is more likely to be irrelevant for organisms whose complexity is high. We could note also that there are two sources of DNA in the eukaryotic cell, nuclear DNA and mitochondrial DNA. The information in the nuclear DNA dominates over the mitochondrial DNA, and we also expect it to dominate over other sources of cellular information. It is possible, however, that the other sources of information approach some fraction (e.g., 10%) of the information in the nuclear DNA, causing a small correction to our estimates.

f. We have implicitly assumed that the development process of a biological organism is deterministic and uniquely determined by the genome. Randomness in the process of development gives rise to additional information in the final structure that is not contained in the genome. Thus, even organisms that have the same DNA are not exactly the same. In humans, identical twins have been studied in order to determine the difference between environmental and genetic influence. Here we are not considering the macroscale environmental influence, but rather the microscale influence. This influence begins with the randomness of molecular vibrations during the developmental process. The additional information gained in this way would have to play a relatively minor functional role if there is significance to the genetic control over physiology. Nevertheless, a complete estimate of the complexity of a system must include this information. Without considering different scales of structure or behavior, on the macroscale we should not expect the microscopic randomness to affect the complexity by more than a factor of 2, and more likely the effect is not more than 10% in a typical biological organism.

g. We have also neglected the macroscale environmental influences on behavior. These are usually described by adaptation and learning. For most biological organisms, the environmental influences on behavior are believed to be small compared to genetic influences. Instinctive behaviors dominate. This is not as true about many mammals and even less true about human beings. Therefore, the genetic estimate becomes less reliable as an upper bound for human beings than it is for lower animals. This point will be discussed in greater detail below.

We can see that the assumptions discussed in (a), (b), (c) and (d) would lead to the DNA length being an overly large estimate of the complexity. Assumptions discussed in (e), (f) and (g) imply it is an underestimate.

One of the conceptual difficulties that we are presented with in considering genome length as a complexity estimate is that plants have a much higher DNA length than animals. This is in conflict with the conventional wisdom that animals have a greater complexity of behavior than plants. We might adopt one of two approaches to understanding this result: first, that plants are actually more complex than animals, and second, that the DNA representation in plants does not make use of, or cannot make use of, compression algorithms that are present in animal cells.

If plants are systematically more complex than animals, there must be a general quality of plants that has higher descriptive and behavioral complexity. A candidate for such a property is that plants are generally able to regenerate after injury. This inherently requires more information than the reliance upon a specific time history for development. In essence, there must be some form of actual blueprint for the organism encoded in the genome that takes into account many possible circumstances. From a programming point of view, this is a multiply reentrant program. To enable this feature may very well be more complex, or it may require a more redundant (longer) representation of the same information. It is presumed that the structure of animals has such a high intrinsic complexity that representation of a fully regenerative organism would be impossible. This idea might be checked by considering the genome length of animals that have greater ability to regenerate. If they are substantially longer than similar animals without the ability to regenerate, the explanation would be supported. Indeed, the salamander, which is the only vertebrate with the ability to regenerate limbs, has a genome of 10^{11} base pairs. This is much larger than that of other vertebrates, and comparable to that of the largest plant genomes.

A more general reason for the high plant genome complexity that is consistent with regeneration would be that plants have systematically developed a high complexity on smaller (molecular and cellular) rather than larger (organismal) scales. One reason for this would be that plant immobility requires the development of complex molecular and cellular mechanisms to inhibit or survive partial consumption by other organisms. By our discussion of the complexity profile in Section 8.3, a high complexity on small scales would not allow a high complexity on larger scales. This explanation would also be consistent with our understanding of the relative simplicity of plants on the larger scale.

The second possibility is that there exists a systematic additional redundancy of the genome in plants. This might be the result of particular proteins with chains of repetitive amino acids. A protein formed out of a long chain of the same amino acid might be functionally of importance in plants, and not in animals. This is a potential explanation for the relative lengths of plant genome and animal genome.

One of the most striking features of the genome lengths found for various organisms is their relative uniformity. Widely different types of organisms have similar genome lengths, while similar organisms may have quite different genome lengths. One explanation for this that might be suggested is that genome lengths have increased systematically with evolutionary time. It is hard, however, to see why this would be the case in all but the simplest models of evolution. It makes more sense to infer that there are constraints on the genome lengths that have led it to gravitate toward a value in the range 10^9–10^{10}. Increases in organism complexity then result from fewer redundancies and better compression, rather than longer genomes. In principle, this could account for the pattern of complexities we have obtained.

Regardless of the ultimate reason for various genome lengths, in each case the complexity estimate from genome length provides an upper bound to the genetic component of organism complexity (c.f. points (e), (f) and (g) above). Thus, the human genome length provides us with an estimate of human complexity.

8.4.3 *Component counting*

The objective of complexity estimation is to determine the behavioral complexity of a system as a whole. However, one of the important clues to the complexity of the system is its composition from elements and their interactions. By counting the number of elements, we can develop an understanding of the complexity of the system. However, as with other estimation methods, it must be understood that there are inherent problems in this approach. We will find that this method gives us a much higher estimate than the other methods. In using this method we are faced with the dilemma that lies at the heart of the ability to understand the nature of complex systems—how does complex behavior arise out of the component behavior and their interactions? The essential question that we face is: Assuming that we have a system formed of N interacting elements that have a complexity C_0 (or a known distribution of complexities), how can the complexity C of the whole system be determined? The maximal possible value would be NC_0. However, as we discussed in Section 8.3, this is reduced both by correlations between elements and by the change of scale from that of the elements to that of the system. We will discuss these problems in the context of estimating human complexity.

If we are to consider the behavioral complexity of a human being by counting components, we must identify the relevant components to count. If we count the number of atoms, we would be describing the microscopic complexity. On the other hand, we cannot count the number of parts on the scale of the organism (one) because the problem in determining the complexity remains in evaluating C_0. Thus the objective is to select components at an intermediate scale. Of the natural intermediate scales to consider, there are molecules, cells and organs. We will tackle the problem by considering cells and discuss difficulties that arise in this context. The first difficulty is that the complexity of behavior does not arise equally from all cells. It is generally understood that muscle cells and bone cells are largely uniform in structure. They may therefore collectively be described in terms of a few parameters, and their contribution to organism behavior can be summarized simply. In contrast, as we discussed in Chapter 2, the behavior of the system on the scale of the organism is generally attributed to the nervous system. Thus, aside from an inconsequential number of additional parameters, we will consider only the cells of the nervous system. If we were considering the behavior on a smaller length scale, then it would be natural to also consider the immune system.

In order to make more progress, we must discuss a specific model for the nervous system and then determine its limitations. We can do this by considering the behavior of a model system we studied in detail in Chapter 2—the attractor neural network model. Each of the neurons is a binary variable. Its behavior is specified by whether it is ON or OFF. The behavior of the network is, however, described by the values of the synapses. The total complexity of the synapses could be quite high if we allowed the synapses to have many digits of precision in their values, but this does not contribute to the complexity of the network behavior. Given our investigation of the storage of patterns in the network, we can argue that the maximal number of independent parameters that may be specified for the operation of the network consists of the neural firing patterns that are stored. This corresponds to $\alpha_c N^2$ bits of information, where N is the number of neurons, and $\alpha_c \approx 0.14$ is a number that arose from our analysis of network overload.

There are several problems with applying this formula to biological nervous systems. The first is that the biological network is not fully connected. We could apply a similar formula to the network assuming only the number of synapses N_s that are present, on average, for a neuron. This gives a value $\alpha_c N_s N$. This means that the storage capacity of the network is smaller, and should scale with the number of synapses. For the human brain where N_s has been estimated at 10^4 and $N \approx 10^{11}$, this would give a value of $0.1 \times 10^4 \times 10^{11} = 10^{14}$ bits. The problem with this estimate is that in order to specify the behavior of the network, we need to specify not only the imprinted patterns but also which synapses are present and which are absent. Listing the synapses that are present would require a set of number pairs that would specify which neurons each neuron is attached to. This list would require roughly $NN_s \log(N) = 3 \times 10^{16}$ bits, which is larger than the number of bits of information in the storage itself. This estimate may be reduced by a small amount, if, as we expect, the synapses of a neuron largely connect to neurons that are nearby. We will use 10^{16} as the basis for our complexity estimate.

The second major problem with this model is that real neurons are far from binary variables. Indeed, a neuron is a complex system. Each neuron responds to particular neurotransmitters, and the synapse between two specific neurons is different from other synapses. How many parameters would be needed to describe the behavior of an individual neuron, and how relevant are these parameters to the complexity of the whole system? Naively, we might think that taking into account the complexity of individual neurons gives a much higher complexity than that considered above. However, this is not the case. We assume that the parameters necessary to describe an individual neuron correspond to a complexity C_0, and it is necessary to specify the parameters of all of the neurons. Then the complexity of the whole system would include $C_0 N$ bits for the neurons themselves. This would be greater than 10^{16} bits only if the complexity of the individual neurons were larger than 10^5. A reasonable estimate of the complexity of a neuron is roughly 10^3–10^4 bits. This would give a value of $C_0 N = 10^{13}$–10^{14} bits, which is not a significant amount by comparison with 10^{16} bits. By these estimates, the complexity of the internal structure of a neuron is not greater than the complexity of its interconnections.

Similarly, we should consider the complexity of a synapse, which multiplies the number of synapses. Synapses are significantly simpler than the neurons. We may estimate their complexity as no more than 10 bits. This would be sufficient to specify the synaptic strength and the type of chemicals involved in transmission. Multiplying this by the total number of synapses (10^{15}) gives 10^{16} bits. This is the same as the information necessary to specify the list of synapses that are present.

Combining our estimates for the information necessary to specify the structure of neurons, the structure of synapses and the list of synapses present, we obtain an estimate for complexity of 10^{16} bits. This estimate is significantly larger than the estimate found from the other two approaches. As we mentioned before, there are two fundamental difficulties with this approach that make the estimate too high—correlations among parameters and the scale of description.

Many of the parameters enumerated above are likely to be the same, giving rise to the possibility of compression of the description. Both the description of an individual neuron and the description of the synapses between them can be drastically simplified if all of them follow a pattern. For example, the visual system involves processing of a visual field where the different neurons at different locations perform essentially the same operation on the visual information. Even if there are smooth variations in the parameters that describe both the neuron behavior and the synapses between them, we can describe the processing of the visual field in terms of a small number of parameters. Indeed, one would guess (an intuition-based estimate) that processing of the visual field is quite complicated (more than 10^2 bits) but would not exceed 10^3–10^5 bits altogether. Since a substantial fraction of the number of neurons in the brain is devoted to initial visual processing, the use of this reduced description of the visual processing would reduce the estimate of the complexity of the whole system.

Nevertheless, the initial visual processing does not involve more than 10% of the number of neurons. Even if we eliminate all of their parameters, the estimate of system complexity would not change. However, the idea behind this construction is that whenever there are many neurons whose behavior can be grouped together into particular functions, then the complexity of the description is reduced. Thus if we can describe neurons as belonging to a particular class of neurons (category or stereotype), then the complexity is reduced. It is known that neurons can be categorized; however, it is not clear how many parameters remain once this categorization has been done.

When we think about grouping the neurons together, we might also realize that this discussion is relevant to the consideration of the influence of environment and genetics on behavior. If the number of parameters necessary to describe the network greatly exceeds the number of parameters in the genetic code, which is only 10^{10} bits, then many of these parameters must be specified by the environment. We will discuss this again in the next section.

On a more philosophical note, we comment that parameters that describe the nervous system also include the malleable short-term memory. While this may be a small part of the total information, our estimate of behavioral complexity should raise questions such as, How specific do we have to be? Should the content of short-term memory be included? The argument in favor would be that we need to represent the human being in entirety. The argument against would be that what happened in the past five minutes or even the past day is not relevant and we can reset this part of the memory. Eventually we may ask whether the objective is to represent the specific information known by an individual or just his or her "character."

We have not yet directly addressed the role of substructure (Chapter 2) in the complexity of the nervous system. In comparison with a fully connected network, a network with substructure is more complex because it is necessary to specify the substructure, or more specifically which neurons (or which information) are proximate to which. However, in a system that is subdivided by virtue of having fewer synapses between subdivisions, once we have counted the information that is present in the selection of synapses, as we have done above, the substructure of the system has already been included.

The second problem of estimating complexity based on component counting is that we do not know how to reduce the complexity estimate based upon an increase of the length scale of observation. The estimate we have obtained for the complexity of the nervous system is relevant to a description of its behavior on the scale of a neuron (it does, however, focus on cellular behavior most relevant to the behavior of the organism). In order to overcome this problem, we need a method to assess the dependence of the organism behavior on the cellular behavior. A natural approach might be to evaluate the robustness of the system behavior to changes in the components. Human beings are believed to lose approximately 10^6 neurons every day (even without alcohol) corresponding to the loss of a significant fraction of the neurons over the course of a lifetime. This suggests that individual neurons are not crucial to determining human behavior. It implies that there may be a couple of orders of magnitude between the estimate of neuron complexity and human complexity. However, since the daily loss of neurons corresponds only to a loss of 1 in 10^5 neurons, we could also argue that it would be hard for us to notice the impact of this loss. In any event, our estimate based upon component counting, 10^{16}, is eight orders of magnitude larger than the estimates obtained from text and six orders of magnitude larger than the genome-based estimate. To account for this difference we would have to argue that 99.999% of neuron parameters are irrelevant to human behavior. This is too great a discrepancy to dismiss based upon such an argument.

Finally, we can demonstrate that 10^{16} is too large an estimate of complexity by considering the counting of time rather than the counting of components. We consider a minimal time interval of describing a human being to be of order 1 second, and we allow for each second 10^3 bits of information. There are of order 10^9 seconds in a lifetime. Thus we conclude that only, at most, 10^{12} bits of information are necessary to describe the actions of a human. This estimate assumes that each second is independently described from all other seconds, and no patterns of behavior exist. This would seem to be a very generous estimate. We can contrast this number with an estimate of the total amount of information that might be imprinted upon the synapses. This can be estimated as the total number of neuronal states over the course of a lifetime. For a neuron reaction time of order 10^{-2} seconds, 10^{11} neurons, and 10^9 seconds in a lifetime, we have 10^{22} bits of information. Thus we see that the total amount of information that passes through the nervous system is much larger than the information that is represented there, which is larger than the information that is manifest in terms of behavior. This suggests either that the collective behavior of neurons requires redundant information in the synapses, as discussed in Section 8.3.6, or that the actions of an individual do not fully represent the possible actions that the individual would take under all circumstances. The latter possibility returns us to the discussion of Eq. (8.3.47) and Eq. (8.3.59), where we commented that the expression is an upper bound, because information may cycle between scales or between system and environment. Under these circumstances, the potential complexity of a system under the most diverse set of circumstances is not necessarily the observed complexity. Both of our approaches to component counting (spatial and temporal) may overestimate the complexity due to this problem.

8.4.4 Complexity of human beings, artificial intelligence, and the soul

We begin this section by summarizing the estimates of human complexity from the previous sections, and then turn to some more philosophical considerations of its significance. We have found that the microscopic complexity of a human being is in the vicinity of 10^{30} bits. This is much larger than our estimates of the macroscopic complexity—language-based 10^8 bits, genome-based 10^{10} bits and component (neuron)-counting 10^{16} bits. As discussed at the end of the last section, we replace the spatial component-counting estimate with the time-counting upper bound of 10^{12} bits. We will discuss the discrepancies between these numbers and conclude with an estimate of $10^{10\pm2}$ bits.

We can summarize our understanding of the different estimates. The language-based estimate is likely to be somewhat low because of the inherent compression achieved by language. One way to say this is that a college education, consisting of 30 textbooks, is based upon childhood learning (nonlinguistic and linguistic) that provides meaning to the words, and therefore contains comparable or greater information. The genome-based complexity is likely to be a too-large estimate of the influence of genome on behavior, because genome information is compressible and because much of it must be relevant to molecular and cellular function. The component-counting estimate suggests that the information obtained from experience is much larger than the information due to the genome—specifically, that genetic information cannot specify the parameters of the neural network. This is consistent with our discussion in Section 3.2.11 that suggested that synapses store learned information while the genome determines the overall structure of the network. We must still conclude that most of the network information is not relevant to behavior at the larger scale. It is redundant, and/or does not manifest itself in human behavior because of the limited types of external circumstances that are encountered. Because of this last point, the complexity for describing the response to arbitrary circumstances may be higher than the estimate that we will give, but should still be significantly less than 10^{16} bits.

Our estimate of the complexity of a human being is $10^{10\pm2}$ bits. The error bars essentially bracket the values we obtained. The main final caveat is that the difficulty in assessing the possibility of information compression may lead to a systematic bias to high complexities. For the following discussion, the actual value is less important than the existence of an estimate.

Consideration of the complexity of a human being is intimately related to fundamental issues in artificial intelligence. The complexity of a human being specifies the amount of information necessary to describe and, given an environment, predict the behavior of a human being. There is no presumption that the prediction would be feasible using present technology. However, in principle, there is an implication of its possibility. Our objective here is to briefly discuss both philosophical and practical implications of this observation.

The notion of reproducing human behavior in a computer (or by other artificial means) has traditionally been a major domain of confrontation between science and religion, and science and popular thought. Some of these conflicts arise because of the supposition by some religious philosophers of a nonmaterial soul that is presumed to animate human beings. Such nonmaterial entities are rejected in the context of science because they are, by definition, not measurable. It may be helpful to discuss some of the alternate approaches to the traditional conflict that bypass the controversy in favor of slightly modified definitions. Specifically, we will consider the possibility of a scientific definition of the concept of a soul. We will see that such a concept is not necessarily in conflict with notions of artificial intelligence. Instead it is closely related to the assumptions of this field.

One way to define the concept of soul is as the information that describes completely a human being. We have just estimated the amount of this information. To understand how this is related to the religious concept of soul, we must realize that the concept of soul serves a purpose. When an individual dies, the existence of a soul represents the independence of the human being from the material of which he or she is formed. If the material of which the human being is made were essential to its function, then there would be no independent functional description. Also, there would be no mechanism by which we could reproduce human behavior without making use of precisely the atoms of which he or she was formed. In this way the description of a soul suggests an abstraction of function from matter which is consistent with abstractions that are familiar in science and modern thought, but might not be consistent with more primitive notions of matter. A primitive concept of matter might insist that the matter of which we are formed is essential to our functioning. The simplest possible abstraction would be to state (as is claimed by physics) that the specific atoms of which the human being are formed are not necessary to his or her function. Instead, these atoms may be replaced by other indistinguishable atoms and the same behavior will be found. Artificial intelligence takes this a large step further by stating that there are other possible media in which the same behavior can be realized. A human being is not directly tied to the material of which he is made. Instead there is a functional description that can be implemented in various media, of which one possible medium is the biological body that the human being was implemented in, when we met him or her.

Viewed in this light, the statement of the existence of a soul appears to be the same as the claim of artificial intelligence—that a human being can be reproduced in a different form by embodying the function rather than the mechanism of the human being. There is, however, a crucial distinction between the religious view and some of the practical approaches of artificial intelligence. This difference is related to the notion of a universal artificial intelligence, which is conceptually similar to the model of universal Turing machines. According to this view there is a generic model for intelligence that can be implemented in a computer. In contrast, the religious view is typically focused on the individual identity of an individual human being as manifest in a unique soul. We have discussed in Chapter 3 that our models of human beings are to be understood as nonuniversal and would indeed be better realized by the concept of representing individual human beings rather than a generic artificial intelligence. There are common features to the information processing of different individuals. However, we anticipate that the features characteristic of human behavior are predominantly specific to each individual rather than common. Thus the objective of creating artificial human beings might be better described as that of manifesting the soul of an individual human.

We can illustrate this change in perspective by considering the Turing test for recognizing artificial intelligence. The Turing test suggests that in a conversation with a computer we may not be able to distinguish it from a human being. A key problem with this prescription is that there is no specification of which human being is to be modeled. Human beings have varied complexity, and interactions are of varied levels of intimacy. It would be quite easy to reproduce the conversation of a mute individual, or even an obsessed individual. Which human being did Turing have in mind? We can go beyond this objection by recognizing that in order to fool us into thinking that the computer is a human being, except for a very casual conversation, the computer would have to represent a single human being with a name, a family history, a profession, opinions and a personality, not an abstract notion of intelligence. Finally, we may also ask whether the represented human being is someone we already know, or someone we do not know, prior to the test.

While we bypassed the fundamental controversy between science and religion regarding the presence of an immaterial soul, we suspect that the real conflict between the approaches resides in a different place. This conflict is in the question of the intrinsic value of a human being and his place in the universe. Both the religious and popular view would like to place an importance on a human being that transcends the value of the matter of which he is formed. Philosophically, the scientific perspective has often been viewed as lowering human worth. This is true whether it is physical scientists that view the material of which man is formed as "just" composed of the same atoms as rocks and water, or whether it is biological scientists that consider the biochemical and cellular structures as the same as, and derived evolutionarily from, animal processes.

The study of complexity presents us with an opportunity in this regard. A quantitative definition of complexity can provide a direct measure of the difference between the behavior of a rock, an animal and a human being. We should recognize that this capability can be a double-edged sword. On the one hand it provides us with a scientific method for distinguishing man from matter, and man from animal, by recognizing that the particular arrangement of atoms in a human being, or the particular implementation of biology, achieves a functionality that is highly complex. At the same time, by placing a number on this complexity it presents us with the finiteness of the human being. For those who would like to view themselves as infinite, a finite complexity may be humbling and difficult to accept. Others who already recognize the inherent limitations of individual human beings, including themselves, may find it comforting to know that this limitation is fundamental.

As is often the case, the value of a number attains meaning though comparison. Specifically, we may consider the complexity of a human being and see it as either high or low. We must have some reference point with respect to which we measure human complexity. One reference point was clear in the preceding discussion—that of animals. We found that our (linguistic) estimates of human complexity placed human beings quantitatively above those of animals, as we might expect. This result is quite reasonable but does not suggest any clear dividing line between animals and man. There is, however, an independent value to which these complexities can be compared. For consistency, we use language-based complexity estimates throughout.

The idea of biological evolution and the biological continuity of man from animal is based upon the concept of the survival demands of the environment on man. Let us consider for the moment the complexity of the demands of the environment. We can estimate this complexity using relevant literature. Books that discuss survival in the wild are typically quite short, 3×10^5 bits. Such a book might describe more than just basic survival—plants to eat and animal hunting—but also various skills of a primitive life such as stone knives, tanning, basket making, and primitive home or boat construction. Alternatively, a book might discuss survival under extreme circumstances rather than survival under more typical circumstances. Even so, the amount of text is not longer than a rather brief book. While there are many individuals who have devoted themselves to living in the wild, there are no encyclopedias of relevant information. This suggests that in comparison with the complexity of a human being, the complexity of survival demands is small. Indeed, this complexity appears to be right at the estimated dividing line between animal (10^6 bits) and man (10^8 bits). It is significant that an ape may have a complexity of ten times the complexity of the environmental demands upon it, but a human being has a complexity of a hundred times this demand. Another way to arrive at this conclusion is to consider primitive man, or primitive tribes that exist today. We might ask about the complexity of their existence and specifically whether the demands of the survival are the same as the complexity of their lives. From books that reflect studies of such peoples we see that the description of their survival techniques is much shorter than the description of their social and cultural activities. A single aspect of their culture might occupy a book, while the survival methods do not occupy even a single one.

We might compare the behavior of primitive man with the behavior of animal predators. In contrast to grazing animals, predators satisfy their survival needs in terms of food using only a small part of the day. One might ask why they did not develop complex cultural activities. One might think, for example, of sleeping lions. While they do have a social life, it does not compare in complexity to that of human beings. The explanation that our discussion provides is that while time would allow cultural activities, complexity does not. Thus, the complexity of such predators is essentially devoted to problems of survival. That of human beings is not.

This conclusion is quite intriguing. Several interesting remarks follow. In this context we can suggest that analyses of animal behavior should not necessarily be assumed to apply to human behavior. In particular, any animal behavior might be justified on the basis of a survival demand. While this approach has also often been applied to human beings—the survival advantages associated with culture, art and science have often been suggested—our analysis suggests that this is not justified, at least not in a direct fashion. Human behavior cannot be driven by survival demands if the survival demands are simpler than the human behavior. Of course, this does not rule out that general aspects or patterns of behavior, or even some specific behaviors, are driven by survival demands.

One of the distinctions between man and animals is the relative dominance of instinctive behavior in animals, as compared to learned behavior in man. It is often suggested that human dependence on learned rather than instinctive behavior is simply a different strategy for survival. However, if the complexity of the demands of survival are smaller than that of a human being, this does not hold. We can argue instead that if the complexity of survival demands are limited, then there is no reason for additional instinctive behaviors. Thus, our results suggest that instinctive behavior is actually a better strategy for overcoming survival demands—because it is prevalent in organisms whose behavior arises in response to survival demands. However, once such demands are met, there is little reason to produce more complex instinctive behaviors, and for this reason human behavior is not instinctively driven.

We now turn to some more practical aspects of the implications of our complexity estimates for the problem of artificial intelligence—or the re-creation of an individual in an artificial form. We may start from the microscopic complexity (roughly the entropy) which corresponds to the information necessary to replace every atom in the human being with another atom of the same kind, or alternatively, to represent the atoms in a computer. We might imagine that the computer could simulate the dynamics of the atoms in order to simulate the behavior of the human being. The practicality of such an implementation is highly questionable. The problem is not just that the number of bits of storage as well as the speed requirements are beyond modern technology. It must be assumed that any computer representation of this dynamics must ultimately be composed of atoms. If the simulation is not composed out of the atoms themselves, but some controllable representation of the atoms, then the complexity of the machine must be significantly greater than that of a human being. Moreover, unless the system is constructed to respond to its environment in a manner similar to the response of a human being, then the computer must also simulate the environment. Such a task is likely to be formally as well as practically impossible.

One central question then becomes whether it is possible to compress the representation of a human being into a simpler one that can be stored. Our estimate of behavioral complexity, $10^{10\pm2}$ bits, suggests that this might be possible. Since a CD-ROM contains 5×10^9 bits, we are discussing $2 \times 10^{\pm2}$ CD-ROMs. At the lower end of this range, 0.02 CD-ROMs is clearly not a problem. Even at the upper end, two hundred CD-ROMs is well within the domain of feasibility. Indeed, even if we chose to represent the information we estimated to be necessary to describe the neural network of a single individual, 10^{16} bits or 2 million CD-ROMs, this would be a technologically feasible project. We have made no claims about our ability to obtain the necessary information for one individual. However, once this information is obtained, it should be possible to store it. A computer that can simulate the behavior of this individual represents a more significant problem.

Before we discuss the problem of simulating a human being, we might ask what the additional microscopic complexity present in a human body is good for. Specifically, if only 10^{10} bits are relevant to human behavior, what are most of the 10^{31} bits doing? One way to think about this question is to ask why nature didn't build a similar machine with of order 10^{10} atoms, which would be significantly smaller. We might also ask whether we would know if such an organism existed. On our own scale, we might ask why nature doesn't build an organism with a complexity of order 10^{30}. We have already suggested that there may be inherent limitations to the complexity that can be formed. However, there may also be another use of some of the additional large number of microscopic pieces of information.

One possible use of the additional information can be inferred from our arguments about the difference between TM with and without a random tape. The discussion in Section 1.9.7 suggests that it may be necessary to have a source of randomness to allow human qualities such as creativity. This fits nicely with our discussion of chaos in complex system behavior. The implication is that the microscopic information becomes gradually relevant to the macroscopic behavior as a chaotic process. We can assume that most microscopic information in a human being describes the position and orientation of water molecules. In this picture, random motion of molecules affects cellular behavior, specifically the firing of neurons, that ultimately affects human behavior. This does not mean that all of the microscopic information is relevant. Only a small number of bits can be relevant at any time. However, we recognize that in order to obtain a certain number of random bits, there must be a much larger reservoir of randomness. This is one approach to understanding a possible use of the microscopic information content of a human being. Another approach would ascribe the additional information to the necessary support structures for the complex behavior, but would not attribute to it an essential role as information.

We have demonstrated time and again that it is possible to build a stronger or faster machine than a human being. This has led some people to believe that we can also build a systematically more capable machine—in the form of a robot. We have already argued that the present notion of computers may not be sufficient if it becomes necessary to include chaotic behavior. We can go beyond this argument by considering the problem we have introduced of the fundamental limits to complexity for a collection of molecules. It may turn out that our quest for the design of a complex machine will be limited by the same fundamental laws that limit the design of human beings. One of the natural improvements for the design of deterministic machines is to consider lower temperatures that enable lower error rates and higher speeds, and possibly the use of superconductors. However, the choice of a higher temperature may be required to enable a higher microscopic complexity, which also limits the macroscopic complexity. The mammalian body temperature may be selected to balance two competing effects. At high temperatures there is a high microscopic complexity. However, breaking the ergodic theorem requires low temperatures so that energy barriers can be effective in stopping movement in phase space. A way to argue this point more generally is that the sensitivity of human ears and eyes is not limited by the biological design, but by fundamental limits of quantum mechanics. It may also be that the behavioral complexity of a human being at its own length and time scale is limited by fundamental law. As with the existence of artificial sensors in other parts of the visual spectrum, we already know that machines with other capabilities can be built. However, this argument suggests that it may not be possible to build a systematically more complex artificial organism.

The previous discussion is not a proof that we cannot build a robot that is more capable than a human being. However, any claims that it is possible should be tempered by the respect that we have gained from studying the effectiveness of biological design. In this regard, it is interesting that some of the modern approaches to artificial intelligence consider the use of nanotechnology, which at least in part will make use of biological molecules and methods.

Finally we can say that the concept of an infinite human being may not be entirely lost. Even the lowly TM whose internal (table) complexity is rather small can, in arbitrarily long time and with an infinite storage, reproduce arbitrarily complex behavior. In this regard we should not consider just the complexity of a human being but also the complexity of a human being in the context of his tools. For example, we can consider the complexity of a human being with paper and pen, the complexity of a human being with a computer, or the complexity of a human being with access to a library. Since human beings make use of external storage that is limited only by the available matter, over time a human being, through collaboration with other human beings/generations extending through time, can reproduce complex behavior limited only by the matter that is available. This brings us back to questions of the behavior of collections of human beings, which we will address in Chapter 9.

9

Human Civilization II:
A Complex(ity) Transition

Conceptual Outline

█ 9.1 █ A danger of thinking about collective human systems is that our perspective on the importance of the individual may be diminished. However, this is only a problem because emergence and interdependence are not generally understood.

█ 9.2 █ By treating human civilization as a complex system, we may go beyond qualitative analogies in our efforts to understand it.

█ 9.3 █ In recent years, human civilization has become manifestly interdependent. Therefore we conclude that it is a complex organism.

█ 9.4 █ There is evidence that a transition in the structure of human organizations is occurring with intriguing consequences. Historical and contemporary evidence suggests that human organizations are undergoing a transition away from hierarchical control. From a complex systems perspective, a hierarchical system implies that the complexity of the behavior of the entire organization (at its own scale) must be less than the complexity of the controlling individual. Thus, the transition away from hierarchical control is consistent with a transition in complexity—previously human organizations behaved in a manner that is simpler than an individual, now they are more complex.

█ 9.5 █ For an individual, the consequences of this transition are manifold and manifest. There is increasing specialization of social and professional contexts. As individuals, we cannot fully understand the social and economic processes that are going on around us. However, as components of a complex organism, we are protected from many dangers.

█ 9.6 █ Our ability to predict the collective behavior of human civilization is limited. Nevertheless, there are a variety of intriguing questions that may be discussed.

9.1 Introduction: Complex Systems and Social Policy

Our objective in this chapter is to consider complex systems that are composed of collections of human beings. There are many such complex systems, ranging from a family unit to the totality of civilization. This endeavor brings us to the domain of a set of fields that we have not yet encountered in this text—social psychology, sociology, anthropology, political science and economics, and to the borders of public policy, social work and social welfare. Once we enter into this societal domain, we must evaluate carefully how to apply scientific methods. One of the central difficulties is ensuring that our desires and concerns don't interfere with our observations. We must strive to understand what is happening, and defer questions of how we would like the society to be, or to become. In order to understand, the scientist must first act as an observer rather than evaluator of good and bad. The questions, What is happening? Why is it happening? and How is it happening? are primary. While there has been a call for scientists to become involved in social policy, there are dangers to this approach. The dispassionate analytic perspective can inform, but is not a substitute for, a compassionate social policy.

Our concern in this section, however, is not to discuss the general problems of the scientific approach in social policy, but rather to discuss a specific way that the study of human civilization in a scientific context may have a negative impact on social policy thinking. There are specific dangers to be avoided. Considering the collective behavior of human beings as a complex organism can, and historically has, led to problems in attitudes that inform social policy when the value of individuals is dismissed in comparison to objectives of the collective. The danger is that we will cause a decrease in respect for the importance of the individual. In the following paragraphs we discuss and clarify this problem as a cautionary preface to our discussion of human civilization as a complex system.

Various forms of collective human systems are taken for granted in anthropology, sociology, politics and economics. In much of recent history the nation-state has been the most prominent political organization. Similarly, the corporation has been the primary economic organization. In Western law corporations are recognized as individuals with rights that are similar to the rights of individual human beings, though there are some distinctions. Other collective human systems of the past and present are the tribe, city-state and community.

In the field of biology, the existence of collective behavior of organisms has been described using the terminology "superorganism." The superorganism terminology expresses the concept that the "actual" system of interest is not the individual biological organism but rather the collective system formed out of many individuals. Applied originally to insect colonies, this term has also been applied more broadly, even to human civilization. However, within the context of complex systems, there are important distinctions that can be made between different kinds of superorganisms. The existence of interactions between insects does not necessarily imply that the collective is the relevant organism rather than the individual insect. We could try to determine the relevant organism by comparing the complexity of the individual to that of the collective. However, it is more important to understand the interdependence of individual and collective organism behavior. The primary significance of the term "superorganism," when applied to a collective, is the implicit suggestion that all of the standard biological concepts of a living organism apply to the collective. Some of these concepts—reproduction, consumption of food and production of waste—can also apply to collections of noninteracting, or decoupled individuals. These may, however, be modified in a collective context. Other concepts, such as interdependence and specialization, which occur in an insect colony, are directly relevant to our discussions of complex systems.

It is not a novel concept to consider human society as analogous to a biological organism. In some elementary biology textbooks the concept of an organism as a collection of interdependent cells is explained by analogy to interdependent human beings in society! The use of analogies, such as the analogy of a biological organism to society, is sometimes helpful in pointing out similarities. However, the limitations of analogies are not often discussed. Analogies can be misleading when they break down, suggesting similarities that are invalid. This leads to a danger of drawing conclusions that are really improper extrapolations. It is the objective of science to develop principles or mathematical models that explicitly capture the commonalties and display the differences between systems, in part so that improper extrapolations are not made. We will discuss specific analogies in Question 9.2.1. Our objective here is to understand possible conceptual implications of a superorganism analogy for human civilization in order to clarify and bound the scientific discourse.

Implications of a superorganism analogy center around relationships between the collective and a part of the collective. In a social context, there are consequences for our understanding of rights and responsibilities. When one person hits another, the hand is not considered responsible for the act. The individual is responsible. Why is this the case? Is it because the hand cannot act by itself, or because the hand is under direct control of the brain? A better answer is that it is due to interdependence of the various parts of the person. What is the level of interdependence at which the part becomes responsible for the act rather than the whole? If an individual is part of a collective, when is the collective responsible for his or her acts? In another type of circumstance, a limb may be amputated to save the individual. When we consider an individual cell, we notice that for the benefit of the collective organism, many individual cells are killed—skin cells are constantly dying to create a protective layer around the body. When can a part of the organism be sacrificed for the benefit of the collective? How much benefit or loss of harm justifies how much sacrifice? We will illustrate these considerations by corporate and societal examples.

The first example pertains to the use of the superorganism concept in limiting both rights and responsibilities of the individual as part of the superorganism. In a corporation, the individual's rights of commerce and communication may be superseded by the rights of the corporation; at the same time, the corporation relieves the individual of responsibilities for certain actions. This is manifest in the protection of employees, including presidents and chairmen of the board, from direct accountability for the consequences of decisions that are made with respect to company policy. This release from accountability has been challenged in recent years. It is enlightening to consider the arguments both pro and con in the context of a complex system framework. Let us say that a president of a corporation makes a decision that causes a faulty product to be manufactured, which leads to the death of some of those who purchase the product. Should the president be held accountable? The problem is that the decision was made in the context of company policies that reflect the history of the company as well as the individual. Other individuals at the corporation would by necessity have to cooperate in order for the product to be manufactured. Moreover, we can ask whether most other people in the same position governed by the same corporate policies would have made the same decision. One could also ask whether production-line employees have the responsibility to evaluate the implications of their work, and thus responsibility for device failure and its consequences. The question to be addressed in this context is whether the corporation and its policies should be punished and through this punishment cause change in the corporate policies that led to the harm to others, or whether the individual who made the decision should be punished to change individual behavior? The answers may require more specific information about a particular case. For us, the questions reveal a balance between the existence of a corporation as a superorganism and the individuals from which it is formed.

Throughout this text we have focused on the interdependence of parts and the whole of a collective complex system. When we think about this relationship in the context of human beings, we can also identify mutual benefit and conflict. Benefit arises when the actions of the individual and the collective are mutually advantageous. Conflict arises when the actions of an individual or the collective do not benefit both individual and collective. Considering the interplay between these is made more difficult when we recognize that collective actions are manifest as actions of individuals, and individuals may misinterpret or misrepresent their actions as collective actions. An extensive discussion of these issues is beyond the scope of this text. However, what is pertinent is that there are many circumstances where the objectives of individuals are subordinated to objectives of the superorganism. This may be illustrated by statements of the following form: Your rights/interests are secondary to the benefit of the society, corporation, or state. Examples include the firing of corporate employees, and the jailing of criminals or of political prisoners.

The need to protect the individual has been recognized. For example, democratic ideals were designed to prevent dominance of the rights of the state over the rights of the individual. The legal system is generally designed to delineate the rights and responsibilities of the individual with respect to society as a whole and with respect to other individuals within the society. Most laws restrict the independence and freedom of individual action. The concept and articulation of human rights (e.g., in the U.S. Bill of Rights) is directly related to ensuring respect for individual goals and benefits in the context of society. Various labor laws are designed to avoid the dominance of corporate rights over those of employees.

Thus, the existence of a balance between the rights of the individual and of the collective can be seen to be necessary. Our cautionary remarks are directed at the process of arriving at this balance.

There is a key distinction between implicit and explicit use of the superorganism concept. Implicit use of the concept means that rights are established by directly considering the benefit to both the individual and the collective. From the ancient times of widespread slavery to the present, the historical progression has often led to greater rights of the individual rather than of the collective. Yet, even in the present context of strengthened individual rights, it is understood that limitations must be placed on the individual in the context of society. The justifications for this are either the protection of the rights of others, or the prevention of substantial financial or other loss to the society as a whole. Such limitations are debated as social policy issues without reference to the superorganism concept. The superorganism concept enters the discussion only through the consideration of collective benefits.

In contrast, explicit use of the superorganism concept invokes the superorganism as a reason for subjugation of the rights of the individual. Claims that the state or corporation has a greater importance than the individual may directly lead to the suspension of individual rights. A telling example is the use by the Nazis of a particular biological superorganism analogy. They described the Jewish people as a cancer to be eradicated from the flesh of Germany. This superorganism concept was used to motivate and justify the involvement of physicians in the design of gas chambers for the Holocaust. The main distinction between the explicit and implicit form of the superorganism is that in the explicit form it is the concept of superorganism itself that is used to justify actions. There is no direct accounting for individual and collective benefits. Aside from the terrible consequences, we may recognize that the biological analogy is inherently ambiguous. It would be impossible to tell if the Nazi actions were an immune response or an autoimmune disease. What is more significant for our concern here is that any collective biological analogy distances us from individual human tragedy.

The preceding paragraph is a cautionary statement about the use of superorganism concepts to direct social policy. In general, science avoids consideration of analogies from physical or biological systems to social or political conditions. This is to be commended, since such analogies have led to abuses and loss of human rights. The advent of the field of complex systems, however, places an additional burden on science—not to ignore the analogies but rather to test and verify or reject them. The use of the organism analogy for the human collective may suggest that once again the rights of individuals are forfeit to the collective. The difficulty will be to keep the use of such models in perspective. In this regard, the most important conceptual tool is recognition of the interdependence in a complex system that gives rise to emergent behavior. This implies that the collective should be concerned about the well-being of its parts. However, there is a further, more specific conclusion that we reach in this chapter that should limit the motivation to utilize complex system models to address social policy matters. In the previous chapter we estimated the complexity of various organisms. In this chapter we will continue this discussion to evaluate the complexity of human civilization. Our analysis will suggest that traditional superorganisms such as states and corporations have been less complex than the individuals of which they are formed, implying the historical importance of individual rights and responsibilities. However, it appears that we are making a transition to a global superorganism that is more complex than an individual human being. Should we conclude that the rights of an individual human being should therefore be diminished in importance? In a sense this might be justified when we consider these rights with respect to the totality of human civilization. However, there is a crucial catch. Our argument is inherently based on the understanding that the superorganism is qualitatively more complex than any human being. This must mean that there is no individual who can understand it. Thus there is no individual who can be trusted to know which, if any, individual rights should be sacrificed. We find that in the context of individuals that are more complex than the superorganism, the rights of the individual are paramount. When the rights of the individual can be said to be secondary, we can at least be assured that no individual has the right to prescribe the nature of this sacrifice. We conclude that it would be unreasonable to base social policy decisions on the benefit or consequence to a system that we as individuals cannot understand.

Before we proceed with the central topic of this chapter—the complexity of the global human superorganism—we discuss a few related issues. One of the recent popular movements has suggested that the biosphere of the earth is in some sense alive. This suggestion is known as the Gaia hypothesis, where Gaia (from the Greek word for Earth) is the name given to the biosphere. The central proposal is that the biosphere is able to react to disturbances and, for example, rebalance itself. Considered in the context of complex system behavior, such a reactive organism is very simple. From the point of view of conventional science, even a chemical equilibrium reacts to disturbances. It would be highly unlikely that the biosphere, when affected on a global scale, does not have similar reactive capability. However, the notion of the collective of life on earth acting in concert is not a conventional view. We will be pursuing this further to explore the complexity of such a global organism, though our focus will be on the human superorganism. It should be understood that there is no clear understanding at this time of the nature of the boundaries of this organism. Should we expand the organism to include the flora and fauna of the earth, or even the earth itself? It may be correct to include all of the biosphere, since at the present time it would be impossible for the "human superorganism" to survive without the rest of the biosphere. This, however, is also true about any animal in its environment. For our purpose, the problem of defining the boundary of the organism is not critical, since we have considered the nervous system as a complex system despite its inseparability from the biological organism that contains it.

The Gaia hypothesis is not generally considered to be within the fold of science. Yet our objective is to pursue the topic of the global economy as a collective human superorganism that is far beyond the Gaia hypothesis in many ways. It is helpful to return to the discussion in the preface to this text, where the question of addressing the origins and destiny of man was briefly mentioned. As pointed out there, these questions have been traditionally within the domain of religion and more recently of science fiction. The field of complex systems is an endeavor to understand a new aspect of our environment as well as of ourselves. There is a natural connection between this field and the subject of the origins and destiny of man. If we did not authorize ourselves to enter into such areas and explore the possibility of scientific inquiry, we would be unduly limiting the field. This is an opportunity "to boldly go" into a new domain of scientific inquiry.

9.2   Inside a Complex System

One of the difficulties we face when discussing human civilization as a complex system is that we know of only one example. The scientific approach inherently does not allow discussion of a single system. An individual system can be discussed as one of a class of systems when principles that apply to the class can be determined. This only works when an appropriate class of systems can be described. For example, Newtonian mechanics enables prediction of the trajectories of planets because there is a broad class of systems that satisfy the same principles. Through observations, the principles could be inferred and then applied to them all. Even though the solar system is, in our experience, unique, it is still part of the class of systems that satisfy Newton's laws, and therefore its dynamics may be predicted.

The question we face is whether human civilization is a completely unique system or whether it is a member of a class of systems. There are nonscientific ways of grouping systems, or describing the similarity between one system and another. These are analogies. Analogies suggest that distinct systems share common properties. When we think about human civilization as a complex system, we can think about it as analogous to other complex systems about which we are more knowledgeable because there are many instances of them. For example, we can think about human civilization as a growing plant, or we can think about it as a colony of cells in a pond, or we can think about it as an animal formed out of various tissues. Such analogies may suggest qualitative similarities and point out features of human civilization. However, they are inherently laden with various assumptions that are not valid. This is apparent in the great variation between the three distinct analogies that have just been mentioned.

Mathematical models are the scientific form of analogies. This kind of analogy shows more precisely how two systems are similar. It may also reveal limitations of the similarities. For example, within every mathematical model are quantitative parameters. The values of these parameters are often different when applied to different systems. The extent to which model parameters are similar, or the degree to which they are different, can inform us about the similarity or difference of the original systems. It should be understood that a mathematical model that is used to capture a particular aspect of two systems does not necessarily capture other aspects. Similar to qualitative analogies, the relevance of mathematical models to describing a system is limited. This is particularly true when we consider the modeling of complex systems where, by their very nature, simplified mathematical models cannot capture the full description or complexity of the system being modeled.

In this context we see how the theory of complex systems has both its most difficult challenge in describing the properties of human civilization, and its greatest opportunity for contributing to our understanding. It is precisely the application of general principles of complex systems that can teach us about human civilization. The class of systems being considered consists of all complex systems, and so human civilization can be included. Moreover, rather than simply rejecting the apparent qualitative analogies between human civilization and other complex systems, the theory of complex systems may reveal both their validity and their limitations. Analogies should not be dismissed out of hand; neither should they be taken beyond their realm of validity.

We thus anticipate that the study of human civilization will be an important application of the study of complex systems. It should be emphasized, however, that there is a realm beyond which science cannot go. The unique aspects of the existence of a single organism cannot be predicted by science. A similar statement applies to an organism's environment. To the extent that the human organism is unique, there will always be aspects of its environment that cannot be predicted—they must only be experienced.

Question 9.2.1 Describe analogies between (1) a corporation and (2) a nation-state and a biological organism. In what ways do the analogies break down?

Solution 9.2.1 Biological organisms have many and varied properties. For example, plants and animals are qualitatively different in their behavior and in many of their attributes. The degree of cooperativity between cells in organisms also varies widely. Thus a discussion of analogies to biological organisms either allows for a broad class of properties, or must be made more specific to capture intended properties. Here we consider some universal biological properties:

1. Corporation
Reproduction—Corporations can split into smaller corporations; individuals from one corporation can leave to start a new one. It is not clear, however, in what way the resulting corporations are reproductions of the original one. Specifically, what are the hereditary traits and how are they transmitted from generation to generation? Characteristic size is typically a hereditary trait among biological organisms, but not among corporations. Generally there is no well-defined equivalent of sexual reproduction among corporations, unless we allow ourselves to consider the formation of a company by several individuals previously working at different corporations as a form of sexual reproduction. Corporations also merge and acquire other corporations. This process seems like the reverse of reproduction. We could try to fit mergers and acquisitions into the analogy by suggesting that they are similar to the consumption of food. However, biological organisms generally decompose

food into molecular components. By contrast, corporate mergers and acquisitions have a wide variety of effects. The previously existing corporate structures may remain largely intact, or they may be completely dismantled. Such variety is not characteristic of consumption in conventional biological organisms.

Growth—Like biological organisms, corporations grow. Corporations grow by increasing net worth, number of employees, sales and net profits. However, they also shrink. We might try to think about this as similar to trees that grow new leaves each year and lose them, or animals adding layers of fat and then consuming them in times of scarcity. However, the processes are quite different. Unlike fat tissue, the growth of corporations is of functional rather than nonfunctional tissue. Unlike trees, what is grown and lost is not manifestly distinct from what is retained.

Food consumption and waste excretion—Corporations consume sources of energy and raw materials. Waste is produced by corporations in the form of used chemicals, smoke, paper or other byproducts of the work being done. Corporations produce products. What is the biological analogy of a product? It is hard to consider the product as excreted waste!

Differentiation of parts—Corporations have significant functional differentiation of parts.

Breakdown of the analogy—The above comments point out some differences between corporations and biological organisms. Other distinctions include the observation that ownership defines a corporation. There is no analog of ownership for biological organisms. In particular, there is no mechanism for a takeover by outside agents. Diseases are not a comparable concept. The mechanisms of reproduction of corporations and biological organisms are quite different, even if we use the concept of reproduction loosely. Corporations can also form spontaneously without being reproduced. This is not the case for biological organisms. Corporations may be directed/guided/owned by a single individual. This is not the case for biological multicellular organisms. Large substructures in complex biological organisms cannot be traded among biological organisms the way people or even corporate divisions can be traded among corporations.

2. State

Reproduction—The primary example of state reproduction is the formation of a colonial settlement followed by independence of the settlement. What are the hereditary properties? The form of governance is a possibility, but it may not persist. The size of the state is not hereditary. Similar to corporations there is no well-defined equivalent of sexual reproduction.

Growth—States grow by increasing territory through wars; populations grow by migrations as well as by biological reproduction. Biological reproduction of populations is similar to the growth of biological organisms by cellular reproduction. However, war and migrations are not similar, because the growth of one state occurs at the expense of shrinking another state. Similar to the discussion of corporations, the possibility that states shrink is not analogous to a property of biological organisms.

Food consumption and waste excretion—States consume resources and produce wastes like biological organisms.

Differentiation of parts—Different parts of the nation may be differentiated in function.

Breakdown of the analogy—We have pointed out several distinctions in discussing reproduction and growth. Among the most dramatic of these is the possibility that part of, or the entirety of, one state will be conquered by another. As with corporate acquisitions, this is not analogous to biological consumption.

We see that analogies between human organizations and biological organisms break down even when we consider quite fundamental biological properties. The limited usefulness of the biological analogies does not carry over to the more general concepts of complex systems that have been developed in this text. As will be discussed in the Question 9.4.1, these concepts continue to be useful in the context of human organizations. ∎

9.3   Is Human Civilization a Complex System?

The reader of this text is, if he or she has followed the discussions of the previous chapters, an expert in the new field of complex systems. As a participant in human civilization, and given information generally known about human interactions and organizations, the reader is in a position to directly address whether we should consider human civilization as a complex system. Questions 9.3.1–9.3.3 are designed to encourage the reader to review various attributes of complex systems and consider their application to human civilization. We rely upon collective knowledge rather than specific references in this discussion.

For quick reference, we briefly review again the central concepts. A complex system is composed out of many elements. These elements interact in such a way as to give rise to collective behaviors on various scales up to that of the entire system. Our principal approach to characterizing the properties of a complex system has been to consider interdependence and substructure. By removing or modifying part of the system and observing the effects of this modification on the rest, we can determine the degree of interdependence of the system. We associate such interdependence with the properties of a complex system. This connection was made more specific by the study of complexity—the length of the description of a system. The complexity of a system on its own scale was shown to be related to the dependence of its behavior on its components' behavior. If we have to specify the state of each of the parts of the system in order to describe the behavior of the whole, then it requires a lot of information to describe. We distinguished between two types of complex systems—complex materials and complex organisms. A complex material has the property that removal or modification of a large part of the system affects a smaller part of it. A complex organism has the property that removal or modification of a small part of the system affects the rest. Thus, we are particularly interested in whether civilization satisfies the properties of a complex organism—whether the collective behavior is affected by removing or modifying part of it.

Question 9.3.1 To illustrate the relevance of the concepts of complex systems in the context of collectives of human beings, discuss the nature of interdependence in corporations.

Solution 9.3.1 In corporations, the degree of interdependence varies tremendously. Some corporations are loose federations of smaller, essentially independent units. Other corporations are tightly knit—interdependent organizations, where a loss of part of the system would cripple the rest. In cases where the corporations are loose federations, a unit (division) may be removed without substantially affecting either the division or the rest of the corporation. This suggests that corporations that satisfy these properties have simple collective behavior. On the other hand, when various parts of a corporation participate in joint manufacture of a product, the interactions and interdependence may be quite complex. When one factory manufactures components that are used by another factory, there are many ways that changing what happens in one will affect what happens in the other. Indeed, this applies whether the two factories are part of the same corporation or part of different corporations. Recognizing the level of interdependence is relevant to various issues pertinent to the functioning and planning of corporations. ∎

Question 9.3.2 Complete both of the following sentences with a list of properties that describe human civilization.

a. Human civilization appears to be a complex system because ...

b. Human civilization does not appear to be a complex system because ...

The objective of this question is not to determine whether human civilization is a complex system, but rather to list some of the necessary or typical features of complex systems that apply to human civilization. Question 9.3.3 addresses more directly whether human civilization is a complex system.

Solution 9.3.2 Human civilization appears to be a complex system because it is characterized by:

1. Many elements: human beings, machines
2. Interactions:

   Communication: oral and written languages, mail, telecommunications

   Economic: buying and selling, borrowing, renting

   Social: meetings, celebrations, gatherings, conferences

   Long-range interactions:

in space (travel and telecommunications) and

in time (books, music, pictures and sculptures preserved over time)

3. Substructure:

Family, community, town, city, state

Company, industry, profession, association, organization

Nationality, religion, race, language

4. Processes of Organization:

Biological evolution, social evolution, history

5. It is interdependent (see Question 9.3.3)

6. It has a complex behavior (see Section 9.4)

Human civilization does not appear to be a complex system because:

1. It does not interact with other complex systems of the same kind.

2. Its response to the environment is not manifestly complex. ∎

Question 9.3.3 Discuss the divisibility/interdependence of human civilization. Consider a few other times in the history of civilization as well as the present. What is the evidence that changes in one part of the world affect other parts of the world? Would the life of people in one place change if dramatic changes happened in another part of the world? When possible, give specific historical events as evidence.

Solution 9.3.3 Our discussions of interdependence (Section 1.3, Chapter 2) were based upon considering the effect of changes in 10%–20% of the system. Geographically this would correspond to subdividing the world into continents: North America, South America, Asia, Africa, Europe, Australia and Antarctica, and considering what would happen to the others if one of them was dramatically affected. In recent years, there has been a general awareness of global interdependence in discussions of the global economy and various geopolitical events. We will place this in a historical context.

Over the course of history there are indications that some areas were interdependent, but other areas were essentially independent over substantial periods of time. For example, human civilization in North and South America was essentially independent of the rest of the world during much of recorded history. Even within the connected continents of Europe, Asia and Africa, there are parts that were almost completely isolated from each other. Great empires of antiquity occupied limited spheres of influence. The Persian Empire and the Roman Empire did not generally affect events in the Far East, including the Chinese Empires. For most purposes, Asia was separated into three regions isolated from each other. These regions are delineated by drainage basins of great rivers: the Tigris and Euphrates Rivers, the Indus River, and the Huang He (Yellow River). There were migrations and cultural transfers that did involve substantial fractions of humanity over the course of centuries—time scales characteristically much longer than a human lifetime. Nevertheless, even as recently as the early 1900s, there were only limited ways in which the disappearance of a substantial fraction of the population in one part of the world, on one of the six populated continents, would impact the others.

One might ask, for example, how the disappearance of North America would have affected the rest of the world. The impact would have been greater after European settlement, but would still be limited to specific major trade items, and migrations from Europe to the Americas. One might also trace the transfer of a particular technology around the world to see the limited degree of influence. An example that comes to mind is the iron plow, invented in the United States and then transferred to other parts of the world in a manner that is slow on time scales that we are used to today. However, the interdependence has increased over time. In recent times it has become manifest. The time scale has become shorter, and the scale of interdependence has reached that of the individual.

The signature of change became apparent through the World Wars, especially World War II, when alliances and battles spread through all major parts of the world and directly involved a significant fraction of the world's economic and social systems. Even World War I was essentially a European conflict. In considering interdependence, we focus on how changes in one part affects the others. The global conflict in World War II arose because of changes that originally occurred in only a few nations. These changes then affected individuals throughout the world. In recent times, global interdependence has been manifest in events that primarily involved individual nations, but which resulted in the attention and involvement of people throughout the world. Some of these are geopolitical, others are geoeconomic in nature.

The following list of keywords is designed to evoke events and concerns that indicate the global interdependence:

Political/Military—governmental changes, civil wars, local wars, nuclear weapons

Economic—trade, depressions, industrialization, global corporations

Environmental—rain forests, polar ice caps, depletion of fish, acid rain

Natural disasters and disaster relief—flooding, famine, hurricanes, earthquakes

Information—publication, invention, software/hardware, global science

For example, the invasion of Kuwait by Iraq in 1990 had a manifest global response despite originally involving only a tiny proportion of the global population. The effects of the oil embargo and OPEC in the 1970s illustrated the global impact of the supply of oil from the Middle East and is reflected in the continued global concerns in that region. The impact on consumers, corporations and economies of the world of the production of automobiles and consumer electronics in Japan is well appreciated, as is the growing impact of the exports of other Pacific Rim nations. A disruption of the supply of products, even a partial disruption, as occurred for example in the wake of the earthquake in Kobe, can have global impact. The potential impact that a small nation can cause through development of nuclear weapons has recently been manifest in the global response to events in North Korea. The widespread destruction that could result from use of nuclear weapons of the arsenals of the nuclear powers is well recognized. The drug production in specific parts of the world, such as in Colombia, has relevance to individuals and the public in many other areas of the world. Various recent occurrences of social disruption and conflict in Somalia, Bosnia and Rwanda illustrate the global response to social disruption in what are considered relatively out of the way places of the world. Since World War II, various local conflicts have attained global significance and attention, e.g., Korea, Vietnam, and the Middle East. Changes of government in diverse countries such as Iran in the 1970s and South Africa in the 1990s occurred in an environment of global influences and consequences. The example of South Africa is of particular interest, since the global influence (the boycott) was directed at internal civil rights rather than external interactions. The global aid in response to famines in Africa, and earthquakes and floods in other parts of the world, are further indications of the global response to local events. The impact of fluctuations of the value of currencies during the 1990s in Italy and England, Mexico, and recently the United States have illustrated the power of global currency markets.

These examples illustrate how, at the present time, events on a national scale can have global effects. However, we can also analyze smaller-scale events that can have a global effect. One of the manifestations of the global interdependence is the wide geographic distribution of product manufacturing and utilization. Manufacturing a product involves raw materials, capital, design, assembly and marketing. Today each may originate or occur in a different part of the world, or even in several. The loss of a factory in any one of tens of countries may significantly affect the production of a corporation. Since individual corporations can be primary suppliers of particular products, this can in turn affect the lives of individuals throughout the world.

In order to consider the effects of the world on a particular individual, we must specialize. We can consider, for example, the influx of students from around the world into universities in the greater Boston area and analyze how this affects faculty, students, and the Boston area economy, as well as how the existence of Boston affects them. We might ask even more specifically how one student from one part of the world can affect another student from another part of the world when both meet in Boston. Or how an individual faculty member affects students that come from many parts of the world, and how students coming from many parts of the world affect a faculty member. Even to ask these questions demonstrates the interdependence at the individual level that now exists throughout the globe. Moreover, we did not yet account in detail for the effects of direct information exchange through the telephone, global mass media, international journals and conferences, and recently the Internet. ∎

Our conclusion from Question 9.3.3 is unambiguous—human civilization is a complex organism. It is clear that the behavior of parts of the system is strongly interdependent. It is also apparent that the behavior of the whole is strongly dependent on its parts. The strength of interdependence is to be measured by the amount of information (bits) needed to describe all of the distinct ways that one part affects the others. Our conclusion is based upon common and well-known phenomena. In this regard we are only echoing many discussions of the global economy, global communications and global interdependence. Yet it is a significant observation. It is also significant that the phenomena of interdependence have become manifest relatively recently. Thus we have an indication that a transition to a manifestly complex organism has occurred during this century. Prior to this time the behavior was not characteristic of a complex organism. In the following section we focus on this transition.

9.4   Toward a Networked Global Economy

In Section 9.4 we used evidence of interdependence to arrive at the conclusion that human civilization is a complex organism. In this section we use a different approach to arrive at the same conclusion. By taking a different route, we will reinforce our conclusion and gain a deeper insight into processes that are taking place in society around us. Our primary tool in this section will be the concept of complexity and the complexity profile discussed in Chapter 8. There is a fundamental connection between the behavior of the complexity profile and interdependence of substructure. We know this because at every level of organization the complexity of the whole arises from correlations in the behavior of the components. However, there are also more direct ways to connect the complexity profile with the functional structure of human organizations, as will become apparent in this chapter.

We begin our discussion with an effort to understand the changes that have occurred in recent years that have led to greater global interdependence. This interdependence led us to conclude that civilization is a complex organism. What is significant is that arriving at this conclusion one hundred years ago, or even fifty years ago, would have been much more ambiguous. Thus, there appears to have been a transition in the behavior of global civilization that is important for us to understand.

9.4.1. *Evidence for decrease in central control*

The history of human civilization has been marked by various stages identified by the nature of social/political/economic structures and tools/technology. One of the more recent transitions is the industrial revolution. From the point of view of technology, the industrial revolution marked a transition to the widespread use of machines powered by coal and oil, which replaced animal and human labor. From the social point of view, it marked the transition from rural to urban life. Economically, it marked the transition from family agriculture to large corporation manufacturing. Politically it strengthened but did not change qualitatively the existence of nation-states, which emerged during the Middle Ages.

In recent years there have been many discussions of the possibility that another significant change in technology, society and the economy is taking place. This has been variously characterized as the information revolution, growth of the service economy, or emergence of the global economy. Other changes that are occurring include geopolitical changes in the significance of national boundaries—the development of trading blocks, global free trade, the end of the cold war, and the emergence of widespread international cooperation in addressing various geopolitical events.

We will first discuss the current change in the global economy as a change in the manner of the exercise of control. Specifically, the hierarchical control structures that have characterized political, economic and other social organizations since antiquity may be disappearing in favor of cooperative networks. Such networks of interacting elements are more characteristic of complex systems we have been considering. Indeed, we have not discussed any specific example of control hierarchies in other complex systems. The change from hierarchies to networked systems is a specific and dramatic indicator of many changes that are taking place. It suggests that the present changes are more significant than those of the industrial revolution. We will show that these changes are related to an increase in complexity of the collective behavior of human beings and the related emergence of civilization as a complex organism. In this section we discuss some of the evidence in historical and current events that a change away from control hierarchies is taking place. In Section 9.4.2 we consider possible reasons for loss of viability of central control that are not satisfactory. In Section 9.4.3 we discuss why the loss of central control is consistent with a transition in complexity. Section 9.4.4 reviews historical phenomena in this light.

In the following paragraphs we review a series of changes that have occurred in recent years, ranging from the nature of governments to the state of interpersonal relations. While no one of these changes could be interpreted to suggest a dramatic change in the structure of civilization, their collective evidence gives some weight to this suggestion. In approaching this discussion it is important to distance ourselves from the notion of proof. Indeed, proof is not possible except in closed mathematical model systems. Our objective is to provide a reasonable case, where counter arguments are possible and to be respected.

1. *Dictatorships in the western hemisphere*—During the early 1980s a series of events occurred in the Americas and in several other countries around the world that decreased significantly the number of nations governed by dictatorships (Table 9.4.1). In many of these countries democracy and dictatorship have come and gone a number of times over the past century. It would be hard to conclude from a single government change that recent events are extraordinary. However, it should be noted that at this time there are no dictatorships in the western hemisphere except Castro's Communist regime in Cuba. Of particular importance is that among the changes of government were revolutions that did not follow the pattern of historical revolutions. Historically, a revolution begins from an attempt to reform the government, then more extreme views and individuals take over; these extreme views lead to a bloody conflict and finally a return to a form

Table 9.4.1 List of mainland Central and South American countries and the date and kind of their most recent major change of government. Until the late 1970s a patchwork of military dictatorships and democracies existed. By the early 1990s a transition to almost universal democratic governments had occurred. A tilde (~) before the word Democracy indicates significant control is still exercised by military leaders within the democratic regime. For countries whose governments have not changed since the early 1970s, no transition is indicated. While not part of the Americas, we added Greece, South Africa and the Philippines at the bottom of the list. Their recent governmental changes were not characteristic of the historical process of revolutions. ■

Country	Before change	After change	Year of change	Manner of change
Argentina	Military Dict	Democracy	1983	Peaceful
Belize	Colony	Democracy	1981	Peaceful
Bolivia	Military Dict	~Democracy	1979	Peaceful
Brazil	Military Dict	Democracy	1985	Peaceful
Chile	Military Dict	Democracy	1990	Peaceful
Colombia	Democracy			
Costa Rica	Democracy			
Cuba	Military Dict			
Ecuador	Military Dict	Democracy	1979	Peaceful
El Salvador	Military Dict	~Democracy	1980-92	Bloody
French Guiana	Possession			
Guatemala	Military Dict	~Democracy	1985	Background violence
Guyana	Democracy			
Nicaragua	Dictatorship	Democracy	1978-90	Bloody
Panama	Military Dict	Democracy	1989	US Military Intervention
Paraguay	Military Dict	Democracy	1989	Peaceful
Peru	Military Dict	~Democracy	1980	Peaceful
Suriname	Military Dict	~Democracy	1985	Peaceful
Uruguay	Military Dict	Democracy	1984	Peaceful
Venezuela	Democracy			
Greece	Military Dict	Democracy	1974	Peaceful
Philippines	Dictatorship	Democracy	1986	Peaceful
South Africa	Apartheid	Democracy	1991	Peaceful

of government structurally similar to that which existed before. This pattern was exemplified by the French and Russian revolutions, but has been realized more recently in the revolution in Iran (1970s). It may be necessary to point out that the American Revolution was actually a war of independence rather than a revolution and did not follow this pattern. The historical pattern of revolutions suggests that there are underlying reasons for a dictatorial form of government. A desire for change does not necessarily eliminate these underlying causes. In contrast, several of the recent revolutions occurred in a peaceful manner and resulted in structural changes in government. Of particular interest was the revolution in the Philippines, where violence was imminent but was averted. While violence did occur in some other revolutions, the pattern of these transitions, and their consistent outcome, may suggest a change in the underlying factors.

2. *Communism*—During the late 1980s the Soviet Union and the Soviet bloc disappeared along with communism as it was known before (Fig. 9.4.1). This dramatic change did not occur in any obvious way as a result of external forces, such as the military ones that characterized geopolitics during most of the twentieth century. Instead it appeared to occur as a result of internal forces. The change occurred peacefully. The change was a great surprise to most observers, as was the lack of violence. The surprise suggests and is consistent with the observation that this process did not fit previous patterns of governmental change. Moreover, once the change occurred, in hindsight it appeared inevitable. Internal weaknesses, and particularly an inability to maintain an effective modern economy, seemed to doom the government. Effectiveness was measured by the ability to supply citizens with products ranging from necessities to advanced technology. The system appeared to simply break down. Since this change, other communist governments around the world, with the exception of Cuba and North Korea, have relinquished control over their economies. This is particularly apparent in China, which still maintains a form of communist government but allows a rapidly growing free market economy.

3. *Privatization in democracies*—Democracies are less centrally controlled than countries with other forms of government. There are still ways in which elected governments exercise control. Control is exercised through government-run services and industries, taxes and purchases, and regulations. These should not all be considered equivalent. One way in which control was recently reduced in democratic countries throughout the world is through privatization of government-run industries. In the United States there were few government-run industries to begin with, so this has not been as manifest. On the other hand, there has been privatization of governmental services. Even garbage collection/recycling has been privatized in many communities.

4. *Decrease in proportion of U.S. government to economy*—The total amount of taxes and the federal budget, as a fraction of the U.S. economy, has not changed significantly in recent years. However, this includes a growing proportion of the budget devoted to social security and interest on the national debt. These are parts of the budget over which little control is exercised. If we measure the size of the government by purchases that are more directly controlled, and that affect the direction of economic activity, then the picture is quite different. Recently the fraction of the economy represented by governmental purchases has declined significantly (Fig. 9.4.2).

5. *Decrease in proportion of large corporations to the economy*—The proportion of the economy that reflects the activity of the largest corporations has decreased in

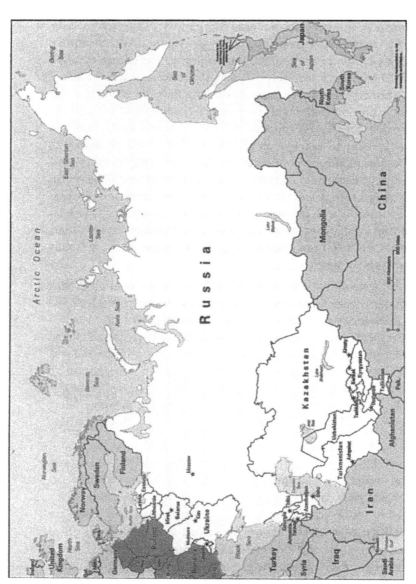

Figure 9.4.1 A map of the nations that resulted from the breakup of the Soviet Union. All of the unshaded nations were part of the Soviet Union prior to 1989. The dark shaded areas, prior to 1989, were part of other countries governed by communist regimes that collapsed just before the Soviet regime. Most of these countries were part of the Soviet bloc of nations whose policies were strongly affected by Soviet policy. The change in the governmental structure of the region that occurred during the years 1989–91 was dramatic, unprecedented, largely unanticipated, and in retrospect appeared inevitable. (Adapted from a map produced in 1995 by the United States Central Intelligence Agency)

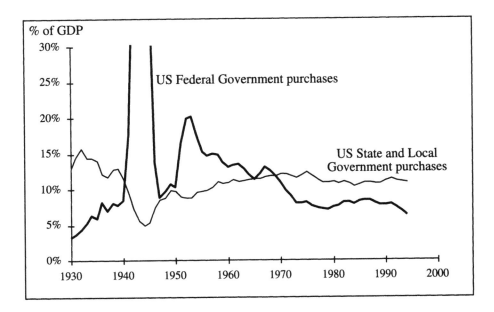

Figure 9.4.2 Size of the U.S. federal government measured by purchases as a fraction of the total U.S. economy (GDP — gross domestic product). By this measure, the federal government has declined in size since the mid-1950s. For comparison the aggregate size of state and local governments is shown (source: Bureau of Economic Analysis, U.S. Department of Commerce). ∎

recent years. One company that for many years was considered to be the basis of the economy is General Motors. It used to be said, "What's good for GM is good for the country." This was not only because this company was large as measured by sales, but also because the number of its employees was a significant fraction of the workforce. The proportion of the workforce employed by Fortune 500 companies as a function of time is shown in Fig. 9.4.3. We note that the changes in corporation size in this and the next two points are only relevant to our argument as long as the companies are centrally controlled. We will address whether they are in point 8.

6. *Systematic downsizing of large corporations*—Since the late 1980s the predominant process in corporation change has been downsizing. More generally, the economy has followed a time-dependent behavior that results in better and worse times, both for the economy as a whole and for individual corporations. These somewhat cyclical variations have been superimposed on a general trend toward increasing value—expansion—of the economic activity. In previous decades, some corporations followed these trends by increasing and decreasing employment when sales increased and decreased. This is to be contrasted with recent trends. During the late 1980s and early 1990s corporations systematically de-

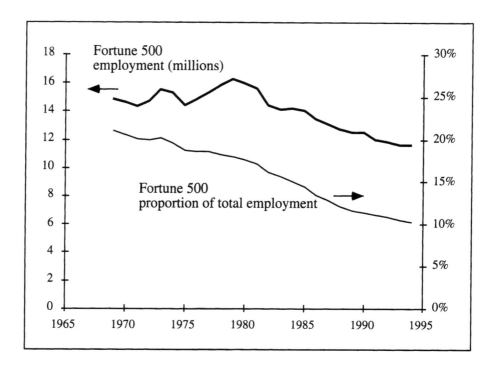

Figure 9.4.3 Total employment of the 500 largest U.S. corporations as compiled by *Fortune* magazine. Since the early 1980s the total employment of the largest companies has declined. Starting much earlier, it has declined as a fraction of the total U.S. employment (source: D. Birch, Cognetics Inc.). ∎

creased the number of their employees, almost independent of whether general expansion or contraction was occurring. This reduction is counter to the previous dominant trend of increasing numbers of employees. In prior times, increased profitability of a corporation was assumed to be based upon increased numbers of employees. This seems natural, since a greater number of employees implies greater production, greater market share and profits. In contrast, at the present time, improved profitability appears to be based on reducing the number of employees. Production appears to be largely unaffected by major cuts in employment. This suggests that changes in the underlying mechanisms of production have occurred.

7. *Growth of small corporations*—While large corporations have systematically decreased in size, certain small corporations have increased in size. In recent years, jobs added by rapidly growing companies have more than compensated for the loss of jobs in large corporations. In the meantime, this suggests a turnover of corporations rather than a change in the nature of corporations. Thus we could interpret the changes in the economy as reflecting a transition to a service or information economy, where the new large corporations are merely different in

their products from the industrial corporations of the past. We will not adopt this approach, because there are more fundamental changes that appear to be occurring in the management of corporations.

8. *Changes in corporate management*—There have been a number of changes in recent years that suggest a detachment of upper-level corporate management from production activities, and a redistribution of decision making within corporations. Upper-level management in many corporations has been active primarily in acquisitions and mergers that often have little to do with company operations. In the past, progressively larger corporate bureaucracies appeared to be an essential part of a corporation. Currently, the downsizing discussed in point 6 is often primarily at the expense of the bureaucracy. Management approaches such as total quality management (TQM) are based on decision making arising from teams of employees rather than directives passed down from upper management. In some cases, individuals or small groups are assigned greater responsibility for the profitability of their own work and consequent decision-making power. This implies that the corporation acts not in the manner of a single entity but more as a collection of individuals interacting in part through the external market system. In other cases, the coordination of employee activities within a corporation are implemented through process-oriented corporate restructuring, which relies upon distributed decision making.

9. *Boundaries of corporations*—A related development that diminishes corporate control is the existence of porous corporate boundaries. A corporation's activities include subcontracting, and hiring consultants and temporary employees. Companies focus on core technologies and "outsource" other aspects of their activities. A single company is also typically formed out of many smaller groupings of individuals. One of these groups may produce a product, while a second group may use a similar product purchased from a different corporation.

10. *Military control restructuring*—Even in the military, generally understood to be a strictly hierarchical structure, there is significant local independence. One example of this is described by General Norman Schwarzkopf in his autobiography. In discussing logistical activities, he writes, "US logistics officers in the field could never tolerate an unresponsive centralized decision-making process. Every unit …[had its own logistics officer]…to take care of his troops." (in H. Norman Schwarzkopf with P. Petre, *H. Norman Schwarzkopf: The Autobiography: It Doesn't Take a Hero* [Bantam Books, New York, 1992], p. 423, see also pp. 358–363). The process of decentralization of control has continued with development of decision teams and military hierarchy flattening—applications of TQM and reengineering within the military.

11. *Individual loss of dominance*—A recent topic of discussion is a change in interpersonal relations both in the context of conventional control hierarchies and elsewhere. This is especially apparent in the relationships between men and women, and parents and children. There are substantial social forces that are

directed to prevent abuse of power, or even the existence of power, in such interpersonal contexts. This has also given rise to the phenomenon of the "angry white male," who according to reports is faced with the loss of power and control.

9.4.2 *Hierarchy versus the individual*

Why is there a change away from hierarchical and centrally controlled structures? We start by considering the effects of technology on the abilities of an individual. We consider the impact of technological change because it is an important driving force in modern civilization, as it was in the industrial revolution. Moreover, individual empowerment is traditionally a natural counterpoint to the control hierarchy. In this context, empowerment is the ability to perform tasks with the aid of technology. We will find, however, that this approach is less than satisfactory.

The effects of technological advance on the abilities of an individual can be attributed to at least seven major interdependent areas of progress:

1. Knowledge—the availability of shared information and tools.
2. Energy—the availability of energy and mechanisms for using it to achieve tasks.
3. Transportation—rapid movement of individuals as well as materials and products.
4. Computation—particularly its decentralization in the form of personal computers.
5. Duplication and storage—mass production, printing, electronic reproduction and storage.
6. Communication—telephone, mass communication, computer networks.
7. Health—well-being through medical knowledge and technology.

How can we quantify the effect of technology on the abilities of an individual? One approach is through the notion of slave-equivalents. It was suggested, as early as the late 1970s, that U.S. citizens could think of themselves as slaveholders owning the equivalent of roughly 10,000 slaves. This figure was based solely on per capita energy consumption, and the corresponding number of slaves that would expend the same amount of energy. By such an estimate today, not for energy consumption but for computations by computers and other tasks facilitated by technology, we would reach a number of slave-equivalents many orders of magnitude higher. This suggests that modern technology greatly empowers individuals to perform tasks through control over the equivalent of large armies of slaves.

What should be the consequences of these advances on human organizations? There would seem to be several possibilities. The first is that the increased abilities could lead to independent and self-sufficient individuals, each providing for his or her own needs. Examples of such behavior do exist, but it is not the dominant trend. The second is that these abilities could enable dictators, CEOs, etc., to control more effectively. This projection was manifest in the dystopian novel *Brave New World*, by Aldous Huxley. However, this projection is counter to the evidence discussed above.

Instead, a third possibility appears to be happening—the formation of networks of interdependent individuals.

A tentative argument for a transition to networks based on technological developments would require several steps, not all of which are obvious. An individual is empowered by the development of tools. These tools allow an individual or small group of individuals to perform tasks that would previously have been possible only for a larger number of people, or would not have been possible at all. As a consequence, individuals can perform complementary and diverse tasks. This results in an increasing complexity of activity. The diverse individual activities are difficult to control because it is impossible for an individual to know how to control and coordinate many diverse activities. At the same time, the coordination of activities through a network becomes possible through advances in communication.

This argument does not withstand detailed scrutiny. However, we can extract from it that the quantity that can be tied most directly to a loss of effectiveness of central control is complexity. Simply stated, the complex behavior of a collection of individuals is impossible for one individual to control. This argument is described more thoroughly in the following section.

9.4.3 *Hierarchy versus network: A complexity transition*

We have argued that a dramatic change is taking place—the hierarchical structures that have been part of human civilization for thousands of years are disappearing. What are the underlying changes that have taken place that might result in this transition? Why is it happening now? What are the primary driving forces? How are they related to the progression in development of civilization? In the following paragraphs we begin to address these questions in the context of our study of complex systems and particularly through the quantitative concept of complexity and the complexity profile developed in Chapter 8.

In order to understand why hierarchical structures are disappearing, we must first understand what the hierarchical structure represents from the point of view of complex systems. Our studies of other complex systems in previous chapters did not reveal such structures. Structural hierarchies were discussed in Chapter 2, but not control hierarchies. The essential point is that the nature of a hierarchically controlled system requires that the behavioral complexity of the controlled group is smaller than the controlling individual. Thus, a hierarchical system implies a limit to the complexity of the collective behavior on whatever scale and in whatever aspect the control is exercised. To understand this further we turn to our discussions of the complexity profile in Section 8.3.

An extreme example of a hierarchical control structure is when a single individual is in direct (absolute) control over the behavior of a large number of other individuals. Biologically, such control structures exist—for example, the collective contraction of the cells of a muscle in response to control by nerve cells. It is apparent that the descriptive complexity of the muscle contraction is not larger than the descriptive complexity of the nerve cell activity that triggers the contraction. However, this analysis is missing the essential discussion of scale. Thus we might consider the complexity profile of a muscle compared to that of a single muscle cell or the nerve cell that is directing it. This is similar to our discussion of the complexity profile of coherent movement in Section 8.3.5. There, we contrasted the complexity profile of coherent motion with that of incoherent motion. Incoherent individuals with a complexity C_0 on their scale L_0 would have a very small collective complexity on the collective scale L_1. The collective complexity was increased in two steps. First, the pattern of behavior of the individual was modified to be simpler on the scale L_0, but fully visible and thus more complex on the scale L_1. This resulted in an individual complexity $C_0' < C_0$ at both scales. Second, the movements of different individuals were made coherent. Under these circumstances, the collective complexity at the scale L_1 was larger, but it was bounded by the simplified individual complexity C_0'. Because the individual behavior must be simplified in order to be visible on the larger scale, the collective behavior on all scales is simpler than the potential behavior of an individual.

Using this model, we can also understand both similarities and differences between two classic forms of human organization associated with the exercise of control: military force and factory production. Conventional military behavior is closer to our discussion of coherent behavior and large-scale motion in the model in Chapter 8. Similar to this model, in the military the behavior of an individual is simplified to follow a limited set of patterns. The behaviors—such as long marches—are designed to be visible on a larger scale. Then, many individuals perform the large-scale behaviors coherently. Consistent with our discussion of changes in the modern military, this model is better used to understand the activities of ancient armies— Roman legions, or even U.S. Civil War armies—than many types of modern military activity.

A conventional industrial production line also simplifies the behavior of an individual. Each individual performs a particular repetitive task. The effect of many individuals performing repetitive tasks results in a large number of copies of a particular product. However, both the simplification of behavior and the coherence is not the same as in the military model. The actions of each individual are not visible on a larger scale, and all individuals do not perform the same actions. Instead, the activities of the individual are coordinated to those of others so that the larger-scale behavior can arise. Thus, there is a relationship between the actions of different individuals that serves in place of direct coherence. As with the military model, the factory model we are describing is more appropriate to early versions of the factory and less appropriate to modern factory production. The differences between the factory and the military model are relevant to our understanding of the role of hierarchical control, which we now discuss.

We must now expand our understanding of complexity profiles in order to describe control hierarchies. It is important to recall that a complexity profile describes the complexity of the entire system, but at different scales of observation. A military force, a corporation, or a country has a collective behavior on various scales, including the scale of the system as a whole. While we have discussed ways to define the scale of observation of behavior in Section 8.3, it is not essential that we use a formal definition to appreciate the concept of collective behavior at the scale of the entire system.

At this scale, many of the details of the behavior of individuals are not apparent. In this context we can understand that a control hierarchy is designed to enable a single individual (the controller) to control the collective behavior, but not directly the behavior of each individual. Indeed, the behavior of an individual need not be known to the controller. What is necessary is that there be a mechanism for ensuring that control over the collective behavior be translated into controls that are exercised over each individual. This is the purpose of the control hierarchy.

We can thus draw a complexity profile for a system controlled by a hierarchy (Fig. 9.4.4). We assume, as is the case in human control hierarchies, that the maximum complexity of any individual in the hierarchy is essentially the same value C_0. There are two reference complexities—the maximum complexity of an individual on his or her scale C_0, and the "ideal" complexity of N individuals NC_0. We can understand the complexity profile of a hierarchy by comparison with the model of coherent behavior—the

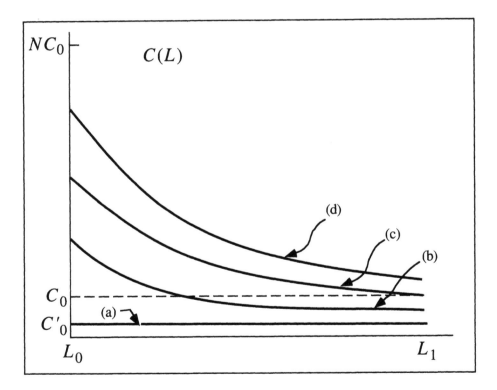

Figure 9.4.4 Comparison of schematic complexity profiles of collective systems that are controlled in distinct manners. The maximum complexity of an individual C_0 is indicated by a dashed line. The scale of an individual L_0 and the scale of the collective L_1 bracket the scales that are shown. The individual curves are as follows. (a) Coherent behavior of simplified individuals, with complexity C_0', whose entire behavior is visible on the collective scale. (b) A system coordinated by a control hierarchy. (c) A system that has the maximum complexity a control hierarchy can achieve. (d) A network which has emergent collective behavior of higher complexity than an individual. ∎

simplest control hierarchy. There are two primary differences between them that pertain to the complexity at the scale of the individual and at the scale of the collective. These differences can be understood by reference to the factory model.

The first difference is that the complexity on the scale of the individual—the complexity of describing the behavior of all of the individuals—can be higher for the control hierarchy. Indeed, the complexity on the scale of the individual can be much larger than C_0. There are two reasons for this. First, since the behavior of each individual need not be manifest on the scale of the collective, it need not be limited by a specific smaller complexity C_0' and may be closer to the maximum complexity C_0. Second, the behavior of different individuals is not the same; therefore describing one individual is not enough to describe what all the individuals are doing. Thus, the complexity of describing all of the individuals on the scale L_0 may be greater than C_0. There is, however, a limit to the complexity at the scale of the individual—it must be significantly smaller than NC_0. This limitation arises because the individual behaviors must be correlated so that the collective behavior can arise. The correlation/coherence/coordination of different individuals is imposed by the hierarchy. The assumption is that lateral communication is not essential for the functioning of the system, and therefore does not play a role in creating the correlations that enable the collective behavior to occur.

The second difference is that the complexity on the scale of the entire system can be higher than C_0'—the complexity of the simplified individual designed for coherent actions. Since the individuals do not act coherently, the complexity of their actions is not directly related to the complexity of the system. What is not changed, by the existence of the hierarchy of control, is that the complexity on the scale of the collective must still be smaller than C_0, because this is the complexity of the controlling individual—a group of individuals whose collective behavior is controlled by a single individual cannot behave in a more complex way than the individual who is exercising the control. This must be true as long as the individual exercises control over the collective behavior. Thus, while the complexity of the whole can be larger than the simplified individual C_0', it cannot be larger than the maximum complexity of an individual C_0. We can now understand why control hierarchies did not appear in our earlier studies of complex systems in previous chapters. In those studies, we were interested in the emergence of complex collective behavior from simple individuals. Hierarchical control structures are symptomatic of collective behavior that is no more complex than one individual.

The limit we have established on the collective complexity of a hierarchy does not yet explain why such hierarchies should disappear. More generally, we would like to understand the forces that cause changes in human organizations over history. To understand this we must understand that corporations and other human systems exist within an environment that places demands upon them. If the complexity of these demands exceed the complexity of a system, the system will fail. Thus, those systems that survive must have a complexity sufficiently large to respond to the complexity of environmental demands. As a result, a form of evolutionary change occurs due to competition between organizations. As discussed in Chapter 6, such competition is a natural process by which complexity may increase. While the detailed process of evolution involving processes of reproduction, variation and selection can be discussed in the context of human organizations, our purposes are served by simply postulating a progressive complexity of the collective behavior of organizations. This is a self-consistent statement, because the environment itself is formed out of organizations of human beings. Thus, there is a self-consistent process of complexity increase where competition between organizations causes the complexity of one organization to serve as the environment in which others must survive. Using the progressive increase of complexity, we can understand the nature of the transition that is under way. To do this we must assume that the complexity of demands upon collective human systems have recently become larger than an individual human being. Once this is true, the hierarchy is no longer able to impose the necessary correlations/coordination on individuals. Instead, interactions and mechanisms characteristic of networks in complex systems like the brain are necessary.

We can now make a powerful connection between the apparent transition toward networked structures from hierarchical structures in the economy and in society and our discussion of human civilization as a complex organism. The transition is consistent with a collective behavior that is more complex than the behavior of an individual. Thus, it implies that various collectives of human beings are now behaving in a manner that is more complex than an individual. This statement could not be made tens or hundreds of years ago. The breakdown of hierarchies at scales up to essentially the scale of civilization as a whole (e.g., the Soviet Union) is consistent with our observation of the recent increase in interdependence of civilization, and the conclusion that civilization is a complex organism. We will pursue this discussion further in the following section.

Question 9.4.1 Consider the properties of a hierarchical organization in response to its environment. How does this contrast with sensorimotor response in an animal?

Solution 9.4.1 In a hierarchical organization, there are various sources of information that might affect the organization's behavior. The information that is obtained about the environment generally flows up the hierarchy. The response to this information may occur at any level of the hierarchy, but this response can only involve the part of the organization that is under the control of the manager that directs the response. If the entire organization must respond to the information, the information must reach the individual who controls the entire organization. Thus the rate of response of the organization is limited by the rate of response of the individual in control, and his or her complexity as indicated above.

The sensorimotor system in an animal also involves a process of filtering of the necessary information. However, the response is dictated by the collective behavior of the network, and is not dependent on a single individual component, i.e., on a single neuron. ∎

9.4.4 *Historical review of the complexity transition*

If we review history, we can see how the development of hierarchies enabled progressively more complex behaviors up until the present time, when this process broke down in favor of networks (Fig. 9.4.5). There are two complementary aspects to the development, complexity at the scale of the individual and at the scale of the collective. In general they do not relate directly to each other. In the context of a control hierarchy, however, there is an association of greater complexity of the individual behaviors with greater complexity of the collective behavior.

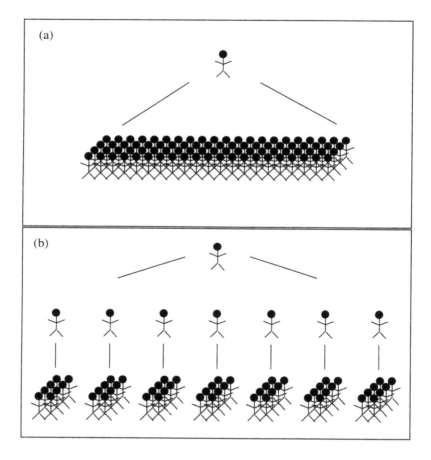

Figure 9.4.5 A brief history of human organizations capturing the effect of increasing collective complexity as illustrated in Fig. 9.4.4. (a) In the first stage a single individual directs the behavior of a large number of other individuals. This coordinates their activities, which are simple when viewed individually and collectively. (b) As the organizations become more complex, intermediate layers of hierarchy are added to the control structure. They filter information about the activities of the workers so that only a simplified picture of the activities reaches higher levels. They also elaborate the directives given by the higher levels so as

Ancient empires replaced various smaller kingdoms that had developed during a process of consolidation of yet smaller associations of human beings. The degree of control in these systems varied, but the progression toward larger more centrally controlled entities is apparent. As per our discussion of the difference between independent individuals and coherent behaviors, this led to a decrease of complexity of behavior of many individuals, but a more complex behavior on the larger scale.

During the time of ancient empires, large-scale human systems executed relatively simple behaviors, and individuals performed relatively simple individual tasks that

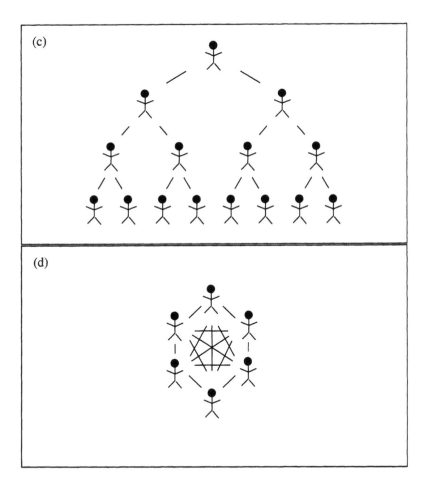

to implement them in the workers' activities. This control structure is effective only if the collective behavior can be meaningfully simplified. (c) The transition occurs when the collective complexity exceeds the maximum complexity of an individual. Then, filtering of information on the way up, and elaboration of directives on the way down, are ineffective. (d) The system structure becomes a network of individuals exerting mutual influence similar to other systems with complex emergent collective behavior. ∎

were repeated by many individuals over time to have a large-scale effect. This applies to soldier armies, as well as slaves working in agriculture, mines or construction. The scale of ancient empires controlled by large armies, as well as the scale of major projects of construction, would be impressive if performed today. However, the activity was simple enough that one individual without much of a hierarchy could direct a large number of individuals. The scale of activity was possible, without modern technology, because of the large number of individuals involved. Thus, hierarchies had a large branching ratio—a large number of controlled individuals for each controller.

As time progressed, the behavior of individuals diversified as did the collective tasks performed by them. Diversity of individuals implies that the behavior of the entire system on the scale of the individual became more complex. This required reducing the branching ratio by adding layers of management that served to exercise local control. As viewed by higher levels of management, each layer simplified the behavior to the point where an individual could control it. The hierarchy acts as a mechanism for communication of information to and from management. In our perspective, the role is also a filtering one, where the amount of information is reduced on the way up. Conversely, commands from the top are elaborated (made more complex) on the way down the hierarchy. As the collective behavioral complexity at the scale of an individual increases, the branching ratio of the control structure becomes smaller and smaller so that fewer individuals are directed by a single manager, and the minimum possible number of layers of management increases. The formation of such branching structures allows an inherently more complex local behavior of the individuals, and a larger complexity of the collective behavior as well.

However, at the point at which the collective complexity is the maximum individual complexity, the process breaks down. Hierarchical structures are not able to provide a higher complexity. We can recognize, however, that a hierarchy serves to create correlations in the behavior of individuals that are similar in many ways to the behavior of a network. The hierarchy serves as a kind of scaffolding for creating a complex system. At the complexity transition, it becomes impossible to exercise control, so the management effectively becomes divorced from the functional aspects of the system. Lateral interactions that replace the control function must be introduced. These interactions act like those of other networks to achieve the correlations in behavior that were previously created by management. As such mechanisms are introduced, layers of management can be removed. Over the course of the transition, the hierarchy exercises control over progressively more limited aspects of the system behavior. Some of the behavior patterns that have been established through the control hierarchy may continue to be effective; others will not, since an increase in system complexity must come about through changes in behavior. Among these changes are the coordination mechanisms themselves, which must be modified to involve lateral interactions. It could be argued that this picture describes much of the dynamics of modern corporations. Upper levels of management have often turned to controlling fiscal rather than production aspects of the corporation. Corporate downsizing has often been primarily at the expense of the middle management, with a subsequent lowering of payroll and little change in production. Hierarchical control has been replaced by decision teams that are introduced by corporate restructuring; and the reengineering of corporations has focused on the development of processes that are task related and do not depend on direct hierarchical control.

Ultimately, the development of greater complexity of collective behavior must continue to involve correlations/coordination of activities of various individuals. Without central control, coordination involves groups of interacting individuals achieving a collective behavior both through external influences and through mutual agreement. Among the many forms of modern corporations discussed are adhocracies, virtual corporations and networked corporations. Some of these structures may act similarly to the networks we used to describe the brain in Chapter 2. However, it is not likely that we understand at this time the various forms that coordination networks may take.

Using this argument we can understand in a straightforward way why control structures ranging from communism to corporate hierarchies could not perform the control tasks required of them in current times. As long as the activities of individuals are uniform and can be simply described—for example, soldiers marching in a row, or manufacturing workers producing a single product by a set of repetitive and simple activities (pasting eyes on a doll, screwing in bolts)—control can be exercised. The individual's activities can be specified once for a long period of time, and the overall behavior of the collective can be simply described. The collective behavior is simple when it can be summarized using a description of a simple product and the rate of its production. In contrast, central control cannot function when activities of individuals produce many products whose description is complex; when production lines use a large number of steps to manufacture many different products; when the products vary rapidly in time; and the markets change rapidly because they themselves are formed of individuals with different and rapidly changing activities.

It is useful to distinguish networks that coordinate human activity from markets that coordinate resource allocation. Markets are a distinct type of system that also results in an emergent collective behavior based upon the independent actions of many individuals. Markets such as the stock exchanges or commodity markets coordinate the allocation of resources (capital, labor and materials) according to the dynamically changing value of their use in different applications. Markets function through the actions of many agents (individuals, corporations and aggregate funds). Each agent acts according to a limited set of local objectives, while the collective behavior can coordinate the transfer of resources across many uses. Markets are distinct from networks in that they assume that the interactions among all agents in regard to a single resource can be summarized by a single time-dependent variable, which is the value of the relevant resource.

To illustrate the problem of central control of a complex economic system, we might consider examples of the problem of resource allocation. An example might be the supply of oil to a country. For an individual to allocate the supply of oil, all of the needs of different users in amounts and times, the capabilities of different suppliers, and the transportation and storage available must be taken into account. Even if one were to suggest that a computer program might perform the allocation, which is recognized as a formally difficult computational problem, the input and output of data would often eliminate this possibility. One of the crucial features of such an allocation problem is that there are both small and large suppliers and small and large users. As the number of independent users and the variation in their requirements increases, the allocation problem becomes impossible to solve. At the same time, a market is effective in performing this allocation with remarkable efficiency.

A more familiar example, which in many ways is more salient, is the problem of food supply to a metropolitan area. The supply of food is not a market, it is a network based upon a market structure. In a metropolitan area, there are hundreds to thousands of small and large supermarkets, thousands to tens of thousands of restaurants, each with specific needs that in the optimal case would be specified by immediate requirements (on demand) rather than by typical or average need over time. The suppliers of foods are also many and varied in nature. We might start by considering general categories of foods—produce, canned goods, baked goods, etc. The transportation and storage requirements of each are subject to different constraints. The many types of vehicles and modes of transportation represent another manifold of possibilities. The market-based system achieves the necessary coordination of food supply without apparent hitch and with necessary margins of error. To consider conceptually the dynamic dance of the supply of food to a city that enables daily availability is awe-inspiring. Even though there are large supermarket chains that themselves coordinate a large supply system, the overall supply system is much greater. When we realize that this coordination of effort relies upon the action of many individuals, it gives meaning to the concept of emergent behavior. We can also understand why in a centrally controlled system, consistent and adequate food supply becomes a problem. In order to have any hope of controlling such a supply problem, it would have to be simplified to allow for only a few products in only a few stores. These were well-known characteristics of food supply in communist regimes. They were seen to reflect the general economic ineffectiveness of such forms of government. In this context we see that the connection is quite direct. While considering the allocation problem in the context of food supply may illustrate the problems associated with central control, the same argument can be applied to various resource allocation and other coordination problems in large and small corporations.

In conclusion, the result of this discussion is that we can understand the implication of the disappearance of central control structures. The implication is that the behaviors of collections of human beings do not simplify sufficiently to be controlled by individuals. Instead of progressive simplification from an individual to larger and larger collections of individuals, we have the opposite—an increasing complexity that is tied to an increasing complexity of the demands of the environment. This makes it impossible for an individual to effectively control collective behaviors. While specific individuals have been faulted for management errors that have led to corporate failures, the analysis we have performed suggests that it is inevitable for management to make errors under these circumstances.

In Chapter 8 we estimated the complexity of various systems by several approaches. The first approach used linguistic descriptions, either imagined or actual, of the systems. The complexity of a human being was estimated to be roughly 30 books (10^8 bits)—the length of an encyclopedia. If we consider the functioning of the global economy and the behavior of its intermediate scale components (corporations, states, etc.) we can readily see that the complexity of its description using language is much larger than the estimate given for a human being. This conclusion may apply to a single product manufactured by a single company. The number of pages of text necessary to describe an airplane, a car, a computer or the processes necessary to produce them would exceed the length of an encyclopedia. It is generally acknowledged that large computer programs exceed the ability of a single person to understand. The UNIX operating system, found on many computers, requires a storage of 4×10^9 bits, which is comparable to our estimate of human complexity. This is only a very small part of the information necessary to describe the operation of civilization. Estimates of complexity of a product or an operating system are relevant to understanding the complexity of the internal functioning of civilization. This does not by itself imply that the complexity of the behavior of collections of human beings is of this size. Thus, more directly relevant to obtaining an estimate are: the inability of one individual to coordinate human activities, the apparent breakdown of central control, and the manifest interdependence of human civilization. As we have argued in the previous chapter, an actual estimate of the complexity of civilization should be impossible for an individual to obtain if the human being is less complex than civilization.

Finally, we can rethink our previous discussion of the global economy and global civilization in this context. In Question 9.3.3 we discussed the growing interdependence of the global system. This interdependence is directly related to increasing complexity. After all, it is precisely the dependence of events in one place on events in another place that leads to much of the complexity that affects all decision making. Thus, we have established a connection between increasing global interdependence, increasing complexity, and the breakdown of hierarchical control in political and economic systems. What is still missing is a realization of the implication that global human civilization is manifestly a complex organism in relation to which we, as individuals, are elementary parts.

9.5   Consequences of a Transition in Complexity

The result of our discussion up to this point is the suggestion that a complexity transition is occurring in human civilization at this time. Prior to the transition, the complexity of various organized structures of human beings was less than the complexity of the individual; now the organized structures have greater complexity. When we say there is a growing complexity to life, this appears to be justified. What are the consequences of such a transition? The disappearance of central control is one that we have discussed and utilized to argue the existence of the transition. There are other important consequences. We will discuss these in two parts (Section 9.5.1 and 9.5.2). The first part is the consequences for an individual human being in the context of an environment that has recently become more complex than himself or herself. The second part reflects the relationship of human civilization as an organism to the individual human being. When we consider an individual in the context of a more complex environment, we find a strong motivation for specialization and for insecurity. When we take into account the relationship of the human organism to the human individual, we find reasons to eliminate the insecurity.

9.5.1 *Consequences for the individual*

We can develop a perspective on the complexity transition by recognizing that until the present, an individual human being was, as far as we know, the most complex organism. We pointed out in Chapter 8 that the demands of survival are much simpler than a human being. How are we to understand the consequences of the existence of a more complex organism which is now the environment of individual human beings? We consider the circumstances of other organisms that are in environments more complex than themselves. Most animals are simpler than the environment in which they live. They survive by limiting their exposure to the environment—restricting themselves to only a limited part of the possible environments that might be found. This results in a substantial simplification. A second strategy is to reproduce rapidly, where the excess reproduction compensates for low probability of individual survival.

The former strategy can be applied to human beings. We can anticipate that individuals will specialize professionally and socially so as to limit their exposure to the complexity of modern civilization. The degree of professional specialization has been increasing. Specialization occurred because of the existence of an increasingly large body of knowledge. This can be understood by comparing the number of books in the Library of Congress, 10^7, with the number of textbooks (courses) in a college education, 30. The existence of a large amount of knowledge does not necessarily mean that all of the knowledge is relevant to the functioning of human civilization. However, for other reasons discussed in this chapter, we see that the functional complexity of civilization has increased as well. This should motivate still more dramatic forms of specialization that relate not only to the information necessary for an individual to know, but also to the nature of his or her interactions with various aspects of the environment.

The complexity of civilization suggests that there are many possible sets of knowledge that an individual might need to know in order to achieve the analog of survival in society—beyond physical survival, this may include other goals such as a successful social and professional life. These sets of knowledge are analogous to ecological niches. In a sense we can consider them to be possible realities. The social and professional reality of one individual may be qualitatively different from the social and professional reality of another individual. This implies, for example, that decision-making strategies cannot be transferred in a simple way from one such reality to another. Moreover, it will be difficult if not impossible for an individual to be suited to more than one such reality. It will be impossible for an individual to address all possible realities. The specific skills inherent in performing a particular task become of crucial relevance to the ability of an individual to perform it. This also implies that education should be directed toward specific and individualized professions, and that these professions must be well suited to the individual's talents in order to enable success.

One oversimplified way to understand specialization is to consider examples where professional specialization is apparent. We might consider singers or athletes as examples. Viewed in an oversimplified way, we can argue that the existence of mass communications, recording and duplication makes it possible for a few singers to perform for a large number of people. This means that fewer singers are able to support themselves, the few that do are wildly successful, and the competition for the attention of the audience increases. Moreover, there are more opportunities for potential singers to try to sing, and the best of these will be the ones selected. In this way only the best of the best are professional singers. The high degree of competition is equivalent to the selection of one from among many. This corresponds (by information theory) to the high complexity of the tasks involved. In order for an individual to be selected, he or she must be well suited in every way, genetically and educationally, to this specific task. Similar statements can be made about the selection of the best athletes in a particular sport, or in a particular competitive event.

The suggestion that only a few—the best of the best—can succeed in a particular profession is not a complete picture. The intensive competition for a single profession is complemented by the increasing existence of diverse professions, including diverse forms of music, and diverse athletic events, in which different individuals can be successful. Thus, while each niche must be filled by a very specific individual, there are many such niches that are to be filled by distinct individuals. Moreover, this oversimplified view does not take into account the nature of collective behavior. We have chosen examples of professions where individual competition is apparent. By virtue of the nature of human civilization as a complex system, the tasks to be performed occur at many levels of organization and involve various numbers of individuals. Thus, while specialization is essential, the nature of competition as a process of selection is not well described by these professions.

A generally recognized feature of the present economy is a dramatic increase in changes of profession by individuals. This is not restricted to changes in employment, but also reflects rapid changes in projects and activities in a single job. We can attribute this to the rapid development of diversification and the rapid changes of technology. We might consider this as symptomatic of economic restructuring, which may resolve itself and result ultimately in a return to stability. This would be similar to the dislocation in employment and changes of profession that occurred during the industrial revolution. However, we can also consider this process in light of the necessity of placing individuals into occupations (niches) that are best suited to their abilities. In a complex system where diversity of professions is a principal property of the system, it may be essential to have such a dynamic flow of individuals until each finds optimal or near optimal suitability to a profession. This process would occur during the transition, and might not continue afterward. On the other hand, an individual in the complex system may also play a number of different roles, requiring various combinations of skills and capabilities. This would be similar to a network of neurons with various collective states, each composed out of a distinct set of activities of individual neurons, as discussed in Chapter 2.

Another implication of the complexity transition is a shift in the objectives and goals of individuals. Since control becomes impossible, the traditional goals of achieving authority, power and control become largely obsolete. For many individuals, as well as entire professions, achieving a position of power and control is the definition of accomplishment and fulfillment. We can already see a significant change in popular literature of the United States away from the traditional descriptions of an individualistic superhero/superachiever and toward the description of team players, networks of interacting individuals, and other more cooperative models for behavior. This is true even in circumstances where control appears to be exercised. A good example may be found in the difference between the original *Star Trek* TV series and the subsequent *Star Trek: The Next Generation* TV series, where the importance of crew members, teamwork, specialization, and complementary functions are more prominent. This change reflects the transition we have been discussing, which must be echoed in a change of personal goals and perspectives on success. While our objective is not to place value on developments, we can see that while some may applaud disappearance of the abuses of central control, the loss of the opportunity to exercise authority may be a disappointment in the context of the individual goals of the past. This is consistent with negative emotional reactions when an individual recognizes his or her inability to control, or even to understand, his or her environment.

When we consider an individual encountering a system of greater complexity, we may ask how the individual will model it. The construction of models by a simple observer of complex systems was discussed briefly in Section 8.3.7, and we continue the discussion here. Our discussion is an effort to gain perspective on how an individual human being will understand his or her environment. Any model developed by the individual must remove some features of the more complex system. One possibility is to ignore all but a limited part of the environment. In this case an individual's model of reality denies the existence of many of its aspects. A second possibility simplifies the complexity to a random process. Events are considered to be random, uncorrelated and thus unpredictable. This reflects our understanding that a random process has a low behavioral complexity. Finally, a model may presume associations or relationships that are overly simplified and therefore inconsistent with reality under all but a limited set of circumstances.

The discrepancy between models of reality and the reality itself has implications for individual actions, decision making and attitudes toward this decision making. Individuals are faced with the necessity for making decisions based upon their models of reality; this is the primary reasons for such models. Models take the form of an expectation that particular actions lead to anticipated outcomes. When the models are incomplete, the anticipated outcomes are not always realized. One of the primary conventional human responses to such inconsistency is to learn and adapt by improving the model. This is the usual process of trial-and-error learning. As long as the complexity of the individual is larger than the environment, adaptation can enable the individual to respond correctly to all circumstances. However, when the complexity of the environment is larger, adaptation becomes less effective.

To understand this point, we consider the behavioral complexity of the individual as a measure of the length of description of his or her pattern of responses to the environment. By our discussion in Question 8.3.8, different responses to distinct environmental conditions must be correlated. The degree of complexity of the individual reflects the extent to which independent responses can be made to distinct conditions. The complexity of the environment is a measure of the complexity an organism needs to survive in the environment. Thus, for an environment with higher complexity, there are more distinct conditions that require independent responses. If a simpler organism adapts to one subset of these conditions, then its responses to others are dictated by this, and are inadequate. Thus, it does not help to adapt to every new condition that arises, since this adaptation causes the individual to lose the ability to respond to conditions that the individual was suited to before. This may explain why simpler animals are not as adaptive as human beings: adaptation is less effective when the organism's complexity is smaller than that of its environment.

There are direct implications for the ability of an individual to perform common and special tasks—to find and retain jobs or conduct interpersonal interactions. We may assume that for many individuals, this inability to develop an effective set of responses to the environment will lead to frustration. Indeed, such frustration has become widespread. We note that in the complex environment, both success and failure are temporary; success at one time does not imply continued success, failure at one time does not imply continued failure.

Another aspect of this problem is the response by one individual to the behavior of another. This has relevance in various aspects of interpersonal and professional interactions. In a complex environment, the reality of one individual may not have a large overlap with the reality of another. We infer that one individual will view another individual as behaving in a random or incomprehensible fashion. Due to the increasing exposure to occurrence of such behavior, individuals may presume that others will not be comprehensible. This may either lead to respect for incomprehensibility or disdain for others. Both are manifest in scientific discourse and are likely to appear in other social and professional contexts.

The increasing specialization of individuals also implies and is consistent with an increasing specialization in sources of information. In this context it might be anticipated that conventional news sources which report on globally important events may become progressively irrelevant to an individual. This occurs because of the general inability of the individual to retain large amounts of information and because of the increasing irrelevance of general news to an individual's decision making. Instead, a system of more individually directed communication is likely to become dominant. In such a system, each individual would be better able to select the nature of information to which he is exposed. This self-consistent process of information exposure and selection may have all of the interesting properties of iterative maps that were discussed in Section 1.1, or self-consistent collective behaviors discussed in Section 1.6.

A better model, however, may be the pattern-formation processes in Chapter 7, in which the pattern of activities of individuals ultimately forms the basis for collective function of the human superorganism.

9.5.2 *Relationship of the individual to civilization*

Thus far our discussion of consequences of the complexity transition has taken the approach of considering an individual human being in the context of an environment whose complexity is greater than him or herself. We now turn toward considering the implications of the relationship between an individual and the complex organism of which he or she is a part. The difficulty in discussing this relationship is the inherent one—that we must assume that we cannot understand the behavior of the collective. Nevertheless, we will attempt to proceed in part by analogy and by assuming that the interdependence of a system and its components has universal implications. We can evaluate the consistency of the conclusions by comparison with observations.

In order to set the stage for this discussion we may note that the number of human beings in the world is of order 5×10^9, roughly comparable to the number of neurons in the brain. No functional analogy between the brain and humanity should be assumed. If we were to adopt a physiological analogy, we might be better off considering the analogy of human beings with mobile cells such as the immune cells in the body. However, there should be no assumption that the physiological analogy can be direct. The main purpose of the numerical analogy is to establish some sense of scale. It suggests that the relationship of an individual to the collective may be much more impressive than we might otherwise assume. The elimination of central control may be only a first step toward the potential complexity of the global system of which we are a part. As long as the human collective did not function as an organism, it played a small role in our perspective on the world, and on our actions. This may change rapidly in upcoming years so that our conscious recognition of this relationship as well as its effects becomes an important part of our existence.

As just described, the various changes that are taking place have led to an increasing sense of insecurity in individuals that are unable to plan for the future in a complex system whose behavior cannot be anticipated. However, when we consider the relationship of a complex organism to its components, rather than an individual in an environment of greater complexity, we see that this insecurity may be only temporary. The complex organisms we know act at least in part to protect and support the existence of their components. We may suggest that the human collective will protect individual human beings. It is likely to protect the individual better than the individual would be able to protect him or herself.

We can test this perspective in the light of historical developments. One way to measure the possibility that the human superorganism will protect individual human beings is through the improvements of life expectancy and quality of life from ancient to modern times. We have argued in the previous chapter that survival of a primitive human was possible because an individual was more complex than his environment. This survival was a statistical one (of order 10%–50% is sufficient) and required only survival to reproductive age. We can contrast this with the ongoing increase in life expectancy and quality of life, particularly during the twentieth century. The improvement in life expectancy occurred first in the West and has been spreading throughout the world. It was achieved through eradication of diseases and other hazards. It originates in technological and social advances that require collective actions of many individuals. This improvement in the human condition does not have as its objective the reproductive success of an individual human being. It is related to collective objectives of societal progress. More recently, collective actions have led to an alleviation of major sources of suffering and death around the world. Famines and natural disasters as well as other forms of social disruption have been addressed by global responses that are historically unprecedented. Moreover, the risk of self-inflicted worldwide cataclysm by nuclear destruction has been dramatically reduced in recent years.

The continued existence of local wars or revolutions in such places as Bosnia and Chechnya may be interpreted as a gap in this argument. The possibility of global conflict may be reduced, but local conflicts appear to continue. This, however, is likely to be temporary, since there is a growing recognition that the main cause of such conflicts—a desire for territory and control—has diminished in importance or practicality. Wealth no longer accumulates from national territory per se. Much of modern wealth is achieved through technological developments in industrial production, services and information. Moreover, from our previous discussion, in many cases control is only possible in name. It is likely that the current local conflicts are a residuum of outdated perspectives. The collapse of the Soviet Union released individuals to act on these perspectives. The individuals involved must interact with the new circumstances in a direct way before they recognize that gain cannot be achieved through military conflict.

At the same time as actions have been taken to alleviate global disease and suffering, there are other developments that increase life expectancy and quality in developed nations. In the United States, deaths from major disease categories, such as heart disease and cancer, have been declining. Deaths from the largest source of accidental death, automobiles, have consistently declined over the last few years. We should contrast the goal of an individual with the goal of the collective in relation to accidental death or death by disease. If we think about the goals of an individual, we realize that it is sufficient to reduce the probability of accidental death to the point where it is unlikely for the individual—say 1 in 100 in a lifetime. From the point of view of the collective, this is unacceptable, because it means that 1 in 100 individuals will die from this cause. We can argue that a new attitude is appearing that the loss of an individual human being has become unacceptable. This is a fundamental change of perspective. A goal of no loss of life is an inherently collective one. Various forms of factory work or building construction are known to have a certain statistical probability of injury or death. These probabilities give rise to a certain number of deaths each year. In the past, this death rate was known and considered to be acceptable. In more recent times goals have been set to reduce the risks to the point where even a single death is improbable. In addition to occupational hazards, this discussion is consistent with standards for product safety (from toys to buildings), where the basic criteria for safety is not just that products are safe under proper use, but that even improper use does not result in death or injury.

If the collective system serves in part to protect its components—individual human beings—then the relationship between the individual and the complexity of the environment changes. Rather than inducing a continuing struggle for survival, which currently appears to be manifest in the struggle for financial well-being, the collective may accommodate individual needs. There is some evidence for this, though the eventual resolution is not yet apparent. The evidence that exists is in the relative lack of dislocation when compared to the magnitude of changes that are taking place. Whether we consider the collapse of the Soviet Union or the job loss in the U.S. economy, the changes have been dramatic. However, the individual dislocations have been relatively mild compared to what can be easily imagined. In particular, there has not been general violence in the former Soviet Union despite several opportunities. In the United States, despite the dramatic reduction in employment at large corporations, it has been possible for small companies to more than compensate for the job loss. Thus it is possible that the collective organism is functioning constructively to transfer individuals from one framework to another in at least a partially effective manner.

In the context of considering human civilization as an organism in relation to individuals, we should revisit the traditional conflict between individual and collective good and rights. This philosophical and practical conflict manifested itself in the conflict between democracy and communism. It was assumed that communism represented an ideology of the collective while democracy represented an ideology of the individual. If we accept the transition to a complex organism, we may consider this conflict to be resolved, not in favor of one or the other, but rather in favor of a third category—an emergent collective formed out of diverse individuals. The traditional collective model was a model that relied upon uniformity of the individuals rather than diversity. Similarly, the ideology of the individual did not view the individual in relation to the collective, but rather the individual serving himself or herself. It should be acknowledged that both philosophies were deeper than their caricatures would suggest. The philosophy of democracy included the idea that the individualistic actions would also serve the benefit of the collective, and the philosophy of communism included the idea that the collective would benefit the individual. Nevertheless, the concept of civilization as an emergent complex organism formed out of human beings is qualitatively different from either form of government.

9.6   Civilization Itself

Our discussions of the relationship of the individual to civilization apply only to the finest scale of civilization as a complex organism formed out of human beings. In this section we turn to discussion of various other aspects of civilization as a complex organism. It is important to accept that there are many matters that we will not be able to describe or predict. This is consistent with the perspective that human civilization is more complex than we are as individuals. When we strive to understand, we expect that this knowledge will, at least in part, enable us to gain additional control. The previous statement makes clear that our knowledge may be limited in its ability to serve this function when applied to the entirety of human civilization.

From this discussion we should realize that there are limits to useful speculation due to lack of predictability. This limit was anticipated in the discussion in Section 9.1. In a sense, it points to a difference between our model and Newtonian models of simple systems with predictable behavior. The study of complex systems is more akin to quantum mechanics, where it is understood that certain questions cannot be answered within the context of science. Moreover, even if we were discussing a phase transition in a thermodynamic system (Sections 1.3 and 1.6), we would find an inherent lack of predictability. In a first-order phase transition, the ability to predict the specific behavior of the system is limited by the properties of nucleation that are sensitive to impurities. In a second-order transition, fluctuations make the local properties of a system inherently unpredictable. The inherent lack of predictability, however frustrating, does not mean that other questions cannot be asked and addressed. Interesting examples of questions follow.

Question 9.6.1 The basis of our discussion of human civilization as a complex system in Section 9.4 was the disappearance of central control in social and economic systems. Do our conclusions about the complexity of these systems mean that we can predict a further decline, or even the complete disappearance of hierarchical structures in human civilization?

Solution 9.6.1 One of the seemingly natural predictions of the model of loss of central control due to increasing complexity is that hierarchical systems or instruments of central control that exist today will continue to disappear over the upcoming years. However, the model of emergence of a collective complex organism suggests that this prediction is not a definite one. Functional segregation in a complex system may lead some parts of the system to retain central control, while others become networks. This is analogous to the existence of a neural network on the one hand, and muscles on the other. Thus, hierarchies may well continue to exist. Without any prior knowledge about the eventual structure that human civilization is to attain, we cannot predict where and in what way. Even though we might expect that dictatorships or centrally controlled economies which still exist in some parts of the world will completely disappear, such predictions may not be valid due to functional segregation.

An example is the relatively centrally controlled economy of Japan. Compared to the U.S. economy, the Japanese economy has a much more hierarchical (centrally controlled) structure. If it is generally true that such systems must fail due to increasing complexity, then we should anticipate that the Japanese economy will experience difficult times. These will occur due to inevitable mistakes made by the central authorities. Eventually the central control will be abandoned. However, a different scenario is possible—that the Japanese economy will continue to be effective and centrally controlled, but that the products of this economy will be limited to those that can be effectively produced in such a system. This is consistent with the model of functional segregation. A second example can be identified in the instruments of central control in the U.S. economy. At present the most powerful instrument of central control over the economy appears to be the Federal Reserve. It may be suggested that this mechanism of control will also fail due to the problem of increasing economic complexity. However, the argument may not apply here as well. In physiology there are glands, such as the adrenal gland, that control various aspects of the overall behavior of the system, such as metabolic activity. By such an analogy, the Federal Reserve may serve its function through controlling the overall level of financial activity even in the complex economy. ∎

Question 9.6.2 Consider global civilization as a single complex organism. What are the implications for the possibility of colonization on other planets?

Solution 9.6.2 Standard scenarios of colonization follow the model of colonization that occurred on Earth. A few individuals are sent to a new location and they independently function as a new society. This scenario does not work in the context of a complex organism. The interdependence of the complex organism implies that we cannot take part of it away and expect the part to function in the same manner as the whole. This is precisely the property of interdependence that we have used to characterize the complex organism.

There are two different models for how colonization might work. One of these is that the colony is not separate from the rest of human civilization but continues to function as a part of it. The second is that the process of colonization follows the same historical process that was followed by human civilization. This would be akin to a process of reproduction that occurs in other complex organisms. In order for the colony to follow the same developmental process, rather than beginning from modern technology it would have to start from a primitive state and develop technology through a similar process to that which occurred on Earth. ∎

Question 9.6.3 (for further thought) Discuss the possible origins of human civilization as a complex organism. Consider the various possible mechanisms for forming complex systems—spontaneous formation by a dynamical process, evolution, and development. Which of these can be relevant to the formation of a single complex organism? What conditions are necessary for it to occur? Which of the mechanisms for forming complex systems might apply to the formation of human civilization as a complex organism?

Question 9.6.4 (for further thought) We have concluded that global civilization (as a collective organism) is more complex than an individual human being. We have also concluded that an individual human being is more complex than the environmental demands upon him or herself. What process would cause an organism to form that is more complex than the environment?

Additional Readings

The following is a list of additional readings rather than a bibliography. The range of topics discussed in this text does not allow for a comprehensive bibliography. Our focus is on the effort to develop concepts and methodologies that enable the study of complex systems in a unified manner. Nevertheless, this effort must be informed by many fields and their phenomenologies. The following list attempts to address this by providing accepted keywords for literature searches as provided by the Library of Congress.

In addition to the keywords, a few references are provided with comments. Many of these texts were obtained from literature searches, and have been checked as relevant to the concepts we have been discussing. These references serve several purposes. First, they provide the student with an opportunity to pursue the phenomenology or theory in greater depth. Second, in a more specific domain, they provide a point of entry into the literature through a bibliography. Third, some references have an approach that is particularly compatible with the material presented in this text, or to the field of complex systems generally.

This list, however, does not serve three conventional purposes. It does not serve to trace the historical origin of concepts presented, or to motivate them from phenomenological grounds, or to prove them using experimental observations. Any of these would be a worthwhile but equally challenging endeavor to the objective of demonstrating the unity of concepts, which is the motivating force behind this text.

As is fitting for concepts that are to be a general underpinning of our understanding of complex systems, points made in this book appear in many contexts in the literature. A stronger statement may be made—the generality of the concepts presented in this text must imply that there are many ways to arrive at them, and many conclusions that may be drawn from them that can be compared with a large body of experimental literature. The effort in this text to draw conclusions from a very small set of assumptions is only a beginning in the effort to understand how widely applicable such concepts can be. In the few cases where we have made a greater effort to make contact with specific phenomenology and thus where support is necessary for material presented in the text (e.g. the discussion of sleep in Chapter 3), we have provided a few more specific references.

Chapter 0

keywords: system theory; autopoiesis; biological systems; chaotic behavior in systems; cybernetics; linear systems; social systems; system analysis; systems engineering; complexity (philosophy)

[0.1] Herbert A. Simon, *The Sciences of the Artificial,* 3d ed. (MIT Press: Cambridge, 1996). See the last chapter for an alternate overview of this text.

There are a remarkable number of popular or semipopular books on various concepts in the study of complex systems. For a number of reasons these books have appeared instead of textbooks. They are of two types: books written by observers of the field, and books written by researchers presenting their ideas to a popular audience.

There are both positive and negative aspects of this approach, the ultimate benefit of which will be judged by others. Here we provide a few references to this literature:

[0.2] James Gleick, *Chaos: Making a New Science* (Penguin: New York, 1987). Concept and personality history, focusing on chaotic and nonlinear dynamics (Section 1.1), and fractals (Section 1.10) but relevant to the study of complex systems in general.

[0.3] Douglas R. Hofstadter, *Gödel, Escher, Bach* (Vintage: New York, 1989). Semipopular and creative romp emphasizing mathematical and philosophical aspects of logic (Section 1.9).

[0.4] Roger Lewin, *Complexity: Life at the Edge of Chaos* (Macmillan: New York, 1992).

[0.5] M. Mitchell Waldrop, *Complexity: The Emerging Science at the Edge of Order and Chaos* (Simon & Schuster: New York, 1992).

[0.6] John L. Casti, *Complexification: Explaining a Paradoxical World through the Science of Surprise* (HarperCollins: New York, 1994).

[0.7] Brian Goodwin, *How the Leopard Changed its Spots: The Evolution of Complexity* (Charles Scribner's Sons: New York, 1994).

[0.8] Stuart A. Kauffman, *At Home in the Universe* (Oxford University Press: New York, 1995).

[0.9] John H. Holland, *Hidden Order: How Adaptation Builds Complexity* (Helix Books, Addison-Wesley: Reading, Mass., 1995).

[0.10] Peter Coveney and Roger Highfield, *Frontiers of Complexity: The Search for Order in a Chaotic World* (Fawcett Columbine: New York, 1995).

[0.11] Per Bak, *How Nature Works: The Science of Self-Organized Criticality* (Copernicus, Springer-Verlag: New York, 1996).

More technical references include some with classic and others with modern approaches:

[0.12] James Grier Miller, *Living Systems* (McGraw-Hill: New York, 1978).

[0.13] George J. Klir, *Architecture of Systems Problem Solving* (Plenum: New York, 1985).

[0.14] Gérard Weisbuch, *Complex Systems Dynamics* (Addison-Wesley, Reading, Mass., 1991).

[0.15] Thomas J. Sargent, *Bounded Rationality in Macroeconomics* (Carendon, Oxford, 1993). Title is inadequate.

[0.16] Mikhail V. Volkenstein, *Physical Approaches to Biological Evolution* (Springer-Verlag: Berlin, 1994). Touches on many classic contributions in complex systems.

A series of books under the collective title "The Santa Fe Institute Studies in the Sciences of Complexity" published by Addison-Wesley, collects various workshops and lectures sponsored by the Santa Fe Institute on subjects relevant to complex systems. This series illustrates the great diversity of concepts and applications of this field. We do not include a full list of these books here. A few volumes in this series are mentioned below as appropriate.

Chapter 1

Section 1.1
keywords: chaotic behavior in systems

[1.1.1] Predrag Cvitanovic, ed. *Universality in Chaos: A Reprint Selection,* 2d ed. (Adam Hilger: Bristol, 1989).

[1.1.2] Robert L. Devaney, *A First Course in Chaotic Dynamical Systems: Theory and Experiment* (Addison-Wesley: Reading, Mass., 1992). Excellent undergraduate-level textbook on basic mathematics of iterative maps and chaos.

[1.1.3] Robert L. Devaney, *Introduction to Chaotic Dynamical Systems,* 2d ed. (Addison-Wesley, Reading, Mass., 1989). More advanced mathematical treatment than [1.1.2].

[1.1.4] Steven H. Strogatz, *Nonlinear Dynamics and Chaos. With Applications to Physics, Biology, Chemistry, and Engineering* (Addison-Wesley: Reading, Mass., 1994). Undergraduate textbook on nonlinear dynamics that arise from differential equations.

[1.1.5] Edward Ott, *Chaos in Dynamical Systems* (Cambridge University Press: Cambridge, 1993).

Section 1.2
keywords: probabilities; combinatorial probabilities; correlation (statistics); distribution (probability theory); games of chance (mathematics); limit theorems (probability theory); random variables; stochastic processes; stochastic sequences; random walks (mathematics)

Probability and statistics is a traditional field of study in many fields with varying emphasis depending on whether it is used for analysis of data, for modeling of systems, or for more abstract formal concepts. A reference that is particularly relevant to our purposes is:

[1.2.1] N. G. van Kampen, *Stochastic Processes in Physics and Chemistry* (North-Holland, Amsterdam, 1981). Outstanding graduate text on the concepts and applications of stochastic processes. Somewhat more formal than this text, while still providing a useful conceptual framework.

Section 1.3
keywords: thermodynamics; statistical physics; statistical mechanics; phase transformations (statistical physics); statistical thermodynamics

Thermodynamics/statistical physics is a traditional field of physics covered by undergraduate and graduate textbooks with various approaches and flavors. Examples include:

[1.3.1] Kerson Huang, *Statistical Mechanics,* 2d ed. (Wiley: New York, 1987). Undergraduate text.

[1.3.2] Edward A. Guggenheim, *Thermodynamics: An Advanced Treatment for Chemists and Physicists* (North-Holland: Amsterdam, 1967). Graduate text with elegant review of basics.

[1.3.3] Lev Davidovich Landau and E. M. Lifshitz, *Statistical Physics* (*Course of Theoretical Physics*, vol. 5) 2d ed. (Pergamon: Oxford, 1969). Classic advanced text.

Section 1.4

keywords: chemical reaction, conditions and laws of; chemical kinetics

The two-state system analysis is based on classic transition-state theory covered in many physical chemistry books as a model for chemical reaction kinetics.

Section 1.5

keywords: cellular automata

[1.5.1] Stephen Wolfram, ed. *Theory and Applications of Cellular Automata* (World Scientific, Singapore, 1983). Many of the original articles are collected in this book. Includes an extensive bibliography.

[1.5.2] Doyne Farmer, Tommaso Toffoli and Stephen Wolfram, eds. *Cellular Automata* (North-Holland: Amsterdam, 1984). A conference proceedings volume.

[1.5.3] Tommaso Toffoli and Norman Margolus, *Cellular Automata Machines: a New Environment for Modeling*, (MIT Press: Cambridge, Mass., 1987). Many useful concepts and methods discussed.

Section 1.6

keywords: phase transformations (statistical physics); ferromagnetism; Ising model

Most books on statistical physics (Section 1.3) include a discussion of the Ising model. See especially:

[1.6.1 H. Eugene Stanley, *Introduction to Phase Transitions and Critical Phenomena* (Oxford University Press: New York, 1971).

[1.6.2] Giorgio Parisi, *Statistical Field Theory* (Addison-Wesley, Reading, Mass., 1988). Advanced formal treatment.

Section 1.7

keywords: simulation methods; mathematical models; computer simulation; Monte Carlo method

[1.7.1] M. P. Allen and D. J. Tildesley, *Computer Simulations of Liquids* (Oxford Science Publications: Oxford, 1987). Readable, practical guide to modern simulation strategies.

[1.7.2] Malvin H. Kalos and Paula A. Whitlock, *Monte Carlo Methods*, vol. 1: *Basics* (Wiley & Sons: New York, 1986).

Section 1.8
keywords: information theory; statistical communication theory

[1.8.1] C. E. Shannon, "A Mathematical Theory of Communication," in *Bell Systems Technical Journal*, July and October 1948; reprinted in C. E. Shannon and W. Weaver, *The Mathematical Theory of Communication* (University of Illinois Press: Urbana, 1963). The original manuscript on this subject is still the best discussion. Note the change in the first word of the title from original to reprint.

Section 1.9
keywords: logic, symbolic and mathematical; machine theory; computer science; Turing machines

[1.9.1] Ira Pohl and Alan Shaw, *The Nature of Computation: An Introduction to Computer Science* (Computer Science Press, Potomac, Md., 1981). Pleasant introduction to concepts and language of computer science.

[1.9.2] John N. Crossley, *What is Mathematical Logic?* (Oxford University Press: Oxford, 1972). Dense overview.

[1.9.3] Elliott Mendelson, *Introduction to Mathematical Logic*, 2d ed. (Van Nostrand: New York, 1979).

[1.9.4] Herbert B. Enderton, *A Mathematical Introduction to Logic* (Academic Press: New York, 1972).

[1.9.5] H. (Hartley) Rogers, *Theory of Recursive Functions and Effective Computability* (McGraw-Hill: New York, 1967; MIT Press: Cambridge, 1987). Formal discussion of the theory of universal computation.

Section 1.10
keywords: fractals; renormalization group; scaling laws (statistical physics); multigrid methods numerical analysis

The subject of fractals has strong overlaps with the topic of chaos (Section 1.1) due to the connection between multiscale phenomena and chaotic dynamics discussed in Chapter 9. Thus, see also the references in Section 1.1.

[1.10.1] Benoit B. Mandelbrot, *The Fractal Geometry of Nature* (W. H. Freeman: New York, 1983). The original source.

[1.10.2] R. J. Creswick, H. A. Farach and C. P. Poole, Jr., *Introduction to Renormalization Group Methods in Physics* (Wiley: New York, 1992). A relatively accessible discussion of renormalization methods.

[1.10.3] Albert-Laszlo Barabasi, H. Eugene Stanley, *Fractal Concepts in Surface Growth* (Cambridge University Press, New York, 1995).

[1.10.4] F. Family and T. Vicsek eds. *Dynamics of Fractal Surfaces* (World Scientific: Singapore, 1991). A collection of articles on the application of scaling ideas to surfaces. The principal relevance to us is the scaling treatment of spatial and temporal properties of these systems. References on the application of scaling to polymers are given in Chapter 5.

[1.10.5] James H. Bramble, *Multigrid Methods* (Longman: Harlow, Essex, 1993; Wiley: New York, 1993).

[1.10.6] William L. Briggs, *A Multigrid Tutorial* (SIAM: Philadelphia, 1987).

Chapter 2

keywords: neurophysiology; brain—localization of functions; neural networks (neurobiology); cognitive neuroscience; artificial intelligence; neural networks (computer science); pattern-recognition systems

Discussions of neural function from a biological perspective:

[2.1] Michael S. Gazzaniga, Richard Ivry and George R. Mangun, *Fundamentals of Cognitive Neuroscience* (W.W. Norton: New York, 1997).

[2.2] Eric R. Kandel, James H. Schwartz, Thomas M. Jessell eds. *Principles of Neural Science*, 3d ed. (Elsevier: New York, 1991).

[2.3] Gordon M. Shepherd, *Neurobiology* (Oxford University Press: New York, 1983).

[2.4] *Scientific American* (September 1992). Issue devoted to the biological approach to brain function, specifically cellular function and brain imaging.

Almost any book on neural networks, of which there are a number, will offer a basic introduction to various types of neural networks including the attractor network and the feedforward network, variations on these networks and other simple models. Unfortunately, the field is polarized, with distinct camps taking different approaches and claiming priority on ideas, realism or other issues. The complexity of biological neural systems enables various approaches to coexist without much more than acknowledging each other. A collection of articles that are central to the development of various threads in the field of neural networks is contained in:

[2.5] J. A. Anderson and E. Rosenfeld eds. *Neurocomputing* (MIT Press, Cambridge, Mass., 1988).

While it is important to respect the value of all approaches, the treatment emphasized in this chapter originates from J. J. Hopfield. This approach emphasizes simplicity of the microscopic components so that collective behavior can be more easily (but still not trivially!) understood. Books expanding on this:

[2.6] Daniel J. Amit, *Modeling Brain Function: The World of Attractor Neural Networks* (Cambridge University Press, Cambridge, 1989). A systematic description of the analysis of attractor networks using techniques developed in statistical mechanics. The early chapters motivate the use of the attractor network from a biological perspective and introduce the models.

[2.7] Marc Mezard, Giorgio Parisi, and Miguel Angel Virasoro, *Spin Glass Theory and Beyond* (World Scientific: Singapore, 1987).

Specific references:

[2.8] P. Baldi and S. Venkatesh, "Number of Stable Points for Spin-Glasses and Neural Networks of Higher Orders," *Phys. Rev. Lett.*, 58, 913 (1987). The maximum number of stored independent bits cannot be greater than $2N^2$, or $2N$ uncorrelated patterns—note that the synaptic matrix is not required to be symmetric.

Chapter 3
Section 3.1
keywords: sleep; sleep—physiological aspects
Discussions of sleep phenomenology and models of its function are contained in:

[3.1] James A. Horne, *Why We Sleep: The Functions of Sleep in Humans and Other Mammals* (Oxford University Press, Oxford, 1988). Excellent review focusing on sleep-deprivation studies. Counters notion that sleep serves physiological restorative function.

[3.2] Andrew Mayes, ed. *Sleep Mechanisms and Functions in Humans and Animals: An Evolutionary Perspective* (Van Nostrand Reinhold (UK): Wokingham, 1983). See particularly Chapter 1 by W. B. Webb for the evolutionary perspective on sleep criticized in this chapter.

[3.3] William Fishbein, ed. *Sleep, Dreams and Memory: Advances in Sleep Research* vol. 6, (Spectrum Publications Medical and Scientific: New York, 1981). This and the next reference suggest that sleep serves a role in memory.

[3.4] David B. Cohen, *Sleep and Dreaming: Origins, Nature and Functions* (Pergamon: Oxford, 1979).

[3.5] J. Allan Hobson, *The Dreaming Brain* (Basic Books: New York, 1988). Dissociation of neural function from sensory information is discussed on pp. 209–210.

[3.6] Ernest L. Hartmann, *The Functions of Sleep* (Yale University Press: New Haven, 1973). Mention of the similarity of aspects of dreams to cognition of postlobotomy patients, pp. 136–138.

Specific references:

[3.7] F. Crick and G. Mitchison, "The Function of Dream Sleep," *Nature* 304, 111 (1983).

[3.8] J. J. Hopfield, D. I. Feinstein and R. G. Palmer, "'Unlearning' Has a Stabilizing Effect in Collective Memories" *Nature* 304, 158 (1983).

[3.9] T. Geszti and F. J. Pázmándi, "Learning within Bounds and Dream Sleep," *J. Phys.* A20, L1299 (1987); "Modeling Dream and Sleep," *Physica Scripta* T25, 152 (1989).

[3.10] L. M. Mukhametov, "Sleep in Marine Mammals," in A. A. Borbély and J. L. Valatx, eds. *Sleep Mechanisms* (Springer-Verlag, 1984) pp. 227–238. Studies of dolphins sleeping half brain at a time. Among other arguments, refutes the evolutionary perspective of ref. [3.2].

[3.11] M. A. Wilson and B. L. McNaughton, "Reactivation of Hippocampal Ensemble Memories During Sleep," *Science* 265, 676 (1994).

Section 3.2

keywords: knowledge, theory of; intellect; perception; human information processing; artificial intelligence; philosophy of mind; cognitive science; memory; psychology of learning

[3.12] Marvin Minsky, *Society of Mind* (Simon and Schuster, New York, 1985).

Specific references:

[3.13] L. Standing, "Learning 10,000 pictures," *Quarterly Journal of Experimental Psychology* 25, 207 (1973). Testing recognition.

[3.14] R. D. Hawkins, T. W. Abrams, T. J. Carew, and E. R. Kandel, "A Cellular Mechanism of Classical Conditioning in Aplysia: Activity-Dependent Amplification of Presynaptic Facilitation" *Science* 219, 400 (1983); E. R. Kandel and R. D. Hawkins, *Scientific American* (September 1992) pp. 78–86, Experimental studies of the biology of neurons showing synapses that couple three neurons; e.g. implementing the logical AND operation.

Chapter 4

keywords: proteins; proteins–conformation; protein folding
Note: the problem of identifying time scale can also be seen in other fields. In computer science see keyword: computational complexity.

[4.1] Jack Kyte, *Structure in Protein Chemistry* (Garland: New York, 1995).

[4.2] Carl Branden and John Tooze, *Introduction to Protein Structure* (Garland: New York, 1991).

[4.3] Alan Ferscht, *Enzyme Structure and Mechanism*, 2d ed. (Freeman: New York, 1985).

[4.4] Thomas E. Creighton, *Proteins: Structures and Molecular Principles* (Freeman: New York, 1983).

Chapter 5

keywords: macromolecules; polymers; polymer solutions; biopolymers
Books on the scaling properties of polymers:

[5.1] Pierre-Gilles de Gennes, *Scaling Concepts in Polymer Physics* (Cornell University Press, Ithaca, N.Y., 1979).

[5.2] Masao Doi and Sam F. Edwards, *The Theory of Polymer Dynamics* (Oxford University Press, Oxford, 1986).

A discussion of parallel-processing simulations is found in:

[5.3] B. M. Boghosian, *Computers in Physics* 4, 14 (1990).

Specific references:

[5.4] P. G. de Gennes, "Kinetics of Collapse for a Flexible Coil," *Journal de Physique Lettres* 46 L639 (1985).

[5.5] B. Ostrovsky, M. A. Smith and Y. Bar-Yam, "Applications of parallel computing to biological problems," *Annual Review of Biophysics and Biomolecular Structure* 24, 239 (1995).

[5.6] Y. Bar-Yam, "Polymer Simulation Using Cellular Automata: 2-d Melts, Gel-Electrophoresis and Polymer Collapse," in G. Bhanot, S. Chen, and P. Seiden eds. *Some New Directions in Science on Computers* (World-Scientific: Singapore, 1996).

Chapter 6

keywords: evolution; evolution (biology); heredity; adaptation(biology); variation (biology); natural selection; genetics; population genetics; cytoplasmic inheritance; egoism; genetic algorithms

[6.1] Charles Darwin, *On the Origin of Species (By Means of Natural Selection)* (a facsimile of the first edition, 1859) (Harvard University Press: Cambridge, Mass., 1964). The original discussion is still enlightening, and shows the strong phenomenological basis for the conceptual developments.

[6.2] Douglas J. Futuyma, *Evolutionary Biology*, 2d ed. (Sinauer Assoc., Sunderland, Mass., 1986). Outstanding discussion of phenomena in biological evolution and their relation to conceptual issues in the theoretical understanding of evolution.

[6.3] Robert N. Brandon and Richard M. Burian, eds. *Genes, Organisms, Populations: Controversies Over the Units of Selection* (MIT Press, Cambridge, Mass., 1984).

[6.4] Richard Dawkins, *The Selfish Gene*, 2d ed. (Oxford University Press: Oxford, 1989). Presents the extreme reductionist view criticized in this chapter.

[6.5] Stephen Jay Gould, *Wonderful Life: The Burgess Shale and the Nature of History* (Norton, New York, 1989). Punctuated equilibria. See also ref. [0.11]

[6.6] Stuart A. Kauffman, *The Origins of Order: Self Organization and Selection in Evolution* (Oxford University Press: New York, 1993). A complex systems perspective.

[6.7] Karl Sigmund, *Games of Life: Explorations in Ecology, Evolution, and Behavior* (Oxford University Press: Oxford, 1993). Many fine points on mathematical treatment of evolution.

[6.8] William Day, *Genesis on Planet Earth: The Search for Life's Beginning*, 2d ed. (Yale University Press: New Haven, Conn., 1984). Molecular to cellular evolution.

Artificial life references include:

[6.9] Christopher G. Langton, ed. *Artificial Life: The Proceedings of an Interdisciplinary Workshop on the Synthesis and Simulation of Living Systems* (Addison-Wesley, Reading, Mass., 1989). See also the sequels in the same series, *Artificial Life II–IV*. See the video for *Artificial Life II*.

[6.10] Christopher G. Langton, ed. *Artificial Life: An Overview* (MIT Press, Cambridge, 1995).

For genetic algorithms, see:

[6.11] John H. Holland, *Adaptation in Natural and Artificial Systems*, 2d ed. (MIT Press: Cambridge, Mass., 1992).

[6.12] Melanie Mitchell, *An Introduction to Genetic Algorithms* (Bradford, MIT Press: Cambridge, Mass., 1996).

For discussions of the philosophy of egoism—self-interest vs. altruism, see:

[6.13] David P. Gauthier, ed. *Morality and Rational Self-interest* (Prentice-Hall: Englewood Cliffs, N.J., 1970).

[6.14] Robert M. Axelrod, *The Evolution of Cooperation* (Basic Books: New York, 1984). Shows the marginal stability of the most primitive form of cooperation (tit for tat); i.e. upon introduction of spatial correlations (correlations in selective forces).

[6.15] David P. Gauthier, *Morals by Agreement* (Clarendon: Oxford, 1986).

Chapter 7

keywords: developmental biology; embryology; pattern formation (biology); chemical reactions

[7.1] Lee A. Segel, *Modeling Dynamic Phenomena in Molecular and Cellular Biology* (Cambridge University Press: Cambridge, 1984).

[7.2] James D. Murray, *Mathematical Biology* (Springer-Verlag, New York, 1989).

[7.3] Hans Meinhardt, *The Algorithmic Beauty of Sea Shell Patterns* (Springer-Verlag: New York, 1994).

[7.4] H. F. Nijhout, "Pattern Formation in Biological Systems," in Lynn Nadel and Daniel L. Stein, eds. *1991 Lectures in Complex Systems* (Addison-Wesley: Reading, Mass., 1992). A brief presentation.

[7.5] Przemyslaw Prusinkiewicz and Aristid Lindenmayer with James S. Hanan, *The Algorithmic Beauty of Plants* (Springer-Verlag: New York, 1990). L-systems.

Chapter 8

keywords: Kolmogorov complexity

[8.1] http://www.fmb.mmu.ac.uk/~bruce/combib/. An extensive bibliography on complexity.

[8.2] Gregory J. Chaitin, *Algorithmic Information Theory* (Cambridge University Press: Cambridge 1987).

[8.3] W. H. Zurek, ed. *Complexity, Entropy and the Physics of Information* (Addison-Wesley: Reading, Mass., 1990).

[8.4] Ming Li and Paul Vitanyi, *An Introduction to Kolmogorov Complexity and Its Applications* (Springer-Verlag: New York, 1993).

Specific references:

[8.5] J. E. Lovelock, *Gaia, A New Look at Life on Earth* (Oxford University Press: Oxford, 1979).

[8.6] E. R. Cohen and B. N. Taylor, "The 1986 Adjustment of the Fundamental Physical Constants," Rev. Mod. Phys. 59, 1121 (1987). Value of the electron magnetic moment.

[8.7] Neville H. Fletcher, *The Chemical Physics of Ice* (Cambridge Univ. Press: London, 1970). Excellent review of basic knowledge of the low temperature properties of ice.

[8.8] Charles H. Bennett, "The Thermodynamics of Computation—A review," *International Journal of Theoretical Physics* 21, 905 (1982).

Chapter 9

keywords: civilization—history, civilization—philosophy; social history; social change; technology and civilization; organization; management; management science; economic history; international economic relations; man—origin

For modeling using system dynamics, a method that was not discussed in this text, see:

[9.1] George P. Richardson and Alexander L. Pugh III, *Introduction to System Dynamics Modeling with DYNAMO* (MIT Press: Cambridge, Mass., 1981).

Books on the structure of corporations and the recent changes in civilization include:

[9.2] Henry Mintzberg, *The Structuring of Organizations: A Synthesis of the Research* (Prentice-Hall, Englewood Cliffs, 1979). Remarkable phenomenologically-driven discussion.

[9.3] David Mitchell, *Control Without Bureaucracy* (McGraw-Hill, London, 1979). Experience-driven discussion.

[9.4] Gregory Stock, *Metaman: The Merging of Humans and Machines into a Global Superorganism* (Simon & Schuster: New York, 1993). Includes an extensive set of references.

Index

Bold numerals identify a key or defining reference.
Numbers followed by the letter "f" indicate figures.

839